TECHNIQUES AND INSTRUMENTATION IN ANALYTICAL CHEMISTRY – VOLUME 24

ACCELERATION AND AUTOMATION OF SOLID SAMPLE TREATMENT

TECHNIQUES AND INSTRUMENTATION IN ANALYTICAL CHEMISTRY

TECHNIQUES AND INSTRUMENTATION IN ANALYTICAL CHEMISTRY – VOLUME 24

ACCELERATION AND AUTOMATION OF SOLID SAMPLE TREATMENT

María Dolores Luque de Castro

José Luis Luque García

Department of Analytical Chemistry, University of Córdoba
Campus of Rabanales, 14071 Córdoba, Spain

2002

ELSEVIER

Amsterdam – Boston – London – New York – Oxford – Paris
San Diego – San Francisco – Singapore – Sydney – Tokyo

ELSEVIER SCIENCE B.V.
Sara Burgerhartstraat 25
P.O. Box 211, 1000 AE Amsterdam, The Netherlands

First edition 2002

Library of Congress Cataloging in Publication Data
A catalog record from the Library of Congress has been applied for.

ISBN: 0-444-50716-7

Typeset by Q3 Bookwork, United Kingdom

♾ The paper used in this publication meets the requirements of ANSI/NISO Z39.48–1992 (Permanence of Paper).
Printed in The Netherlands.

Contents

CHAPTER 6. HIGH-PRESSURE, HIGH-TEMPERATURE SOLVENT EXTRACTION

CHAPTER 7. ANALYTICAL SUPERCRITICAL FLUID EXTRACTION

CHAPTER 8. DEVICES FOR SOLID SAMPLE TREATMENT PRIOR TO INTRODUCTION INTO ATOMIC SPECTROMETERS: ELECTROTHERMAL DEVICES AND GLOW-DISCHARGE SOURCES

CHAPTER 9. LASER-ASSISTED SOLID SAMPLING

Foreword

Writing a book is always an audacious act, but writing a book such as this one on sampling and sample pretreatment is even more than that: it is a real challenge. The book encompasses the entire field of analytical chemistry and oozes a profound knowledge of the concepts involved.

The role of the analytical chemist has changed profoundly over the past few decades. The levels of analytes that can be determined routinely have decreased by several orders of magnitude and the sample sizes that need to be handled have dropped significantly in recent years. In parallel, the overall number of determinations that need to be performed has grown enormously. Also, the advent of intelligent instruments and laboratory automation has help analytical chemistry become a real information science. This significant, steady progress in analytical science, however, has been brought about as much by gradual advances in the preliminary steps of the analytical process (e.g. sample handling) as by a number of leaps in technological progress.

Often, in discussing analytical science, chemical measurements are reduced to fundamental processes such as detection, identification or quantitation. Although analytical and data processing techniques are dealt with in many books, sampling and sample pretreatment have so far been the subject matters of only a few books and review articles. This sustained lack of attention is surprising if one considers the overall time spent on these preparatory stages leading to the final analytical act and when one realizes how intricately the quality of the results depends on the initial, preparatory steps of an analysis. The development of clean sampling and handling techniques, which involves substantial investments of time and resources, has always been a mandatory step towards obtaining accurate, reliable final analytical data.

Scientific journals are rarely considered to be appropriate forums for discussing the guiding principles behind sampling and sample pretreatment. A book like this, which brings together all relevant wisdom and knowledge on this particular field, is therefore a more than welcome addition to the analytical chemist's library.

This particular book is designed to provide the analytical chemist with the background required to understand the concepts inherent in automated solid sample pretreatment. Its chapters provide a logical, systematic introduction to all aspects involved in the different methods of sample pretreatment and their automation. Topics such as freeze-drying, ultrasounds, microwave treatment, high pressure, high temperature solvent extraction, superfluid extraction, laser ablation and laser-induced plasma spectroscopy, the use of electrothermal devices and glow-discharge processes, are dealt with in sequence in all their relevant aspects. One chapter is devoted to specific sample pretreatment methods involving the removal of volatile species and provides a comprehensive discussion of hydride generation, headspace sampling, pervaporation and solid-phase micro-extraction. Automated and computer-controlled methods are given proper emphasis throughout, and the role of workstations and robots in the analytical process is discussed in a separate chapter.

Many of the ideas presented in this book have matured through the years from the wide experience gathered by the authors in the pursuit of their research on analytical chemistry and practical analysis at the University of Córdoba, where Professor Luque de Castro leads a research group specializing in "Innovations in Continuous and Discrete (Robotic) Systems for the Automation of Analytical Methods" at the Department of Analytical Chemistry. Her research interests include the combined use of discrete (robotic) and continuous (flow injection and completely continuous flow) systems for the development of fully automated methods; the design of flow-through (bio)chemical sensors; immunoassays; chromatographic and non-chromatographic continuous separation techniques; solid-sample pretreatment by subcritical and supercritical fluid extraction; focused and multimode microwaves and ultrasounds; laser-induced processes; applications to environmental, clinical and food analysis; and on-line process monitoring. She has co-authored several books on subjects such as Flow Injection Analysis, Automatic Methods of Analysis, Non-Chromatographic Continuous Separation Techniques, Analytical Supercritical Fluid Extraction and Flow-Through (Bio)chemical Sensors.

Freddy Adams
University of Antwerp, Belgium

Preface

Analytical Chemistry has traditionally been largely ignored by society and even by the scientific community, which, at most, has used it in the service of the other sciences. At present, Chemistry itself appears to have been reduced to a similar status. For example, the cover story of the August 20, 2001 issue of Time magazine was entitled "America's Best Science and Medicine" and, in it, the editors focused on the most exciting fields of research and then looked for the men and women who were doing the most cutting-edge research in such fields. The eighteen individuals honoured by the magazine — none of whom was a chemist — were conducting work in fields such as cell biology, the human origin, child psychology, pediatrics, genomics, cardiology, oncology, climatology, ecology, AIDS research, astrophysics, paleontology, biomedical engineering, neurobiology, cell death, spinal cord repair and molecular mechanics. In this issue, Time reflected present social perceptions in positing the following two destructive equations: science = biology and cutting edge = medical progress. Clearly, however, all these sciences rely on Chemistry and, ultimately, on Analytical Chemistry — which, invariably, provided the data that led to the achievements distinguished by Time.

We analytical chemists have always been aware of such unfair ignorance and yet have continued to work in pursuance of our goals. This book is a modest contribution intended to help those aiming to meet one of the most pressing needs of Analytical Chemistry: automation. A need that has arisen from the increasing demand for analyses in various fields of social interest including industry, health and the environment. Because sample pretreatment is the single step of the analytical process in greatest need of automation, this book compiles the wide range of tools available for this purpose with a view to assisting current and future users in choosing the best tool for each problem. Also, because full automation is almost always impossible, the primary goal — and also the main achievement in some cases — is to expedite this analytical step. Hence the book's title.

Rather than a comprehensive discussion of available choices for expediting or automating sample pretreatment, this book provides a description of the most widely favoured choices at present and those with a high potential which, however, remains largely unexplored. Its ten chapters revolve around this criterion.

The first chapter introduces both general aspects of sample preparation and the main problems encountered in automating sample treatments. The second provides a brief discussion of the underexploited potential of freeze-drying for delivering samples in forms that facilitate their subsequent analysis. Chapter 3 is devoted to a kind of energy that has also received inadequate attention from analysts: ultrasounds.

Those analytes that are either volatile or easily converted into volatile products are good candidates for separation from solid matrices by use of classical techniques such as hydride generation and dynamic or static headspace sampling, or more recent alternatives such as solid-phase microextraction and analytical pervaporation — with which virtually the whole research group headed by the authors is acquainted; all are dealt with in Chapter 4. The growing use of microwaves in both batch and continuous systems is the subject matter of Chapter 5, where the authors have strived — whether successfully or not readers will tell — to avoid overemphasizing their own work. The controversial, sparse applications of high-pressure, high-temperature solvent extraction, a widely used technique of a high potential, are systematically discussed in Chapter 6.

Chapter 7 describes the rapid expansion of supercritical fluid extraction but also how its scope has been restricted by differences in the behaviour of analytes naturally occurring in samples and those added to them for the purpose of analysis. Chapters 8 and 9 deal with various atomic techniques in which the authors are no experts, so any omissions or overpraising are totally unintentional. Chapter 8 is concerned with the use of electrothermal vaporizers and atomizers, and glow-discharge sources, with solid samples; it also provides a brief description of alternative techniques with a weaker impact that include arc nebulization and direct solid sample insertion. Chapter 9 describes the most salient laser-based techniques for solid sampling, namely: laser ablation and laser-induced breakdown spectroscopy; this chapter and the previous one examine the advantages and disadvantages of coupling the respective devices to analytical instruments.

Finally, Chapter 10 focuses on robotic treatment, the only currently available choice for completely automated solid sampling by virtue of its ability to have samples weighed without human intervention. Because of their high cost, robotic stations and, especially, workstations, are inaccessible to many users, so their capabilities are only outlined in the book. Would-be users should thus seek more comprehensive information — which continues to grow and change day by day, particularly as regards workstations — before they embark on the costly venture of setting up a robotic configuration.

The authors wish to express their gratitude to Antonio Losada, MSc, for his linguistic revision of the manuscript, Francisco Doctor for drawing the artwork and José Manuel Membrives for producing the tables.

The authors
Córdoba, December 2001

Chapter 1

Introduction to solid sample pretreatment

1.1. CONVENTIONAL PROCEDURES FOR SAMPLE PRETREATMENT AND THEIR STANDARDIZATION

Analytical processes typically consist of several discrete steps. While some steps such as measurement and transduction of the instrumental signal, and data collection and processing, are common to all analytical processes, those preceding the insertion of the treated sample into the instrument for measurement (viz. sampling, dissolution, clean-up, preconcentration, individual separation, derivatization), which are critical with a view to ensuring the obtainment of accurate, reproducible results, vary markedly from process to process. Virtually all analytical methods include a sample preparation step. Despite the dramatic advances in analytical equipment and microcomputer technology, many sample preparation practices continue to rely on 19th-century technologies. For example, the ubiquitous *Soxhlet extraction* was developed more than one hundred years ago; also, many of the sample preparation techniques currently in use have been around for decades with little or no improvement over the years.

The most common methods for analyte removal or sample dissolution are as follows:

(a) *Leaching*, which involves a solid–liquid extraction with an appropriate solvent (water for ionic compounds or a more or less polar solvent for non-ionic substances).
(b) *Sample digestion*, which is usually done with pure or mixed concentrated acids.
(c) *Fusion*, in which the sample is mixed with an appropriate flux material and heated until melting, after which the mixture is allowed to cool and the fusion cake dissolved in a suitable solvent.
(d) *Dry ashing*, which is especially common as a preliminary step in the analysis of organic samples for metallic and non-metallic elements. The sample is heated at 400–500°C in a muffle furnace for several hours and the target elements are either converted into gaseous species and collected into an appropriate absorber for determination or kept as a residue and dissolved into a suitable acid for analysis.

One of the main constraints on conventional procedures for sample preparation is their lack of uniformity. In fact, many users introduce specific modifications in the general procedures that eventually result in enormous differences in the way they are implemented. Such a lack of uniformity precludes comparison of results obtained using what was seemingly the same procedure and validation of the ensuing methods. Comparability of the results can thus only be achieved by using actually identical procedures; this justifies the efforts currently in progress at standardizing sample preparation procedures.

While some authors argue that standardization may "fossilize" progress in analytical science in a wide range of cases, it is widely accepted that the only way to ensure comparability when using operationally defined procedures is to standardize such procedures and apply them according to very strict protocols. In response to this need, the Standards, Measurements and Testing (SM&T) programme (formerly BCR) of the European Union developed a number of protocols for standardizing procedures over the past two decades, prominent examples of which are the operationally defined extraction procedures for soil and sediment analysis [1].

Standardizing a given procedure involves the following actions:

(a) Feasibility studies or examination of the literature, consultation with experts and selection of a few methods.
(b) Interlaboratory studies using the selected methods and reference materials. From the results, changes in the method to be used in the last step may be decided upon.
(c) Validation through interlaboratory studies, in which the participating laboratories receive natural samples for analysis following a strict protocol. The results thus obtained lead to either adopting the method or restarting the standardization process by reselecting methods in the light of the outcome of the failed attempt.

Standardization studies are slowly being superseded by sample preparation equipment that comes with bundled cookbooks where the optimum operating conditions for each procedure are clearly described. Some time is required in most cases, however, for universal adoption of automated procedures following extensive application by users.

1.2. THE STATE OF THE ART IN SAMPLE PREPARATION

Usually, even simplest types of samples are unsuitable for direct analysis; in most cases, they are too dilute, too concentrated or simply incompatible with normal instrument operation. The solution to these and other problems has been sample preparation using a wide range of approaches — many of which, as noted in the previous section, have been in use for decades. Sample preparation (SP) remains one of the more time-consuming, error-prone aspects of analytical chemistry, so much so that it is increasingly being recognized as a legitimate area of specialization in this science [2]. In fact, SP technology is gaining increasing momentum with the rapid development of specialized equipment, much in the same manner as was formerly the case with chromatography and spectroscopy. While the last two are by now mature, well-established techniques that are even recognized as analytical subdisciplines, sample preparation is still consolidating, even though it is expanding and growing at an incredible rate. Thus, SP has rapidly emerged as a specialized field in response to increasing developments and reliance on instrumental analysis.

Analytical chemistry as we know it today is the result of the major shift in nature it experienced over the second half of the 20th century. Previously, single-analyte wet chemical analysis procedures relied on hot-plate sample preparation. These procedures were gradually replaced by sophisticated instrumental methods that required equally sophisticated sample preparation [3,4]. A scientist from the 1950s would be amazed at the progress made in analytical instrumentation but would feel quite at home with many of the sample preparation techniques in contemporary use [5].

Efforts at introducing novel sample preparation concepts have frequently been ignored by practitioners, who were explicitly required by law to use the approved traditional methods. Problems associated with these traditional sample preparation methods such as those arising from the use of toxic organic solvents and of multi-step procedures that often result in the loss of analytes during the process, frequently make sample preparation the greatest source of error in an analysis and preclude integration with the rest of the analytical process. This picture is about to change for many reasons, not all of which are driven by science. Thus, legislation will restrict or ban the use of several common solvents such as chlorinated hydrocarbons; also, the costs associated with the use, transport and disposal of solvents in general these days provide an economic incentive to minimize solvent consumption. Renewed awareness of the pollution and hazards caused by hydrocarbons, and carcinogenic effects, have promoted international initiatives to eliminate the production and use of organic solvents on which many current sample preparation methods rely. This phasing out of solvents is poised to induce a major change in analytical methodology and also provides an opportunity to formulate practical alternatives to existing sample preparation methods.

Recycling pure solvents for trace analysis causes more problems than it solves and is generally inviable — recycling analytical solvents for re-use in less demanding areas of chemistry such as organic synthesis could be viable if integrated solvent management programmes were introduced. The current fashion, in any case, is the development of new, or the improvement of old, sample preparation methods that reduce solvent consumption significantly.

The attention given to sample preparation technology has brought it into the limelight much more so than in the past, such that it has become an acceptable area of academic interest — but rarely of funding. Users currently know much more about traditional sample preparation procedures than they did not long ago; also, they are developing an increasing awareness of the influence of sample matrix characteristics on, for example, analyte extractability in terms of a fundamental understanding of extraction kinetics and thermodynamic analyte–matrix interactions.

The greatest hurdle that keeps practitioners from adopting new analytical methods is the expense, in terms of both capital cost and training requirements, involved. An ideal sample preparation technique should be solvent-free (or, at least, use as little solvent as possible), straightforward, inexpensive, efficient, selective, easily automated and compatible with a wide range of matrices — and applications as a result; it should allow simultaneous separation and concentration of the target components, and be amenable to on-site use.

New sample preparation technologies have been slow on the uptake despite their proven advantages over some of the older technologies. In spite of many strong driving forces such as increased sample loads, decreasing skilled labour force, worker safety and less exposure to chemical hazards, the need for enhanced productivity, better quality data with increasing regulatory constraints, and the greater need for information management, automation of sample pretreatment and integration of information management into the analytical process have been accepted with reluctance. Notwithstanding all these adverse impacts, new sample preparation techniques are on the horizon and are being introduced with strength. Automation of sample preparation is a clear present trend meeting the needs of current analytical chemistry and implicitly included among the research priorities of the 21st century [6].

Automation was recognized as the buzzword in ACHEMA 2000 (the 26th Exhibition Congress, International Meeting on Chemical Engineering, Environmental Protection and Biotechnology, held in Frankfurt, Germany, from May 22–27, 2000). This fashion has reached both the process industry and the laboratory. Connection to the Internet or an Intranet was recognized as essential during the meeting. The new developments in laboratory devices placed special emphasis on cost-cutting, low maintenance and automation. This also applied to sample preparation having a major influence on the quality of the results. Even in small laboratories, attempts are being made at automating the often laborious sample preparation process. Flexible structure and easy-to-use software are indispensable if an analyser is to become established in practice.

As an ongoing experiment, the SamplePrep Web, a central resource site on the Internet, was created at Duquesne University in the past decade with a view to taking advantage of the immediacy and global nature of the information superhighway. The resources held on the SamplePrep Web and other, related Internet sites, were intended to provide support for sample preparation in aspects such as sample dissolution, sample extraction or leaching, reaction chemistry, applications, synthesis, literature, standards and clean chemistry. Eventually, the site would be linked to an electronic journal on sample preparation that would help focus effort on this important area. Such support was planned to be an on-line electronic library responding to the perceived need for rapid communication of ever-changing ideas about sample preparation. Clearly, sample preparation technologies would hopefully mature more rapidly and with more stability if support was provided early, and new, global information resources could be the perfect vehicles for dissemination. To this end, scientists were invited to contribute and participate in order to help in the rapid development of this area. Unfortunately, this call had little response and the SamplePrep Web has not been revisited since 1998.

1.3. PROBLEMS ENCOUNTERED IN AUTOMATING SAMPLE PREPARATION

Despite the progress, sample preparation continues to be the bottleneck of the analytical process as, only in the past 10–15 years, has its automation been actively addressed — by contrast, advances in automated chromatographic and spectroscopic equipment started one or two decades before. The reasons why SP remains the bottleneck of the analytical process vary among experts [7], but the most widely endorsed are as follows:

(a) As a rule, SP has always been viewed as a manual technology, often regarded as "low-tech", and usually assigned to the least trained staff.
(b) The addition of instrumentation to a traditionally non-instrumental step requires chemists' conversion to new techniques and chemists are somewhat reluctant to quickly accept new technologies.

(c) Sample preparation has little academic appeal and is rarely recognized as an analytical chemical area by academe. As a result, many universities have shied away from investigating it.
(d) In industry, the chemist's role is thought to be more "high-tech": to support the use of the analytical instruments as well as analyte detection. Industrial chemists spend most of their time evaluating and implementing new analytical technologies and generating and interpreting data. In addition, analytical techniques are more generic in nature; by contrast, SP can be done in many ways, a number of which are sample-specific.
(e) There are not many new concepts for the automation of SP, and early attempts tended to mimic manual operations. These approaches were unsuccessful and expensive as many of the manual procedures were not amenable to automation.
(f) Many of the major instrument manufacturers do not identify SP automation as important and it is mainly the smaller firms that have become niche players and provided the required instrumentation. For this reason, most of the early instruments were stand-alone SP modules that could not be integrated with analytical equipment as parts of the analytical process. The large firms, however, are slowly beginning to introduce automated equipment for SP.
(g) Robotics has been a revolutionary development in the context of SP automation; the implementation of robotic technology, however, is quite difficult and poses strong technological demands. Clearly, robotics and laboratory unit operations constitute the first insight into flexible automation of SP.
(h) The acceptance time of new technologies is a major constraint in SP. Widespread acceptance of a new technology usually requires a long time (often 10 years or more). Also, the time needed to validate the ensuing methods further restricts acceptance of the new technology. Regulatory agencies (particularly in the USA) are very slow to adopt new technologies, mainly as a result of most regulated methods being prescription- rather than performance-based. The inertia of most workers, usually reluctant to change and new training, is one other reason for slow acceptance.
(i) Financial payback is another major factor in the acceptance of the newer technologies, which are usually expensive. Whether the purchase of laboratory automation equipment is justified is largely dependent on the specific intended application and is actually a combination of both financial and strategic benefits.

All these constraints have restricted the development of automated sample preparation devices.

1.4. BATCH VERSUS SERIAL APPROACHES TO AUTOMATED SAMPLE PREPARATION

Automated sample preparation can be accomplished by using a batch or a serial approach. In the batch mode, multiple samples are prepared and then transferred to the analytical instrument for measurement. In the serial mode, samples are prepared one at a time and the SP device is connected (integrated) with the analytical instrument only.

From a temporal viewpoint, it is assumed that, when the SP time is equivalent to or less than the analysis time, then a serial method is preferable. Serial methods are also more useful with samples where the assays are time-dependent or when there is a stability question of the prepared analyte (e.g. with labile samples or matrices). Examples of better application of serial techniques are OPA-amino acid derivatives for high pressure liquid chromatography (HPLC) analysis, where some derivatives have very short half-lives, and for the determination of moisture, where batching of assays could expose the samples to moisture in the environment and give rise to spurious results. When the sampling cycle time is very long, then serial SP makes more sense. Two extreme examples are environmental early warning analysis of contaminated water streams, where samples are collected every 4–6 h, and fast sample preparation procedures such as microwave digestion. In the latter case, collecting the sample directly from the microwave vessel on-line makes a great deal of sense — some microwave digesters, however, are integrated with the analytical instrument to perform that task.

From the viewpoint of sample availability, if limited sample is available, the user may feel more confident in preparing one sample at a time; by contrast, batch preparation is to be preferred for abundant, complex samples.

Regarding the equipment subsequently used, off-line SP makes sense when the instrument is very expensive (e.g. a GC–MS, which one would not want to "tie up" during routine use). In general, the serial mode will require the development of low-cost analysers. Where this mode has its greatest potential is in feedback control led systems, which enable real-time (or *pseudo* real-time) process monitoring. Batch systems are more easily implemented in laboratories with routine loads and less so in laboratories that require a high flexibility or mixing and matching SP and instruments.

One other situation where the batch mode might make sense is when a wide variety of SP techniques feed a wide range of analytical instruments. The type of SP procedure involved dictates whether the batch or serial approach is required. For example, centrifugation and microwave-assisted digestion tend to require batch processes. On the other hand, new technologies such as those of axial centrifuges and the new flow-through microwave digesters have revealed that even traditionally batch steps can be made serial.

The batch environment is often an accommodation of the human operator in the laboratory rather than the most appropriate way for a given chemical–instrumental system. Frequently, batch methods are not science-based but developed by the constraint of an eight-hour working day, so phrases such as "let stand overnight to settle" or "allow to digest for 1 h" are common in these procedures. Automated systems are not so strongly constrained and should not be penalized with human limitations. The human element also dictates the use of the batch mode because of our limited ability to keep track of too many events.

In short, automated sample preparation is changing most SP protocols and giving a new face to this step of the analytical process, which was formerly envisaged as a clumsy, unproductive step.

1.5. DEFINITIONS AND RESTRICTIONS OF TERM MEANINGS

This section discusses some terms that are very common in the context of solid sample treatment but are not always properly used or understood.

1.5.1. Sample preparation and sample pretreatment

"Sample preparation" is a term in wide use at present. There is, however, some confusion with "sample pretreatment" as no clear-cut boundary between the two exists, nor has where sample preparation ends or what precedes sample treatment been clearly established. To the authors' minds, sample preparation includes rendering the sample — or, rather, the target analytes contained in the original sample — ready for insertion into the measuring instrument and can involve one or more steps such as dissolution or leaching (with solid samples), clean-up, preconcentration, derivatization, etc. Sampling is seemingly excluded from sample preparation as this designation assumes that the sample already exists as such before preparation starts. On the other hand, sample pretreatment can be envisaged as the first step of sample preparation or as a step preceding one other specific action (e.g. a pretreatment for subsequent liquid–liquid extraction or insertion of a solution into a chromatograph), which, in the case of solid samples, can be sample dissolution or leaching — in a more or less selective manner — of the target analytes. It is unclear when such a step finishes and when detection starts in the vaporizer or atomizer in electrothermal treatments, and so is in glow-discharge and laser-induced breakdown methods. When one of these approaches is coupled to, for example, ICP–MS, is the ionization step in the plasma also a part of the pretreatment or should it be considered a part of the detection step? In many cases, the boundaries between steps are blurred and the names such steps receive within the framework of the overall analytical process are confusing.

In dealing with the subject matter of this book, solid samples, the term "sample pretreatment" is used to refer to the step whereby the analytes are isolated from the whole matrix and placed in — usually — a solution.

1.5.2. Solid sampling

One other terminological controversy arises from the ordinary usage of "solid sampling" in relation to solid samples delivered to an electrothermal device, acting as electrodes in glow-discharge applications or being "shot" by a laser. Is it correct to use this designation after stating that "the sample is weighed and placed in the graphite tube or L'vov platform", "the sample is placed in the glow-discharge chamber" or "the sample is positioned on the XYZ translation stage for subjection to laser shots"? When does sampling take place in these techniques, before or after the sample is introduced into the specific device used? Because no universal consensus exists as regards the use of "solid sampling" to designate the particular step involved in each case, this book uses it consistently throughout.

1.5.3. Digestion versus leaching

The word *digestion* has several, rather different meanings. Thus, the Oxford English Dictionary (OED) defines it as "the operation of dissolving a substance by the action of heat and moisture" [8], whereas Webster's Third New International Dictionary (WTNID) provides the definitions "the process by which a material is softened or decomposed by heat or moisture or a chemical, often under pressure" and "the process by which soluble ingredients are extracted from plant or animal materials by warming a liquid" [9]. The last definition is fairly close to that for *leaching* in both WTNID ("the process or an instance of separating the soluble components from some material by percolation") [9] and the OED ("subjecting to the action of percolating water with a view to removing the soluble constituents") [8]. In order to set a boundary between both terms and adhere to the most frequent usage patterns in the literature, "digestion" is used in this book when all the components of the solid sample are dissolved under drastic conditions (viz. with the aid of heating and an acid or base) and "leaching" when only partial, selective dissolution is sought, using milder conditions.

1.5.4. Use of a leaching step versus a digestion step

Based on the most widely accepted meanings of leaching and digestion, the former is always desirable when possible. Partial dissolution of the sample with complete removal of the target analytes provides a less complex liquid and the possibility to avoid interferences. It should be noted, however, that complete dissolution of the analytes must be ensured if the target analytes are to be quantified with adequate precision. Special care must be exercised in applying a method developed with spiked matrices to natural samples — rather than with real samples as opposed to prepared or artificial samples since the opposite of "real samples" would be "imaginary samples".

Complete dissolution, which does not always require digestion, is usually easier to accomplish than leaching; however, the increased complexity of the latter is usually offset by the fact that it provides analyte solutions which are easier to process.

1.5.5. Leaching versus extraction

One other controversial term in wide use is *extraction*, which is defined as "the act to separate or otherwise obtain (as constituent elements or juices) from a substance by treating it with a solvent, distilling, evaporating, subjecting to pressure or centrifugal force, or by some other chemical or mechanical process" by WTNID [9] and as "the action to obtain (elements, juices, etc.) from a thing or substance by any chemical or mechanical operation" by the OED [8]. Extraction is thus used as a synonym for, and virtually always in place of, leaching. Thus, it is rather unusual to find the designation "supercritical fluid leaching" even though "supercritical fluid extraction" is much more general and less precise. In this book, "extraction" is used mainly to refer to steps in which a solid sample is partially dissolved to remove target species, but just because this is the more common term of the two; in fact, the authors greatly favour the use of "leaching" as it is more accurate (it reflects the fact that it is a solid sample that is treated to dissolve a portion thereof). The term "lixiviation" is synonymous with "leaching".

1.6. SOME NOTES ON THE BOOK CONTENTS

Clearly, the variety of solid samples that can be subjected to some pretreatment precludes the development of a universal device for processing any type of sample whatever its nature. Substantial efforts have been made, however, at adapting commercially available equipment to as many types of samples as possible (liquids included) with a view to expanding their scope. Two cases in point are supercritical fluid extraction with CO_2 in the presence of a modifier for removal of polar, even ionic, compounds; and radiofrequency-assisted glow-discharge sampling, which enables sputtering of non-conducting materials without the need to "deceive" the technique by mixing the sample with a conducting support.

The primary aim of this book is to provide readers interested in solid sample pretreatment with an overview of available techniques for development of this step of the analytical process. The title of the book is intended to reflect that it is mainly concerned with the dissolution or removal of target analytes from solid samples. Once they have selected the technique most closely fitting their intended purpose, readers can obtain a deeper knowledge about the technique of choice in the specialized literature — in fact, providing a thorough description of each of the wide variety of sample pretreatment techniques available at present was obviously outside the scope of a book like this. In fact, only those aspects that can be illustrated with reasonable concision are dealt with specifically in it. For identical reasons, the book does not touch on the subsequent steps of the analytical process. The authors therefore assume that the reader will be acquainted with the general principles of chromatography in its different variants, as well as with those of commonplace molecular optical and electroanalytical techniques, and atomic and mass spectrometries.

Although no explicit separation has been made, the devices and techniques dealt with in this book have been grouped in two different, broad categories, namely:

(a) Those that allow samples to be pretreated in such a way that the target analytes are released from the matrix (e.g. those based on ultrasounds, microwaves, superheated liquids and supercritical fluids, which are discussed in Chapters 3 to 7) or help release the analytes for a subsequent treatment such as dissolution or leaching prior to instrumental measurement (e.g. freeze-drying, which is the subject matter of Chapter 2).
(b) Those using electrothermal vaporizers or atomizers, a glow-discharge, laser ablation or laser-induced breakdown (with which Chapters 8 and 9 are concerned), where the pretreatment is integrated with detection, whether directly or via a plasma (ICP or DCP) source.

Because of their special features, workstations and robotic stations are dealt with separately (in Chapter 10) from the techniques in the previous two groups.

Obviously, not all techniques involving solid samples are included in this book. Thus, such techniques as neutron activation analysis (INNA) and X-ray photoelectron spectroscopy (XPS), which require equipment inaccessible to many laboratories, are only considered in comparisons with other techniques based on more affordable instrumentation but yielding similar results. Some readers may argue that, for example, laser ablation or laser-induced breakdown spectroscopy are uncommon in analytical laboratories at present; however, the authors have decided to cover them on the grounds of their high potential. Also, the short chapter devoted to freeze-drying is intended to draw the readers' attention to a technique that is rarely regarded as a form of sample pretreatment but deserves special recognition on account of its extensive use.

The authors have avoided the inclusion of exhaustive application tables, which would have unduly lengthened the book. Rather, only those references considered to be essential to illustrate some theoretical or practical aspect of the different techniques, and others representative of the state of the art in each, have been included. For comprehensive tables compiling applications or some other aspect, interested readers are referred to the reviews periodically published in specialist journals.

Although sample preparation may sound archaic and alchemy-related in a time where the buzzwords are terms such as genomic or proteomics, both updated and emergent sciences and technologies continue to rely on analyses and, ultimately, on sample pretreatment.

REFERENCES

1 P. Quevauviller, *Trends Anal. Chem.*, 17 (1998) 289.
2 J. O'Donnell, *Internat. Lab. News*, December (1999) 10.
3 B. Rosenkranz, P. Quevauviller, J. Bettner, *Internat. Lab.*, May (2000) 20.
4 S. Slaets, F. Adams, *Anal. Chim. Acta*, 414 (2000) 141.
5 J.L. Gómez-Ariza, D. Sánchez-Rodas, I. Giráldez, E. Morales, *Talanta*, 51 (2000) 257.
6 Technical Report, *Environ. Sci. Technol.*, 31 (1997) 20 A.
7 R.E. Majors, *LC–GC*, 13 (1995) 742.
8 *Oxford English Dictionary*, 2nd edn, Oxford University Press, Oxford (1989).
9 *Webster's Third New International Dictionary*, Merriam–Webster, Inc., Chicago, IL (1981).

Chapter 2

Analytical freeze-drying

2.1. INTRODUCTION

Freeze-drying, called "lyophilization" by the pharmaceutical industry, is a drying technique by which a product is solifidied by freezing and the solvent that contains it (usually water) is evaporated by sublimation (a chemical phenomenon) upon heating. The transformation of the solid phase into the gas phase takes place with neither the water nor the solvent passing through the liquid state.

During the process, the product loses more than 90% of the water it initially contains. One feature of freeze-drying that singles it out from other dehydration techniques is that dehydration takes place from a frozen — and hence less chemically active — product and in vacuum — which minimizes the effects of oxidation. These assets have made freeze-drying a widely accepted technique for processing heat-labile products such as enzymes and proteins for long-term storage.

One other salient feature of freeze-drying is that the product retains its original texture and regains its initial morphology upon addition of an appropriate amount of a solvent (usually water). In fact, freeze-drying avoids the denaturation caused by the heating procedures used in conventional drying methods.

The advantages of freeze-drying have been exploited since ancient times. Thus, the Peruvian Incas of the Andes increased the stability of their food reserves by storing them at the summits of Machu Picchu. There, the low prevailing temperature froze foods and water, in the form of ice, gradually evaporated by virtue of the low atmospheric pressure. However, the process was not made commercially available until World War II, where it gained widespread use as a means of preserving the blood plasma needed at front lines. Shortly afterwards, the process was applied to penicillin and freeze-drying was recognized as a major scientific technique for preserving biological samples.

Freeze-drying has a broad scope of application, particularly in the food and pharmaceutical industries. Freezing immobilizes the sample, thus allowing it to retain its original form; also, the absence of water prevents microbial growth and chemical changes leading to deterioration of the sample. Because water sublimates so readily, the conditions needed to freeze-dry a food do not remove most other constituents (e.g. acetaldehyde, which gives citrus fruits part of their flavour).

There are, however, many other possible uses for freeze-drying that include the preservation of microbial cultures, the restoration of books and other objects damaged by water, the concentration and recovery of reaction products, and the preservation of whole animal specimens as a form of taxidermy.

The freeze-drying technique is also used in the analytical laboratory. Two additional benefits make it attractive to analysts, namely: (a) because the removal of water allows samples to be subsequently reconstituted in a minimal volume of liquid, a preconcentration effect is achieved; and (b), even more important, because of their texture, freeze-dried materials are very easily attacked by solvents. The process can be optimized for a specific purpose by using an appropriate temperature, pH, reagent addition sequence and time. One major disadvantage of freeze-drying, however, is that some volatile compounds may be removed from the sample matrix during the process.

The conditions leading to the freeze-drying state can only be established with special equipment. Such equipment is currently available in a wide variety of models and sizes that range from laboratory-scale apparatus to industrial production systems.

This chapter deals with the different aspects of the freeze-drying process, beginning with the steps involved and the conditions influencing them. The state of the art in freeze-drying technology and equipment is then reviewed, and the aims and major analytical chemical applications of the technique (particularly as regards the pretreatment of solid samples) are discussed.

2.2. STEPS OF THE FREEZE-DRYING PROCESS

The freeze-drying process consists essentially of three steps, namely:

(a) *Freezing*, by which most of the water contained in the sample is converted into ice by cooling. A small fraction of the water, however, is tightly bound to the sample matrix and is not frozen, the amount of water that remains in the liquid state depends on the initial composition of the sample and on the temperature used.
(b) *Sublimation*, by which the ice formed in the previous step is converted into vapour by input of energy. The passage of solid to gas takes place with no intermediate conversion of water to the liquid state.
(c) *Desorption*, by which the water tightly bound to the matrix (viz. sorbed water) is removed from it when the water pressure of the solid-associated water equals the water pressure of the condenser.

2.2.1. The freezing step

The first step in freeze-drying a product involves converting it to the frozen state. Any alterations of the sample during this step should be avoided. Freezing has three major effects on the sample with strong implications on the outcome of the freeze-drying process; thus, it causes partial dehydration of the sample while preserving its initial structure and determining its morphology.

During freezing, most of the water in the sample is converted into ice; as a result, the solids in the remaining liquid become more concentrated — or, conversely, the solids are partially dehydrated. The ice thus formed remains in the sample, so freezing only causes a physical separation between the water transformed into ice and that remaining in the liquid state and containing dissolved solids. The ice is effectively separated from the sample matrix during the sublimation step. If the freezing step is conducted in a stand-alone apparatus rather than in the freeze-dryer proper, then the temperature of the sample should be maintained in transferring it from the freezing device to the dryer.

The extent of solid dehydration achieved with freeze-drying depends on the particular type of material and on the temperature reached by the sample during the freezing step. In any case, some water (containing dissolved solids) will always remain unfrozen in the sample.

Effectively preserving the initial structure of the frozen sample relies on a combination of stiffness in the frozen and unfrozen regions. Ice is quite resistant to deformation, so it acts as a support for the unfrozen region — the stiffness of which is more widely variable. The amount of ice formed and that of material which remains unfrozen depend on the sample temperature and the initial solid concentration in the sample, which are thus the two variables most markedly influencing the efficiency of ice as a support.

Sample morphology in this context refers to the spatial distribution of the sample components following freezing. Too rapid cooling produces a large number of crystals oriented in different directions, whereas too slow cooling yields increased amounts of elongated crystals oriented in the same direction and less restrictive channels in the matrix during the drying process.

The small, disoriented ice crystals formed in viscous samples by virtue of the restricted water motion, hinder mass transfer during the sublimation step, thereby resulting in less efficient drying and hence in samples more liable to structural changes.

The freezing temperature also influences crystal size and hence the rate at which sublimation occurs. It is therefore essential to establish the best freezing conditions for each product. The determination of freezing points or eutectic points is probably the most complex task in the freezing protocol. The presence of any liquid constituent other than water in a liquid matrix lowers its freezing point. It is thus very important to pre-freeze the product to below the eutectic temperature prior to freeze-drying proper. In fact, any small pockets of unfrozen material remaining in the product may expand and compromise the structural stability of the freeze-dried product.

The freezing points of biochemical products containing hygroscopic molecules (e.g. sugars, protein complexes) are well below zero by effect of some water molecules being tightly bound to molecules of the dissolved product. Eutectic points are usually determined from electrical resistivity measurements.

As a rule, products should be frozen below −20°C for no less than 3–4 h.

In some cases, freezing or freeze-drying a highly concentrated solution may prove difficult and require diluting the sample prior to freezing so that an adequate amount of ice is formed in the freezing step.

2.2.2. The sublimation step

The sublimation step removes the water molecules that were separated from the solids as ice crystals in the freezing step. Sublimation is the process by which solid water is directly transformed into vapour water. When sublimation occurs during the freeze-drying step, it is commonly referred to as the "primary drying process". The sublimation step does not affect unfrozen water associated to the solids (sorbed water), which is removed during the desorption step.

TABLE 2.1

VARIATION OF THE WATER VAPOUR PRESSURE WITH TEMPERATURE

Temperature (°C)	Water vapour pressure (mbar)
0	6.104
−10	2.599
−20	1.034
−30	0.381
−40	0.129
−50	0.036
−60	0.011
−70	0.0025
−80	0.0005

Sublimation always starts at an open surface and then moves inwards into the sample. After some of the ice is sublimated, the sample exhibits two distinct regions, namely: the dry layer (from which ice crystals have sublimated) and the frozen layer (where ice crystals are still present). These two regions meet at the so-called "ice interface", "sublimation interface", "freeze-drying interface" or, simply, "interface".

The sublimation step requires very careful control of two of the key variables of the freeze-drying process: temperature and pressure.

While sublimation can occur at atmospheric pressure, the process is rather slow because the gas molecules from the ice must find their way through the atmospheric gases that are bombarding the surface of the ice. This slow process by which the water molecules leave the ice surface is known as "diffusion".

The rate of sublimation of ice from the frozen product depends on the difference in vapour pressure between the product and the ice collector. It can be increased by decreasing the pressure over the ice surface. This can be accomplished by placing the frozen material in an evacuated chamber, where molecules will migrate from the sample to an area of lower pressure. Although the extent to which the pressure is reduced increases with increasing vacuum in the chamber, there is relatively little change in sublimation rate of the water from the ice surface. Only when the pressure in the chamber becomes less than the pressure of the ice is there a marked increase in sublimation rate. The vapour pressure of the ice is dictated by the ice temperature. A comprehensive list of temperatures and vapour pressures is given in Table 2.1.

It is extremely important that the temperature at which a product is freeze-dried is balanced between the level that maintains the frozen integrity of the product and that which maximizes its vapour pressure. Such a balance is a key to optimal drying. The typical phase diagram of Fig. 2.1 illustrates this point. Most products are frozen well below their eutectic or glass transition point, and then the temperature is raised to just

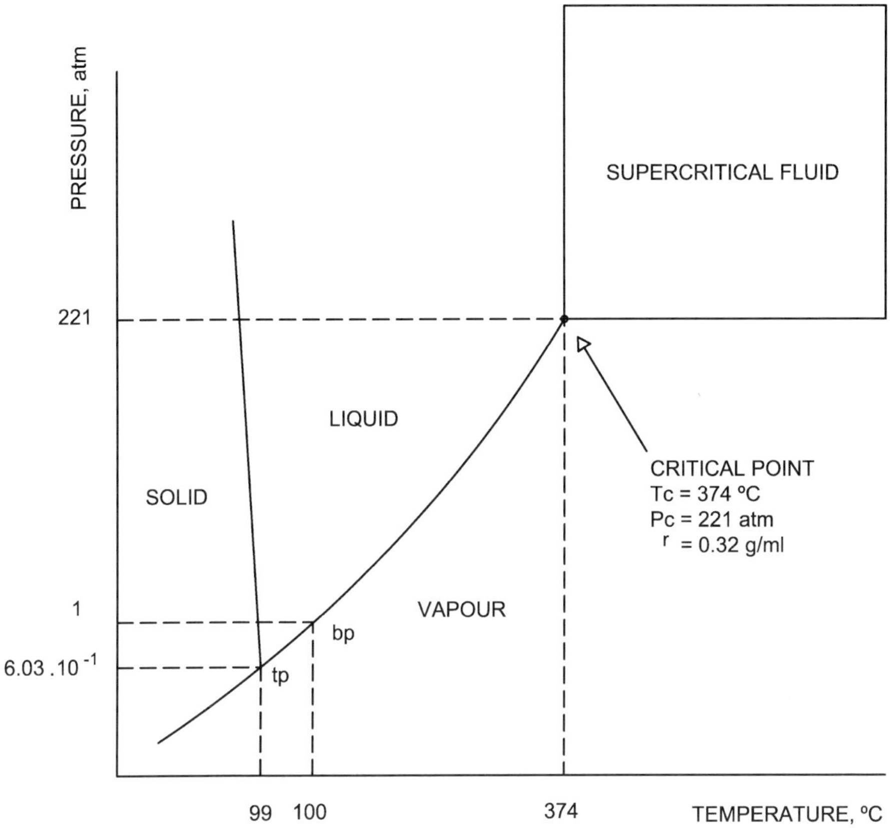

Fig. 2.1. *p-T* diagram showing phase equilibrium lines, the triple point (tp), boiling point (bp) and critical point for water.

below this critical temperature, the products being subjected to a reduced pressure. At that point, the freeze-drying process is started.

Some products such as aqueous sucrose solutions can undergo structural changes during drying that result in so-called "collapse". Although the product is frozen below its eutectic temperature, warming during the freeze-drying process can affect the structure of the frozen matrix at the boundary of the drying front. This causes the matrix structure to collapse. In order to prevent collapse of sucrose-containing products, the temperature must be maintained well below a critical collapse threshold during primary drying. For example, the collapse temperature for sucrose is −32°C (Table 2.2 shows the temperatures for miscellaneous compounds).

One problem arising from collapse of the matrix structure is that, when a diluent (e.g. water) is added to reconstitute the product, the solutes may require an extended period to return into the solution, so the product will not dissolve so rapidly as in a freeze-dried solution. Additional problems associated to matrix collapse arise from excessive

TABLE 2.2

COLLAPSE TEMPERATURES FOR SELECTED PRODUCTS AND SUBSTANCES

Substance	Collapse temperature (°C)
Dextran	−9
Ovalbumin	−10
Coffee extract	−20
Orange juice	−24
Inositol	−27
Lactose	−32
Maltose	−32
Sucrose	−32
Lemon juice	−36
Glucose	−40
Apple juice	−42
Sorbitol	−45
Fructose	−48

moisture, which may result in decreased stability, a loss of potency or even chemical reactions between solids yielding a turbid solution.

Energy, supplied in the form of heat, is the third key variable in the freeze-drying process. The transformation of ice into water vapour requires energy. In fact, almost as much energy is required to sublimate one gram of water from the frozen to the gaseous state as is needed to freeze one gram of water. Sublimating one gram of ice entails using about 675 calories.

All other conditions being appropriate, heat must be applied to the product to facilitate the removal of water in the form of vapour from the frozen product. The amount of heat used must be carefully controlled as applying more than evaporative cooling in the system can remove will warm the product above its eutectic or collapse temperature. Heat can be supplied in various ways (e.g. directly through a thermal conductor or from the environment).

2.2.3. The desorption step

The sublimation step separates the ice crystals formed during the freezing step. When an ice crystal forms, that which remains is the concentrated solute phase called the "dry layer". This will become the freeze-dried material at the end of the process. Immediately following passage through the interface, however, the solids in the dry layer still contain a substantial amount of water (about 25–30 g per 100 g of solids), which continues to be strongly bound to the solids. Most sample materials will not be structurally or chemically stable unless most of this water (called "sorbed water") is removed. The process

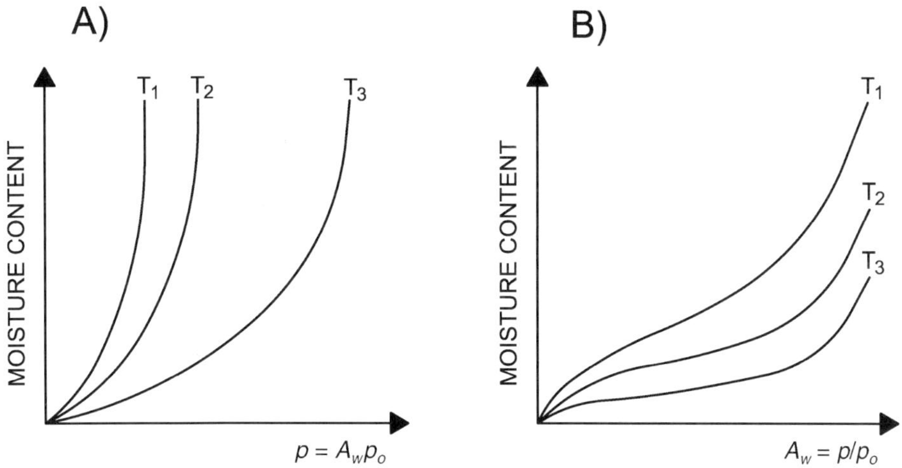

Fig. 2.2. Typical water sorption isotherms ($T_3 > T_2 > T_1$). (A) Based on water pressure. (B) Based on water activity. A_w water activity, p water pressure exerted by the sorbed water molecules in the solid at a given temperature, p_0 water pressure over pure water at the same temperature.

by which this more strongly associated water is removed is called "desorption" (or "secondary drying").

Sorbed, associated or bound water is water that exerts a lower pressure than pure liquid water at the same temperature. This pressure reduction results from binding of the water to the solids, the binding strength dictating to what extent the pressure is reduced.

Sorption isotherms (Fig. 2.2) expose the relationship between the water pressure and water content of a material at a given temperature. The sorption isotherm for the material varies with temperature in such a way that, at each sample moisture level, the water pressure increases with increase in sample temperature.

The most severe problem encountered in this step of the freeze-drying process arises from the need to supply the solid-bound water molecules with energy (heat) in order to effectively evaporate them. One way of solving the problem is by raising the pressure in the system — provided the stability of the product is not compromised — which can be accomplished by injecting an inert gas for some time and then re-evacuating. The inert gas provides the system with molecules that drive the energy from the heat source to the bulk product and act as brushes by transferring the vapour molecules from the product to the condenser.

2.3. VARIABLES INFLUENCING THE FREEZE-DRYING PROCESS

The two variables most strongly affecting the freeze-drying process are pressure and temperature. Temperature (or, rather, a temperature difference) is the driving force for heat transport. Energy moves from locations of higher temperature to locations of lower temperature. Similarly, pressure differences are the driving force for mass transport. The influence of these two variables on the different steps of the freeze-drying process, and their relationship to other influential variables, are discussed below.

The *freezing step* is conducted at atmospheric pressure. It is crucial to ensure that the product will be thoroughly frozen by the end. This entails maintaining the temperature below the eutectic or collapse point of the product.

In order to freeze the product, the shelf temperature is reduced to a level well below the collapse temperature. Depending on the dimensions of the container and on the fill volume of the product, the product temperature is maintained at shelf level for a specified period of time to ensure the formation of a completely frozen matrix.

In general, the term "shelf temperature" as applied to freeze-drying refers to the shelf fluid rather than to the shelf surface temperature. The shelf temperature may represent the fluid temperature in a reservoir, the inlet or outlet temperature of the fluid to the shelves or the fluid temperature in the inlet or outlet of the shelf manifold. To avoid ambiguity, it is preferable to use "shelf temperature" to refer to a measured shelf fluid temperature.

In the absence of a product temperature monitor, the formation of a completely frozen matrix relies on the length of time the product is on a shelf at a given fluid temperature. For vials loaded into the dryer on trays, additional time must be allocated to compensate for the variation in the heat transfer properties of the trays.

The *sublimation step* begins after the freeze-dryer attains the required condenser temperature and chamber pressure. The shelf temperature is increased such that sublimation of the ice in the product matrix occurs and the product temperature [1] is significantly lower than the collapse temperature to ensure a completely frozen matrix throughout the sublimation step. Ideally, the product temperature during this second step of the process will be directly dependent on the shelf temperature and chamber pressure. Hence, in order to maintain a given product temperature, one cannot alter the chamber pressure without introducing a compensating change in shelf temperature. This relationship was demonstrated when the temperature of ice in a stainless steel tray increased from −47°C to −24°C as the pressure in the chamber was raised from 55 mTorr to 500 mTorr at a shelf-surface temperature of −16°C [2]. Also, a change in the shelf-surface temperature at a given, constant pressure, produces a concomitant change in product temperature. For example, lowering the shelf-surface temperature decreases the product temperature at a fixed chamber pressure.

Validation of the freeze-drying process has been found to rely upon the establishment of a range of product temperatures yielding a product with properties falling within a set of prescribed limits [3–5]. It is therefore essential to ensure that the product temperature is maintained within such limits during the sublimation step.

One way to ascertain that the sublimation step has completed is by monitoring the pressure rise in the drying chamber as a result of temporarily isolating it from the pumping system [6]. An increase in chamber pressure at a given shelf pressure will result in an increase in product temperature. Such a temperature increase may exceed the collapse temperature and result in collapse or, under extreme conditions, in meltback of the product. In the absence of monitoring of the product temperature, this method is only applicable if the pressure in the chamber does not exceed the vapour pressure of ice at the collapse temperature. Because the shelf temperature during primary drying will exceed the collapse temperature, ensuring that the product temperature does not exceed the collapse temperature can only be accomplished by monitoring the product temperature. The pressure rise test has been found to alter the drying rate [6].

The characteristics of the dryer can also affect the product temperature if spurious pressure values are measured. Depending on the thickness of shelf-to-shelf spacing, pressure differentials may exist across the shelf. Such differentials mean that the product temperature in the middle of the shelf may be significantly higher than that on its edge near the inlet to the condenser chamber. The magnitude of the pressure differential increases with decreasing distance between shelves. In addition, the differential varies from process to process depending on the rate of sublimation of the product. Differences in sublimation rate across the shelf during the sublimation step have indeed been observed [7]: the drying rate was found to be lower in the middle and back of the shelf than in the front. These results suggest that a substantial amount of radiant energy might be transferred from the dryer door to the product. Product monitors placed at the front of the dryer would indicate product temperatures not representative of the general population of the product on the shelf.

The *desorption step* serves to reduce the residual moisture content in the product such that it will no longer support biological growth or chemical reactions [1]. The final residual moisture content in the product will depend on its adsorption and desorption isotherms [8,9]. The equilibrium moisture content of the product depends on its temperature, the vapour pressure on its surface and its potential chemical interactions with water vapour molecules. For a given product, the final moisture content can be reduced by increasing the shelf temperature at a constant water vapour pressure or by decreasing the water vapour pressure on the product surface at a constant temperature.

The completion of the desorption step is generally signalled by the product temperature approaching the shelf temperature for some specified length of time. In the absence of significant errors in chamber pressure potentially affecting the flow of water vapour from the container, pressure will not affect the product temperature as much as the shelf temperature. The main difficulty during the desorption step is that arising from changes in shelf temperature.

2.4. FREEZE-DRYING METHODS

Freeze-drying can be implemented in various ways the most common of which are manifold drying, batch drying and bulk drying. Each is used for a different purpose, the most suitable choice in each case depending on the specific initial and end products.

2.4.1. The manifold method

This method uses a turret with multiple ports (a manifold) to which flasks and vials are fitted via suitable valves (see Fig. 2.3). The product is either frozen in a freezer (by direct submersion in a low-temperature bath) or shelf-frozen, depending on the nature of the initial and end product, and also on the volume to be freeze-dried. The pre-frozen product must be immediately attached to the drying chamber or manifold to prevent

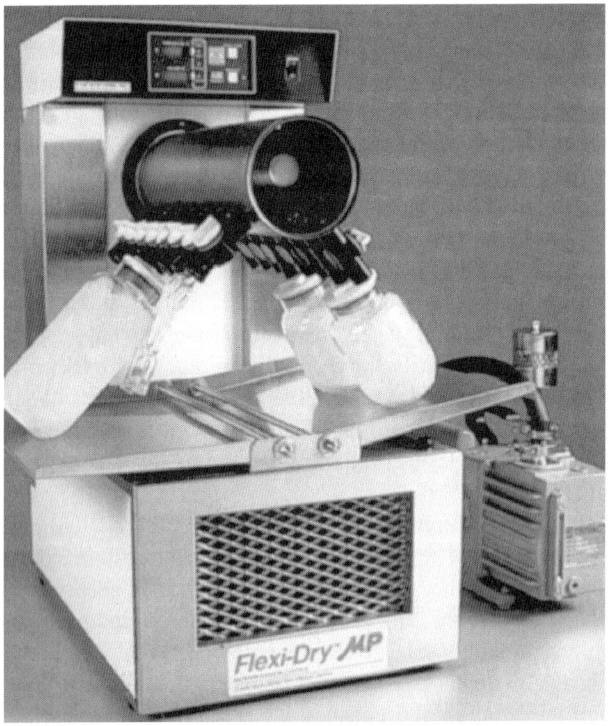

Fig. 2.3. Manifold-based freeze-dryer. (Reproduced with permission of FTS Systems.)

warming. The product container should be rapidly evacuated and the operator must rely on evaporative cooling to maintain the product temperature. This procedure is only applicable to products with high eutectic and collapse temperatures, and to small volumes at a time.

Manifold drying has several advantages over batch tray drying. Because vessels are attached to the manifold separately, each vial or flask has a direct path to the collector. This eliminates some of the competition for molecular space created in a batch system and is most ideally realized in a cylindrical drying chamber where the distance between each product and the collector will be identical. In a "tee" manifold, the water molecules leaving the product in vessels farthest from the collector experience some traffic congestion as they travel past ports of other vessels.

Heat input can be effected by exposing the vessels to room temperature or by using a recirculation bath. Manifold drying may be unsuitable for some products requiring strictly controlled temperatures.

Vessels of variable size and closing mechanism can be attached to the same manifold in order to dry various types of product simultaneously.

As the products and their volumes may differ, each vessel can be removed from the manifold separately as its drying cycle is completed. Also, the close proximity to the collector creates an environment that maximizes the drying efficiency.

2.4.2. The batch method

The vessels used in this drying method are all of the same size, contain the same product and are placed together in a tray dryer. The product is usually frozen on the dryer tray.

The product drying temperature and the amount of heat applied must be kept constant throughout the process. Usually, all the vessels in a batch are processed similarly, even though variations in the system can reasonably be expected. Thus, the amount of heat applied to different zones of the tray may vary slightly. Also, vessels located in the front portion of the shelf may be radiantly heated through the clear door. These slight variations can result in small differences in residual moisture.

The batch method allows one to stopper all vessels in a lot at the same time, under identical atmospheric conditions. The vessels can be stoppered in vacuum or after back-filling with an inert gas. Stoppering all the vessels at once ensures a uniform environment in each vessel and uniform product stability during storage.

Batch drying is the method most frequently used by the pharmaceutical industry to prepare large number of flasks or vials of the same product for marketing.

2.4.3. The bulk method

Like batch drying, bulk drying is performed in a tray dryer. However, the product is poured into a bulk pan and dried as a single unit. Although the product is spread throughout the surface area of the shelf and may be the same thickness as products dried in vials, the lack of empty spaces within the product mass changes the rate of heat input, which is restricted primarily to that provided by contact with the shelf.

Unlike manifold-based or batch drying, bulk drying allows no control of the conditions to which the product is exposed. Usually, the product is removed from the freeze-drying system prior to closure and then packaged in air-tight containers. Bulk drying is usually restricted to stable products scarcely vulnerable to oxygen and moisture.

2.5. COMPUTER-ASSISTED ANALYTICAL FREEZE-DRYING

The high performance of currently available computer hardware and software has inevitably led to their application in the freeze-drying process. A carefully compiled data library can be used to determine the optimum values of such crucial variables as the pressure, temperature and time required in each step simply by inputting the composition of the sample to be processed.

Below are briefly discussed the inputs one would need to enter or that should be present in the software in order to perform the three steps of the freeze-drying process as implemented by Jennings [8].

2.5.1. Computer-assisted freezing

In this first step, the computer software will require an input or a function that will compute the following minimum thermal properties of the sample: degree of crystallization, collapse or eutectic temperature and degree of supercooling. This last thermal property depends not only on the composition of the sample but also on the nature of its container (a vial or tray).

2.5.2. Computer-assisted sublimation

The sublimation step should be started when the matrix is in a completely frozen state. In order to ensure that the matrix will preserve its frozen state throughout the sublimation step, the computer should be provided with the collapse or eutectic temperature of the sample. Since the vast majority (99%) of samples will have a collapse rather than a eutectic temperature, the software will need to base the product temperature on some knowledge of, or means of computing that for the sample concerned. Such a frequency distribution of collapse temperatures arises from the fact that the glassy state, formed in the interstitial region of the matrix, is non-stoichiometric in composition, so the collapse temperature can vary from container to container depending on the particular frequency distribution.

Based on the size (sample volume and number of vials) of the lot, the software must take into account that the calculated ideal product temperature may exceed the collapse temperature.

The chamber pressure during the sublimation step (i.e. the primary drying process) has been found to be related to the product and shelf-surface temperatures [8]; however, determining the shelf temperature required is more difficult as it depends on the nature of the heat transfer fluid used to control the shelf temperature and also on the particular design of the freeze-dryer.

The sample container rests on a boundary layer at the top of the shelf surface. Such a layer is a region where the flow of heat transfer fluid is minimal or zero (i.e. the fluid is stationary). As a result, the sublimation step, which involves the transfer of heat from the fluid to the shelf surface, creates a temperature gradient across the boundary layer that depends on the thermal load exerted by the sublimation process and on the nature (viscosity, thermal conductivity and flow across the shelves) of the heat transfer fluid. The temperature gradient also depends on the number of shelves, their design and build, and on the pumping capacity of the circulation pump. These variables in turn depend on the size and particular manufacturer of the freeze-dryer, so the software used should include an input of data for the materials used, and for the dryer's design and build.

The computer must determine the rate at which this step develops from the heat of sublimation of the ice in the sample and the uncrystallized water present in the interstitial region. Also, it should take into account the heat transfer coefficient of the sample container — in the case of a vial, not just a single container, but the frequency distribution of an entire lot of containers. From this statistical information, the computer can then determine the highest shelf-surface temperature where, based on knowledge of the batch size, the chances that the heat transfer coefficient of a container would result in a product temperature exceeding the collapse temperature and yielding a defective product are minimal. The frequency distribution of the heat transfer coefficients of the containers would also provide the computer with the information needed to extend the primary drying process long enough to allow containers with coefficients falling on the lower portion of the frequency distribution to complete the sublimation step.

2.5.3. Computer-assisted desorption

This step is intended to reduce residual moisture to levels allowing no microbial growth or chemical reactions of the end product. The amount of residual moisture present in a product depends on its desorption isotherms. Such isotherms in turn depend on various factors including the product temperature, pressure chamber, partial vapour pressure in the container and nature of the interaction of the water vapour with the interstitial material formed in the freezing step. The computer should be fed with information on the target sample component. For example, if the component of interest is a protein, then overdrying may alter its configuration and decrease the potency of the end product. Consequently, the computer should control not only the final product temperature but also the partial water vapour pressure and the duration of the desorption step.

2.6. LABORATORY-SCALE FREEZE-DRYERS

2.6.1. Basic components of a laboratory freeze-dryer

All freeze-dryers are constructed from the same basic components, even though they may vary greatly in size and appearance.

A freeze-dryer for food production of tons-per-day capacity (e.g. one for coffee) operates on essentially the same principles as a laboratory freeze-dryer used to prepare grams-per-day amounts of tissue samples for microscopic analysis.

Basically, a freeze-dryer consists of a port for inserting the product, a drying chamber, a vacuum pump, a condenser, and vacuum and temperature gauges and controls.

The product insertion port is usually a tray or a manifold where the frozen product is placed for freeze-drying.

The drying chamber in a shelf freeze-dryer provides a thermally insulated vessel to accommodate the dryer shelves, which are specially constructed so that their temperature may be changed by passing a heat transfer fluid or gas. The gas or liquid is used to cool or heat the product. For freeze-drying in vials, the shelves can be moved so as to stopper the vials prior to opening the dryer door. The drying chamber should be equipped with a vacuum gauge in order to accurately monitor the pressure during the drying process.

The freeze-dryer is equipped with a mechanical pumping system that removes non-condensable gases. With oil sealed mechanical pumps, one should be careful to operate the dryer so that no hydrocarbon vapours from the pump can backstream into the drying chamber.

The condenser's main purpose is to remove the water vapour that is either sublimated or desorbed from the product. The latter is accomplished by directing the flow of vapours from the product over a cool surface. In an "internal" condenser, the condensing surfaces are located in the drying chamber; in an "external" one, the surfaces are contained in a separate, insulated chamber. The principal advantages of an internal condenser are that the dryer occupies less floor space and is less expensive. On the other hand, an external condenser can be isolated from the drying chamber during backfilling of the chamber with a dry gas (e.g. nitrogen). In this way, the possibility of water vapour being transported from the condensed ice surface back to the product is avoided.

The dryer should possess temperature sensors to measure the shelf and product temperatures throughout the freeze-drying process. Most freeze-dryers can be equipped with a computer system to control the pressure in the chamber and the shelf temperature as a function of time. The product temperature allows one to confirm that the shelf temperature and chamber pressure are within their preset limits.

2.6.2. Evolution of laboratory freeze-dryers

The mechanical design of freeze-dryers has evolved slowly since the 1940s. Many users of freeze-drying systems realized the need for special designs; in response, they altered their laboratory equipment themselves or had the required changes made on a commercial scale. In the mid-1990s, manufacturers started to incorporate the allegedly most universally demanded variations. This introduced new control equipment and mechanical designs in freeze-dryers for research purposes. Below are comments on the most salient shortcomings of the original freezer-dryers and the ways they were circumvented.

Pressure and temperature controls

Tray-based freeze-dryers were formerly heated by means of electrical wires, the temperature being controlled by using a thermostat. The temperature of the shelves (or the product itself) was measured with thermocouples furnished with analogue indicators.

Pressure measurements were read off a mercury gauge. These obsolete devices were not precise enough for control and measurement purposes or were cumbersome to use.

Freeze-dryer controls have experienced the most dramatic developments in recent years. The original mercury gauge and thermostatically controlled heaters were initially replaced with analogue vacuum gauges, thermocouples and controls. Eventually, the latter two were superseded by digital devices.

Digital controls were initially non-automatic. The operator was required to set the shelf temperature to be used to freeze the product (in tray-based systems) and, once such a temperature was reached, to activate the condenser cooling mechanism. When the condenser attained a temperature of −40°C, the operator started the vacuum pump and allowed the system to stabilize; a new intervention was subsequently required to raise the shelf temperature for the sublimation step. A thermal delay in time between the temperature set by the operator and the product temperature usually existed as a result of the low pressure in the system. In order to maximize the freeze-drying efficiency, the operator had to go through many trial-and-error cycles until the amount of heat supplied was balanced with that of vapour migrating outside the unmelt product; this led to unnecessarily long freeze-drying cycles. In addition, the most widely used method for detecting the end of the sublimation step involved isolating the tray chamber from the condenser module by using a vacuum pump. A rise in chamber pressure indicated that the process was still in the primary drying step. If the valve was shut when the temperature of the product was close to its melting point, the resulting pressure drop caused the product to melt down.

The thermostatically controlled heaters formerly used in tray-based freeze-dryers were gradually abandoned as they resulted in inconsistent product temperatures. Most freezers of this type currently use hollow trays through which a heat transfer liquid is passed at a controlled temperature.

Operating currently available freeze-dryer models is less labour-intensive; also, modern equipment affords the implementation of some initial production stages. Once the few sample (melting point) and protocol data (final tray temperature) required is input by the operator following insertion of the product, the apparatus conducts an entire freeze-drying cycle.

Pushing control buttons is all the operator has to do in order to have the system operate — there is no need to continually return to the freeze-dryer during the work cycle. The fact that the dryer pressure is strictly controlled avoids a thermal gap between the tray and product temperatures.

For safety reasons, the pressure is set in such a way that the temperature is 5°C below the melting point of the product. On an hourly basis, the freeze-dryer automatically uses a drain valve to allow the pressure within the system to drop. A simultaneous fall in product temperature is indicative of the presence of frozen solvent that should thus be extracted during the sublimation or primary drying step. Detecting any residual solvent in this way avoids the risk of the product melting down. Before the desorption step is started, this pressure control cycle is repeated until no further drop in product temperature is observed.

The vacuum pump

The vacuum pumps initially used in freeze-dryers were of the belt-drive type and, as such, extremely bulky and noisy by current standards. At present, most research-scale freeze-dryers use direct-drive pumps, which are much quieter than the previous models and also, frequently, very small. Because they are constructed from corrosion-proof materials, they are highly flexible in operation.

The low-pressure zone and the vapour trap

In the earliest commercially available freeze-dryers, the functions of the low-pressure zone and the vapour trap were both served by a chemical or ice trap.

A major change in the design of the low-pressure zone was introduced in 1955. Dry ice used as coolant was replaced with mechanically cooled low-temperature condensers, which proved much easier to use and maintain. The earliest condensers were made of stainless steel. As the scope of freeze-drying expanded, a need for a chemically resistant type of trap arose. In response, titanium traps that proved much more resistant to corrosion were developed.

The closing system

The devices initially used to stopper flasks on the shelves used a manually actuated screw-type mechanism that pushed the trays down. The procedure was very cumbersome and prone to introducing contaminants in the product, so it was eventually superseded by one based on air diaphragms. A bag is filled with atmospheric air to act on a plate that forces stoppers into flasks and vials. Depending on its shape, the bag can cause stoppers to seat misaligned in the vial mouths and the vials to break. If the bag bursts while being filled, the product will be contaminated. Built-in electromechanical closing devices have made air bags redundant. They avoid contamination of the product with external agents during closure and ensure simultaneous stoppering of all flasks or vials without breaking them.

Additional safety systems

Modern laboratory-scale freeze-dryers include additional devices for improved safety. Some avoid condenser overload or power outages.

During freeze-drying, the condenser temperature creates a low-pressure zone in the dryer for vapour to migrate. Such a zone should therefore be maintained at a preset, low level. If such a level is exceeded, then a pressure rise in the system can cause the product to melt down. Freezer-dryers equipped with condenser overload controls automatically cool the shelves until the emergency is solved.

A freeze-dryer that has gone through a power outage will continue to operate until the pressure within the system is so large that no sublimation or vapour migration will occur. Some freeze-dryers check the condenser and system pressures each time the system is reset. If both fall within the prescribed ranges, the freeze-dryer automatically continues on to the next step. However, if the conditions established during the power outage do not allow the process to continue, the freeze-dryer will automatically refreeze the product and the condensate but will not start the evacuation system.

Changes and advances in mechanical design of freeze-drying equipment and control systems have had a strong impact. Modern freeze-dryers are easier to use, require less operator intervention and are applicable to a wide variety of products. Current, automated freeze-dryers allow the initial steps of the protocol to be implemented, thereby providing the operator with more data of interest; also, they are safer and easier to operate than previous models.

2.6.3. Laboratory-designed freeze-dryers

A variety of small devices has been designed to meet specific functions not effectively served by commercially available freeze-dryers. Such devices were preceded by a number of customized dryers that were developed for various analytical purposes. Thus, Nakaguchi *et al.* [10] designed a new drying apparatus equipped with a device for trapping evaporated substances that was used to condition biological samples prior to determining trace elements.

Auxiliary devices for freeze-dryers have been designed for various purposes. Thus, a tray apparatus containing several wells was developed to freeze-dry biological substances (particularly anticoagulants used in blood gas analyses). This apparatus is suitable for producing a "pledget" or single-unit dosage of predetermined values of freeze-dried biological material that can be stored for extended periods in a sterile syringe to be used as blood anticoagulant. The tray has an upper plate laminate (covered by an insulator during freezing) with rows of wells for holding a biological solution of a preset concentration and a lower laminate in the form of a sheet of conducting material for establishing a temperature gradient across the height of the tray to assist in freezing the tray contents as the stackable trays sit on the cooling or heating shelves of the freeze-dryer [11].

An instrument performing measurements of D2 (viz. the ratio of the resistivity of ice to that of a product at a given temperature) and implementing DTA (differential thermal analysis) was developed to control freeze-drying processes; the principal components are a test chamber, a cooling and heating unit, a digital computer system and a printer. It was used to evaluate the thermal features of a 20% sucrose solution [12].

Rigid and semi-rigid disposable reaction containers for measuring absorbance changes have been reported [13]. A windowed reaction chamber was adapted to transmit light and connected to an auxiliary chamber via a small opening that allowed the reagent to be forced through to trigger a reaction. The container may be used in automated assays. Also, components of a freeze-dried reagent can be stored in compartments for later analysis.

2.6.4. Commercially available freeze-drying equipment

A wide range of apparatus exists that meets most laboratory needs as regards freeze-drying. Below are described some of the more widely used designs.

Manifold-based designs

These are best exemplified by the Cryodos-50 and Cryodos-80 models from Telstar (Fig. 2.4), which consist essentially of the following elements:

(a) A *vacuum pump* furnished with a gas ballast and a cut-off valve. The valve isolates the body of the freeze-dryer from the pump when the latter is stopped and effectively maintains a high vacuum in the dryer while atmospheric pressure is restored within the pump, thereby avoiding oil backstreaming. The pump also contains an eject filter.
(b) A *cooling system* that lowers the condenser temperature to the working level. The condensation surface is sized in such a way as to ensure condensation of the water vapour released by the product during the freeze-drying process and prevent it from reaching the vacuum pump.
(c) A *stainless steel cylinder* located at the top of the system and equipped with eight stopcocked ports each of which is furnished with a three-way valve that allows each flask to be isolated and atmospheric pressure restored within it on an individual basis

Fig. 2.4. Manifold-based freeze-dryer (Cryodos 6 model). (Reproduced with permission of Telstar.)

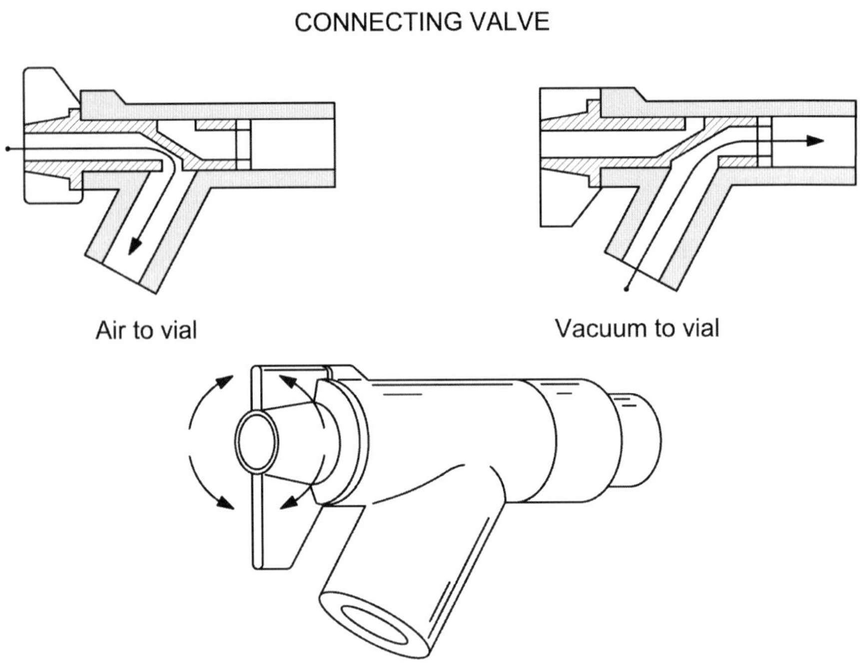

Fig. 2.5. Three-way valve for isolating each individual flask and restoring atmospheric pressure with it. (Reproduced with permission of Telstar.)

(Fig. 2.5). The valves allow ground-mouthed flasks to be fitted via a conical adaptor. By removing the adaptor, tubes and vials can be directly fitted to the valve.

(d) The *control panel*, which includes an alphanumeric LCD display that shows the menu for the freeze-drying cycle (and the condenser temperature and vacuum pressure throughout the process).

(e) *Ancillary devices* that allow small volumes of product contained in various packages, vials, ampoules (or even bulk products) to be processed.

Tray-based designs

This type of design is aptly exemplified by the Telstar models Lioalfa 6–50 and 6–80 (Fig. 2.6). Both are compact, benchtop apparatuses that accommodate the vacuum pump and cooling system. Also, both use the same type of vacuum pump as the Cryodos models.

Fig. 2.6. Tray-based freeze-dryer (Lioalfa model). (Reproduced with permission of Telstar.)

The cooling system consists of one or two air-tight, one-step compressors and uses air as coolant.

Like the previous ones, the Lioalfa models can be equipped with ancillary devices for processing bulk products and samples in flasks, vials or ampoules.

Miscellaneous designs

Depending on the number of containers to be used at once, some freezer-dryers use a multiport turret (a manifold) or a tray module. The former is suitable for containers of varying shape or size that need not be processed simultaneously; the latter is to be preferred when a large amount of product or number of containers is to be processed in a simultaneous manner. The tray module can be equipped with an electromechanical device to close the containers after the products are freeze-dried in order to avoid external contamination. One example of this type of freeze-dryer is the model from FTS Systems.

Some multi-purpose models such as Telstar's Liogamma can be used as research plants with a view to studying or storing sensitive biotechnological products or as pilot plants for the accurate reproduction of entire freeze-drying processes or the examination of freeze-drying curves of industrial use. They allow small production lots to be freeze-dried with a view to assessing the marketing potential of new products.

2.7. GENERAL AIMS OF ANALYTICAL FREEZE-DRYING

A product can be freeze-dried for various purposes depending on the particular features the resulting material is required to possess. Below are discussed some of the more salient uses of the freeze-drying technique.

2.7.1. Increasing long-term storage stability

The very essence of the freeze-drying process is that it is a stabilizing process. In foods and biological samples such as enzymes, blood fractions, urine, etc., freeze-drying can effectively avoid unwanted chemical or biochemical reactions and hence deterioration of the material. It is especially important — particularly with biological samples — that any inherent biological activity be preserved throughout the process and recovered upon rehydrating the material.

The stability of freeze-dried materials is dictated by various factors two of the most important of which are moisture and oxygen.

All freeze-dried products contain a small amount of water called "residual moisture". The exact amount that remains in the material depends on the nature of the product and on the length of the secondary drying step. Residual moisture can be measured by chemical, chromatographic, manometric or gravimetric means. It is expressed as a weight percentage of the total weight of dried product. Residual moisture values typically range from < 1% to 3%.

By nature, freeze-dried products are hygroscopic, so exposure to moisture during storage can destabilize them. Packaging used for freeze-dried materials must be impermeable to atmospheric moisture. Storing products in low-humidity environments can reduce the risk of degradation by exposure to moisture. Oxygen is detrimental to the stability of most freeze-dried materials, so the packaging used should also be impermeable to air.

The detrimental effects of both oxygen and moisture are temperature-dependent. The higher the storage temperature is, the faster a product degrades. Most freeze-dried products can be stored at refrigerator temperatures (i.e. 4–8°C). Storing freeze-dried products at lower temperatures extends their shelf life, however. The shelf life of a freeze-dried product can be predicted by measuring its rate of degradation at a high temperature. This is called "accelerated storage". By choosing the proper time and temperature relationships at high temperatures, the rate of product degradation at lower temperatures can be predicted.

Freeze-drying slows down the kinetic clock (viz. the rate at which a substance undergoes change). One example of a substance undergoing a slow change is rusting iron; one of a substance experiencing a rapid change is burning gun powder. While freeze-drying is not used to prevent rusting of iron or burning of gun powder, it is used by the health care industry to slow down the kinetic clock of pharmaceuticals. For example, a pharmaceutical formulation will retain at least 90% of its original potency for two weeks when stored at 4–8°C in a refrigerator. After two weeks, the potency is found to be less than 90% and the product must be discarded. This means that the pharmaceutical product would have to be prepared, bottled, shipped — while being maintained at 4–8°C — and administered by physician to patient all in less than two weeks. If the manufacturer shipped too much product then it would experience a financial loss as a result of the unused product being discarded.

Freeze-drying converts one second at refrigeration temperature into one minute at room temperature. By changing the rate of the kinetic clock, the formulation that was only stable for 2 weeks at 4–8°C is now stable for 120 weeks (2.3 years). In slowing down the clock, the manufacturer now has time to test the product before releasing it to the public. In addition, the product no longer requires refrigeration at 4–8°C, which makes worldwide distribution much easier and safer. Also, there is less chance that the product will not be available for a patient or that it will expire and need to be discarded.

2.7.2. Preparing tissues for microscopic examination

Properly conditioning tissue samples for examination under a microscope usually entails their prior drying. This can be done by freeze-drying, provided the tissue preserves its structure throughout the process.

2.7.3. Preparing samples as an intermediate step in (bio)chemical analytical procedures

Some types of chemical analysis require that the sample be in dry powder form, either because of restrictions imposed by the analytical system to be used or in order to concentrate the sample and boost the sensitivity of the analytical method applied. The necessary condition for freeze-drying to be applicable with this purpose is that it should not alter the sample composition, which will depend exclusively on the particular freeze-drying conditions used.

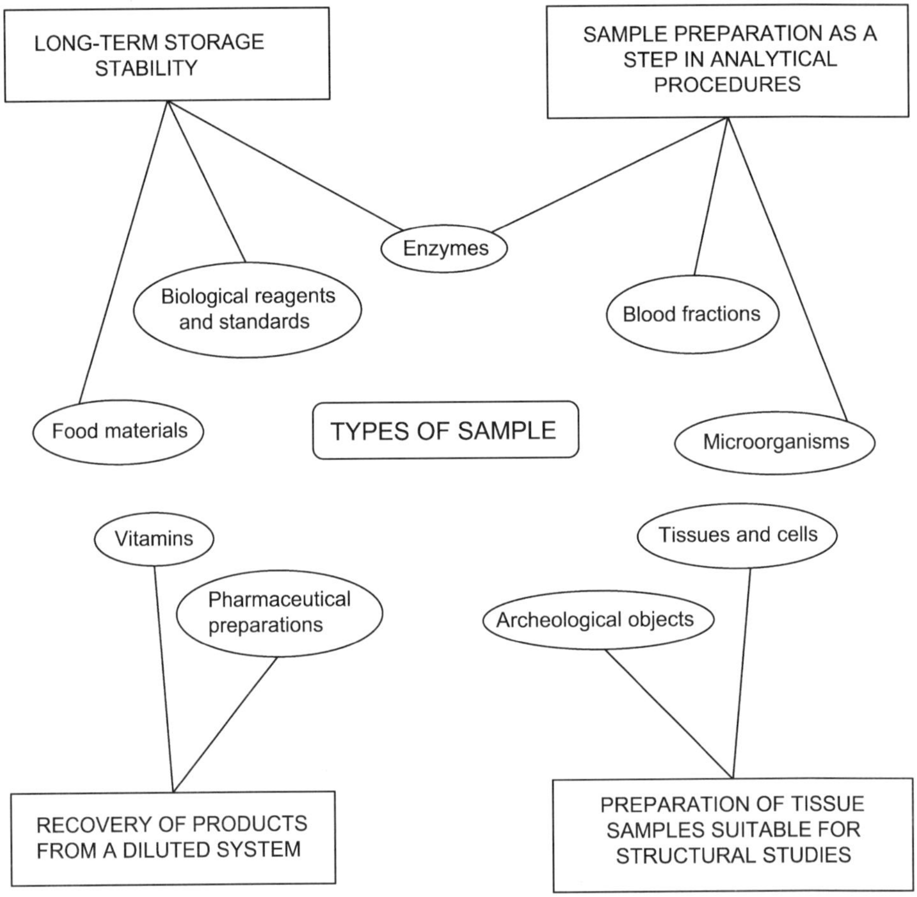

Fig. 2.7. Selected materials typically subjected to freeze-drying and aim of the process in each case.

2.7.4. Recovering products from a reaction mixture

Freeze-drying is a convenient, advantageous alternative to recovering reaction products from dilute media provided it does not alter the chemical or biological activity of the product of interest. Figure 2.7 shows selected products typically subjected to freeze-drying and the main purpose of using this drying technique in each case.

2.8. ANALYTICAL USES OF FREEZE-DRYING

Freeze-drying has rarely been considered an effective choice with a view to automating sample pretreatment, possibly because of the contradictory results obtained so far. However, there is solid evidence that this technique is suitable for many purposes.

The analytical uses of freeze-drying can be classified into three broad categories according to target: reagents, standards or samples. This section describes the principal applications of analytical freeze-drying, particularly in relation to solid samples; some mention of liquid samples is also made as serum and urine are among those most frequently used to develop analytical standards.

2.8.1. Reagents

Storing reagents poses special problems, particularly in clinical analysis. One major difficulty encountered in evaluating the results of immunochemical tests is the absence of affordable reagent standards [14]. This shortcoming has been partly circumvented by using freeze-dried reagents. The large number of patents granted to applications in this field in the 1980s testifies to the high potential of the freeze-drying technique for immunochemical analysis.

Freeze-drying a reagent will only be justified if (a) the process improves its stability and (b) the reagent can be readily reconstituted. In some cases, the stability of the final, freeze-dried reagent can be improved by using a suitable additive.

Most immunoassay reagent standards possess improved stability upon freeze-drying. Thus, freeze-dried IgG aggregates stored at 4°C for 4 months exhibit minimal changes in biological reactivity and physical integrity with respect to identical preparations frozen at −20°C [15]. This allows IgG aggregates to be used as reagent standards for the comparison and standardization of complex immunoassays.

Freeze-dried reagents can be reconstituted by rapidly injecting the diluting solution onto the vial walls or by slowly or rapidly injecting it onto the surface of the freeze-dried material. The properties of the resulting reagent solution will depend on the particular method used. Thus, rhGH preparations reconstituted using the previous three methods exhibit a varying degree of gelling [16]. The method of choice in each case will be that providing the best results with the reagent concerned.

In many cases, obtaining a stable freeze-dried reagent requires using an additive. For example, the horseradish peroxidase-conjugated monoclonal antibodies (mAb-HRP) used in diagnostic serology are severely altered by freeze-drying; however, supplying the sample with 0.3 M trehalose or a mixture containing 0.8% lactalbumin and 3.2% sucrose substantially increases the long-term stability of the dried product with minimal loss of biological activity upon reconstitution [17].

2.8.2. Standards

Freeze-drying for preparation and storage of standards has found especially widespread application in the clinical field, particularly in relation to material standards containing a large variety of analytes.

The sera used as control samples and standards in clinical chemistry should remain stable over relatively long periods of time if they are to be useful and provide reliable results.

Freeze-dried serum provides no advantage over serum frozen at −30°C in the determination of species such as glucose, urea, nitrogen, sodium, potassium, total protein and serum glutamic oxalacetic transaminase [18]. Although freeze-drying is widely assumed to alter the protein matrix of serum through changes in the ternary structure of proteins, some results appear to contradict this widespread belief. Thus, isotachophoretic bands suggest that, while some changes in ternary structure result from freeze-drying, treatments such as dialysis and the addition of chemicals or tissue extracts during the manufacturing process can also substantially alter the protein matrix. The storage conditions, the presence of proteolytic bacteria and "aging" of serum can also induce changes in protein constituents [19].

The preparation of control serum containing lipo- and apolipoproteins is specially difficult. Thus, lipoproteins in plasma cannot be stored in a liquid state for long. Also, some lipoproteins become insoluble following freeze-drying; the resulting serum is turbid and provides spurious spectrophotometric measurements. Consequently, lipoproteins are usually removed from quality control materials in order to substantially reduce their cholesterol and triglyceride contents. Most lipoproteins are affected by freeze-drying, so they cannot be related to the original lipoproteins; this hinders their determination with electrophoretic and immunoassay methods.

The denaturation of serum lipoproteins during freeze-drying can be prevented in various ways. One simple manner is by supplying the serum or control plasma with sucrose prior to freeze-drying [20]. The mechanism by which sucrose protects lipoproteins from denaturation during freeze-drying remains obscure. Possibly, the sugar displaces water from the hydration sphere during the process. The material will easily accept water during the reconstitution step, thereby facilitating reconstruction of the hydration sphere. The methods most strongly affected by the physico-chemical conditions of the target analyte are the electrophoretic determination of lipoproteins and the nephelometric determination of apolipoprotein B. With both methods, the results for the control serum cannot be distinguished from those of fresh whole serum. The ability to prepare freeze-dried human serum containing stepped levels of apolipoproteins A_1 and B for use as reference material is of great interest on account of the prominent roles of these substances as indicators of heart disease risk [20,21]. One way of increasing the thermal and long-term stability of these lipoproteins is by adding ethyl alcohol and acetic/acetate buffer to precipitate fractions enriched with them that will withstand both the passage of time (up to 3 years) and high temperatures (up to 56°C) [22]. This technology can thus be used as a preparative method for reference materials of apolipoproteins at concentrations spanning the broad ranges required by clinical applications.

Apolipoprotein E, which has a direct influence on the development of cardiovascular and neurodegenerative diseases, exhibits no change in immunoreactivity upon freeze-drying provided a bicarbonate buffer is used as the matrix [23]. As can be seen from Fig. 2.8, the immunological behaviour of various dilutions of reconstituted freeze-dried recombinant apo E_3 (one of the isoforms of apolipoprotein E) in relation to the frozen recombinant apo E_3, the human VLDL (very low density lipoprotein) purified apo E_3 and fresh human serum samples, as determined using turbidimetric and ELISA methods, is similar. The stability of freeze-dried recombinant apo E_3 is acceptable, without any major alterations of its immunological reactivity.

Freeze-dried serum certified for the determination of trace elements is used to check the accuracy and precision of analytical procedures for quantifying low levels of trace

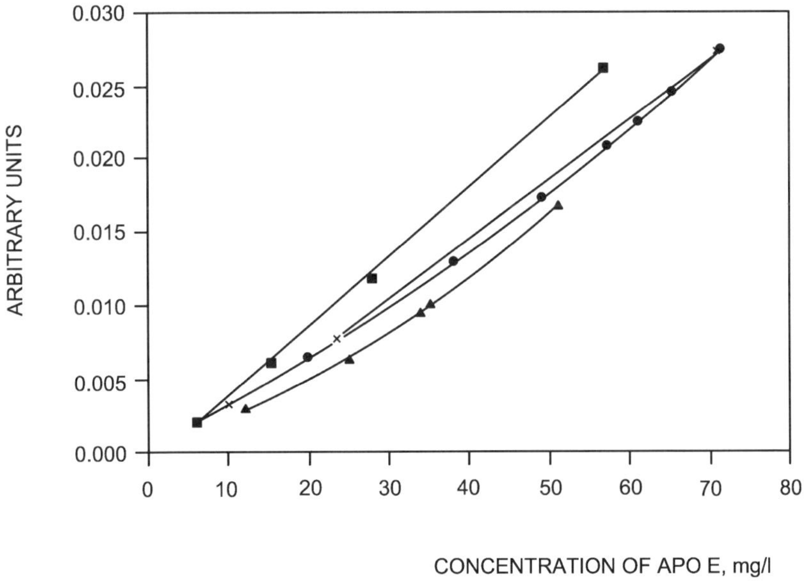

Fig. 2.8. Immunoturbidimetry. Immunological behaviour of the freeze-dried (▲) and the frozen recombinant apo E3 (■) in comparison with human VLDL purified apo E3 (×) and human fresh serum (●). (Reproduced with permission of Springer-Verlag.)

elements and to rapidly identify errors in reference materials containing higher levels of the trace elements [24].

Freeze-dried materials for urine samples are nearly as commonplace as serum samples. These materials possess long-term stability and are used for the determination of pentachlorophenol [25] and metals such as Pb [26] and Cd [27], among others. In most cases, the results for the freeze-dried reference material are indistinguishable from those for fresh urine.

2.8.3. Samples

Freeze-drying is also used as a sample pretreatment method (mainly for preservation purposes but also for preconcentration and to facilitate subsequent dissolution). Whether freeze-drying can be used with a given sample depends on the type of analyte (organic or inorganic) to be determined following reconstitution of the sample — which ultimately dictates the usefulness of the pretreatment.

Inorganic species

Traces and subtraces of inorganic compounds have been the most frequent targets of freeze-dried reagents determined with radiochemical [28–30], X-ray [31,32] and voltammetric techniques [33,34], among others. Obviously, the loss of species present at such low concentrations should be avoided at any rate. This is not always easy when, for example, samples are dried in an oven and ashed in oxygen plasma. Freeze-drying is safer for this type of sample as it avoids significant losses of trace elements in most cases.

The endorsed procedures for dehydrating biological materials can result in losses of trace elements even at relatively low temperatures. Thus, elements such as Cr, Zn and Se, which are accumulated by the oyster *Carssostrea gigas* and determined on account of the high activity of their radioisotopes, exhibit major losses relative to freeze-drying when the samples are oven-dried at 50, 90, 105 or 120°C [28]. In some cases, the magnitude of the losses depends not only on temperature but also on the type of tissue that contains the target analytes. Thus, kidney and fecal samples dried in an oven at 80°C exhibit no significant losses of Cr(III); such losses start to be appreciable at 105°C and become significant at 120°C. On the other hand, Cr(VI) losses are negligible at any of the previous temperatures [29].

One other important issue frequently ignored in the use of freeze-drying with biological samples to be analysed for trace materials is the need to physically condition biological tissues in such a way as to facilitate the subsequent destruction of organic matter. This entails optimizing the freeze-drying step in order to maximize the amount of dry residue obtained without loss or gain of the target analytes through volatilization and contamination, respectively. The variables to be optimized are the freeze-drying time, heating time and heating temperature. The determination of trace elements in bovine liver tissue by anodic stripping voltammetry (ASV) requires a prior digestion which is easier if the sample is previously freeze-dried. The three variables should be experimentally optimized before the freeze-drying process is started. Initially, the freeze-drying time is varied over a broad range and the time yielding the highest proportion of dry residue is selected for subsequent use. Then, the influence of the heating time is examined, using a constant temperature in the region of 40°C. Finally, the optimum heating temperature is determined [34].

The use of freeze-drying to preconcentrate inorganic species has so far been concerned primarily with aqueous samples. The main difficulty encountered in this context is to retain the original species. One critical operation in ensuring proper retention — particularly of volatile species — is to maintain the sample frozen throughout the freeze-drying process [35]. Freeze-drying has proved as efficient as solvent extraction and ion-exchange with chelating resins for the preconcentration of metal elements including Fe, Mn, Ni, Al, Cr, Cu and Cd in freshwater samples [36]. However, some species such as Hg may be completely lost during the freeze-drying of acid and neutral aqueous samples [37–42].

Freeze-drying, freezing replacement and cryomicrotomy are the three most frequently used techniques prior to cellular detection of labelled species (viz. complexes that diffuse across plant tissues). However, using microautoradiographs to process large batches of samples for cryomicrotomy or freezing replacement is impractical. In fact, thin frozen slides are difficult to handle with cryomicrotomy. Freezing replacement is also inconvenient as it requires maintaining anhydrous, especially cold conditions and fixing samples immediately upon processing. By contrast, it is fairly easy to freeze-dry a large number of samples in a single run and store them in an anhydrous state at room temperature for indefinitely long periods of time prior to fixing. Freeze-dried tissue is usually fixed by infiltration with a non-polar solvent such as xylene or diethyl ether, followed by alternate use of various solvents and resins in epoxy resin. One modified infiltration technique uses pressurized epoxy resin containing no solvent for long incubation periods to facilitate the processing of the mass of wood samples from cotton plants for microautoradiography; it provides good molecular retention of ^{14}C-labelled water-soluble compounds with a high reproducibility [43].

The conditioning of samples for electron microprobe analysis in studies of the soil/root interface [44] or electron microscopy research based on the use of critical drying point to prepare solid samples [45], the extraction of soluble acid phosphates from plants during fixation of the material with liquid nitrogen [46], and the comparison of the performance of thermal processes and freeze-drying as storage methods for food products prior to the determination of metal traces [47], are but a few selected applications that testify to the improvements provided by freeze-drying the sample prior to the determination of inorganic species.

Organic species

The effects of freeze-drying on various properties of organic compounds present in biological matrices and aqueous solutions have been studied by several authors. Such effects are discussed separately for plants and foods below on account of their special features.

Biological matrices Some lipid compounds behave in a contradictory manner when freeze-dried. For example, freeze-drying phospholipids with a view to their subsequent determination by ^{31}P NMR spectroscopy compares unfavourably with simply drying the tissue with acetone [48]. The sterol contents of some freeze-dried samples stored at 4°C was found to decrease within 6 weeks and to fall to only 50% after 6 months. The deviation in sterol content from unlyophilized samples relative to lyophilized ones was 2%; after 2–2.5 months, the largest deviation was 5.1%, which confirmed that freeze-drying provided no significant advantage in this case [49]. In others such as the supercritical fluid extraction of liquids, the freeze-drying technique provides higher recoveries with lyophilized samples than it does with hydromatrix or untreated samples [50].

Freeze-drying is used in a single-step process to concentrate biogenic amines and their metabolites in biological fluids containing low levels of proteins and other interfering substances. This rapid, convenient, inexpensive method, provides very good results: recoveries in the region of 89% for standard monoamines and their metabolites in artificial cerebrospinal fluid solutions following freeze-drying in the presence of glutathione and EGTA [51]. An extraction step must be conducted when concentrating plasma compounds containing large amounts of proteins and other electrochemically active materials.

One effective method for extracting drugs from urine uses freeze-drying and a subsequent liquid–liquid extraction. Urine samples are supplied with acetic acid and freeze-dried, the resulting residue being extracted and analysed by thin layer chromatography. This method provides much higher recoveries than standard extraction procedures for all types of compounds [52].

Freeze-drying in combination with resin infiltration and UV light-induced polymerization at a low temperature under vacuum is an effective alternative to the preparation of tissues for microprobe analysis. Samples can be immobilized in a non-polar resin at a low temperature, which affords infiltration and polymerization at temperatures around −50°C. Slices of the immobilized material can be dry-cut at −60°C or at room temperature. Slicing at a low temperature is an effective choice for samples that are difficult to handle at room temperature. Samples preserve their appearance to an adequate extent for mitochondrial structures, lysosomes and various types of endoplasmic reticula in liver cells to be identified. This allows significant differences in element composition among freeze-dried or cold-replaced tissues to be detected prior to immobilization. The freeze-drying technique is the most expeditious choice. Because it avoids contact with organic solvents, it decreases the risks of ion losses and redistribution. In contrast with thin freeze-dried cryosections, immobilized material at a low temperature can be sliced using fine microscopy and areas of interest containing specific cell populations selected. The initial preparative step, cryofixation, ensures efficient morphological preservation of freeze-dried, fixed tissues [53–55]. Slides that were freeze-dried prior to their microscopic examination may undergo rehydration, which can be avoided by exposing them to fixative vapours of, for example, formaldehyde or osmium tetroxide. Fixation with dry vapours preserves the integrity of samples for microanalysis and allows one to direct immunocytochemical reactions in serial slices [56].

Plant and food matrices Freeze-drying is seemingly the most suitable procedure for storing and preconcentrating plant material and foods for subsequent analysis. Some materials, however, do not benefit from the use of this technique. In some cases, simply heating the material in question is much more practical than freeze-drying it (especially with bulky samples). Nevertheless, the freeze-drying technique provides substantial advantages in many instances. Thus, while heat-drying at 71°C for 12 h is more practical than freeze-drying samples to determine total nitrogen, nitrate, soluble and insoluble nitrogen, α-amino nitrogen, nicotine, reducing sugars, acids and pH in green tobacco leaves, the latter technique is much more efficient when the metabolic process has been characterized [57]. Freeze-drying also provides good results as regards accuracy and reproducibility. Thus, the accuracy achieved in the determination of carotenes in forage and grass following freeze-drying and grinding of samples was found to be much higher than with alternative pretreatment methods: the largest oscillation was less than 8 mg/g and nearly one half of all duplicate determinations yielded the same result [58].

Freeze-drying is also highly useful as a preconcentration method. For example, using it to concentrate the sodium salts of free fatty acids in cheese following extraction with ether substantially improves the recovery of volatile acids in general and acetic acid in particular [59].

The main problem encountered in applying freeze-drying to this type of matrix is the occasional loss of species. A conventional freezing procedure will be more efficient than freeze-drying if it results in smaller losses of volatile elements — a combination of the two may be the ideal choice [60]. In some cases, however, freeze-drying is the best choice; for example, it results in the smallest losses of amino acids and amine nitrogen [61].

Whether freeze-drying is the technique of choice for a given matrix should only be decided upon after the stability and retention of the target species have been examined as the results for specific samples and compounds can rarely be extrapolated to others.

Aqueous matrices The main use of freeze-drying with aqueous matrices is preconcentrating the target analytes. The risk of species being lost during drying is also present here. Losses in this case arise from transformations in the media containing the species of interest but can be minimized by using suitable conditions such as those provided by controlled acidification [62], or by applying a chemical treatment or adding an excipient [63]. This ensures accurate results for compounds such as organophosphorus pesticides, the loss of which during freeze-drying is due to partial hydrolysis under the highly alkaline conditions provided by bicarbonate solutions [64,65]. Freeze-drying losses here can also arise from changes in the vapour pressure or sublimation pressure of the compounds concerned with temperature, as well as from variations in their heats of vaporization or sublimation. Such is the case with the global losses observed in some metabolically significant aliphatic acids present in aqueous solutions [66,67].

One potential problem to be considered in dealing with this type of matrix is that derived from the salt content and the potential presence of humic material. In fact, a high salt content considerably hinders freeze-drying; also, humic acid complexes are structurally altered — and often rendered either insoluble or difficult to dissolve — upon freeze-drying, which makes reconstituting the sample especially difficult [68].

The disparate behaviour of organic substances such as nitrosamines [69], cobalamines [70], aldehydes and ketones [71], carbohydrates [72], peptides [73], dibenzodioxins [74,75], non-specific organic matter in river water [76,77], organic pollutants [78,79] and sludge [80] reveals that, as with the previous matrices, no universal conclusions as regards the applicability of freeze-drying as a pretreatment method can be made.

Broader, more systematic studies need to be conducted with a view to unequivocally elucidating the behaviour of compound families towards the freeze-drying technique and establishing its true potential, which has so far been underexploited in analytical chemistry.

References

1 T.A. Jennings, H. Duan and J. Parenter, *Sci. Tech.*, 49 (1995) 272.
2 T.A. Jennings and J. Parenter, *Sci. Tech.*, 40 (1986) 95.
3 T.A. Jennings and Marcel Dekker, Inc., New York (1986) 595.
4 E.H. Trappler, *MD&MI IVD Technology* (1995) 20.
5 T.A. Jennings, *1st European Conference* at Wishaw, North Warwickshire, England (1995).
6 S.L. Nail and W. Johnson, *Develop. Biol. Stand.*, 74 (1990) 137.
7 D. Grief and J. Parenter, *Sci. Tech.*, 44 (1990) 118.
8 T.A. Jennings, *Lyophilization Seminar Notes* published by Phase Technologies, Inc., Conshohocken, PA (1994).
9 M.J. Pikal and S. Shan, *Develop. Biol. Stand.*, 74 (1990) 165.
10 Y. Nakaguchi, M. Morita, O. Fujino, T. Aono and K. Hiraki, *Kinki Daigaku Kenkyu Hokoku*, 22 (1986) 87.

11 C.D. Williams, *US 4,521,719 (Cl. 422–102; B01L3/00)*, (1985).
12 T.A. Jennings, *Med. Device Diagn. Ind.*, 2 (1980) 49.
13 R.C. Trivedi, H.R. Bungay III and P.J. Cuccolo, *US 4,040,786 (Cl. 23–230R; G01N33/16)*, (1977).
14 P.H. Lambert, *J. Chem. Lab. Inv.*, 1 (1978) 1.
15 D.A. McCarthy, M. Field, P. Mumford, S.R. Moore, E.J. Holborrow and R.N. Maini, *J. Immunol. Met.*, 82 (1985) 155.
16 M. Kawashige, T. Sendo, K. Otsubo, T. Aoyama and R. Oishi, *Biol. Pharm. Bull.*, 18 (1995) 1793.
17 K. Nielsen, *J. Immunoassay*, 16 (1995) 183.
18 M.D. Roland and E. Berry, *Tech. Bull. Regist. Med. Technol.*, 38 (1968) 304.
19 P.M.S. Clark, T.P. Whitehead and L.J. Kricka, *Anal. Chem. Symp. Ser.*, 6 (1981) 109.
20 H. Wieland and D. Seidel, *Clin. Chem.*, 28 (1982) 1335.
21 J. Brunzel, A. Sniderman, J. Albers and P. Kwiterovich Jr., *B. Apoproteins, Atherosclerosis*, 4 (1989) 79.
22 L.O. Henderson, J.S. Kazlehurst, L. Taylor and W.H. Hannon, *Clin. Biochem.*, 21 (1988) 219.
23 F. Schiele, A. Barbier, A. Visvikis, L. Aggerbeck, M. Rosseneu, L. Havekes, M. Huttinger, C. Profilis and G. Siest, *Fresenius J. Anal. Chem.*, 360 (1998) 501.
24 J. Versieck, L. Vanballenberghe, A. de Kesel, J. Hoste, B. Wallaeys, J. Vandenhaute, N. Baeck, H. Steyaert, A.R. Byrne and F. William Sunderman Jr., *Anal. Chim. Acta*, 204 (1988) 63.
25 R.G. Treble and T.S. Thompson, *Chemosphere*, 33 (1996) 693.
26 P.J. Parsons and W. Slavin, *Spectrochim. Acta*, 54B (1999) 853.
27 R.F.M. Herber, M. Stoeppler and D.B. Tonks, *Fresenius J. Anal. Chem.*, 338 (1990) 279.
28 H.O. Fourte and M. Peischa, *Radiochem. Radioanal. Lett.*, 26 (1976) 277.
29 G.V. Iyengar, K. Kasperek and L.E. Feinendegen, *Anal. Chim. Acta*, 138 (1982) 355.
30 Z.L. Xue and Y.X. Wang, *J. Radioanal. Nucl. Chem.*, 119 (1987) 425.
31 K. Zierold and D. Schaefer, *Verh. Dtsch. Zool. Ges.*, 80 (1987) 111.
32 F.D. Ingram and M.J. Ingram, *Sci. Biol. Specimen. Prep. Microsc. Microanal.*, Proc. Pfferkorn Con. 2nd (1984) 167, 74.
33 A. Izquierdo, M.D. Luque de Castro and M. Valcárcel, *Electroanalysis*, 6 (1994) 764.
34 A. Izquierdo, M.D. Luque de Castro and M. Valcárcel, *Electroanalysis*, 6 (1994) 894.
35 D.F. Schutz and K.K. Tureklan, *Geochim. Cosmochim. Acta*, 29 (1965) 259.
36 A. Hall and M. Casimiro Godinho, *Anal. Chim. Acta*, 113 (1980) 369.
37 M. Carison and R. Litman, *Radiochem. Radioanal. Lett.*, 35 (1978) 161.
38 V.A. Vizhenskii, E.I. Andrievskii, T.P. Molchanova, R.T. Panova, S.V. Trofimenko and V.M. Byr'ko, *Yad. Fiz. Metody Anal. Kontrole Okruzh. Sredy*, 129 (1985) 34.
39 S.H. Harrison, P.D. LaFleur and W.H. Zoller, *Anal. Chem.*, 129 (1975) 1685.
40 K.H. Lieser, W. Calmano, E. Heuss and V. Neitzert, *J. Radioanal. Chem.*, 37 (1977) 717.
41 A. Foldzinka and W. Zmijewski, *Radiochem. Radioanal. Lett.*, 27 (1976) 225.
42 X. Qian and X. Li, *Huanjing Huaxue*, 5 (1986) 60.
43 T.C. Volgelmann and R.E. Dickson, *Plant Physiol.*, 70 (1982) 606.
44 W. Shi, M. Xu and Z. Liu, *Turang Xuebao*, 24 (1987) 286.
45 R. Tippkoetter, *Z. Pflanzenermaehr. Bodenkd.*, 148 (1985) 92.
46 O.A. Sokolov, *Agrokhimiya*, 9 (1973) 133.
47 A.P. Tereshchenko, E.I. Pokrovskaya, L.A. Filatkina and O.G. Livanskaya, *Izv. Vyssh. Uchebn. Zaved., Pishch. Tekhnol.*, 6 (1977) 146.
48 P. Meneses, P.F. Para and T. Glonck, *J. Lipid Res.*, 30 (1989) 458.
49 A.A. Jekel, H.A.M.G. Vaessen and R.C. Schothorst, *Fresenius J. Anal. Chem.*, 360 (1998) 595.
50 J.M. Snyder, J.W. King and M.A. Jackson, *J. Am. Oil Chem. Soc.*, 74 (1997) 585.
51 Q. Cheng Meng, Y.F. Chen and S. Oparil, *Life Sci.*, 44 (1989) 1207.
52 J.R. Broich, D.B. Hoffman, S.J. Goldner, S. Andryauskas and J. Umberger, *J. Chromatogr.*, 63 (1971) 309.
53 K. Zierold, *Scanning Electron Microsc.*, 2 (1983) 809.
54 R. Wroblewski, J. Wroblewski, M. Anniko and L. Edstrom, *Scanning Electrom.*, 1 (1985) 447.

55 M.S. Burns, J. Chabala, P. Hallegot and R. Levi-Setti, *Microbeam Anal.*, 23 (1988) 445.
56 P.M. Frederik, P.H.H. Bomans, W.M. Busing and R. Odselius, *J. Phys. Colloq.*, C2 (1984) 451.
57 P.J. Gous, W. Kroontje and T.R. Terrill, *Tob. Sci.*, 14 (1970) 63.
58 J.T. Gillingham, *J. Assoc. Off. Anal. Chem.*, 50 (1967) 828.
59 E. Hote-Baudart, *Bull. Rech. Agrom. Gembloux*, 3 (1968) 689.
60 R.J. Norstrom and H.T. Won, *J. Assoc. Off. Anal. Chem.*, 68 (1985) 129.
61 G. Basarova and I. Cerna, *Kvasny Prum.*, 18 (1972) 78.
62 H. Bargnoux, D. Pepin, J.L. Chabard, F. Vedrine, J. Petit and J.A. Berger, *Analusis*, 6 (1978) 107.
63 D. Pepin and H. Bargnoux, *J. Français d'Hydrologie*, 9 (1978) 125.
64 H. Bargnoux, D. Pepin, J.L. Chabard, F. Vedrine, J. Petit and J.A. Berger, *Analusis*, 4 (1977) 170.
65 H. Bargnoux, D. Pepin, J.L. Chabard, F. Vedrine, J. Petit and J.A. Berger, *Analusis*, 8 (1980) 117.
66 R.A. Chalmers and R.W.E. Watts, *Analyst*, 97 (1972) 224.
67 A. Jart, *Acta Polytech. Scand. Chem. Incl. Metall. Ser.*, 121 (1974) 116.
68 D.E. Wells, *Analyst*, 123 (1998) 983.
69 L.S. Du Plessis and J.R. Nunn, *Anal. Dorm. Proc. Work. Conf.*, 55 (1951) 63.
70 R.A. Beck and J.J. Brink, *Environ. Sci. Technol.*, 10 (1976) 173.
71 L.H. Sadowski, J.C. Harris, *ADL-82480–19, EPA-600/2–84–012* (1984).
72 E.C. To and J.M. Flink, *J. Food Technol.*, 13 (1978) 551.
73 S.A. Margolis and P.J. Longenbach, *Role Pept. Neuronal Func.*, 49 (1980) 67.
74 L. Stieglitz, G. Zwick and W. Roth, *Chemosphere*, 15 (1986) 1135.
75 G. Orcharova, T. Vitanov and V. Angelova, *Izv. Durzh. Inst. Kontrol Lek. Sredstra*, 11 (1978) 49.
76 I.S. Sirotkina, N.S. Zagudaeva and G.M. Varshal, *Gidrokhim. Mater.*, 53 (1972) 147.
77 H. Kunte and E.H. Pfeiffer, *Bull. Environ. Contam. Toxicol.*, 34 (1985) 650.
78 K. Urano, S. Ch. Yao and K. Hayashi, *Bull. Chem. Soc. Jpn.*, 56 (1983) 1271.
79 M. Uno, Y. Onji, H. Nakahira and K. Tanigawa, *Nara-ken Eisei Kenkyusho Nenpo*, 18 (1983) 73.
80 I. Bauer and E. Thomanetz, *Gas-Wasserfach Wasser Abwasser*, 118 (1977) 538.

Chapter 3

Analytical uses of ultrasounds

3.1. INTRODUCTION

Ultrasound is simply sound pitched above human hearing. Ultrasounds have found a myriad of uses in many areas. At home, they are typically used for dog whistles, burglar alarms and jewellery cleaners. In hospitals, doctors use them to remove kidney stones without surgery, to treat cartilage injuries and to image fetal development during pregnancy. Ultrasonic scalpels are used by surgeons where they want to cut without exerting any pressure. In industry, ultrasounds constitute an effective tool for emulsifying cosmetics and foods, welding plastics, cutting alloys and large-scale cleaning. None of these uses, however, takes advantage of the potential effects of ultrasounds on chemical reactivity.

In 1927, Alfred Loomis first noticed the unusual chemical effects of cavitation. Dr Karl Dussik, a psychiatrist at the hospital in Bad Ischl, Austria, was the first person to report a medical use of diagnostic ultrasound in 1942; he was trying to locate brain tumours. The first application of ultrasounds to a reacting system was reported in 1946 and involved measuring the extent of polymerization in a condensation or radical process [1]. Over the following years, several other chemists explored sonochemistry but the field remained fairly quiet until the 1980s, when appropriate laboratory equipment became available.

In just a few decades of existence, sonochemistry has found applications ranging from biomedical diagnosis to materials chemistry to environmental clean-up. For example, tiny, air-filled balls have helped cardiologists around the world visualize the flow of blood through patients' hearts. Smaller than red blood cells, the balls are protein microspheres produced by ultrasonic cavitation. Upon injection into the bloodstream, the microspheres increase contrast in echocardiograms (ultrasound images of the heart) and allow cardiologists to better diagnose their patients' conditions.

Analytical chemistry is one of the chemical areas where ultrasonic radiation has been only infrequently used. This auxiliary technique could be a powerful aid in the acceleration of various steps of the analytical process.

Although analytical chemists have shown little interest in the use of ultrasound, its potential usually surpasses that of other, conventional auxiliary energies. Thus, ultrasound is of great help in the pretreatment of solid samples as it facilitates and accelerates operations such as the extraction of organic and inorganic compounds [2,3], slurry dispersion [4], homogenization [5] and various others [6–8].

After a brief description of the cavitation phenomenon and commercially available devices for the production of ultrasound, this chapter discusses its principal applications to sample treatment, with special emphasis on those involving solid samples — liquid–liquid systems are also dealt with here, however, as they also benefit from the practical advantages of ultrasounds.

3.2. GENERAL ASPECTS OF THE CAVITATION PHENOMENON

Sound waves are mechanical vibrations in a solid, liquid or gas. Ultrasound is the same, but at a frequency higher than the range audible to humans (viz. 1 Hz to 16 kHz). The lowest ultrasonic frequency is normally taken to be 20 kHz (i.e. 20 000 cycles per second). The top end of the frequency range is limited only by the ability to generate the signals; frequencies in the gigahertz range (upwards of 1 billion cycles per second) have been used in some applications.

Sound waves are intrinsically different from electromagnetic waves. Thus, while the latter (radio waves; infrared, visible or ultraviolet light; X-rays; gamma rays) can travel in vacuum, sound waves must be contained in some form of matter as they involve expansion and compression cycles travelling through a medium. Compression cycles push molecules together, whereas expansion cycles pull them apart.

In a liquid, the expansion cycle produces negative pressure that pulls molecules away from one another. If the ultrasound is strong enough, the expansion cycle can create bubbles or cavities in the liquid. This is the case when the negative pressure exerted exceeds the local tensile strength of the liquid, which varies depending on its nature and purity. The process by which vapour bubbles form, grow and undergo implosive collapse is known as "cavitation". The steps involved are illustrated in Fig. 3.1. As can be seen, the whole process takes place within about 400 μs. Normally, cavitation is a nucleated process, i.e. it occurs at pre-existing weak points in the liquid such as gas-filled crevices in suspended particulate matter or transient microbubbles from prior cavitation events. Most liquids are contaminated with sufficiently large amounts of small particles for cavitation to be readily initiated at fairly low negative pressures [9].

Sir John Thornycroft and Sydney W. Barnaby, in 1894, were the first to describe the cavitation phenomenon. During trials of a new high-speed British Navy ship, they ascribed the strong vibrations found and the erosion observed in the propeller to strong turbulence leading to the formation of cavitational bubbles. Although the source of cavitation was turbulence, high-intensity ultrasound has similar effects.

The significance of cavitation to sonochemistry is not so much how the bubbles form but rather what happens when they collapse. At some point, a bubble can no longer efficiently absorb the energy from the ultrasound so it implodes. Rapid adiabatic compression of gases and vapours within the bubbles or cavities produces extremely high temperatures and pressures. Suslick *et al.* [10] estimated the temperature of these hot spots to be about 5000°C — similar to the surface of the sun. The pressure is roughly 1000 atm, equivalent to that at the Marianas Trench — the deepest point in the ocean. The size of the bubbles is very small relative to the total liquid volume, so the heat they produce is rapidly dissipated with no appreciable change in the environmental conditions. The cooling following collapse of a cavitational bubble has been estimated to be in the region

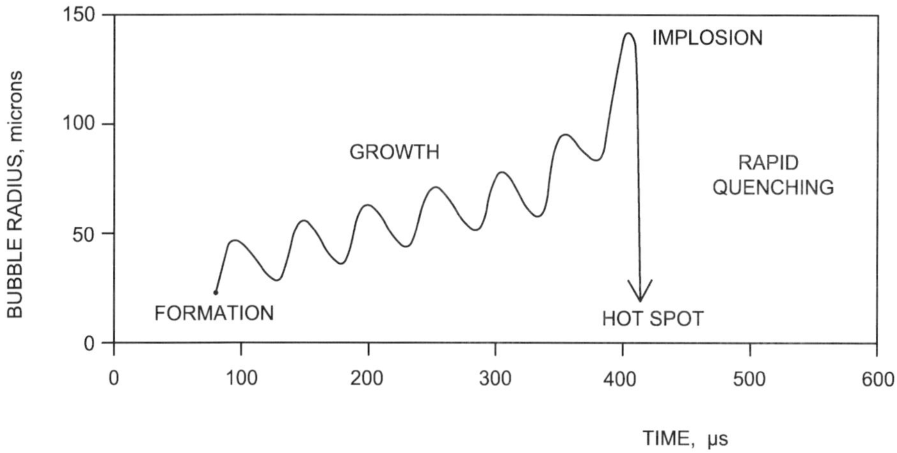

Fig. 3.1. Bubble growth and implosion in a liquid irradiated with ultrasound. (Reproduced with permission of Scientific American.)

TABLE 3.1

MAIN FREE RADICALS FORMED DURING ULTRASONIC IRRADIATION OF WATER AND THEIR REACTIONS

Argon sparge
$H_2O \rightarrow \cdot OH + \cdot H$
$\cdot OH + \cdot H \rightarrow H_2O$
$2\ \cdot OH \rightarrow H_2O_2$
$2\ \cdot H \rightarrow H_2$
$2\ \cdot OH \rightleftarrows \cdot O + H_2O$
$2\ \cdot O \rightarrow O_2$
Additional reactions in the presence of oxygen
$O_2 \rightarrow 2\ \cdot O$
$O_2 + \cdot O \rightarrow O_3$
$O_2 + \cdot H \rightarrow \cdot OOH$ (or $\cdot OH + O$)
$\cdot O + \cdot OOH \rightleftarrows \cdot OH + O_2$
$\cdot O + H_2O \rightarrow 2\ \cdot OH$
$2\ \cdot OOH \rightarrow H_2O_2 + O_2$

of 10 billion °C/s (a million times faster than that of a red-hot iron rod plunged into water). Acoustic cavitation thus provides a unique interaction of energy and matter.

The high temperatures and pressures produced lead to the formation of free radicals and other compounds; thus, the sonication of pure water causes its thermal dissociation into H atoms and OH radicals, the latter forming hydrogen peroxide by recombination [9,11,12]. Table 3.1 shows the main reactions occurring in water irradiated with ultrasounds. If the water contains some salt such as potassium iodide, iodine free radicals are also released in addition to the previous species [4].

One major phenomenon derived from ultrasonic cavitation is sonoluminescence, by which a tiny flame is formed in a cool liquid. This emission of light results from the high-temperature formation of reactive chemical species in excited electronic states. Emitted light from such states provides a spectroscopic probe for the cavitation event.

When cavitation occurs in a liquid close to a solid surface, the dynamics of cavity collapse change dramatically. In pure liquids, the cavity retains its spherical shape during collapse as its surroundings are uniform. Close to a solid boundary, however, cavity collapse is rather asymmetric and produces high-speed jets of liquid. The potential energy of the expanded bubble becomes kinetic energy for a liquid jet that moves through the bubble's inside and penetrates the opposite bubble wall. Liquid jets driving into the surface at speeds close to 400 km/h have been observed. The impact of the jets on the solid surface is very strong. This can result in serious damage to impact zones and produce newly exposed, highly reactive surfaces. Distortions of bubble collapse depend on a surface several times larger than the resonant size of the bubble.

3.3. TYPES OF ULTRASONIC DEVICES

Ultrasounds can be applied to chemical systems by using ultrasonic baths or probes. Although baths are more widely used, probes are more efficient as a result of (a) the lack of uniformity in the transmission of ultrasounds (in baths, only a small fraction of the total liquid volume in the immediate vicinity of the ultrasound source experiences the effects of cavitation) and (b) the decline in power with time, which leads to exhaustion of the energy applied to baths. Both phenomena result in substantially decreased experimental repeatability and reproducibility. For this reason, the use of baths should be restricted to cleaning operations and removal of dissolved gases, their intended applications. A wide variety of commercially available ultrasonic baths exists ranging from laboratory to industrial-scale models.

By contrast, ultrasonic probes (also called "sonotrodes") focus their energy on a specific sample zone, cavitation in which is thus enormously boosted. Also, they are subject to no exhaustion restrictions, so they are much more suitable for use in analytical chemistry than are ultrasonic baths.

One other question to be taken into account with ultrasonic probes is that they are more flexible as regards construction, which allows them to be designed for specific purposes. What effects ultrasounds have on a process depends on a number of factors including the direction, amplitude and frequency of the vibrations at the point of application, and the method of clamping the workpiece. All of these should be specified with a view to maximizing the effect on the process within the constraints of the ultrasonic system.

Having chosen what vibrations are required at the "working end", an appropriate sonotrode (ultrasonic tool) must be designed to generate them. The required movement must be a part of a mode of vibration which has a natural frequency equal to that of the rest of the system. Also, there must be a position somewhere on the sonotrode where the forcing vibrations can be applied (losses in the system and energy transfer to the work-piece will always demand that the sonotrode be driven to continue vibrating, even though it is a resonant system). At the point of application, the vibrations should be precisely normal to the surface since any out-of-axis movement of the transducer will cause it to self-destruct. For anything but the simplest system, finite element analysis will be essential to produce a satisfactory sonotrode design.

The principal components of an ultrasonic probe are as follows:

(a) A *power supply* that converts mains electrical power to the frequency, voltage and current required by the ultrasonic system. All three must be continually monitored and adjusted automatically to keep the system operating properly.
(b) A *transducer* (or *converter*) that converts electrical power into mechanical vibrations. This is a tuned system, resonant at the operating frequency.
(c) A *transducer cover*. A cover is normally fitted around the ceramic discs, electrodes and electrical connections. This protects the user from high voltages and the transducer from accidental damage.
(d) *Piezoelectric ceramic discs*, which are the heart of the transducer. As a voltage is applied, the discs expand and contract along the transducer axis. Several discs (usually 2 or 4) are used to increase the movement generated. Using an even number of discs ensures that high voltage is applied only within stack.
(e) *Electrodes*, which are used on both sides of each ceramic disc to apply the voltage that causes it to expand and contract.
(f) A *machine screw* (normally high-tensile steel) through the centre of the transducer clamps the ceramic discs together. The screw should be kept under compressive stress — even when the transducer is stretched to its maximum — in order to prevent the discs from cracking.
(g) The *front-end* of the transducer (titanium or high-strength aluminium alloy), which is used to transmit its motion to the rest of the system.
(h) A *back-block* (normally steel or titanium alloy), which provides a mass behind the ceramic discs to balance the motion of the transducer.
(i) An optional *booster* or interstage (titanium or high-strength aluminium alloy) fitted between the transducer and the ultrasonic tool. This is also resonant at the operating frequency. It alters — usually increases — the vibration amplitude and may also be used to mount all the mechanical parts of the ultrasonic system.
(k) The *sonotrode*, which is the sole part of the system that can be changed between processes. It comes in all shapes and sizes, depending on the particular application, but, like other components, should be resonant at the operating frequency. The material used to construct it (steel, stainless steel, titanium alloys, ceramics) is a compromise between the needs of ultrasounds and those of the application.

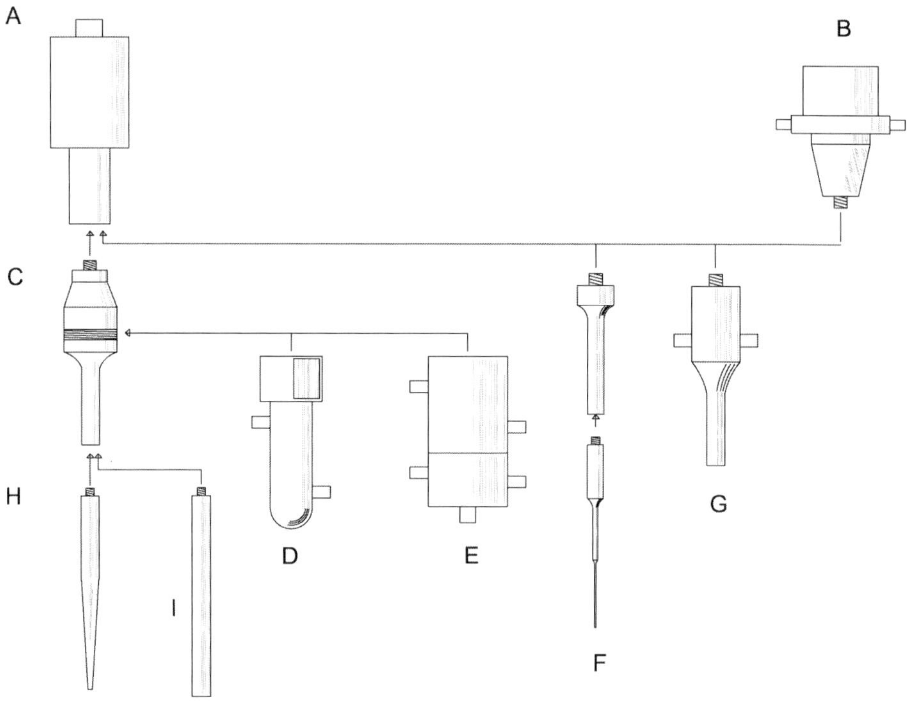

Fig. 3.2. The Branson Sonifier® II and accessories. A converter, B cup horn, C standard horn, D sealed atmosphere chamber with water jacket, E continuous-flow attachment, F special microtip with coupling section, G flow-through horn, H standard microtip, I extender. (Reproduced with permission of Branson Ultrasonics Corporation.)

Standard ultrasonication systems are commercially available in a variety of frequency and power ranges from a number of manufacturers. One salient example is the Branson Sonifier® II, which is a flexible laboratory unit available in two power levels for disintegration and homogenization. Each model consists of a power supply, converter and mechanically tuned horn. The power supply converts conventional 50 Hz line voltage to high-frequency electrical energy at 20 000 Hz. Figure 3.2 depicts some of the accessories available for this model, which range from cup horns with a water jacket to a continuous-flow attachment.

3.4. ULTRASOUND-ASSISTED LEACHING

Ultrasound-assisted leaching is an effective way of extracting a number of analytes from different types of samples. The influence of extremely high effective temperatures — which result in increased solubility and diffusivity — and pressures — which favour penetration and transport — at the interface between an aqueous or organic solution subjected to ultrasonic energy and a solid matrix, combined with the oxidative energy of radicals created during sonolysis of the solvent (hydroxyl and hydrogen peroxide for water), results in a high extractive power.

In many situations, ultrasound-assisted leaching is an expeditious, inexpensive, efficient alternative to conventional extraction techniques and, in some cases, even to supercritical fluid and microwave-assisted extraction. A number of applications to both organic and inorganic analytes in a wide variety of samples exist. Most are conducted by hand. As with automatic extractions, applications involving continuous systems are still very scant and hence one possible target for future research.

3.4.1. Batch systems

As noted earlier, most applications of ultrasound-assisted leaching involve discrete systems using a bath or an ultrasonic probe. Although ultrasonic baths are more common, ultrasonic probes have the advantage that they focus their energy on a localized sample zone, thereby providing more efficient cavitation in the liquid.

This section discusses various applications of discrete ultrasonic systems according to the organic or inorganic nature of the leached species, and to sample type.

Inorganic species

Ultrasonic extraction is an effective method for the extraction of a number of heavy metals from *environmental* and *industrial hygiene samples*. In many cases, it provides a means for quantitative recovery of metals and replacing drastic preparation procedures requiring the use of concentrated acids and the application of high heat and/or pressures (i.e. hot plate and/or microwave extraction).

The potential of ultrasonic extraction for field-based extractions has been put into use in the industrial hygiene and environmental single-element analysis of, for example, lead from glass fibre filter ambient air samples [13,14] or from lead-based paint, urban dust and river sediment [15]; hexavalent chromium from coal fly ash and paint chips [16]; and strontium from river sediment [17]. Ultrasonic extraction has also proved effective as a prior step in multi-element determinations of heavy metals.

The growing interest in acquiring information about the mobility and bioavailability of heavy metals present in environmental samples [18] has led to the widespread use of sequential extraction schemes. These provide highly relevant information about samples but involve time-consuming experimental work that includes a large number of slow, tedious steps. The use of ultrasounds has allowed some sequential extraction schemes to be expedited; such is the case with the three-stage sequential extraction procedure proposed by the European Community Bureau of Reference (BCR) for the speciation of copper, chromium, nickel, lead and zinc in sludge samples collected from urban wastewater treatment plants [19]. In this case, ultrasonic energy is an effective alternative to the conventional shaking system and substantially reduces the operation time of the BCR's sequential extraction method. The extraction process using focused ultrasound energy reduces the time for the overall sequential extraction method from 51 h to only 22 min [19]. The Tessier sequential extraction scheme [20] used for metal fractionation in sewage sludge [21] and for the speciation of particulate trace elements in river sediments [22] also benefits from ultrasonication. A comparison of the two sequential extraction schemes (viz. conventional and ultrasound-accelerated) in terms of extraction efficiency, precision, treatment time and partitioning patterns of metals revealed the efficiency of the accelerated process to depend strongly on the particular metal to be leached. Thus, while both extraction procedures provide similar fractionations for Ni and Pb in sewage sludge, the extraction of Cr from the oxidizable fraction is virtually zero with both types of sample in the ultrasound-assisted procedure. The low recovery of Cr has been ascribed to a poor extraction capability of the oxidizing reagent (H_2O_2) with the ultrasonic treatment. Also, the recovery of some metals such as Zn has been found to decrease when the sonication time is extended beyond the optimum value owing to re-adsorption of the metal on the matrix.

Ultrasonic extraction is especially efficient with environmental and industrial hygiene samples; however, in addition to the inapplicability to the extraction of some metals and the inability to quantitatively extract heavy metals from very large bulk environmental samples [14,15], ultrasounds occasionally produce ionic species that were absent from the original sample. The new species give unidentified signals that yield spurious analytical results; such is the case with the extraction of ionic species from airborne particulate matter [23], where the new ions formed prevent accurate determination of those initially present in the sample.

The determination of metals in *biological samples* is an important part of analytical chemistry as many are toxic agents; hence the need to establish a safe concentration range for each and elucidate its roles as a part of the food chain. Metals may be encapsulated within cell walls that must be broken down by the combined effects of dilute acid attack and ultrasonication in order to bring the metals into the liquid extractant. Some authors have successfully used ultrasonic extraction to isolate metals from biological samples [24–26]; quantitative extraction usually requires the use of high intensity probe sonication. However, as with environmental samples, not all metals behave identically, so maximizing the extraction yield entails optimizing the variables of the process for each specific situation. With this type of sample, the efficiency of the extraction process is essentially governed by four variables, namely: acid concentration, particle size, sonication time and amplitude. For example, in the extraction of cadmium and lead from mussels, Cd allowed the use of mild sonication conditions (viz. minimum sonication time and amplitude), whereas lead called for more stringent conditions (viz. maximum amplitude, sonication time and acid concentration, together with minimum particle size) [24]. Occasionally, the effect of temperature and the presence of matrix modifiers must also be considered in order to avoid volatilization of the analytes and minimize the effect of interferents. Such is the case with the determination of selenium by electrothermal atomic absorption spectroscopy, where an optimized temperature–time programme and a carefully chosen amount of nickel [27–30] or palladium [31–35] modifier minimize interferences arising from a phosphate-rich matrix. The leaching of selenium by diluted nitric acid as a liquid medium accelerated by ultrasounds allows the analyte to be separated from the matrix and provides a simple, rapid, relatively inexpensive "first action" method of sample pretreatment [26].

Organic species

Ultrasound-assisted leaching is an effective method for the extraction of *organic pollutants* from various types of sample. One of the principal groups of organic pollutants is that of polycyclic aromatic hydrocarbons (PAHs), which are widely distributed in the environment. These pollutants result from incomplete combustion of fossil fuels, as well as from diagenetic processes during fossil fuel formation and, in smaller amounts, from forest fires and, possibly, microbiological synthetic processes. Some PAHs possess carcinogenic and/or mutagenic properties, and are included on the US Environmental Protection Agency and European Community lists of priority pollutants. The solid samples usually analysed in connection with the determination of these compounds are soils and sediments [36–39], and also particulate matter [40–42].

Ultrasound-assisted leaching is used as a rapid, inexpensive separation step prior to the determination of PAHs in the previous types of sample. However, the efficiency with which PAHs are extracted depends largely on the particular congener; usually, the recovery of a given PAH is inversely proportional to the number of rings it contains, so obtaining results comparable with those of other extraction techniques entails careful optimization of the extraction variables. With some types of sample such as stack ash, the recovery of PAHs containing many rings additionally depends on factors such as sample composition, extraction time and the ratio of applied ultrasonic power to mass of sample extracted. Under these conditions, the recovery of large-ring PAHs such as benzo[a]pyrene

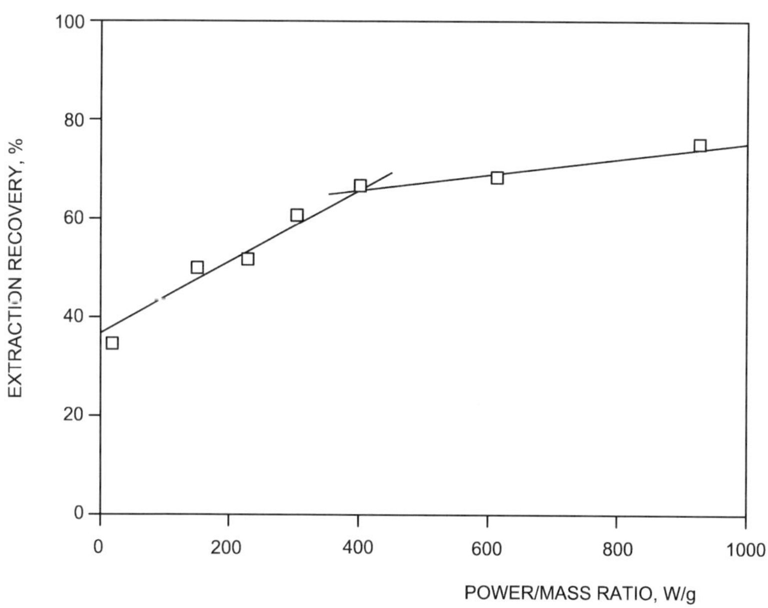

Fig. 3.3. Variation of the ^{14}C-BaP extraction recovery from stack ash as a function of the power/mass ratio. (Reproduced with permission of the American Chemical Society.)

(BaP) can be significantly improved by increasing the power/mass ratio as exemplified in Fig. 3.3 for 7,10-carbon-14-labelled BaP [40].

One other major group of pollutants for which ultrasound-assisted leaching is an effective extraction method is that of polychlorinated biphenyls (PCBs) [43–45]. Their persistence and accumulation in the environment has led to their inclusion in monitoring programme analyses. In most cases, ultrasonic extraction is recommended as a fast, efficient, straightforward choice for these compounds, particularly in routine analyses [44].

Ultrasound-assisted leaching has also been used, to a lesser extent, for the extraction of nitrophenols [46] and pesticides [47] from soil, and of alkylbenzene sulphonate from plant tissue among other materials [48], as well as for organotin speciation analysis in molluscan tissue [49]. The sonication procedure for the extraction of organotins from bivalve tissues, based on the use of non-acid reagents, allows the extraction of butyltin species with recoveries similar to those yielded by acid treatment and extraction. The use of methanol under sonication as extractant is simpler and faster than the traditional acid treatment and extraction, which, in addition, can degrade organotin species. Not always does ultrasound-assisted extraction avoid the degradation of organic compounds typical of other extraction techniques, however; in fact, ultrasounds themselves induce the degradation of some organic species. This poses a serious problem in some analytical determinations, the results of which do not accurately reflect the original composition of the sample as a consequence. Solving this problem entails performing a stability study in each specific situation as degradation arises from long exposure to ultrasounds during sample preparation [50] in some cases and from the particular solvent used in others [51]; for example, the use of CH_2Cl_2 as solvent results in the detection of various chlorinated and oxygenated derivatives of acenaphthylene [52]. Figure 3.4 depicts the set-up used to examine the stability of acenaphthylene against ultrasonic irradiation in various solvents. The titanium probe used was directly immersed in the sample (*ca.* 1 cm from the bottom of the beaker), held in a glass vessel furnished with a gas inlet and outlet to adjust the gaseous atmosphere. The vessel was closed during tests and immersed in an ice bath to maintain a constant temperature.

The degradation of some organic compounds by the action of ultrasounds under specific conditions has also found favourable uses. Thus, the use of ultrasounds for the treatment of chemical contaminants with a view to their removal has gained significance over the past few years. The principal application of this type reported so far is the sonolytic destruction of toxic substances in liquid samples (e.g. alkylphenol ethoxylate surfactants [53] and methyl *tert*-butyl ether in water [54]). The effects of ultrasounds are monitored in a continuous manner, using the dynamic approach illustrated in Fig. 3.5. The reaction volume is contained in a cylindrical, double-walled (i.e. water-cooled) vessel furnished with four sampling ports at the top that are used for venting of gases, withdrawal of aqueous samples and insertion of background gases. The presence of ozone inside the sonolytic system enhances the overall rate of destruction through thermal decomposition in cavitation bubbles, which yields $^{*}OH$ species, and facilitates direct reaction with some substrates.

The unique properties of some plastics can only be fully realized by using *additives* including stabilizers, pigments, lubricants and various others. The types of additives used and their contents influence the processing and durability of the end-product by providing

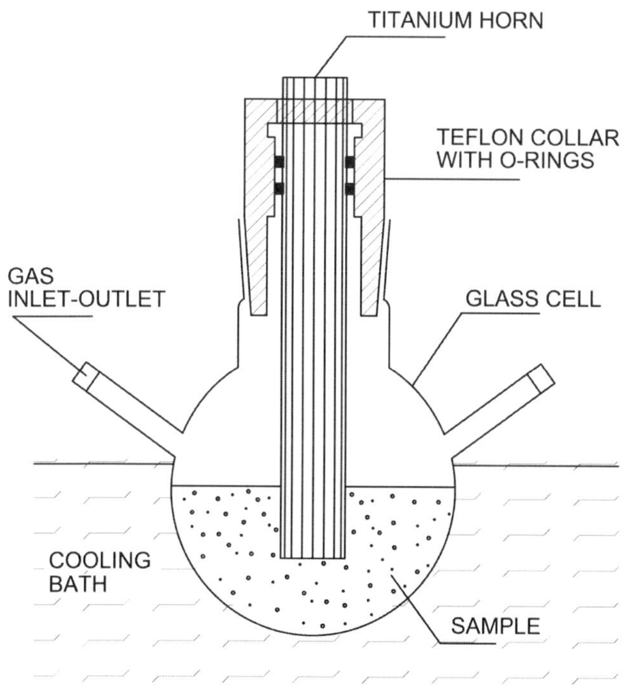

Fig. 3.4. Schematic diagram of a set-up for studying the stability of acenaphthylene under ultrasonic irradiation. (Reproduced with permission of the American Chemical Society.)

the required properties during manufacturing and use. The extraction, and subsequent separation and quantitation, of polymer additives in poly(olefins) has proven a major challenge for analytical chemists [55]. Extracting and recovering poly(olefins) additives in proportions above 90% within a reasonable time is rather difficult. One of the main problems encountered in extracting these additives is degradation upon heating. Thus, a temperature rise can cause the polymer to undergo a transition from the glassy to the rubbery form at the glass transition temperature. The ultrasonic extraction of additives from polymers provides fast, quantitative recovery with no appreciable degradation or the need for additional evaporation of the solvent or redissolution of the additive [56]. The structure of the target additives affects the extraction kinetics, which entails careful selection of the extraction temperature and time in each situation.

Ultrasound-assisted leaching has also been used to extract *natural compounds* such as vitamins A, D and E from feeds [57], paclitaxel and related taxoids from leaf tissue of *Taxus* [58], opiates from hair samples [59] and antioxidants from rosemary [60]. Ultrasounds have so far had much more restricted application in this field than in the previous ones, possibly as a result of the technique being at a disadvantage with respect to alternatives such as microwave-assisted extraction [57] or supercritical CO_2 extraction [60].

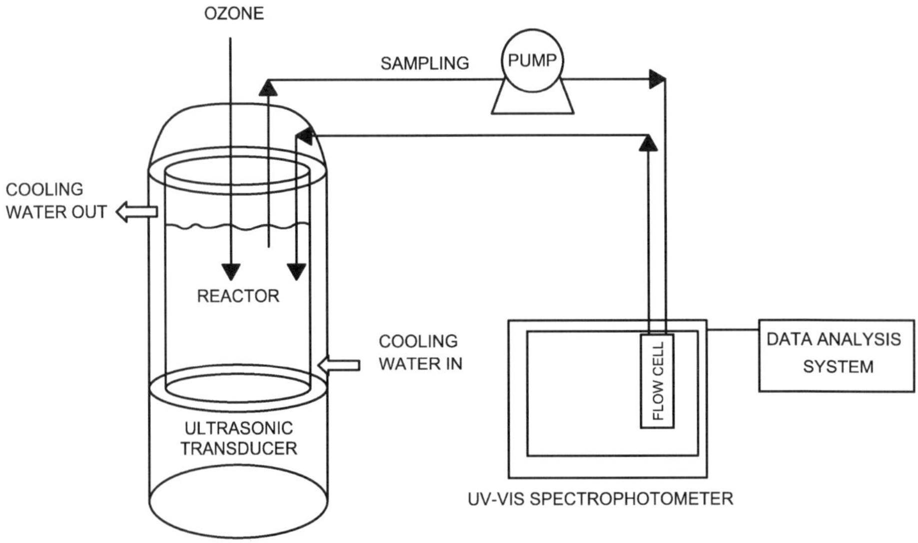

Fig. 3.5. Schematic diagram of a sonochemical reaction system for the sonolytic destruction of methyl *tert*-butyl ether. (Reproduced with permission of the American Chemical Society.)

3.4.2. Continuous systems

As stated above, the use of ultrasound as auxiliary energy for leaching has been widely exploited for the extraction of a variety of analytes from different types of solid samples. While ultrasonic irradiation has been advantageously used over other methods to accelerate manual leaching of solid samples, this type of energy has scarcely been employed for pretreatment in continuous systems, even though it expedites the sampling process considerably.

The *sample container* to be used for ultrasound-assisted continuous leaching should allow both passage of the leacher and tight closing to hinder leakage of the solution — which is favoured by the action of ultrasounds. Two types of sample container have so far been used in this context. The one-piece container of Fig. 3.6A consists of a small chamber with inlet and outlet orifices furnished with connectors for placing the container in-line with the dynamic manifold. The insertion of solid samples into the chamber is highly time-consuming owing to its small size. The dismountable container of Fig. 3.6B is similar to the one-piece device but consists of two halves, the lower of which is used to weigh the sample. Following weighing, the two parts are tightly joined together by screws that also act as connectors. Both types of cell can be made either in stainless steel, or methacrylate, the last material having the advantage that it allows one to visualize the inside of the cell. After connection to the dynamic manifold, the cell containing the sample is plunged into a bath of transmitting liquid and the tip of the ultrasonic probe is placed above the cell, at an appropriate, optimized distance. When a programmable peristaltic pump is used to move the leacher go forward and back through the sample,

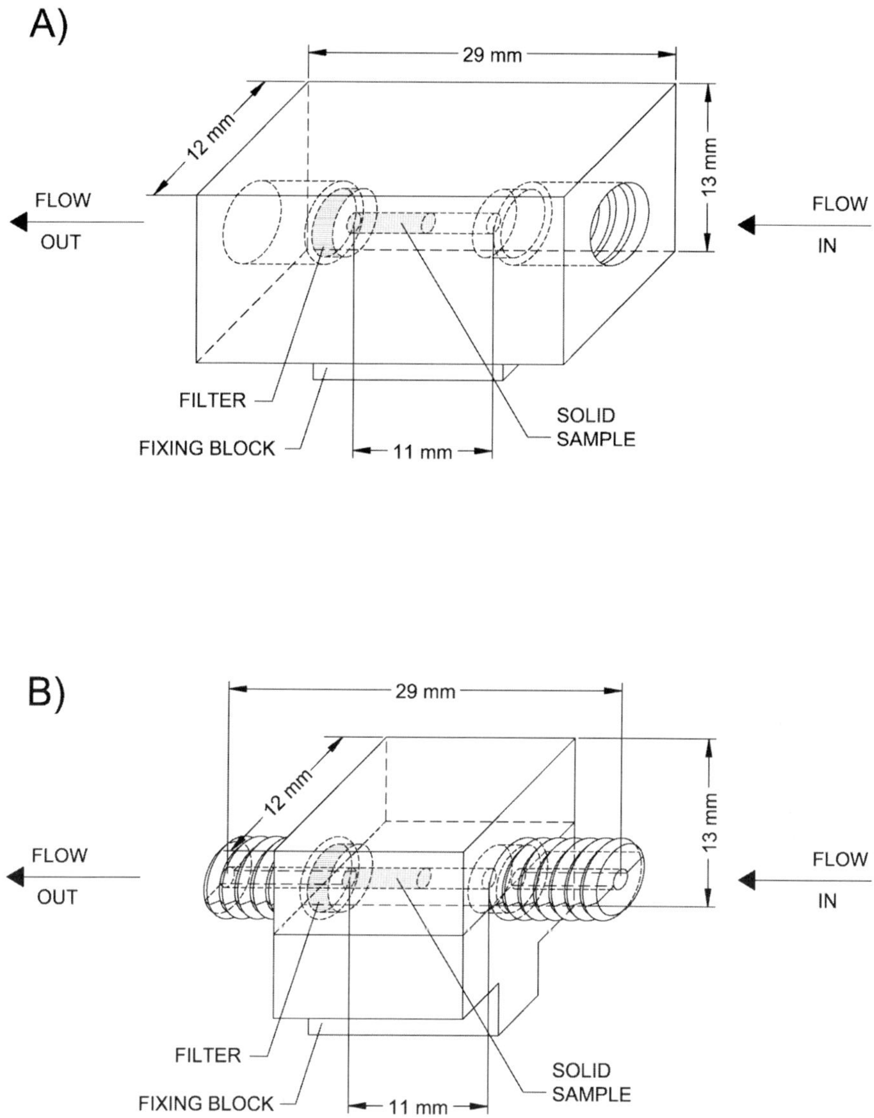

Fig. 3.6. Sample containers for continuous ultrasound-assisted leaching. (A) One-piece container. (B) Dismountable container. (Reproduced with permission of Elsevier.)

each end of the container is fitted with an appropriate filter. This dynamic approach avoids undesirable compaction in the container and hence pressure changes in the flow system.

There are three different *dynamic approaches* to ultrasound-assisted leaching in continuous-flow systems, namely:

(a) The most simple, based on the manifold of Fig. 3.7A, can only be used when the leached species respond to a specific type of detector placed in-line with the close circuit. The leacher is circulated in one direction, the leaching step can be continuously monitored and a rising signal is obtained, the slope of which reflects the kinetics of the leaching process. Attainment of the extraction equilibrium is signalled by a plateau in the detector response, after which SV in the figure is switched to its open position and the extract driven to waste.

(b) The manifold of Fig. 3.7B is used when the leached species must be derivatized in order to obtain a detector response. In this case, the leacher is recycled as in the previous approach and, after an optimal leaching interval, SV is switched to its open position and the carrier drives the circuit contents to a point of merging with a reagent stream, the derivatization reaction taking place between this point and the detector. Valve SV is in fact an injection valve in whose loop the recycling system is mounted; it operates as such in the filling position (Fig. 3.7B1). The contents of this macroloop are injected into the carrier solution in the derivatization–detection step (Fig. 3.7B2).

(c) When the derivatization reaction is required after leaching, the dynamic manifold can be simplified by programming the functioning of the peristaltic pump in such a way that the leacher is driven go forward and back through the sample container at preset intervals. After a number of cycles, the extract is merged with the derivatization reagent for reaction and detection as in the previous approach.

When necessary, a debubbler is connected in-line with the closed circuit to remove occluded air in the solid, which can give rise to parasitic signals at the detector.

The variables to be optimized in a continuous ultrasound-assisted leaching approach can be classified into two groups, namely: those concerning the relative position of the probe and sample cell and their surroundings, and those related to the sonicator settings.

The variables concerned with the relative position of the probe and sample cell and their surroundings are the depth of the probe in the transmitting liquid, the probe–sample cell distance, the size and shape of the reservoir holding the liquid transmitter and the nature of the liquid. The depth of the probe in the liquid also containing the sample cell is considered because if the probe is only slightly immersed, it causes foaming at the liquid surface, which in turn results in the loss of ultrasonic energy; on the other hand,

Fig. 3.7. Approaches to implementing continuous ultrasound-assisted leaching. (A) With direct monitoring of the process. (B) and (C) With a derivatization step before detection. (B1) With leaching in the recirculation mode. (C1) With leaching in the go forward and back mode. (B2) and (C2) Derivatization–detection steps. P peristaltic pump, UP ultrasound probe, SS solid sample, DB debubbler, D detector, SV switching valve, LC leaching carrier, R reagent, W waste. (Reproduced with permission of Elsevier.)

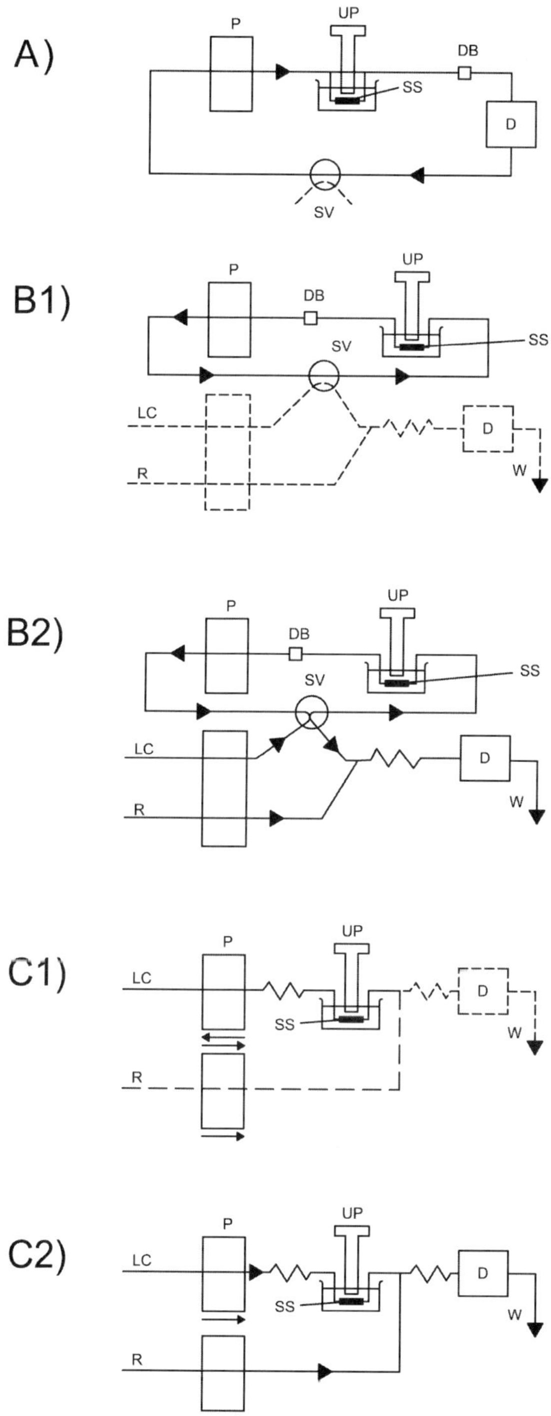
A)
P
UP
DB
SS
D
SV
B1)
P
UP
DB
SV
SS
LC
R
D
W
B2)
P
UP
DB
SV
SS
LC
R
D
W
C1)
P
UP
LC
SS
D
W
R
C2)
P
UP
LC
SS
D
W
R

TABLE 3.2

COMPARISON OF THE DETERMINATION OF IRON BY AAS AND PHOTOMETRY FOLLOWING COMPLEXATION WITH *o*-PHENANTHROLINE, WITH AND WITHOUT THE AID OF ULTRASONIC IRRADIATION

Iron found, µg/g			
Sample	AAS	Photometry	
		Without ultrasounds	With ultrasounds
Beech leaves	463.0 ± 3.1	281.2 ± 6.2	472.5 ± 3.6
Spruce needles	127.4 ± 2.9	96.1 ± 5.1	134.2 ± 2.1

TABLE 3.3

COMPARISON OF CONTINUOUS ULTRASONIC-ASSISTED LEACHING AND MANUAL LEACHING FOR THE EXTRACTION OF BORON FROM SOIL SAMPLES *

Sample	Boron found, µg/g	
	Ultrasonic leaching	Manual leaching
A	12.86	13.37
B	34.29	33.90
C	9.40	9.03
D	14.55	12.88
E	7.01	7.96
F	17.73	17.39
G	17.92	15.42

* The leaching time was reduced from 40 to 5 min in the presence of ultrasounds

if the probe is immersed to an excessive depth, the energy supplied is not adequately transmitted through the liquid and the efficiency of the leaching solution suffers. The distance between the probe and the sample cell, and the nature of the transmitting liquid, are two other influential variables. Small volumes make direct impingement of the ultrasonic probe on the sample microcell redundant. With large volumes, the efficiency of ultrasounds decreases relative to that obtained with small reservoirs and the same probe–cell distance. This is a result of the ultrasonic energy dispersing in the bulk solution.

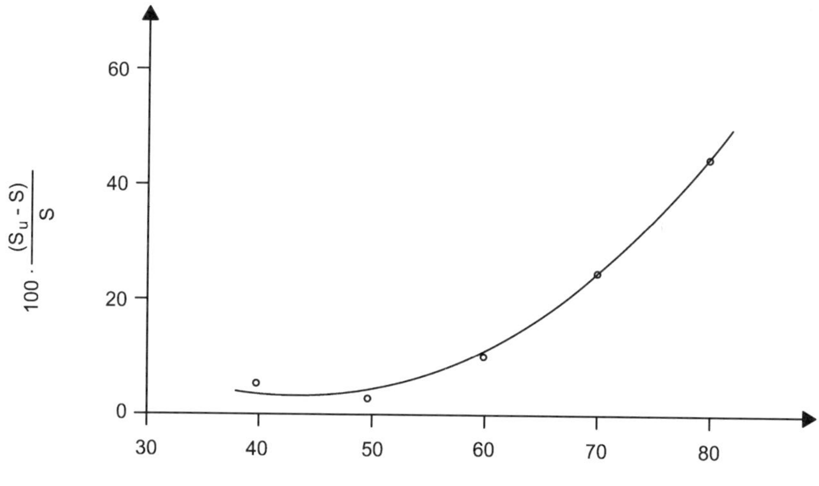

Fig. 3.8. Effect of temperature on the leaching of boron in soil in the presence (Su) and absence (S) of ultrasounds. For details, see text. (Reproduced with permission of Elsevier.)

Usually, the sonicator system allows adjustment of the sonication interval, the radiation amplitude and its continuous or intermittent functioning. The optimum values of these three variables depend directly on the particular analyte to be leached and the type of matrix that contains it.

The variables involved in the reaction step, if there is one, and the detection step, are those typical of flow-injection work (viz. the composition of the carrier and reagent streams, and their flow-rates).

The usefulness of continuous ultrasound-assisted leaching has been demonstrated with applications such as the determination of iron in plant material [61] and that of boron in soil [62]. In the former case, formation of the well-known complex between the analyte and 1,10-phenanthroline provided results consistent with those of the conventional method using atomic absorption spectrometry following hot-acid treatment (see Table 3.2). The determination of boron in soil was based on the use of azomethine-H as derivatizing reagent. The leaching time was reduced from 40 to 5 min by applying ultrasonic irradiation while the leaching solution was passed through the solid, with similar results as shown in Table 3.3. The temperature was found to have a crucial influence on the leaching step. The effect of temperatures between 40 and 80°C in the presence and absence of ultrasound, using 0.1 M HCl as leacher, is shown in Fig. 3.8. As can be seen, the dissolving power increased with increasing temperature, whether or not ultrasounds were used; however, the effect was more obvious when the ultrasonic probe was employed. The analytical signal obtained after derivatization increased by almost 50% from 40 to 80°C with the ultrasonic treatment. Clearly, increased temperatures favour dissolution of boron and increase the efficiency of ultrasounds. The latter effect can be easily understood by considering that elevated temperatures increase the distance between molecules in both solids and liquids, thereby decreasing attenuation of ultrasonic irradiation and strengthening its effect at the interface between the circulating liquid and the static sample.

The principal advantages of continuous ultrasound-assisted leaching are modest consumption of sample and reagents, the need for few or none of the chemicals required for dissolution in the manual method and the expeditiousness of analyses (about 25 soil samples per hour can be processed in the determination of boron).

3.4.3. Ultrasound-assisted leaching versus other leaching techniques

Ultrasound-assisted extraction has rapidly evolved from a novel extraction technique [63] to a part of various reference methods that have superseded previous ones based on alternative extraction techniques [44,64]. Below are discussed the advantages and disadvantages of ultrasound-assisted leaching in relation to the more common leaching alternatives.

Ultrasound-assisted extraction versus conventional Soxhlet extraction

Methods based on Soxhlet extraction have traditionally been used as references to assess the performance of methods based on other principles (ultrasound included).

The most salient advantages of ultrasound-assisted leaching over conventional Soxhlet extraction are as follows:

(a) The cavitation phenomenon increases the polarity of the system (extractants, analytes and matrix included); this increases the extraction efficiency, which can be similar to [63] or greater than [43] that provided by conventional Soxhlet extraction.
(b) Ultrasound-assisted leaching allows the addition of a co-extractant to further increase the polarity of the liquid phase [65,66].
(c) It also allows the leaching of thermolabile analytes, which are altered under the working conditions of Soxhlet extraction.
(d) The operating time is invariably shorter than with Soxhlet extraction [67].

However, ultrasound-assisted leaching is at a disadvantage with respect to Soxhlet extraction in the following respects:

(a) In batch methods, which are the most widely used, the solvent cannot be renewed during the process, so its efficiency is a function of its partitioning coefficient. The need for filtering and rinsing after extraction lengthens the overall duration of the process and increases solvent consumption and the risk of losses and/or contamination of the extract during handling.
(b) Soxhlet extraction is usually more reproducible than both ultrasound-assisted modes.

Ultrasound-assisted leaching versus supercritical fluid extraction

The ultrasound-based technique surpasses supercritical fluid extraction in respects such as the following:

(a) The equipment needed is much more simple, so the overall cost of the leaching process is much lower.
(b) Ultrasound-assisted leaching allows the extraction of a wide variety compounds, whatever their polarity, as it can be used with any solvent. On the other hand, supercritical fluid extraction uses virtually exclusively CO_2 as extractant (with or without the presence of co-solvents), so its scope is restricted to non-polar analytes.

On the other hand, supercritical fluid extraction (SFE) surpasses ultrasound-assisted leaching in the following respects:

(a) The SFE technique is more simple and expeditious than are some liquid solvent sonication methods [60].
(b) Unlike the solvents typically used for sonication (e.g. cyclohexane, tetrahydrofuran and binary mixtures such as that of dichloromethane and acetone) [64], supercritical CO_2 is not environmentally hazardous.
(c) Usually, SFE methods are more precise than their ultrasound-assisted counterparts [44].

Ultrasound-assisted extraction versus microwave-assisted leaching

Ultrasound-assisted extraction provides several interesting advantages over microwave-assisted leaching, namely:

(a) Ultrasound-assisted extraction is occasionally faster than microwave-assisted leaching [57].
(b) In acid digestions, the ultrasonic procedure is safer as it requires no high pressure or temperature [68].
(c) In many cases, the whole procedure is simpler as it involves fewer operations and is thus less prone to contamination.

On the other hand, ultrasound-assisted extraction has the following shortcomings relative to microwave-assisted extraction:

(a) Particle size is a critical factor in ultrasound-assisted applications [68].
(b) Ultrasound-assisted procedures are usually less robust than microwave-assisted ones since, as suggested by Cencic-Kobda and Marcel [45], ageing of the ultrasonic probe surface can alter the extraction efficiency.

3.5. ULTRASOUND-ASSISTED SAMPLING

Ultrasounds have been used to assist in the insertion of samples into solid–liquid analytical systems for some time. The earliest ultrasonic nebulizers and automatic slurry sampling systems were reported in the 1980s. However, the actual potential of ultrasounds for increasing the efficiency of sampling systems remains to be explored. Thus, a recently developed technique based on acoustic levitation has been found to substantially improve microanalytical results. This section deals with the two most widely used types of ultrasound-based sample insertion systems and discusses the most salient features of the ultrasonic levitation technique, which is especially suitable for liquid samples but also applicable to solid ones.

3.5.1. Ultrasonic nebulizers

Direct nebulization of an aqueous or organic phase containing extracted analytes has been widely used in flame atomic absorption spectroscopy [69–72], inductively coupled plasma atomic emission spectrometry [73–76], microwave induced plasma atomic emission spectrometry [77–80] and atomic fluorescence spectrometry [81], as well as to interface a separation step to a spectrometric detection [82–85].

Ultrasonic nebulizers for plasma-based spectrometry

The pneumatic nebulizer has for many years been the most universal sample insertion device for plasma-based spectrometry. The inherent lack of transport efficiency, coupled with the continuing need for increased sensitivity, has promoted research into the use of ultrasonic nebulizers to boost detection capabilities. Such research has focused on various aspects including fundamental aerosol properties [86–88], instrument development [89], nebulizer comparisons [90,91], desolvation effects [92,93], direct nebulization applications [94,95] and speciation [96].

Ultrasonic nebulization in combination with a desolvation system has been found to improve the limits of detection provided by conventional pneumatic systems [80]. When a desolvation tube is used, its temperature is one crucial operating parameter for the ultrasonic nebulizer. Figure 3.9 illustrates the dependence of the ion signal on the desolvation tube temperature in the determination of selenium compounds in urine. If the temperature of the desolvation tube is too high, the water and the organic solvent evaporate completely and the aerosol particulate may be too small to be carried by the gas flow, thereby decreasing the transport efficiency. If the tube temperature is too low, too much water and organic vapour may be introduced and the plasma cooled as a result [83]. The use of a membrane desolvator substantially improves the removal of water and volatile organic solvents. Solvent vapour permeates through a heated, porous PTFE membrane and is removed to vent by a flow of argon sweep gas. Figure 3.10 shows the modular equipment U-6000AT+, from CETAC Technologies, which includes a membrane desolvation system of this type.

The insertion of volatile organic solvents and metal organic complexes into an inductively coupled plasma (ICP) is made difficult by overloading and pyrolysis effects. These include carbon build-up in the torch and spectral interferences. These shortcomings have restricted the use of solvent extraction as a method for preconcentrating trace metals for their ICP-based determination but can be circumvented by using an ultrasonic nebulizer–separation membrane interface for the direct insertion and separation of organic solvents using ICP atomic emission spectrometry or ICP mass spectrometry. Usually, the amount of aerosol produced and that of analyte transferred across the membrane are

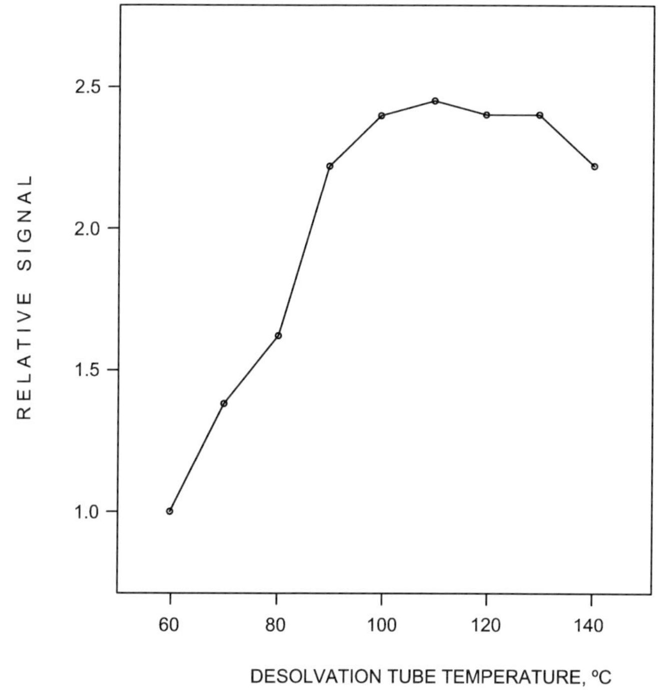

Fig. 3.9. Effect of the desolvation tube temperature on ion signals. (Reproduced with permission of Elsevier.)

greater than those provided by a pneumatic nebulizer. This results in improved limits of detection relative to pneumatic devices [97].

In many cases, coupling a separation technique to a spectrometric one to facilitate insertion of the sample provides a number of advantages. However, the problem arises that transfer of the analyte across the liquid chromatograph/ICP interface is low. This problem can be minimized by using a microscale separation technique. However, coupling a separation procedure to an inductively coupled plasma technique raises many instrumental problems. Thus, the very small size of the sample exacerbates the already marginal response of the system and the low liquid flow-rates inherent in separation techniques make direct coupling to conventional ultrasonic nebulizers difficult. Microflow ultrasonic nebulizers may solve these problems as they can operate in a steady manner with total liquid flow-rates as low as 5 μl/min and no need for a make-up solvent. The system of Fig. 3.11 allows direct insertion of the sample and coupling of techniques such as capillary electrophoresis and micro-liquid chromatography to inductively coupled plasma atomic emission spectrometry, as well as to inductively coupled plasma mass spectrometry [98].

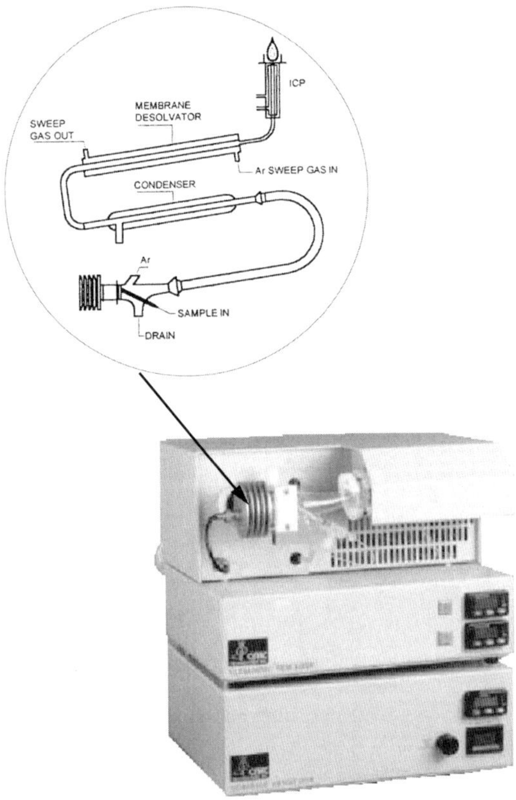

Fig. 3.10. CETAC U-6000AT+ ultrasonic nebulizer/membrane desolvator. (Reproduced with permission of CETAC Technologies.)

Ultrasound-assisted electrospray formation

Electrospray ionization has rapidly emerged as a promising technique for the interfacing of liquid chromatography and capillary electrophoresis to mass spectrometry. Originally developed in Fenn's laboratory at Yale, this interface has the ability to transfer compounds such as proteins, peptides, nucleic acids, carbohydrates and other delicate or thermally labile species from the liquid phase to the gas phase while ionizing them as well. The electrospray technique has been successfully used with virtually all types of mass spectrometer including quadrupole, magnetic, Fourier transform, ion trap and time-of-flight models. The ion spray mode, which uses a high-velocity nebulizing gas in the same direction as the liquid flow, is an effective alternative to liquid nebulization.

Exciting though the potential of electrospray ionization may seem, the use of electrosprays for liquid chromatography/mass spectrometry has been limited in practice by the severe restrictions on solvent composition and volumetric flow-rate that are amenable

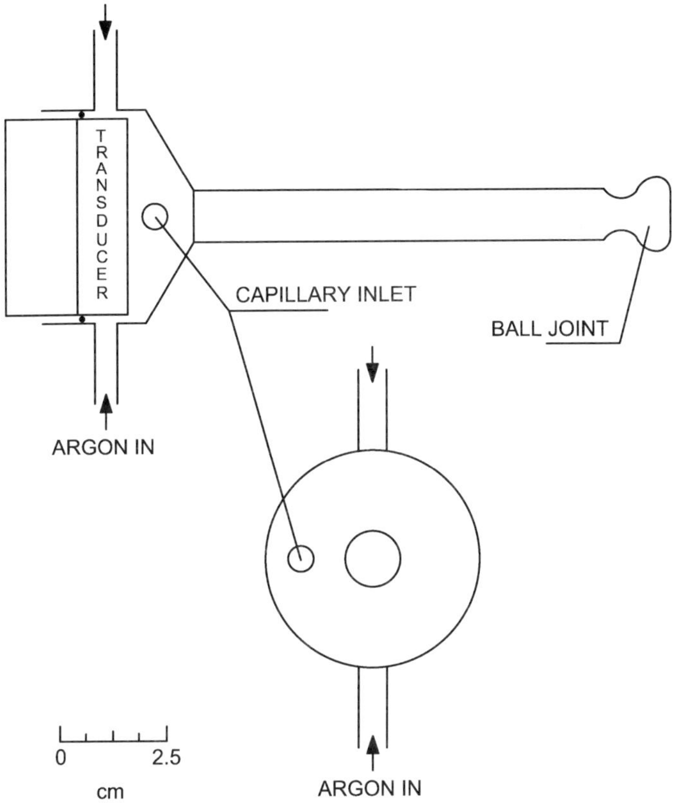

Fig. 3.11. Microflow ultrasound nebulizer spray chamber design. All glass-made. (Reproduced with permission of the American Chemical Society.)

to the electrospraying process. The incompatibility between the liquid chromatography mobile phase and the electrosprayable solution composition imposes stringent limits on the types of liquid chromatography applications that can be used with electrospray ionization and is the single most significant hindrance at this time, restricting wider acceptance of the technique. Electrospray ionization cannot be used with the following types of chromatographic mobile phases: (a) those whose volumetric flow-rate exceeds 5 μl/min; (b) those that possess a high conductivity; and (c) those exhibiting a high surface tension.

The previous restrictions are not trivial and affect most liquid chromatography applications of current interest. The flow-rate restrictions make this technique amenable to capillary liquid chromatography — which uses rates of a few microlitres per minute — only. Alternatively, some type of post-column flow splitting can be used in conjunction with larger columns. The conductivity restriction precludes the use of ion-exchange and reversed-phase liquid chromatography with highly conductive mobile modifiers. Finally, the surface restriction excludes the use of most aqueous mobile phases (e.g. those used for the separation of polar compounds on reversed-phase columns) [84].

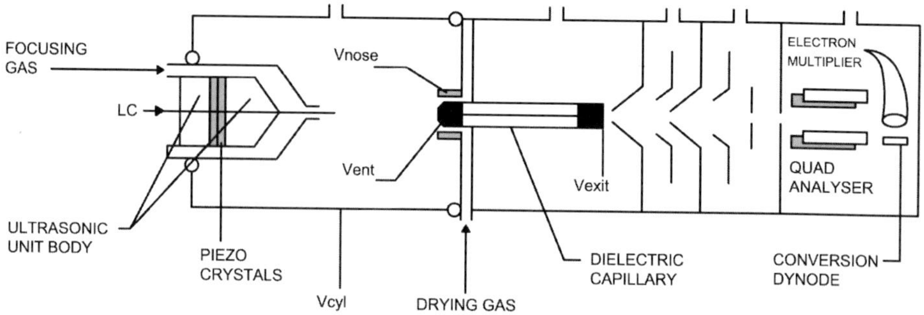

Fig. 3.12. Schematic diagram of electrospray source with ultrasonic nebulizer. (Reproduced with permission of the American Chemical Society.)

In order to improve the efficiency of the electrospray technique, researchers have sought a means to decouple as much as possible the spray formation and ionization processes so as to facilitate production of the fine spray of droplets required for successful ion evaporation in the source. The primary requirement here is that spray formation should be relatively independent of the physical properties of the chromatographic mobile phase and its flow-rate. This challenging goal suggested that the spray should be produced by some mechanical means (e.g. ultrasonic vibration instead of application of an electric field [99]). Figure 3.12 depicts an electrospray source equipped with an ultrasonic nebulizer. This device is capable of producing the fine spray of droplets required for electrospray ionization by purely mechanical means, which is less dependent on the influence of applied electrical fields than is unassisted electrospray ionization. For these reasons, liquid chromatography mobile phases, which are normally not amenable to electrospraying, can be used with liquid chromatography/electrospray ionization/mass spectrometry, thus extending the range of suitable liquid chromatography mobile phase compositions to include high flow-rates, high electrical conductivities and high-surface tension solutions. The bulk of liquid chromatography applications that are currently impossible or very difficult to accomplish with conventional unassisted electrospray ionization is thus encompassed.

3.5.2. Ultrasonic slurry sampling

Graphite-furnace atomic absorption spectrometry (GF–AAS) has proved a useful method for the direct analysis of solids [100]. Solid sampling has a number of advantages over conventional sample preparation procedures such as acid digestion, namely: (a) decreased sample preparation times; (b) a decreased risk of analyte losses through volatilization or other phenomena prior to analysis; (c) reduced analyte losses relative to retention by an insoluble matrix in the case of leaching; (d) decreased likelihood of contamination by reagents, containers, etc.; (e) increased sensitivity resulting from the need for no dilution; (f) avoidance of hazardous acid, toxic or environmentally unfriendly solvents; and (g) the ability to selectively analyse microamounts of solids. GF–AAS suffers from none of the particle size effects on nebulization techniques [101]. Although solids contain particles in variable sizes, which suggests problems arising also from variable atomization efficiencies, the use of integrated absorbances with GF–AAS, which extends residence times, ensures accurate determination of trace metals in solids. Among the most common problems with direct solid sampling are the difficulty of inserting samples automatically in order to avoid the poor precision obtained with manual sampling and those arising from the lack of calibration standards.

Slurry sampling has been deemed a highly suitable approach to solid sampling that incorporates the benefits of solid–liquid sampling. The practical advantages derived include (a) the ability to use an autosampler for unattended operation; (b) that to process several replicates from a single aliquot and hence to better establish the precision of the method; (c) that to handle increased masses of sample to improve representativeness; (d) that to alter the slurry concentration simply by changing the solid–liquid ratio to fit the analytical signal within the best interval of the calibration curve; (e) better analytical performance by virtue of the slurry technique combining the benefits of solid and liquid sampling; and (f) improved repeatability and reproducibility [102]. Ultrasonic probes for homogenizing slurries provide additional advantages such as increased throughput, the ability to automate slurry sample preparation and insertion into the furnace, improved reproducibility and analyte extraction, and increased representativeness of the aliquot assayed. In 1988 [103], the concept of manual ultrasonic mixing of slurries was introduced and in 1989 [104] an automated ultrasonic mixing accessory for slurry GF–AAS was reported. The development by Perkin–Elmer of the USS-100 Ultrasonic Slurry Sampler, which is based on this technology [105,106], has been followed by a series of successful applications of this approach [107–109]. Figure 3.13 depicts the USS-800 Ultrasonic Slurry Sampler, one of the latest, improved versions of the original apparatus. This model is fully automated and can be coupled to a suitable atomic absorption detector by using a software command to change the autosampler operation from conventional liquid sampling to slurry sampling. The power level and the mixing time can be programmed for each specific application. The ultrasonic probe is made of solid titanium for corrosion resistance and is automatically flushed between samples to reduce carry-over and contamination. The operational sequence of the apparatus is as follows: the USS-800 probe and autosampler capillary are rinsed, and the capillary is moved to a stand-by position. The mixing probe is then inserted into the appropriate sample cup. The probe ultrasonically mixes the sample for the user-specific time. After mixing, the ultrasonic probe moves quickly out of the cup. A fraction of a second later, before the slurry particles have the time to settle, the autosampler capillary takes an aliquot of the slurry from the cup and transfers it directly to the furnace. When the graphite furnace analysis is complete, the cycle is repeated for the next sample or replicate.

Two particularly important considerations in ultrasonic slurry GF–AAS are that method development and optimization can be based on conventional liquid sample protocols and that, frequently, only limited changes are needed to facilitate direct insertion of solids. The recommended overall procedure for ultrasonic slurry GF–AAS determinations [101] involves six steps that are conducted in the following sequence:

(1) Grinding the sample to produce a powder. Particle sizes up to 300–500 μm may be acceptable. The grinding modes used for this purpose include mortar grinding, Teflon

Fig. 3.13. The USS-800 Ultrasonic Slurry Sampler. (Reproduced with permission of Perkin–Elmer Corporation.)

beads and polyethylene bottles (for biological and botanical samples); grinding, pulverizing, planetary, centrifugal or jar mills; and cryogenic grinding. Stainless steel containers can introduce contamination, so their use should be avoided.

(2) Planning to optimize slurry preparation. The slurries must have analyte concentrations that are appropriate for the analyte line selected. The factors of interest include homogeneity of the solid, distribution of the analyte in the solid, density, particle size and analyte partitioning in the slurry. If the analyte distribution in the solid is heterogeneous, one must strive for very small (< 10 μm) particles. The minimum mass required for analysis based on particle size and density should be computed. The "volume-to-volume" ratio (solid volume/liquid volume ratio) should be computed in order to ensure that it is lower than 0.25.

(3) Preparing slurries for analysis. Microweighing should be done on an electronic microbalance. An amount of 1–50 mg of ground material can be weighed directly in a Teflon autosampler cup and supplied with 1.0 ml of diluent (5% sub-boiling distilled HNO_3 containing 0.004% Triton X-100).

(4) Establishing the analytical conditions. Wavelength selection will depend on analyte concentrations. Less sensitive non-resonance lines may be useful. Quantitation is then accomplished by using aqueous standards for peak area measurements. The use of a matrix modifier and a char step may be unnecessary.

(5) Ultrasonic slurry mixing. The power output to the ultrasonic probe should be adjusted to provide good mixing (typically 40–80%). The usual mixing time is in the region of 20–25 s.

(6) Establishing the number of determinations. Typically, 5 readings of each slurry preparation are adequate when analysing an unknown sample. The data thus obtained should be reviewed to see if the determined concentrations suggest a dependence on sample weight or heterogeneity.

Ultrasonic slurry sampling is also discussed in Chapter 8 alongside other modes of solid sample insertion into atomic spectrometers.

3.5.3. Acoustically levitated droplets

One problem frequently encountered in microanalyses arises from contact of the sample with container walls. This may lead to alterations of the sample composition through adsorption of the analyte at walls, or desorption of either analyte or interfering substances from them. The best way of minimizing these alterations is by avoiding contact between the sample and container walls, at least during some steps of the analytical process (particularly during sample pretreatment). One way of handling small samples without contacting container walls is by using acoustic levitation [110].

In contrast to most levitation techniques such as levitation in electrostatic or magnetic fields, acoustic levitation requires no specific properties of the sample, so almost every substance, whether solid or liquid, can be acoustically levitated. The maximum possible diameter for a levitated sample is a function of the ultrasonic wavelength and turns out to be about one-half the wavelength under ambient atmospheric conditions. Usually, levitators are operated by ultrasound frequencies between 15 and 100 kHz resulting in wavelengths from 2.2 to 0.34 cm.

If the wavelength of the ultrasound exceeds 8 mm, the maximum volume of a liquid drop is no longer dictated by the above-described ratio but by that between hydrostatic pressure and the capillary pressure due to surface tension inside the drop. If this ratio, which is referred to as the Bond number of the levitated drop, exceeds 1.5, the drop disintegrates. Therefore, the maximum volume of a drop is a function of the surface tension and specific density of the liquid. For example, the maximum volume of a water drop is 155 μl; for most organic solvents, it is about 40 μl [111].

A number of techniques have been developed for contactless sample positioning [112]. Table 3.4 summarizes their analytically most relevant features [113–123].

Easy positioning of fluids with reproducible volumes is an essential requirement for experiments with levitated drops. The most effective way of achieving it is by using a gas chromatography syringe. For better detachment of the water drop from the syringe, it is advisable to coat the needle with a hydrophobic substance such as paraffin or silicon. One other way to position samples in the nodal points of a standing ultrasonic wave is to let them lift off from a small (in the case of water) hydrophobic coated wire loop by the attractive force of the nodal point. Accurate positioning of samples (particularly liquid samples) requires some practice and would thus benefit from automation.

In the absence of an appropriate technique for measuring analyte contents directly in the drop, this must be transferred to a detection device. There are two different ways of transferring samples from an acoustic levitator to an analytical detector, namely: by transferring the whole sample to an integral detector (e.g. a graphite-furnace atomic absorption

TABLE 3.4

COMPARISON OF SELECTED LEVITATION METHODS IN TERMS OF SUITABILITY FOR INCLUSION IN MICROANALYTICAL PROCESSES

Method	Levitating force	Levitable volume	Gradients in the levitating force	Samples	Applications	References
Optical levitation	Radiation force of an intense light source (e.g. a laser)	A few millilitres	Yes (vertical) and horizontal)	Transparent, slightly absorbing samples may be accelerated towards the light source	Positioning of fusion targets, laser forceps	113–115
Electrostatic levitation	Electrostatic fields	A few nanolitres to 100 μl	No	Capable of carrying electrostatic charge	Investigations in single drops	116–118
Aerodynamic levitation	Aerodynamic force of a vertical gas stream	A few nanolitres to 100 μl	Depending on design of the device	Almost any type. Difficulties may arise with liquids of low viscosity and/or surface tension	Atmospheric chemistry experiments	112
Acoustic levitation	Acoustic radiation pressure of an ultrasound source	A few nanolitres to 100 μl	Yes (vertical and horizontal)	Almost any type. Difficulties may arise with liquids of low viscosity and/or surface tension	Micro and trace analysis. Investigation of physical properties	111,112 119–123

spectrometer) or by transferring only an aliquot. Using a sample aliquot for analysis is more difficult because evaporation introduces some uncertainty in the drop volume.

Samples can be removed from the levitator with the same instruments that are used to position them in the nodes of the standing ultrasonic wave (viz. syringes, microlitre pipettes and glass capillaries). Microlitre pipettes are the most efficient for removing whole samples.

The efficiency with which substances from the surrounding atmosphere are absorbed and distributed in the drop depends on the degree of mixing within the drop. Movement inside the drop can be observed in two ways, namely: by suspending solids (e.g. aluminium oxide particles) in the levitated drop and by spiking the drop with a highly concentrated dye solution. In both cases, the speed and the pattern of the distribution inside the drop can be visually observed.

One experimental system for the determination of analytes in ultrasonically levitated samples is an automated set-up that can be used in conjunction with spectrometric detection using the method of standard additions and/or microtitration with spectroscopic detection of the end-point. Although the potential of this combined approach remains to be explored, acoustic levitation appears to be a highly promising choice for microanalyses [124].

3.6. ULTRASOUND-ASSISTED ELECTROANALYSIS

The coupling of power ultrasound with well-characterized electrochemical experiments has opened the door to a new field of sensitive, reliable, diversely applicable electroanalytical techniques [125]. This combination of technologies has allowed the realization of the promise of such techniques, which have so far been relatively scarcely used owing to passivation, reliability and sensitivity problems [126].

This section describes the principal uses of ultrasounds in relation to voltammetric and, to a lesser extent, polarographic [127] and potentiometric techniques [128]. Although most applications of ultrasounds in combination with voltammetric techniques have so far involved liquid samples, they are discussed in this chapter on account of the growing interest they are arousing among analytical chemists.

Electrochemical stripping analysis is an extremely sensitive analytical technique that has traditionally allowed the determination of a wide range of species at concentrations down to 10^{-10} M under favourable conditions, all with fairly straightforward, inexpensive equipment. Anodic stripping voltammetry (ASV), adsorptive stripping voltammetry (AdSV) and, to a lesser extent, cathodic stripping voltammetry (CSV), have all been used to varying degrees for the detection of a wide range of organic and inorganic species including Pb, Sn, Cu, Zn, Cd, Bi, Sb, Ga, Mn, cardiac and anticancer drugs, vitamins and pesticides. All three stripping techniques involve two distinct steps. In the initial, preconcentration step, the target analyte is accumulated from solution onto a suitable working electrode substrate. Depending on the nature of the analyte, this can be accomplished either by electrochemical deposition or by physical adsorption. The potential of the working electrode is then adjusted in such a way as to release the analyte from the electrode, with the resulting Faradaic current allowing direct quantitation of the amount of analyte present in the sample. A number of uses of the three stripping techniques in this context have been reported. However, each of these classical techniques has its drawbacks. Thus, ASV is commonly plagued by the formation of intermetallic species, which can complicate its analytical integrity. On the other hand, AdSV can be very slow, with the time required for the accumulation step having a strong effect on sensitivity. Finally, CSV has historically been the most technically difficult of the three stripping techniques and has deterred many workers in favour of ASV and AdSV.

Mercury thin film electrodes (MTFEs) are well-established electrode materials for ASV. These electrodes provide a high surface area-to-volume ratio and, as a result, better sensitivity and resolution of neighbouring signals than those obtained with a mercury drop electrode. Glassy carbon is commonly used as substrate for MTFEs by virtue of its advantages, which include ease of preparation, high sensitivity, a wide potential window and low interaction with the mercury. However, the weak adherence to mercury often results in poor mechanical properties. Other electrode materials such as platinum, copper and silver as substrates give rise to mercury amalgams, the resulting formation of intermetallic compounds with analyte metals deposited onto the film restricting their use. Environmental and health issues concerning mercury and its toxicity have prompted the development of mercury-free solid electrodes [129,130], especially for on-line monitoring of foodstuff, clinical and field tests.

One problem frequently encountered in ASV is deactivation of the electrode surface. All classical voltammetric techniques are generically hindered by electrode fouling, which occurs in analytical applications — particularly with microelectrodes. Many real-world analytical environments encountered outside the laboratory contain surface-active substances that can passivate electrodes; as little as monolayer coverage often suffices to perturb the path of electrode processes. The presence of organic substances in the sample matrix can cause peak depression or even complete fouling of the electrode. Thus, the electrochemical determination of copper ion in beer is rendered impossible by passivation of the electrode by adsorbed organic species. Samples containing proteins, soluble polymers, sugars and blood plasma, for example, also tend to cause electrode fouling. The result is low sensitivity and poor reproducibility. As a consequence, the sample must often be mineralized to avoid these interferences. The treatment involves acidification or UV irradiation in the presence of an oxidant. This complicates and lengthens the analytical protocol, and introduces contamination risks. One way of preventing electrode fouling is by electrochemically activating or modifying the electrode. More efficient, however, is the use of ultrasound-assisted electrochemical methods.

The use of ultrasounds in electrochemistry has provided a powerful electroanalytical tool. Thus, ultrasound-enhanced voltammetry possesses some attractive features such as the following:

(a) Ultrasonic irradiation ensures very high mass transport of electroactive species and their products to and from the electrode surface [131,132].
(b) Ultrasounds can alter the mechanisms of chemical and electrochemical reactions via the action of highly reactive radicals such as $^{*}OH$ and $^{*}H$ formed during the sonolysis of water [133,134].

(c) The adsorption of species involved in an electrochemical reaction can be reduced by the use of ultrasounds [135].
(d) Ultrasonic irradiation facilitates in situ cleaning [136,137] and activation of the electrode surface [138,139].

Among the variety of experimental approaches to introducing ultrasounds into an electrochemical cell, direct-immersion sonicator probes are highly practicable. A suitable cell configuration contains the working electrode inserted from the bottom and placed directly below the tip of a vibrating titanium horn. One of the most attractive features of this experimental configuration is the extremely enhanced mass transport resulting from the high ultrasonic power intensity in the vicinity of the electrode. In the configuration of Fig. 3.14, ultrasound acts not only by facilitating mass transfer, but also by allowing coupling of a leaching step in the electrolytic cell itself. The figure illustrates the extraction of Pb from waste cathode ray TV tubes (CRTs). Aqueous Pb^{2+} ions leached from the CRT glass accumulate on a boron-doped diamond electrode at a potential of +0.27 V versus saturated calomel and form PbO_2. Lead(II) ions in the dilute nitric acid leachate constitute the ideal form for electrochemical recovery of metal lead at an appropriate cathode [140].

One of the most serious problems posed by ultrasound-enhanced electrochemical systems using a mercury film electrode as working electrode is the need to preserve its structural integrity. One way of ensuring stability of the electrode is by depositing the mercury in situ via a film of the perfluorosulphonated ionomer Nafion, which provides a reliable electrode assembly that is resistant to ultrasonic irradiation. For practical applications, the Nafion coating has the additional advantage of reducing interferences from complex matrices. This approach is especially attractive for direct study of ultrasound-enhanced ASV towards biological and environmental samples by using other effects of ultrasounds such as rapid cell disruption and homogenization of the sample material [135,141].

As electrochemical stripping analysis has developed, new techniques have been introduced with a view to overcoming the problems of lengthy deposition times and improving sensitivity.

Rotating disc electrodes are used to decrease the time required for the accumulation step in stripping analysis. Extremely high rates of mass transport can be achieved by using this type of electrode in solutions subjected to ultrasonic irradiation, which allows the observation of steady-state currents at scan rates as high as 25 V/s [131]. Figure 3.15 shows an experimental device used to couple ultrasonic irradiation to a rotary disc electrode. The electrochemical cell was constructed in such a way that the surface of the working electrode was held securely through a threaded bushing directly below and parallel to the titanium amplifying horn and was therefore positioned in the most intense region of the cavitational plume. This configuration provides stability for the reference electrode and prevents detection at the working electrode of products formed at the auxiliary electrode.

Microelectrodes (viz. electrodes with dimensions in or below the micrometer range) are also employed in this context on account of their ability to increase mass transport;

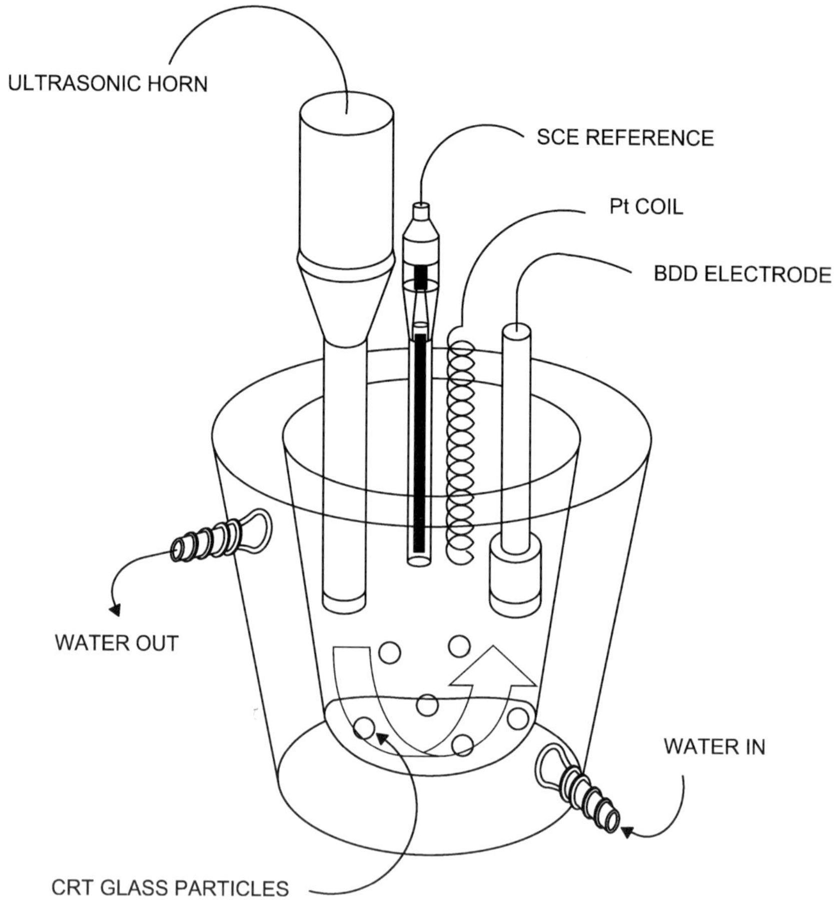

Fig. 3.14. Schematic diagram of a jacketed glass cell used for sonoleaching and sonoelectroanalysis of lead in cathode ray TV tubes. BDD boron-doped diamond, CRT cathode ray TV tubes, SCE saturated calomel electrode. (Reproduced with permission of the Royal Society of Chemistry.)

steady-state currents can thus be observed as a result of radial diffusion. Mass transfer can be boosted by subjecting the microelectrode to ultrasonic irradiation. The effect can be ascribed to two transient processes, namely: (a) bubble collapse at or near the solid–liquid interface and the consequent formation of a high-speed liquid microjet directed at the electrode surface; and (b) bubble motion near or within the electrode's diffusion layer [142].

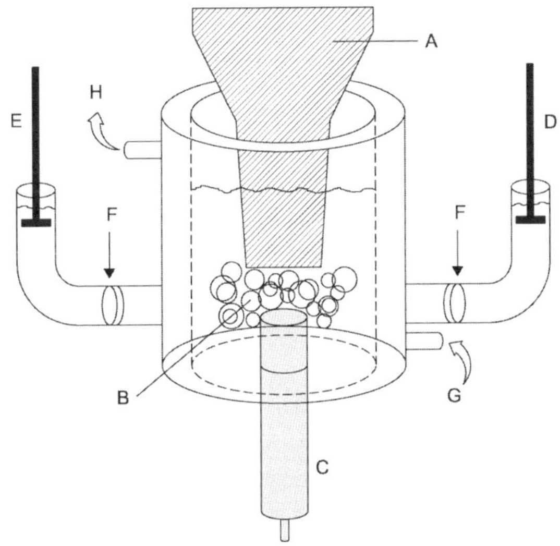

Fig. 3.15. Sonoelectrochemical cell used to ultrasonicate a rotating disc electrode. A titanium amplifying horn, B cavitational plume, C Pt disc working electrode, D Ag/AgCl reference electrode, E auxiliary electrode, F fine porosity glass frit, G coolant inlet, H coolant outlet. (Reproduced with permission of the American Chemical Society.)

3.7. OTHER ANALYTICAL USES OF ULTRASOUNDS

This section completes the discussion of the uses of ultrasounds in connection with solid–liquid systems and electroanalysis dealt with in the previous ones with brief comments on the most salient applications of ultrasonic energy to liquid–liquid systems.

3.7.1. Ultrasonic degassing

Occasionally, a dissolved gas must be removed from a liquid solution (i.e. the solution must be degassed). Degassing is also a natural consequence of the application of high-intensity ultrasounds to a solution. Cavitation has been called "cold boiling" because dissolved gases are extracted at room temperature in a vacuum system at nearly water vapour pressure while ultrasonic energy is applied to the sample. Figure 3.16 depicts one experimental system used to degas water that allows the gas removed to be collected for subsequent analysis by gas chromatography. The degassing line consists of a sample bottle (A) that is connected via a valve to a gas burette (B). An evacuated gas container (C) is attached on top of the gas burette. Gas for on-board gas chromatography analysis is sampled by a syringe via a silicon septum (D). At the lower end of the sample bottle (A), the water sample is connected by a glass tube and a valve via a second glass bottle (E, water reservoir) to the vacuum system (protected by a water trap bottle, F).

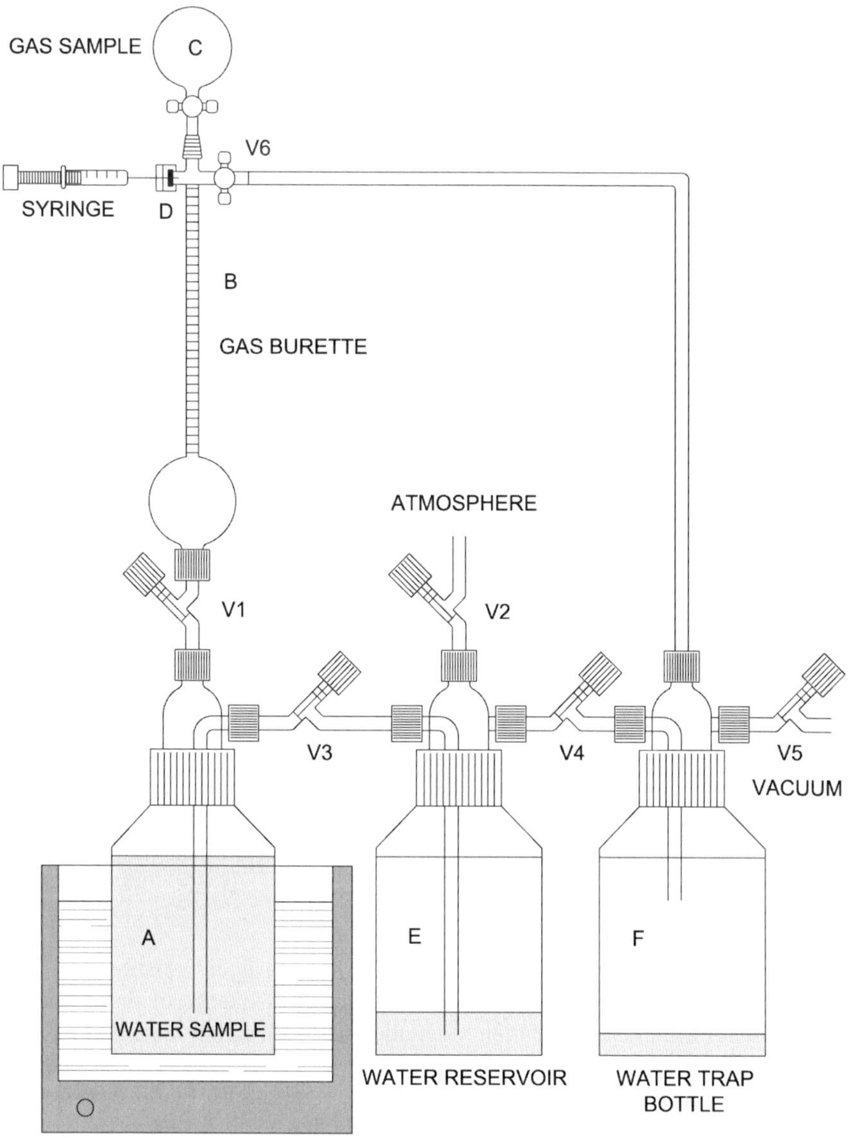

Fig. 3.16. Ultrasonic vacuum degassing line used for the extraction of methane from sea water (for details, see text). (Reproduced with permission of the American Chemical Society.)

For sampling, bottle A is disconnected from the line, evacuated and, via V1, filled up with water. For analysis, bottle A is attached to the burette. Some water is pumped off into bottle E until bottle A contains approximately 1 litre of water. The vacuum of the head space in bottle A at this time is close to the water vapour pressure. Ultrasonic energy is applied over a period of 5 min. Once gas extraction is finished, the vacuum-degassed water in bottle B is added to the water sample in bottle A. Via V1, the water level rises into the burette until atmospheric pressure is reached. The volume of gas can be read on the burette scale [143].

3.7.2. Ultrasonic filtration

An on-line sampling device is an important part of any analytical system for process analysis. A number of techniques and devices have been developed and tested in the past for on-line sampling. Most such devices are membrane-based and suffer from drawbacks including short lifetimes, and high mechanical complexity and cost. There is also the problem of membrane fouling, which reduces the efficiency through blocking of the pores during operation. However, ultrasonic devices provide an excellent dynamic solution to these problems for on-line sampling, and also an efficient method for obtaining clarified samples for analysis.

Particles (either solids or cells) suspended in an ultrasound standing wave can be rapidly aggregated and separated in the pressure nodes located at half-wavelength intervals along the wave [144]. Recently, this idea was used to design a filter for the separation of particles from various media by combining ultrasound aggregation and gravity sedimentation. Such ultrasound filtration provides a new technology for use in on-line sampling. The fouling problems associated with other techniques are avoided because ultrasound separation does not rely on a physical barrier to particle separation; this makes the devices ideal for use in a process environment. The system is very robust, with no moving parts, and can be stream-sterilized (which makes it highly suitable for fermentation and clinical applications).

Only three types of ultrasound filters have so far been designed, for as many types of samples, that can be coupled to continuous analytical systems [145]. Each type of filter uses an ultrasound resonance field generated by a solid-state transducer to cause particle aggregation without any physical filtering surface. The filter for mammalian cell culture samples (Fig. 3.17A) is built into a system for the periodic removal of samples from a fermenter. This is accomplished without breaching the sterile integrity of the vessel, thereby allowing samples to be taken for on-line monitoring. When the sampling procedure is started, the medium containing the cells is pulled into the chamber. The ultrasound waves force the cells to align in vertical bands, where they aggregate and then settle back into the fermenter under gravity. The clarified medium is then pumped into the injection port through the sample loop for on-line analysis.

Clarification of whole blood in the clinical application is achieved by using a batch step (Fig. 3.17B). The sample tube is held central to the transducer by a dip in the centre of the base. The tubular transducer is fixed to the top section of the filter. With the tube in the filter, the blood level is set to coincide with the top of the transducer. A pulsing

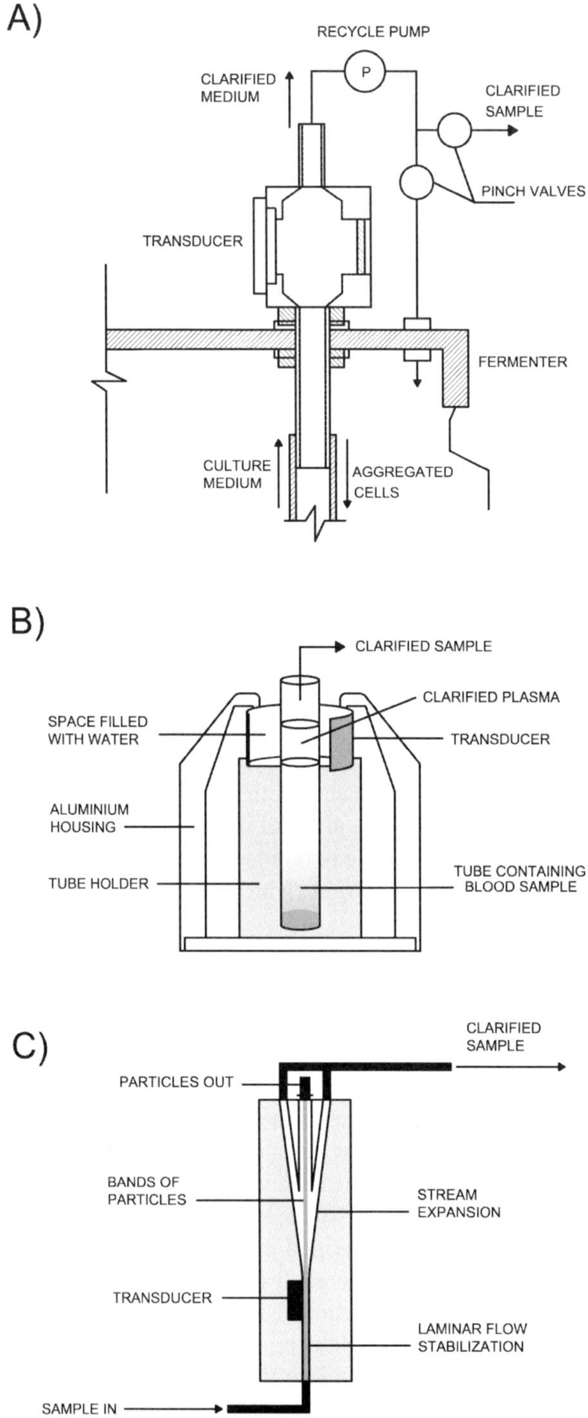
A)
RECYCLE PUMP
P
CLARIFIED MEDIUM
CLARIFIED SAMPLE
PINCH VALVES
TRANSDUCER
FERMENTER
CULTURE MEDIUM
AGGREGATED CELLS
B)
CLARIFIED SAMPLE
CLARIFIED PLASMA
SPACE FILLED WITH WATER
TRANSDUCER
ALUMINIUM HOUSING
TUBE HOLDER
TUBE CONTAINING BLOOD SAMPLE
C)
CLARIFIED SAMPLE
PARTICLES OUT
BANDS OF PARTICLES
STREAM EXPANSION
TRANSDUCER
LAMINAR FLOW STABILIZATION
SAMPLE IN

regime is developed for optimized separation containing ten cycles with "on/off" periods. The overall time needed for the filtration process is 5 min 40 s. The first ten cycles accumulate blood cells in bands formed by the ultrasonic field. The next 15 cycles encourage sedimentation of the cells in the bottom of the tube. During the "off" period, sonication is maintained at a very low level, giving the effect of a break in sonication of the cells.

Clarification of surface water samples is achieved by using a laminar flow filtration chamber (Fig. 3.17C). The concentrated particles and clarified water are removed through separate outlets. The filter operates vertically, with a continuous flow of water sample being fed upward from the base of the system into the ultrasonic field. Particles concentrate along the central axis of the chamber and allow the clarified material to be drawn out of the filter at the edges.

References

1 J. Workman, D.J. Veltkamp, S. Doherty, B.B. Anderson, K.E. Creasy, M. Koch, J.F. Tatera, A.L. Robinson and L. Bond, *Anal. Chem.*, 71 (1999) 121R.
2 J.L. Gómez-Ariza, E. Morales, R. Beltrán, I. Giráldez and M. Ruiz-Benítez, *Analyst*, 120 (1995) 1171.
3 J. Mierzwa, Y.C. Sun and M.H. Yang, *Anal. Chim. Acta*, 355 (1997) 277.
4 G. Wibetoe, D.T. Takuwa, W. Lund and G. Sawula, *Fresenius J. Anal. Chem.*, 363 (1999) 46.
5 K. Vinodgopal, J. Peller, O. Makogon and P.V. Kamat, *Water Res.*, 32 (1998) 3646.
6 B.W. Pack, S.J. Ray, R.A. Potyrailo and G.M. Hieftje, *Appl. Spectrosc.*, 52 (1998) 1515.
7 P.L. Buldini, A. Mevoli and J.L. Sharma, *Talanta*, 47 (1998) 203.
8 B. Fransson and U. Ragnarsson, *J. Chromatogr.*, 827 (1998) 31.
9 K.S. Suslick, *Scient. Amer.*, Feb (1989) 80.
10 K.S. Suslick, *The Year Book of Science and Future, Encyclopedia Britannica*: Chicago (1994) 138.
11 L.R.F. Carvalho, S.R. Souza, B.S. Martinis and M. Korn, *Anal. Chim. Acta*, 317 (1995) 171.
12 C.A. Wakerford, R. Blackburn, P.D. Lickiss, *Ultrason. Sonochem.*, 6 (1999) 141.
13 S.J. Long, J.C. Suggs and J.F. Walling, *J. Air Pollut. Control Assoc.*, 29 (1979) 28.
14 *US EPA, 40CFR Part 50: Reference method for the determination of lead in suspended particulate matter collected from ambient air (Appendix G)*; Federal Register 44, p. 564, US Government Printing Office, Washington, DC (1979).
15 K. Ashley, *Electroanalysis*, 7 (1995) 1189.
16 J. Wang, K. Ashley, E.R. Kennedy and C. Neumeister, *Analyst*, 122 (1997) 1307.
17 M. Akcay, A Elik and S. Savasci, *Analyst*, 114 (1989) 1079.
18 G. Rauret, R. Rubio, J.F. López-Sánchez and E. Casassas, *Int. J. Environ. Anal. Chem.*, 35 (1989) 89.
19 B. Pérez-Cid, I. Lavilla and C. Bendicho, *Anal. Chim. Acta*, 360 (1998) 35.
20 A. Tessier, P.G.C. Campbell and M. Bisson, *Anal. Chem.*, 51 (1979) 844.
21 B. Pérez-Cid, I. Lavilla and C. Bendicho, *Fresenius J. Anal. Chem.*, 363 (1999) 667.
22 B. Pérez-Cid, I. Lavilla and C. Bendicho, *Environ. Anal. Chem.*, 73 (1999) 79.
23 L.R.F. Carvalho, S.R. Souza, B.S. Martinis and M. Korn, *Anal. Chim. Acta*, 317 (1995) 171.

Fig. 3.17. Cross-sectional views of ultrasound filters for mammalian cell cultures (A), separation of whole blood (B) and use with ground water (C). (Reproduced with permission of Elsevier.)

24 I. Lavilla, J.L. Capelo and C. Bendicho, *Fresenius J. Anal. Chem.*, 363 (1999) 283.
25 E.A. Chmilenko and L.V. Baklanova, *J. Anal. Chem.*, 53 (1998) 784.
26 J. Mierzwa, S.B. Adáåoju and H.S. Dhindsa, *Anal. Sci.*, 13 (1997) 189.
27 R.D. Ediger, *At. Absorption Newslett.*, 14 (1975) 127.
28 K. Julshamn, O. Ringdal, K.E. Slinning and O.R. Braekkan, *Spectrochim. Acta*, 37B (1982) 473.
29 G.R. Carnrick, D.C. Manning and W. Slavin, *Analyst*, 108 (1983) 1297.
30 J. Bauslaugh, B. Radziuk, K. Saeed and Y. Thomassen, *Anal. Chim. Acta*, 165 (1984) 149.
31 L. Ping, K. Fuwa and K. Matsumoto, *Anal. Chim. Acta*, 171 (1985) 279.
32 L.M. Voth Beach and D.E. Shrader, *Spectroscopy*, 1 (1986) 49.
33 G. Schlemmer and B. Welz, *Spectrochim. Acta*, 41B (1986) 1157.
34 B. Welz, G. Schlemmer and J.R. Mudakawi, *J. Anal. At. Spectrom.*, 1 (1986) 119.
35 E.H. Larsen and J. Ekelund, *Analyst*, 114 (1989) 915.
36 O.P. Heekem, N. Theobald and B.W. Wenclawiak, *Anal. Chem.*, 69 (1997) 2171.
37 F.S. Sun, D. Littlejohn and M.D. Gibson, *Anal. Chim. Acta*, 364 (1998) 1.
38 C.H. Marvin, L. Allan, B.E. McCarry and D.W. Bryant, *Int. J. Environ. Anal. Chem.*, 49 (1992) 221.
39 M.N. Kayali-Sayadi, S. Rubio-Barroso, C.A. Díaz-Díaz and L.M. Polo-Díez, *Fresenius J. Anal. Chem.*, 368 (2000) 697.
40 W.H. Griest and B.A. Tomkins, *Anal. Chem.*, 60 (1988) 2169.
41 H.A. Claessens, M.M. Rhemrev, J.P. Wevers, A.A.J. Janssen and L.J. Brasser, *Chromatographia*, 31 (1991) 569.
42 M.N. Kayali and S. Rubio-Barroso, *J. Liq. Chromatogr.*, 18 (1995) 1617.
43 A. Beard, K. Naikwadi and F.W. Karasek, *J. Chromatogr.*, 589 (1992) 265.
44 R. Fuoco, M. Perla-Colombini and A. Ceccarini, *Mikrochim. Acta*, 123 (1996) 175.
45 Z. Cencic-Kobda and J. Marcel, *Chromatographia*, 49 (1999) 21.
46 Z. Voznakova, J. Podehraska and M. Kohlickova, *Food Chem.*, 56 (1996) 285.
47 S. Babic, M. Petrovic and M. Kastelan-Macan, *J. Chromatogr.*, 823 (1998) 3.
48 Z.Q. Qu, L.Q. Jia, H.Y. Jin, A. Yediler, T.H. Sun and A. Kettrup, *Chromatographia*, 44 (1997) 417.
49 J.L. Gómez-Ariza, E. Morales, R. Beltrán, I. Giráldez and M. Ruiz-Benítez, *Analyst*, 120 (1995) 1171.
50 G.M. Cao and T. Hoshino, *Anal. Sci.*, 14 (1998) 835.
51 D. Drijvers, H. Van Langenhove and M. Beckers, *Water Res.*, 33 (1999) 1187.
52 E. Leonhardt and R. Stahl, *Anal. Chem.*, 70 (1998) 1228.
53 H. Destaillats, H.M. Hung and M.R. Hoffmann, *Environ. Sci. Technol.*, 34 (2000) 311.
54 J.W. Kang, H.M. Hung, A. Lin and M.R. Hoffmann, *Environ. Sci. Technol.*, 33 (1999) 3199.
55 H.J. Vandenburg, A.A. Clifford, K.D. Bartle, J. Carroll, I. Newton, L.M. Garden, J.R. Dean and C.T. Costley, *Analyst*, 122 (1997) 101R.
56 N. Haider and S. Karlson, *Analyst*, 124 (1999) 797.
57 C.L. Chen, D.X. Yuan and M. Chen, *Fenxi Kexue Xuebao*, 15 (1999) 36.
58 R.E.B. Ketchum, J.V. Luong and D.M. Gibson, *J. Liq. Chromatogr. Relat. Technol.*, 22 (1999) 1715.
59 M. Rothe and F. Pragst, *J. Anal. Toxicol.*, 19 (1995) 236.
60 M.T. Tena, M. Valcárcel, R.J. Hidalgo and J.L. Ubera, *Anal. Chem.*, 69 (1997) 521.
61 F. Lázaro, M.D. Luque de Castro and M. Valcárcel, *Anal. Chim. Acta*, 242 (1991) 283.
62 D. Chen, F. Lázaro, M.D. Luque de Castro and M. Valcárcel, *Anal. Chim. Acta*, 226 (1989) 221.
63 U. Hechler, J. Fischer and S. Plagemann, *Fresenius J. Anal. Chem.*, 351 (1995) 591.
64 J.D. Besert, M. Ejem, R. Holzer and P. Lischer, *Anal. Chim. Acta*, 383 (1999) 263.
65 V. López-Avila, R. Young and N. Teplitsky, *J. AOAC. Int.*, 79 (1996) 142.
66 S. Soporstoel, N. Gjoes and G.E. Carlberg, *Anal. Chim. Acta*, 151 (1983) 231.
67 T.F. Jenkins and M.E. Walsh, *J. Chromatogr. A.*, 662 (1994) 178.
68 A.V. Filgueiras, J.L. Capelo, I. Lavilla and C. Bendicho, *Talanta*, 53 (2000) 433.
69 Y. Ren, W. Shi and Q. Jin, *Guangpuxue Yu Guangpu Fenxi*, 10 (1990) 32.
70 J. Stupar and W. Frech, *J. Chromatogr.*, 541 (1991) 243.

71 J. Stupar, *Vestn. Slov. Kem. Drus.*, 36 (1989) 75.
72 E.A. Chmilenko, A.N. Baklanov and V.T. Chuiko, *Khim. Tekhnol. Vody.*, 13 (1991) 62.
73 R.I. Botto, *J. Anal. At. Spectrom.*, 8 (1993) 51.
74 Q. Jin, C. Zhu, K. Brushwyler and G.M. Hieftje, *Appl. Spectrosc.*, 44 (1990) 183.
75 B. Budic, *J. Anal. At. Spectrom.*, 13 (1998) 869.
76 Y. Nakamura, K. Ide and R. Hasegawa, *Bunseki Kagaku*, 48 (1999) 339.
77 J. Alvarado, M.G. Wu and J.W. Carnahan, *J. Anal. At. Spectrom.*, 7 (1992) 1253.
78 J.S. Zhang, L.H. Li, H.Q. Zhang, D.M. Ye and Q.H. Jin, *Guangpuxue Yu Guangpu Fenxi*, 16 (1996) 56.
79 K. Jankowski, D. Karmasz, A. Ramsza and E. Reszke, *Spectrochim. Acta*, 52B (1997) 1813.
80 Y. Madrid, M. Wu, Q. Jin and G. Hieftje, *Anal. Chim. Acta*, 277 (1993) 1.
81 A. Woller, Z. Mester and P. Fodor, *J. Anal. At. Spectrom.*, 10 (1995) 609.
82 Z. Mester and P. Fodor, *Anal. Chim. Acta*, 386 (1999) 89.
83 K.L. Yang and S.J. Jiang, *Anal. Chim. Acta*, 307 (1995) 109.
84 J.F. Banks, S. Shen, C.M. Whitehouse and J.B. Fenn, *Anal. Chem.*, 66 (1994) 406.
85 R. Straub, M. Linder and R.D. Voyksner, *Anal. Chem.*, 66 (1994) 3651.
86 M.A. Tarr, G. Zhu and R.F. Browner, *Anal. At. Spectrom.*, 7 (1992) 813.
87 M.A. Tarr, G. Zhu and R.F. Browner, *Anal. Chem.*, 63 (1993) 1689.
88 R.H. Clifford, H. Tan, H. Liu and A. Montaser, *Spectrochim. Acta*, 48 (1993) 1221.
89 P.D. Goulden and D.H. Anthony, *J. Anal. Chem.*, 56 (1984) 2327.
90 E. Michaud-Poussel and J.M. Mermet, *Spectrochim. Acta*, 41 (1986) 49.
91 K.E. LaFreniere, G.W. Rice and V.A. Fassel, *Spectrochim. Acta*, 40 (1985) 1495.
92 A. Montaser, H. Tan, I. Ishii, S.H. Name and M. Cai, *Anal. Chem.*, 63 (1991) 2660.
93 D.R. Wiederin, R.S. Houk, R.K. Winge and A.P. D'Silva, *Anal. Chem.*, 62 (1990) 1155.
94 I.B. Brenner and E. Dorfman, *Anal. At. Spectrom.*, 8 (1993) 833.
95 L.C. Alves, L.A. Alen and R.S. Houk, *Anal. Chem.*, 65 (1993) 2468.
96 C.W. Huang and S.J. Jiang, *J. Anal. At. Spectrom.*, 8 (1993) 681.
97 I.B. Brenner, J. Zhu and A. Zander, *Fresenius J. Anal. Chem.*, 355 (1996) 774.
98 M.A. Tarr, G. Zhu and R.F. Browner, *Anal. Chem.*, 65 (1993) 1689.
99 J.F. Banks, J.P. Quinn and C.M. Whitehouse, *Anal. Chem.*, 66 (1994) 3688.
100 C. Bendicho and M.T.C. Loos-Vollebregt, *J. Anal. At. Spectrom.*, 6 (1991) 353.
101 N.J. Miller-Ihli, *Fresenius J. Anal. Chem.*, 345 (1993) 482.
102 M.J. Cal-Prieto, A. Carlosena, J.M. Andrade, S. Muniategui, P. López-Mahía, E. Fernández and D. Prada, *J. Anal. At. Spectrom.*, 14 (1999) 703.
103 N.J. Miller-Ihli, *J. Anal. At. Spectrom.*, 3 (1988) 73.
104 N.J. Miller-Ihli, *J. Anal. At. Spectrom.*, 4 (1989) 295.
105 N.J. Miller-Ihli, *US Patent 4,930,898* (1990).
106 G.R. Carnrick, G. Daley and A. Fotinopoulos, *At. Spectrosc.*, 6 (1989) 170.
107 W. Klemm and G. Bombach, *Fresenius J. Anal. Chem.*, 353 (1995) 12.
108 C. Santos, F. Álava-Moreno, I. Lavilla and C. Bendicho, *J. Anal. At. Spectrom.*, 15 (2000) 987.
109 E.C. Lima, F. Barbosa Jr., F.J. Krug, M.M. Silva and M.G.R. Vale, *J. Anal. At. Spectrom.*, 15 (2000) 995.
110 E. Welter and B. Neidhart, *Fresenius J. Anal. Chem.*, 357 (1997) 345.
111 E.G. Lierke, *Acoustica*, 82 (1996) 220.
112 E.G. Lierte, *Forsch. Ing. Wes.*, 61 (1995) 201.
113 H.G. Hoffman, E. Lentz and B. Schrader, *Rev. Sci. Instrum.*, 64 (1993) 823.
114 A. Ashkin, *Phys. Rev. Lett.*, 24 (1970) 156.
115 A. Ashkin, *Science*, 210 (1980) 1081.
116 W.B. Written, J.M. Ramsey, S. Arnold and B.V. Bronk, *Anal. Chem.*, 63 (1991) 1027.
117 K.C. Ng, W.B. Written, S. Arnold and J.M. Ramsey, *Anal. Chem.*, 64 (1992) 2914.
118 M.D. Barnes, K.C. Ng, W.B. Written and J.M. Ramsey, *Anal. Chem.*, 65 (1993) 2360.
119 Bücks and H. Müller, *Z. Phys.*, 84 (1933) 75.
120 A.R. Hanson, E.G. Domich and H.S. Adams, *Rev. Sci. Instrum.*, 35 (1964) 1031.
121 W.A. Oran, L.H. Berge and H.W. Parker, *Rev. Sci. Instrum.*, 51 (1980) 626.

122 E.H. Trinh and H. Chaur-Jian, *J. Acoust. Soc. Am.*, 79 (1986) 1335.
123 E.H. Trinh, *Rev. Sci. Instrum.*, 56 (1985) 2059.
124 O. Rohling, C. Weitkamp and B. Neidhart, *Fresenius J. Anal. Chem.*, 368 (2000) 125.
125 M. Mohammad, *Bull. Electrochem.*, 6 (1990) 806.
126 A.J. Saterlay and R.G. Compton, *Fresenius J. Anal. Chem.*, 367 (2000) 308.
127 A.J. Reviejo, A. González, J.M. Pingarrón and L.M. Polo, *Anal. Chim. Acta*, 264 (1992) 141.
128 V.K. Chebotarev, I.V. Voronkina, N.N. Artyukhova and Y. Kraev, *Zh. Anal. Khim.*, 49 (1994) 989.
129 C. Agra-Gutiérrez, J.L. Hardcastle, J.C. Ball and R.G. Compton, *Analyst*, 124 (1999) 1053.
130 A.J. Saterlay, J.S. Foord and R.G. Compton, *Analyst*, 124 (1999) 1791.
131 C.R.S. Hagan and L.A. Coury Jr., *Anal. Chem.*, 66 (1994) 399.
132 D.J. Walton, S.S. Phull, A. Chyla, J.P. Lorimer, T.J. Mason, L.D. Burke, M. Murphy, R.G. Compton, J.C. Eklund and S.D. Page, *J. Appl. Electrochem.*, 25 (1995) 1083.
133 R.G. Compton, J.C. Eklund and S.D. Page, *J. Phys. Chem.*, 99 (1995) 4211.
134 R.G. Compton and F.M. Matysik, *Electroanalysis*, 8 (1996) 218.
135 F.M. Matysik, S. Matysik, A.M. Oliveira Brett and C.M.A. Brett, *Anal. Chem.*, 69 (1997) 1651.
136 Y.C. Tsai, J. Davis and R. Compton, *Fresenius J. Anal. Chem.*, 368 (2000) 415.
137 J.L. Hardcastle, G.G. Murcott and R.G. Compton, *Electroanalysis*, 12 (2000) 559.
138 R.G. Compton, J.C. Eklund, S.D. Page, G.H.W. Sanders and J. Booth, *J. Phys. Chem.*, 98 (1994) 12410.
139 H. Zhang and L.A. Coury Jr., *Anal. Chem.*, 65 (1993) 1552.
140 A.J. Saterlay, S.J. Wilkins and R.G. Compton, *Green Chem.*, *in press*.
141 C. Agra-Gutiérrez, M.F. Suárez and R.G. Compton, *Electroanalysis*, 11 (1999) 16.
142 P.R. Birkin and S. Silva-Martínez, *Anal. Chem.*, 69 (1997) 2055.
143 M. Schmitt, *Anal. Chem.*, 63 (1991) 529.
144 E.N. Andrade, *Phil. Trans.*, 230 (1932) 413.
145 R.Y. Wang, J.A. Jarratt, P.J. Keay, J.J. Hawkes and W.T. Coakley, *Talanta*, 52 (2000) 129.

Chapter 4

Solid sample treatments involving the removal of volatile species

4.1. INTRODUCTION

The techniques described in this chapter are subject to a major constraint: they can only be used with volatile analytes or those that can be converted into a volatile product using a variably complex, labour-intensive and/or expensive procedure. This constraint, however, results in a favourable feature as regards selectivity: only volatile species and those that can be converted into volatile products are isolated from the matrix, which provides some selection by itself.

One other salient feature of the techniques dealt with in this chapter is that they are applied mostly to liquid samples; hence, the equipment needed to implement them — most often a continuous system, whether laboratory-made or commercially available — can only be used with liquids. Solid samples require dissolution or leaching, which is usually done by using a (frequently) labour-intensive standard procedure. However, the typical designs for liquid samples can usually be adapted for use with solid samples, with (often) slight changes. This chapter places special emphasis on this aspect, i.e. on the advantages and disadvantages of inserting the sample, whether in its original form or following slight treatment, into a unit capable of isolating the analytes as volatile species — the optimum form for detection with an atomic spectrophotometer (for hydrides and cold vapours), a gas chromatograph (with the headspace and solid-phase microextraction techniques) or, indifferently, a gas chromatograph or any type of detector (with pervaporation).

The techniques discussed in this chapter vary in automatability and frequency of use. Thus, while automatic hydride and cold mercury vapour generation are implemented in laboratory-constructed or commercially available dynamic equipment that is straightforward, easy to operate and inexpensive, automating laboratory headspace modes and solid-phase microextraction is rather complicated and commercially available automated equipment for their implementation is sophisticated and expensive. Because of its fairly recent inception, analytical pervaporation lacks commercially available equipment for any type of sample; however, its high potential and the interest it has aroused among manufacturers is bound to result in fast development of instrumentation for both solid and liquid samples. This technique, which is always applied under dynamic conditions, has invariably been implemented in a semi-automatic manner to date; also, its complete automatization is very simple.

The techniques dealt with in this chapter are discussed in terms of similarities. For this reason, hydride generation and cold mercury vapour generation are addressed first, notwithstanding the limited scope of this technique as regards analytes — scope that can be expanded by formation of organometals but will always be restricted to inorganic analytes. The headspace technique, whichever its mode, is more widely used than the previous ones, particularly in organic analysis — as such, it is a complement to them. Because of its similarity, and also for easier comparison, it is discussed immediately after analytical pervaporation. Despite its scant use with solid samples, solid-phase microextraction is commented on in a separate section.

4.2. HYDRIDE AND COLD MERCURY VAPOUR GENERATION

4.2.1. Introduction

Separating devices for the removal of a gas from either a liquid or a solid sample were formerly clearly distinguished according to whether the separation was based on gas diffusion or gas expansion. The main difference between the two types of separators was that a membrane was almost invariably used for gas diffusion from one channel to another, whereas gas expansion was conducted in substantially larger separation chambers that usually contained no membrane. At present, however, a growing tendency exists to using membranes in expansion chambers, so the two types of separators only differ in the previous geometric aspect. Both hydride and cold mercury vapour generation belong to the gas expansion category; also, they use no separation membrane in most cases.

Because the hydride and cold vapour generation techniques have so far been used mainly with liquid and dissolved samples, all aspects related to the variables influencing the process and the characteristics of the ensuing methods, among others, have been established in the light of a liquid entering the separator. Most such aspects also affect solid samples and are thus worth some comment, as are those that are exclusive to them — all briefly as this technique has scarcely been applied to solid samples.

4.2.2. Samples, analytes and reagents

Samples

Although candidate samples for hydride and cold mercury vapour generation can be liquid or solid, the latter must be dissolved or leached prior to having the analytes contact the reagents. Alternatively, solid samples can be mixed with an acid solution of the reagent in the separator in order to obtain a suspension or slurry that will facilitate analyte–reagent contact. Such contact can be improved by stirring or by mixing the solid with inert glass beads in order to prevent compaction of the sample.

Analytes

The species most often determined following hydride generation include Se, As, Sb, Bi, Pb, Te, Sn and Ge [1–3]. Although Hg is regarded as a hydride-forming element, some authors have expressed doubts about the nature of the resulting volatile species and considered the potential partial formation of a hydride at the beginning of the derivatization step that may subsequently evolve in rapid manner to the elemental, also volatile form [4]. Only Se, As and Sb among the previous elements can readily form hydrides at their lower oxidation states; species in higher states must thus be previously reduced if a hydride is to be formed. All inorganic species of these elements eventually form the same hydride (SeH_2, AsH_3 and SbH_3, respectively); speciation analysis can thus be enabled by generating the hydride in two steps with an intervening reduction of the most oxidized forms. Some organocompounds of these analytes can be discriminated in one run upon hydride generation; such is the case with methyl and dimethylarsonic (or stibonic) acids, which produce volatile $MeAsH_2$ and Me_2AsH [4,5] (or $MeSbH_2$ and Me_2SbH [6]), respectively. Trialkyllead species form stable hydrides whereas dialkylleads are non-reactive [7]. Mercury and methylmercury seemingly form hydrides amenable to gas chromatography [8]. Hydride generation has aroused the greatest interest for the speciation of organotin analytes since, although it can easily be interfered with by natural samples, it allows the simultaneous speciation of ionic methyl and butyl species in one run, using fairly simple set-ups [9–12].

Reagents

The reagent most commonly used for hydride generation is sodium tetrahydroborate, in an acid (HCl) medium for inorganic analytes but in the presence of an organic solvent such as hexane [13] or isooctane [14] when organometals are involved. The generic reaction can be written as follows:

$$M^{n+} + n\ NaBH_4 + 3n\ H_2O = MH_n + n\ H_3BO_3 + 3n\ H_2 + n\ Na^+$$

With mercury, the end-product of the reaction with $NaBH_4$ depends on the organic or inorganic nature of the species [15]. Thus:

$$Hg^{2+} + 2\ NaBH_4 + 6\ H_2O = Hg^0 + 7\ H_2 + 2\ H_3BO_3 + 2\ Na^+$$

$$MeHg^+ + NaBH_4 + 3\ H_2O = MeHgH + 3\ H_2 + H_3BO_3 + Na^+$$

Cold mercury vapour is usually produced by using an acid stannous ion solution. The reaction for mercuric ion is as follows:

$$HgCl_2 + SnCl_2 = Hg^0 + SnCl_4$$

Similarly to hydride and cold mercury vapour generation, alternative reagents can be used to form volatile species for subsequent transfer to a gas chromatograph. The use of such reagents is intended to circumvent the shortcomings of hydride generation, which is extremely prone to interference (especially from transition metals) that detract from expeditiousness and analytical precision [16]. Tetraalkylborates (particularly tetraethylborate, $NaBEt_4$) provide substantial improvements in the formation of some volatile (ethylated) compounds relative to hydride formation (e.g. they enable speciation [17,18]).

Grignard reagents (viz. methyl-, ethyl-, propyl-, butyl-, pentyl-, hexyl- and phenylmagnesium chlorides and bromides [19]) have also been used to obtain volatile species. The main shortcoming of the Grignard reaction is that it requires an aqueous-free medium. In addition, unreacted reagent must be decomposed prior to injecting the derivatized extract onto a column. Acetonates, trifluoroacetonates and dithiocarbamates are finding increasing use as alternatives to hydrides in analyses for volatile species [20,21].

4.2.3. The separator and its associated equipment

The characteristics of the separator and its associated equipment used for hydride and cold mercury vapour generation are dictated by the state of the sample.

Separator

Ideally, the unit used to isolate the gas formed from the liquid or suspended phase (whether a hydride or metal vapour) should meet several requirements, namely:

(a) It should operate smoothly and regularly in order to avoid irreproducibility — the particular sources of which vary depending on the state of the sample. With solid samples, grain size and the type of matrix involved can affect the ease of formation or removal of the gas phase; with liquid samples, irreproducibility arises mainly from disaggregation units included in the continuous-flow system.
(b) It should induce minimal dispersion or dilution of the gaseous analyte, i.e. void volumes should be as low as possible — which will ensure adequate sensitivity — but not so low as to result in incomplete separation of the gas from the sample.
(c) It should prevent the liquid or solid from reaching the detection system, which would give rise to major precision and selectivity problems, and might also deteriorate the detector (e.g. a quartz atomizer in AAS or a plasma torch in ICP–AES).

With *liquid samples*, the separator is a continuous-flow unit intended to collect one or several streams containing a liquid mixed with some gases (volatile analytes, hydrogen, stripping gas) or the stripping gas alone and separate the two phases. The gas phase is then transferred to the detector while the liquid phase is driven to waste, either directly or by adjusting the flow-rate of the emerging stream via a channel of the peristaltic pump. There are two general types of separators for liquid samples that differ in the way isothermal separation is accomplished.

Separators based on diffusion of the gas phase across a permeable membrane, which are known as "dual-phase gas diffusion separators", can be of two types, namely:

(a) Sandwich cells (Fig. 4.1A), which consist of two Perspex, Teflon or aluminium blocks with identical internally engraved conduits that constitute the inner chamber (the

Fig. 4.1. Hydride and cold mercury vapour separators. (A) Sandwich-type. (B) Tubular-type. (C) Modified Vijan-type. (D) Commercial-type (Perkin–Elmer FIAS-200 module). (E) With direct introduction of the stripping gas into the separator. (F) Dismountable separator for solid samples. (Reproduced with permission of the Royal Society of Chemistry and Perkin–Elmer.)

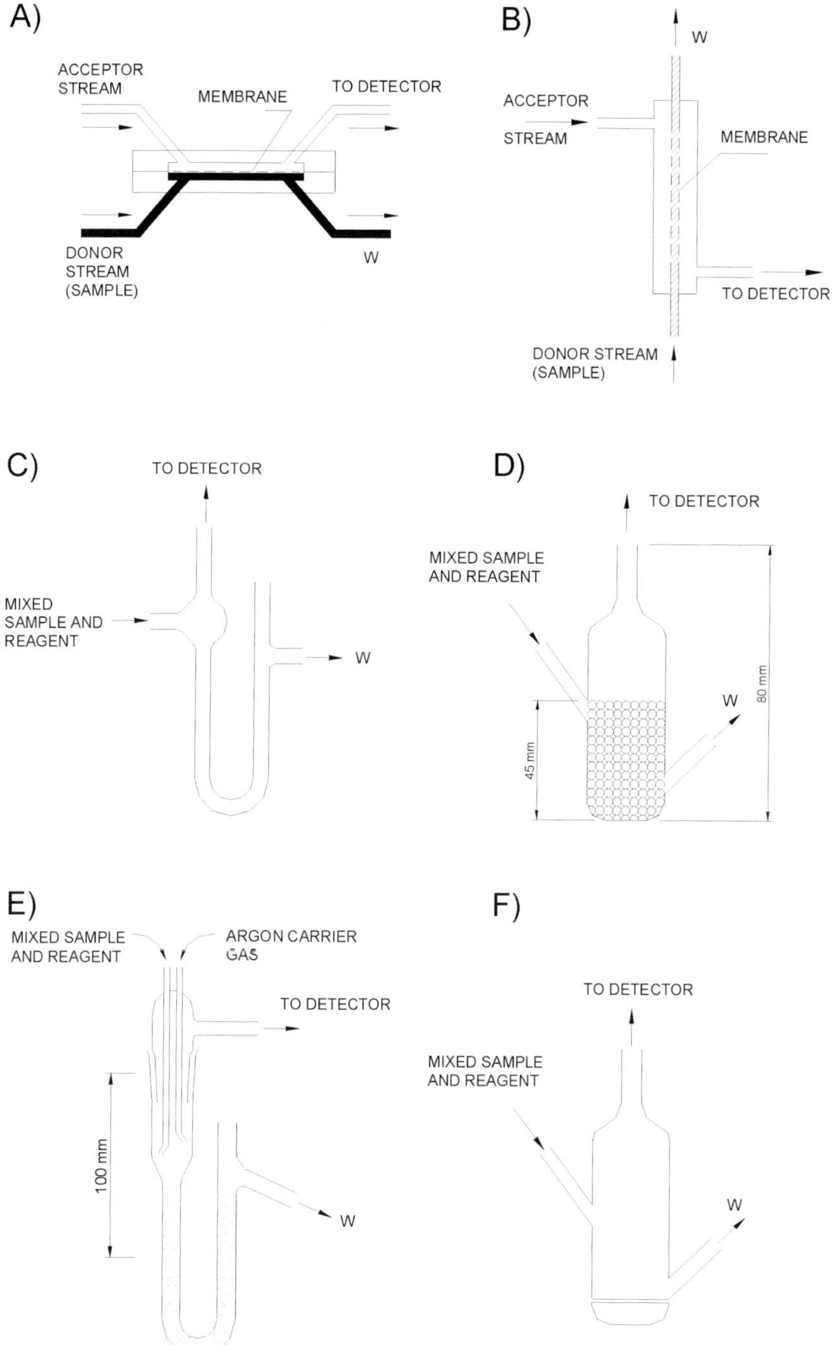
A)
ACCEPTOR STREAM
MEMBRANE
TO DETECTOR
DONOR STREAM (SAMPLE)
W
B)
W
ACCEPTOR STREAM
MEMBRANE
TO DETECTOR
DONOR STREAM (SAMPLE)
C)
TO DETECTOR
MIXED SAMPLE AND REAGENT
W
D)
TO DETECTOR
MIXED SAMPLE AND REAGENT
80 mm
W
45 mm
E)
MIXED SAMPLE AND REAGENT
ARGON CARRIER GAS
TO DETECTOR
100 mm
W
F)
TO DETECTOR
MIXED SAMPLE AND REAGENT
W

geometry of which varies among models). The membrane is placed between the two blocks, which must be joined tightly in order to avoid leakage. Each engraved micro-conduit has two holes on its ends that connect it with the manifold tubing. One variant of this design (Fig. 4.1B) uses a Gore-Tex (tubular) membrane.

(b) Gas-sample expansion units, which are usually glass minichambers. The wide variety of designs available can be classified according to the zone through which the stripping gas is inserted. The most popular type of unit for gas–liquid mixing prior to the detector is that based on the U-shaped design of Vijan [22], which has been modified to improve specific aspects — particularly to reduce dispersion through miniaturization, which decreases the amount of analyte (sample) and reagents needed. Figures 4.1C and 4.1D show the basic, custom design of Vijan and that of the Perkin–Elmer FIAS-200 module, respectively. The commercial design is packed half-full with glass beads about 3 mm in diameter, which reduces aerosol formation and foaming. An additional precaution can be adopted to prevent migration of liquid or solid particles into the heated quartz cell by using a length of microporous PTFE tubing (with one end blocked) connected to the separator outlet, the tube protruding into a secondary outer tube of a larger diameter [23]. In gas–liquid expansion separators, the stripping gas is inserted in a direct manner. In that of Fig. 4.1E, originally developed by Thompson *et al.* [24], the mixed sample–reagent and argon streams enter the minichamber; the liquid is drained out by a communicating-vessel effect through a U-shaped tube along the bottom of the expansion zone, while the gas is directly driven to the detector.

The separator for *solid samples* depicted in Fig. 4.1F is a dismountable device where the lower part is lightweight as the sample is weighed in it. The lower part is connected to the rest and the reagents are inserted through a septum (simultaneously with the start of stirring and passage of the stripping gas).

In principle, the devices used for hydride removal can also be employed with cold vapour — which in addition requires a non-heated quartz-cell atomizer. In any case, research interests have focused on systems for specific use with mercury vapour.

Ancillary equipment

Once hydrides are formed, they must be driven to the detector (usually an atomic spectrometer but also, occasionally, a gas chromatograph) under optimal conditions as regards concentration and the absence of species unfriendly to the chromatographic column.

A U-tube filled with a GC sorbent (e.g. polymethoxysilane coated silica beads) and maintained in liquid nitrogen (where hydrides are trapped) allows the product to be concentrated prior to insertion, either into the detector or onto the column. A thermal desorption system is mandatory for proper, fast removal of retained, preconcentrated species. A water trap (another U-tube maintained at −15°C or filled with anhydrous $CaCl_2$), a Nafion tube and various other devices are often placed in between the separator and detector.

4.2.4. Variables influencing vapour generation

Gas-forming reagent

The reagent to be used to aid in the formation of volatile species depends on whether the sample is solid or liquid. Solutions containing sodium tetrahydroborate (or some derivative) must be merged with a hydrochloric acid stream prior to meeting a liquid sample if a dynamic approach is used to drive the ingredients to the separator. With cold mercury vapour, the acid stannous solution is also merged with the sample stream prior to reaching the separator. With solid samples, both sample and reagent are directly placed in the separator, where the gas-forming reaction takes place. When both solid and liquid samples are to be processed, the concentration of $NaBH_4$ used dictates whether or not an additional H_2 stream should be employed in order to maintain the flame in flame atomizers [25].

Stripping gas

The release and transfer to the detector of a hydride or cold mercury vapour can be facilitated by using a nitrogen, argon or hydrogen stream at a flow-rate of 50–200 ml/min. Although the flow-rate is not critical, it should be kept as constant as possible throughout. The release of large, variable amounts of hydrogen in the reaction zone results in over-pressure of also variable magnitude on the inside of the narrow-bore tubing. In order to ensure uniform flow, a throttle in the form of a small copper disc with a tiny aperture (*ca.* 0.2 mm) is fitted to the outlet of the pressure gauge of the gas container. A back-pressure of more than 3 kg/cm^2 may result from such a low gas flow-rate as 150 ml/min; however, the flow-rate is not affected by small back-pressure changes in the flow system.

Accessories for implementing a hydride or cold mercury vapour-based method

Although the starting samples typically subjected to hydride generation or cold mercury vapour formation are usually solid, they are most often dissolved using a conventional or accelerated procedure prior to mixing with the reagents. For this reason, most approaches to the partial or complete automatization of the process from generation of the volatile molecular or atomic species have been designed for working with liquids. A number of commercial flow injection devices for hydride and/or cold vapour generation exist (e.g. those from Perkin–Elmer or P.S. Analytical) that can be coupled to atomic detectors; developing custom devices for this purpose is fairly easy, however. The manifold in Fig. 4.2 illustrates a dynamic approach to the formation of volatile species and their transfer to an atomic detector. The sample is inserted into a carrier stream (usually water) by means of an injection valve. An acid — most often HCl — stream is merged with the carrier and with one or more reagent streams in order to establish an appropriate pH for the subsequent reactions. Cleavage of bonds in organometals and analyte discrimination (speciation) are the usual aims of using an additional reagent prior to that employed to form the volatile species (sodium tetrahydroborate or tin chloride).

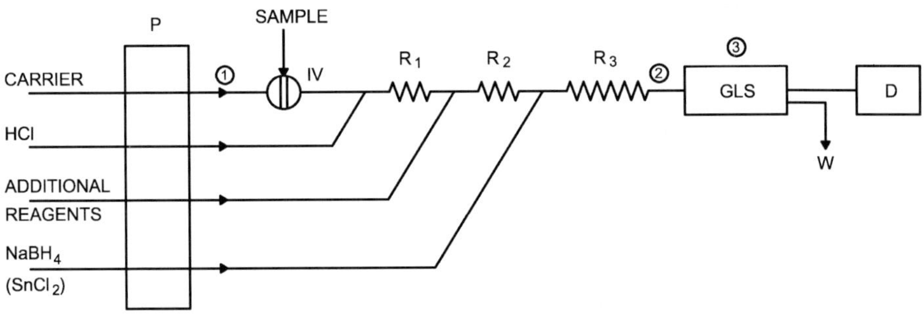

Fig. 4.2. Flow injection manifold for automatic hydride generation. 1-3 Points at which the stripping gas may be inserted into the dynamic system, P peristaltic pump, IV injection valve, R reactor, GLS gas–liquid separator, D detector, W waste.

The stripping gas can be inserted prior to the sample injection valve, between the reaction coil and the gas–liquid separator, or directly through this unit. The gas phase formed (a hydride or mercury vapour, and the hydrogen resulting from the reaction when tetrahydroborate or a similar reagent is used) is driven to the detector by the stripping gas. With flame atomizers, the flame can be fed only by the hydrogen produced in the hydride formation reaction or a mixture of an inert gas (e.g. argon) and hydrogen can be used as stripping gas to save reagent [25].

Detectors for volatile hydrides and cold mercury vapour

The physico–chemical properties of the analytes and the way they reach the detector have made atomic spectroscopy the detection technique of choice in most instances. A heated quartz cell or a similar device is connected directly to the gas outlet of the separation cell [26]. The use of an atomic fluorescence detector has provided methods for selenium [25,27] and mercury [28,29] that possess excellent analytical features and use inexpensive instruments. On a less affordable level are ICP emission [30] and atomic emission cavity spectrometers [31].

4.2.5. Features of methods based on hydride or cold mercury vapour generation

Steady-state and kinetic methods of hydride and cold vapour generation

When a preconcentration unit is placed behind the separator, the hydride-forming reaction usually develops to completion, which results in increased sensitivity — at the expense of throughput.

Shortening the collection time is recommended when other hydride-forming elements present in the sample react more slowly than the analyte forming the volatile species. This fixed-time kinetic approach is less sensitive than its steady-state counterpart but provides better selectivity and shorter analysis times [32,33].

Tolerance to interferences

One salient advantage of fixed-time kinetic methods is significantly increased selectivity. Tolerated concentrations of interfering metal ions are often reported to be one to two orders of magnitude higher than in equilibrium methods. The increased tolerance to interferences of kinetic methods may be the result of the following facts:

(a) The reaction time for the hydride or cold vapour generation step can be precisely controlled via the flow-rate and line lengths in order to favour the main reactions — which are usually fast. Slower interfering reactions are frequently suppressed by using shorter reaction coils. Such kinetic discrimination is impossible in batch processes.
(b) The high flow-rate of sample and reagent through the reaction conduit and separator leaves little opportunity for reduced metal or metal boride deposits accumulating therein. Such deposits are often considered to be major sources of interference with these methods.
(c) The small sample volumes used introduce reduced amounts of interferents into the hydride generation system; in fact, the absolute amount of interferent has frequently been shown to be more influential than its relative concentration in competing with the analyte product for free radicals in the atomization process. The favourable effect of using a decreased sample volume in flow injection systems is therefore self-obvious.

4.2.6. Applications

Most of the applications of hydride generation and cold mercury vapour formation to solid samples involve a previous step in which the sample is treated with an appropriate solution (usually an acid solution) with the aid of some type of energy for accelerating dissolution [25,26]. In a number of applications, however, the solid sample is directly introduced into the evaporator, which provides a simple, expeditious method for treating samples prior to insertion into an atomic detector. As early as the 1980s, Kruse developed an apparatus for fully automated sample insertion and changeover in cold vapour and hydride generation AAS. Both liquids (less than 5 ml) and solids (less than 500 mg) were inserted into the evaporator in a sequential manner and provided intermittent signals. Absolute limits of detection of 1, 1 and 5 ng were obtained for Hg(II), As(III) and As(V), respectively, and up to 60 samples per hour were processed in this way [34]. The English Department of the Environment has reported several methods for determining the previous analytes at levels below 20 mg/kg [35]. Environmental samples (soil, sewage sludge, fly ash and waste incineration ash) of very small grain size (90 μm) were mixed with 10 ml of 4 M HCl under stirring, the solution then being supplied with 1 ml of 1% $NaBH_4$ for 0.5 min and the resulting arsine determined by AAS. The influence of interfering elements present in the sample matrix was investigated and found to be generally weaker than that observed in the hot acid extraction method. Arsenic recoveries were better than those provided by such a method, and the relative standard deviation was 7% [36].

Table 4.1 [37–43] shows the figures of merit of typical hydride generation and cold mercury vapour formation methods using atomic absorption or fluorescence detection.

TABLE 4.1

PERFORMANCE AND CHARACTERISTIC DATA OF REPRESENTATIVE FI HYDRIDE AND COLD VAPOUR GENERATION ATOMIC DETECTION METHODS

Detection	Analyte	Sample	Separator	Sample volume, µl	f, h^{-1}	Limit of detection, µg/l	RSD, %	Ref.
AAS	Bi	Aqueous	U-tube	700	180	0.08	0.2–0.8	37
	Se	Wheat, orchard leaves, soils, coal fly ash	U-tube	400	250	0.06	1.6	38
	As	Soils, rice, wastewater	U-tube	400	220	0.10	1.5	39
	Se	Geological	U-tube	200	50	0.2	—	40
	As, Se, Te, Sb, Bi	Wheat, orchard leaves, coal fly ash	U-tube	500	120	0.08–0.6	—	41
	As	Waters	Tubular membrane	1000	150	0.07	2.5	42
	As, Se, Sb, Te, Bi, Sn	Soil, sludge, sea water, blood, urine, etc.	W-tube	500	180	0.06–0.27	—	43
AFS	Se (IV), Se (VI)	Tap water, hemodialysis fluids	U-tube	600	50	0.11	4.5	25
	Se	Nutritional supplements, shampoos	U-tube	100	50	0.4	2.9	27
	PhHgAc	Sewage sludge	U-tube	0.1 g	5	1 µg/kg	1.8	28
	Hg	Cosmetics	U-tube	Continuous 15 min	4	15 ng/l	5.9	29

4.3. HEADSPACE SAMPLING

4.3.1. Introduction

Sample injection in gas chromatography often seems deceptively simple: a microlitre aliquot is rapidly injected into an inlet system, and elution and detection follow. Samples containing substantial amounts of non-volatile material, however, require one or more preparation steps in order to isolate volatile analytes from non-volatiles that would otherwise contaminate the inlet system and column, eventually leading to impaired chromatographic performance. Examples of such procedures include liquid–liquid extraction, solid-phase extraction and filtration. The use of a pre-column (viz. a retention gap or a guard column) is often required, even if prepared samples are used.

The headspace technique, which is considered to be either a sample treatment technique or a sampling system for volatiles, was introduced in the late 1950s to early 1960s. In its static and equilibrium modes, it provides a clean, convenient way around preparation requirements for a wide variety of materials. In its oldest version, the headspace sampling is done in the static mode: the sample, liquid or solid, is placed in a sealed vial and heated at a predetermined temperature for a preset length of time. As a result, volatiles partition between the gas and the liquid or solid phase(s) in the vial. At the end of the thermostatting time, a portion of the volatiles in the gas phase (the headspace) is removed and injected onto the column. Non-volatiles are left behind in the sample vial, so they cannot contaminate the chromatograph. Headspace treatment is not useful for analysing high-boiling solutes (viz. those with boiling points higher than approximately that for n-C_{18}) as the analyte concentrations that can be gathered in the headspace at normal temperatures is usually inadequate.

Originally, static headspace gas chromatography (HS–GC) was used to measure the solubility of anaesthetics and to determine gases, alcohols and solvents in biological samples. The equipment and procedure used were quite simple, but were subject to a number of practical drawbacks. One was a lack of precision. Through improved instrumentation and further theoretical development, the technique was refined to such an extent that it currently provides accurate quantitative analyses. In addition, new, more efficient modes such as multiple headspace (MHS), dynamic headspace (DHS), and purge and trap (PT) have provided improved means for removing volatile and semi-volatile substances from both liquid and solid samples, thus increasing its use. Basically, all four modes rely on the presence of an air-gap phase in between the sample and the septum of the sample container where the analytes evaporate, so they can be labelled as headspace modes.

4.3.2. Equipment and procedures

The basic set-up for headspace analysis comprises an HS element — the characteristics of which depend on the particular mode used for pretreatment and a gas chromatograph or, less often, an alternative detector for measurement. Static and dynamic headspace (purge and trap included) differ in the type of equipment required; multiple headspace uses the same automated device as static headspace.

Static or equilibrium headspace

In its static version, the headspace element can be either manual or automated. In the manual version, it consists of a vaporization container (Fig. 4.3) where equilibrium is reached, a heating unit that keeps the container at a constant temperature and an injection device that transfers the vapour phase from the HS container into the GC column. Automating the technique dramatically complicates the equipment needed. By way of example, the headspace equipment commercially available from Perkin–Elmer consists of an autosampler with a 16, 40 or 110-sample capacity that performs overlapping thermostatting of consecutive vials; thus, after the first sample is injected, the unit heats and equilibrates the next sample so it can be injected as soon as the gas chromatograph is ready for the next run. The headspace uses a pneumatic pressure-balanced system — a proprietary technique that provides better sensitivity than its manual counterpart. Each thermostatted glass vial containing the liquid or solid sample is closed by a septum and a heated moveable needle with a hollow part permits flow in either direction. This system, having the required time, temperature and pressure control, is connected to the inlet part of the chromatograph. The headspace sampling procedure essentially consists of the following three steps:

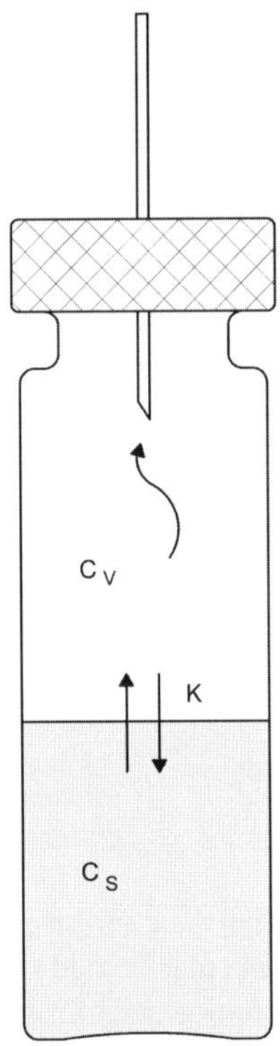

Fig. 4.3. Basic static headspace device (for details, see text).

(a) Equilibration (Fig. 4.4A), where the vial is thermostatted and the carrier gas is directly driven to the chromatographic column in the usual manner.
(b) Pressurization (Fig. 4.4B). After equilibrium is established, the needle penetrates the septum into the vial headspace. Simultaneously, a valve array (simplified in the figure as a three-way valve) allows part of the carrier gas to flow into the sample vial to build up its pressure.
(c) Sample transfer (Fig. 4.4C). The flow of carrier gas to the column is temporarily disconnected. As the sample vial is now open to the column through the needle, a flow of the headspace gas reaches the column and transfers an aliquot of the gas into it. The aliquot volume is controlled either through the time of transfer or by inserting a loop between the sample loop and the vial.

At the end of the sampling period, the valve is re-opened and the normal carrier gas flow resumed. In this way, the sample is introduced onto the column without using a gas syringe, thus avoiding fractionation due to pressure changes in the syringe. Since the system is closed, no headspace is lost during the transfer. The equipment requires no multiport valves, which reduces the number of components that come into contact with delicate samples.

After a preset time, the needle is removed from the vial and the system is made ready for sampling from the next vial. Although the equilibration time is the basis for static headspace, attainment of equilibrium is not mandatory when fully automated equipment is used provided every instrumental parameter including the (incomplete) thermostatting time is accurately reproduced [44].

Multiple headspace

The multiple headspace (MHS) mode emerged in the early 1980s [45] as a method for circumventing the shortcomings of quantitative analyses by conventional HS when the sample was not homogeneous, so the usual internal, external and standard addition methods failed. MHS provides an effective solution as it is an absolute quantitative method where the determination is independent of the sample matrix. The automated equipment typically used in HS also allows implementation of MHS simply by introducing a straightforward modification in the control software. In simple terms, MHS involves dynamic gas extraction conducted stepwise. In each step, equilibrium conditions are established in the vial between the sample and its headspace. The analyte concentration in the headspace decreases exponentially with each consecutive extraction step; by proper mathematical extrapolation, the total peak area — which will be proportional to the overall amount of analyte present in the original sample — can be calculated. Also, calibration under the same conditions allows the relationship between peak area and the amount of analyte to be established.

The overall MHS process involves the three above-described steps for conventional automated HS (A to C in Fig. 4.4) and the two additional steps labelled D and E in the same figure. Thus, after the sample is transferred, the system is vented and a re-equilibration step is performed — these steps can be repeated as many times as required.

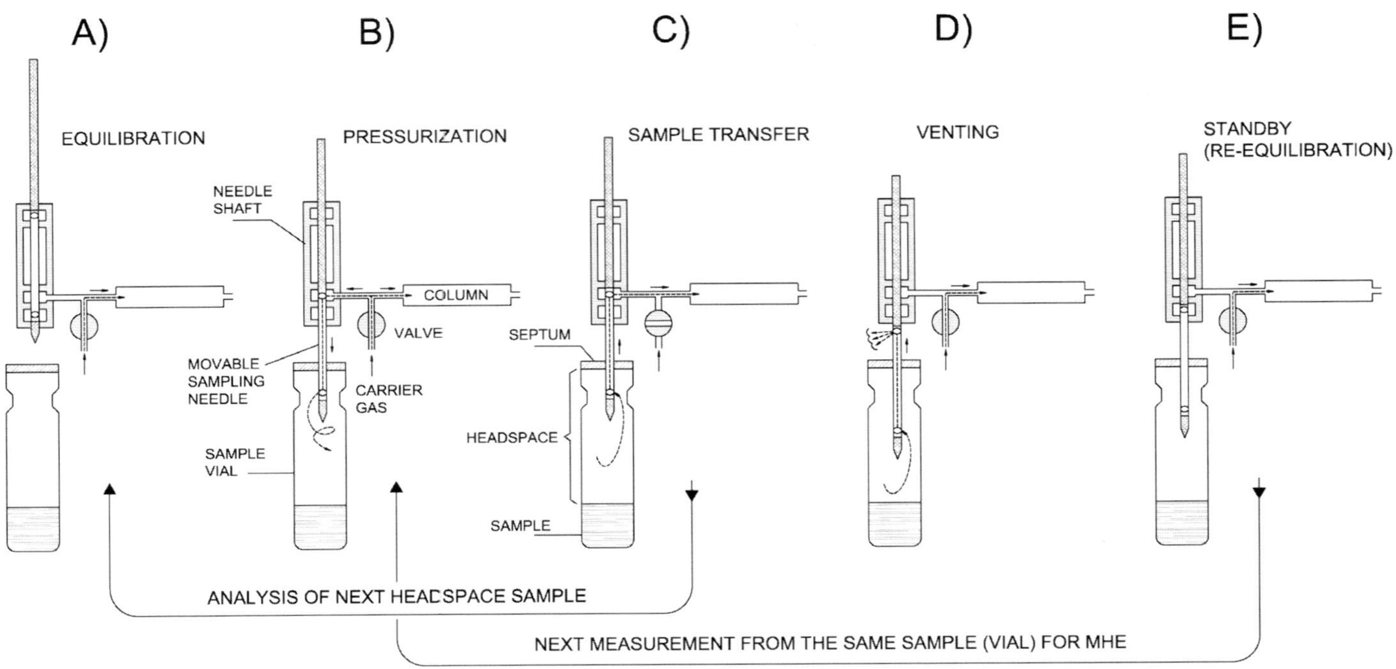

Fig. 4.4. The three basic steps (A–C) of the pressurized headspace technique and two additional steps (D and E) involved in the multiple-headspace mode: (A) equilibration, (B) pressurization, (C) transfer of an aliquot of headspace gas to the chromatographic column, (D) venting to the atmosphere and (E) re-equilibration of the vial for the next sample. (Reproduced with permission of Springer-Verlag.)

Purge and trap

In its simplest form, a purge and trap procedure requires the use of a sample-holding container, a trapping device and a series of sample lines and valves. Many custom-made purge and trap units have been reported that, in form, contain the elements needed for concentration. In function, however, commercially available equipment has greatly improved the reliability and reproducibility of tests. Analyses are started by inserting an aliquot of sample into the sample-holding or sparging container. In most cases, a 25 ml volume may be needed to achieve the minimum detection limits (MDLs) specified in some methods. The sample is purged with either ultrapure (> 99.998%) helium or nitrogen at a specified flow-rate, temperature and time (Fig. 4.5A). This extracts the volatile analytes from the sample matrix and transfers them to an ambient temperature trap (viz. a short, 1/8-in i.d. packed column) containing from as few as one to as many as four different adsorbents. Following the purging and trapping step, volatiles are thermally desorbed (injected) onto the gas chromatographic column (Fig. 4.5B). This is accomplished by rapidly heating the He-swept trap to a high enough temperature in order to transfer the analytes in a narrow injection band — in an attempt to simulate the effect of a direct syringe injection. Then, chromatographic separation and detection are performed as required. The final step in the analysis is the preparation of the sample concentrator for the next sample. This involves baking the adsorbent trap at a high temperature with flow through the trap in the reverse direction of the purge extraction flow (Fig. 4.5C). The baking step and the chromatographic separation are, for the most part, carried out simultaneously.

It should be noted that most users of these dynamic techniques, and also equipment manufacturers, make no distinction between dynamic headspace (DHS) and purge and trap (PT). The difference, when considered [46], is based on bubbling of the auxiliary gas inside the sample in PT and sweeping it over the sample surface in DHS. All other

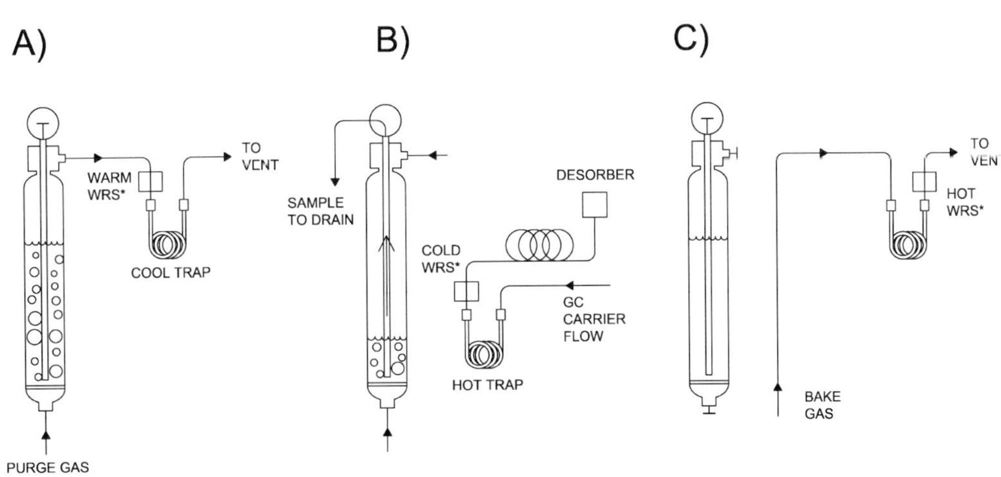

Fig. 4.5. Steps of the purge and trap process: (A) purge, (B) desorption–injection and (C) bake. WRS water removal system. (Reproduced with permission of Preston Publications.)

steps are the same in both modes. Consequently, the two are dealt with as a single mode (purge and trap) here. Wherever analytes are removed without purging — a rare occurrence—it is specifically noted. Purge and trap methods are the most frequently and widely used HS methods. The units typically employed in their implementation are described in detail below.

Trapping system The trapping efficiency in the PT technique depends on various factors including the vapour pressure of the compound, the surface area of the trapping material and thermodynamic interactions between the analyte and the adsorbent. Trapping adsorbents are usually more efficient at low temperatures. In order to minimize breakthrough, the concentrator trap should be near 25°C or within 1–2°C of ambient temperature — which should not exceed 30°C. Some sub-ambient temperature trap cooling devices are available; however, the modest gain in performance and the need to use a liquid nitrogen or carbon dioxide coolant with standard SW-846 methodology have deterred potential users.

The earliest traps specified in EPA's purge and trap methods contained equal amounts of Tenax, silica gel and coconut-based charcoal. Tenax absorbs compounds that are liquid, and silica gel substances that are gaseous, at room temperature. Charcoal is used to trap dichlorodifluoromethane (Freon 12), the boiling point of which is −29°C. The superior adsorption and desorption efficiencies of these adsorbents for purge and trap analysis are supported through their continuous use today; however, some alternative adsorbents have lately become popular for specific applications [47].

Water removal systems In the mid-1980s, the interference of water arising from the use of silica gel and charcoal became apparent through the increased use of some mass spectrometers and selective conventional detectors such as ECD, which were affected by the presence of moisture. In PT and DHS, water vapour is carried along with the volatiles onto the trap and is trapped mainly on silica gel, which is a highly hydrophilic adsorbent. The presence of water has little effect on the adsorbent's trapping or desorption efficiency; however, if desorbed to the GC, water does affect separation and detection. Problems resulting from excessive water transfer include decreased column efficiency (especially for early eluting compounds), quenching of the detector response, increased limits of detection, and also increased maintenance costs. In the late 1980s, the use of alternative hydrophobic sorbents coincided with the development of commercial purge and trap water removal systems. The adverse effects of water promoted the use of molecular sieves consisting of synthetic carbon (e.g. Carbosieve, a trademark of Supelco, Bellefonte, PA) in traps. These sieves are effective hydrophobic alternatives to silica gel and charcoal as they adsorb less water and thus result in an acceptable extent of water transfer from the trap. VOCARB, a synthetic adsorbent bed also from Supelco [48], has proved highly successful for many of the same reasons. Efficient purge and trap water removal systems have all but eliminated water retention as a criterion for adsorbent selection. The choice of trapping adsorbent should now rely mainly on breakthrough volumes for a target analyte list.

In the absence of components affected by water, the need for water removal systems in PT analysis would never have arisen. However, the increased use of mass spectrometry

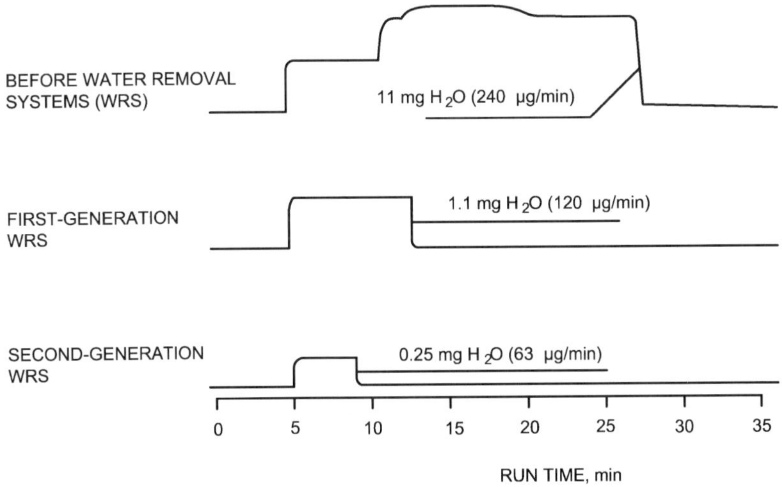

Fig. 4.6. Improvements in water-removal systems (WRS). EPA Method 524.2. Operating conditions: 10 ml/min He carrier; 0.53 mm i.d. column; TCD responding to water blank. (Reproduced with permission of Preston Publications.)

(MS) alone provided incentive for the development of built-in water removal devices for PT units. Of the two techniques used for water removal in PT work, viz. permeation and condensation, the former is highly efficient in the removal of both moisture and polar analytes. Early methods for the analysis of drinking water specified few polar analytes; with the addition of polar analytes such as ketones, however, permeation was rapidly abandoned. Condensation is now the most common choice for removing moisture without adversely affecting the efficiency for polar analytes. Since 1988, both thermoelectric cooling and ordinary fans have been used to cool the water removal area in the PT unit when coupled to MS detectors. Improvements in water removal efficiency have led to the transfer of as little as 0.063 µl/min of water during the trap desorption step [49]. Figure 4.6 shows the evolution of condensation-based water removal systems. Further reduction in water transfer would likely result in diminishing returns. As it is, the level of water removal is sufficient for MS, which tends to be more sensitive to the insertion of moisture. Nafion driers are also used to remove water after purging; Fig. 4.7 depicts a custom-made design for sample insertion into a multi-capillary gas chromatograph prior to transfer to a microwave-induced plasma atomic emission spectrometer for the speciation of mercury [50]. All these devices are based on water separation. One alternative choice for water removal is the use of hydrophobic membranes across which volatile species can diffuse. The principle behind the use of membranes in pervaporation (see Section 4.4) can be applied to the removal of water, not by eliminating it, but by hindering its passage to the trapping system.

Desorption systems Desorption, like extraction, is dependent upon flow-rate and temperature. Its aim is to ensure instantaneous sample introduction, thereby minimizing band broadening caused by the sample transfer step. When the PT technique was developed,

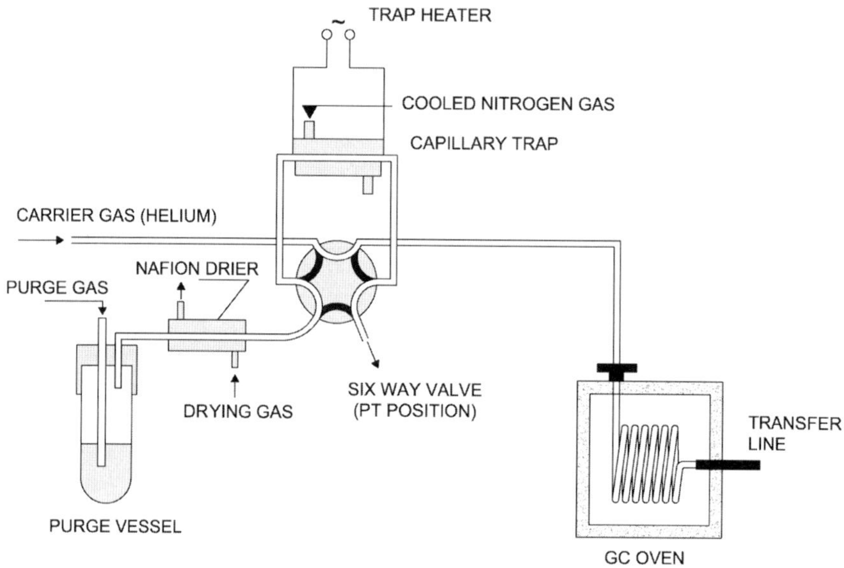

Fig. 4.7. Custom-made device for interfacing a PT module to a multi-capillary gas chromatograph using a Nafion drier for water removal. (Reproduced with permission of the Royal Society of Chemistry.)

packed column flow-rates (greater than 30 ml/min) were sufficient for expeditious analyte transfer from the trap and trap heating rates were as low as 200°C/min. With the transition to capillary columns and lower carrier flow-rates (1–10 ml/min), trap desorption became more challenging. Increasing the heating rate of the trap was a major improvement in capillary column chromatography by PT. Inductive trap heating is achieved by applying current directly to the concentrator trap (e.g. without a heater sleeve), which provides heating rates in excess of 800°C/min. The faster the trap is heated, the quicker the analyte (within a small amount of carrier gas) is released from the adsorbent. This rapid heating technique results in compounds (especially the early eluting gases) being flash-desorbed in the narrowest injection band to the column, even under capillary column flow-rates. Even faster trap heating rates can be expected as usage of smaller bore capillary columns grows (especially in GC–MS). Irrespective of how it is accomplished, a current trend exists in trap desorption to reducing the time required to transfer the analytes to the gas chromatograph. Such a reduction in transfer time provides the additional benefit of reducing the amounts of water (irrespective of whether a water-removal system is used) and carbon dioxide that are transferred to the gas chromatograph during the desorption step, thereby improving peak shape, narrowing peak widths and increasing sensitivity.

Baking systems Trap baking is mainly a function of the trapping adsorbent used, as well as of the application concerned. The bake times required to prepare the trap for the next sample (i.e. to minimize carry-over) are decreasing. Because the GC run begins at the start of desorption, the bake step occurs simultaneously with separation. In the recent past, the trap could bake for at least 20 min and the bake time still be completed before the end of the separation and GC cool down. With the increasing usage of thinner columns, the reduction in separation times has led to a reduction in bake times to prevent the PT unit from limiting productivity. For this reason, the important issues concerning trap baking are choosing adsorbents that can be baked quickly and efficiently, and pre-screening suspicious samples that might overload the system. Time spent pre-screening samples has led to significantly decreased PT downtimes in many laboratories [51]. Trap baking can be accomplished either by back-flushing the trap in the same direction of the desorption flow or forward-flushing it in the same direction as the purge flow. Baking by back-flushing the trap is more efficient because any remaining analyte is baked only through weaker adsorbents to vent. Forward-flushing a multi-sorbent trap requires higher bake temperatures (above 220°C) and longer times because volatiles remaining on the weakest adsorbent must be forced through stronger adsorbents. A minimum bake time (back-flushing) of 5 min has been determined for gasoline-range organics up to C_8 [52].

Commercial PT modules Commercial purge and trap equipment from various manufacturers including Tekmar, Perkin–Elmer and Varian is currently available. The Dynamic Thermal Stripper (DTS) from Dynatherm Analytical Instruments, Inc., provides a choice between two purging methods: conventional dynamic headspace, where the sample is neat (dry), and steam distillation, where the sample is immersed in water. In both cases, the purge gas percolates through the sample and adsorbs the components of interest as they are released from the sample matrix or extracted by the water. Once absorbed in the gas stream, these compounds exit through the top of the sample vial, pass through a heated interface for water removal and are deposited on an adsorbent packed tube. Upon conclusion of the stripping process, the sorbent may be removed from the DTS, sealed in a storage container and ultimately thermally desorbed onto an analytical GC column for component separation and analysis.

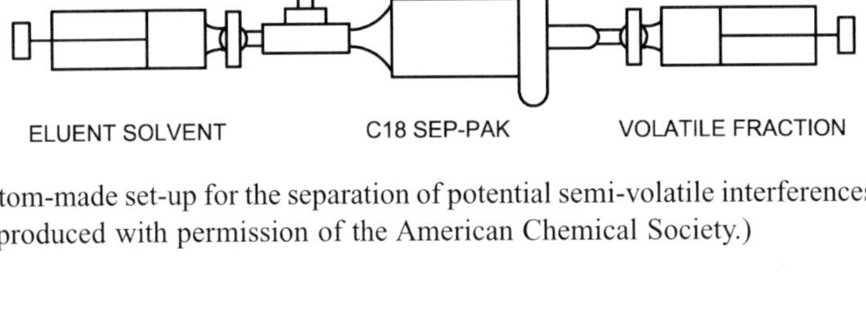

Fig. 4.8. Custom-made set-up for the separation of potential semi-volatile interferences from volatile analytes. (Reproduced with permission of the American Chemical Society.)

Auxiliary devices

In addition to the headspace module, headspace-based methods use three different types of ancillary units, namely: devices for conducting preliminary steps such as dilution, suspension or concentration; an instrument for detection (and, usually, also individual separation) of the target species; and an interface between the headspace module and the detector or chromatograph.

The most common preliminary steps involved in analytical methods (e.g. dilution and slurry preparation) are conducted by using typical manual or automated devices developed for this purpose. Some units, however, have been specially conceived for use in headspace work. Such is the case with the approach to the separation of potential semivolatile interferents from volatile analytes illustrated in Fig. 4.8, which was built with three 5-ml gas-tight Hamilton syringes and a three-way stainless steel valve. Two of the syringes are Luer-lok attached to the valve and the remaining one is Luer-lok connected to a C_{18} Sep-Pak cartridge that is in turn connected to the three-way valve. The sample is inserted via the sample syringe, through the cartridge and into the volatile fraction receiving syringe while the eluent solvent syringe valve is closed. The sample syringe is then closed, the eluent solvent syringe valve opened and the eluent (80:20 methanol/water) passed from the eluent solvent syringe into the receiving volatile fraction syringe to form a final volume of 5 ml. The valve for the volatile fraction syringe is closed and the contents are made ready for the purge and trap step. This clean-up approach has been applied to radioactive, hazardous mixed waste samples containing normal paraffin hydrocarbon from a Hanford single-shell tank [53].

Although some authors have used microwaves as an auxiliary form of energy in HS — the ensuing methods being labelled "microwave-assisted purge and trap" — microwaves are in fact used in a preliminary, separate sample dissolution step that is followed by cooling and insertion of the sample into the HS module [54].

As noted earlier, the most widely used piece of equipment after the headspace module is a gas chromatograph, which is in turn connected to a suitable (flame ionization, electron capture, mass spectrometric, atomic absorption, atomic emission) detector. Some high-resolution detectors including mass spectrometers have been directly connected to the HS module.

Refining the procedures by which the sample analytes are transferred to the gas chromatograph or detector is as important as refining HS devices themselves. Initially, interfacing in HS applications involved connecting parts via a 1/8-in i.d. packed column. The most common way of connecting the HS module into the GC plumbing was by cutting the carrier gas inlet line approximately 2 inches from the injection port. Then, using a low-volume stainless steel union, the carrier gas was routed back to the injection port via the heated transfer line. Splicing into a packed injection port is not suitable for capillary columns. Because of the decreased flow-rates used (1–10 ml/min), desorbed analytes have longer residence times in the injector, which exposes them to a relatively hostile environment of hot metal surfaces. Sample losses are a usual result of such long exposures. Even more detrimental is the large dead volume in the injector cavity.

The most useful and effective interfaces between HS devices and megabore columns are probably the low-dead volume (LDV) interface and the PT–LDV injector. The former is implemented by installing a special fitting in a pre-existing GC injector body and associated plumbing. Like other GC injectors, the PT injector is a stand-alone injection port with its own plumbing and heater assembly designed specifically for use with an HS device. In either case, the sample transfer path is either deactivated glass or fused silica-lined tubing reduced to a minimum volume, making it conducive to the lower flow-rates associated with 0.53 and 0.45-mm i.d. columns. Syringe injections can be performed with both interfaces and each type is well-suited to GC systems or GC–MS ones using a jet separator.

The jet separator is a bridge between the high-desorption flow of the PT module and the vacuum of the ion source on the MS. The trend away from using jet separators is mainly because they require high maintenance and sensitivity suffers as a result of a discriminatory portion of the sample potentially being lost to the separator's vacuum system. For these reasons, interfacing 0.25-mm i.d. and thinner capillary columns directly into the MS has become a more popular interfacing choice. Two different interfaces (viz. the cryogenic injector and the split–splitless injector) are used for this purpose. Cryofocusing is necessary when the desorption rate of the carrier gas is less than about 5 ml/min as peak distortion occurs for those analytes not refocused in the column stationary phase under these conditions. This type of interface is usually connected to an existing GC injector that is cooled as required. Once desorption from the trap begins, a cryoinjector thermally traps the analytes at liquid nitrogen temperatures (−160°C) in a short section of capillary tubing (usually of deactivated fused silica). Higher refocusing efficiency is achieved when the fused silica is coated with a temperature-stable polymer phase [55]. When trap desorption is complete, the injector is rapidly heated (at a rate greater than 1000°C/min) to efficiently inject the sample onto the head of the analytical column in a tight band.

One PT–MS interface that has received much attention is the capillary split injection interface, which makes both cryofocusing and jet separation redundant for interfacing PT to a mass spectrometer. Figure 4.9 shows the flow path for the desorption step in PT. A variable total gas flow is supplied through the sample concentrator's six-port valve from a constant-flow controller. This gas is used to desorb the analyte(s) from the trap to the injector via the sample concentrator's heated transfer line, which is connected via an LDV union directly into the carrier gas inlet of a split GC injector. With a 0.25-mm i.d., or smaller, capillary column connected to the inlet, 0.5–1.5 ml/min of column flow is allowed into the column, the remainder being diverted to a variable split vent. The most salient advantages of this interface are that desorption and the chromatographic separation take place at their optimum flow-rates and that either the total flow or the split vent flow can be adjusted for low- and high-level analyses.

4.3.3. Theoretical background

Although the static and multiple headspace modes use similar equipment, the two rely on rather different principles. On the other hand, purge and trap, and dynamic headspace, possess the same foundation, the only difference between them being the location of the tubing used to transfer the carrier gas to the sample container.

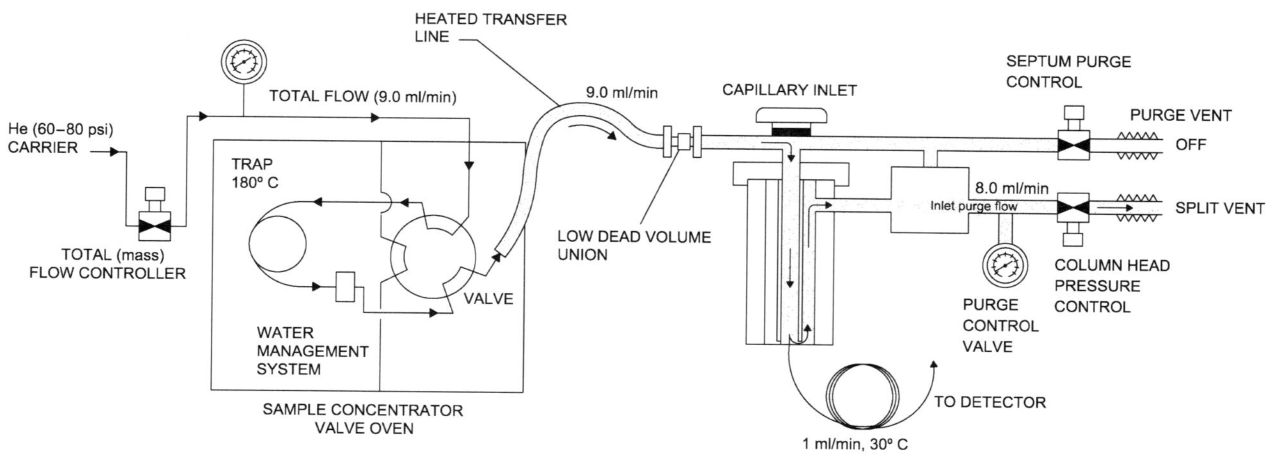

Fig. 4.9. Split injection interface for a PT–MS coupled system. (Reproduced with permission of Preston Publications.)

Static or equilibrium headspace: theory

The vapour–liquid distribution (partition) coefficient, K, is defined as the ratio of the equilibrium concentration of the solute in the sample (C_s) to that in the gas phase (C_g):

$$K = \frac{C_s}{C_g} = \frac{m_s}{m_g} \cdot \frac{V_g}{V_s} \tag{4.1}$$

where m_s and m_g are the amount of volatile compound in the sample and gas phase, respectively, and V_g and V_s the volumes of the gas and sample phase, respectively.

The mass balance for a volatile analyte between two phases can be expressed with reference to one of the states as follows [56]:

$$m_s^0 = m_s + m_g \quad \text{or} \quad C_s^0 V_s = C_s V_s + C_g + V_g \tag{4.2}$$

where C_s^0 denotes the analyte concentration in the sample phase before headspace equilibrium is reached; C_s and C_g are the equilibrium concentrations of the sample and gas phase, respectively; and V_s and V_g their respective volumes.

If the phase ratio, β, is defined as V_g/V_s, then eq. (4.2) can be transformed into

$$C_s^0 = C_g (K + \beta) \tag{4.3}$$

The peak area A obtained after chromatographic separation and detection is proportional to the analyte concentration in the gas phase:

$$A \approx C_g = \frac{C_s^0}{(K + \beta)} \tag{4.4}$$

Partition coefficients can be determined by vapour-phase calibration (VPC) [54], by the phase-ratio variation (PRV) method [also known as the "vapour–liquid equilibrium" (VLE) method] [57] for many solvents in their aqueous solutions, and by VLE for ethanol in water. If two sample vials of different volume are both filled with the same sample, the partition coefficient, K, will be the same. In order to determine the solute's partition coefficient, K, each vial, at equilibrium, is subjected to headspace analysis in order to derive the slope of the linear equation (4.1). The concentrations of a solute in the two vials can be written as

$$C_s^0 = C_{g,1}(K + \beta_1) \tag{4.5}$$

and

$$C_s^0 = C_{g,2}(K + \beta_2) \tag{4.6}$$

where $\beta_1 = V_{g,1}/V_{s,1}$ and $\beta_2 = V_{g,2}/V_{s,2}$.

The solute concentration in the vapour phase, C_g, is proportional to the peak area derived from GC measurements. Thus, the following correlation can be obtained from two consecutive measurements:

TABLE 4.2

PARTITION COEFFICIENTS, *K*, FOR SELECTED COMPOUNDS IN WATER

Compound	40°C	60°C	80°C
Tetrachloroethene	1.5	1.3	0.9
1,1,1-Trichloroethane	1.6	1.5	1.2
Toluene	2.8	1.6	1.3
o-Xylene	2.4	1.3	1.0
Cyclohexane	0.07	0.05	0.02
n-Hexane	0.14	0.04	⩾0
Ethyl acetate	62.4	29.3	17.5
n-Butyl acetate	31.4	13.6	7.6
Isopropyl alcohol	825	286	117
Methyl isobutyl ketone	54.3	22.8	11.8
Dioxane	1618	642	288
n-Butanol	647	238	99

$$\frac{C_{g,1}}{A_1} = \frac{C_{g,2}}{A_2}$$

The partition coefficient can be derived from eqs (4.5) and (4.6), following rearrangement, and calculated from

$$K = \frac{A_1\beta_1 - A_2\beta_2}{A_2 - A_1} \tag{4.7}$$

Table 4.2 lists the partition coefficients for selected compounds.

Multiple headspace: theory

This HS mode uses a mathematical model developed by McAuliffe [58,59] that is applied in a stepwise manner.

Step 1 corresponds to the end of the first equilibration process, when equilibrium between the sample in the vial and the headspace above it has been reached. Similarly as before, the thermodynamic equilibrium is described by the partition coefficient, but subscripts now include a number to identify the particular step. Thus:

$$K = C_{s1}/C_{g1} \tag{4.8}$$

$$m_1 = m_{s1} + m_{g1} \tag{4.9}$$

$$C_{s1} = m_{s1}/V_s \tag{4.10}$$

$$C_{g1} = m_{g1}/V_g \tag{4.11}$$

Note that all concentrations in the previous equations are amounts per volume.

Because equilibrium is followed by pressurization and sample transfer steps, and then by separation of the volatile compounds in the column,

$$A_1 = \text{const} \cdot C_{g1} \tag{4.12}$$

It should be noted that, during sample transfer to the column, the total amount of analyte present in the vial's headspace is reduced by the amount transferred. The change is neglected here because the amount transferred is usually small relative to the total amount present in the vial. In the case of splitless sampling into an open-tubular (capillary) column, the total volume of gas taken from the headspace is about 50 μl. If a vial with a volume of 22 ml is assumed, 50% of which will be the headspace ($\beta = 1$), then the transferred aliquot will account for only 0.45% of the total headspace. Compared to the vented volume (see next step), this is significant. In the case of split sampling or when packed columns are used, the total volume of gas taken from the headspace is higher (about 0.5 ml, which accounts for 2.3% of the total headspace). Regardless of its volume, it is still reproducible and simply adds to the vented volume. Thus, at the end of step 1, the vial contains the respective headspace and sample volumes of V_g and V_s; the amounts and concentrations of the analyte in these phases are m_{g1}, m_{s1}, C_{g1} and C_{s1}. At this point, the pressure in the vial is p_h, which is higher than the atmospheric level.

Step 1/2 is a transitional step corresponding to the venting step that follows the first headspace analysis (Fig. 4.4D). The gas phase in the vial is vented to the atmosphere; as a result, its pressure, p_h, is reduced to atmospheric or near-atmospheric level (p_0). This step can be visualized as a gas expansion: the gas, having a volume V_g, expands to $V_{g1/2}$ as its pressure is reduced to p_0. Note that p_0 is not always identical with atmospheric pressure. Pressure release (venting) in the pressurized vial should be very rapid so as to avoid any change in the sample concentration (C_{s1}). Consequently, p_0 may be slightly higher than atmospheric pressure as a result of the partial pressures of the matrix compound(s). This is particularly the case with aqueous samples when the pressure p_0 in the vial is dictated by the saturated vapour pressure. Gas expansion during venting is instantaneous and, initially, only affects the gas phase. In other words, the amount and concentration of the analyte in the sample remains unchanged (the analyte has not time to repartition), so

$$C_{s1/2} = C_{s1} = \frac{m_{s1}}{V_s} \tag{4.13}$$

Also, the amount m_{g1} originally present in V_g will now be present in the expanded volume $V_{g1/2}$, so the concentration of this expanded volume will be

$$C_{g1/2} = C_{s1} = \frac{m_{g1}}{V_{g1/2}} \tag{4.14}$$

and

$$C_{g1/2} < C_{g1} \tag{4.15}$$

From the basic gas laws, the relationship between steps 1 and 1/2 can be described as

$$P_h \cdot V_g = p_0 \cdot V_{g1.2} \tag{4.16a}$$

$$\frac{V_g}{V_{g1/2}} = \frac{p_0}{p_h} \tag{4.16b}$$

We shall henceforward use ρ to refer to the ratio of the two pressures:

$$\rho = p_0/p_h \tag{4.17}$$

Solving eqs (4.11) and (4.14) for the respective volumes, and substituting the resulting expressions into eq. (4.16b) yields

$$C_{g1/2}/C_{g1} = \rho \tag{4.18}$$

The gas volume $V_{g1/2}$ in this step consists of two parts, namely: one, V_{v1}, that is vented, and the other, V_g, which corresponds to the headspace volume above the sample in the vial:

$$V_{g1/2} = V_{v1} + V_g \tag{4.19}$$

Similarly, the amount of analyte originally present in the vial headspace under equilibrium conditions, m_{g1}, can be split into two parts, namely: one, $m_{v1/2}$, that is vented and the other, $m_{g1/2}$, that remains in the vial headspace:

$$m_{g1} = m_{v1/2} + m_{g1/2} \tag{4.20}$$

In this step, the analyte concentration in the headspace will be $C_{g/12}$; the same as established for the expanded volume [*cf.* eq. (4.14)]. Such a concentration can be expressed in terms of $m_{g1/2}$ and V_g (the vial's headspace volume):

$$C_{g1/2} = m_{g1/2}/V_g \tag{4.21}$$

Returning to the total amount m_1 present in the vial in step 1, the following material balance can be written [*cf.* eqs (4.9) and (4.20)]:

$$m_1 = m_{g1/2} + m_{v1/2} \tag{4.22}$$

Since $m_{v1/2}$ is vented, the total amount of analyte remaining in the vial in this transition step will be

$$m_{1/2} = m_{s1} + m_{g1/2} \tag{4.23}$$

In *step 2*, the needle is closed and the vial is re-equilibrated at the same temperature as in step 1. At equilibrium, the distribution coefficient will be the same as in step 1, which means that the ratio of the equilibrium concentrations will also be the same; however, the individual concentrations will now be different:

$$K = C_{s2}/C_{g2} \tag{4.24}$$

With these new concentrations, the respective amounts in the sample and gas phase at equilibrium will be m_{s2} and m_{g2}, respectively; because the volumes of sample and gas phase remain constant,

$$C_{s2} = m_{s2}/V_s \tag{4.25}$$

$$C_{g2} = m_{g2}/V_g \tag{4.26}$$

The total amount of analyte now present in the vial, m_2, will be the same as that at the end of the transition step, which, however, is less than m_1 (the amount of sample present in step 1):

$$m_2 = m_{1/2} \tag{4.27}$$

Consequently,

$$m_2 = m_{s2} + m_{g2} \tag{4.28}$$

Substituting eqs (4.23) and (4.28) into (4.27) yields

$$m_{s2} + m_{g2} = m_{s1} + m_{g1/2} \tag{4.29}$$

Solving eqs (4.10), (4.21), (4.25) and (4.26) for the respective amounts allows the following material balance to be written:

$$C_{s2} \cdot V_s + C_{g2} \cdot V_g = C_{s1} \cdot V_s + C_{g1/2} \cdot V_g \tag{4.30}$$

Substituting $\beta \cdot V_s$ for V_g in the previous equation gives

$$C_{s2} + C_{g2} \cdot \beta = C_{s1} + C_{g1/2} \cdot \beta \tag{4.31}$$

By solving eqs (4.8), (4.24) and (4.18) for C_{s1}, C_{s2} and $C_{g1/2}$, respectively, and substituting them into (4.31), one obtains

$$C_{g2} \cdot K + C_{g2} \cdot \beta = C_{g1} \cdot K + C_{g1} \cdot \rho$$

$$\frac{C_{g2}}{C_{g1}} = \frac{K + b \cdot \rho}{K + \beta} = \frac{K/(\beta + \rho)}{K/(\beta + 1)} \tag{4.32}$$

Substituting C_{g1} for $C_{g1/2}$ is legitimate because, at the end of step 2, the pressure in the vial will be the same as that at the end of step 1.

If equilibrium is followed by pressurization and sample transfer, and by another GC analysis from the new headspace aliquot, then the peak area obtained will be A_2. By analogy with eq. (4.12), the following expression can be written:

$$A_2 = \text{const} \cdot C_{g2} \tag{4.33}$$

Based on the proportionalities between the equilibrium headspace concentrations and the peak area [*cf.* eqs (4.12) and (4.33)], the concentrations in eq. (4.32) can be substituted by the respective peak areas:

$$\frac{A_2}{A_1} = \frac{K/(\beta + \rho)}{K/(\beta + 1)} \tag{4.34}$$

Further steps: if the MHS procedure is repeated several times, the peak area obtained in the i-th step can be described by the following relation:

$$\frac{A_i}{A_1} = \left(\frac{K/(\beta + \rho)}{K/(\beta + 1)}\right)^{i-1} \tag{4.35a}$$

or

$$\frac{A_i}{A_1} = \left(\frac{K/(\beta + \rho)}{K/(\beta + 1)}\right)^{i-1} \tag{4.35b}$$

Equation (4.35b) describes a geometric progression the quotient, Q', of which is

$$Q' = \frac{K/(\beta + \rho)}{K/(\beta + 1)} \tag{4.36}$$

The mathematical condition for a convergent geometric progression is that the value of its quotient, Q', is between -1 and $+1$. Since $p_0 < p_h$, ρ will always be less than unity. For this reason [*cf.* eq. (4.36)], Q' will also be less than unity (between 0 and $+1$). This fulfils the condition for a convergent geometric progression. Q' can be calculated from a regression analysis of the linearized form of eq. (4.35b):

$$\ln A_i = (i - 1) \cdot \ln Q' + \ln A_1 \tag{4.37}$$

The sum of the geometric progression expressed by eq. (4.35b) is

$$\sum_{i=1}^{n} A_i = \frac{A_i}{1 - Q'} \tag{4.38}$$

Substituting eq. (4.36) for Q' in eq. (4.38) yields

$$\sum_{i=1}^{n} A_i = A_i \frac{K/(\beta + 1)}{1 - \rho} \tag{4.39}$$

This means that, if all three parameters (K, β and ρ) are known, the total peak area can also be calculated from a single area measurement. However, this is not recommended for two reasons: first, the parameter values are usually not readily available and their accurate determination would take at least as much time as the three or four consecutive steps of MHS; second, determining a single peak area provides no information about the existence of equilibrium conditions, which is a pre-requisite for MHS [60].

Recently, a new headspace mode was developed that allows complete or partial sampling of the vapour phase from a pre-equilibrated headspace vial, using a dual needle system. The theoretical treatment of this system has shown that the maximum mass (M_{max}) of an analyte in the gaseous phase (M_g) is generated when the gas–condensed phase ratio (V_g/V_c) in the equilibrated headspace vial is equal to the square root of the distribution coefficient [61].

Multiple headspace and solid samples

The most important criticism concerning MHS is that it is very difficult to implement with solid samples [62]. Were this true, it would obviously severely restrict its applicability, but it is not. Proper conduct of an MHS procedure requires fulfilling the following five conditions: (a) the relationship between peak area and the concentration (amount) of analyte in the headspace should be linear throughout the concentration range of interest; (b) the coefficient of distribution of the analyte between the headspace and the sample should be constant throughout the range examined; (c) before the headspace is sampled, it should reach equilibrium with the sample in the vial; (d) pressure release (venting) from the pressurized vial against the atmosphere should be rapid enough to avoid any change in the analyte concentration in the sample during this step; and (e) ρ ($= p_0/p_h$) should be the same in every MHS measurement step. Fulfilment of these conditions results in constancy of Q', which in turn results in a high correlation coefficient for the linear regression analysis of eq. (4.37) or, expressed differently, in a linear relationship between ln A_i and $(i-1)$. In other words, if a linear relationship is found, the conditions are fulfilled, and vice versa.

Condition (a) above is not exclusive to headspace analysis; in fact, it is a pre-requisite for quantitative analysis of any sample in gas chromatography. Essentially the same is true for condition (b). Condition (d) is primarily a design problem. Finally, constancy of ρ is assured by proper automation of the system (i.e. by exact repetition of the operational parameters) and by the fact that the calibration standard is carried through the same MHS steps as the sample itself. Therefore, the greatest problem is posed by the need to ensure equilibrium between the two phases in the vial.

With liquid samples, the vial containing the two phases (gas and liquid) represents a partitioning system. Here, equilibration will sooner or later be accomplished and the time needed for it will depend primarily on the temperature and volume of the liquid. At increased temperatures, equilibration is faster than with larger volumes because, in the latter case, the analyte molecules take longer to diffuse in and out of the sample. Establishing the minimum time needed for equilibration in an unknown sample is one of the basic measurements recommended in headspace analysis [45,63].

With solid samples, equilibrium can only be reached under certain conditions or after specific changes. One major advantage of MHS is that it can be used as a preliminary approach to establish whether regression analysis will expose linearity in the ln A_i vs $(i-1)$ relationship. This allows the analyst to determine whether the solid sample represents a partition system or an adsorption system.

Typical solid samples representing a partition system include polymers heated above their glass transition point. These samples are amenable to application of the usual HS modes (e.g. the standard-addition technique). The standard can be added to the gas phase in the closed vial by injection through the septum, equilibrium being established by diffusion from both sides. If the regression analysis reveals the absence of linearity, then either some instrumental parameter (e.g. the sample temperature or equilibration time) will have to be adjusted or the system is not a partition but rather an adsorption system.

If the solid sample represents an adsorption system, then the distribution of the volatile analyte(s) will be dictated by an adsorption coefficient. Usually, adsorption coefficients are a function of concentration or are constant over a short concentration range only. Obviously, such a non-linear system cannot be handled by the MHS extrapolation technique. There are, however, a number of alternatives for adapting the system to MHS or MHS-like determinations, three of which are described below.

If the volatile analyte is present at a very low concentration and is mostly adsorbed on the sample surface, one can physically measure the peak area for the volatile analyte, step by step, until no further peak is found. Usually, only a few extraction steps are needed until the whole amount of analyte is depleted. In this case, the sum of peak areas will be proportional to the total amount of analyte present in the sample.

Alternatively, one can convert the adsorption system into a partition system by adding a displacer to the sample in the vial headspace. One typical example is the determination of residual halocarbon solvent in decaffeinated instant coffee using an excess of water as displacer [45]. For analytes adsorbed on activated charcoal — which is used in personal monitoring tubes — benzyl alcohol is an effective choice for desorbing the volatile analyte [49,64].

Finally, one can use the "suspension approach" [65] to determine water in solid samples that would otherwise act as adsorption systems. The sample is suspended in a water-miscible dry polar solvent such as ethylene glycol monomethyl ester (EGME). The second solvent displaces water from the solid sample and dissolves it; if the water concentration in EGME is low enough (less than 5% v/v), then the peak area is linearly related to the concentration.

With some solid samples, even the previous modifications fail and no apparent equilibrium is reached. Such is the case with coarse or bulky materials (e.g. large granules or solid pieces). Occasionally, the situation can be effectively improved by milling granules while cooling with dry ice, as in the determination of residual styrene monomer in large polystyrene granules [45].

It should be noted that the problem of very slow diffusion in and out of a bulky solid matrix makes any type of gas extraction impractical and the results dubious. In other words, these restrictions apply not only to the MHS approach but also to dynamic headspace processes in general.

Purge and trap: theory

In a continuous gas extraction system such as purge and trap and dynamic headspace, the volatile analyte is continuously purged or removed from the sample by means of an inert gas system and the process can continue until the analyte is completely eliminated. The removed analyte is trapped from the gas stream; hence, at the end of the continuous gas extraction, the whole amount of analyte is available for determination using standard GC techniques, with no interference from the excess matrix compound.

The efficiency of the separation process is controlled by the rate of diffusion of the analyte from the liquid or solid sample phase to the gaseous phase. An infinitesimal variation in concentration of the sample phase, $\mathrm{d}C_s$, can be described by the equation

$$\mathrm{d}C_s = \frac{KA}{V}\,\mathrm{d}t \tag{4.40}$$

where K is the flux of analyte from the sample to the gas phase, expressed in mol/cm$^2 \cdot$ s; A is the interface area, in cm^2; and V is the solution volume, in millilitres. Based on Fick's first law and the Nernst approximation, eq. (4.40) can be rewritten as

$$\frac{\mathrm{d}C_s}{C} = -\frac{\mathrm{D}}{h\delta}\,\mathrm{d}t \tag{4.41}$$

where D is the diffusion constant of the analyte, δ the thickness of the diffusion layer and h the height of the sample in the vessel.

Under the boundary condition that $C_l = C_s^0$ at $t = 0$, integration of the previous equation yields a first-order kinetic-like function:

$$C_s = C_s^s \cdot \exp(-bt) \tag{4.42}$$

where $b = \mathrm{D}/h\delta$.

The efficiency with which a given analyte will be removed can be estimated from the following ratio for an equivalent mass of analyte processed by both purge and trap (PT), and direct injection (DI):

% Extraction = Area of analyte (by PT)/Area of analyte (by DI)

The extraction efficiency for an analyte can be predicted using Henry's law at constant extraction volume, sample volume and extraction temperature [66]. The extraction of most analytes follows a first-order kinetics irrespective of the effects of hydrogen bonding or physical properties such as boiling point or water solubility [67].

4.3.4. Variables affecting performance

This section deals not only with the variables pertaining to the headspace process (i.e. those affecting the removal of volatile species) but also with those involved in the passage of such species to the gas phase and the insertion of the compounds removed from the sample into the chromatographic column.

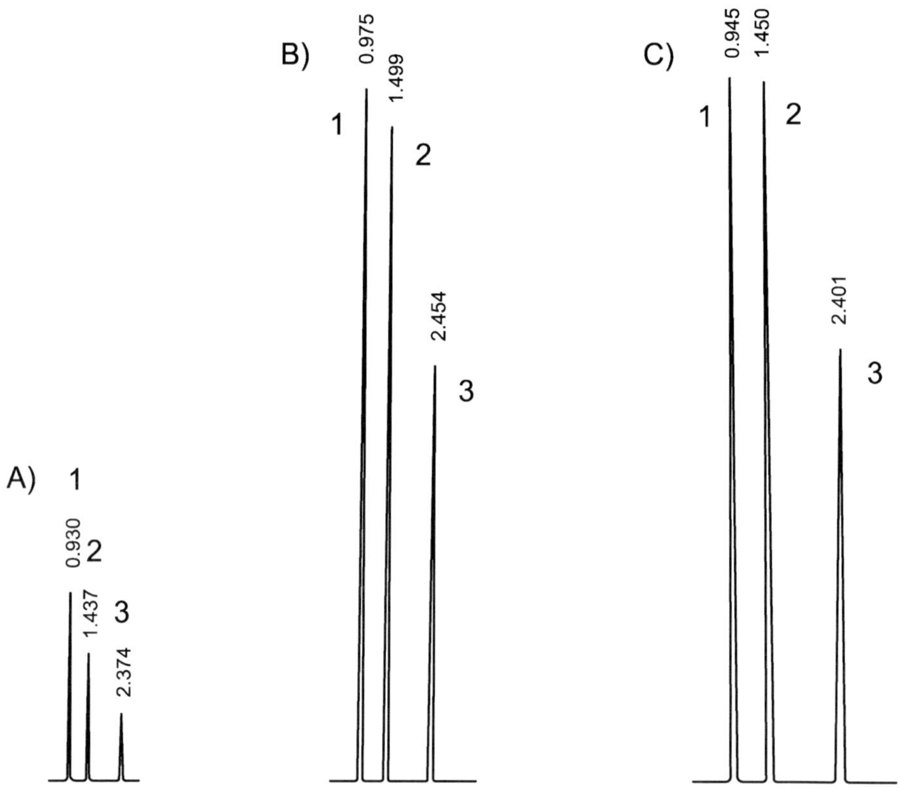

Fig. 4.10. Chromatograms for solvent residues in pharmaceutical base material after equilibration at (A) 40°C, (B) 70°C and (C) 80°C. 1 benzene, 2 toluene, 3 *o*-xylene. All chromatograms were obtained by using a 15 m × 0.32 mm i.d. × 5 μm SBP™-1 fused-silica column. (Reproduced with permission of Springer-Verlag.)

Variables common to all headspace modes

The *partition coefficient* is the driving force for removal of the volatile species to the gas phase in establishing equilibrium, attainment of which depends on the particular procedure. This coefficient depends on the analyte concentration: it decreases with increasing concentration but remains constant over a concentration range that depends on the particular type of compound. Thus, the range of virtual constancy for BTX (benzene, toluene and *o*-xylene) is 0.1–10 μg/ml; as a result, these compounds can only be determined reliably at concentrations below 10 μg/ml [68].

Temperature has a dramatic effect on the separation efficiency of conventional HS with both solid and liquid samples. The partition coefficient decreases slightly with increasing temperature. As a result, the headspace technique is more sensitive at high temperatures. Figure 4.10 illustrates the effect on equilibrium HS for liquid samples at three different temperatures [68]. Users tend to decrease the distribution coefficient

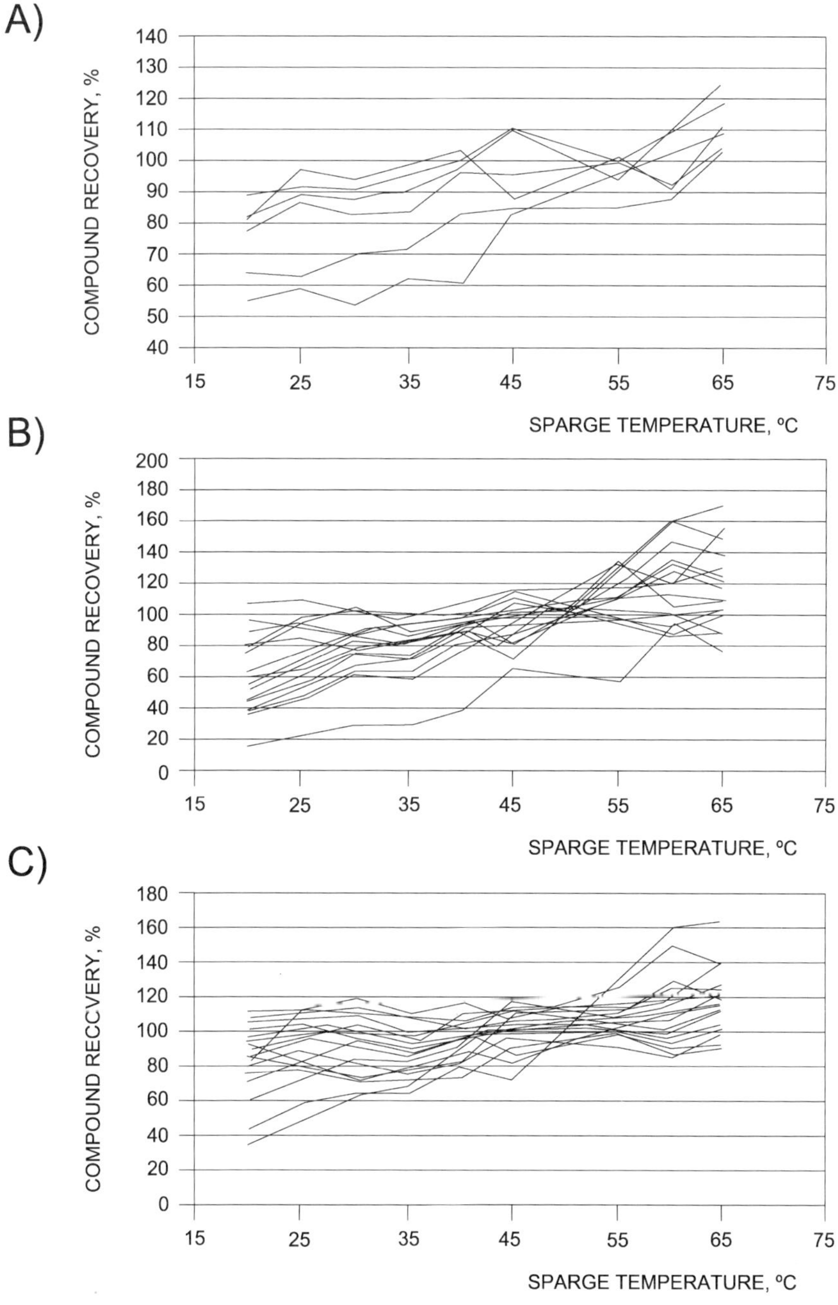

Fig. 4.11. Effect of the sparge temperature on compound recovery. (A) Gases. (B) Polar compounds. (C) Non-polar compounds. (Reproduced with permission of BPA International.)

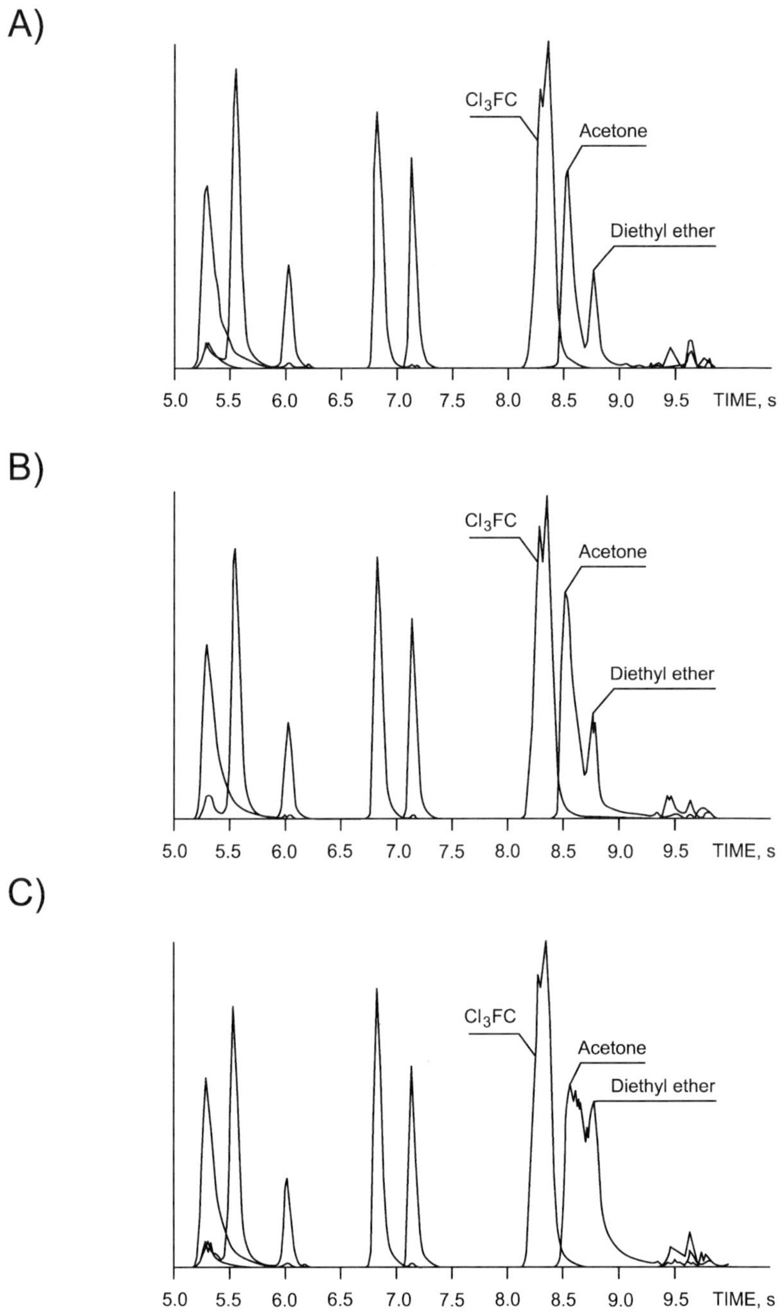

Fig. 4.12. Extracted ion chromatograms for compounds 1–8 on the list of US EPA Method 524.2, rev. 4, at a sample sparge temperature of (A) 45°C, (B) 55°C and (C) 65°C. (Reproduced with permission of BPA International.)

(i.e. to raise the concentration of the target compound in the headspace) by increasing the thermostat temperature. However, the matrix compound should also be considered if it is volatile as its concentration in the headspace should not increase beyond a certain limit, particularly if its peak elutes from the column close to and/or before the analyte peak — which will have an unfavourable effect on the GC separation. Thus, when ethanol is determined from aqueous solutions (e.g. blood), samples are most often thermostatted at 60°C, where the distribution coefficient for the alcohol is still high (about 550). Using a somewhat higher thermostat temperature would result in a lower distribution coefficient for ethanol (e.g. *ca.* 250 at 80°C). At this temperature, however, the vapour pressure of water increases to an undesirably high level and its large peak, which emerges fairly close to that for ethanol, interferes with it. On the other hand, halocarbons have much higher vapour pressures at lower temperatures, so a near-unity distribution coefficient provides highly favourable conditions for their separation.

The effect of temperature on PT depends strongly on the particular type of compound. Thus, a PT recovery study involving the analysis for VOCs using EPA Method 524.2, rev. 4 [69], which involved analysing replicate 5 ml aliquots of a 4.0 ng/ml standard at purge temperatures from ambient level (*ca.* 19°C) to 65°C, revealed the effect of temperature to depend on the specific analyte. As can be seen in Fig. 4.11, the increase in extraction efficiency with increasing temperature was more marked for the polar compounds, most of which were quantitatively recovered at 45°C. Temperatures above this level resulted in little additional improvement in compound recovery, and, as can be seen from Fig. 4.12, also in decreased chromatographic resolution in some instances (primarily with polar compounds). The detrimental effect of increased temperatures arose from the increased amount of water that was transferred to the trap along with the compounds. The additional water occupied adsorptive sites in the trap that would otherwise have been taken by the target compounds. The additional water forced the target compounds to migrate further into the trap, thus having adverse effects on the desorption and chromatographic profiles [70–72].

The *presence and nature of salts* also affects — to a lesser extent than temperature — the partition coefficient, and hence the separation efficiency. The effect is illustrated in Fig. 4.13 for BTX at a constant concentration of the target salt; the effect of the salt concentration on K is shown in Table 4.3. It should be noted that the phase ratios in the vials for the liquid samples were kept constant at $\beta_1 = 3$ and $\beta_2 = 1$.

The *addition of solvents* influences the separation of volatiles from samples by altering their partition coefficients. The sign of the effect depends on the relative polarity of the compounds and solvents involved. Table 4.4 illustrates the effect of the presence of an additional polar solvent such as an alcohol or acetone in an aqueous solution of BTX. As can be seen, the partition coefficient for the three analytes increases in the presence of a polar organic additive [e.g. 5% v/v alcohol (methanol, ethanol, *n*-propanol, *n*-butanol)] or acetone in the aqueous solution. Consequently, the additives aid dissolution of polarizable aromatics in water and ultimately decrease the sensitivity by a factor of 2–3.

The *sample volume* has rather a different influence depending on the particular HS mode. For equilibrium HS, Ettre and Kolb introduced the concept "sample phase fraction", mathematically defined as $\phi_s = V_s/V_g$, and showed that, provided this parameter

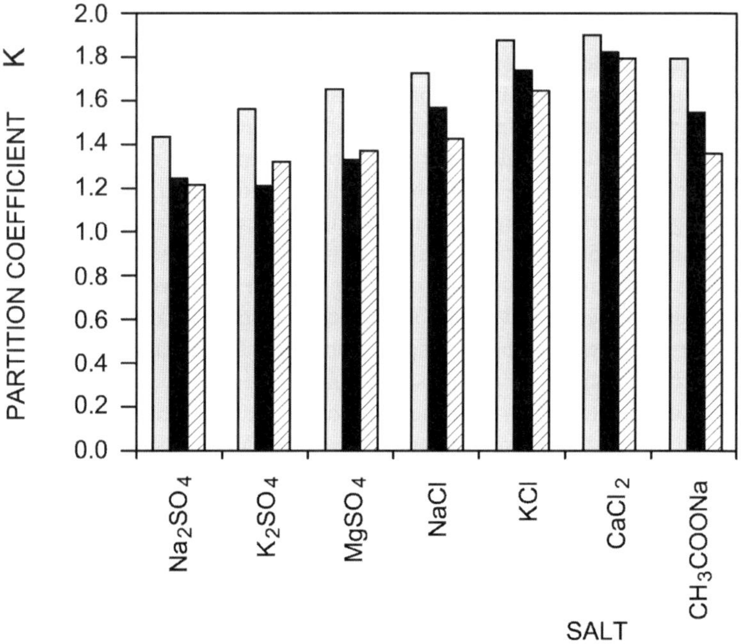

Fig. 4.13. Effect of salts on the partition coefficients of three analytes in aqueous solutions in a gas–water system. The salt concentration was 20% w/w. (Reproduced with permission of Springer-Verlag.)

is kept constant, the vial's volume has no influence on the results as the analyte concentration in the headspace will be almost constant except with low distribution coefficients [73]. In PT, the influence of the sample volume is dramatically affected by other variables including the sample temperature and total extraction volume. The above-mentioned study of VOC recovery using EPA Method 524.2, rev. 4 [69], revealed that changing the sample volume from 5 to 25 ml lowered the minimum detection limits (MDLs) for 93% of the 86 target analytes. However, MDLs were not reduced by a factor of 5 as might be expected; rather, the MDLs for all types of compounds (gaseous, polar and non-polar) were improved by a factor of only 2–3 relative to a 5 ml sample.

Variables pertaining to purge and trap

The PT extraction efficiency depends essentially on two variables, viz. the total extraction volume and the sample temperature. The total extraction volume, or the total amount of purge gas that is passed through the sample during the extraction step (or over the sample in DHS) is determined by multiplying the flow-rate of the carrier gas by the time it is circulated. Since the earliest PT methods, the purge time has been specified to be 11 min at a flow-rate of 40 ml/min, which provides a total extraction volume of 440 ml. In theory, these flow-rates have been specified to optimize the extraction of volatiles (especially

TABLE 4.3

EFFECT OF SALTS ON PARTITION COEFFICIENTS IN AQUEOUS SOLUTIONS AT 40°C. PHASE RATIOS: $\beta_1 = 3$, $\beta_2 = 1$

Salt	Salt concentration (%, w/w)	Partition coefficient		
		Benzene	Toluene	*o*-Xylene
Na_2SO_4	5	2.56	2.35	2.00
	10	1.87	1.72	1.62
	20	1.44	1.25	1.22
	40	1.21	1.17	1.15
K_2SO_4	5	2.66	2.47	2.29
	10	1.95	1.82	1.75
	20	1.56	1.21	1.32
	40	1.30	1.24	1.22
$MgSO_4$	5	2.72	2.52	2.31
	10	2.05	1.97	1.78
	20	1.65	1.32	1.37
	40	1.35	1.29	1.20
NaCl	5	2.60	2.41	2.15
	10	1.92	1.80	1.70
	20	1.72	1.56	1.42
	40	1.54	1.40	1.31
KCl	5	2.86	2.52	2.45
	10	2.32	2.12	2.00
	20	1.86	1.74	1.63
	40	1.62	1.52	1.48
$CaCl_2$	5	2.95	2.60	2.50
	10	2.42	2.23	2.11
	20	1.89	1.81	1.78
	40	1.75	1.69	1.70
$NaCH_3COO$	5	2.52	2.50	2.35
	10	2.01	1.85	1.75
	20	1.79	1.54	1.35
	40	1.62	1.42	1.27

TABLE 4.4

PARTITION COEFFICIENTS IN AQUEOUS SOLUTIONS CONTAINING VARIOUS ORGANIC SOLVENTS AT A 5% CONCENTRATION AT 40°C

Solvent	Partition coefficient		
	Benzene	Toluene	*o*-Xylene
Methanol	4.55	3.98	3.58
Ethanol	9.56	8.25	7.85
Propanol	6.25	5.25	4.56
Butanol	7.52	5.87	5.12
Acetone	8.45	6.65	6.12

relatively non-polar analytes) from the sample matrix and trapping of the gaseous analyte on the concentrator's adsorbent trap. In practice, the purge flow is usually optimized over a series of standard analyses on a given system and is rarely, if ever, measured again. The effect of the purge flow-rate is becoming apparent as more polar analytes are being added to existing purge and trap methods. The analytes typically analysed by PT can be classified into the four categories of Fig. 4.14. The extraction of analytes that are difficult to purge can be improved by increasing the total extraction volume that is passed through the sample during the purge step. This can be done by increasing either the purge time at a constant flow-rate or the purge flow-rate. Additional purge flow, however, is likely to improve the extraction of inefficiently removed compounds only at the expense of the light gases. In general, over the specified purge time of 11 min, the extraction efficiency of all but the heaviest compounds — even those characterized as well-behaved — decreases as a result of trap breakthrough across the total extraction volume range 200–600 ml. The new, more polar analytes added to EPA Method 542.4 (revised version 4) [73] have further challenged the purge efficiency tradeoff. In contrast to the analyte list in revision 3, the efficiency of these newly added analytes is generally improved over the 200–600 ml extraction volume range. Therefore, the trend towards reducing the purge time as a way of reducing the overall analysis time is likely to be reversed because of the addition of new purge-inefficient compounds. The breakthrough volumes for most volatile organics have been determined [74–76] and may constitute major reference material as the list of analytes processed using the purge and trap technique expands.

Other steps in the PT process potentially influenced by temperature are adsorption and desorption. As noted in Section 4.3.3., the optimum temperature in the former depends both on the type of compound to be retained and on the trapping material used.

As shown in Section 4.3.3, desorption performance is dependent upon the rate at which the trap is heated, the final trap temperature and the total desorb flow (i.e. the amount of gas that is passed through the trap). One other aspect of temperature to be considered is desorb preheat. Preheating the adsorbent trap immediately prior to desorption without

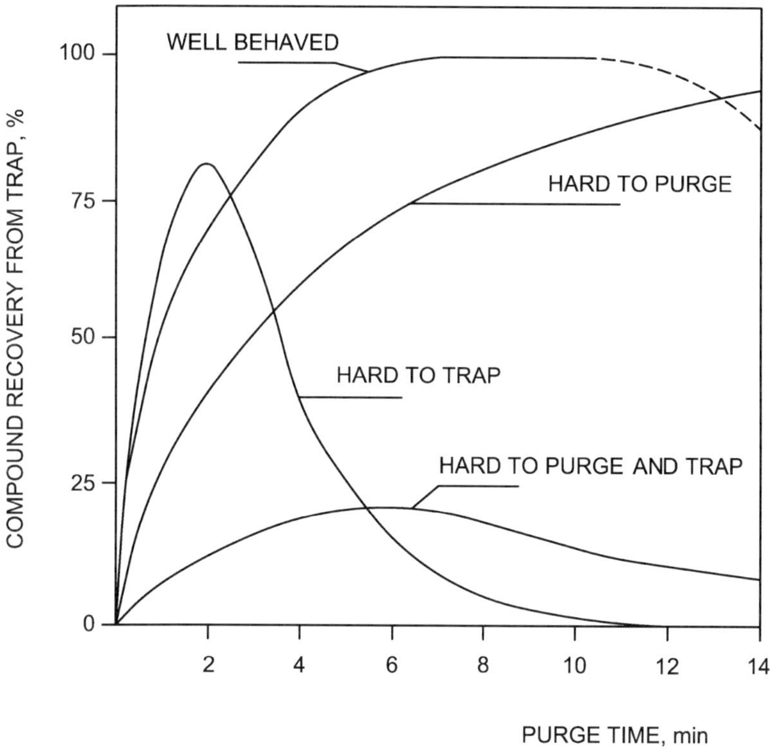

Fig. 4.14. Different types of analyte behaviour towards purge and trap. (Reproduced with permission of Preston Publications.)

a carrier flow is usually redundant when rapid direct-resistance heating is used. However, the technique is occasionally useful to narrow the desorption bandwidth when the release of selected compounds from the trap is slow. Under these conditions, slowly released compounds spread out in the transfer gas stream and fail to be inserted into the column as a discrete plug. By heating the trap with no gas flow, the compounds can be released in a narrower band with reduced linear diffusion.

4.3.5. Calibration and analytical features

Calibration

Calibration in headspace work is usually done by using one of four different techniques, namely: external-standard, internal-standard, standard-addition and vapour-phase calibration.

External-standard calibration can be implemented by using either a simulated matrix or a pure matrix, the similarity between both being a function of the ease of preparing a simulated matrix closely resembling the pure matrix

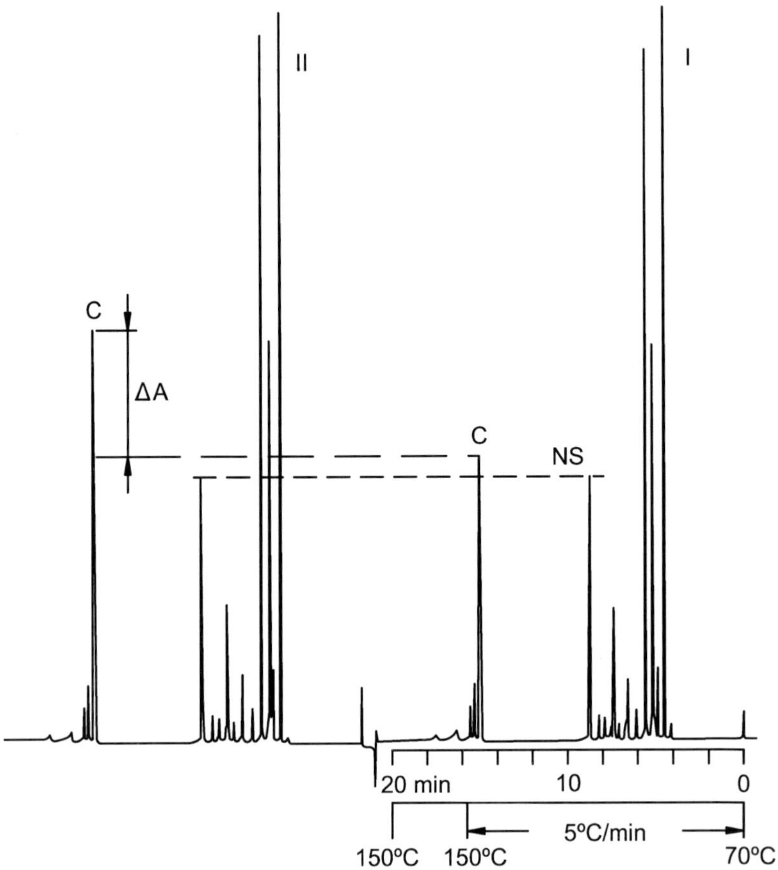

Fig. 4.15. Standard addition with a normalization standard (NS). Chromatogram obtained after headspace separation of 1.1% camphor (C) in a massage cream. (Reproduced with permission of Perkin–Elmer.)

Single or multiple linear regression standard-addition calibration can be performed by using a single liquid or gas phase addition — the latter being more appropriate for solid samples; however, it can also be implemented in conjunction with additional standards for normalization (see Fig. 4.15).

Vapour-phase calibration (VPC) is based on the principle that the concentration of the volatile analyte in the gas phase can be determined by external-standard calibration. If the total amount present in the vial is known, the concentration in the sample phase at equilibrium is calculated from the difference. This technique, where the distribution of a volatile compound between two phases in a headspace vial is determined by using a pure vapour as reference, was originally implemented by Kolb using automated headspace equipment to determine distribution coefficients in gas–liquid [77] and gas–solid systems [78], and later by Schoene *et al.* [79] to determine solubility coefficients in vapours of both solid and liquid polymers. Although these investigations focused on non-linear distribution as found both in concentrated solutions and in gas–solid adsorption systems, the same technique can obviously be applied equally well to dilute solutions, where the partition coefficient is constant and independent of the concentration. A description of the fundamentals of this technique can be found in a paper by Kolb *et al.* [80].

Sensitivity

The sensitivity of headspace-based methods can be modulated by altering a number of variables, namely:

(a) *The distribution coefficient and distribution ratio.* Basically, the headspace sensitivity for a given concentration of a volatile compound in the sample depends on K and the phase ratio, β — which includes the sample volume, V_s. Whether an increased sample volume increases headspace sensitivity depends on K. Figure 4.16 illustrates the effect for dioxane (K = 642 at 60°C) and cyclohexane (K = 0.05 at 60°C); the

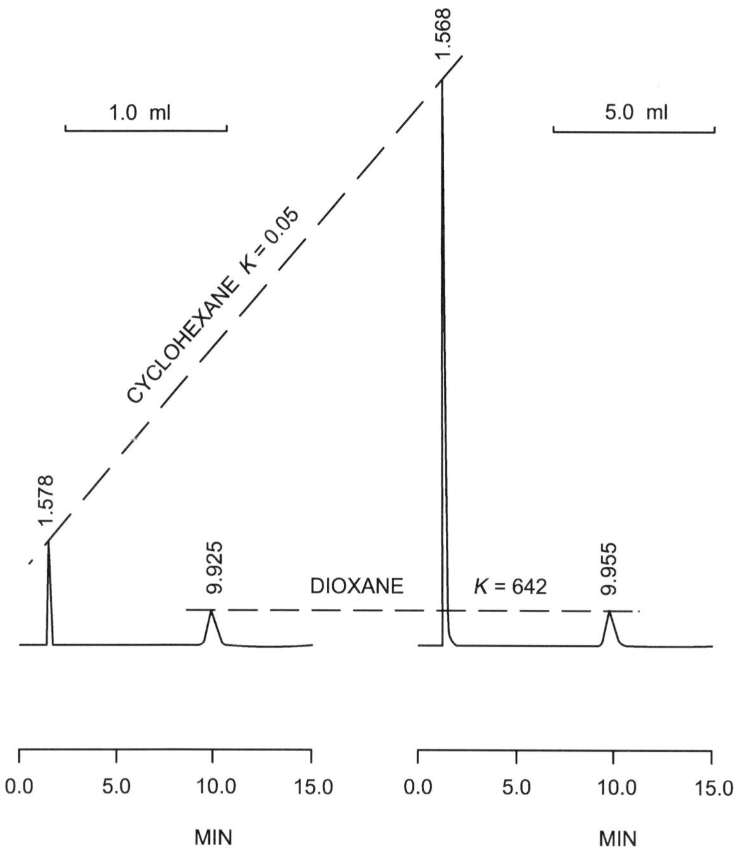

Fig. 4.16. Influence of K and β values on the sensitivity of the headspace technique. (Reproduced with permission of Perkin–Elmer.)

TABLE 4.5

HEADSPACE SENSITIVITY DEPENDING ON THE TEMPERATURE FUNCTION OF K

	Sensitivity = $f(K)$				
Solvent	40°C	50°C	60°C	70°C	80°C
n-Hexane	0.16	0.07	0.03	0.01	0
Toluene	2.8	2.2	1.8	1.5	1.3
2-Butanone	166	108	73	50	35
Ethanol	1370	826	515	330	216

phase ratio, β, was 21.3 and 3.46 at a sample volume (V_s) of 1 and 5 ml, respectively. Also, based on their relationship to the area of the chromatographic peak, $A \cong C_s^0/(K + \beta)$. As noted in Section 4.3.4, the influence of the sample volume is only significant with low distribution coefficients.

(b) *Temperature.* The principal effect of this variable on sensitivity is exerted through the distribution coefficient. Table 4.5 shows the changes in K associated to temperature. As can be seen, the temperature can dramatically affect the sensitivity of HS methods.

(c) *Addition of salts.* The addition of an inert salt to an analyte solution can also affect the sensitivity of an HS method. The effect of the salt increases with increasing analyte solubility. The nature of the salt plays no crucial role in increasing the sensitivity. The effect is described in more detail in Section 4.3.4.

(d) *Presence of polar solvents.* As noted earlier, the sensitivity of an HS method decreases when a polar solvent is added to an aqueous solution of low-polar compounds.

(e) *Variables pertaining to the chromatographic and/or detection steps.* Such variables are keys to both the selectivity and sensitivity of HS methods, and must be adjusted in accordance with the number, nature and concentration of the volatile compounds that are to be removed from the matrix. Additional factors such as cryofocusing or the injection mode also affect the performance of HS methods. A detailed discussion of their effect is beyond the scope of this book, however.

Precision

The precision of HS-based methods relies strongly on strict control of the operating conditions, as well as on sample uniformity, the effect of which is more marked on solid than on liquid samples. The relative standard deviation of HS methods ranges from as low as 0.5% to as high as 50% [69,80,84,94].

4.3.6. Applications

The use of the headspace technique (particularly in its dynamic modes) for sampling volatile compounds in solids has by now reached most fields of industrial, environmental and social interest.

In *environmental analysis*, the headspace technique (particularly in the purge and trap mode) has been extensively used in the determination of VOCs in soils, sediments and sludges [81–92].

In the *industrial field*, this separation technique has been used for purposes as diverse as the analysis of odour and taste problems in high-density polyethylene [93], contaminants in recycled high-impact polystyrene [94], benzene residues in recycled polyethylene terephthalate [95] and denture adhesives [96], residual toluene in medical plasters [97], and lipstick (for assignation of products to manufacturers) [98,99].

The *analysis of natural compounds in foods* is also assisted by the use of the purge and trap technique in methods for distinguishing strawberry varieties [100]; the aroma of unprocessed foods including gherkin [101], durian [102], garlic [103] and meat [104–107]; cheese [108,109] and other dairy products [110]; tobacco, tea and coffee [111,112]; and peanuts [112]. *Food additives* including sulphur dioxide [113,114] and *food contaminants* such as VOCs [115–126], have been recovered by PT, particularly from table-ready foods. Animal [127,128] and plant tissues [129,130] have also been subjected to PT for separation of volatile compounds.

The analysis of volatile organic compounds (VOCs) by PT is perhaps the most widely used method of trace analysis in environmental organic chemistry. The primary reasons for this massive use are that the method is applicable to a wide variety of analytes in virtually any type of sample matrix and that it is still unsurpassed in terms of sensitivity. The purge and trap technique has been adapted for the analysis of VOCs in a variety of matrices including human blood [131], consumer products (e.g. floor coverings and air deodorizers) [132], offshore waste gas in crude oil production [133], and on surfaces. The technique has also been used for the determination of residual solvents in pharmaceutical products [134], and is recommended for the analysis of bottled drinking water [135]. On the other hand, the equilibrium HS mode has been used for the determination of partition coefficients [80]. Good proof of the present maturity of headspace modes (especially PT) is the procedure developed by Supelco for the determination of BTX in soil and sediments [136] using a specially designed trap material. One uncommon application of PT is the monitoring of volatile hydride and methylated arsenic species produced by anaerobic microbes in gases released from sewage treatment facilities and municipal landfills [137]. Some commercially available units have been altered for specific applications; such is the case with an improved sparger for PT concentration that was used to determine volatile organics in landfill leachates and septic samples [138]. Losses of VOCs were minimized by using the same tube for sample collection, storage and sparging. Also, the technique of diffusive gradients in thin films [139] has been used as a step prior to PT for sampling sulphide species through a polyacrylamide hydrogel, accumulated sulphide being measured by PT followed by colorimetry [140]. The usefulness of HS methods is clearly reflected in their growing adoption by EPA [69].

Static and dynamic HS have been used in combination to identify aroma compounds in yoghurt. Thus, overall, 91 compounds were identified by MS after HS–GC, nonen-3-one being detected for the first time (at a concentration of 9 pg/kg) [141].

The performance of methods using various HS modes has been compared and found to depend on how strong the matrix–analyte bond is.

Soil spiked with trichloroethylene and toluene was analysed using a gas chromatograph equipped with a PT concentrator that was found to be replaceable by a headspace unit in order to simplify the overall assembly. The headspace analysis of soil samples was found to be restricted by incomplete desorption of the contaminants in soil–water mixtures; this shortcoming, however, was effectively overcome by the addition of methanol. Henry's law constants for volatile organics in methanol must previously be determined if the method is to be applied to soils [142]. A comparison of the performance of static and dynamic (PT) headspace modes in the determination of nine VOCs in five different soils revealed poor PT recoveries in relation to those of static headspace (which ranged from 68 to 88%); the latter, however, required longer development times [143].

As stated above, most users of the headspace technique make no distinction between dynamic HS and PT. In one of the few publications that distinguished and compared these two HS modes, dynamic HS and PT were assessed as steps preceding high-resolution GC–electron capture detection for the determination of nitrous oxide in sea water. The process was found to exhibit a first-order kinetics in both cases and the matrix to exert a significant effect that was proportional to the nitrous oxide concentration in bidistilled water, as well as in synthetic and natural sea water. As expected, PT provided better extraction recovery, sensitivity and limits of detection — which fell in the picomole-per-millilitre range [46].

Based on available results, dynamic and static HS mode is complementary rather than competitive. The better choice in each case will depend on the sample–analyte interaction. Thus, the dynamic modes are better suited to semi-volatile compounds in water, while the static modes are to be preferred for solid and semi-solid samples.

In any case, HS modes are not the panacea for separating volatiles from all types of sample; in fact, comparisons with other techniques have revealed HS to be no better choice in many cases — particularly with strongly bound analytes.

The primary US EPA-recommended methods VERVSW-846.5030 and 8260 A for measurement of VOCs in soils, which involves PT followed by GC–MS, only appears to be useful for measuring readily desorbed organic contaminants from soil pore spaces and external soil interfaces. It does not, however, measure contamination that has diffused into internal micropores of the soil matrix, as shown by Askari *et al.* [144]; thus, the purge and trap method measures only a small fraction of total soil contaminants, especially in long-contaminated soils, where about 90–99% of contamination is within the soil matrix. These authors compared three methods for determining VOCs in aged field samples, viz. purge and trap, methanol immersion and hot methanol extraction. As can be seen from Table 4.6, hot solvent extraction proved much more effective than the EPA-approved PT method. The results also reflect the inefficacy of PT as a method for assessing vapour extraction remediation technology and suggest that EPA should revise the use of PT methodology for measuring VOC concentrations in soils as the efficiency of the methods is strongly dependent upon analyte–matrix interactions.

TABLE 4.6

COMPARISON OF THREE METHODS FOR MEASURING VOC IN SOILS

Soil type (component)	Compounds	Purge and trap[a], mg/kg	Methanol immersion[a], mg/kg	Hot methanol extraction[a], mg/kg
Kentucky (clay)	Trichloroethylene	100 ± 57	150 ± 15	240 ± 31[b]
Louisiana (silty loam)	*cis*-1,1-Dichloroethylene	3070 ± 351	26 000 ± 4359[b]	41 700 ± 2082[b]
	Trichloroethylene	5900 ± 1210	85 000 ± 13 115[b]	121 700 ± 11 719[b]
Florida (silty, fine to very fine sand)	Methylene chloride	240 ± 62	530 ± 31[b]	630 ± 58[b]
	Benzene	2 ± 1	110 ± 5[b]	150 ± 29[b]
	Toluene	190 ± 41	240 ± 11[c]	270 ± 37[b]
	Chloroform	2 ± 0	110 ± 5[b]	130 ± 25[b]

[a] Average of three soil sample measurements
[b] Significantly higher mean than purge and trap at 95% confidence level
[c] Significantly higher mean than purge and trap at 90% confidence level

4.4. ANALYTICAL PERVAPORATION

4.4.1. Introduction

Analytical pervaporation emerged in the late 1980s as a non-chromatographic continuous separation technique advantageously competing with other membrane-based separation techniques thanks to the absence of sample–membrane contact, which makes it useful for dirty, aggressive and particle-containing samples [145–148]. The later integration of steps of the analytical process other than separation (viz. derivatizing chemical [149,150] and biochemical reactions either preceding [151,152] or following [153,154] the pervaporation step) increased its potential and miniaturizability. Finally, its use as a tool for treating solid samples has further expanded its scope.

One of the most salient, favourable aspects of pervaporation is its ability to integrate many steps of the analytical process, thereby facilitating miniaturization. On the other hand, the greatest shortcomings of this technique are its still incipient development and the resulting absence of commercially available equipment to implement it.

4.4.2. Principles of pervaporation

Industrial pervaporation

Industrial pervaporation [155–157] selectively separates a liquid feed mixture by partial vaporization through a non-porous polymer membrane. The separation is not based on relative volatilities, as in the case of distillation or evaporation, but rather on the relative rates of permeation through the membrane. The mixture is circulated through the membrane and the permeating component diffuses through it. A vacuum (vacuum pervaporation) or sweeping gas (sweeping-gas pervaporation) is applied to the membrane on the permeate side and the permeating component desorbs from the membrane as a vapour that can be collected or released, as required. The driving force for the separation is the chemical potential on both sides of the membrane. The permeating component is transported through the membrane because its partial pressure on the permeate side is lower than that in the saturated vapour [158].

Separation from mixtures is achieved because the membrane transports one component more readily than the others, even if the driving forces are equal. The effectiveness of pervaporation is measured by two parameters, namely: flux, which determines the rate of permeation; and selectivity, which measures the separation efficiency of the membrane (controlled by the intrinsic properties of the polymer used to construct it). The coupling of fluxes affecting the permeability of a mixture component can be divided into two parts, namely: a thermodynamic part expressed as solubility, and a kinetic part expressed as diffusivity. In the thermodynamic part, the concentration change of one component in the membrane due to the presence of another is caused by mutual interactions between the permeates in the membrane in addition to interactions between the individual components and the membrane material. On the other hand, kinetic coupling arises from the dependence of the concentration on the diffusion coefficients of the permeates in the polymers [155].

For a binary mixture of components A and B, the flux can be expressed for the entire permeate (J_T, total flux) or each component (J_A and J_B, the flux of component A and B, respectively), having dimensions of mass/(area × time). The flux can be calculated provided the mass of the permeating component, the membrane area and the time of measurement are known. To this end, the following expression can be used:

$$J_i = -\frac{L_i \, \Delta \mu i}{l} \tag{4.43}$$

where L_i is the phenomenological diffusion coefficient, $\Delta \mu i$ the chemical potential driving force across the membrane and l the membrane thickness.

The selectivity of pervaporation is given by the ratio of the mass fractions of components A and B for the permeate and feed. In the case of selective permeation of component A, the following equation applies:

$$\alpha_{\mathrm{B}}^{\mathrm{A}} = \frac{y_{\mathrm{A}}/y_{\mathrm{B}}}{x_{\mathrm{A}}/x_{\mathrm{B}}} \tag{4.44}$$

where x_A and x_B are the mass fractions of components A and B, respectively, in the feed, and y_A and y_B those in the permeate. A value greater than unity indicates selective permeation of A over B, whereas a value smaller than unity indicates selective permeation of B over A. The selectivity is dimensionless and occasionally described as an enrichment factor β that is the ratio of the concentration of one component on the permeate to that in the feed [158].

The process is favoured by increased temperatures, which result in a vapour pressure difference that increases the permeability of the substances through the membrane by allowing the diffusion of permeated molecules and decreasing interactions between permeates. The morphology of the membrane (rubber, glassy polymer) also influences the separation efficiency.

Although pervaporation appeared at the beginning of the 20th century, research aimed at expanding industrial applications of this technique did not start until the 1970s, where the world energy crisis promoted the development of three different separation approaches, namely: (a) the dehydration of aqueous–organic mixtures by using water-selective hydrophilic membranes (e.g. the dehydration of ethanol produced by fermentation of sugar in refineries); (b) the permeation of organics from an aqueous–organic mixture using a hydrophobic membrane for purposes such as solvent recovery, hazardous waste treatment, water purification or beverage processing; and (c) the permeation of specific organic compounds from an anhydrous mixture (e.g. aromatics–paraffin, branched hydrocarbons–paraffin, olefins–paraffin, isomer mixtures), which is unfeasible with currently available membranes.

Pervaporation surpasses conventional industrial separation in several respects; thus, it makes more efficient use of energy, allows the ready separation of azeotropic mixtures and dehydration of multicomponent mixtures, avoids contamination of the product with entrained compounds and the environmental pollution usually resulting from treatment of entrained substances, uses little space and is easy to implement and install on-site as the pervaporator is skid-mounted [155].

Analytical pervaporation

Analytical pervaporation is the process by which volatile substances in a heated donor phase evaporate and diffuse through a porous hydrophobic membrane, the vapour condensing on the surface of a cool acceptor fluid on the other side of the membrane. Surface tension forces withhold the fluids from the pores and prevent direct contact between them. A temperature difference that results in a vapour pressure difference across the membrane provides a strong driving force for the separation, which also occurs in the absence of a temperature gradient. Evaporation will occur at the sample surface if the vapour pressure exceeds that at the acceptor surface. One important feature of pervaporation modules used for analytical purposes is the air gap between the donor phase and the hydrophobic membrane, which avoids any contact between them and reduces the problems associated with fouling of the membrane.

The mechanism of transport by pervaporation can be described in the light of the sample diffusion model [159], which comprises the following steps: (a) evaporation of the analyte into the air gap; (b) sorption into the membrane on the sample side; (c) diffusion of the sorbed component through the polymer matrix; and (d) desorption into a liquid or gas phase on the acceptor side. The last three steps are also included in industrial pervaporation processes.

Analytical pervaporation is a very mild process that can be operated at the required temperature and needs no high pressure or cross-flow velocity, and no additional chemicals. Because of its short life, the theoretical principles of analytical pervaporation have not yet been established except for liquid samples and gas-phase acceptors [160]; there is, however, research in progress on solid and liquid samples processed with both liquid and gaseous acceptors [161].

4.4.3. The analytical pervaporator and auxiliary units

So far, laboratory-scale pervaporators have been user-designed and built. Although pervaporation can be applied to liquid, solid and slurry samples, the basic separation unit is identical whichever the sample type, the sole difference as regards equipment requirements being the use of appropriate accessory units. An analytical pervaporator consists of two essential parts, namely: the body of the separation module (including the devices for hindering gas losses) and the membrane.

Pervaporator body

Basically, an analytical pervaporator consists of the elements shown in Fig. 4.17A, namely: an upper acceptor chamber (a) with inlets and outlets through which the acceptor stream is circulated and in which the gaseous analyte (or its reaction product if the analyte is not volatile) is collected; a lower, donor chamber (d) that contains the solid sample or through which the feed stream of liquid or slurry sample is circulated; a thin (*ca.* 1 mm) membrane support (b) made of polytetrafluoroethylene (PTFE) or metal; and spacers (c) of variable thickness (2–10 mm) that can be placed below or above the membrane support in order to increase the volumes of the corresponding chambers.

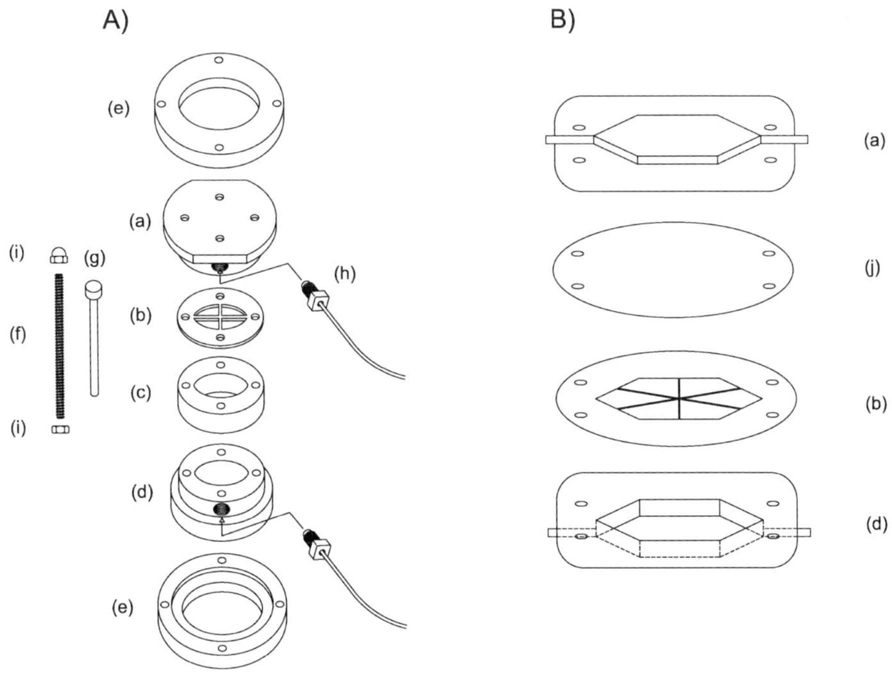

Fig. 4.17. (A) Parts of a conventional pervaporation module: (a) acceptor chamber; (b) membrane support; (c) spacer; (d) donor-sample chamber; (e) aluminium supports; (f) and (g) rods for screwing and aligning the module, respectively; (h) connectors; (i) screws. (Reproduced with permission of Wiley & Sons.) (B) Most significant parts of a hexagonal pervaporation module: (j) membrane. (Reproduced with permission of Elsevier.)

The whole module can be made of methacrylate, which enables continuous checking during experiments. Some Perspex pervaporators have also been reported [162]. The upper and lower chamber, and the membrane support, are aligned by inserting metal rods in the four orifices (g); closer contact is achieved by screwing them with four screws (f) between two aluminium plates (e). The temperature is adjusted either by circulating a fluid at an appropriate temperature [163], by immersing the module in a thermostatted water bath [150], by inserting an electronic temperature controller [164] or by using focused microwaves [28]. Usually, only the lower chamber is heated and no temperature control system is used for the upper chamber. The efficiency of the pervaporation step can be boosted by cooling the acceptor chamber; however, special precautions must be adopted as the temperature gradient may result in condensation of water vapour present in the air gap upon contact with the sample side of the membrane, thus decreasing its active surface.

The hexagonally shaped unit of Fig. 4.17B was designed with the aim of decreasing void volumes and increasing throughput through faster removal of both the sample and acceptor solutions containing the pervaporated species from the donor and acceptor chamber, respectively.

The analytical pervaporator can be used in combination with a flow-injection manifold, either in the upper chamber when the pervaporated species must be derivatized for adaptation to the detector and/or in the lower chamber for the pervaporation of analytes from liquid samples or slurries. Alterations of either the auxiliary dynamic manifold or the pervaporator itself are required when the pervaporation step is assisted by focused microwaves, the separation step assists in the continuous monitoring of an evolving system, untreated solid samples are used or pervaporation is integrated with detection.

The module used for microwave-assisted pervaporation [28] is similar to that shown in Fig. 4.17, but smaller so that it can be accommodated inside the microwave vessel. Since no metal parts are admitted in the microwave chamber, the cell is constructed entirely in PTFE/methacrylate and the screws in PEEK® – poly(ether–ether ketone).

When the pervaporation unit is used for the continuous monitoring of a fermentation process with a view to determining analytes in samples containing suspended particles, or in slurries, the module is altered by increasing the diameter of both channels (inlet and outlet) of the donor chamber, which might otherwise be clogged.

For solid samples, a septum is placed at the inlet of the lower chamber through which the reagents, if necessary, are injected with a syringe furnished with a hypodermic needle. The outlet of the lower chamber is shut with a screw in order to avoid leakage and the resulting loss of analytes.

When separation and detection of volatile species occur simultaneously, a hole is drilled at the centre of the upper chamber top of the pervaporation cell in order to accommodate the sensor [149,165] with the aid of appropriate adaptors. This module can be used with both liquid [149,152] and solid samples [165]. The flow of the acceptor stream is stopped during measurement so as to allow the accumulation of the analyte released from the matrix during the process and hence to increase the sensitivity [166].

Membranes

As a rule, permeability in glassy polymers (e.g. cellulose) is lower than in rubbery polymers (e.g. polydimethylsiloxane, PDMS); on the other hand, selectivity is dictated by the molecular dimensions of the permeating species [167]. The polymers used as membranes in analytical pervaporation are similar to those employed for gas separation and possess a dense, non-porous macroscopic structure. The difference between the two lies in the transport mechanism and arises mainly from a large affinity difference between the permeating molecules and the polymer membrane.

The membranes used for analytical pervaporation are hydrophobic membranes of the types usually employed in ultrafiltration and gas-diffusion processes. In practice, PTFE is the most frequently used membrane material, followed by hydrophobic polyvinylidene-fluoride (PVDF). Ultrafiltration membranes are very thin, which, in combination with the large surface area of both the donor and acceptor chamber, leads to their easy bending. This results in changes in the flux of the permeating component through an altered membrane area and hence in changes in the efficiency of the process. As a result, membranes must be replaced fairly often. Because of their thickness, gas-diffusion membranes are not so easily bent, so the same membrane can be used over long periods. The pore size of the

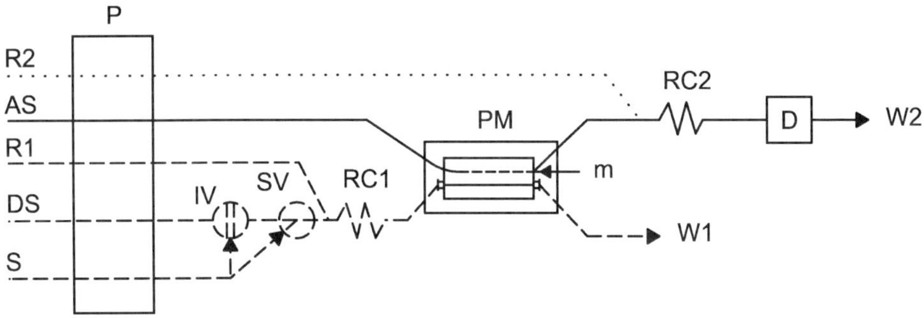

Fig. 4.18. Continuous–discrete approach to implementing analytical pervaporation of solid samples. The dotted line corresponds to a potential derivatization reaction of the pervaporated species and the dashed lines represent the continuous manifold used for automatic insertion of liquid samples. P peristaltic pump, AS acceptor stream, IV injection valve, SV switching valve for changing between continuous and discrete insertion of sample into the pervaporator, R reagent, DS donor-sample stream, S sample, RC reaction coil, PM pervaporation module, M membrane, D detector, W waste.

membrane (or its directly proportional thickness) influences the efficiency of the process and hence its sensitivity. This facilitates application to samples with variable analyte concentrations.

Auxiliary manifold

Figure 4.18 depicts a typical flow-injection pervaporation assembly. Solid lines represent fixed parts and the dashed lines, parts included when liquid samples are to be processed.

For solids, an appropriate amount of sample is weighed in the donor chamber, which is then connected to the rest of the pervaporator. The reagents, when necessary, are injected through the septum (at the inlet of the donor chamber) or septa (at both the inlet and outlet of the donor chamber) when several separate reagents are required. Then, the donor chamber is heated and the analytes pervaporate through the membrane and are accepted by the fluid on the other side of the membrane.

The manifold into which the upper chamber is inserted does not depend on the initial state of the sample, but only on the characteristics of the pervaporated analytes, the type of detector used and its position along the manifold. Depending on the particular type of detector used, auxiliary channels will have to be included to bring the pervaporated species into contact with appropriate reagents in order to obtain products to which the detector will respond. Integrated detection and pervaporation requires altering the pervaporator but simplifies the overall manifold. As shown below, preconcentration units, solid-phase reactors (mainly enzyme reactors) and various other devices can also be connected in-line in the manifold when required.

Pervaporation modes

The qualifiers "continuous" and "discrete" as applied to pervaporation refer to different aspects of the process. In fact, analytical pervaporation is a continuous technique because, while the sample is in the separation module, mass transfer between the phases is continuous until equilibrium is reached. "Continuous" also refers to the way the sample is inserted into the dynamic manifold for transfer to the pervaporator. When the samples to be treated are liquids or slurries, the overall manifold to be used is one such as that of Fig. 4.18 (dashed lines included). The sample can be continuously aspirated and mixed with the reagent(s) if required (continuous sample insertion). "Discrete" sample insertion is done by injecting a liquid sample, either via an injection valve in the manifold (and followed by transfer to the pervaporator) or by using a syringe furnished with a hypodermic needle [directly into the lower (donor) chamber of the separation module when no dynamic manifold is connected to the lower chamber]. In any case, the sample reaches the lower chamber and the volatile analyte (or its reaction product) evaporates, diffuses across the membrane and is accepted in the upper chamber by a dynamic or static fluid that drives it continuously or intermittently, respectively, to the detector — except when separation and detection are integrated.

The flow of donor (sample) solution and acceptor fluid (lower and upper chamber, respectively) can be continuous or stopped at any time during the separation step in order to increase the efficiency of the process.

With solid samples, the lower chamber always operates in a discrete manner since, as noted earlier, the sample is weighed in it. In this case, the separation module is also used as a leaching reaction vessel, which also can incorporate the detection step, thus operating as a mini-autoanalyser integrating all of the steps of the analytical process through an appropriate computer connected to the detector via a passive interface [165].

4.4.4. Pervaporation efficiency

The efficiency of a given pervaporation process must be evaluated in order to act on the overall system so as to either boost or reduce mass transfer to the acceptor chamber as required. Such evaluation can be done either in (a) relative terms by comparing signals provided under different conditions by the analyte (or its reaction product), previously collected in the upper chamber and driven to the detector, or in (b) absolute terms by comparing the signal obtained under the working conditions with that corresponding to 100% mass transfer.

The relative procedure is typically used in optimization experiments in order to accommodate the sensitivity (usually by maximizing it), but also to ensure the best possible conditions for derivatization reactions (prior and/or subsequent to pervaporation) and dispersion along the continuous system, among others. No special alterations of the manifold other than those resulting from the optimization process are required in this case.

The absolute procedure requires an auxiliary channel in the upper subsystem, through which the acceptor solution is also passed, furnished with an auxiliary injection valve (AIV in Fig. 4.19) of the same characteristics as the main injection valve (MIV), which is used to inject the sample for analyte separation and monitoring. Two sequential sample

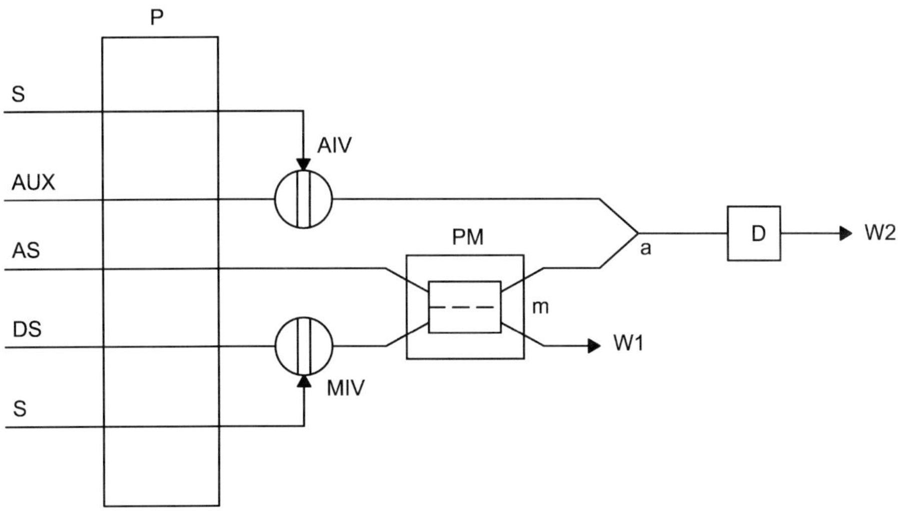

Fig. 4.19. Simplified scheme of a hydrodynamic manifold for evaluation of pervaporation efficiency. Note that, when a derivatization reaction is required prior to or after analyte separation, one or more additional channels for the reagent solution(s) must be included in the donor and acceptor stream, respectively. AIV auxiliary injection valve, MIV main injection valve, PM pervaporation module, m membrane, a merging point, D detector, W waste, AUX auxiliary channel containing acceptor solution, S sample, AS acceptor solution, DS donor solution. (Reproduced with permission of Wiley & Sons.)

injections — the sequence is not influential — from both valves are needed to evaluate the efficiency of the mass-transfer process. First, a volume of sample containing a given analyte concentration is injected via MIV; the analyte evaporates to the air gap, diffuses across the membrane, and is absorbed by the acceptor stream and driven to the detector. The signal produced is proportional to the amount of analyte transferred from the donor to the acceptor stream. Second, an identical volume of the same sample is injected through AIV and driven to the detector. A signal corresponding to the whole amount of injected analyte is obtained. The difference between the two signals provides the amount of analyte transferred to the separation process relative to that in the sample. Note that the dispersion undergone by the monitored species in the two channels is different, so the two must be compared in terms of the areas of the transient signals (representative of the total amount of monitored species that has passed by the detection point) rather than their heights, which are representative of the concentration. Obviously, the comparison is not valid if the flow-rate in the upper manifold changes during the process [168].

4.4.5. Variables influencing analytical pervaporation

Most often, the aim of manipulating the variables that influence a method is improving its efficiency, both by lowering the limit of determination and by obtaining a steeper slope for the calibration curve. When the concentration of the target analyte in the sample varies widely, a real or *pseudo* dilution step is occasionally mandatory in order to fit the analyte concentration within the linear range of the calibration curve. Some ways of changing the mass transfer through the membrane in analytical pervaporation are obvious and can be inferred from the nature of the separation process; others, however, are sophisticated or ingenious approaches. The variables can be characteristic of either the donor or the acceptor chamber (or manifold) of the system.

Variables pertaining to the donor chamber

An increase in *temperature* in the donor chamber increases the vapour pressure of the analyte and boosts mass transfer across the membrane. The usual way of heating this chamber is by immersing it into a water bath. When pervaporation is assisted by microwaves focused on the donor chamber, higher temperatures can be reached with closer control [28]. Different working temperatures result in calibration curves that encompass different analyte concentration ranges (see Fig. 4.20), which can be of interest when the concentration of the target species in the sample ranges over wide bounds.

Sample agitation is strongly influential on pervaporation. Magnetic stirring facilitates the removal of gases from the donor chamber, thus accelerating attainment of the separation equilibrium. Ultrasounds have a similar, but stronger, effect compared to magnetic stirring; however, this type of energy favours leakage and hence losses of the gas phase. The presence of *chemically inert beads* of a suitable size in the donor chamber boosts the transfer of volatiles to the air gap, both with liquid and with solid samples. In the

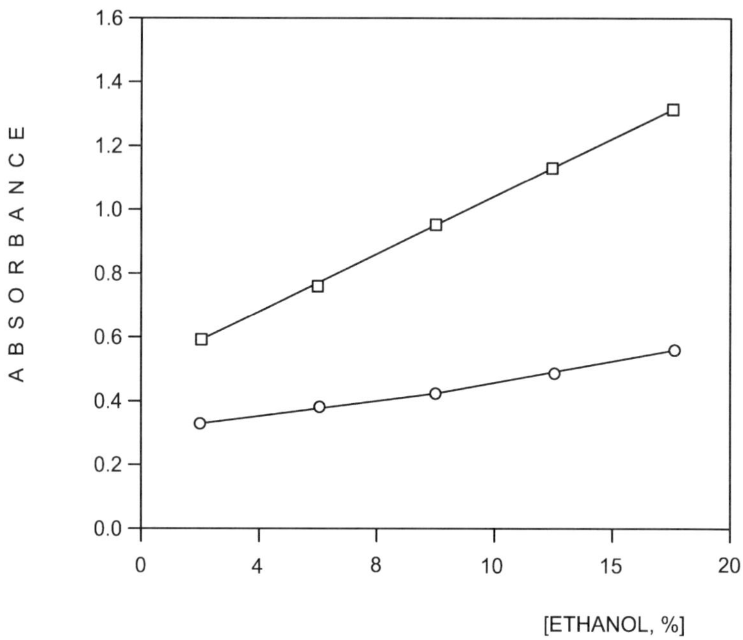

Fig. 4.20. Calibration curves for ethanol obtained at a pervaporation time of (□) 8 min and (○) 4 min.

former case, the time the sample remains in the donor chamber is increased by effect of the beads creating a tortuous path for the liquid [159]. With solids, the beads separate the particles, thus creating a larger surface for transfer of the gas to the air gap.

The *volume of the air gap* over the sample is crucial for efficient pervaporation as it contains a large enough amount of evaporated species to force passage through the membrane by gradient. The smaller the air gap is, the smaller is the amount of analyte required to start mass transfer through the membrane. The best and most controllable way of decreasing the air gap volume in the case of liquid samples (viz. with a continuous manifold connected to the lower chamber) is by lengthening the waste line. This channel creates a pressure in the lower chamber that is proportional to its length. By increasing the length, the overpressure created forces up the level in the donor chamber, thereby decreasing the amount of evaporated analyte needed to establish equilibrium in the gas phase, which results in faster mass transfer through the membrane. Thus, the efficiency obtained by using a waste tube of 280 cm rather than one of 20 cm (both of the same inner diameter) is 14 times higher [147].

Reducing the air gap space also increases the likelihood of sample–membrane contact, so a compromise between sensitivity and selectivity must be adopted.

With liquid samples, the *flow-rate* strongly influences the pervaporation process. The lower the flow-rate is, the longer will be the residence time in the donor chamber of a given volume of sample and the closer to equilibrium will be the analyte mass transfer. The flow can be stopped when the sample is in the donor chamber to allow steady-state mass transfer. This situation is not desirable because it lengthens the separation unduly. Instead, a fresh sample can be passed through the donor chamber as often as required provided the volume of sample available is not limited.

The *concentration of reagents* (pH included) can have dramatic effects on pervaporation if the reagents take part in the reaction that produces the volatile species. In any case, mass transfer can be modulated as required by altering such a concentration.

Variables pertaining to the acceptor chamber or auxiliary manifold

Mass transfer through the membrane can be adjusted by changing two major variables of the acceptor (upper) chamber. Thus, the *flow-rate* of the acceptor fluid is a key to improved pervaporation; this can be achieved through a displacement in the mass transfer equilibrium resulting from the continual use of fresh acceptor fluid — which maximizes the concentration gradient between both chambers. Keeping the fluid static facilitates attainment of an equilibrium or near-equilibrium state for mass transfer.

There are two ways of keeping the acceptor fluid static during pervaporation. One is by stopping the propulsion unit for this stream. This entails using separate propulsion units for the lower and upper streams (if both are required), and disturbs the detector baseline through the start-and-stop sequence, which in turn causes part of the fluid to return to the propulsion unit. These shortcomings can be avoided by using the approach illustrated in Fig. 4.21A, where the upper chamber of the separation module is accommodated in the loop of an injection valve. In the filling position, the acceptor stream passes through the bypass to the detector at the preset flow-rate. After the predetermined collection time has elapsed, the valve is switched to its load position and the stream

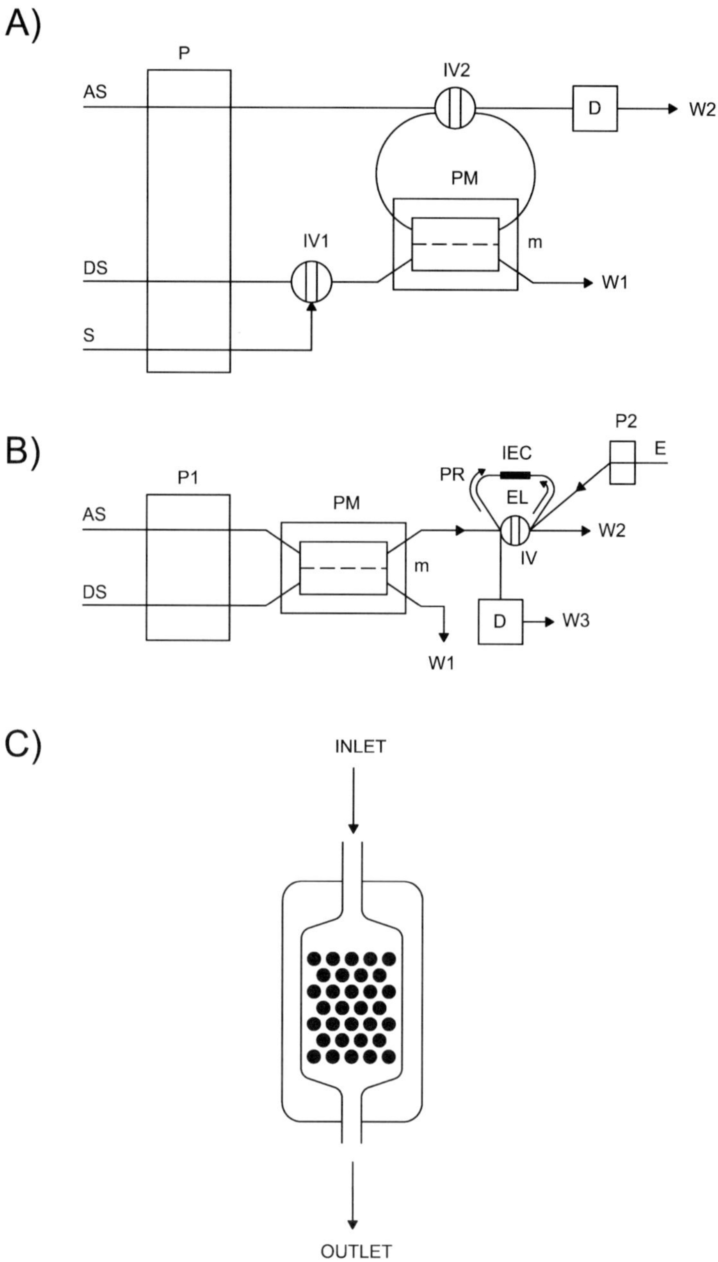

Fig. 4.21. Different ways of improving pervaporation efficiency. (A) By halting the flow in the acceptor chamber without disrupting the overall dynamic system. (B) By on-line retention of transferred volatile species and elution in the opposite direction. (C) By use of a packed flow-cell to integrate reaction and detection. E eluent, IEC ion-exchange column, PR preconcentration, EL elution. (For other abbreviations, see previous figures.) (Reproduced with permission of Wiley & Sons.)

passed through the loop to sweep its contents to the detector. In this way, a situation as close to mass transfer equilibrium as required can be established without disturbing the detector baseline.

Preconcentration of pervaporated analytes is the most effective way of favouring mass transfer. The equilibrium between phases can be efficiently displaced by continuous removal of the transferred species from the upper chamber, followed by concentration (either in a minicolumn placed before the detector or in the flow-cell itself).

A minicolumn packed with an ion-exchange sorbent can be placed between the pervaporation module and the detector, either in the transport tubing or, better, in the loop of an injection valve (Fig. 4.21B). The latter approach allows the retention and elution steps to take place in opposite directions in the column, thus avoiding increased compaction and hence an increased pressure in the dynamic system [169].

By retaining the pervaporated species in the detection zone, the previous two steps can be integrated and the kinetics of the separation process monitored as the volatile species will reach the detector at the same rate as they cross the membrane (Fig. 4.21C). Obviously, this approach is only viable with non-destructive detectors. Both approaches (Figs 4.21B and 4.21C) allow fresh portions of the acceptor fluid to be brought into contact with the membrane in such a way that equilibrium of the separation process is never reached.

Characteristics of the membrane such as pore size and, especially, thickness, have a strong influence on pervaporation. A comparison of three PTFE membranes of pore size 0.5, 2.0 and 5.0 μm, and thickness 0.16, 0.18 and 0.20 mm, respectively, in terms of permeability to hydrogen cyanide revealed the membrane of 2.0 μm pore size and 0.18 mm thickness to be the most permeable [162]. The membrane of 5.0 μm pore size was supposed to exhibit the highest permeability; however, its relatively large pores only allowed better diffusion of the larger molecular gases and not of the smaller ones such as HCN. In this situation, the effects of additional membrane properties including thickness and pore density must be considered. Permeability is inversely proportional to the thickness of the membrane and directly proportional to its pore density. An increased membrane pore size can be expected to require a decreased pore density to ensure acceptable mechanical stability. As a result, the overall area of the pores through which gas diffusion takes place could even decrease with increasing pore size and lead to decreased permeability.

4.4.6. The physics of analytical pervaporation: kinetics of the mass transfer process

Monitoring mass transfer kinetics is of paramount importance in optimization experiments with a view to both identifying the factors influencing the process and establishing an appropriate pervaporation time to ensure a high throughput without significantly detracting from sensitivity. The kinetics of the pervaporation process can be monitored in two ways, namely: (a) by integrating pervaporation and detection; and (b) by integrating a retention step with detection (following pervaporation).

Continuous monitoring of mass transfer across the membrane can be accomplished by placing the active surface of a sensor opposite the membrane (see Fig. 4.22A). In this

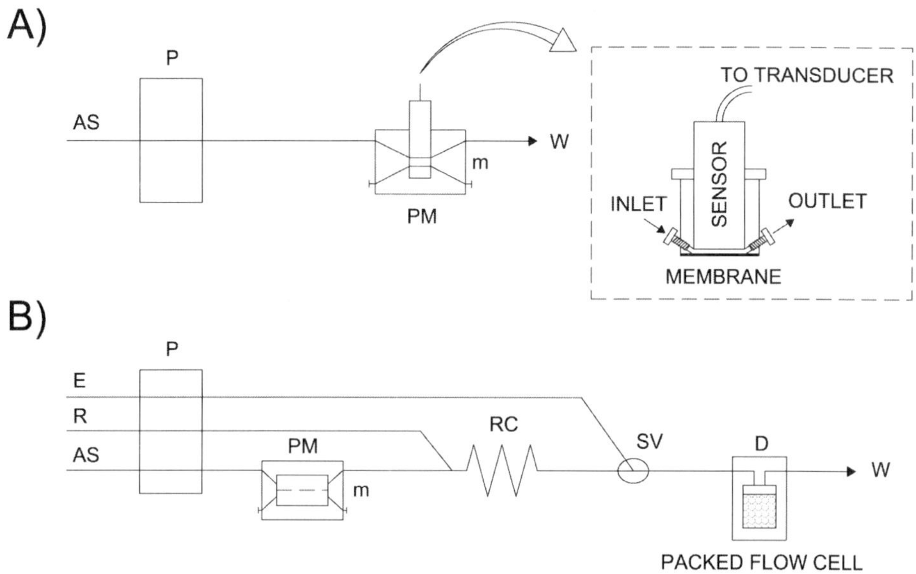

Fig. 4.22. Monitoring of pervaporation kinetics. (A) By integrating separation and detection. (B) By coupling pervaporation with a retention step integrated with detection. For abbreviations, see previous figures. (Reproduced with permission of Wiley & Sons.)

way, a rising signal is obtained that reflects the enrichment of the static acceptor solution with the pervaporated species. The signal evolves to a plateau when the partitioning equilibrium of the analyte between the sample, air gap and acceptor fluid is reached. Electroanalytical (potentiometric, voltammetric) probes and fibre optic-assisted optical detectors have been used for this purpose.

A flow-cell packed with a suitable support (e.g. an ion-exchanger or sorption material) accommodated in a non-destructive detector can be used for continuous retention of the pervaporated species or its reaction product (Fig. 4.22B). In this way, the content of the upper chamber is continuously removed and driven to the detector, and the fresh acceptor stream that reaches the chamber displaces the mass transfer equilibrium, thus increasing the efficiency of the pervaporation process. A rising recording is thus obtained, the slope of which evolves to a plateau but never becomes flat since the concentration of the volatile species in the lower chamber becomes increasingly lower but is never zero. After the measurement interval — the duration of which depends on the quantitativeness to be achieved — has elapsed, a switching valve (SV) located prior to the detector allows the acceptor stream to be replaced with an eluting solution that removes the retained species and drives them to waste. On switching the valve again, the flow of the acceptor solution through the detector is restarted, the baseline re-established and the system made ready for a fresh sample.

4.4.7. Analytical features of pervaporation methods

This section deals with the basic analytical properties (sensitivity, selectivity and precision) of pervaporation-based methods.

Sensitivity

The sensitivity of a method involving pervaporation essentially depends on the efficiency of the separation step. The efficiency of pervaporation can be adjusted in order to fit the signal provided by the analyte or its reaction product to the linear portion of the calibration curve for a given method, thus avoiding the need for a dilution or concentration step. A number of approaches to increasing and decreasing pervaporation efficiency exist.

The analytical signal can be optimized for quantitation by increasing the temperature of the donor chamber, stirring the sample, using chemically inert beads, employing optimal reagent concentrations, increasing the pervaporation time, using an efficient procedure for concentrating the volatile species and increasing the amount (weight or volume) of sample used.

When the concentrations of the target analytes in the sample exceed the upper limit of the linear range of the calibration curve, a dilution or *pseudo* dilution step is mandatory in order to fit the unknown concentration to this portion of the calibration curve, and hence to increase the precision of measurements. A number of approaches can be used in addition to the usual previous dilution of the sample or the weighing of a smaller amount to avoid the errors involved in a dilution step and in the weighing of small amounts of a solid. Such approaches involve (a) the use of a lower pervaporation temperature; (b) changing the chemical conditions to a less favourable situation, if a derivatization reaction is required, so as to reduce the yield of the volatile or monitored species (with the reaction preceding or following pervaporation, respectively); (c) expanding the air gap between the sample and membrane by intercalating as many spacers as required; (d) using a thicker membrane; (e) using a smaller loop for the injection valve (if the sample is a liquid); or (f) rising the flow-rate of the donor and/or acceptor stream.

Selectivity

The fact that pervaporation can only be used to separate volatile analytes may initially seem a shortcoming rather than an advantage; however, it results in increased selectivity for the target analytes and enables their determination in complex samples such as biological fluids and environmental specimens [170] with no prior treatment, using pervaporation as a clean-up step. If the target analyte is not volatile but can be converted into a volatile species via a selective (bio)chemical reaction, it can also be separated by pervaporation. The analyte is made volatile by adding an appropriate reagent to the sample, either in a continuous manner or by direct injection into the matrix. In any case, a derivatization reaction is often required to form the species to be eventually detected. The possibility of employing the derivatization reaction prior to or after the separation step has extended the scope of pervaporation to a variety of analytes and allowed pervaporation units to be coupled to various detector types. Pervaporation can also be used for speciation as it allows the selective determination of different forms of the target analyte.

In multideterminations, and also when the target volatile analytes possess different boiling points, the pervaporation unit can be used at different temperatures to ensure selective separation of the species of interest. Use of a heating system allowing reproducible control of time and temperature is obviously mandatory here.

Discrimination between polar and non-polar volatile analytes can be accomplished by using microwave irradiation of the donor chamber in order to facilitate the evaporation of polar species. Non-polar ones can then be evaporated by conventional heating (whether electrical or by use of a water bath).

Selectivity can also be boosted by changing the membrane used in the separation process. As noted earlier, the permeability of glassy polymers is usually lower than that of rubbery polymers, while selectivity is primarily dictated by the molecular dimensions of the permeated species [167].

Precision

The fact that placement or passage of the sample in or through the donor chamber, respectively, is made quite reproducible by the use of stirring or chemically inert beads results in highly reproducible signals following pervaporation. Experiments conducted in the presence and absence of stirring and chemically inert beads revealed that significantly improved precision, as relative standard deviation, was obtained from the use of glass beads (0.79%) and magnetic stirring (1.78%) in relation to the absence of both (11.29%) [159].

Repeatability can also be improved by aspirating the detector waste at a flow-rate identical with that of the acceptor stream or, if a chemical reaction with an additional reagent is required, to the combined individual flow-rates [149].

When the concentrations of the target analytes are relatively low, a low precision is obtained that can be ascribed to the fact that equilibrium conditions in the gas phase held in the donor chamber are not readily reached. This shortcoming can be circumvented by previously saturating the air gap with the volatile species (i.e. by injecting the required volume of a standard solution containing an appropriate volatile concentration and driving it to the lower part of the separation module). The flow of both the donor and the acceptor stream is then stopped for a preset time and, once resumed, the experiment can be continued with fairly good results (the precision thus obtained in the determination of sulphide was 5.46% as RSD) [145].

The use of long waste tubes in the lower part of the pervaporation module can also improve the precision of the ensuing method since the pressure in the module is thus increased and small variations in pressure caused by both internal and external noise yield a smaller overall effect when the length of the waste tubing changes from 20 to 280 cm [147]. In this way, the RSD decreased from 5.46% to 2.85% for injection of the sample, and from 7.03% to 1.61% for continuous introduction of sample. Halting the flow of both the acceptor and the donor stream resulted in a repeatability as RSD of 2.51% at the normal waste tube length (20 cm) and 1.21% for longer lengths (280 cm) [147].

The use of gas-diffusion membranes instead of ultrafiltration membranes can also improve the precision as the former are not easily bent; this avoids changes in permeate flux resulting from both irreproducible membrane areas and increased volumes of the upper chamber [153,171].

4.4.8. Scope of application of analytical pervaporation

Because the full potential of pervaporation remains to be realized, its applications are still small enough in number to be easily summarized. Table 4.7 shows such applications classified according to whether the sample was liquid or solid. Within each category, applications are classified according to field: environmental, food and clinical. A theoretical application to the calculation of Henry constants is illustrated at the end. The applications of pervaporation can also be classified according to whether they involve any derivatization reaction. Such reactions can be used prior to the separation step in order to convert the analyte into a volatile compound or subsequent to the mass transfer step (in the upper manifold) in order to obtain a product that can be measured by the detector. The reaction in question can be of chemical or biochemical nature. One other possible classification of pervaporation applications is according to whether a single or many parameters are determined. The feasibility of multideterminations relies on the separation process (viz. on the use of different temperatures for analytes with different vapour pressures), a reaction preceding or following pervaporation, the use of a discriminating separation technique or the characteristics of the detector. Whether the separation step is integrated with detection also gives rise to two types of methods. One of the most salient applications of pervaporation is no doubt its use as an alternative to both head-space modes (static and dynamic or purge and trap), which it surpasses in some respects. Below are briefly described various aspects of pervaporation applications.

Sample types

Although this discussion focuses on solid samples, pervaporation is applicable to liquids, solids and even slurries. As stated in Section 4.4.3, both the manifold used and the way the sample is inserted vary depending on whether it is liquid or solid. Also, as noted in the same section, slight alterations to the manifold are required to process slurries. The most salient difference between the treatment of liquid and solid samples is that, with the former, the whole process from injection of the sample to delivery of the results can be fully automated. By contrast, some human intervention is required in sample weighing and tight fastening of the pervaporation module for solid samples. Only partial automation is possible in this case as the rest of the process (viz. derivatization — if required — pervaporation, transport to the detector, monitoring, data acquisition and processing, and delivery of the results) can be performed in an automated fashion.

Derivatization reactions

When a derivatization reaction is developed in either the lower manifold (to form a volatile product) or the upper one (to adapt the monitored product to the characteristics

TABLE 4.7

GENERAL APPLICATIONS OF ANALYTICAL PERVAPORATION

Analyte	Sample type	Derivatization reaction	Detection	Special features	Ref.
Solid samples					
F^-	Tree leaves, fertilizers, ceramic industry wastewater	Chemical	Integrated potentiometry	Continuous–discrete manifold	165
Dichlorvos	Natural water, soil	Biochemical	Fluorimetry	Enzymic derivatization by inhibition	177
Metrifonate	Natural water, soil	Biochemical	Fluorimetry	Triple integration: chemical hydrolysis–pervaporation–enzymic derivatization by inhibition	178
Ammonia/Urea	Soil	Biochemical	Integrated potentiometry	Sequential determination. Speciation analysis	173
Mercury compounds	Soil (CRM)		Atomic fluorimetry	Microwave-assisted pervaporation. Speciation analysis	28
Mercury	CRM		Atomic fluorimetry	Alternative to headspace, speciation	169
VOCs	Soil		ECD(**)	Alternative to headspace	171
Acetaldehyde	Fruit juice, yoghurt, bread	Chemical	Photometry	Food samples	175
Trimethylamine	Fish	Chemical	Photometry	Pretreatment integrated with pervaporation	176

Liquid samples					
F^-	Fertilizers, ceramic industry wastewater	Chemical	Potentiometry	Stopped-flow preconcentration	149
S^{2-}	Kraft liquors	Chemical	Photometry	Improved precision and sensitivity	145,147
Cyanide	Samples from the mining industry	Chemical	Amperometry	Determination of cyanide in the presence of sulphide. Interference study	162
Phenol	Water	Chemical	Amperometry	Stopped-flow preconcentration	185
Ammonia	Industrial effluents	Chemical	Photometry	Determination in the presence of surfactants	191
COD(*)/Inorganic carbon	Bleaching liquors, domestic sewage, river water	Chemical	Photometry/ integrated potentiometry	Simultaneous determination of COD(*) and inorganic carbon	166
Ethanol	Wine	Biochemical	Fluorimetry	Immobilized enzyme on-line monitoring fermentation	174
Ethanol	Fermentation media	Chemical	Photometry	On-line monitoring	193
Ethanol	Wine	Biochemical	Photometry	Immobilized enzyme	188
Acidity	Wine	Chemical	Photometry	Sequential determination of total and volatile acidity	183
Acidity	Wine	Chemical	Photometry	Determination of volatile acidity	186
SO_2	Wine	Chemical	Photometry	Speciation of free and bound SO_2	150
CO_2/SO_2	Wine	Chemical	Photometry/ potentiometry	Simultaneous determination of CO_2/SO_2	182

TABLE 4.7 (*cont.*)

Analyte	Sample type	Derivatization reaction	Detection	Special features	Ref.
Liquid samples (cont.)					
Ethanol/glycerol	Wine	Chemical/ biochemical	Photometry/ fluorimetry	Simultaneous determination of ethanol/glycerol, immobilized enzyme	184
Volatile species	Wine		FID(***)	Alternative to headspace	187
Ethanol/diacetyl	Yeast extract	Biochemical	Fluorimetry	Immobilized enzyme	188
Ethanol	Bakers' yeast culture, beer	Biochemical	Voltammetry, fluorimetry	Electronically controlled temperature, immobilized enzyme	188
Diacetyl	Beer	Chemical	Photometry	Stopped-flow preconcentration	195
Diacetyl	Beer	Chemical	Adsorptive stripping	During fermentation and in final product	192
Acetaldehyde	Fermentation media	Biochemical	Photometry	Immobilized enzyme, continuous monitoring	175
Ammonia/Urea	Serum, urine	Biochemical	Conventional potentiometry	Sequential determination, immobilized enzymes	150
Oxalate	Urine, serum	Biochemical	Conventional potentiometry	Continuous–discrete manifold, immobilized enzyme	151
Formaldehyde	Pharmaceuticals and cosmetics	Chemical	Conventional photometry	Stopped-flow preconcentration	189
F^-	Pharmaceuticals and cosmetics	Chemical	Integrated potentiometry	Continuous–discrete manifold	190
Phenols	Water		Amperometry	Theoretical studies, Henry constants	194

(*) Chemical oxygen demand (**) Electron capture detector (***) Flame ionization detector

of the detector), the basic dynamic manifold assembly may have to be altered depending on the particular reaction to be used. Actions on the lower chamber intended to obtain volatile species involve injecting appropriate reagents for solid samples or merging a stream thereof with the channel that contains the injection valve at a point in between the valve and the pervaporator. When post-pervaporation reactions are needed, the most effective way of implementing them is by using the solution containing the reagents as acceptor solution. In this way, the pervaporated analytes react on the acceptor side of the membrane, thus displacing the mass transfer equilibrium. Solid reagents, which represent the most usual situation in enzyme-catalysed reactions, are packed in a solid reactor (a tube) placed between the injection valve and the separation module when the analyte in the liquid sample is non-volatile in order to obtain a volatile product amenable to separation; however, the solid biocatalyst–support conjugate can also be placed in the donor chamber for two main purposes, namely: (a) to increase the yield of the biocatalysed reaction (because the product is removed from the solution into the air gap and then into the acceptor solution, the reaction equilibrium is gradually displaced to quantitativeness); and (b) to avoid dispersion of the sample in the carrier solution when the analyte concentration is low and the sample scant (and hence inadequate for insertion in a continuous manner) [152]. When the sample is a solid, the use of the biocatalyst bonded to a solid support in the lower chamber is unfeasible unless the support consists of magnetic particles [172], which can be removed and reused after the analysis has been completed. Thus, enzymes are usually employed in dissolved form when a volatile product is to be obtained from a solid sample [173]. When the purpose of the biocatalysed reaction is to obtain a product amenable to detection, the reactor, packed with the support–enzyme conjugate, is placed between the acceptor chamber and the detector. The enzyme has also occasionally been immobilized in the upper chamber of the pervaporation module in order to integrate separation and an enzymatic reaction with an immobilized biocatalyst. This approach has been used to monitor ethanol production during wine fermentation and found to result in longer durability of the immobilized biocatalyst [174].

Individual determinations

The simplest use of pervaporation is for the separation and subsequent determination of a single analyte; as a result, most reported methods using this technique involve individual determinations. The determination of acetaldehyde in semi-solid and solid food samples is one salient example of single-analyte determinations where pervaporation avoids time-consuming sample preparation steps such as filtration, removal of dyestuffs and turbidity from the Carrez solution, and centrifugation [175]. One other example is the determination of trimethylamine (an objective parameter for fish quality evaluation that correlates well with sensory evaluation) [176]. The method is based on pervaporation of the analyte and monitoring of the colour change in Bromothymol Blue caused by the basic character of the amine; the results are consistent with those provided by the official method for trimethylamine.

One typical example of the high potential of pervaporation for boosting the selectivity of processes such as enzymic inhibition-based reactions is the method for the selective determination of pesticides in environmental samples. The method, which is only applicable to volatile pesticides or their volatile degradation products, has been used in the determination of diclorvos (boiling point 35°C) [177] and metrifonate (a compound yielding diclorvos as hydrolysis product) [178]. Diclorvos, both directly and after hydrolysis, has been determined in soil following pervaporation by its inhibitory action on acetylcholinesterase oxidase catalysis. The decrease in the fluorimetric signal obtained after a two-step derivatization reaction involving choline oxidase and horseradish peroxidase is proportional to the inhibition caused by the analyte and hence to its concentration in the sample.

The determination of pectinesterase activity in fruit and vegetables by monitoring the methanol released and pervaporated [179] is one other example of the potential of pervaporation for individual determinations in solid samples.

Multideterminations

Multideterminations involving pervaporation [180] can be implemented either by altering parts or parameters affecting the overall approach or by coupling the dynamic manifold of the upper chamber to a highly discriminating separation technique such as gas chromatography.

Multideterminations involving solid samples Determinations of this type using pervaporation have so far focused on speciation analysis [181]. The two most salient examples reported to date are the speciation of inorganic and organic forms of mercury (Fig. 4.23A) and that of nitrogen as ammonia and urea (Fig. 4.23B).

The speciation analysis of mercury in soil was based on microwave-assisted pervaporation/atomic fluorescence detection in the straightforward, discrete–continuous manifold depicted in Fig. 4.23A. Focused microwaves proved highly efficient when the analytes were polar enough to interact with this type of energy; such was the case of $HgNO_3$ and PhHgOAc, for example. In the manifold used, the sample was weighed in the donor chamber of the pervaporation cell and the cell was tightly closed and connected to the loop of the injection valve accommodating the pervaporator's upper chamber. This allowed the carrier stream (water or argon) to pass either through the acceptor chamber and then into the detector, or via a bypass to the detector while the flow in the acceptor chamber was stopped. Although the pervaporator flow was stopped, the baseline did not change because the carrier stream proceeded to the gas–liquid separator via the bypass loop at the same rate as it was circulated through the pervaporator. Once the sample was in the donor chamber, the reagents were injected via the donor chamber inlet. Septa similar to those used in GC allowed the syringe needle to penetrate the donor chamber, inject the reagents and be removed with no loss of either sample or reagents. One minute after the reagent (acidic $SnCl_2$) was injected, the microwave device was switched on at a preset power for a given time, after which the cell was allowed to cool for 1 min before the valve was actuated to divert the carrier through the acceptor chamber to collect one of the mercury compounds to the gas–liquid separator if the acceptor carrier was liquid or directly to the atomic fluorescence detector if it was gaseous. Once the detector baseline was restored, the valve was switched again to isolate the pervaporator, the reagents (viz. $KBr/KBrO_3$ as oxidant and $SnCl_2$ as reductant) were sequentially injected and the

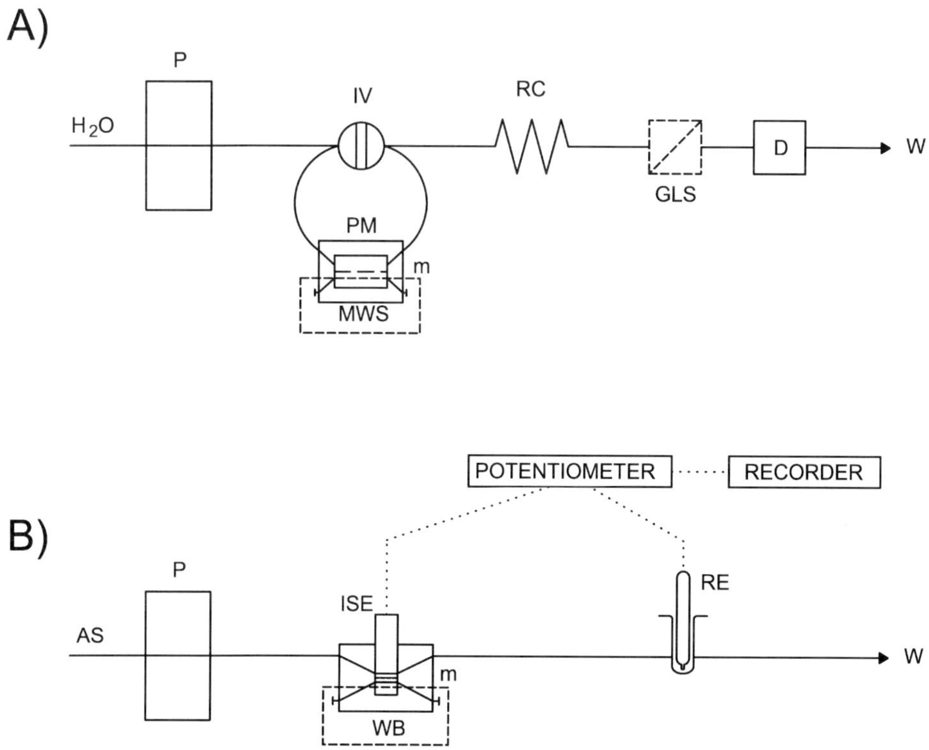

Fig. 4.23. Manifolds used for speciation analysis of solid samples. (A) Mercury speciation (as Hg^{2+} and PhHgH). (B) Nitrogen speciation (as ammonia and urea). GLS gas–liquid separator, MWS microwave system, ISE ammonia-selective electrode, RE reference electrode, WB water bath thermostat. Other abbreviations as in previous figures. (Reproduced with permission of Elsevier.)

microwave digester was switched on again for another preset time at an appropriate power setting for the other mercury — now organic — compound. After the peak was recorded, the valve was actuated to isolate the cell and the system was thus made ready for weighing the next sample [28]. The method was validated by using a certified reference material (viz. CRM 145R from the BCR).

The speciation analysis of nitrogen (as ammonia and urea) in soils was based on an enzymatic reaction and integrated pervaporation/potentiometric detection using the manifold of Fig. 4.23B. The procedure also involved weighing the sample in the donor chamber, shutting the pervaporator and propelling the acceptor stream through the upper chamber until a steady baseline was obtained. Then, an appropriate volume of NaOH solution was injected into the donor chamber through the septum, the flow in the upper chamber being stopped and the signal provided by the selective electrode used (corresponding to the ammonia present in the sample) recorded until the steady state value was reached. The flow was then restored and the sample neutralized by injecting HCl, followed by Tris buffer at pH 8.8 and an appropriate volume of urease solution in the same buffer.

The flow was stopped again and the signal corresponding to the urea concentration in the sample was recorded as before. During measurements, the pervaporation module was kept in a thermostatted water bath at the required temperature [173].

Multideterminations involving liquid samples These are easier to implement and hence more common than the previous ones. Some involve speciation while others have a different analytical purpose.

Pervaporation-based speciation analyses of liquid samples have provided methods for determining two volatile species or one volatile species (determined after passing into the acceptor chamber) and a non-volatile one (determined either in the effluent from the donor chamber or by using another sample aliquot injected simultaneously with that driven to the pervaporator). The method for the determination of free and bound SO_2 in wine [150], and that for ammonia and urea in serum and urine [151], are typical examples of sequential speciation analyses of two volatile species. On the other hand, the methods for measuring chemical oxygen demand and inorganic carbon in bleaching liquors and domestic sewage [166] are examples of volatile and non-volatile speciation analyses.

Monitoring wine fermentation and determining the final product pose special challenges as they involve analyses of liquid samples containing a number of volatile compounds and a matrix that is complex enough to require the use of pervaporation. Typical examples testifying to the high potential of pervaporation in this context include a method for determining two volatile species by using either a single photometric detector (for free and bound SO_2) [150] or two serially arranged detectors (a potentiometric one for CO_2 and a photometric one for SO_2) [182]; one for the determination of a volatile species and a non-volatile one with a single photometric detector (for volatile and total acidity) [183]; and one employing a photometric detector and a fluorimetric one arranged in parallel for ethanol and glycerol, respectively [184].

Coupling a pervaporator to a gas chromatograph: an alternative to headspace sampling

The use of pervaporation as an alternative to the headspace technique is worth separate discussion. This is, in fact, one of the most promising uses of this approach, as revealed by two existing methods for mercury speciation and VOC analysis in solid samples that exemplify the advantages of pervaporation over static and dynamic headspace modes. Both methods were developed by using the overall assembly depicted in Fig. 4.24A, by which the analytical process was developed in the following four steps:

(1) Pervaporation of the analytes. An amount of *ca.* 0.5 g of sample was placed in the pervaporator's donor chamber, which was then closed, connected to the system and placed either in a water bath or in a vessel of a focused microwave device depending on the polar or non-polar nature of the target analytes. These were evaporated into the gas layer above the sample and then diffused through the membrane to the argon acceptor stream.
(2) Preconcentration/desorption. A Tenax minicolumn accommodated in the loop of an injection valve — which was placed in an ice bath — was used to preconcentrate

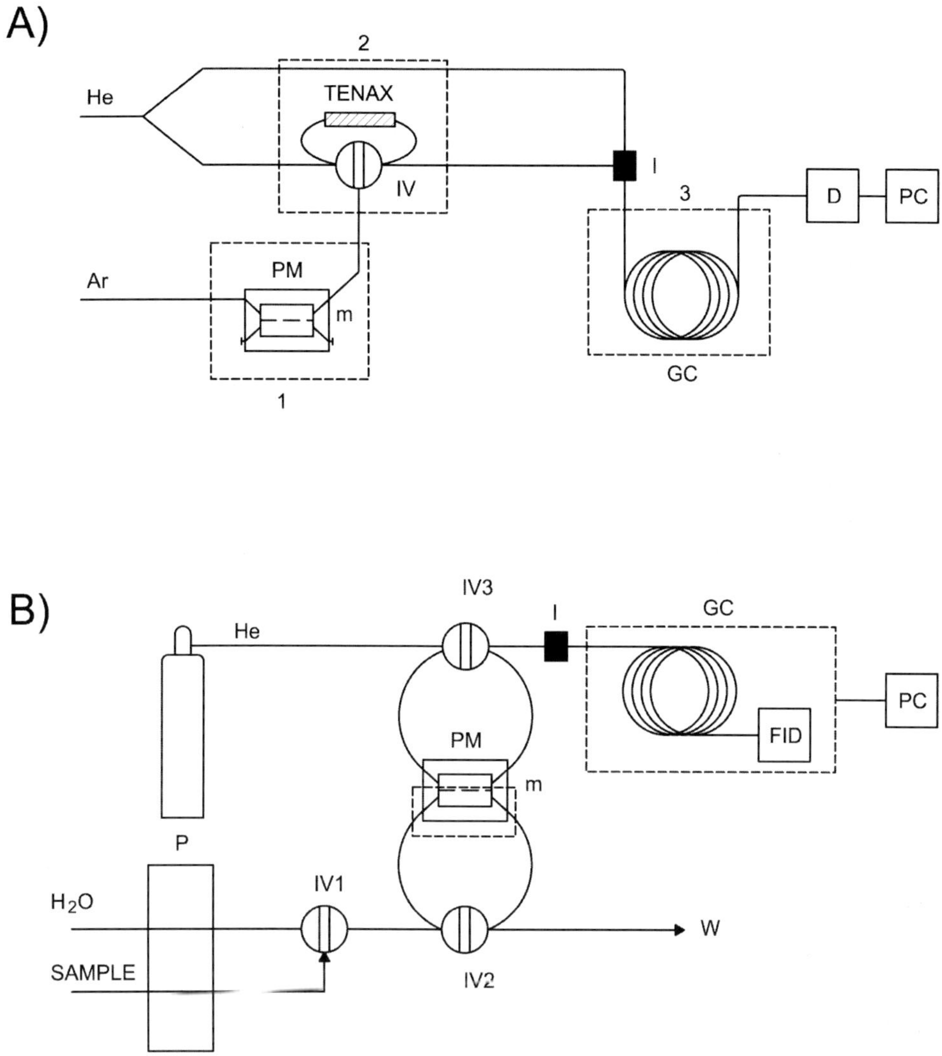

Fig. 4.24. Coupling a pervaporator to a gas chromatograph for (A) solid and (B) liquid samples. I chromatograph injector, GC gas chromatograph, FID flame ionization detector, PC personal computer. 1, 2 and 3 denote pervaporation, retention and chromatographic steps, respectively. Other abbreviations as in previous figures. (Reproduced with permission of Elsevier and Springer-Verlag, respectively.)

the volatile species passing through the membrane in the pervaporation unit. In order to desorb the retained species and inject them onto the chromatographic column, the flow of argon from the pervaporation unit was diverted so that the valve loop could be shut off with no gas flow passing through it. The Tenax minicolumn was removed from the ice bath and placed in a muffle oven at an appropriate temperature for non-polar analytes or in a microwave device for polar analytes, following which the valve was switched to the inject position to allow a helium stream to pass through the column and flush the desorbed species from the loop of the injection valve through the connector between the injection valve and the chromatograph, and onto the chromatographic column.

(3) Chromatographic separation. The helium stream containing the desorbed compounds entered the injector in a continuous manner via a hypodermic needle fitted to the injector inlet. This stream was then merged with an additional one flowing through the normal carrier gas inlet into the injector. The chromatograph oven was programmed as required and the species separated.

(4) Detection of the isolated species. As the individual mercury species emerged from the chromatographic column, they were passed through a pyrolysis unit at 800°C in order to break the compounds down to elemental mercury. The resulting stream was then mixed with a make-up gas on its way to the detector. A second stream of the same gas (argon) was used as sheath gas, which increased the reproducibility as a result of the He/Ar/Hg flow being maintained in the beam of the atomic fluorescence detector. VOCs required no pyrolysis unit, so the isolated analytes were driven directly to the electron capture detector (ECD). No hydrophilic membranes could be used as they allowed the passage of water, which then reached the GC column. PTFE membranes were found to provide increased sensitivity and thermal stability relative to other hydrophobic membranes such as those of the PVDF type [171].

The coupling of an analytical pervaporator to a gas chromatograph for the determination of enological parameters in wines is an excellent example of the usefulness of this approach for the analysis of liquid samples. The FI–pervaporation–GC assembly of Fig. 4.24B differs from that used with solid samples in the need to insert the sample by injection and the fact that no preconcentration/desorption step is required. The process involves the following steps:

(1) Pervaporation/preconcentration of the analytes. First, the loop of the injection valve (IV1 in Fig. 4.24B) is filled with sample while IV2 and IV3 are in their injection and filling positions, respectively. In this way, the donor stream (H_2O) is continuously circulated through the lower chamber of the pervaporation unit and the helium stream is continuously sent to the gas chromatograph; meanwhile, a portion of the helium remains static in the acceptor chamber to collect the volatile species. Then, IV1 is switched to the injection position and the sample is transported by the H_2O stream through the lower chamber of the pervaporation unit, which is immersed in a thermostatic bath at 80°C.

(2) Injection, chromatographic separation and detection. Five minutes after the sample is injected — this is the optimum preconcentration time — IV2 is switched to its

filling position and, immediately, IV3 to its injection position for 8 s. During this interval, the evaporated species are carried out to the top of the chromatographic column and, after separation, driven directly to the flame ionization detector. The upper chamber of the pervaporation unit requires flushing before a fresh sample is injected. To this end, IV2 and IV3 are switched to the filling and injection position, respectively, for 10 s.

Pervaporation provides a number of advantages over the headspace technique that can be summarized as follows:

(a) The thin air gap layer above the sample requires very small amounts of the analytes to establish equilibrium with the solid or liquid sample phase and mass transfer across the membrane.
(b) Continuous removal of the volatilized analytes across the membrane from the air gap displaces the mass transfer equilibrium and increases the separation efficiency.
(c) Continuous removal of the pervaporated analytes to the preconcentration column can be used to allow fresh portions of acceptor gas to continuously come into contact with the diffused species, thus displacing the mass transfer equilibrium.
(d) The separation process can be easily automated for laboratory use with minimal purchase and maintenance costs.
(e) Unlike purge and trap, no water vapour condenser is required, nor is a hydrophobic sorbent as no water-crossing through the (hydrophobic) membrane occurs.

Integrated pervaporation and detection

The best way of monitoring pervaporation kinetics is by inserting a probe-type sensor in the acceptor chamber, with its active side facing the membrane. A number of methods based on this principle have been reported. The determination of fluoride in liquid (ceramic industry wastewater, dissolved fertilizers) and solid samples (e.g. tree leaves) by formation of a volatile product with hexamethyldisiloxane using a manifold such as that of Fig. 4.23B is an excellent example of this approach [165]. The most salient advantages of integrating pervaporation and detection are as follows: (a) the response time is shortened with respect to the conventional location of the sensor behind the separation module as the need to transport the target species from the module to the detector is avoided; (b) the kinetics of mass transfer across the membrane can be monitored in order to obtain a better understanding of the separation step and easier, in-depth optimization; and (c) the size of the set-up required for the determination of a volatile analyte or reaction product in solid samples can be dramatically reduced by integrating the leaching, derivatization, separation and detection steps.

Table 4.7 includes additional analytical uses of pervaporation [185–194].

Detector types

Pervaporators are amenable to coupling to any type of detector, whether molecular or atomic, via an appropriate interface such as a transport tube, a microcolumn packed with adsorptive or ion-exchange material, or a gas–liquid separator. The acceptor carrier stream can be either liquid or gaseous depending on the characteristics of the detector. The detectors most frequently used in conjunction with pervaporation are of the spectroscopic [atomic (fluorimetric, absorptiometric) or molecular (photometric, fluorimetric)], electroanalytical (potentiometric, voltammetric), electron capture and flame ionization types. The low selectivity of some detection techniques such as photometry, ion-selective electrode-based potentiometry and flame ionization is offset by that of the pervaporation step, which endows the overall analytical process with the selectivity required to process highly complex matrices. The potential of the pervaporation technique for sample insertion into water-unfriendly detectors such as mass spectrometers or devices such as those based on microwave-induced plasma remains an unexplored field for pervaporation.

4.4.9. Prospects for analytical pervaporation

As shown above, pervaporation is a useful analytical tool that features simplicity, and automation and miniaturization capabilities [195–197]. The actual potential of this technique for the pretreatment of solid, liquid and slurry samples can readily be inferred from its intrinsic features and from available methods using a pervaporator.

Consolidation of this technique as an effective, widespread analytical tool is bound to rely on future, sequential developments such as the following: (a) the gathering of sufficient research experience to compile a cookbook of methods where potential users can search for solutions to specific problems; and (b) the commercialization of inexpensive pervaporation modules meeting the requirements of a variety of samples and analytes. The development of such methods and theoretical studies should turn pervaporation into a useful tool for routine environmental, clinical, food and industrial analyses.

4.5. SOLID-PHASE MICROEXTRACTION

4.5.1. Introduction

The idea of using a sorbent material to extract trace organic compounds from, in principle, an aqueous sample, emerged as sorbent extraction was developed a few years ago [198]. At present, sorbents are used to extract organics from matrices such as water, air and soil. One of the most salient features of this technique is that it allows analytes to be concentrated by the sorbent. In fact, by using a sorbent with a strong affinity for the target analytes, these can be concentrated from an otherwise highly dilute sample.

Solid-phase extraction (SPE) is one of the most commonly used sorbent extraction techniques. The analytes are extracted together with interfering compounds with some similar characteristics by passing a usually aqueous sample through a plastic cartridge containing dispersed sorbent on a particulate support. A selective organic solvent is normally used to remove interferences and a different solvent is then employed to elute the retained analytes. This technique has a number of attractive features as compared with traditional solvent extraction. Thus, it is quite simple and inexpensive, can be automated and used in the field, and uses relatively small amounts of solvents. Particle-load membranes known as Empore extraction discs have improved extraction efficiency, decreased solvent consumption and reduced plugging in SPE since their introduction in 1990 [199]. However, SPE still has some limitations such as the low recoveries that result from interactions between the sample matrix and analytes, and plugging of the cartridge or blocking of the pores in the sorbent by solid or oily components, which results in low breakthrough volumes and also low capacities. Although Empore discs have to some extent overcome these drawbacks [200], some problems still exist, along with high blank values and batch-to-batch changes in the sorbents. Because SPE is a multi-step approach involving concentration, it is restricted to semi-volatile compounds with boiling points substantially above those for the solvents.

One solution to the shortcomings of SPE is to improve the sorbent geometry by coating it on a fine rod such as a fused-silica fibre or a wire of an appropriate material. The cylindrical geometry of the resulting system ensures rapid mass transfer during extraction and desorption, prevents plugging, and facilitates handling and insertion into analytical instruments [201]. This approach is known as "solid-phase microextraction" (SPME) [202–204].

Because SPME is a static extraction technique, the need for a large surface area is no longer as critical as in SPE. Smooth liquid coatings can be used that avoid plugging. Also, by sampling from headspace, SPME can extract analytes from highly complex matrices such as sludge.

4.5.2. Principles, devices and theoretical aspects of SPME

Principles and devices

A solid-phase microextraction process involves two steps, namely: partitioning of the analytes between the coating and the sample, and desorption of the concentrated species into an analytical instrument. In the first step, the coated fibre is exposed and the target analytes are extracted from the sample matrix into the coating. In the second step, the fibre with the concentrated analytes is transferred to an instrument for desorption. A third, clean-up step can also be incorporated by using selective solvents, as in SPE.

SPME applications have focused on the extraction of organics from various matrices including air, water and soils, followed by direct transfer of the retained species into a GC injector, desorption from the injector, separation on the chromatographic column and measurement by the detector. A fused-silica fibre coated with a gas chromatographic stationary phase such as polydimethylsiloxane is typically used for the microextraction. This type of fibre and its coating are obtained by using well-established techniques for manufacturing optical fibres. The fused-silica fibre itself is chemically inert and very stable, even at high temperatures. The small size and cylindrical geometry of the fibre allow it to be incorporated into a syringe-like device such as that shown in Fig. 4.25, which can be operated like an ordinary syringe and easily accommodated in a GC injector. As shown in the figure, the fused-silica fibre is connected to stainless steel tubing that is used to increase the mechanical strength of the fibre assembly for repeated sampling. The stainless steel tube is contained in a specially designed syringe. During SPME, the fibre is first withdrawn into the syringe needle and then lowered into the vial by pressing

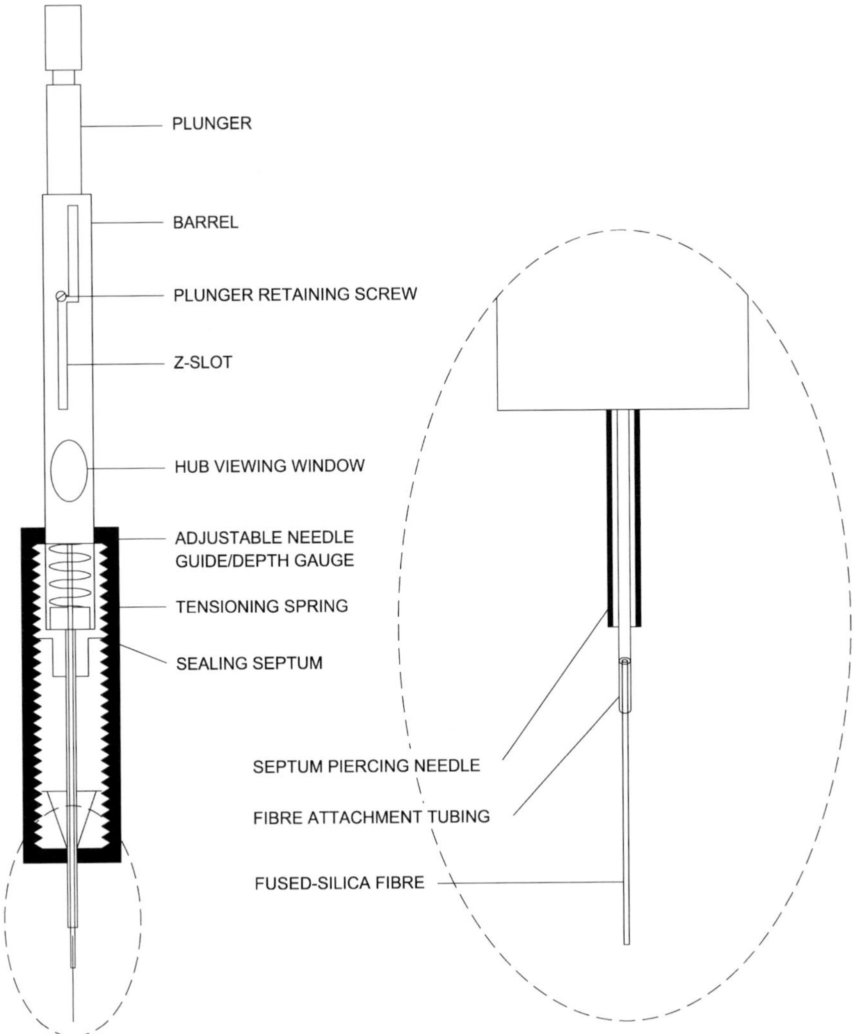

Fig. 4.25. SPME device and detail of the fibre and needle. (Reproduced with permission of the American Chemical Society.)

down the plunger. The fibre coating is exposed for a preset time to extract the analytes from their matrix. Once the sampling is completed, the fibre is directly transferred into a gas chromatographic injector. The analytes are thermally desorbed from the fibre coating and quantitatively analysed by GC. The steps involved in an SPME process can be developed either manually or automatically (with the aid of an autosampler). Because the SPME device is basically a syringe, anyone who can use a syringe properly can implement this extraction technique. The main difference between a typical autosampler and those used in SPME is that the plunger movement and timing must be carefully controlled to ensure correct adsorption and desorption. When SPME is used for sampling in the field and subsequent analysis in the laboratory, special care should be exercised to avoid analyte losses during transport. The needle opening of the SPME device can be sealed with a piece of septum and/or by cooling the needle.

Because increased temperatures have an adverse effect on adsorption, the thermal desorption of analytes from an SPME coating is highly effective in most cases. As the temperature increases, coating/gas partition coefficients decrease and the ability of the coating to retain the analytes drops very rapidly. In addition, the constant flow of carrier gas within a GC injector facilitates the removal of analytes from the coating. For volatile and semi-volatile compounds, analytes can be desorbed from the coating at a temperature between 150 and 250°C in a fraction of a second and then focused at the front of the GC column [205]. The injection band can be further shortened by using an injector that can generate a heating pulse instead of constant heating; by using an internal heating device inside the fibre; or by directly passing the current through a fibre made of metal (a wire), as shown in Fig. 4.26. With compounds of a high molecular weight such as

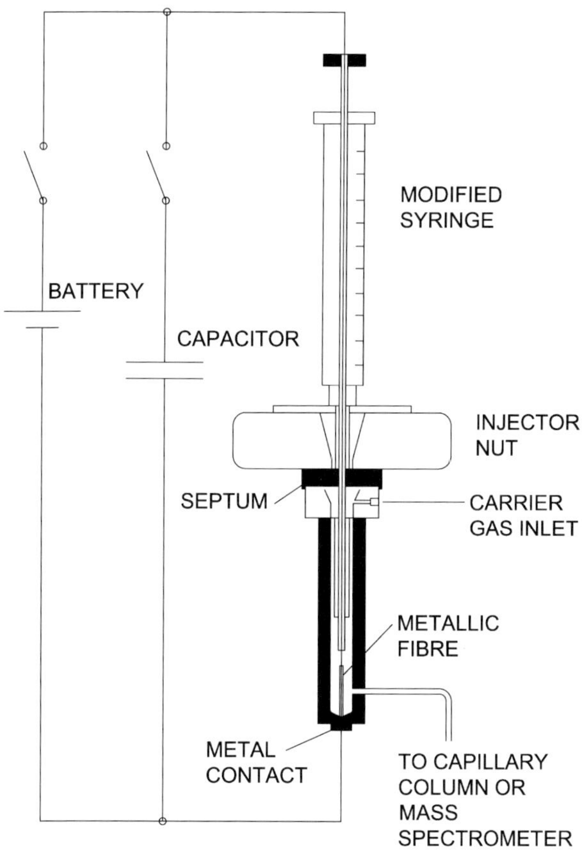

Fig. 4.26. Device for electrically heating an SPME fibre made from a metallic material. (Reproduced with permission of the American Chemical Society.)

perylene, however, carry-over may be a problem. No special thermal desorption module or alteration of the gas chromatograph is needed. The combination of SPME and thermal desorption completely eliminates organic solvents from extraction and injection, and integrates both processes into a single step.

Desorption from the fibre can also be accomplished by using an appropriate solvent, a condensed medium such as compressed CO_2 or even an aqueous solution. For example, the selective extraction of bismuth from a fibre coated with an ion-exchange polymer can be accomplished by using an acidic potassium iodide solution as a complexing reagent [206].

The most important analytical parameters (viz. throughput, sensitivity, accuracy and precision) of SPME are largely determined during the extraction step. Desorption, which strongly influences the efficiency of the chromatographic separation and the precision of the quantitation, largely dictates the quality of the results and the scope of SPME.

Theoretical aspects

The principle behind SPME is the partitioning of the analytes between the sample matrix and the extraction medium. When a liquid polymeric coating is used, the amount of analyte absorbed by the coating at equilibrium is directly related to its concentration in the sample:

$$m = \frac{K_{fs} V_t C_0 V_s}{K_{fs} V_t + V_s} \tag{4.45}$$

where m is the mass of analyte adsorbed by the coating; V_f and V_s are the volumes of the coating and sample, respectively; K_{fs} is the partition coefficient of the analyte between the coating and the sample matrix; and C_0 is the initial concentration of the analyte in the sample. Equation (4.45) establishes a clear-cut linear relationship between the amount of analytes adsorbed by the fibre coating and their initial concentration in the sample.

Because the coatings used in SPME have strong affinities for organic compounds, the K_{fs} values for the target analytes are quite large; this means that SPME exerts a very strong concentrating effect and hence results in good sensitivity. In many cases, however, K_{fs} values are not large enough to ensure thorough extraction of most analytes in the matrix. Rather, SPME, like the static headspace mode, is an equilibrium sampling method and, through proper calibration, can be used to accurately determine the concentrations of target analytes in a sample matrix. As can be inferred from eq. (4.45), if V_s is very large (i.e. if $V_s \gg K_{fs} V_f$), the amount of analyte extracted by the fibre coating,

$$m = K_{fs} V_f C_0 \tag{4.46}$$

is not related to the sample volume. This feature, combined with its straightforward geometry, makes SPME suitable for field sampling. Because the fibre can be exposed to air or dipped directly into a well, lake or river, for example, SPME reduces field analysis time by combining sampling, extraction, concentration and injection into a single, uninterrupted process.

The speed of extraction is controlled by the mass transport of the analytes from the sample matrix to the coating. This process involves convective transport in an air or liquid sample, desorption of the analytes from the solid surface when particulate matter is present, and diffusion of the analytes in the coating [205]. In direct SPME sampling, the mass transfer rate is determined by the diffusion of analytes in the coating provided the sample matrix is thoroughly agitated.

When the rate of mass transfer is determined by the diffusion of the analytes in the coating, equilibrium is reached within less than 1 min for most analytes [205]. The small thickness of the coating (typically between 10 and 100 μm) ensures rapid extraction. In practice, this limit can only be achieved for gaseous samples, which possess large diffusion coefficients. With aqueous samples, equilibration requires the use of vigorous agitation methods such as sonication [207]. With more common agitation methods such as magnetic stirring, the equilibration time is much longer and determined by diffusion through a thin static aqueous layer adjacent to the fibre. Such a thin layer of water around the fibre is very difficult to remove even when water is rapidly stirred to enhance the mass transfer of analytes; the analytes must diffuse through the water before they can be adsorbed by the fibre coating.

Placing a fibre directly into a sample to extract organic compounds works well with gaseous samples and relatively clean water samples. Sampling analytes from a solid matrix of wastewater containing grease, oil or high-molecular weight humic acids is not so easy with SPME, however, and requires using the headspace above the sample matrix. The SPME technique can thus be used to extract organics from virtually any type of matrix provided the target compounds can be released from the matrix into the headspace [208].

For volatile compounds, releasing analytes into the headspace is relatively easy because they tend to vaporize once they are dissociated from the matrix. For semi-volatile compounds, the low volatility and relatively large molecular size may slow down the mass transfer from the matrix to the headspace and, in some cases, the kinetically controlled desorption or swelling process can also slow down the extraction.

When the matrix absorbs the analytes more strongly than the extracting medium does, the analytes partition poorly into the extraction phase. Because of the limited amount of the extraction phase in SPME, the extraction will be subject to a thermodynamic constraint: the partition coefficient, K_{fs}, will be too small and result in poor sensitivity. If the coating has a stronger ability to adsorb the analytes than the matrix does, it will only be a matter of time for a substantial amount of the analytes to be extracted by the fibre coating and only kinetics will play a significant role during extraction. One of the most efficient ways of overcoming the kinetic constraints is by heating the sample, which increases the vapour pressure of the analytes, provides the energy required for analytes to dissociate from the matrix and results in more expeditious transport of the analytes.

Headspace SPME involves three phases (viz. coating, headspace and matrix), the chemical potential difference of the analytes between the phases being the driving force that moves them from the matrix to the fibre coating. With aqueous samples, headspace/water partition coefficients (K_{hs}) are directly related to the Henry constants for the analytes, which are in turn determined by their volatility and hydrophobicity.

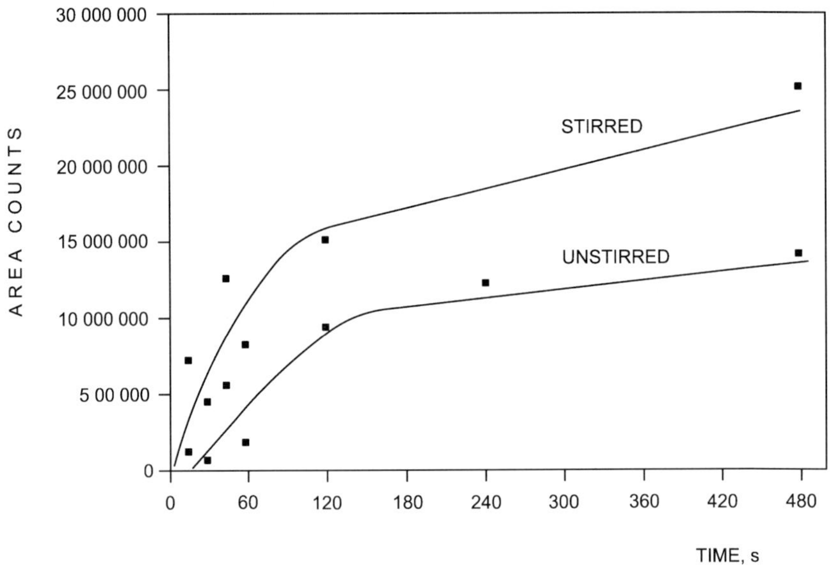

Fig. 4.27. Adsorption–time profile for 0.75 μg/ml trichloroethene in both stirred and unstirred solutions. (Reproduced with permission of the American Chemical Society.)

In aqueous matrices, most compounds have quite small K_{hs} values (< 0.25), so the headspace has a low analyte trapping capacity. As a result, the sensitivity of headspace SPME is almost the same as that of direct SPME. The sensitivity loss is only significant when the target analytes partition well into the headspace (i.e. when they possess large K_{hs} values) and/or a large headspace volume is used. For aqueous VOCs, headspace SPME sampling is faster than direct SPME. In headspace SPME, mass transfer from water to the headspace can be rapidly expedited by continuously stirring the water sample to generate a continuously renewed surface (see Fig. 4.27). The mass transfer of volatile compounds from the headspace to the fibre coating is very fast because of the large diffusion coefficients of the analytes in the gas phase, and volatile compounds are more efficiently transferred from water to headspace to coating than from water directly to coating [208]. When the rate-determining mass transfer step is the mass transport of the analytes from the matrix to the headspace, the extraction time profile curve exhibits an initial steep rise (associated to partitioning of the analytes originally present in the gaseous headspace) followed by a portion with a gentler slope (resulting from slower mass transport of the analytes from the matrix). The rate of mass transfer is dictated by convective and diffusive transport of the analytes in the sample matrix [208] and/or by the slow desorption kinetics as in SFE [209].

Because partition coefficients are temperature-dependent, there is usually an optimum temperature for headspace SPME. As the temperature rises, more analytes are released from the matrix to the headspace, a process that results in high analyte concentrations in the headspace and favours extraction. At high temperatures, however, coating/headspace partition coefficients decrease and the matrix-dependent optimum extraction temperature is determined by interactions among the analytes, coating and matrix. With an appropriate coating, headspace SPME can be used to extract both volatile and semi-volatile compounds from their matrices.

Although SPME is mainly an equilibrium extraction technique, it has the ability to provide thorough extraction. If the coating/matrix partition coefficient, K_{fs}, is very large (i.e. $K_{fs}V_f >> V$), the amount of analyte desorbed by the coating will be

$$m = C_0 V_s \tag{4.47}$$

and thorough extraction achieved.

Coating types

Because different groups of analytes can be extracted by different types of sorbents, SPME has been used with a variety of sorbents. For organics, the basic principle of "like dissolves like" applies. Polar coatings such as polyacrylate and Carbowax® extract polar compounds such as phenols and carboxylic acids very efficiently, whereas non-polar coatings such as polydimethylsiloxane retain hydrocarbons very well. A gold coating works for extraction of mercury and an ion exchanger extracts metal ions. As shown in Table 4.8, SPME can be used to extract a wide range of compounds from various matrices by adjusting the operating conditions.

4.5.3. Variables affecting SPME

The variables that influence SPME performance can be classified according to whether they relate to the retention step or the desorption step.

TABLE 4.8

CHARACTERISTICS AND SCOPE OF SPME MODES

Mode	Treatment	Analytes	Matrices
Direct SPME	Routine	Most compounds	Gaseous and liquid
	In situ chemical derivatization	Polar compounds	
	In situ redox reaction	Inorganic ions	
Headspace SPME	Routine	Volatile and semi-volatile compounds	Any
	Heating/cooling	Volatile and semi-volatile compounds with small partition coefficients	
	In situ chemical derivatization	Polar compounds	

Retention-related variables

Using an appropriate *stationary phase* is the key to ensuring efficient removal of the target analytes. Both the type of coating material and its thickness influence performance here. Stationary phases for SPME are commercially available in different thicknesses and polarities for a variety of analytes. As noted earlier, a polydimethylsiloxane phase is preferable for non-polar analytes and a polyacrylate phase for polar ones. For example, the extraction efficiency for triazine pesticides can be increased by a factor as high as 10–20 simply by using a polyacrylate phase rather than a polydimethylsiloxane one. Also, polydimethylsiloxane, which is a non-polar coating, extracts non-polar compounds such as BTX (benzene, toluene, ethylbenzene and xylene isomers) and PAHs (polycyclic aromatic hydrocarbons) from water very effectively but cannot extract polar compounds such as phenol or its derivatives (see Fig. 4.28A). On the other hand, polyacrylate, a more polar coating, extracts phenol and its derivatives quite well, but is relatively inefficient with BTX compounds (see Fig. 4.28B). The coating/water partition coefficients of BTX for polydimethylsiloxane are much greater than those for polyacrylate, whereas the opposite is true for phenol and its derivatives. For compounds with a large molecular size such as PAHs, large K_{fs} values can be obtained by using the polydimethylsiloxane coating. In fact, quantitative extraction of these analytes from aqueous matrices can be achieved because of their very large coating/water partition coefficients [210]. Two recently introduced phases (viz. Carbowax®–divinylbenzene for alcohols and divinylbenzene-polydimethylsiloxane for volatile amines) are more selective. The latter material is available in various thicknesses. A thick stationary phase is more suitable for volatile compounds, whereas a thinner phase is more efficient with larger molecules.

Testing of various types of *needles* has revealed good manufacturer control of the stationary phase thickness that results in good reproducibility (3–9% as RSD) [211]. The repeatability of extraction with the same fibre depends on phase thickness and is better with thicker phases, but is adversely affected by stirring (particularly when extracting by hand).

One recent development in this context is an HPLC stationary phase of 5 μm particle size glued to the metal needle. Silica, C_8 and C_{18} reversed phases have been tested for this purpose. Adsorption of the analytes was found to be faster and fibre capacity to increase as a result of the larger active surface of the stationary phase. The mechanical stability of the needle was also increased as a result [212].

The *extraction time* is dependent on the partition coefficient of the analyte and on agitation of the sample, and is generally shorter for extractions from the headspace. With sonication, the equilibration time is close to 1 min [207]. With other agitation methods such as magnetic stirring, the equilibration time is typically 2–60 min, depending on the agitation rate and partition coefficient. Ensuring that the maximum possible amount of analyte is extracted entails reaching the equilibration time; this, however, is too long and impractical for many compounds. Usually, an absorption diagram (viz. an extraction–time profile curve such as that of Fig. 4.29) is constructed for the target analyte that shows the dependence of the amount of analyte absorbed on the extraction time. From such a curve, the shortest acceptable time as regards analyte detection is chosen.

The *sample temperature* has a twofold effect when working with liquid samples: diffusion coefficients in water increase and extraction times decrease with increasing

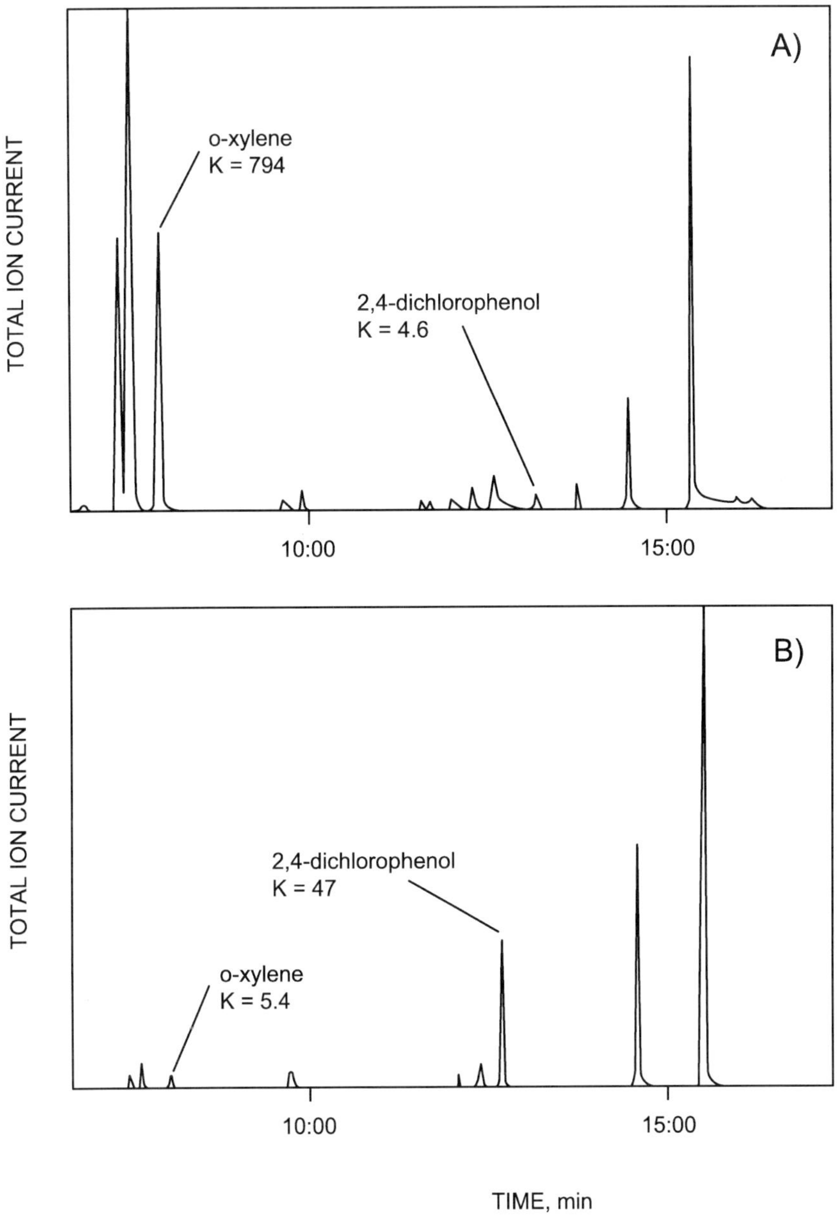

Fig. 4.28. Total ion current GC–MS chromatogram for BTX and phenol mixtures in water. (A) Extracted with a polydimethylsiloxane coating. (B) Extracted with a polyacrylate coating. *K* is the coating/water partition coefficient of the analyte, which drops by more than 2 orders of magnitude for *o*-xylene and increases by more than 1 for 2,4-dichlorophenol upon changing the coating from non-polar polydimethylsiloxane to polyacrylate. (Reproduced with permission of the American Chemical Society.)

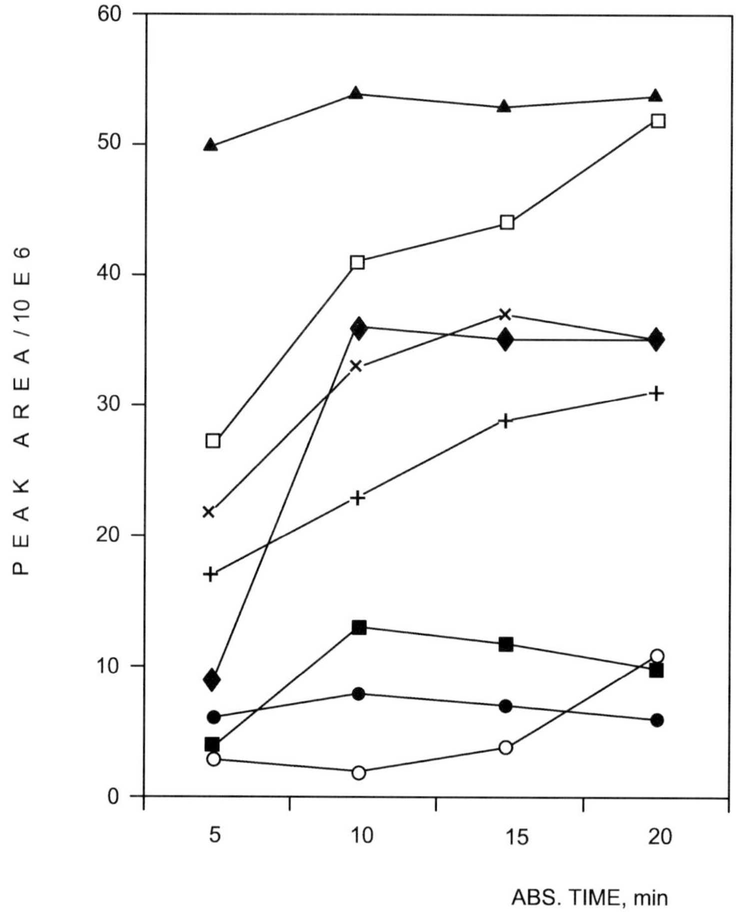

Fig. 4.29. Extraction–time profile curves (signal height versus absorption time) for organochlorine pesticides in aqueous solutions. Conditions: spiking level 0.2–1.2 ng/ml; polydimethylsiloxane fibre length 100 μm; magnetic stirring rate 250 rpm. Although equilibrium was not reached for some analytes, the extraction time was chosen to be 15 min. (◆) α-hexachlorocyclohexane, (■) β-hexachlorocyclohexane, (▲) dieldrin, (×) aldrin, (○) 4,4′-DDT, (●) 4,4′-DDE, (+) 4,4′-DDD and (□) endosulfan sulphate. (Reproduced with permission of Elsevier.)

temperatures, which, however, also result in decreasing partition coefficients. Heating the sample is therefore inadvisable except to expedite the release of analytes into the gaseous phase — particularly for extraction from the headspace above solid samples. Improved transport of compounds from the sample to the gaseous phase can be achieved by heating the sample or by adding water or some other surface-active compound [208,213]. Microwave heating can also generate convective currents in samples that improve transfer of analytes from the matrix to the extracting phase.

The *sample pH* is a crucial variable for slightly acidic or basic compounds. Extraction

TABLE 4.9

EFFECT OF THE ADDITION OF NEUTRAL SALTS ON SPME EFFICIENCY

Compound	Peak area/10^6, arbitrary units	
	No salt added	Saturated with NaCl
α-BHC	8.6	20.9
γ-BHC	3.4	10.0
Dieldrin	12.3	17.8
Aldrin	17.2	23.2
4,4′-DDT	5.6	3.1
4,4′-DDE	35.8	23.2
4,4′-DDD	39.2	37.3
Endosulfan sulphate	17.9	23.1

is more effective if such compounds are kept undissociated, as is well-known from liquid–liquid extraction and SPE procedures. The fibres should be handled with care as, for example, the polydimethylsiloxane phase is not resistant to media with pH below 4 or above 10 [214].

The *ionic strength* of the solution, when working with liquid samples, also affects the extraction step. Because the partition coefficients of the analytes (K_{fs}) are partly determined by the interaction between the target analytes and the matrix, the nature of the matrix can be altered to change the coating/matrix partition coefficients of the analytes. The salts commonly used for this purpose include NaCl and Na_2SO_4, which raise the ionic strength of water, thereby favouring the partitioning of polar organics — but not ions — into the polymer coating. As can be seen from Table 4.9, one potential drawback of using a salt to increase the ionic strength is a loss of fibre selectivity.

The presence of *organic solvents* in the sample suppresses the analyte absorption onto the fibre. Thus, K_{fs} for benzene is decreased by 20% in the presence of 3% methanol [214]. Other well-sorbing hydrophobic substances such as dissolved organic carbon in natural water can have a similar effect.

Air moisture also has some influence on the effectiveness of extraction from the headspace; in fact, at a relative humidity above 90%, absorption of analytes onto the fibre can be reduced by as much as 10% [215].

Derivatization can be used to reduce the polarity of polar compounds such as phenols and carboxylic acids, and also to increase their partition coefficients and improve the chromatographic separation. When working with liquid samples in SPME, polar analytes can be derivatized in their aqueous matrix (e.g. by converting phenols into their acetate derivatives) and then SPME applied [211]. With solid samples, derivatization is accomplished by doping the fibre coating with an appropriate reagent [216,217]; during the partition, polar analytes are simultaneously extracted and derivatized to less polar analogues that are amenable to GC analysis. Because the amount of reagent in the coating is much higher than the equilibrium concentration of the acids, the reaction follows a *pseudo* first-order kinetics and the amount of derivative accumulated by the fibre is proportional to the integral of the acid concentration in air during the time of exposure. For example, the highly polar propionic and butanoic acids can be derivatized within the fibre coating into the pyrenyl ester, which has a high affinity for the coating and can be easily analysed by GC or HPLC.

Problems such as low reaction recoveries or the formation of side products are to be expected in some cases, however. Also, the usual in-solution derivatization procedures are not always applicable to on-fibre derivatization because factors such as pH or the presence of certain catalysts are difficult to control.

Special aspects of extraction from solid samples

Solid samples pose special problems in SPME as they cannot be extracted in a direct manner; rather, analytes must first be released into the headspace and then concentrated in the fibre coating. Some soil matrices such as sand exhibit relatively weak analyte–matrix interactions, which allows the ready headspace SPME sampling of volatile compounds. Others such as clay with a high metal content and large surface area exhibit strong analyte–matrix interactions and make headspace sampling difficult. With thermally stable analytes, heating the sample is an effective, convenient manner of releasing the analytes from the matrix to the headspace and improving sensitivity. A small amount of water and/or some other surface-active compound can also be added to soil samples in order to facilitate the release of analytes. For example, the amount of volatile hydrocarbon compounds extracted from soil matrices by SPME increases significantly when the sampling is conducted at high temperatures and a 10% moisture content [218].

Heating solid samples helps to release analytes from the headspace and facilitates extraction during SPME sampling. As the temperature increases, however, the fibre coating gradually loses its ability to absorb analytes. An optimum temperature therefore usually exists for SPME. If sampling can be done at a high temperature and a low coating temperature can be maintained at the same time, K_{fs} and sensitivity are dramatically increased as a result. By altering the SPME device and using liquid CO_2 as coolant, the sample can be heated to 250°C while maintaining the coating temperature below 20°C. By using a high temperature, additional water, and cooled fibre, quantitative extraction of toluene, ethylbenzene and xylene isomers — which possess small partition coefficients and are impossible to extract quantitatively under room-temperature conditions — can be achieved [210].

SPME from solid samples does not necessarily require the use of the headspace technique. Thus, aqueous suspensions of samples have proved effective for analytes not too strongly bonded to the matrix [219]. When the suspension is too thick, the fibre coating can be damaged during agitation. One alternative approach involves combining SPME with some other extraction technique such as superheated water extraction [220] or conventional liquid–liquid extraction (LLE) [221]. In the latter case, SPME is used as a fast alternative to time-consuming clean-up procedures and has proved especially effective for rapid screening. Quantitativeness in the determinations is hindered by strong analyte–matrix interactions, which can only be overcome by thorough extraction.

Desorption-related variables

The efficiency with which analytes are thermally desorbed from the SPME fibre is dictated by temperature, the desorption time and the position of the needle in the GC injector. The injectors in most modern gas chromatographs are amenable to direct insertion of the fibre, the only condition being that the liner for capillary GC should contain no packing. Although it has not been checked, desorption from the fibre and subsequent GC analysis are probably unfeasible with packed columns owing to the shape difference from the detector.

The *liner volume* affects the shape of the chromatographic peak (larger volumes cause peak tailing). Split–splitless injectors should be operated in the splitless mode.

The *desorption temperature* usually ranges from 150 to 250°C for semi-volatile compounds. Usually, the optimum desorption temperature is roughly equal to the boiling point of the least volatile analyte [214]. In order to prevent broadening of the chromato-

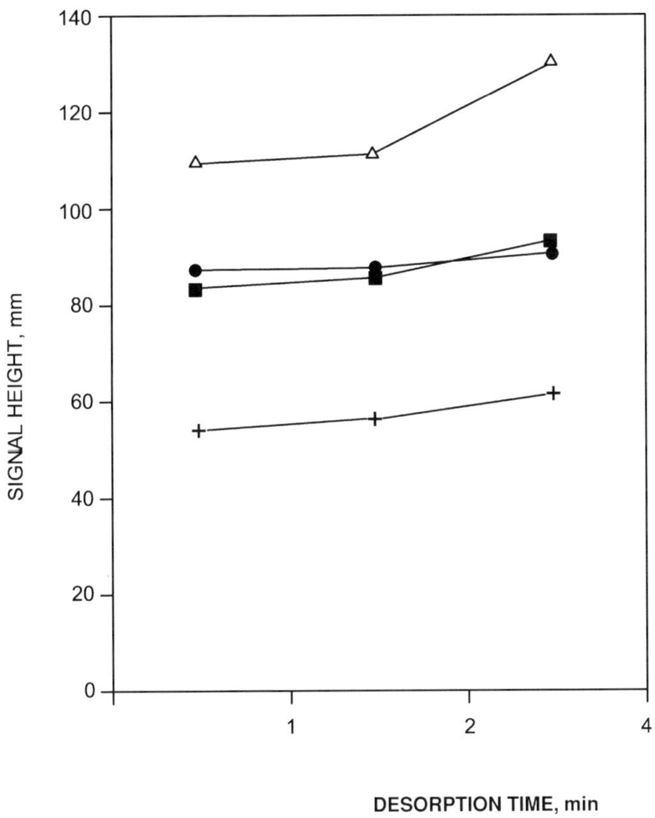

Fig. 4.30. Effect of desorption time on signal height. Sample spiked with chlorophenols and narrow-bore liner. (●) 3-chlorophenol, (■) 2,5-dichlorophenol, (△) 2,3,5-trichlorophenol, (+) 2,3,4,6-tetrachlorophenol. (Reproduced with permission of Elsevier.)

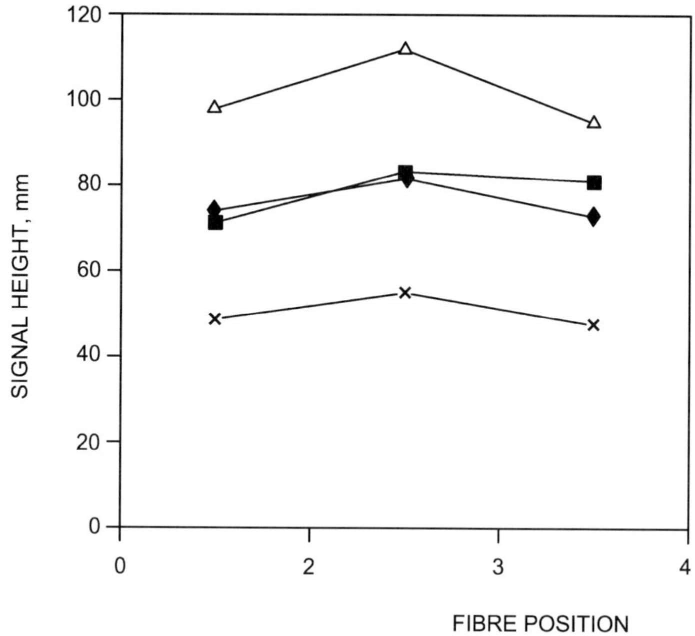

Fig. 4.31. Effect of fibre position (graduation on fibre holder) in the GC injector (narrow-bore liner) during desorption. Sample: water spiked with chlorophenols. The reduction in signal height between the two extreme positions was less than 10%. (♦) 3-chlorophenol, (■) 2,5-dichlorophenol, (△) 2,3,5-trichlorophenol, (×) 2,3,4,6-tetrachlorophenol. (Reproduced with permission of Elsevier.)

graphic peaks, the GC column should be kept at a low initial temperature or even cooled (cryofocused). An initial temperature 80–100°C below the boiling point of the most volatile analyte is recommended [214]. This ensures efficient concentration of the analytes at the column head. SPME and desorption can be repeated several times from the same sample and the analytes concentrated at the front of the cooled column. This, however, is rather time-consuming. The constant flow of carrier gas within a GC injector also facilitates the removal of analytes from the coating.

The *desorption time* depends on the desorption temperature; in general, however, it is less than 1 s, so 1 minute's desorption in the injector should be more than adequate [214] even though considerably longer times are occasionally used (see Fig. 4.30). In any case, it is better to leave the fibre in the GC for a period of time (5–10 min) in order to avoid memory effects.

The *position of the needle in the injector* has proved crucial for some analytes, probably as a result of the injector not being uniformly heated [214]. Although, as can be seen from Fig. 4.31, this factor has little effect on some analytes, it is always preferable to keep the needle in the same position.

As an alternative to thermal desorption, analytes can be desorbed from the fibre by using organic solvents at a special interface and subsequently analysed by HPLC [222].

For compounds amenable to HPLC, a combination of this technique and SPE is the preferred choice on account of its operational simplicity and favourable limits of detection.

4.5.4. Analytical features of SPME

This section deals with the basic analytical properties enhanced by the use of SPME (viz. selectivity and sensitivity), in addition to other major analytical properties including accuracy and precision.

Sensitivity

The amount of analytes that is extracted by SPME depends mainly on the nature of the coating, its volume, whether and how the target analytes are derivatized and the matrix modified, and whether the sample is heated or the coating cooled.

The amount of analyte that is absorbed, and hence the limit of detection for the analyte, can be controlled via the phase thickness. Based on eq. (4.45), the amount of analyte extracted by the fibre coating is directly proportional to the coating volume; also, sensitivity improves as the coating volume is increased (by increasing the coating thickness, the fibre length or both). With non-polar and moderately polar organic compounds, conventional coatings are capable of extracting at least a few picograms of analytes from low part-per-trillion samples, a level that constitutes the typical limits of detection specified by regulatory agencies. This amount is sufficient for identification and quantitation with an ion-trap mass spectrometer. Sensitivity in SPME peaks at equilibrium; for practical purposes, however, the extraction time can be shortened to an extent depending on the particular sensitivity requirements.

Selectivity

As noted earlier for the extraction of BTX compounds and shown in Fig. 4.28, coating/sample partition coefficients depend largely on the type of coating for a given analyte. Derivatization and temperature also have some influence on selectivity and must be adjusted as a function of the analytes, matrix and state of the sample.

Accuracy and precision

Because SPME is primarily an equilibrium technique, it requires calibration when used for quantitation purposes. With air samples, calibration is quite easy [223]. Because sampling occurs either in an open space or in a large volume of air, the amount of analytes extracted by the fibre coating is, based on eq. (4.46), linearly related to the partition coefficient, K_{fs}, which in turn is determined by moisture and temperature. The concentration of analytes in air can thus be determined from the response of the GC detector after correcting for moisture and temperature. For relatively clean aqueous samples (< 1% organics) such as drinking water or other well-defined matrices, external calibration works very well and is usually carried out by spiking a known amount of the target analytes into a clean matrix and then performing SPME. The analyte concentration in unknown samples can be accurately determined by comparing the detector signal with the calibration curve. Multi-level validation for drinking water revealed a high consistency between the results of SPME and the purge and trap technique [224].

With complex soil samples, external calibration may not work well because of matrix effects, and standard additions or internal standards may have to be used instead [206,210,225]. For internal standards to work well, their partition coefficients must be similar to those of the target analytes. Isotopically labelled analogues of target analytes are the best internal standards for SPME because their chemical and physical properties are similar to those of their unlabelled counterparts [210]. Because spiked analytes may not interact with the matrix as strongly as native ones, the experimental conditions must be carefully adjusted to ensure the effective release of native analytes into the extracting media [222]. Analytes with very large partition coefficients — because of their strong affinity for the coating or by effect of the sample being heated simultaneously with cooling of the coating — can be thoroughly extracted. Although calibration is no longer necessary, it is usually done anyway in order to confirm the extraction efficiency.

Because SPME is a single-step technique — which minimizes sources of random errors associated with transfer of the analytes — it exhibits very good precision with clean, model samples. In fact, the precision of SPME is typically in the region of 5% (as RSD) with manual operation and can be as low as 1% with model samples processed using an auto-sampler. On the other hand, complex matrices and carry-over effects can result in RSD values as high as 92% [226,227].

4.5.5. Advantages and shortcomings of SPME

Advantages

As a sample preparation technique, SPME is well-suited to the analysis of volatile and semi-volatile compounds in aqueous, gaseous and solid matrices. Because it dispenses with the use of solvents — not only in extraction but also during injection — it greatly improves the efficiency of chromatographic separations [228]. Solvent-free injection facilitates the use of thinner columns. The cylindrical geometry and small diameter of the fibre can fit into an injector with a very small volume or directly into a column. Analytes can be thermally desorbed very rapidly, thus generating a very narrow band in the column. The use of a column with a small inner diameter and a thin stationary phase improves the separation efficiency and facilitates expeditious GC separations in highly compact instruments. SPME is thus a highly useful sampling tool for on-site environmental analyses when coupled to a portable high-resolution isothermal gas chromatograph. Combined with a smaller injector capable of pulse heating, SPME can be used to insert an extremely narrow band of analytes into a GC column and facilitate separation in a very short time (a few seconds) [229]. All these features make SPME a highly suitable, fast screening technique (especially for field work).

SPME is also an excellent tool for extracting compounds amenable to GC, which provides an alternative to well-established sample preparation methods for polar compounds based on on-line SPE–HPLC.

Shortcomings

One of the factors governing SPME performance is the quality of the needles used, which varies among manufacturers. One major problem arises from the fact that the fibres are fragile and readily broken. The GC injector temperature for desorption of the analytes is also a crucial variable and should be maintained below the maximum operating temperature of the coating (viz. 260–320°C depending on the particular phase). Such a temperature should never be exceeded while conditioning the fibre — a mandatory operation for each new fibre and for fibres that have not been in use for some time. The time required for thermal conditioning is usually specified by the manufacturer. Even with careful conditioning of the fibre, some bleeding of the stationary phase can occur, however.

One other potential problem arises from carry-over of the fibre, which may occur with some analytes (e.g. organochlorine pesticides) and is difficult to suppress — even at high desorption temperatures. If the problem exists, the best solution is to perform blank GC runs with a fibre between successive samplings.

Samples containing large amounts of suspended matter can result in damage to the fibre coating during agitation. High-molecular weight compounds such as humic acids and proteins can be irreversibly absorbed to the fibre, thus altering the properties of the stationary phase and rendering it useless. This can be avoided by protecting it with a semi-permeable membrane [230]. Irreversible absorption is one possible reason for the poor reproducibility and linearity occasionally encountered in SPME, and so is the formation of gas bubbles on the fibre surface, which is difficult to avoid and affects mass transfer rates. Both problems can be circumvented by using headspace SPME (or the above-mentioned isotopically labelled internal standards) in conjunction with mass spectrometric detection. The former solution is simpler; the latter is certainly more reliable, but also rather expensive.

4.5.6. Applications of SPME

The applications of SPME have so far focused on the extraction of organic compounds from various types of matrix including air, water and soil, usually followed by direct transfer into a GC injector for thermal desorption, individual separation on the column and quantitation. In combination with headspace sampling, SPME can also be used to extract organic compounds from wastewater, sludge, soil and other complex matrices [231]. Sub-part-per-billion limits of detection and good precision have been achieved when sampling at room temperature from such complex samples [218]. Even with thin coatings (e.g. 15 μm polydimethylsiloxane), SPME affords the extraction of semi-volatile compounds such as PAHs and PCBs [210].

SPME can also be used to extract target analytes from food and drug samples. Thus, it has been employed for the extraction of caffeine from coffee and tea [225], and for that of volatile impurities in drugs. Headspace SPME has also been tested for flavour analysis in foods. Thus, the SPME/GC/TOF–MS tandem was successfully used for the rapid analysis of volatile flavour compounds in apple fruit. The sample (300–450 g of apple) was subjected to static headspace sampling for 4–6 h in order to allow the volatiles to reach the steady state, after which the headspace was sampled with an SPME fibre 1 cm long coated with a 100 μm thickness of polydimethylsiloxane. The subsequent desorption from the fibre, separation and detection allowed the thorough characterization of the unknown compounds. Throughput, variability, linearity, the saturation kinetics and the matrix effects on sampling were examined in depth using aroma standards. Although the total analysis time was claimed to be 10 min — and hence much shorter than the typical 100–120 min required for purge and trap sampling and subsequent GC–FID — one should also consider the 4–6 h needed for the headspace to reach the steady state [232]. With no steady state allowed, but only 4 min for creating the headspace distribution, the analysis time was dramatically shortened [233].

A comparison of SPME and the purge and trap technique with both thermal desorption and the direct headspace mode for the removal of volatile organic compounds in environmentally hazardous waste provided similar chromatograms for VOCs processed by SPME and PT (46 compounds were detected in both cases with > 90% spectral mass) but poorer results with direct headspace sampling (only 11 compounds were identified in this case). It should be noted that the direct headspace mode was implemented without the aid of cryogenic cooling, which assists in refocusing the analytes at the column head; by contrast, the PT device was equipped with an electrical cold trap in the desorption chamber that refocused the analytes prior to GC separation. The consistency between the qualitative results of SPME and PT, and the faster performance of the former, led to SPME being recommended for screening of VOCs [234]. Table 4.10 compares the results provided by SPME and the headspace modes for petrochemical wastewater. As can be seen, the headspace-based method was more sensitive (responses were 1.6–7.8 times higher); both, however, were similarly reproducible. Compounds with an increased RSD may require a longer equilibration time. As the molecular size of the analyte increases, the speed of convection of the analyte through the matrix to the fibre decreases and diffu-

TABLE 4.10

COMPARISON OF IMMERSION AND HEADSPACE SPME ANALYSIS OF PETROCHEMICAL WASTEWATER

Compound	Immersion response (area × 10^6)	% RSD (n = 5)	Headspace response (area × 10^6)	% RSD (n = 5)
Benzene	83.3	3.1	147	6.1
Toluene	343	5.5	560	5.3
Ethylbenzene	12.9	2.0	101	8.6
m/p-Xylenes	259	12	605	3.3
o-Xylene	138	11	331	5.1
1-Ethyl-2-methylbenzene	34.9	13	115	13
1,3,5-Trimethylbenzene	53.5	23	191	17
1,2,4-Trimethylbenzene	15.6	12	57.5	15
Naphthalene	30.2	22	89.9	16

sion into the fibre coating becomes slower. The addition of a salt to improve extraction and sensitivity tends to delay equilibration. A compromise between sensitivity and equilibration time may be reached by using a fibre with a thinner coating [234].

The characterization of water-soluble components in slurries is one use of SPME with mixed solid–liquid samples. In one application, dried homogenized solid samples (10 mg of sewage sludge or sediment) were slurried in 4 ml of H_2O saturated with NaCl and adjusted to pH 2 with HCl for extraction for 1–15 h, which was followed by desorption into 4:1 methanol/ethanol over 2 min. The extracted compounds were either injected into a liquid chromatograph or fed directly via an electrospray ionization interface to a mass spectrometer with 1 s *m/z* scans from 50–700 or selected-ion monitoring. The major components extracted included phthalates, fatty acids, non-ionic surfactants, chlorinated phenols and carbohydrate derivatives [235].

Although SPME is usually employed with organic compounds, it has also been applied to inorganic ones (e.g. in the extraction of inorganic lead and tetraethyllead from water [236]). Also, an interesting method for the determination of haloacetic acids in water based on in situ derivatization, headspace separation, SPME and GC/ion-trap MS was recently reported [237] whereby the methyl esters of the analytes, formed by treatment with dimethyl sulphate, were subjected to headspace separation and then to extraction using a Carboxen-polydiethylsiloxane fibre. The method was linear over 2 orders of magnitude, the limits of detection were compound-dependent and ranged from 10 to 450 ng/l, and the results were consistent with those of the EPA recommended method for the analytes.

References

1 J. Dedina, *Prog. Anal. Atom. Spectrosc.*, 11 (1988) 251.
2 A.D. Campbell, *Pure Appl. Chem.*, 64 (1992) 227.
3 T. Nakahara, Hydride Generation, in: *Sample Introduction in Atomic Spectroscopy*, J. Sneddon Ed., Elsevier, Amsterdam (1990).
4 L. Ebdon, S. Hill and R.W. Walton, *Analyst*, 113 (1987) 1159.
5 P. Michel, B. Avertt and V. Colandini, *Mikrochim. Acta*, 109 (1992) 35.
6 M.O. Andreae, J.F. Asmodé, P.Foster and L.Van't Dack, *Anal. Chem.*, 53 (1981) 1766.
7 J.S. Blais and W.D. Marshall, *J. Anal. At. Spectrom.*, 4 (1989) 641.
8 M. Fillippelli, F. Baldi, F.E. Brinckmann and G.J. Olson, *Environ. Sci. Technol.*, 26 (1988) 1457.
9 S. Clark and P.J. Craig, *Appl. Organomet. Chem.*, 2 (1988) 33.
10 O.F.X. Donard, S. Rapsomanikis and J.H. Weber, *Anal. Chem.*, 58 (1986) 772.
11 L. Randall, O.F.X. Donard and J.H. Weber, *Anal. Chim. Acta*, 184 (1986) 197.
12 R.J.A. Van Cleuvenbergen, W.E. Van Mol and F.C. Adams, *J. Anal. At. Spectrom.*, 3 (1988) 169.
13 D. Dyne, B.S. Chana, N.J. Smith and J. Cocker, *Anal. Chim. Acta*, 246 (1991) 351.
14 S. Chiavarini, C. Cremisini, T. Ferri and A. Perini, *Sci. Total Environ.*, 101 (1991) 217.
15 J.L. Capelo, I. Lavilla and C. Bendicho, *Anal. Chem.*, 72 (2000) 4979.
16 W. Dirks, R. Lobinski and F.C. Adams, *Anal. Chim. Acta*, 286 (1994) 309.
17 S. Rapsominikis and O.F.X. Donard, *Anal. Chem.*, 58 (1986) 35.
18 L. Liang, N.S. Bloom and M. Horvat, *Clin. Chem.*, 40 (1994) 602.
19 J.S. Lobinska, C. Witte, R. Lobinski and F.C. Adams, *Fresenius J. Anal. Chem.*, 351 (1995) 351.
20 K. Robards and P.J. Worsfold, *Trends Anal. Chem.*, 10 (1991) 121.
21 G. Schwedt, *Chromatographic Methods in Inorganic Analysis*, Hüthig, Heidelberg (1981).
22 P.N. Vijan and G.R. Wood, *At. Absorpt. Newslet.*, 13 (1974) 33.
23 B. Welz and M. Schubert-Jacobs, *At. Spectrosc.*, 12 (1991) 91.

24 M. Thompson, B. Pahlavanpour, S.J. Walton and G.F. Kirbright, *Analyst*, 103 (1978) 568.
25 D.W. Bryce, A. Izquierdo and M.D. Luque de Castro, *J. Anal. At. Spectrom.*, 10 (1995) 1059.
26 W.F. Chan and P.K. Hon, *Analyst*, 115 (1990) 567.
27 L. Gámiz-Gracia and M.D. Luque de Castro, *Talanta*, 50 (1999) 875.
28 D.W. Bryce, A. Izquierdo and M.D. Luque de Castro, *Anal. Chim. Acta*, 324 (1996) 69.
29 L. Gámiz-Gracia and M.D. Luque de Castro, *J. Anal. At. Spectrom.*, 14 (1999) 1615.
30 X. Wang, M. Viczian, A. Lasztity and R.M. Barnens, *J. Anal. At. Spectrom.*, 3 (1988) 821.
31 M. Burguera and J.L. Burguera, *Analyst*, 111 (1986) 171.
32 L. Linag, P. Danilchik and Z. Huang, *At. Spectrosc.*, 15 (1994) 151.
33 O.E. Van Schoor, H.J. Robberecht, I. De Leeuw and H.A. Deelstra, *Z. Lebensm Unters Forsch.*, 180 (1985) 26.
34 R. Kruse, *Levensmittelchem. Gerichtl. Chem.*, 40 (1986) 103.
35 Department of the Environment (UK), *Methods Exam. Waters Assoc. Mater.*, p. 43 (1987).
36 S.J. Haswell, J. Mendham, M.J. Butler and D.C. Smith, *J. Anal. At. Spectrom.*, 3 (1988) 731.
37 O. Anstrom, *Anal. Chem.*, 54 (1982) 190.
38 X. Wang and Z.L. Fang, *Fenxi Huaxue*, 14 (1986) 837.
39 X. Wang and Z.L. Fang, *Fenxi Huaxue*, 16 (1988) 912.
40 C.C.Y. Chan, *Anal. Chem.*, 57 (1985) 1482.
41 M. Yamamoto, M. Yasuda and Y. Yamamoto, *Anal. Chem.*, 57 (1985) 1382.
42 M. Yamamoto, M. Yasuda and Y. Yamamoto, *J. Flow Inject. Anal.*, 2 (1985) 134.
43 B. Welz and M. Schubert-Jacobs, *At. Spectrosc.*, 12 (1991) 91.
44 A. Shinohara, A. Sato, N. Ishii and N. Onda, *Chromatographia*, 32 (1991) 357.
45 B. Kolb, P. Pospisil and M. Auer, *Chromatographia*, 19 (1984) 113.
46 M. Careri, G. Mori, M. Musci and P. Viaroli, *J. Chromatogr.*, 848 (1999) 327.
47 R.E. Shirely and S.B. Cole, *The Supelco Reporter*, 12(1) (1993) 21.
48 S. Hazard, *The Supelco Reporter*, 13(1) (1994) 22.
49 R. Westendorf, *Environ. Lab.*, Aug/Sep (1992) 36.
50 I. Rodríguez Pereiro, A. Wasik and R. Lobinski, *Anal. Chem.*, 70 (1998) 4063.
51 R.S. Rodgers, *Am. Lab.*, Dec. (1993) 20K.
52 S. Abeel and R. Magon, *Pittcon* (1994) Chicago, IL, February 28-March 3.
53 R.B. Lucke, J.A. Campbell, G.A. Ross, S.C. Goheen and E.W. Hopp, *Anal. Chem.*, 65 (1993) 2229.
54 R. Lobinski, I. Rodríguez Pereiro, H. Chassaigne, A. Wasik and J. Szpunar, *J. Anal. Atom. Spectrom.*, 13 (1998) 859.
55 R. Westendorf, G. Mercier and A.K. Vickers, *Pittcon* (1992) New Orleans, LA, March 9–13.
56 Y. Seto, *J. Chromatogr. A.*, 674 (1994) 25.
57 B. Kolb, C. Welter and B. Bichler, *Chromatographia*, 34 (1992) 5.
58 C. McAuliffe, *Chem. Technol.* (1971) 46.
59 C. McAuliffe, *US Pat. 3,759,086* (1973).
60 B. Kolb and L.S. Ettre, *Chromatographia*, 32 (1991) 505.
61 M. Markelov and O. Bershevits, *Anal. Chim. Acta*, 432 (2001) 213.
62 A. Venema, *J. High Resolut. Chromatogr. Commun.*, 13 (1990) 537.
63 L.S. Ettre, E. Jones and B.S. Todd, *Chromatogr. Newslett.*, 12 (1984) 1.
64 B. Kolb and P. Pospisil, *Chromatogr. Newslett.*, 8 (1980) 35.
65 B. Kolb and M. Auer, *Fresenius J. Anal. Chem.*, 336 (1990) 297.
66 J.F. Pankow and M.E. Rosen, *Environ. Sci. Technol.*, 22 (1988) 398.
67 D.P. Lin, C. Falkenberg, D.A. Payne, J. Thakkar, C. Tang and C. Elly, *Anal. Chem.*, 65 (1993) 999.
68 A. Naddaf and J. Balla, *Chromatographia*, 51 (2000) S-283.
69 L. Chambers, *Internat. Lab.*, Nov. (2000) 24.
70 J.F. Pankow, *Environ. Sci. Technol.*, 25 (1991) 123.
71 N.H. Mosesman, L.M. Sidisky and S.D. Corman, *J. Chromatogr. Sci.*, 25 (1987) 351.
72 S. Abeel, *Pittcon* (1992) New Orleans, LA, March 9–13.
73 L.S. Ettre and B. Kolb, *Chromatographia*, 32 (1991) 5.

74 R.E. Shirey, S. Hazard and S.B. Cole, *Pittcon* (1991) Chicago, IL, March 4–8.
75 S.J. O'Doherty, P.G. Simmonds, G. Nickless and W.E. Betz, *J. Chromatogr.*, 630 (1993) 265.
76 J.J. Manura, *Pittcon* (1994), Chicago, IL, February 28–March 3.
77 B. Kolb, *J. Chromatogr.*, 112 (1975) 287.
78 B. Kolb, in: *Applied Headspace Gas Chromatography*, B. Kolb Ed., Heyden & Sons Ltd., New York, (1980) 1.
79 K. Schoene, J. Steinhanses and A. König, *J. Chromatogr.*, 455 (1988) 57.
80 B. Kolb, C. Welter and C. Bichler, *Chromatographia*, 34 (1992) 235.
81 P. Kuran, R. Kubinec, I. Ostrovsky and L. Sojak, *J. Chromatogr.*, 665 (1994) 133.
82 M.M. Minnich, J.H. Zimmerman and B.A. Schumacher, *J. AOAC Int.*, 79 (1996) 1198.
83 F. Hui, J. Xie and R. Rosset, *Analusis*, 24 (1996) 267.
84 J.W. Munch and J.W. Eichelberger, *J. Chromatogr. Sci.*, 30 (1992) 471.
85 J.L. Booker, *J. Chromatogr. Sci.*, 23 (1985) 415.
86 Y. Yokouchi and M. Sano, *J. Chromatogr.*, 555 (1991) 297.
87 A.D. Hewitt and C.L. Grant, *Environ. Sci. Technol.*, 29 (1995) 769.
88 A. Robbat, S. Smararson and Y. Gankin, *Field Anal. Chem. Technol.*, 2 (1998) 253.
89 V. López-Ávila, N. Heath and A. Hu, *J. Chromatogr. Sci.*, 25 (1987) 356.
90 J.V. Headley, *Biomed. Environ. Mass Spectrom.*, 14 (1987) 275.
91 T. Saeed and R. Abu-Tabanja, *J. Assoc. Off. Anal. Chem.*, 68 (1985) 61.
92 P.H. Hemberger, J.E. Alarid, A.D. Cameron, C.P. Leibman, T.M. Cannon, M.A Wolf and R.E. Kaiser, *Int. J. Mass Spectrom. Ion Proc.*, 15 May (1991) 106299.
93 K. Villberg, A. Veijanen, I. Gustafsson and K. Wickstroem, *J. Chromatogr.*, 791 (1997) 213.
94 J. Salafranca, J. Cacho and C. Nerín, *Chromatographia*, 51 (2000) 615.
95 V. Komolprasert, W.A. Hargraves and D.J. Armstrong, *Food Addt. Contam.*, 11 (1994) 605.
96 T.L. Barry, G. Petzinger and G. Lehr, *J. AOAC Int.*, 78 (1995) 413.
97 C. Cai, L.E. Mo and J.H. Guan, *Fenxi Ceshi Xuebao*, 17 (1998) 63.
98 Y. Ehara and Y. Marumo, *Forensic Sci. Int.*, 96 (1998) 1.
99 Y. Ehara, N. Oguri, S. Saito and Y. Marumo, *Bunseki Kagaku*, 46 (1997) 733.
100 M.D.R. Gomes da Silva and H.J. Chaves das Neves, *J. High Resolut. Chromatogr.*, 20 (1997) 275.
101 V. Pinnel and J. Vandegasn, *J. High Resolut. Chromatogr.*, 20 (1997) 343.
102 Q. Liu, J. Zhou, M.D. Xie, Y.F. Zhu and W.Y. Huang, *Fenxi Ceshi Xuebao*, 18 (1999) 58.
103 S.M. Kim, C.M. Wu, A. Kobayashi, K. Kubota and J. Okumura, *J. Agric. Food Chem.*, 43 (1995) 2951.
104 C.Y. Ang, F. Liu and T. Sun, *J. Agric. Food Chem.*, 42 (1994) 2493.
105 D.U. Ahn, C. Jo and D.G. Olson, *J. Agric. Food Chem.*, 47 (1999) 2776.
106 J.R. Vercellotti, J.C.W. Kuan, M.G. Legendre, A.J. StAngelo and H.P. Dupuy, *J. Agric. Food Chem.*, 35 (1987) 1030.
107 D.U. Ahn, C. Jo and D.G. Olson, *J. Food Sci.*, 64 (1999) 230.
108 R.G. Mariaca, T.F.H. Berger, R. Gauch, M.I. Imhof, B. Jeangros and J.O. Bosset, *J. Agric. Food Chem.*, 45 (1997) 4423.
109 J.O. Bosset, U. Buetikofer, T. Berger and R. Gauch, *Mitt. Geb. Lebensmittelunters Hyg.*, 88 (1997) 233.
110 G. Piraprez, M.F. Herent and S. Collin, *Food Chem.*, 61 (1998) 119.
111 T.J. Clark and J.E. Bunch, *J. Chromatogr. Sci.*, 35 (1997) 206.
112 W.M. Coleman, *J. Chromatogr. Sci.*, 30 (1992) 159.
113 W. Holak and J. Specchio, *J. Assoc. Off. Anal. Chem.*, 72 (1989) 476.
114 M. Nagase, *Bunseki Kagaku*, 37 (1988) 30.
115 D.L. Heikes, S.R. Jensen and M.E. Fleming-Jones, *J. Agric. Food Chem.*, 43 (1995) 2869.
116 D.L. Heikes, *J. Assoc. Off. Anal. Chem.*, 68 (1985) 431.
117 D.L. Heikes, *J. Assoc. Off. Anal. Chem.*, 70 (1987) 215.
118 S. Odake, J.P. Roozen and J.J. Burger, *Nahrung*, 42 (1998) 385.
119 M.C. Kuo, Y. Zhang, T.G. Hartman, R.T. Rosen and C.T. Ho, *J. Agric. Food Chem.*, 37 (1989) 1020.

120 J.M. Levy, R.A. Cavalier, T.N. Bosh, A.F. Rynaski and W.E. Huhak, *J. Chromatogr. Sci.*, 27 (1989) 341.
121 J.H. Phillips, C.A. Potera, P.M. Michalko and J.H. Frost, *J. Res. Natl. Bur. Stand.*, 93 (1988) 292.
122 M. Filippelli, F. Baldi, F.E. Brinckman and G.J. Olson, *Environ. Sci. Technol.*, 26 (1992) 1457.
123 E.D. Lund, *J. AOAC Int.*, 77 (1994) 416.
124 D.H. Steele, M.J. Thornburg, J.S. Stanley, R.R. Miller, R. Brooke, J.R. Cushman and G. Cruzan, *J. Agric. Food Chem.*, 42 (1994) 1661.
125 C. Nerín, C. Rubio, J. Cacho and J. Salafranca, *Food Addit. Contam.*, 15 (1998) 346.
126 T.P. McNeal, P.J. Nyman, G.W. Diachenko and H.C. Hollifield, *J. AOAC Int.*, 76 (1993) 1214.
127 S. Holm and E. Lundgren, *Anal. Biochem.*, 136 (1984) 157.
128 J.A. Thompson, B. Ho and S.L. Mastovich, *Anal. Biochem.*, 145 (1985) 376.
129 C.Y. Ye, Y.Q. Feng and J.C. Teng, *Yaowu Fenxi Zazhi*, 12 (1992) 159.
130 S.L. Taylor, J.W. King and J.M. Snyder, *J. Microcolumn Sep.*, 6 (1994) 467.
131 D.L. Ashely, M.A. Bonin, F.L. Cardinaly, J.M. McCraw, J.S. Holler, L.L. Needham and D.G. Patterson, *Anal. Chem.*, 64 (19992) 1021.
132 C.W. Bayer, M.S. Black and L.M. Galloway, *J. Chromatogr. Sci.*, 26 (1988) 168.
133 T.C. Sauer, *Eviron. Sci. Technol.*, 15 (1981) 917.
134 U.S.P. Method 467; *Organic Volatile Impurities*. Pharmacopeial Forum, 19 (1993) 4917.
135 B.D. Page, H.B.S. Conacher, J. Salminen, G.R. Nixon, G. Riedel, B. Mori, J. Gagnon and R. Brousseau, *J. Assoc. Off. Anal. Chem.*, 76 (1993) 26.
136 *Supelco Report*, Note 29 (1994) 1.
137 E.B. Wickenheiser, J. Michalke, C. Drescher, A.V. Hirner and R. Hensel, *Fresenius J. Anal. Chem.*, 362 (1998) 498.
138 D.J. Chichester-Constable, M.E. Barbeau, S.L. Liu, S.R. Smith and J.D. Stuart, *Anal. Lett.*, 20 (1987) 403.
139 W. Davison and H. Zhang, *Nature*, 367 (1994) 546.
140 P.T. Teasdale, S. Hayward and W. Davison, *Anal. Chem.*, 71 (1999) 2186.
141 A. Ott, L.B. Fay and A. Chaintreau, *J. Agric. Food. Chem.*, 45 (1997) 850.
142 S.G. Pavlostathis and G.N. Mathavan, *J. Environ. Technol.*, 13 (1992) 23.
143 T.C. Voice and B. Kolb, *Environ. Sci. Technol.*, 27 (1993) 709.
144 D.F. Askari, M.P. Maskarinec, S.M. Smith, P.M. Beam and C.C. Travis, *Anal. Chem.*, 68 (1996) 3431.
145 I. Papaefstathiou, M.D. Luque de Castro and M. Valcárcel, *Fresenius J. Anal. Chem.*, 354 (1996) 442.
146 M.D. Luque de Castro and I. Papaefstathiou, *Trends Anal. Chem.*, 17 (1997) 41.
147 I. Papaefstathiou and M.D. Luque de Castro, *Anal. Lett.*, 28 (1995) 2063.
148 M.D. Luque de Castro and L. Gámiz-Gracia, Analytical Pervaporation, in: *Encyclopedia of Analytical Chemistry*, R.A. Meyers Ed., John Wiley & Sons, NY (2000) pp. 3084–3101.
149 I. Papaefstathiou, M.T. Tena and M.D. Luque de Castro, *Anal. Chim. Acta*, 308 (1995) 246.
150 E. Mataix and M.D. Luque de Castro, *Analyst*, 123 (1998) 1547.
151 P. Cañizares and M.D. Luque de Castro, *Quím. Anal.*, 15 (1996) 239.
152 P. Cañizares and M.D. Luque de Castro, *Fresenius, J. Anal. Chem.*, 357 (1997) 777.
153 I. Papaefstathiou, U. Bilitewski and M.D. Luque de Castro, *Anal. Chim. Acta*, 330 (1996) 265.
154 F. Delgado-Reyes, I. Papaefstathiou, J.M. Fernández-Romero and M.D. Luque de Castro, *Analyst*, 123 (1998) 2367.
155 R.Y.M. Huang Ed., *Pervaporation Membrane Separation Processes*, Elsevier, Amsterdam (1991).
156 W. Nell, *Pervaporation*, Tec & Doc Lavoisier, Cachan (1998).
157 M.D. Luque de Castro and I. Papaefstathiou, Analytical Pervaporation, in: *Encyclopedia of Environmental Analysis and Remediation*, R.A. Meyers Ed., John Wiley & Sons, NY (1998) pp. 3462–3465.
158 P.J. Hickey and C.S. Slater, *Purif. Methods*, 19 (1990) 93.
159 I.L. Mattos, M.D. Luque de Castro and M. Valcárcel, *Talanta*, 42 (1995) 755.

160 J. Galván, I. Sanz-Vicente, J.R. Castillo and M.D. Luque de Castro, *Anal. Chim. Acta*, 434 (2001) 81.
161 S.D. Kolev, M.D. Luque de Castro, *unpublished results*.
162 H. Sulistyarti, T.J. Cardwell, M.D. Luque de Castro and S.D. Kolev, *Anal. Chim. Acta*, 390 (1999) 133.
163 C. Cámara and R. Muñoz, *private communication*.
164 I. Ogbomo, A. Steffl, W. Schumann, U. Prinzing and H.L. Schmidt, *J. Biotechnol.*, 31 (1993) 317.
165 I. Papaefstathiou and M.D. Luque de Castro, *Anal. Chem.*, 67 (1995) 3916.
166 I. Papaefstathiou and M.D. Luque de Castro, *Int. J. Environ. Anal. Chem.*, 66 (1997) 107.
167 C.M. Bell, F.J. Gerner and H. Strathmann, *J. Membr. Sci.*, 36 (1988) 315.
168 I.L. Mattos and M.D. Luque de Castro, *Anal. Chim. Acta*, 298 (1994) 159.
169 D.W. Bryce, A. Izquierdo and M.D. Luque de Castro, *Anal. Chem.*, 69 (1997) 844.
170 M.D. Luque de Castro and I. Papaefstathiou, *Rev. Anal. Chem.*, 17 (1998) 191.
171 I. Papaefstathiou and M.D. Luque de Castro, *J. Chromatogr.*, 779 (1997) 352.
172 A. Ishida, I. Kameshita and H. Fujisawa, *Anal. Biochem.*, 254 (1997) 152.
173 I. Papaefstathiou and M.D. Luque de Castro, *Anal. Chim. Acta*, 354 (1997) 135.
174 F. Delgado-Reyes, I. Papaefstathiou, J.M. Fernández-Romero and M.D. Luque de Castro, *Analyst*, 123 (1998) 2367.
175 I. Papaefstathiou, U. Bilitewski and M.D. Luque de Castro, *Fresenius, J. Anal. Chem.*, 357 (1997) 1168.
176 J. A. García-Garrido and M.D. Luque de Castro, *Analyst*, 122 (1997) 663.
177 F. Delgado-Reyes, J.M. Fernández-Romero and M.D. Luque de Castro, *Anal. Chim. Acta*, 408 (2000) 209.
178 F. Delgado-Reyes, J.M. Fernández-Romero and M.D. Luque de Castro, *Talanta*, 53 (2001) 961.
179 F. Delgado-Reyes, J.M. Fernández-Romero and M.D. Luque de Castro, *Anal. Lett.*, 34 (2001) 2277.
180 M.D. Luque de Castro and M.M. Jiménez-Carmona, *Quím. Anal.*, 18 (1999) 263.
181 M.D. Luque de Castro and I. Papaefstathiou, *Spectrochim. Acta*, 53B (1998) 311.
182 E. Mataix and M.D. Luque de Castro, *Analyst*, 123 (1998) 1547.
183 E. Mataix and M.D. Luque de Castro, *Anal. Chim. Acta*, 381 (1999) 23.
184 E. Mataix and M.D. Luque de Castro, *Talanta*, 51 (2000) 489.
185 S.Y. Sheikheldin, T.J. Cardwell, R.W. Cattrall, M.D. Luque de Castro and S.D. Kolev, *Anal. Chim. Acta,* 419 (2000) 9., in press.
186 J. González-Rodríguez, J. Pérez-Juan and M.D. Luque de Castro, *Analyst*, 126 (2001) 1177.
187 E. Mataix and M.D. Luque de Castro, *Chromatographia*, 52 (2000) 205.
188 U. Prinzing, I. Ogbomo, C. Lehn and H.L. Schmidt, *Sens. Actuators*, B1 (1990) 542.
189 L. Gámiz-Gracia and M.D. Luque de Castro, *Analyst*, 124 (1999) 1119.
190 L. Gámiz-Gracia and M.D. Luque de Castro, *J. Pharm. Biomed. Anal.*, 22 (2000) 909.
191 L. Wang, T.J. Cardwell, R.W. Cattrall, M.D. Luque de Castro and S.D. Kolev, *Anal. Chim. Acta*, 416 (2000) 177.
192 P.G. Rodrigues, J.A. Rodrigues and A.A. Barros, *Flow Analysis VIII*, June, Poland (2000) Proceedings, p. 78.
193 I.L. Mattos and E.A.G. Zagatto, *Anal. Sci.*, 15 (1999) 63.
194 S.Y. Sheikheldin, T.J. Cardwell, R.W. Cattrall, M.D. Luque de Castro and S.D. Kolev, *J. Environ. Sci. Technol.*, 35 (2001) 178.
195 M.D. Luque de Castro and J.M. Fernández-Romero, *J. Chromatogr.*, 819 (1998) 25.
196 M.D. Luque de Castro and L. Gámiz-Gracia, *Chromatographia*, 52 (2000) 265.
197 M.D. Luque de Castro and L. Gámiz-Gracia, *Anal. Chim. Acta*, 351 (1997) 23.
198 C.F. Poole and S.A. Schuette, *J. High Resolut. Chromatogr.* 6 (1983) 526.
199 D.R. Hagen, C.G. Markell, G. Schmitt and D.B. Blevins, *Anal. Chim. Acta*, 236 (1990) 157.
200 A. Kraut-Vass and J. Thoma, *J. Chromatogr.*, 538 (1991) 233.
201 C.L. Arthur and J. Pawliszyn, *Anal. Chem.*, 62 (1990) 2145.
202 Z. Zhang, M.J. Yang and J. Pawliszyn, *Anal. Chem.*, 66 (1994) 844 A.

203 H. Prosen and L.Z. Zupancic-Kralj, *Trends Anal. Chem.*, 18 (1999) 272.
204 J. Pawliszyn, *SPME, Theory and Practice*, Wiley, New York (1997).
205 D. Louch, S. Mothalgh and J. Pawliszyn, *Anal. Chem.*, 64 (1992) 1187.
206 E. Otu and J. Pawliszyn, *Mikrochim. Acta*, 112 (1993) 41.
207 S. Motlagh and J. Pawliszyn, *Anal. Chim. Acta*, 284 (1993) 265.
208 Z. Zhang and J. Pawliszyn, *Anal. Chem.*, 65 (1993) 1843.
209 J. Pawliszyn, *J. Chromatogr. Sci.*, 31 (1993) 31.
210 D. Potter and J. Pawliszyn, *Environ. Sci. Technol.*, 28 (1994) 298.
211 K. Buchholz and J. Pawliszyn, *Anal. Chem.*, 66 (1994) 160.
212 Y. Liu, Y. Shen and M.L. Lee, *Anal. Chem.*, 69 (1997) 190.
213 A. Fromberg, T. Nilsson, B.R. Larsen, L. Montanarella, S. Facchetti and J.O. Madsen, *J. Chromatogr. A*, 746 (1996) 71.
214 C.L. Arthur, L.M. Killam, K.D. Buchhlolz, J. Pawliszyn and J.R. Berg, *Anal. Chem.*, 64 (1992) 1960.
215 P.A. Martos and J. Pawliszyn, *Anal. Chem.*, 69 (1997) 206.
216 L. Pan and J. Pawliszyn, *Anal. Chem.*, 69 (1997) 196.
217 S. Velikonja Bolta, L. Zupancic-Kralj and J. Marsel, *Chromatographia*, 48 (1998) 95.
218 S. Magdic, A. Boyd-Boland, K. Jinno and J. Pawliszyn, *J. Chromatogr.*, 736 (1996) 219.
219 A.A. Boyd-Boland and J. Pawliszyn, *J. Chromatogr.* 704 (1995) 163.
220 H. Daimon and J. Pawliszyn, *Anal. Commun.*, 33 (1996) 421.
221 H. Prosen and L. Zupancic-Kralj, *Acta Chim. Slov.*, 45 (1998) 1.
222 J. Chen and J. Pawliszyn, *Anal. Chem.*, 67 (1995) 2530.
223 M. Chai, C. Arthur, J. Pawliszyn, R. Belardi and K. Pratt, *Analyst*, 118 (1993) 1501.
224 B. MacGillivary, P. Fowlie, C. Sagara and J. Pawliszyn, *J. Chromatogr. Sci.*, 32 (1994) 317.
225 S. Hawthorne, D. Miller and J. Pawliszyn, *J. Chromatogr.*, 603 (1992) 185.
226 A.A. Boyd-Boland, S. Magdic and J. Pawliszyn, *Analyst*, 121 (1996) 929.
227 R. Young, V. López-Ávila and W.F. Beckert, *J. High Resolut. Chromatogr.*, 19 (1996) 247.
228 T.L. Schumacher, *Supelco Report*, 16(1) (1997) 8.
229 C.L. Arthur, M. Chai and J. Pawliszyn, *J. Microcolumn Sep.*, 5 (1993) 51.
230 J. Pawliszyn and S. Liu, *Anal. Chem.*, 59 (1987) 1475.
231 Z. Zhang, J. Poerschamann and J. Pawliszyn, *Anal. Commun.*, 33 (1996) 219.
232 J. Song, B.D. Gardner, J.F. Holland and R.M. Beaudry, *J. Agric. Food Chem.*, 4 (1997) 1801.
233 R.F. Mindrup, *Supelco Report*, 17(3) (1998) 4.
234 K.J. James and M.A. Stack, *Fresenius J. Anal. Chem.*, 358 (1997) 833.
235 M. Moeder and J. Pawliszyn, *J. Microcolumn Sep.*, 10 (1998) 225.
236 T. Górecki and J. Pawliszyn, *Anal. Chem.*, 68 (1996) 3008.
237 M.N. Sarrión, F.J. Santos and M.T. Galcerán, *Anal. Chem.*, 72 (2000) 4865.

Chapter 5

Microwave-assisted solid sample treatment

5.1. INTRODUCTION

Whether simply for heating up a meal or preparing a sample for analysis via a multi-step process, microwave technology is of considerable significance to both homes and analytical laboratories. In fact, it can help reduce the time taken to develop many processes, so it is of high value with many types of samples. The way in which microwaves enhance some processes is not fully understood, however. The main factors to be considered here include improved transport properties of molecules, molecular agitation, the heating of solvents above their boiling points and, in some cases, product selectivity.

Analytical chemists began using microwaves in 1975 to digest biological samples [1]. A domestic microwave oven was then used to rapidly heat a mixture of sample and digestion acids to its atmospheric boiling point in an Erlenmeyer flask. Microwave-assisted methods have been widely used in analytical chemistry ever since their first application was reported. The latest advances in the use of microwaves in various analytical chemical fields include sample digestion for elemental analysis [2], solvent extraction [3], sample drying [4], moisture measurement [5], analyte adsorption and desorption [6], sample clean-up [7], chromogenic reactions [8], solid-phase retention [9], elution [10], distillation [11], microwave plasma atomic spectrometry [12] and synthetic reactions [13], among others.

This chapter describes the principal applications of microwaves to the pretreatment of solid samples, with special emphasis on digestion and extraction, which are their two main uses in analytical chemistry. The description is preceded by a discussion of the fundamentals of microwave energy and its interaction with matter, and also of the equipment used by analytical laboratories, which can be of the open or closed type depending on whether they operate at atmospheric pressure or a higher level and whether they use multi-mode or focused microwaves. Selected designs developed for specific purposes are also commented on.

Before applications are dealt with, the main variables governing microwave-assisted processes and the parameters characterizing specific microwave treatments are examined. The applications discussed include not only microwave-assisted digestion and extraction — which are the two most widely implemented and hence those with the highest potential interest to readers — but also others of special significance to solid sample treatment such as microwave-assisted drying, distillation and protein hydrolysis. Finally, some safety recommendations on the use of microwave equipment are made.

5.2. FUNDAMENTALS OF MICROWAVE ENERGY AND ITS INTERACTION WITH MATTER

Microwave radiation consists of electromagnetic waves and hence of an electric and a magnetic field that are normal to each other. Microwave energy is a non-ionizing type of radiation that causes molecular motion through migration of ions and rotation of dipoles without altering molecular structure.

The microwave region of the electromagnetic spectrum lies between infrared light and radio frequencies, and spans the frequency range from 300 MHz to 300 GHz, which corresponds to wavelengths from 1 m to 1 cm. The wavelengths from 1 to 25 cm are extensively used for RADAR (RAdio Detection And Ranging) transmissions and the remaining wave-length range is used for telecommunications. Four specific frequencies, viz. 915 ± 25, 2450 ± 13, 5800 ± 75 and 22 125 ± 125 MHz, are used for industrial and scientific microwave heating. These frequencies were reserved for industrial, scientific and medical use by the US Federal Communications Commission (FCC) and conform to the International Radio Regulations adopted in Geneva in 1959. Most domestic microwave ovens operate at 2.45 GHz. The typical energy output of a microwave system is 600–700 W. This allows approximately 43 000 cal to be supplied within 5 min to the microwave cavity for sample heating [14].

There is a clear difference between microwave spectroscopy and microwave dielectric heating effects. Thus, in microwave spectroscopy, molecules are examined in the gas phase and the microwave spectrum for a molecule exhibits many sharp bands [15] over the frequency range 3–60 GHz. Such sharp bands arise from transitions between quantized rotational states of the molecules. Microwave spectroscopy provides an excellent fingerprinting method for identifying molecules in a gas phase and has been used, for example, to confirm the presence of a wide range of molecules in outer space.

The lifetime of the excited state produced by a particular rotational state in a gas phase at a low pressure is long. However, by virtue of Heisenberg's uncertainty principle, at pressures in the region of 10^{-1} mm Hg, the lifetime is reduced by collisions and spectral peaks broaden as a result. In liquids and solids, where molecules are generally not free to rotate separately, spectra are too broad to be observed. It is to these phases that microwave dielectric loss heating effects are relevant as distinguished from spectroscopic effects of microwave energy. A material can be heated by applying energy to it in the form of high-frequency electromagnetic waves. The origin of the heating effect is the ability of an electric field to exert a force on charged particles. If the particles present in a substance can move freely through it, then a current has been induced. On the other hand, if the charge carriers in the substance are bound to certain regions, they will move until a counterforce balances them, the net result being dielectric polarization. Both conduction and dielectric polarization are sources of microwave heating. The heating effect of microwaves depends on their frequency and on the applied power; unlike microwave spectroscopy, however, the effect does not result from well-spaced discrete quantized energy states.

5.2.1. Dissipation factor

The dielectric properties of materials are defined by two different parameters, namely: the dielectric constant and the dielectric loss. The dielectric constant, ε, describes the ability of a molecule to be polarized by the electric field. At low frequencies, ε reaches a maximum as the maximum amount of energy can be stored in the material. The dielectric loss, ε', measures the efficiency with which the energy of the electromagnetic radiation can be converted into heat. The dielectric loss goes through a maximum as the dielectric constant falls [16]. The dissipation factor (tan δ) is the ratio of the dielectric loss of the sample, also called "loss factor", to its dielectric constant: $\tan \delta = \varepsilon'/\varepsilon$.

When microwave energy penetrates a sample, the energy is absorbed by the sample at a rate dependent on its dissipation factor. Penetration is considered to be infinite in materials that are transparent to microwave energy and zero in reflective materials (e.g. metals). The dissipation factor is a finite amount for absorptive samples. Because the energy is rapidly absorbed and dissipated as microwaves pass into the sample, the greater the dissipation factor of the sample is, the less microwave energy at a given frequency will penetrate into it. One effective way to characterize penetration is via the half-power depth or the distance from the sample surface at which the power density is reduced to one-half that at the surface [17]. The half-power depth varies with the dielectric properties of the sample and also, roughly, with the inverse of the square root of the frequency.

5.2.2. Transformation of microwave energy into heat

Microwave energy is lost to the sample via two mechanisms: ionic conduction and dipole rotation. Both occur simultaneously in many practical applications of microwave heating.

Ionic conduction is the conductive migration of dissolved ions in the applied electromagnetic field. This ion migration is a flow of current that results in I^2R losses (heat production) due to resistance to ion flow. All ions in a solution contribute to the conduction processes; however, the fraction of current carried by any given species is determined by its relative concentration and its inherent mobility in the medium. Therefore, the losses due to ionic migration depend on the size, charge and conductivity of the dissolved ions, and are subject to the effects of ion interaction with the solvent molecules [18].

Ion conduction is influenced by the ion concentration, ion mobility and solution temperature. Every ionic solution contains at least two ionic species and each species will conduct current according to its concentration and mobility. The dissipation factor of an ionic solution changes with temperature, which affects ion mobility and concentration.

Dipole rotation refers to the alignment, by effect of the electric field, of molecules in the sample that have permanent or induced dipole moments. As the electric field of microwave energy increases, it aligns polarized molecules. As the field decreases, thermally induced disorder is restored. In fact, applied microwave fields cause molecules, on average, to temporarily spend very slightly more time pointing in one direction than in others. Associated with that very small fraction of preferred orientation there is another very small fraction of molecular order imposed and hence a tiny bit of energy. When the field is removed, thermal agitation returns the molecules to disorder, during relaxation time, and thermal energy is released. At 2450 MHz, the alignment of the molecules followed by their return to disorder occurs 4.9×10^9 times each second and results in very rapid heating. However, the efficacy of heating by dipole rotation depends upon the characteristic dielectric relaxation time of the sample, which in turn is a function of its temperature and viscosity. The dielectric relaxation time is the time taken by the molecules in the sample to achieve 63% of their return to disorder. The maximum energy conversion per cycle for many materials (dielectric loss due to dipole rotation) will occur when $\omega = 1/\tau$, where ω is the angular frequency of the microwave energy (in rad/s) and τ the dielectric relaxation time of the sample [19]. A non-ionic polar sample with a $1/\tau$ value close to the angular frequency of the input microwave energy will possess a high dissipation factor. By contrast, when $1/\tau$ for the sample is substantially different from the microwave frequency, the sample will have a low dissipation factor.

Temperature dictates to a great extent the relative contribution of each of the two energy conversion mechanisms (dipole rotation and ionic conduction). With small molecules such as water and other solvents, the dielectric loss to a sample due to the contribution of dipole rotation decreases as the sample temperature increases. By contrast, the dielectric loss due to ionic conduction increases as the sample temperature increases. Therefore, as an ionic sample is heated by microwave energy, the dielectric loss to the sample is initially dominated by the contribution of dipole rotation. As the temperature increases, the dielectric loss is dominated by ionic conduction.

The relative contribution of these two heating mechanisms depends on the mobility and concentration of the sample ions and on the relaxation time of the sample. If the ion mobility and concentration of the sample ions are low, then sample heating will be entirely dominated by dipole rotation. On the other hand, as the mobility and concentration of the sample ions increase, microwave heating will be dominated by ionic conduction and the heating time will be independent of the relaxation time of the solution. As the ionic concentration increases, the dissipation factor will increase and the heating time decrease. The heating time depends not only on the dielectric absorptivity of the sample but also on the microwave system design and the sample size.

5.2.3. Microwave heating

There is a clear difference between conductive and microwave heating: because the vessels used in conductive heating are usually poor conductors of heat, they take time to heat and transfer such heat to the solution. Also, because the liquid vaporizes at the surface, a thermal gradient is established by convection currents so only a small portion of the fluid is at the temperature of the heat applied to the outside of the vessel. Therefore, with conductive heating, only a small portion of the fluid is above the boiling point of the solution.

On the other hand, microwaves heat all of the sample fluid simultaneously without heating the vessel. Therefore, with microwave heating, the solution reaches its boiling point very rapidly. Because the rate of heating is so much more rapid, substantial localized superheating can occur [14]. Table 5.1 shows the effect of microwave heating on several solvents [20].

TABLE 5.1

TEMPERATURE REACHED BY VARIOUS SOLVENTS UPON HEATING FROM ROOM TEMPERATURE USING MICROWAVE ENERGY OF 2450 MHz AND 560 W FOR 1 MIN

Solvent	*T*, °C	Boiling point,°C
Water	81	100
Methanol	65	65
Ethanol	78	78
1-Propanol	97	97
1-Butanol	109	117
1-Pentanol	106	137
1-Hexanol	92	158
1-Chlorobutane	76	78
1-Bromobutane	95	101
Acetic acid	110	119
Ethyl acetate	73	77
Chloroform	49	61
Acetone	56	56
Dimethylformamide	131	153
Dimethyl ether	32	35
Hexane	25	68
Heptane	26	98
Carbon tetrachloride	26	77

5.3. MICROWAVE EQUIPMENT

Microwave equipment used for sample treatment can be classified into two groups according to the way microwave energy is applied to the sample, namely: *multi-mode systems*, where the microwave radiation is allowed to randomly disperse in a cavity, so every zone in the cavity and the sample it contains is evenly irradiated; and *single-mode* or *focused systems*, where the microwave radiation is focused on a restricted zone where the sample is placed for subjection to a much stronger electrical field than in the previous case. Usually, multi-mode systems are of the closed-vessel type, and focused systems of the open-vessel type [21], the former being defined as systems where the microwave treatment is conducted at a high pressure and the latter as those where microwaves are applied at atmospheric pressure, so the system is subject to no overpressure. This mutual association, however, is incorrect as some commercial devices that operate at a high pressure use focused microwaves [22,23] and domestic ovens have been used to couple microwave treatment to detection by using multi-mode sample irradiation but at atmospheric pressure [24,25]. The latter application has aroused much confusion as many authors consider these systems, which are usually flow-injection manifolds, to be of the closed-vessel type [26]. In a different context, this definition is quite correct and causes no confusion as the sample, once inserted into the flow-injection system, actually

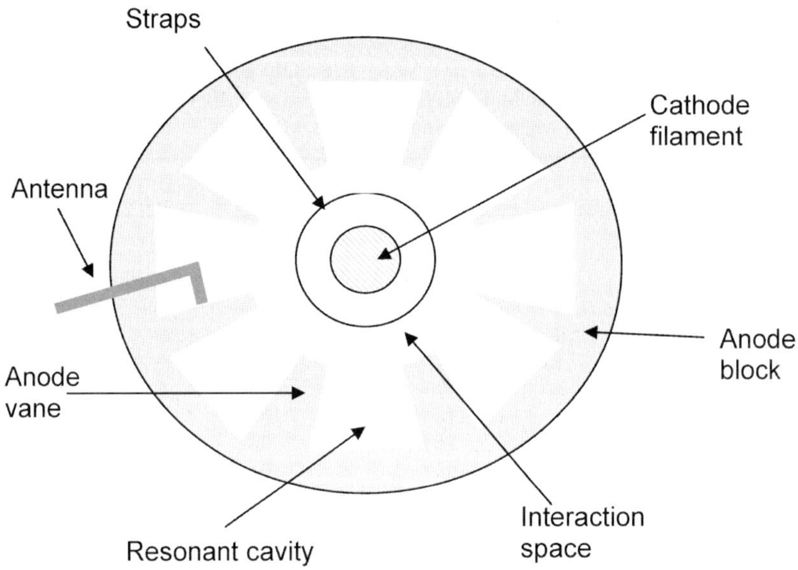

Fig. 5.1. Schematic of a magnetron.

circulates through a series of tubes and devices isolated from the outside. However, in order to avoid confusion, we shall use the designation "closed-vessel" in this chapter only to refer to systems operating above atmospheric pressure.

5.3.1. Main components of a microwave device

Both multi-mode and focused microwave devices consist of four major components, namely:

(a) The microwave generator, usually called the "magnetron", where the microwave energy is produced;
(b) The waveguide, which is used to propagate the microwaves from the magnetron to the microwave cavity;
(c) The applicator, where the sample is placed;
(d) The circulator, which allows microwaves to pass only in the forward direction.

Basically, the *magnetron* consists of a number of identical resonators (small cavities) arranged in a cylindrical pattern around an also cylindrical cathode (see Fig. 5.1). A permanent magnet is used to produce a strong magnetic field normal to the cross-section. The anode, which consists of a series of circuits tuned to oscillate at a specific frequency (or its overtones) is kept at a high positive voltage relative to the cathode. Electrons emitted from the cathode are accelerated towards the anode block, but the presence of the magnetic field produces a force in the azimuthal direction that causes the electron trajectory to be deflected in the same direction. As the deflected electrons pass through the resonator gaps, they induce a small charge into the circuit that causes the resonator to oscillate. Alternate cavities are linked by two small wire straps (mode straps) that ensure the correct phase relationship. This process continues until the oscillation has reached a sufficiently high amplitude. At that point, the radiation is taken off the anode via an antenna enclosed in the vacuum envelope of a tube.

In microwave systems used for sample preparation, the power output of the magnetron is controlled by "cycling" the magnetron to obtain an average power level. The duty cycle is the time the magnetron is "on" divided by the time base. For example, a time "on" of 5 s with a time base of 10 s is a duty cycle of 0.5, and so is a time "on" of 0.5 s with a time base of 1 s. The time base for the magnetron in microwave systems used for sample preparation is 1 s. Thus, in order to obtain one-half the rated output of 600 W, the magnetron would be "on" for 0.5 s and "off" for 0.5 s. Appliance-grade microwave ovens typically have a time base of 10 s. This relatively long time is not desirable for heating analytical samples because heat losses can be significant during the long "off" time (5 s for a duty cycle of 0.5).

The power output of a magnetron can be affected by overheating resulting from reflected microwaves. Reflected power occurs when the travelling electromagnetic waves are reflected and the flow of energy is partly in the reverse direction: towards the magnetron. Methods that transfer travelling waves from one medium to another without reflectance are called "impedance matching methods". If the microwaves travel from the magnetron to the sample and are not reflected, then the system is perfectly matched. If there is reflection, then the system is said to be mismatched. Mismatching may cause overheating of the magnetron, a loss in output power or even destruction of the magnetron.

The *waveguide* is a rectangular or circular channel constructed of reflective material such as sheet metal and is designed to direct the microwaves generated by the magnetron to the cavity without a mismatch. Waveguides with a rectangular cross-section are the more widely used. They are available in sizes for use at frequencies from 320 MHz to 333 GHz. The minimum frequency that can be propagated through the waveguide is related to the dimensions of the rectangular cross-section by the expression $c/f = 2d$, where c is the speed of light, f the cut-off frequency and d the larger of the dimensions of the rectangular cross-section.

The *applicator* is the zone in which microwaves are applied to the sample. It can be a multi-mode cavity where microwaves are randomly dispersed or the waveguide itself, in which case the sample vessel is placed directly inside it to focus the microwave radiation onto the sample.

The walls of a microwave cavity should be reflectant in order to prevent leakage of radiation and increase the efficiency of the oven. The pathway followed by the microwaves is well defined into recognizable patterns or modes that have a beginning and an end. The direction and number of excitation modes in a cavity of a certain size are two factors to be considered with a view to ensuring uniform microwave heating. There is rarely a perfect match between the frequency used and the resonant frequency of the load, so if the energy is reflected by the walls, the absorbance is increased because the energy passes through the sample more often and can be partially absorbed on each passage. A sample in the microwave cavity that has a high dissipation factor could be a model of a perfectly matched microwave system by absorbing nearly 100% of the input power. If too much

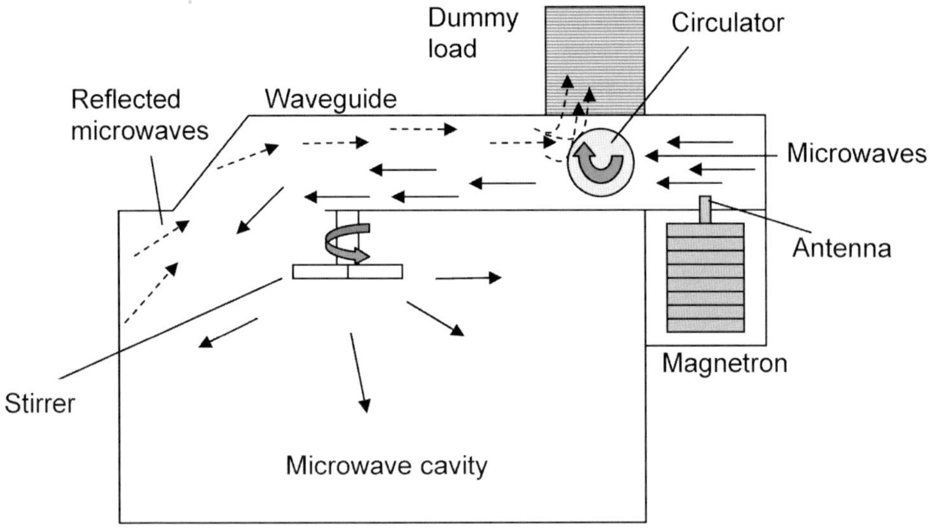

Fig. 5.2. Major components of a typical multi-mode microwave system.

energy is reflected back into the waveguide, a mismatch between the magnetron and the microwave occurs that can alter the output and result in damage to the magnetron through overheating. Uniform distribution of microwave radiation in a microwave cavity can be achieved by using a mode stirrer, which is a fan-shaped blade that is used to reflect and mix the energy entering the microwave cavity from the waveguide. A mode stirrer assists in distributing the incoming energy so that heating of the sample will be more independent of position. Figure 5.2 depicts a typical multi-mode microwave system.

Unlike microwave cavity systems, in focused microwave systems a single vessel is placed directly in a microwave waveguide that acts as the applicator (see Fig. 5.3). The bottom few inches of the vessel are directly exposed to the microwaves, whereas the upper region of the vessel remains cool. This results in an effective condensing mechanism inherent in the design. While vessels are open to the atmosphere, the refluxing action minimizes losses of solvent and some volatiles. Vessel openings have been designed to allow automated reagent addition and restrict contamination from the atmosphere.

Most commercial ovens, whether single-mode or multi-mode models, are protected from potential damage caused by reflected energy by an automatic cut-off mechanism; also, some ovens provide additional protection in the form of a *circulator* that directs reflected energy into a dummy load. A circulator is essentially a device that uses ferrites and static magnetic fields to allow microwaves to pass in the forward direction but diverts reflected waves into a dummy load where the energy is harmlessly dissipated as heat.

5.3.2. Closed-vessel microwave systems

The earliest microwave systems for analytical purposes were closed-vessel systems with a multi-mode cavity. The primary improvement over domestic units was the addition

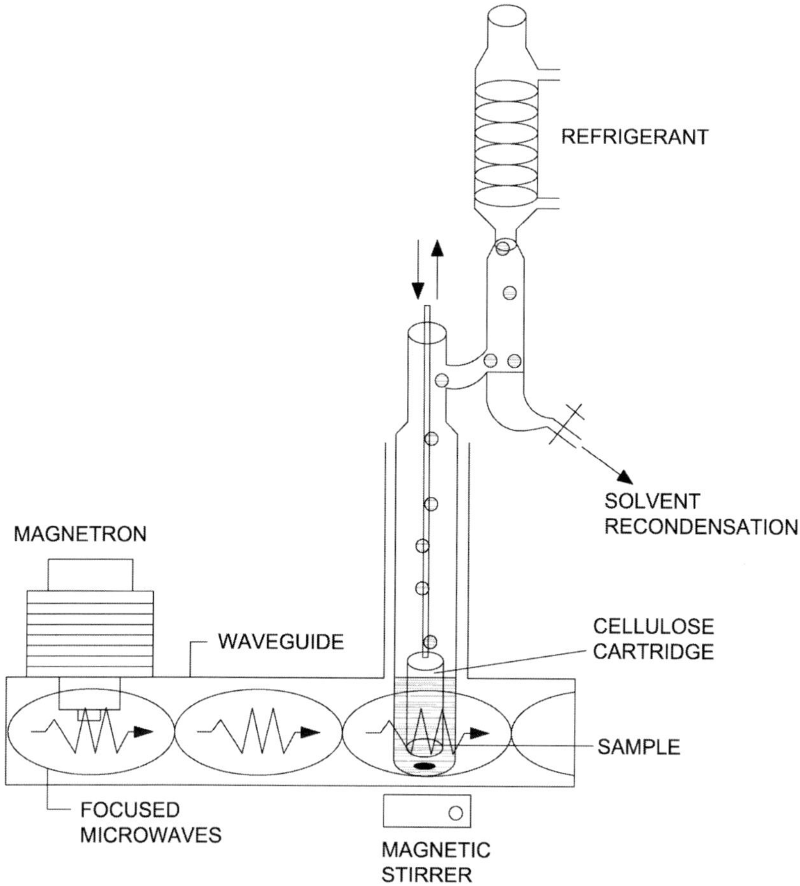

Fig. 5.3. Scheme of a focused microwave system. (Reproduced with permission of Springer-Verlag.)

of safety mechanisms. The early units, though built from domestic-oven cavities and doors, isolated and ventilated the cavity to prevent acid fumes from attacking the electronics. Ever since the first commercial laboratory microwave unit (Model MDS-81 D from CEM Corporation) was introduced in 1985 [27], a number of manufacturers have delivered gradually improved versions with more uniform microwave fields, the ability to control the microwave power and, most importantly, increased safety.

In the mid-1980s, some researchers started to build or adapt temperature- and pressure-monitoring equipment for use inside a microwave cavity. The primary challenges were to develop probes that would not disturb the microwave field and to construct wavelength attenuator cut-offs for the probes so that they could enter the microwave region while preventing microwaves from leaving the microwave cavity. Monitoring temperature and/or pressure during digestions or extractions started the age of controlled microwave treatments, the study of the mechanisms of microwave-assisted processes and the development of transferable standard microwave sample preparation methods. These developments promoted an outgrowth of microwave use that was reflected in a substantial increase in microwave sample preparation publications.

The earliest commercial laboratory microwave units with pressure and temperature feedback control, which were developed in 1989 and 1992, respectively, afforded a more rigorous design and control of microwave sample preparation procedures.

Concurrent with these developments, the microwave vessel also evolved significantly. The vessels in microwave systems are relatively transparent to microwave radiation and inert to the reagents used. The materials in contact with the sample are usually of fluoropolymer, quartz or glass composition. Closed, pressurized vessels are frequently multi-layered, with a structural outer casing of a microwave-transparent polymer such as polyetherimide. This polymer is commonly used for its tensile strength, rigidity and ability to contain accumulated pressure.

Closed microwave vessels exist in two different forms. One encompasses uninsulated, relatively thin, single-walled fluoropolymer vessels and fluoropolymer vessels of the same material with a thin outer layer made of polyetherimide or rigid composite plastic. These vessels have minimal insulating characteristics and allow large amounts of heat to escape. The other type of closed vessel is a well-insulated container, usually of very thick-walled fluoropolymer, or one with a very thick outer layer or casing (or both). These vessels retain heat in a highly efficient manner, so they do not allow rapid cooling by ambient air forced over them within the microwave cavity [16].

Commercially available closed vessels can be chronologically divided into three generations. The first generation was an all-Teflon [polytetrafluoroethylene (PTFE)] design with low-pressure limits that decreased as the vessel aged through stress from previous treatments The second generation was that of jacketed vessels, which typically consisted of a Teflon liner and cap on a polymer case (usually of polyetherimide) that increased the pressure limit of the vessel up to 20 atm. The third generation of vessels, also lined, are thoroughly redesigned models that can withstand pressures as high as 60–110 atm. Figure 5.4 depicts one of the most recent (third-generation) designs of high-pressure closed vessels, together with a typical high-pressure module where up to ten vessels can be placed for simultaneous sample pretreatment. This type of high-pressure vessel was designed by OI Analytical Corporation for maximum safety by incorporating a dual pressure-relief mechanism. This design includes a safety rupture membrane for over-pressure venting at more than 600 psi and a break-away safety disc for rapid pressure relief at more than 1000 psi.

The type of design and, especially, the material used to construct closed vessels, should be chosen as a function of the type of treatment to be applied and the subsequent type of analysis to be performed on the sample. For example, materials such as PFA (a perfluoroalkoxy resin) and TFM [a copolymer of tetrafluoroethylene and a small amount of perfluoro(propyl vinyl ether)] should be used in the microwave digestion of samples prior to ultratrace analysis, where the material should be inert to hot concentrated acids and contain minimal amounts of leachable contaminants [28]. Raster electron microscopy has shown PFA to be the most suitable material for this purpose; in fact, its surface was found to remain virtually unaltered even after 100 digestion cycles [29].

A)

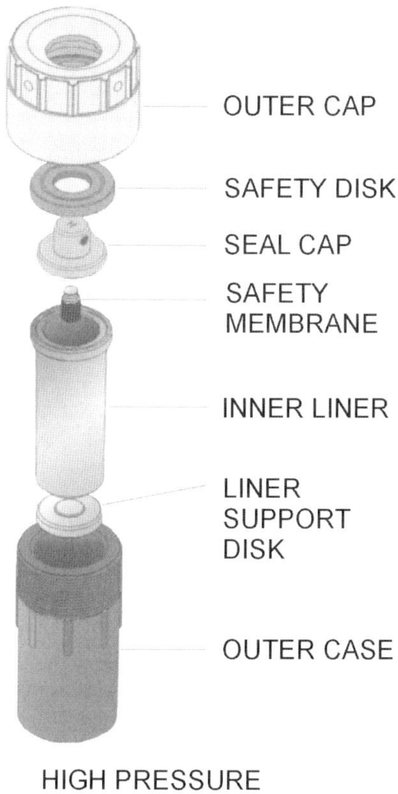

B)

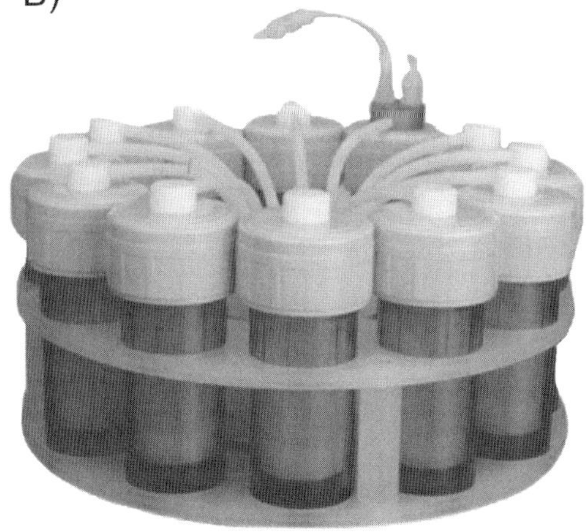

Fig. 5.4. (A) Elements of a high-pressure vessel. (B) Arrangement of the vessels in a high-pressure module.

A)

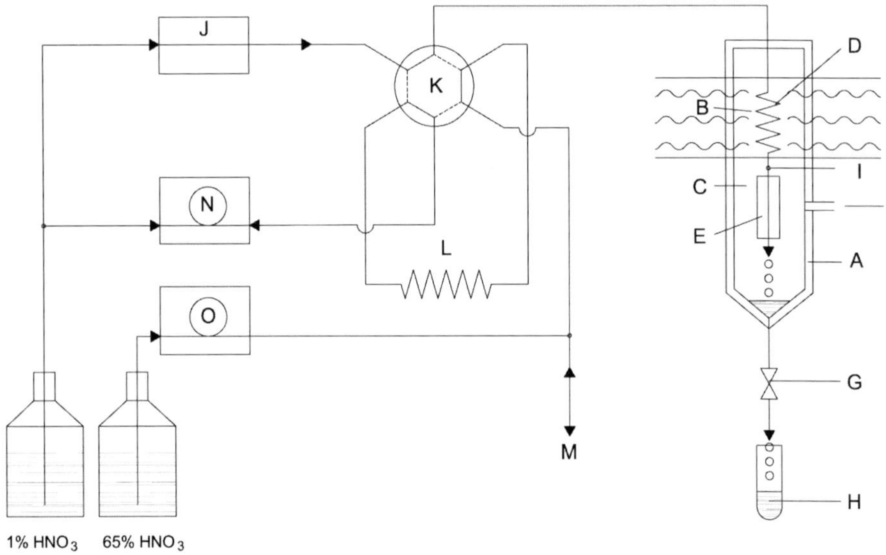

B)

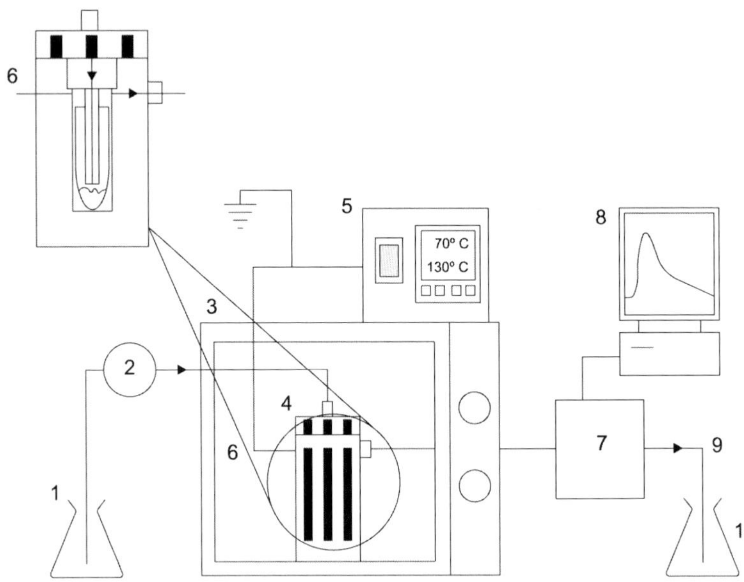

Focused closed-vessel microwave devices

Most commercially available closed-vessel microwave systems are based on multi-mode microwaves; however, the advantages of high-pressure vessels and focused microwave heating have led to the development of systems that combine both assets. These so-called "focused high-pressure, high-temperature microwave systems" [30,31] consist of an integrated closed vessel and a focused microwave-heated system operating at a very high pressure and temperature. One particular system uses a rectangular brass waveguide to direct and focus the microwaves by means of the antenna coupler and a polymer bomb to contain the sample and solvents, and also a special configuration of antennas is used: internal and external antenna systems are carefully sized to achieve efficient coupling of microwave energy between the magnetron and the sample and solvents. This design provides highly efficient (> 90%) energy transfer from the microwave generator to the sample and solvents, and can be used at both low and high powers.

Recently, an extension of the above-described focused single-closed vessel microwave-heated system was developed that allows the simultaneous treatment of up to six samples in a gas-pressurized metal chamber while controlling and measuring the pressure in one vessel. Pressures up to 130 bar and temperatures up to 320°C can thus be reached [32].

Dynamic closed-vessel microwave systems

Dynamic systems for high-pressure microwave treatment were developed much later than open-vessel systems. Operating under a high pressure reduces the flexibility afforded by working at atmospheric pressure. However, some recently developed devices allow microwave-assisted high-pressure digestion and extraction in a dynamic manner [33,34].

In 1999, a microwave-assisted flow digestion system was developed from a novel pressure equilibration system that affords temperatures up to 250°C and pressures in the region of 35 bar with previously slurried solid samples. Figure 5.5A depicts the flow microwave digestion system and illustrates the principle of pressure equilibration. The equipment consists of a pressurized reactor vessel that can be made of PTFE or PFA. The heating zone, located in the microwave field of a waveguide, is made of alumina ceramics and the cooling zone is made of poly(ether–ether ketone) (PEEK®). Both parts are connected in a gas-tight manner. The reaction coil, which consists of PTFE or PFA tubing, is placed within the heating zone of the reactor. The length of the digestion tube is restricted by the dimensions of the waveguide: the longer the digestion tube is, the higher is the reactive volume within the microwave field and the higher the sample

Fig. 5.5. (A) Scheme of a flow digestion system and the principle of pressure equilibration: A pressure reactor, B heating zone, C cooling zone, D digestion coil, E cooling device, F connection for gas supply, G restrictor tube, H collector vial, I temperature sensor, J high-pressure pump, K injection valve, L sample loop, M sample, N and O peristaltic pumps. (Reproduced with permission of the American Chemical Society.) (B) Manifold for dynamic microwave-assisted extraction: 1 solvent, 2 pump, 3 microwave oven, 4 extraction chamber, 5 temperature set-point controller, 6 thermocouple, 7 fluorescence detector, 8 recording device, 9 restrictor, 10 extractor. (Reproduced with permission of Elsevier.)

throughput as a result. Directly outside the microwave field, the temperature of the reaction mixture is measured by means of a Pt sensor. The cooling device consists of a water-cooled titanium rod wrapped with Teflon tubing. The reaction tube ends open at the funnel-shaped bottom of the reactor vessel, which is connected to a nitrogen bomb and supplied with N_2 at a constant pressure up to 5 bar. A restrictor tube is connected gas-tight to the bottom of the reactor vessel, so there is a small continuous gas flow through it. The high pressure within the reactor causes the boiling temperature of the reaction mixture following passage through the digestion coil to rise without exerting mechanical strain on the tube walls. This is the result of the pressure balance inside and outside the reaction tube. It is necessary to supply enough microwave energy to reach the boiling point inside the reaction tube. At the boiling stage, vapour bubbles are formed that can no longer absorb microwave energy; the vapour condenses and microwave energy is absorbed again, so a boiling equilibrium is maintained; in that way, the temperature is controlled only by the applied pressure. The most salient advantage of this pressure equilibrium system is that, even with extremely fast oxidation reactions (e.g. the digestion of glucose), the pressure inside and outside the reaction tube remains identical and the tube does not burst [33].

Recently, Ericsson and Colmsjö [34] developed a dynamic high-pressure microwave-assisted extraction device that allows the treatment of solid samples while avoiding the principal drawback of the previous approach (viz. the relative complexity of the assembly and the inability to directly introduce unslurried solid samples). The experimental set-up used is shown in Fig. 5.5B. The extraction cell is made of PTFE and has an inner volume of 9 ml. It is designed to fit a Soxhlet-type extraction thimble made of cellulose. The tubes and the fingertight fittings are made of PEEK®. During extraction, fresh solvent is continuously pumped through the extraction cell, which is maintained at a slight overpressure by using a fused silica restrictor in order to keep the solvent in the liquid state. This system allows the extraction to be coupled to a subsequent stage of the analytical process by connecting the extractor outlet to the unit concerned. Figure 5.5B illustrates one possible coupling: connecting the extractor to a fluorimetric detector allows the kinetics of the extraction of PAHs from sediments to be monitored [34].

5.3.3. Open-vessel microwave systems

The first completely re-engineered laboratory focused microwave system was introduced by Prolabo in 1986. Most commercial open-vessel microwave systems manufactured since then are of the focused-microwave type, i.e. they use the waveguide as a single-mode cavity [35]. In these systems, the waveguide is closed at both sides and an opening (typically 42 mm wide) is used to facilitate the incorporation of a flask into it. The opening is located in such a way that the axis of the flasks corresponds to an antinode of the waves transmitted by the waveguide. Precise positioning of the wave antinode is achieved by carefully placing the operating and length selection of the waveguide along with an adjustment of a tuning device (e.g. two metal rods). The electromagnetic energy and the sample are thus highly efficiently coupled and a high power density obtained as a result. According to Prolabo, the coupling efficiency is increased by a factor of 10 relative to the use of a multi-mode cavity. Propagation of microwaves outside the waveguide is easily prevented by using a tubular cylindrical stack around the portion of the flask. This arrangement allows the user to add reagents or fresh solvent during the process or to connect a device such as a reflux system.

Even though the principle of focused microwaves is efficient in terms of energy transfer, it has allowed for the use of only one flask at a time in most designs. Automation in some systems can enable the sequential use of flasks, but this characteristic can be viewed as a constraint. One recent development involves the use of four flasks at a time by symmetrically splitting the microwave energy among the flasks at the end of each waveguide.

In these open devices, a continuously adjustable percentage of the maximum power is used over a given period of time. This result differs from a multi-mode cavity, where pulsed power is predominantly used. The maximum power is in the 200–800 W range, depending on the type of microwave system used [16].

A study about the energy transfer in a focused microwave system revealed that it is possible to accurately predict the temperature of different liquids, whether alone or in the presence of small amounts of samples in an open microwave device. The ensuing model shows that the sample receives much more energy than the useful energy, E_u, which is the energy used by the sample to increase its temperature. It also provides an indication of the relative significance of conduction, convection and radiation heat transfer mechanisms. While conduction flow can be considered to be a source of loss for microwave heating, the inverse mechanism is observed in conventional heating techniques. Convective and radiative losses are more important for glass than for the sample because the glass temperature always remains below the sample temperature, whereas glass has no "limiting" temperature such as a boiling point. These results can be used to improve the waveguide and the glass vessel configuration in order to minimize heat losses. The model also provides a numerical estimation of the microwave energy lost in relation to the applied power. This energy loss includes both waveguide loss and glass absorption. The different reagents studied seemingly absorb different amounts of energy under identical operating conditions (given power and time) because of their differential heat capacities and densities. However, different relative heat flows remain similar because they are related to the geometry of the thermal system, which is constant. It has been possible to build abacus for predicting the useful time for a given temperature. Highly precise operating procedures can thus be developed by selecting appropriate time and power settings. Similar abacus could be established for other reagents or organic solvents in order to develop new sample preparation procedures or control product degradation in organic synthesis reactions. Such abacus can also be used to calibrate or check the magnetron [36].

In general, all non-metal vessels used in open-vessel treatments involving conductive heating (e.g. glass beakers, boiling flasks, refluxing containers) can be used in open-vessel treatments with microwave heating. The development of new vessels for specific applications in microwave pretreatments is only limited by the analytical chemist's ingenuity.

In most open-vessel treatments, it is desirable to achieve as much refluxing as possible so that the need for continual additions of solvent to maintain the volume is decreased. Refluxing is generally better in open-vessel microwave treatments because the vessel is secondarily heated as heat from the solution is dissipated and remains cooler. Therefore, with good air flow through the cavity, as is the case with analytical microwave systems, refluxing is more efficient.

As noted earlier, not all open-vessel systems (viz. those that operate at atmospheric pressure) are of the focused type. A number of reported applications use a domestic multi-mode oven to process samples for analytical purposes, usually with a view to coupling the microwave treatment to some other step of the analytical process (generally the determination step). Below are described the most common on-line systems used so far, including domestic ovens (multi-mode systems) and open-vessel focused systems, which operate at atmospheric pressure and are thus much more flexible for coupling to subsequent steps of the analytical process. On the other hand, the increased flexibility of open-vessel systems has promoted the design of new microwave-assisted sample treatment units based on focused or multi-mode (domestic) ovens adapted to the particular purpose. Examples of these new units include the microwave–ultrasound combined extractor, the focused microwave-assisted Soxhlet extractor, the microwave-assisted drying system and the microwave-assisted distillation extractor, which are also dealt with in this section. Finally, the usefulness of the microwave-assisted sample treatment modules incorporated in robot stations is also commented on, albeit briefly as such devices are discussed in greater detail in Chapter 10.

Open-vessel microwave-assisted on-line sample treatment

Most on-line procedures involving microwaves that are conducted with a view to coupling a microwave treatment (usually digestion) with a detector (usually of the atomic type) for the determination of analytes use either a domestic oven [37–40] or a commercial focused system [41–43] plus appropriate connections. Usually, the coupling is done by inserting a Teflon coil in the oven in order to circulate the suspended solid sample to be subjected to the microwaves [44]. Some systems use domestic or commercial focused systems where the solid sample is directly placed in the sample vessel [45] and an aspiration system is used after the microwave treatment to transfer the extract to the determinative instrument used [37,46] or to an apparatus employed in some other step of the analytical process [40,43].

Ever since Burguera *et al.* reported the first instance of coupling of a microwave oven to a flow injection (FI) system for solid sample treatment in 1988 [37], a host of applications have been developed to exploit the many advantages of these on-line systems over off-line methods; such advantages include automatability, reductions on expensive time delays between sample delivery and analysis, protection of samples from ambient agents and improved personal safety. Basically, most such applications use an FI system where the sample is introduced as a stabilized slurry; the system includes a reaction coil that is inserted in a microwave oven to perform a sample treatment which, as noted earlier, is usually a digestion [24,41,44,47–52] but can also be a leaching operation [40,53]. The FI system allows the solution or the resulting extract to be transferred to an appropriate detector in order to determine the target species after the microwave treatment. The sample digestion throughput can be improved by using several coils simultaneously in the oven [48]. Microwave ovens have so far been coupled to detectors based on spectrophotometry

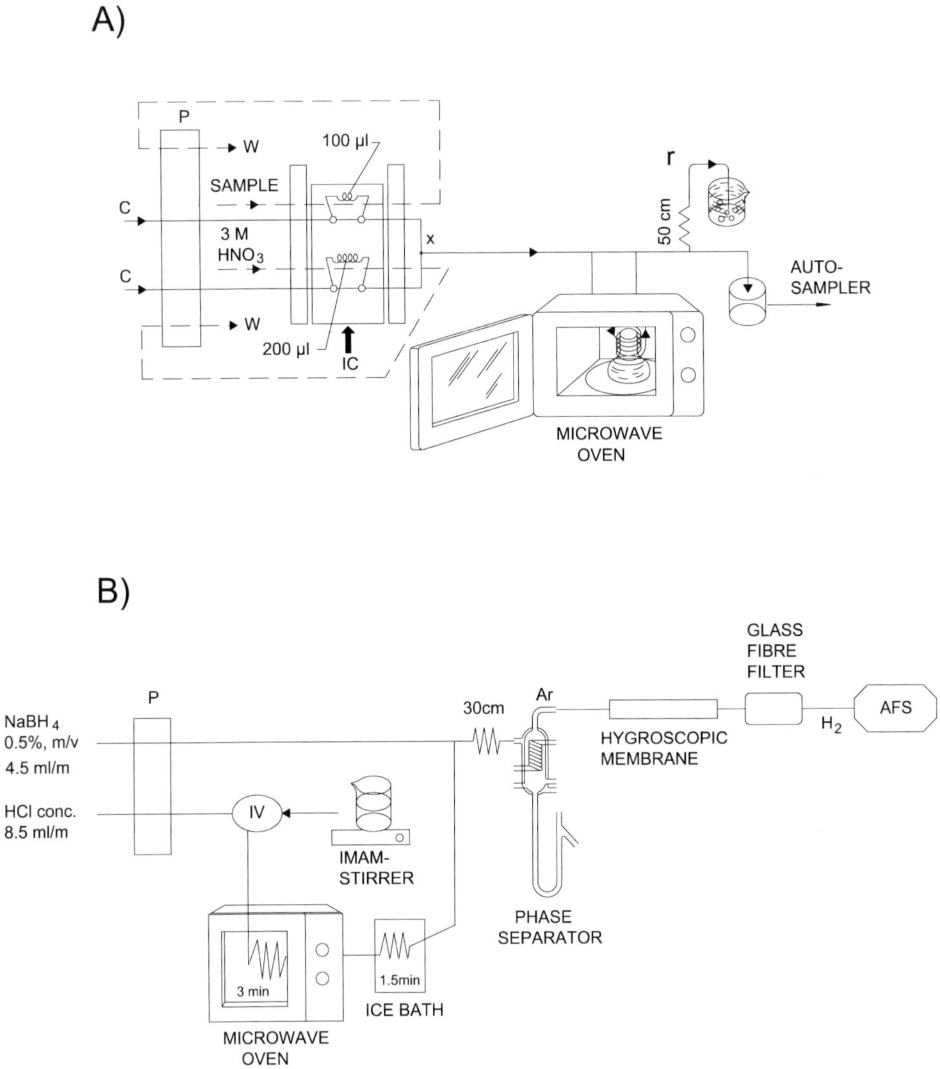

Fig. 5.6. (A) Flow-injection manifold for the wet digestion of shellfish slurry samples: C carrier solution, P peristaltic pump, IC injector commutator, W waste, r restrictor. (B) Dynamic manifold for the analysis of sludge slurries in an on-line digestion–reduction system by FI–HG–AFS: IV injection valve, AFS atomic fluorescence spectrometer, P peristaltic pump. (Reproduced with permission of the Royal Society of Chemistry.)

[38,48,50], flame atomic absorption spectroscopy (FAAS) [46,54], electrothermal atomic absorption spectroscopy (ETA–AAS) [39], inductively coupled plasma–atomic emission spectroscopy (ICP–AES) [41,51] and inductively coupled plasma–mass spectrometry [49]; and with derivatization reactors in techniques such as hydride generation–atomic absorption spectrometry (HG–AAS) [47], hydride generation–atomic fluorescence spectroscopy (HG–AFS) [44] and cold vapour-atomic fluorescence spectroscopy (CV–AFS) [52]. In each coupled system, an FI manifold was used to merge the sample with the required reagents (usually acids) prior to the microwave treatment or to add the reagents, separation devices or components needed to derivatize the sample prior to analysis. Figure 5.6 illustrates two examples of coupled systems. That in Fig. 5.6A is a straightforward system where the FI manifold serves the sole purpose of allowing the sample and acid to merge prior to reaching the microwave oven and where the solution resulting from the digestion process is transferred to the autosampler of an ETA–AAS assembly [39]. The system of Fig. 5.6B, which is much more complex than the previous one, includes a phase separator, a hygroscopic membrane and a fibreglass filter to enable the determination of species by AFS following digestion of the sample [44].

The main problem with these systems is the inherent inhomogeneity of the microwave power distribution within the oven cavity: with only a small area occupied by the reaction coil, a high proportion of the microwave power within the cavity is not absorbed. Thus, the positioning and distribution of the reaction coil within the cavity has been identified as an important parameter in achieving uniform heating of the sample and reagents with a view to ensuring efficiently controlled digestion. In addition to being inefficient, the high proportion of unabsorbed power probably reduces the lifetime of the oven. Any resulting heating of the magnetron may in fact reduce the microwave output.

Some methods use a domestic or commercial focused microwave oven into which the sample is not introduced as a slurry but rather placed in an appropriate vessel that is then introduced into the oven [37,40,43,45,46]. Figure 5.7 shows two such systems using a domestic microwave oven where, as in the previous case, the microwave treatment was coupled to a subsequent step via an FI system. The system of Fig. 5.7A allows the simultaneous treatment of several samples. It consists of a series of test tubes closed in a glass container that is coupled to a water pump connected to an FI manifold via a selective aspiration system. As can be seen in the figure, while the system affords the simultaneous processing of several samples, it requires up to 14 valves to couple digestion with AAS determination, which results in an extremely complicated set-up [37,46]. On the other hand, the system of Fig. 5.7B is much simpler but can only process one sample at a time. The filter at the extractor outlet prevents clogging of some elements of the FI system by the colloid fraction of the solid sample. As can be seen, these systems possess the same flexibility as the previous ones in addition to that inherent in the FI system — which, as noted earlier, is only limited by the user's ingenuity. This system includes some elements such as a membrane phase separator and a mini-sorption column that enable partial automation of the process [40].

In some commercially available focused microwave systems, the sample is placed directly in the vessel [45] or in an extraction cartridge that is in turn placed in the vessel [43]. Figure 5.8 depicts two such systems. In the example of Fig. 5.8A, the microwave system was not coupled to a subsequent step of the analytical process; rather, a dynamic

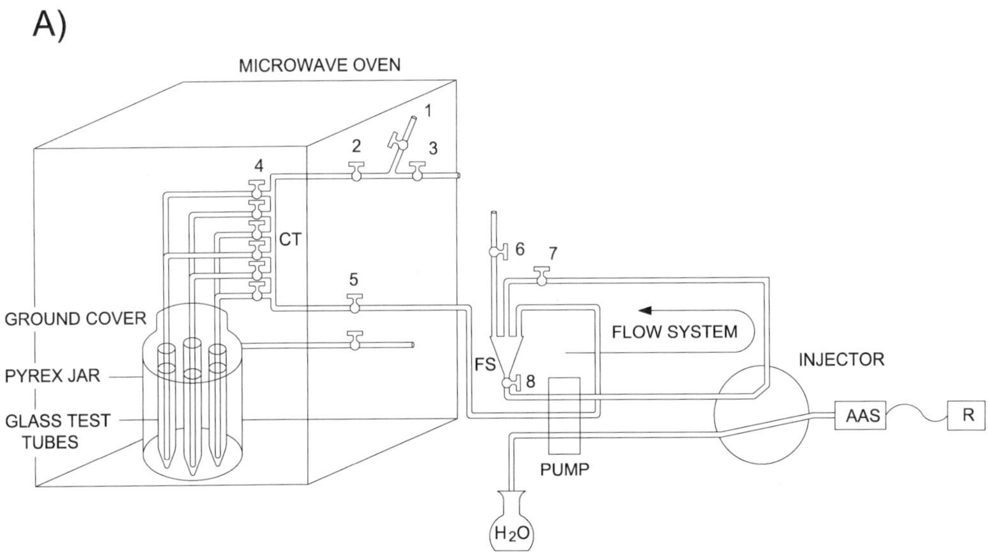

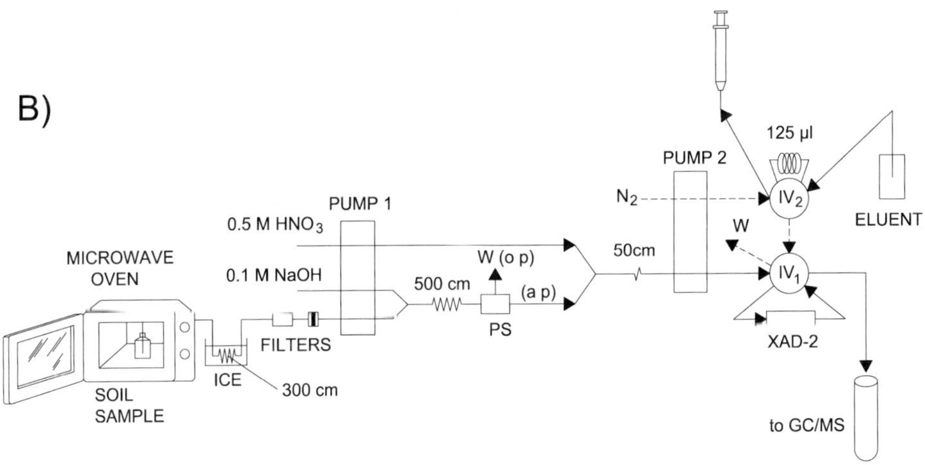

Fig. 5.7. (A) Assembly used in the determination of cadmium and zinc in kidney and liver tissue by simultaneous microwave-assisted acid digestion of 6 samples, FI and AAS: CT collector tube, FS flowing sample collector, R recorder. (B) Experimental design for the continuous microwave-assisted leaching, liquid-liquid extraction, sorption/clean-up of phenol compounds in soil samples: IV, injection valve, PS membrane phase separator, o.p. and a.p. organic and aqueous phase, W waste. (Reproduced with permission of Elsevier.)

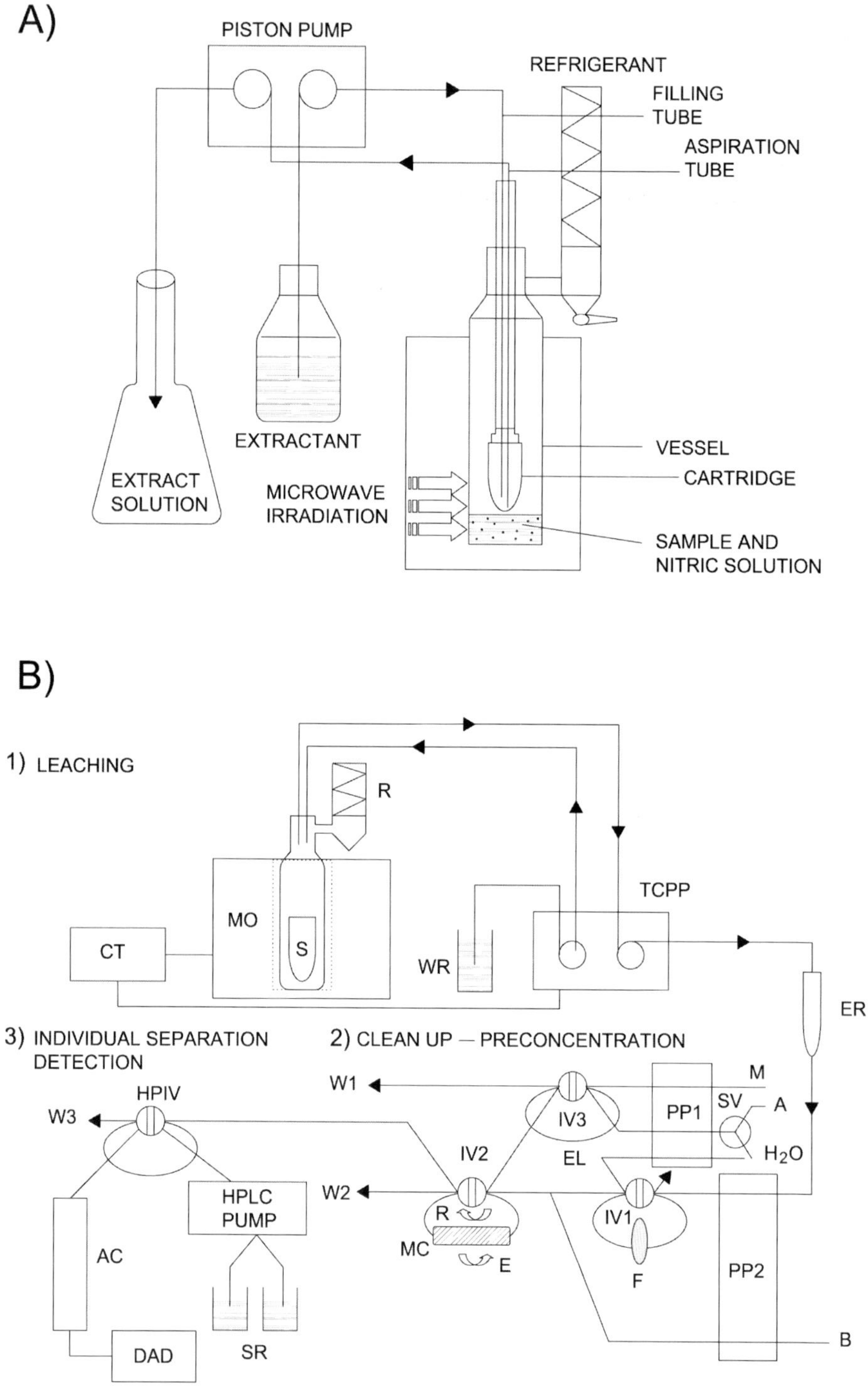
A)
PISTON PUMP
REFRIGERANT
FILLING TUBE
ASPIRATION TUBE
EXTRACTANT
EXTRACT SOLUTION
MICROWAVE IRRADIATION
VESSEL
CARTRIDGE
SAMPLE AND NITRIC SOLUTION
B)
1) LEACHING
R
MO
CT
S
WR
TCPP
ER
3) INDIVIDUAL SEPARATION DETECTION
2) CLEAN UP — PRECONCENTRATION
W1
M
IV3
PP1
SV
A
H2O
EL
HPIV
W3
IV2
W2
R
MC
E
IV1
F
HPLC PUMP
AC
DAD
SR
PP2
B

system was designed with a view to allowing several consecutive extraction cycles to be performed in order to ensure quantitative extraction of the analytes [45]. Addition of the extractant (a nitric acid solution) and aspiration of the extract after each cycle were done by means of a two-channel piston pump. The implementation of successive cycles afforded quantitative extraction of the analytes, which proved impossible in a single cycle. The sample was placed directly in the extraction vessel and the extract provided by each cycle was aspirated through a cellulose filter to retain solid particles and obtain a clean extract. The use of the pump itself as the interface between the extractor and an FI manifold (Fig. 5.8B) allowed microwave-assisted extraction to be coupled to filtration, preconcentration, individual chromatographic separation and UV detection steps with a view to automating the whole analytical process [43]. Extraction was done by supplying an appropriate extractant (water) in an automatic manner by means of the piston pump. Then, the sample was irradiated with microwaves for a preset time and the extract aspirated through the other pump channel to a vessel connected to the FI manifold, where the analytes where filtered, preconcentrated and chromatographically determined in an automatic manner.

Special open-vessel microwave systems

Microwave–ultrasound combined extractor The microwave–ultrasound combined system was constructed from a 2.45 GHz, 0–300 W Prolabo Maxidigest 350 single-mode microwave oven; the sample solution was held in a double-walled open 20–150 ml borosilicate reactor with a reflux column at the top. A cup-horn Branson Sonifier 250 at the base of the microwave oven was used for indirect ultrasonic agitation of the sample at 20 kHz via 200 ml of decalin held between the reactor walls. The ultrasonic probe was placed distant enough from the electromagnetic field to avoid interactions and short-circuits. The microwave system was used to dissolve 0.5 g of Co_3O_4 by charring with 20 ml of 16 M HNO_3 and oxidation with 2 ml of H_2O_2 for 3–10 min at a microwave power of 150 and 290 W, respectively. The digestion time in the presence and absence of ultrasonic agitation was 1 and 3 h, respectively. The system was also used to digest 1 g of olive oil for the subsequent determination of Cu by charring with 20 ml of HNO_3 for 27 min and oxidation with 3 ml of H_2O_2 for 3 min at a microwave power of 180 and 290 W, respectively. The digestion time with and without ultrasonic agitation was 30 and 45 min, respectively [55].

Focused microwave-assisted Soxhlet extractor The first prototype of a focused microwave-assisted Soxhlet extractor was designed by Luque de Castro *et al.* and

---

Fig. 5.8. (A) General scheme of a dynamic focused microwave-assisted extractor. (B) Experimental set-up used to integrate microwave-assisted extraction with the subsequent steps of the analytical process. (1) Leaching step: CT controller, MO microwave oven, S sample, R condenser, WR water reservoir, TCPP two-channel piston pump, ER extract reservoir, SV switching valve. (2) Clean-up/preconcentration step: M methanol, A air, B buffer, PP peristaltic pump, F filter, EL elution loop, MC mini-column, R retention direction, E elution direction, IV1–IV3 injection valves, W waste. (3) Individual separation–detection step: HPIV high-pressure injection valve, AC analytical column, DAD diode array detector, SR solvent reservoirs.

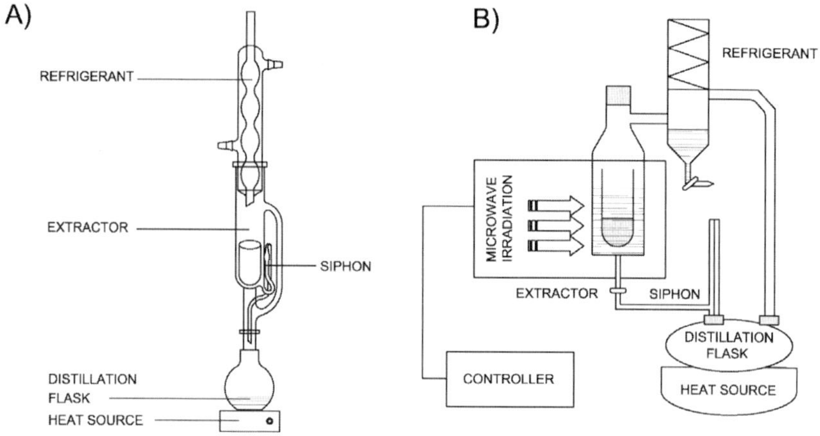

Fig. 5.9. (A) Conventional Soxhlet extractor. (B) Focused microwave-assisted Soxhlet extractor from Prolabo. (Reproduced with permission of Elsevier.)

constructed by Prolabo (Paris, France) in 1997 [56]. This prototype combines the advantages of the Soxhlet extractor with those of the microwave-assisted process. The extractor design is based on the same principles as a conventional Soxhlet extractor (Fig. 5.9A) modified to facilitate accommodation of the sample cartridge compartment in the irradiation zone of a microwave oven. The latter was also modified by making an orifice at the bottom of the irradiation zone, thus enabling connection of the cartridge compartment to the distillation flask through a siphon as illustrated in Fig. 5.9B. The use of this device has provided better results than conventional Soxhlet extraction in much shorter periods of time [57–60]. The device, which enables focused microwave-assisted Soxhlet extraction (FMASE), retains the advantages of conventional Soxhlet extraction while overcoming restrictions such as the long extraction time and non-quantitative extraction of strongly retained analytes due to the easier cleavage of analyte–matrix bonds by interactions with focused microwave energy, unavailability for automation and the large volumes of organic solvents wasted — unlike a conventional Soxhlet extractor, the microwave-assisted system allows recycling of up to 75–85% of the total extractant volume. In addition, solvent distillation in the FMAS extractor is achieved by electrical heating, which is independent of the extractant polarity, thus avoiding the principal problem of commercial focused microwave devices such as those of the Soxwave series from Prolabo.

Notwithstanding their performance similarities, there is a substantial difference between FMASE and Soxhlet extraction. Thus, the FMAS extractor is usually equipped with glassware of a design similar to that of the Soxhlet extractor, which can be substituted by a pump serving the functions of the glassware with additional advantages such as the ability to couple the FMAS extractor with other dynamic devices for on-line clean-up and detection. For example the device was coupled to a fluorimeter as shown in Fig. 5.10 in order to monitor the extraction kinetics of fluorescent analytes such as PAHs and

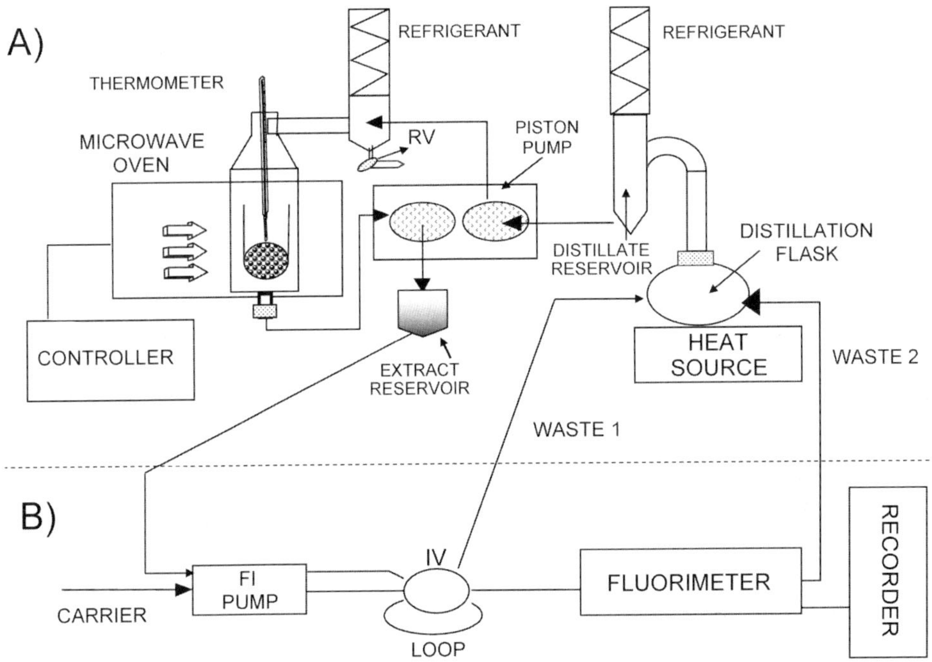

Fig. 5.10. Approach to the matrix-independent removal of fluorescent compounds. (A) Controlled FMASE: RV recycling valve. (B) Flow-injection manifold: IV injection valve. (Reproduced with permission of the American Chemical Society.)

identify the end of the leaching step irrespective of the sample matrix concerned, thus avoiding exceedingly long extraction times [61].

The main drawback of the FMASE prototype was the difficulty of using water or other high-boiling point solvents as extractants owing to the length of the glassware required for evaporation, condensation and dropping of the solvents on the samples, which considerably slowed the process. In order to enable the use of water as extractant, as well as to improve automatability and operability, a new prototype based on the same principles as the previous FMAS extractor was designed by Luque de Castro *et al.* and constructed by SEV (Puebla, Mexico). The new prototype, called MIC V (Fig. 5.11), consists of a single unit where the shortened glassware pathway allows reception of the water vapour on a condenser connected to the top of the sample-cartridge vessel with minimal heat losses on the way to its subsequent condensation and dropping on the solid sample [62]. The system operates with two extraction units, which allows the simultaneous processing of two samples. Also, it includes an optical sensor that is positioned at the desired syphon height to have the magnetron start irradiation of the sample when the solvent reaches the preset level. A solenoid valve in the bottom of the syphon is switched on at the end of the irradiation process to empty the sample vessel. The equipment operates similarly to a conventional Soxhlet extractor (i.e. the sample is continuously brought into contact with fresh, recycled solvent), the sole difference being that the sample receives microwave

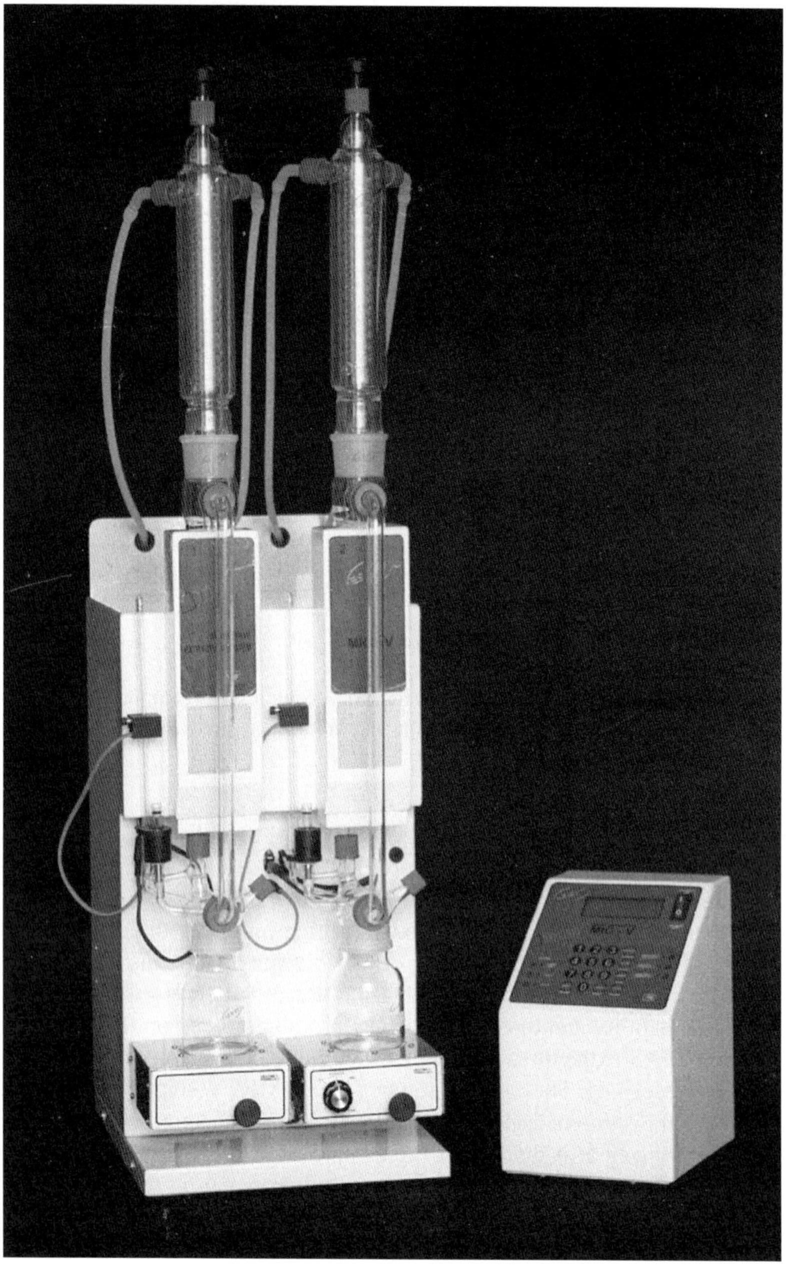

Fig. 5.11. Novel focused microwave-assisted extractor design. (Reproduced with permission of SEV, Mexico.)

irradiation over a preset period. This prototype is being tested at present and is expected to be marketed shortly.

Microwave-assisted drying system A special drying rotor has been used to dry samples using microwaves [4]. The system consists of a cylindrical chamber through which an air stream is circulated after introduction via 6 filter discs located at the cylinder's top plate. This allows the chamber to be steadily purged with clean air, which drives off the moisture released by the sample. Samples, placed in Petri dishes made of borosilicate or quartz glass, rest on a porcelain desiccator plate. Porcelain and the borosilicate glass walls of the cylinder are both slightly self-heating under microwave irradiation, which avoids condensation of moisture even when large amounts of water evaporate (e.g. from sewage sludge samples).

Microwave-assisted distiller Microwave energy has also been used to facilitate the distillation of various compounds from solid samples [11,63–65]. Figure 5.12A depicts a typical microwave-assisted distillation apparatus. It consists of a laboratory-made polytetrafluoroethylene (PTFE) vessel of 120 ml inner volume and 10 mm wall thickness, and a 60 ml collector flask. The screw cap of the vessel includes a drilled PTFE disc located on its upper part that avoids sample ejection during distillation. A hole made in the vessel cap provides a means for evacuating volatile compounds via a PTFE tube that is passed through the vent holes of the microwave cavity and is then adapted to a glass line finished by a bubbler which is inserted in the collector flask. A domestic microwave oven equipped with a 2450 MHz magnetron is normally used for this purpose, altered only as required to connect the sample vessel with the collector flask.

A more complicated system (Fig. 5.12B) was developed by Conte *et al.* [66,67] with a view to coupling distillation with a solid-phase absorbent trapping step. The microwave oven (A) was modified as follows: an opening was drilled in the oven for insertion of a probe and another one in its outer casing for introducing a Teflon gas transfer line (I). The Teflon sparge gas tube (E) was inserted through an opening in the bottom of the outer case and then through a small opening in the inner compartment. The flow of helium sparge gas was regulated by a rotometer (H). Tube E was connected to a check valve (G) to prevent reversed flow of liquids during and after distillation. The sparge gas tube (E) was fitted through a Teflon male tee (C) and sealed in place with a Teflon reducing ferrule (F). Component C was connected to a 250 ml thick-wall hydrogenation flask (B) via a Teflon nut (D) furnished with a Teflon O-ring. Argon was used to purge the distillate formed from the microwave radiation through tube I and the flow was directed from I into the condenser (N) via a tee-adaptor (K) and a Teflon union (M). The tee adaptor was attached to I and a Teflon rod (L) by means of a Teflon bushing and O-ring (J). Component L was used as an opening for the addition of the elution solvent. The temperature of the condenser (N) was controlled by a benchtop recirculating cooling unit (S). A support disc (O) was attached to the bottom of the condenser by means of an adaptor (P). The condenser accommodated a stainless steel union having a male Luer lock fitting epoxy-glued to one side (Q). An appropriate solvent material was secured to the Luer fitting.

A)

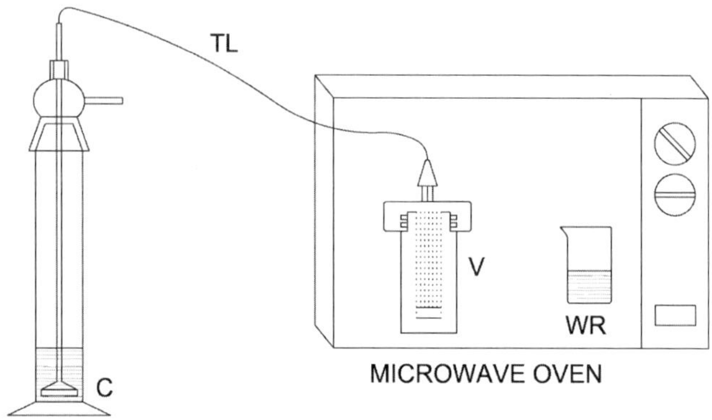

B)

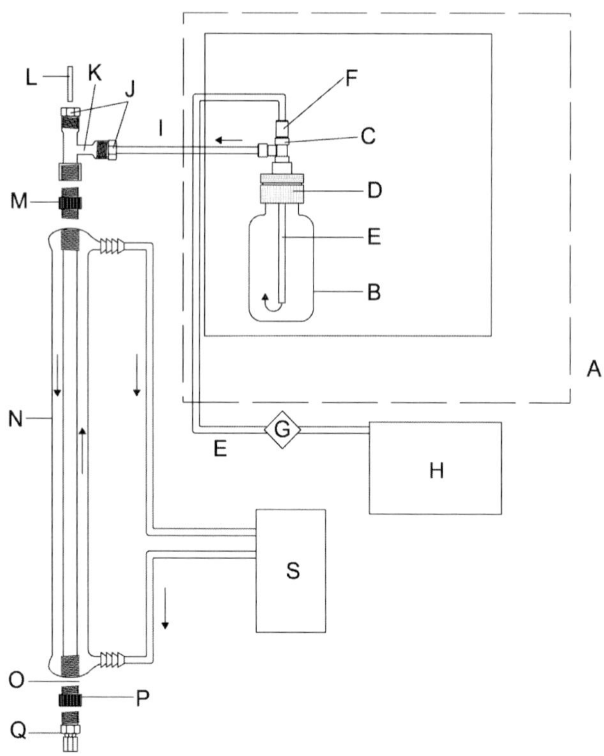

This design affords shorter analysis times with minimal solvent consumption; also, no preconcentration step is required as preconcentration is achieved simultaneously with distillation by retaining distilled analytes on the solid-phase material. This device features a broad range of uses in the analysis of semi-volatile compounds in solid samples with high moisture contents.

Microwave-assisted robotic methods

The high flexibility of a robot station affords the insertion and/or exchange of modules. Such is the case with the system designed by Torres *et al.* [68], who incorporated a microwave treatment module with a view to developing a method for the determination of trace metals in soils where such steps as sample weighing, reagent and solvent addition, sample digestion and insertion into the measuring equipment would be done in a fully automatic manner. Special robotic stations have also been used for the automatic simultaneous or sequential microwave treatment of several samples. Thus, the system developed by Ciommer *et al.* consists of a robot hand with special fingers for grasping digestion vessels, a vibrator hand, a pipetting hand, a rotation hand, opening and dosing stations, a sample balance module, a sample rack for 90 samples and 5 digestion vessel racks with 8 positions each. The microwave section was modified so that the digestion vessels and the sealing from the automatic handling section could be accommodated and its operation fully automated [69]. In both cases, the inclusion of the microwave digester in the robotic station dramatically decreased the time required for sample treatment as compared to conventional stirring-based methods, with a concomitant increase in sample throughput.

5.3.4. Closed versus open microwave systems

The earliest microwave systems adapted for laboratory work were of the closed-vessel type; however, many were superseded by open-vessel systems that were developed to circumvent their shortcomings. In any case, each type of system has its own advantages and disadvantages, so neither can be said to be the better choice for all microwave-assisted treatments of solid samples. The advantages of closed-vessel systems can be summarized as follows:

(a) They can reach higher temperatures than open-vessel systems because the boiling points of the solvents used are raised by the increased pressure inside the vessel. The higher temperatures in turn decrease the time needed for the microwave treatment.

(b) Losses of volatile substances during microwave irradiation are virtually completely avoided by virtue of the absence of vapour losses.

Fig. 5.12. (A) Microwave-assisted distillation set-up – C: collector; TL: tansfer line; V: vessel; WR: water reservoir. (Reproduced with permission of Elsevier.) (B) Microwave-assisted distillation/solid-phase extraction apparatus. The parts designated by capital letters are described in the text. (Reproduced with permission of the American Chemical Society.)

(c) Less solvent is required. Because no evaporation occurs, there is no need to continually add solvent to maintain the volume. Also, the risk of contamination is avoided as a result.
(d) There is little or no risk of airborne contamination.
(e) The fumes produced during an acid microwave digestion are contained within the vessel, so no provision for handling potentially hazardous fumes need be made.

On the other hand, the use of closed-vessel systems is subject to several shortcomings, namely:

(a) The high pressures used pose safety (explosion) risks derived from the generation of hydrogen in the acid digestion of metals and alloys.
(b) The amount of sample that can be processed is limited (usually less than 1 g for inorganic samples and 0.5 g for organic samples).
(c) The usual constituent material of the vessel, PTFE, does not afford high solution temperatures.
(d) These systems are rarely suitable for organic compounds.
(e) The single-step procedure excludes the addition of reagents or solvents during operation.
(f) The vessel must be cooled down before it can be opened after the treatment.
(g) The use of porous PTFE can result in memory effects.
(h) Unlike digestion, leaching involves a mass-transfer equilibrium of the analytes between the sample matrix and the leaching agent that hinders quantitative removal of the target species.

Atmospheric pressure (open-vessel) microwave sample preparation can be as effective a method as closed-vessel microwave sample preparation methods or even more so. The use of atmospheric pressure provides substantial advantages over pressurized vessels, namely:

(a) Increased safety resulting from operating at atmospheric pressure with open vessels containing, for example, gas-forming species.
(b) The ability to add reagents at any time during the treatment, which enables sequential acid attacks in the case of digestion.
(c) The ability to use vessels made of various materials including PTFE, glass and quartz.
(d) The ability to operate at high temperature with quartz material when using sulphuric acid near its boiling point to destroy organic compounds in digestion methods.
(e) The ease with which excess solvent can be removed to ensure complete dryness of the digest or extract.
(f) The ability to process large samples.
(g) No need for cooling down or depressurization.
(h) The low cost of the equipment required.
(i) The ability to develop several leaching cycles until quantitative removal of the target species is achieved.

In addition to these advantages, the following specifications would make open-vessel systems even more useful:

(a) Highly efficient transfer and precise control of the energy deposited into the sample.
(b) Fully automatic operation.
(c) The digestion of large (3–10 g) samples — especially those with a high carbon content — produces large amounts of gas and vapours. In open vessels, these are released by the reaction mixture and continuously swept from the headspace above the sample. Thus, in contrast to closed vessels, completion of gas-forming reactions is favoured as per Le Chatelier's principle.
(d) Open-vessel operation is more suitable with thermolabile species (e.g. organometals) as it uses low temperatures relative to closed-vessel systems.

Despite their many advantages, open-vessel systems are also subject to several shortcomings, namely:

(a) The ensuing methods are usually less precise than those developed using closed-vessel systems.
(b) The sample throughput is lower as most open systems cannot process many samples simultaneously — closed-vessel systems can handle 8–14 samples at once.
(c) The operation times required to obtain results similar to those of closed-vessel systems are usually longer.
(d) Digestion is especially cumbersome with some samples owing to the difficulty of reaching the drastic conditions they require.

5.4. VARIABLES GOVERNING MICROWAVE-ASSISTED PROCESSES

Performance in microwave-assisted processes depends on a number of variables including the microwave power output, exposure time, solvent and sample size used. However, specific treatments such as digestion are additionally dependent upon factors such as the digesting acid, pressure and its relationship to temperature in closed-vessel systems, the residence time in flow systems, the number of cycles used in an FMAS extractor, etc., all of which should be optimized for each specific situation.

5.4.1. Microwave power and exposure time

The principal variables to be considered in developing a microwave-assisted analytical method are the applied power and the irradiation (exposure) time. The power and the corresponding time will depend on the type of sample and reagent used. Usually, these two variables have opposing effects: for a given process, the use of a high microwave power affords a decreased exposure time; on the other hand, the use of a low power entails irradiating the sample for a longer time to apply the same amount of energy.

In theory, one should use a high microwave power to reduce the exposure time as much as possible. This is the rule to be followed in most digestion processes, i.e. to use the highest power afforded by the microwave equipment in order to minimize the digestion time [2]. Occasionally, however, using a very high power is inadvisable; such is the case with digestion processes involving acid mixtures, abrupt heating of which can cause them to react explosively (e.g. the sulphuric–nitric acid mixture in the presence of organic material) or in extraction processes where complete dissolution of the sample should be avoided in order to minimize the effect of potential interferences. In some cases, using a very high microwave power decreases the extraction efficiency through degradation of the sample or its analytes [59,60], or rapid boiling of the solvent in open-vessel systems (which hinders contact with the sample). In the extraction of herbicides from acid soils using an open-vessel focused microwave system, the application of microwaves over short, intermittent periods resulted in substantially increased extraction efficiency relative to the continuous application of microwaves, which caused the solvent to boil and hindered sample–extractant contact [43].

In dynamic digestion systems where the sample is inserted as a slurry in a Teflon coil, the microwave exposure time is directly related to the residence time of the slurry inside the oven, which in turn depends on the length of the digestion coil and the flow-rate of slurry through the coil. With focused microwave-assisted Soxhlet extraction, the process develops in cycles where the sample is brought into contact with volumes of fresh extractant, thereby favouring displacement of the analytes from the matrix to the extractant. The sample is irradiated with microwaves for a preset time during each cycle. The number of cycles and the irradiation time used in each cycle are thus two variables to be optimized in using a focused microwave-assisted extractor.

5.4.2. Temperature and pressure

Temperature is a key variable in most analytical processes. In microwave-assisted processes, it plays a prominent role and affects the rate of some reactions, the degradation of thermolabile species and the solubilization of some substances, among others. A number of devices have been developed for monitoring or even controlling the temperature, some of which are commented on in Section 5.3.

Pressure is a highly influential factor in closed-vessel systems. The development of vessels capable of withstanding pressures as high as 41 bars has enabled digestion at very high temperatures with pure or mixed liquid acids, thereby dramatically increasing the digestion efficiency and decreasing the exposure time (see Section 5.5). Some devices also discussed in Section 5.3 allow the pressure to be both monitored and controlled.

5.4.3. Type of solvent

The choice of solvent in a microwave-assisted sample treatment is a direct function of the type of treatment used. Thus, digestion and hydrolysis are best done with aqueous solutions of pure or mixed acids, whereas selective extraction of organic compounds from a sample matrix usually requires an organic solvent.

The microwave-assisted extraction of non-polar compounds usually entails the adoption of a compromise since, although these compounds are more readily dissolved in non-polar solvents such as *n*-hexane, the interaction of microwaves with the solvent depends on its dielectric constant (the greater ε, the stronger the interaction); this requires the use of a mixture of solvents of different polarity in many cases. For example, the most suitable solvent for the microwave-assisted extraction of organic compounds such as PAHs and PCBs is a mixture of hexane and acetone. In specific cases such as the extraction of organometal compounds, which is discussed in greater detail in Section 5.5, the use of methanol acidified with acetic acid allows organometals to be extracted in a rapid, efficient manner with no degradation.

The solvents of choice for digestion (microwave-assisted digestion included) are solutions of both oxidizing (e.g. HNO_3, H_2SO_4) and non-oxidizing inorganic acids (e.g. HCl, HF). The acid or mixture of acids should be chosen for its efficiency in decomposing the matrix. Thus, knowledge of the sample matrix and its major elements and compounds is essential with a view to selecting the appropriate acid to ensure complete sample dissolution. Combinations of acids are frequently used to digest a sample matrix. For example, hydrofluoric acid alone is inappropriate for the decomposition of plant material. However, if a siliceous component is present, hydrofluoric acid is added to nitric acid in order to release trace elements that would otherwise remain trapped with the silica. These combinations must be chosen on the basis of the chemistry of the sample matrix [70,71].

The high significance of microwave-assisted digestion and its wide application warrants some comments on the characteristics of the different acids usually employed in the process and also of hydrogen peroxide, not an acid itself but widely used for digestions on account of its oxidizing power.

Nitric acid

Nitric acid is the primary choice here on several grounds. It is an oxidizing acid that can dissolve most metals to form soluble metal nitrates. It has poor oxidizing strength below 2 M but is a powerful oxidizing acid in concentrated form. Its oxidizing power can be enhanced by addition of chlorate, permanganate, hydrogen peroxide or bromine, as well as by increasing its temperature and pressure. Most metals and alloys are oxidized by nitric acid; however, gold and platinum are not oxidized and some metals are passivated when attacked by concentrated nitric acid. Such metals can be dissolved by using a combination of acids or a dilute nitric solution. At temperatures in the region of 180°C, nitric acid can decompose most organic molecules, converting hydrocarbons into carbon dioxide and water.

Hydrochloric acid

Hydrochloric acid is a non-oxidizing acid that exhibits weak reducing properties during digestion. Concentrated HCl is an excellent solvent for some metal oxides and for metals that are oxidized more easily than hydrogen. At a high pressure and temperature, many silicates and numerous other refractory oxides, sulphates and fluorides are attacked by HCl to produce generally soluble, chloride salts.

Because it is not an oxidizing agent, concentrated hydrochloric acid is normally not used to digest organic materials. Nevertheless, it is an effective solvent for basic compounds such as amines and alkaloids in aqueous solutions, as well as for some organometallic compounds. The hydrolysis of natural products with HCl is a routine preliminary procedure for the analysis of amino acids and carbohydrates.

Hydrofluoric acid

Hydrofluoric acid is a non-oxidizing acid the reactivity of which is based on its strong complexing capacity. It is most commonly used in inorganic analysis because it is one of the few acids that can dissolve silicates. Its strong complexation capabilities prevent the formation of sparingly soluble products of various metals and increases their solubility and stability. Digestion with HF produces mainly soluble fluorides — by exception, the fluorides of alkaline earths, lanthanides and actinides are insoluble or sparingly soluble at most. In order to improve digestions, HF is routinely combined with another acid such as HNO_3. Insoluble fluorides may frequently be redissolved by removing the HF after digestion.

Phosphoric acid

Hot phosphoric acid has been successfully used to digest ion-based alloys when HCl would have volatilized specific trace constituents. Phosphoric acid can also digest a wide range of aluminium slags, iron ores, chromium and alkali metals. The temperature and pressure profiles for H_3PO_4 indicate that temperatures of 240°C can be attained with just 3 atm. Because of its low vapour pressure, relatively high temperatures can be reached without disrupting the digestion.

Sulphuric acid

While dilute sulphuric acid exhibits no oxidizing properties, in concentrated form the acid is capable of oxidizing many substances. In fact, concentrated H_2SO_4 is an effective solvent for a wide range of organic tissues, inorganic oxides, hydroxides, alloys, metals and ores. However, concentrated sulphuric acid (98.7%) has a boiling point of 339°C, which exceeds the working range of all Teflons. As a result, quartz is the material of choice for the vessels used in H_2SO_4 digestions.

Sulphuric acid is commonly used with other acids and reagents. Two of the most frequent combinations are with perchloric acid or hydrogen peroxide. Sulphuric acid can act as a dehydrating agent and dramatically increase the oxidizing power of perchloric acid; however, the mixture can react explosively with organic matrices in closed vessels or if heated too rapidly.

Perchloric acid

Hot $HClO_4$ is a strong oxidant which attacks metals that are unresponsive to other acids. Perchloric acid also effects the thorough decomposition of organic materials.

Because of its oxidizing capacity, the hot acid is frequently used to take elements to their highest oxidation state. Cold concentrated and hot diluted H_3PO_4 pose little hazard; however, too concentrated $HClO_4$ is potentially explosive when in contact with organic materials and easily oxidized inorganics. Extreme safety precautions are required when using this concentrated acid at high temperatures. Because of this potential hazard, expensive acid hoods, special scrubbers and duct work are needed to handle this acid.

Hydrogen peroxide

Typically, concentrations of about 30% hydrogen peroxide are used for digestion; more recently, however, 50% solutions have become available. Hydrogen peroxide alone can react explosively with many organics, especially in the more concentrated form. Hydrogen peroxide is usually combined with an acid because its oxidizing power increases as the acidity increases. The combination of H_2O_2 and sulphuric acid forms monoperoxosulphuric acid (H_2SO_5), a very strong oxidant. Because of its oxidizing power, hydrogen peroxide is frequently added after the primary acid has completed a pre-digestion of the matrix. The hydrogen peroxide can complete digestion and the potential safety hazards are minimized. In this respect, H_2O_2 is used similarly to $HClO_4$.

5.4.4. Influence of sample viscosity and sample size on microwave heating

The viscosity of a sample reflects its ability to absorb microwave energy because it affects molecular rotation. The effect of viscosity is best illustrated by considering ice water. When water is frozen, the water molecules become locked in a crystal lattice. This greatly restricts molecular mobility and makes it difficult for the molecules to align with the microwave field. Thus, the dielectric dissipation factor of ice is low (2.7×10^{-4} at 2450 MHz). When the temperature of the water is increased to 27°C, the viscosity decreases and the dissipation factor rises to a much higher value (12.2).

The input microwave frequency also affects the penetration depth of the microwave energy. As noted in Section 5.2.1, at a given input frequency, the greater the dissipation factor of a sample is, the less it will be penetrated by microwave energy. In large samples with high dissipation factors, the heating that occurs beyond the penetration depth of the microwave energy is due to thermal conductance through molecular collisions. Therefore, temperatures at or near the surface will be higher. Because boiling and other agitation increase the rate of thermal conductance, surface heating is not a problem unless penetration is very low and sample heating extremely superficial. In such a case, heat losses through the vessel walls can be significant and increase the sample heating time.

Although the small sample size used in most analytical dissolution processes has some advantages, it is also subject to a least one disadvantage: the amount of microwave energy absorbed decreases with decreasing sample size. With small sample sizes, a substantial amount of energy is not absorbed but reflected. Reflected energy can damage the magnetron, so, in using small samples for analytical work, it is advisable to employ microwave systems designed to protect the magnetron from reflected power.

5.5. APPLICATIONS OF MICROWAVES TO SOLID SAMPLE TREATMENT

Microwave-assisted sample pretreatment is being increasingly used in analytical laboratories all over the world. This section describes the principal applications developed since the inception of microwave devices for solid sample treatment in the analytical laboratory. Owing to the large number of papers published on the topic since then (in excess of 2000), only the most salient aspects of the body of information they contain are discussed here in order to provide readers with a global picture of the potential of this auxiliary energy for treating solid samples. Readers wishing to go deeper into any specific aspects are thus referred to the corresponding publications. For consistency with the subject matter of this book, applications of microwaves to liquid samples, microwave-assisted elution [10], thermospraying [72], ozonization [73] and desolvation [74–76] are not dealt with here.

Microwave-assisted digestion and extraction are discussed extensively as they constitute the two most relevant areas and those producing the largest numbers of analytical applications of microwaves. Digestion applications are classified according to the type of matrix involved as, in most cases, digestion is followed by the determination of metal species, so the conditions that govern the digestion efficiency depend on the type of matrix concerned. Microwave-assisted Kjeldahl methods are also commented on, at the end of this section, because of the good results they have provided to date. On the other hand, extraction applications are classified according to the analytes involved as the efficiency of the extraction step is dictated by the particular species to be leached.

Finally, other salient uses of microwaves for treating solid samples such as microwave-assisted drying, distillation and protein hydrolysis are also briefly described.

5.5.1. Microwave-assisted digestion

Sample digestion methods must efficiently decompose the sample matrix so that the target analytes can be thoroughly released from it and solubilized in a form compatible with the determination method of choice.

Ever since Abu Samra reported in 1975 the earliest use of microwaves as a heat source for wet digestion [1], microwave digestion has gradually gained widespread acceptance as an effective means of sample treatment. With the aid of microwaves, not only have digestion times been dramatically reduced (by a factor of 2–5), but also other benefits such as reduced contamination and reagent and sample usage, and improved safety, have been derived. The advantages of microwave-assisted digestion have led to its wide application as an effective sample pretreatment method, mainly for the determination of metals in a wide range of sample matrices; in fact, many of the ensuing methods have been adopted as references by several bodies [77–81]. An increasing number of laboratories are replacing conventional methods with others based on this technology, as reflected in the ever increasing amount of material published on the subject [2,82]. The reports on solid sample digestion involved geological, biological (zoological, botanical, food), soil, sediment and sludge samples, coal and ash, metals and synthetic materials.

As stated in Section 5.3, there are two main types of microwave units, namely: closed-vessel systems, which use pressurized containers, and open-vessel systems, which use focused microwaves at atmospheric pressure. Both types of system have been used to digest solid samples and each has its own advantages and disadvantages, so it is impossible to recommend either as the more suitable overall.

Geological samples

Geological materials such as ores, rocks and minerals often contain immutable matrices resistant to acid attack. Many such matrices consist of refractory minerals with high melting points or other compounds that may require several hours of heating in acids, or fusion processes, for successful digestion. The introduction of microwave-assisted digestion has been a major breakthrough in geochemical analysis, as shown by the many studies involving the use of microwave energy for the digestion of geological samples [83–93].

Regarding microwave equipment, most workers have used modified domestic ovens [84,89,91,94,95] for digestion. The most common modification is the incorporation of a device for purging the oven cavity of fumes to a hood using either compressed air, CO_2 gas or fan-forced air. Most microwave-assisted digestions have been performed using sealed vessels and a pressure release mechanism [85,96] plus pressure and/or temperature monitors [96–98] in some cases. Some special systems have been designed for specific purposes [96–98] or for safer use of closed vessels and microwave energy [84,89]. Also, some workers have placed closed vessels inside containers in order to prevent leaked fumes from destroying the oven cavity [84,86,89,91,96].

Validation of microwave-assisted digestion methods has relied on the use of common certified reference materials and comparisons have also been made with other, conventional procedures such as prolonged heating with acids with a view to testing the reliability, accuracy and precision of microwave-based methods [83,85,88,92,93,96, 99–105]. By judicious adjustment of the different variables involved (e.g. the choice of acid mixture, power setting and exposure time), it has been found possible to digest most types of geological samples using microwave-assisted digestion. Precision and accuracy are generally as good as or even better than those obtained with conventional digestion methods [82].

Simple nitric or hydrochloric acid digestions suffice for the determination of some elements. For example, good Fe recoveries from iron ore samples were obtained using a simple hydrochloric acid digestion; by contrast, limestone rocks required the additional use of hydrofluoric acid [106]. Other acid mixtures such as HNO_3–HCl [97,98,102], aqua regia–HCl [90,96,105] and HNO_3–$HClO_4$ [86,95,107] have also been used for the efficient digestion of geological samples.

Biological samples

The usual aim of digesting biological samples is the subsequent determination of metal traces potentially acting as nutrients or toxins, or even causing diseases such as urolithiasis [108]. One of the major problems in determining trace metals in biological (zoological, botanic, food) samples arises from the need to dissolve organic matter [109]. There are many reports on the use of microwave digestion with biological samples [110–142]. The high pressure generated from the use of closed vessels raises the boiling point of the acid(s) employed and substantially increases the dissolution rate of these organic matrices. For example, nitric acid normally boils at 120°C; however, in a closed vessel at 5 atm, it boils at 176°C [14]. This is 56°C above its normal boiling point and increases its oxidation potential and accelerates its reactions. Under classical hot-plate conditions, the organic matrices of biological samples would remain intact due to the low boiling point of nitric acid. Under these circumstances, reagents such as perchloric acid would have to be used to ensure complete destruction of the organic matter.

The closed-vessel microwave approach dispenses with the use of perchloric acid, which is potentially explosive upon contact with readily oxidized inorganic or organic materials, especially at high temperatures. Great care should be exercised when heating perchloric acid under pressure, even by itself: it has been shown that closed vessels heated with only 5 ml of $HClO_4$ remain pressurized even after cooling with liquid nitrogen. Instead of vaporizing and condensing like other acids, perchloric acid at high temperatures generates chlorine gas via an irreversible decomposition reaction. The chlorine gas is not easily frozen and maintains a high pressure within the vessel. Despite the hazard, some authors have used perchloric acid mixtures under pressure to decompose biological samples [33,143–148]. Fortunately, pressure digestion involving other acid mixtures can be used with acceptable results [149]. However, even when other acids are used, the substantial change in internal pressure generated by the production of CO_2 and NO_2 as organic matrices are decomposed must be anticipated and controlled to prevent vessel rupture, oven damage and possible personal injury. As a safety precaution, vessels with some kind of pressure release mechanism are recommended.

Biological (zoological, botanical, food) samples are primarily composed of three basic constituents, namely: carbohydrates, proteins and lipids. Carbohydrates are the first constituents to decompose, at approximately 140°C, followed by proteins at *ca.* 150°C and lipids at about 160°C [14,150]. At each of these temperatures, significant increases in pressure are generated in the vessel with virtually no increase in temperature. Thus, for safety reasons and for determining the most efficient decomposition temperature, the analyst should have some knowledge of the constituents of the sample matrix.

Although biological samples are most often digested in closed-vessel systems, focused microwaves have also been used for this purpose in some instances [151–155]. However, the use of microwave-transparent, inert, closed vessels appears to be the method of choice for destroying organic matter [156]. The loss of analytes through volatilization or adsorption onto the vessel walls is negligible despite the high pressures and temperatures associated with the digestion. In most instances, such high pressures and temperatures result in complete dissolution of the organic matter that often remains after digestion with conventional techniques [82].

Much of the research on biological sample digestion has included results from standard reference materials for validation purposes. Also, much of the development work has been conducted in parallel with a more traditional method such as conventional heating [157,158] and also with new methods such as pressurized ashing [117].

The results provided by microwave digestion have been found to be comparable or even better in most cases, with advantages such as shorter operational times, higher safety concerning the use of acids and higher precision as a result of the use of automated equipment [158].

There are some contradictions as regards the efficiency of the different acids used for digestion in the determination of the same element. For example, a simple nitric acid digestion was successfully used for the determination of iron in cocoa [126], horse kidney [127], total diet [117], bovine liver [128,133,139], mussels [128], chlorella [128], sargasso [114] and pepperbush [114] samples. However, Mingorance *et al.* [132] obtained low, imprecise results using a similar method with botanical samples. A slight improvement was obtained with nitric acid and hydrofluoric acid digestion, but the results were still outside the certified range. Whether digestion with hydrofluoric acid is necessary for the determination of aluminium is a controversial subject. Thus, low aluminium recoveries from citrus leaves and pine needles were obtained using closed-vessel HNO_3–HF digestion [129] that were improved by employing both acids in combination with hydrogen peroxide. Low recoveries were also obtained from spruce needles after HNO_3–H_2O_2 digestion [123], as well as from shellfish [39], total diet samples [140], apples and peach leaves [135] following closed-vessel digestion with nitric acid alone. On the other hand, the addition of hydrofluoric acid to the HNO_3–H_2O_2 mixture provided improved results. However, evidence also exists which suggests that digestion with hydrofluoric acid is unnecessary for some samples. For example, good results have been obtained for aluminium in total diet [117] and citrus leaves [142] after a simple nitric acid digestion. A combination of nitric acid and hydrogen peroxide was successfully used for the digestion of wheat flour, but the results for rice flour where slightly high [134]; recoveries were not increased, however, by including hydrofluoric acid in the procedure.

Similar inconsistencies have been encountered in the determination of other elements such as selenium [125,133], arsenic [118,130], copper [119,123] and zinc [120,122]. On the other hand, methods for the determination of mercury are slightly more consistent, with many workers using a closed-vessel nitric acid digestion procedure. Good results have thus been obtained with pig kidney [124,141], mussel tissue [138], cod muscle [137], citrus leaves [136], pine needles [136], albacore tuna [131] and fish tissue [124], among others.

There are various approaches to the on-line digestion of biological samples. The development of these flow-through digestion systems for coupling to a variety of instruments is a further step towards complete automation. Virtually all the designs described in Section 5.3, whether of the closed-vessel [32,33] or open-vessel type [33,37,39,46–48, 50,146,159–161], have been used for the on-line digestion of biological samples — particularly open-vessel systems where the sample is inserted as a stabilized slurry. This has probably been the result of the ease with which these systems can be coupled to a detector (especially an atomic detector) and of their low cost and complexity in relation to high-pressure systems for on-line microwave-assisted digestion.

The greatest advantage of on-line digestion methods over batch methods is that they minimize the problem of acid fumes produced during the digestion of biological samples in the traditional Teflon vessels used in batch treatments. Gas evolution in the on-line methods, however, can lead to problems resulting in pressure build-up and disturbances of the carrier flow. In some cases, air carrier streams have been used in order to sweep the gases formed during digestion [48].

Notwithstanding the above-described advantages of on-line open-vessel systems and their more extensive development, high-pressure on-line systems such as that of Fig. 5.5A are especially effective for digesting biological samples, which requires strong, fast oxidation of the organic matter. The main advantage of these high-pressure on-line systems is the substantially increased digestion temperature achieved, which significantly raises the oxidation potential of acids such as HNO_3. This enables the digestion of tough organic compounds and accelerates the oxidation reaction in general, thus shortening the digestion time [33].

Sediment, soil and sludge samples

Soil and sediments are highly variable, complex matrices; the latter are considered to be the ultimate sink of heavy metals released into the environment [162] and also reservoirs for other environmental pollutants such as organochlorine insecticides [163]. The relative ease with which metals are released from, or bound to, soils and sediments is poorly understood. With increasingly stringent environmental and health regulations, the demand for environmental analysis and the range of sample types have grown enormously over the past decade. The most important type of sludge is that resulting from the treatment of sewage. Although there are appreciable benefits in the application of this sludge to farmland [164,165], there is also the risk of toxic elements and bacteria accumulating in the soil if the sewage is not properly treated. The accumulation can lead to contamination of the soil, ground water, lakes and streams, and eventually of the food chain (e.g. fish, crops and livestock), so the analysis of the metal content in sludge is also mandatory for environmental and health regulations. Microwave-assisted digestion is gaining wide acceptance as the method of choice for the pretreatment of these types of samples [93,162,164, 166–185]. In fact, most official microwave-assisted digestion methods are used prior to the determination of metal species in soil, sediments and sludges [77–81], and employ nitric acid, alone or in mixture with hydrochloric or hydrofluoric acid, as the main reagent.

Most on-line microwave-assisted digestions of these types of samples have been done by introducing them as acid slurries and coupling the manifold to a suitable atomic detector [24,49,51,52,54].

Metallic samples

The determination of metal purity and the elemental composition of alloys is of utmost importance to the metallurgical industry. Microwave-assisted digestion is often well-suited to metals and metallurgical samples that pose no difficulty and dissolve readily and safely with the aid of microwaves [148,186–196]. For example, hydrofluoric acid can be used in closed vessels to digest silicate matrices and stop the hydrolysis of refractory elements without loss of volatile fluorides or passivation. After cooling, boric acid can be added to complex unreacted hydrofluoric acid [14]. The solid sample itself may absorb microwave radiation, thus creating a heated surface on which the acid or acids can react. Microwave muffle furnaces are commercially available [197] based on oven linings made out of the highly efficient microwave absorber silicon carbide. This microwave absorber rapidly heats up the oven cavity to ashing or fusion temperatures. Other solid samples may reflect microwaves and create high sample/acid interface microwave densities. This can result in very fast heating of the acid(s) used and create turbulence and sample agitation, both of which sweep sample surfaces clean of acid-resistant coatings and expose fresh, unexposed sample surfaces to acid attack.

The use of microwave energy in the presence of metallic materials is not recommended in all circumstances. Specifically, the interaction of some metals with microwave energy can result in uncontrolled sparking and electrical arcing, possibly causing damage to electrical components and the magnetron. This can be a problem and is a safety hazard when performing closed-vessel digestion. Some metallic materials (especially iron-based alloys) spark when subjected to microwave energy. Such sparking can ignite the hydrogen gas generated in the digestion of metals [14,194].

Miscellaneous samples

Microwave-assisted digestion has been applied to other types of samples such as ceramics [198,199], air particulate matter [200–202], polymers [203,204], lithographic materials [205], coal and ash [206], among others. As with the previous samples, the main purpose was the determination of their metal contents. The results provided by microwave-assisted digestion are similar to those obtained with conventional methods.

Microwave-assisted Kjeldahl methods

Monitoring total nitrogen in soil, which is of special interest to those concerned with plant growth, has traditionally been a cumbersome procedure. The most common method for extracting total nitrogen is Kjeldahl digestion, a procedure developed by Johan Kjeldahl in 1882 and reviewed by McKenzie in 1994 [207]. A Kjeldahl digestion simply involves boiling the sample in sulphuric acid for several hours in the presence of a mercury or copper catalyst. This extraction yields the nitrogen as ammonium sulphate. The time required for the extraction of nitrogen from samples in a conventional Kjeldahl step ranges from 2 to 4 h, which makes it the rate-determining step of the overall analysis. When a large number of samples are to be analysed, this time requirement can prove very costly and an enormous drain on resources and manpower.

In 1988, Alvarado *et al.* [208] investigated the use of closed test-tube microwave Kjeldahl digestion in a domestic microwave oven for the determination of the total nitrogen content in various foods. Some very encouraging results were obtained in the form of improved precision and reduced digestion times. For example, the digestion of cassada leaves took 130 min with the conventional method and 14.6 min with the aid of microwaves. The method did, however, require continuous attention from the operator, turning the microwave power on and off so as not to allow the pressure to exceed the pressure that the tubes can tolerate. Later developments, again in investigating the digestion of food, were reported by Feinberg *et al.* [209], with the introduction of a more sophisticated open-vessel focused microwave digestion system. They reported that using a metal catalyst was unnecessary if hydrogen peroxide was employed. This procedure, at around 40 min, was still quite lengthy and provided little advantage over thermal methods. In 1996, Collins *et al.* [210] used a focused microwave system and achieved similar results. However, the system still only had the ability to handle one sample at a time. More recently, Mason *et al.* [211] reported a method that combines the microwave Kjeldahl system and a flow-injection manifold with microwave-enhanced colour development; the new system greatly reduces the time needed to determine total nitrogen and increases throughput from 36 samples/h on the thermally heated system to 60 samples/h on the microwave-heated system [211].

5.5.2. Microwave-assisted extraction

Microwave-assisted extraction (MAE) is the process by which microwave energy is used to heat solvents in contact with solid samples and to partition compounds of interest from the sample into the solvent. Extractions are done in closed or open microwave-transparent vessels where the solvent and sample are combined and then uniformly exposed to microwave energy. Partitioning may occur by any one or several of the following three heating mechanisms: (a) a single solvent or mixture of solvents possessing high dielectric loss coefficients; (b) solvent mixtures of high and low dielectric loss; (c) susceptible samples with a high dielectric loss in a solvent of a low dielectric loss.

Although the traditional Soxhlet and solvent extraction techniques are widely accepted, they have inherent limitations and problems. Thus, Soxhlet extraction requires 12–24 h in most cases and uses high volumes of organic solvents (hundreds of millilitres). In contrast to conventional methods, microwave-assisted extraction can reduce the extraction time to less than 30 min and solvent consumption to under 50 ml [12]. Moreover, the recoveries obtained with microwave-assisted extraction are comparable with those provided by alternative extraction methods [7].

Unlike digestion, microwave-assisted extraction is used mainly with soil and sediments rather than with biological samples. As regards analytes, microwave-assisted extraction is applied primarily to organic pollutants such as PAHs, PCBs and pesticides rather than to metal species.

The use of MAE is a continuously expanding area of research at present. As a result, it is difficult to provide a completely updated overview. Rather, selected groups of compounds of interest that have been subjected to MAE are discussed here. Unless otherwise stated, the microwave systems used in these applications were of the open-vessel type, which require none of the drastic pressure and temperature conditions employed with closed-vessel systems.

Pesticides

The increasing risks to human health posed by the widespread use of pesticides in our environment is well established [212,213]. Thus, the determination of pesticides in water, plants, soil, sediments, foodstuff, etc., is of major concern today [214,215]. Microwave-assisted extraction provides an efficient, reproducible alternative to classical methods based on Soxhlet extraction or sonication for the extraction of pesticides from environmental samples.

A wide range of pesticides including organochlorine [216,217] and organophosphorus compounds [218], triazines [219–222], herbicides [43,62] and imidazolinones [223] have been extracted with the aid of microwaves, with recoveries close to 100% in most cases.

Sample ageing is an important factor in evaluating a new extraction method as it allows some degree of analyte–matrix interaction to occur and, therefore, provides a more realistic environmental situation. This consideration is very important in pesticide analysis, so ageing studies of the samples to be tested must be performed, as shown by the study on organochlorine pesticides listed in EPA Method 8081. The compounds in question included hexachlorocyclohexane (HCH) isomers, DDT and DDD analogues, as well as endrin and dieldrin [224]. Sand was found to be the cleanest matrix (i.e. it provided the highest recoveries and precision). Co-extractable materials from the topsoil and composted materials are believed to interfere with the analysis and lead to high recoveries of endrin and other compounds. When 45 organochlorine pesticides listed in EPA Method 8081 were freshly spiked onto a series of soils that were aged for a variable time, 38 were recovered to an extent of 80–120%. Most of the compounds were recovered at similar levels after ageing for 24 h — by exception, captafol, captan and dichlone were poorly recovered. Recoveries also decreased with increased ageing time, which can be ascribed in part to microbial degradation. In any case, they were higher than those obtained for comparable Soxhlet-extracted or sonicated samples by about 7% [225].

In another study, 47 organophosphorus pesticides listed in EPA Method 141A were extracted from freshly spiked and variously aged soil, 35 of them being recovered virtually quantitatively. A few including dichlorvos and fensulfathion were recovered at 50–80%. Spiked samples aged for 14 days fared better: 37 compounds were recovered at > 80%, a few at 50% and some at > 120%. Recovery of poorly extracted compounds from freshly spiked soils always improved in briefly aged topsoils. Continued ageing reversed the overall trend because recoveries of the same analytes decreased after 21 days. This anomalous behaviour was attributed to microbial digestion [226].

Because pesticides persist in the environment and are lipophilic, they accumulate along the food chain, so food, zoological tissue and plant analyses for them are of paramount importance. Monitoring programmes tracking this activity have recently reported the use of MAE to improve sample pretreatment activities required in connection with pesticide analysis and assessment of accumulations. Recoveries of organochlorine pesticides higher than 90% were obtained with 6 samples within less than 40 min for grey seal rubber and spiked pork fat [227], and close to 100% within 20 min for medicinal plants [228]. Recoveries of atrazine and organophosphorus pesticides from oranges ranged from 93% to 101%, with advantages over traditional shake-out extraction methods such as a short extraction time (20 min versus 1.5–2 h), low solvent consumption (10 ml versus 150–300 ml) and automatability for routine sample treatment [218].

Polycyclic aromatic hydrocarbons

Polycyclic aromatic hydrocarbons (PAHs) constitute an extensive group of compounds distributed in the environment as a result of natural processes or anthropogenic inputs [229,232]. Traditionally, the extraction of such compounds from solid samples prior to analysis has been conducted using the Soxhlet method, which, while being efficient, usually takes more than 6 h and uses large volumes of solvents. Microwave-assisted extraction is an advantageous alternative to the conventional extraction of these compounds [53,233,234].

Several reports indicate that PAHs can be extracted from solid matrices such as soil, plants and animal tissues relatively easily by using simple microwave-assisted extraction [235]. The most effective solution used for leaching PAHs is the acetone–hexane mixture, usually in a 1:1 ratio. Barnabas *et al.* [236] investigated the effect of varying this ratio on the recovery of PAHs from a native contaminated soil. The combinations ranged from 80:20 to 0:100 hexane–acetone in steps of 10. It was found that, as the solvent mixture became more polar (viz. at a hexane-to-acetone ratio of 10:90 or higher), it was possible to recover greater amounts of PAHs from the native soil. The use of water as extractant was also tested with a view to developing clean methods using no organic solvent. Because of the high hydrophobicity of PAHs, however, the recoveries were very low in spite of the strong interaction between the microwaves and the water. One way of using water effectively for the microwave-assisted extraction of non-polar compounds (PAHs included) is by micelle formation. The ensuing methods provide recoveries close to 100% [3,237,238].

Most studies about the microwave-assisted extraction of PAHs from solid samples have been conducted using closed-vessel systems [12,214,226,236,239–246] and only a few with open-vessel focused microwave devices [57,247–252]. Because open-vessel systems operate at atmospheric pressure, the extraction vessel can be used as a reactor in order to perform on-line purification pretreatments of the total extracts (reagents can be readily added to the medium) [53] or directly introduce the extract into the determination instrument, as in the focused microwave-assisted extractor with on-line fluorescent monitoring of Fig. 5.10, which provides a matrix-independent approach to the extraction of PAHs [61].

Polychlorinated biphenyls

Polychlorinated biphenyls (PCBs) were produced from 1930 to 1983 in the form of complex mixtures for a variety of uses (e.g. dielectric fluids in transformers and capacitors, and plasticizers in paint and rubber sealants). Because of their physical and chemical stability, and also of their lipophilic affinity, they are highly persistent and tend to accumulate in sediments and biota in the aquatic environment [253].

Microwave-assisted extraction (MAE) has become a major choice for the extraction of PCBs from solid matrices. In fact, this technique has been used to extract PCBs from a wide range of samples including soil, sediments and animal tissues [254–257], as well as certified reference materials (CRMs) [258]. Normally, the extractant used is the same as that employed with PAHs, viz. a 1:1 hexane–acetone mixture [246]; however, organized media [256] have also provided results similar to those obtained with conventional methodologies such as Soxhlet extraction [259]. For example, the results for a range of Aroclors (1254, 1260, 1016 and 1248) were quite consistent with their certified values. The microwave-assisted extraction of PCB Aroclors 1248, 1254 and 1260 followed by enzyme-linked immunosorbent assay (ELISA) was shown by López-Ávila *et al.* [225] to compare favourably with traditional extraction methods. No degradation of the PCBs was found to occur during heating of solvent–soil suspensions with microwave energy [224,225].

Organometallic compounds

Synthetic compounds combining the special properties of metals with a hydrocarbon moiety as a vehicle for their introduction into chemical formulations such as arsones and selenones, arsines, organotins and organomercuries have become widespread in the environment since the early 1960s. For example, roxarsone, a disubstituted phenylarsonic acid, has been used as a growth promoter in animal food production. Triphenyl tin (TPT) and tributyl tin (TBT) are added to marine antifouling paints as growth inhibitors. Recently, several MAE methods were developed for the extraction of organometallic compounds such as 3-nitro-4-hydroxy-phenylarsonic acid [260] and methylmercury [261–263] from sediments and animal tissues. The MAE recoveries were consistently higher than those achieved with sonication or Soxhlet extraction. All species recoveries were matrix-dependent, as previously found with other pollutants, but using methanol acidified with acetic acid under the optimal MAE conditions afforded rapid, efficient extraction of all analytes.

Microwave-assisted extraction has also been used as a solid sample treatment prior to speciation analysis [264–266], leaving the organometallic compound moiety intact. This is a prerequisite for a successful extraction procedure to be applied prior to speciation analysis and can be met by careful optimization of the conditions of the microwave attack. Open-vessel treatment is preferred to pressurized bomb systems commonly used in the analysis for total metals because it offers milder reaction conditions — the increase in temperature is governed to a great extent by the boiling point of the solvent — and easier control of process variables [266].

Miscellaneous organic compounds

Other organic compounds such as phenols [40], hydrocarbons [267], polymer additives [268,269] and natural compounds [270] have also been extracted from a wide range of solid matrices with good results. This confirms that, under optimal conditions of solvent, microwave power and exposure time, etc., the MAE technique is suitable for the extraction of organic compounds from solid samples.

One of the most interesting fields of application of microwave-assisted leaching of organic compounds is probably food analysis for lipids. This step, traditionally developed by conventional Soxhlet extraction, has recently been performed with the focused microwave-assisted Soxhlet prototype of Fig. 5.9B. The extractions of oil from olives [58], and sunflower seeds and soybean [271]; those of the lipid fraction of dairy products (milk [272] and cheese [273]); and those of fatty acids from precooked [59] and sausage foods [60] have significant advantages over conventional methods including dramatically reduced extraction times, lower degradation of thermolabile analytes and acceleration of other analytical steps such as hydrolysis in milk samples, in addition to completeness of analyte extractions — which is not always achieved with conventional methods.

Metals

The extraction of metals from solid samples has also been assisted by microwaves [68,274–277], most often with a view to speciation analysis [44,278–282].

Metals such as Fe, Cu, Zn, Pb, Hg, Se, As, Co, Cd, Cr and Ni have been extracted using an acid solution (usually HNO_3) in a focused open-vessel system for periods ranging from 3 min for soil samples [68] to 50 min for coal (where metals are much more strongly retained) [46]; the system of Fig. 5.8A allowed the sample to be brought into contact with fresh solvent in four cycles, which resulted in the relatively short extraction time achieved. One clear example of the dramatically reduced extraction times afforded by microwaves is the extraction of Pb, Zn and Cu from calcareous soil [277], where the sequential procedure used proved considerably faster than conventional extraction (2 h versus 20 h).

Microwave-assisted extraction has been used both as a pretreatment for chemical speciation (mainly to discriminate As [278,282] and Se species [44,279]) and as a means for accelerating the fractionation of metals in sewage sludge for operational and/or functional speciation [280,281]. The former type of application has usually been developed using open-vessel systems coupled to various types of equipment for individual separation and/or determination (e.g. hydride generation–atomic fluorescence detection [44], electrochemical detection [279] and high performance liquid chromatography–hydride generation–inductively coupled plasma mass spectrometry [278]) following leaching.

5.5.3. Microwave-assisted sample drying

Drying is defined as the removal of volatiles from a sample. One common problem associated with the process of assaying in analytical laboratories is obtaining precise, reproducible methods of sample drying. This step should result in identical sample compositions irrespective of differences in starting volatile content and sample history. Usually, the major volatile component to be removed from a sample is water. Different sample types absorb variable amounts of water and therefore must be dried to an equilibrium weight. When a plot of weight loss versus time reaches a plateau, equilibrium weight has been achieved. Many samples are hygroscopic and must be kept in a moisture-free environment (e.g. that provided by a desiccator) after drying. Variation in water content is one of the most common sources of irreproducible sample weights. This can result in decreased assay precision and accuracy, and possibly give biased analytical data. In order to avoid or at least minimize the problem, samples are usually dried before weighing. Classical conventional methods of drying include the use of convection ovens, vacuum ovens and desiccators. All three methods are effective and can provide samples with reproducible composition. However, some samples may contain different temperature-dependent volatile components or exhibit variable degrees of hydration. This can lead to varying equilibrium weights and should be of concern to the analyst. Generally, the least aggressive method yielding reproducible results and stable sample composition will be the pretreatment of choice. The least aggressive method minimizes the risk of chemical alteration of the sample.

Microwave-assisted drying provides a simple, expeditious alternative to conventional drying methods. Specially designed microwave ovens are now commercially available for this purpose [197,283–286]. One example is described in Section 5.3. Because solvent molecules are directly heated by the microwave radiation, they can be volatilized without heating the container that holds them. This reduces the cooling time required in a desiccator before accurate weight measurements can be made. In concert with drying samples, moisture determinations can be performed [4,7,287,288]. One important advantage of microwave-assisted drying methods is that, under reduced pressure, water in the sample can be removed at lower temperatures, so the decomposition or loss of volatile organic compounds and the formation of local hot spots are avoided to some extent [5].

5.5.4. Microwave-assisted distillation

Distillation has long been used to separate volatile chemical components. The technique is still widely employed today in a variety of industries and for a range of separations. While conventional distillation methods rely on conduction for heating, microwave heating results from direct interaction of the energy with the sample, thus allowing energy transfer to the entire sample as opposed to this taking place via convection currents as with conductive heating. This advantage makes microwave heating an excellent candidate for replacing distillations and refluxing procedures limited by long heating periods. This microwave-assisted step has been used in the distillation of inorganic arsenic from soil [11] and seafood products [63,65], as well as in that of organic off-flavour compounds from animal tissues [66,67,289–292] (using a coupled solid-phase extractor such as that of Fig. 5.12B to allow the retention and preconcentration of distilled analytes). In both cases, microwave-assisted distillation afforded a significant reduction in sample preparation time relative to conventional distillation and also reduced the amount of sample to be handled.

5.5.5. Microwave-assisted protein hydrolysis

Acid hydrolysis of protein samples has been the rate-determining step in amino acid analysis ever since commercial HPLC amino acid analysers became available in the late 1960s. Microwave-assisted hydrolysis is an alternative to conventional hydrolysis that has changed this situation dramatically. Thus, protein hydrolysates can now be prepared in less time than a single chromatographic run without compromising accuracy or precision. Microwave-assisted protein hydrolysis is particularly advantageous in cases where two or more hydrolyses are performed to achieve optimum recoveries of labile and refractory amino acids. For example, microwave hydrolysis runs of 20 and 40 min duration can be performed without altering the chemistry of amino acids instead of one and two days of conventional hydrolysis.

Early studies in microwave-assisted hydrolysis of proteins showed the recoveries of amino acids after microwave hydrolysis to be quite consistent with the results provided by conventional hydrolysis [293,294]. Engelhardt *et al.* used on-line microwave hydrolysis in an HPLC system followed by derivatization with OPA to improve the detection sensitivity for proteins [295].

The initial work on microwave-assisted hydrolysis of proteins was conducted using commercial domestic microwave devices. Different laboratory microwave ovens featuring various methods of operation are now commercially available as well. The microwave-assisted hydrolysis of proteins or peptides is usually performed in sealed containers [296] where the sample is in direct contact with the concentrated acid used for hydrolysis. In vapour-gas microwave hydrolysis [297], the acid is evaporated and only the vapour contacts the sample, which prevents contamination of the samples by small amounts of amino acids potentially present in the acid used for analysis.

Microwave-assisted hydrolysis systems equipped with temperature regulators suitable for quality control of protein products are now available. In fact, microwave hydrolysis is the sole technique in which the acid temperature and vapour pressure are measured directly, in situ. Other devices measure the temperature of a heating unit such as metal heating block or oven surrounding a hydrolysis chamber.

One major concern in the hydrolysis process is the potential racemization of amino acids. Peter *et al.* [298] used both microwave and conventional hydrolysis to determine the amino acid composition of three synthetic peptides and found microwave-assisted hydrolysis to result in reduced racemization and in higher recovery of sensitive amino acids than hydrolysis by conventional heating with hydrochloric acid [299].

5.6. SAFETY CONSIDERATIONS ON THE USE OF MICROWAVE ENERGY

Standards, limits and tolerance ranges have been established for microwave radiation exposure in most of the industrial world. The United States, the former Soviet republics, the United Kingdom, the former Czechoslovakia, Canada, Australia and the European Community, military and governmental bodies in these countries, as well as international organizations, have all established safety standards [300,301]. An underlying reason for the large number of exposure standards in existence today is the manner in which they are defined (e.g. by the electromagnetic energy frequency, duration of exposure, body mass and time or periodicity of exposure).

Research into the biological effects of microwave radiation exposure has been extensively discussed in several reviews dealing with scientific, industrial and medical applications [300–302]. Overall, the effects on human tissue are thermal in nature and relate to overheating of exposed tissue. The underlying protective principle of several standards is derived from data on the amount of energy required to raise human skin and tissue temperatures to biologically significant levels. Exposure to energy such as sunlight is basically a surface phenomenon; however, microwave energy penetrates the skin into the subcutaneous tissue and therefore also raises the temperature level of tissue and blood.

Proper usage of laboratory microwave equipment is the responsibility of laboratory personnel. It is possible to render the safety devices of many instruments and vessels ineffective by carelessness or misuse. It is the responsibility of the analyst to follow good laboratory practices and the manufacturer's instructions when assembling, using and maintaining microwave equipment. For example, by placing a microwave system inside a fume hood, where exhausted acid fumes may get circulated around the unit, the designed physical isolation of the electronics from the cavity is defeated. Accelerated corrosion of the electronics, including the safety interlock mechanisms and control circuits, can result. Chemical vapours should always be transported away from the unit, or the cavity air swept away to an exhaust hose, a fume extraction or neutralization system, or a hood. Deterioration of the waveguide, door seals or cavity walls can provide leak paths for the escape of microwave radiation and result in degradation of the equipment.

The hazards associated with inappropriate use of microwave equipment cannot be entirely prevented by interlocking devices and other safeguards. However, the risk can be minimized if the analyst continually inspects the system to ensure that the equipment is maintained in safe working order. If any portion of the microwave unit such as door seal or vessel casing becomes damaged by a serious event such as an acid spill, prolonged wear or impact, the safety of the equipment should be re-assessed before it is returned to service.

In addition to compliance with microwave energy leakage standards, safety interlock devices are required to prevent accidental exposure on all commercial and consumer microwave equipment. These interlocks protect against initiating or continuing the emission of microwave energy into the cavity if the microwave system door is open or misaligned. Safety devices should never be removed or comprised on any microwave equipment, but especially on laboratory systems. Other important components for safe operation such as wavelength attenuators in atmospheric-pressure systems, door seals or waveguides should be inspected and tested for microwave leakage if corrosion is observed or a vessel vents and reagents have prolonged contact with non-resistant parts of the equipment.

Other considerations such as the constituent materials of closed vessels and the cautions to be exercised in handling hazardous acids such as $HClO_4$ have been addressed in detail throughout this chapter.

References

1 A. Abu Samra, J.S. Morris and S.R. Koirtyohann, *Anal. Chem.*, 47 (1975) 1475.
2 K.J. Lamble and S.J. Hill, *Analyst*, 123 (1998) 103R.
3 V. Pino, J.H. Ayala, A.M. Afonso and V. González, *J. Chromatogr.*, 869 (2000) 515.
4 B. Maichin, P. Kettisch and G. Knapp, *Fresenius J. Anal. Chem.*, 366 (2000) 26.
5 P.A. Tanner and L.S. Leong, *Anal. Chim. Acta*, 342 (1997) 247.
6 I. Kubrakova, T. Kudinova, A. Formanovsky, N. Kuz'min, G. Tsysin and Y. Zolotov, *Analyst*, 119 (1994) 2477.
7 Q. Jin, F. Liang, H. Zhang, L. Zhao, Y. Huan and D. Song, *Trends Anal. Chem.*, 18 (1999) 479.
8 H. Wang, Y. Zhou, Y. Zhao, Y. Ning, X. Chen and Z. Hu, *Anal. Lett.*, 33 (2000) 2075.
9 K.K. Chee, M.K. Wong and H.K. Lee, *Anal. Chim. Acta*, 330 (1996) 217.
10 K.K. Chee, M.K. Wong and H.K. Lee, *Mikrochim. Acta*, 126 (1997) 97.

11 C.M. Barra, M.L. Cervera, M. de la Guardia and R.E. Santelli, *Anal. Chim Acta*, 407 (2000) 155.
12 K.K. Chee, M.K. Wong and H.K. Lee, *J. Chromatogr.*, 723 (1996) 259.
13 A. Gourdenne and Q. LeVan, *Polym. Prepr.*, 22 (1981) 125.
14 H.M. Kingston and L.B. Jassie, *Introduction to Microwave Sample Preparation*, ACS Professional Reference Books, Washington, DC (1998).
15 G. Herzberg, *Infrared and Raman Spectra*, Van Nostrand, New York (1945).
16 H.M. Kingston and S.J. Haswell, *Microwave-Enhanced Chemistry*, ACS Professional Reference Books, Washington, DC (1997).
17 R.D. Smith, *Elec. Power Res. Inst.*, EM-3645 (1984) A-8.
18 R.V. Decareau, *Microwaves in the Food Processing Industry*, Academic Press, New York (1985).
19 A.R. Von Hipel, *Dielectric Materials and Applications*, John Wiley, New York (1954) 301.
20 R.N. Gedye, F.E. Smith and K.G. Can, *J. Chem.*, 66 (1988) 17.
21 C. Banks, *Lab. News*, 581 (1998) 18.
22 D. Constable, K. Raner, P. Somlo and C. Strauss, *J. Microwave Power Electromagnetic Energy*, 27 (1992) 195.
23 H. Matusiewicz, R.E. Sturgeon and S.S. Berman, *J. Anal. At. Spectrom.*, 6 (1991) 283.
24 M. Domínguez de Almeida, K.C. Leandro, C. Vidal da Costa, R.E. Santelli and M. de la Guardia, *J. Anal. At. Spectrom.*, 12 (1997) 1235.
25 A. Cuesta, J.L. Todolí, J. Mora and A. Canals, *Anal. Chim. Acta*, 372 (1998) 399.
26 M.Y. Pérez-Jordán, A. Salvador and M. de la Guardia, *Anal. Lett.*, 31 (1998) 867.
27 The 100 Most Significant New Products of the Year, *Ind. Res. Dev.*, 29 (1987) 55.
28 G.M. Kimber and S. Kokot, *Trends Anal. Chem.*, 9 (1990) 203.
29 G. Giebenhain, U. Sengutta and H.J. Staerk, *GIT Labor Fachz.*, 42 (1998) 390.
30 H. Matusiewicz, *Anal. Chem.*, 66 (1994) 751.
31 *Report 12/93*, Plazmatronika Ltd, Wroclaw, Poland (1993).
32 H. Matusiewicz, *Anal. Chem.*, 71 (1999) 3145.
33 U. Pichler, A. Haase and G. Knapp, *Anal. Chem.*, 71 (1999) 4050.
34 M. Ericsson and A. Colmsjö, *J. Chromatogr.*, 877 (2000) 141.
35 R. Commarmot, D. Didenot and J.F. Gardais, *US Patent 4,681,740* (1987).
36 C. Suard, R.M. Mourel, B. Cerdan, G. Bart and M.H. Feinberg, *Anal. Chim. Acta*, 318 (1996) 261.
37 M. Burguera, J.L. Burguera and O.M. Alarcón, *Anal. Chim. Acta*, 214 (1988) 421.
38 E. Luque-Pérez and S.J. Haswell, *Anal. Proc.*, 32 (1995) 85.
39 M.A.Z. Arruda, M. Gallego and M. Valcárcel, *J. Anal. At. Spectrom.*, 10 (1995) 501.
40 M.A. Crespín, M. Gallego and M. Valcárcel, *J. Chromatogr.*, 897 (2000) 279.
41 J. Flock, F. Michael and K.D. Ohls, *Fresenius J. Anal. Chem.*, 363 (1999) 306.
42 Z. Bouhsain, S. Garrigues, A. Morales-Rubio and M. de la Guardia, *Anal. Chim. Acta*, 330 (1996) 59.
43 J.L. Luque-García, S. Morales-Muñoz and M.D. Luque de Castro, *Chromatographia*, 55 (2002) 117.
44 M.E. Moreno, C. Pérez-Conde and C. Cámara, *J. Anal. At. Spectrom.*, 15 (2000) 681.
45 V. Fernández-Pérez, L.E. García-Ayuso and M.D. Luque de Castro, *Analyst*, 125 (2000) 317.
46 M. de la Guardia, A. Salvador, J.L. Burguera and M. Burguera, *J. Flow Injection Anal.*, 5 (1988) 121.
47 C. Cabrera, Y. Madrid and C. Cámara, *J. Anal. At. Spectrom.*, 9 (1994) 1423.
48 E.R. Pereira-Filho and M.A.Z. Arruda, *Analyst*, 124 (1999) 1873.
49 E.S. Beary, P.J. Paulsen, L.B. Jassie and J.D. Fassett, *Anal. Chem.*, 69 (1997) 758.
50 E.R. Pereira, F. Jarbas, J.R. Rohwedder and M.A.Z. Arruda, *Analyst*, 123 (1998) 1023.
51 L. Bordera, V. Hernandis and A. Canals, *Fresenius J. Anal. Chem.*, 355 (1996) 112.
52 A. Morales-Rubio, M.L. Mena and C.W. McLeod, *Anal. Chim. Acta*, 308 (1995) 364.
53 S.L. Cresswell and S.J. Haswell, *Analyst*, 124 (1999) 1361.
54 V. Carbonell, M. de la Guardia, A. Salvador, J.L. Burguera and M. Burguera, *Anal. Chim. Acta*, 238 (1990) 417.
55 A. Lagha, S. Chemat, P.V. Bartels and F. Chemat, *Analusis*, 27 (1999) 452.

56 *PCT application nºPCT/FR97/00883 (Published under no. WO 97/44109)*, Societé Prolabo, M.D. Luque de Castro and L.E. García-Ayuso (1997).
57 L.E. García-Ayuso, M. Sánchez, A. Fernández de Alba and M.D. Luque de Castro, *Anal. Chem.*, 70 (1998) 2426.
58 L.E. García-Ayuso and M.D. Luque de Castro, *Anal. Chim. Acta*, 382 (1999) 309.
59 J.L. Luque-García, J. Velasco, M.C. Dobarganes and M.D. Luque de Castro, *Food Chem.*, 76 (2002) 241.
60 E. Priego-López, J. Velasco, M.C. Dobarganes, G. Ramis and M.D. Luque de Castro, *Anal. Bioanal. Chem.*, submitted.
61 L.E. García-Ayuso, J.L. Luque-García and M.D. Luque de Castro, *Anal. Chem.*, 72 (2000) 3527.
62 J.L. Luque-García and M.D. Luque de Castro, *Anal. Chem.*, 73 (2001) 5903.
63 O. Muñoz, D. Vélez, M.L. Cervera and R. Montoro, *J. Anal. At. Spectrom.*, 14 (1999) 1607.
64 K. Barkács, A. Varga, K. Gál-Solmos and G. Záray, *J. Anal. At. Spectrom.*, 14 (1999) 577.
65 J.C. López, C. Reija, R. Montoro, M.L. Cervera and M. de la Guardia, *J. Anal. At. Spectrom.*, 9 (1994) 651.
66 E.D. Conte, C.Y. Shen, P.W. Perschbacher and D.W. Miller, *J. Agric. Food Chem.*, 44 (1996) 829.
67 E.D. Conte, C.Y. Shen and D.W. Miller, *Anal. Chem.*, 68 (1996) 2713.
68 P. Torres, E. Ballesteros and M.D. Luque de Castro, *Anal. Chim. Acta*, 308 (1995) 371.
69 B. Ciommer and N. Muller, *GIT Fachz. Lab.*, 40 (1996) 389.
70 W.M. Johnson and J.A. Maxwell, *Rock and Mineral Analysis*, Wiley & Sons, New York (1981).
71 J. Dolezal, P. Povondra and Z. Sulcek, *Decomposition Techniques in Inorganic Analysis*, Elsevier, New York (1968).
72 L. Ding, F. Liang, Y.F. Huan, Y.B. Cao, H.Q. Zhang and Q.H. Jin, *J. Anal. At. Spectrom.*, 15 (2000) 293.
73 W. Jiang, S.J. Chalk and H.M. Kingston, *Analyst*, 122 (1997) 211.
74 J. Mora, A. Canals, V. Herrandis, E.H. Van Veen and M.T.C. de Loos-Vollebregt, *J. Anal. At. Spectrom.*, 13 (1998) 175.
75 L. Gras, J. Mora, J.L. Todoli, A. Canals and V. Hernandis, *Spectrochim. Acta*, 54B (1999) 469.
76 L. Gras, J. Mora, J.L. Todoli, A. Canals and V. Hernandis, *Spectrochim. Acta*, 52B (1997) 1201.
77 SW-846 EPA Method 3031: "Acid Digestion of Oils for Metals, Analysis by FLAA or ICP Spectroscopy", in: *Test Methods for Evaluating Solid Waste*, 3rd edn, 3rd update, US Environmental Protection Agency, Washington, DC (1995).
78 SW-846 EPA Method 3052: "Microwave Assisted Acid Digestion of Siliceous and Organically Based Matrices", in: *Test Methods for Evaluating Solid Waste*, 3rd edn, 3rd update, US Environmental Protection Agency, Washington, DC (1995).
79 *Method V 03–100: Kjeldahl Nitrogen*, French Standard.
80 *NIEA Meted C303. O1T: Acid Digestion of Fish and Shellfish*, NIEA, Republic of China (1994).
81 *AST Method D5513–94. Standard Practice for Microwave Digestion of Industrial Furnace Feedstreams for Trace Element Analysis*, The American Society for Testing and Materials, Philadelphia, PA (1994).
82 F.E. Smith and E.A. Arsenault, *Talanta*, 43 (1996) 1207.
83 F. Smith, B. Cousins, J. Bozic and W. Flora, *Anal. Chim. Acta*, 177 (1985) 243.
84 P.J. Lamothe, T.L. Fries and J.J. Consul, *Anal. Chem.*, 58 (1986) 1881.
85 A.J. Kemp and C.J. Brown, *Analyst*, 115 (1990) 1197.
86 L.B. Fisher, *Anal. Chem.*, 58 (1986) 261.
87 T. Suzuki and M. Sensui, *Anal. Chim. Acta*, 245 (1991) 43.
88 M. Betinelli, U. Baroni and N. Pastorilli, *J. Anal. At. Spectrom.*, 2 (1987) 485.
89 P.R. Klock and P.J. Lamothe, *Talanta*, 33 (1986) 495.
90 R.A. Nadkarni, *Anal. Chem.*, 56 (1984) 2233.
91 W.R. Alexander and T.M. Shimmield, *J. Radioanal. Nucl. Chem.*, 145 (1990) 301.
92 W. Xu, M. Zou, J. Zhao and Z. Li, *Fenxi Huaxue*, 20 (1992) 1291.
93 J. Alvarado and A. Petrola, *J. Anal. At. Spectrom.*, 4 (1989) 411.
94 R.A. Davidson, D.D. Harbuck and D.D. Hammargren, *At. Spectrom.*, 11 (1990) 7.

95 S.A. Matthes, R.F. Farrell and A.J. Mackie, *Tech. Prog. Rep. US Bur. Mines*, TRP 120 (1983) 9.
96 R.T.T. Rantala and D.H. Loring, *Anal. Chim. Acta*, 220 (1989) 263.
97 F.E. Smith, B.G. Cousins and J.Y. Maillet, *Educ. Chem.*, 24 (1987) 13.
98 J. Bozic and F.E. Smith, Ores and Concentrates-Special Considerations, in: *Handbook on the Analysis of Geological Materials*, C. Riddle Ed., Marcel Dekker, Inc. (1993).
99 T. Paurket, *Chem. Listy*, 86 (1992) 143.
100 Z. Zhang and J. Cai, *Fenxi Huaxue*, 18 (1990) 1078.
101 B. Li, Z. Yu, K. Han, *Guangpuxue Yu Guangpu Fenxi*, 11 (1991) 60.
102 H. Ma, S. Wu, Y. Chen, C. Chen and H. Hu, *Yankuang Ceshi*, 12 (1993) 28.
103 J.L. Bouvier and G.E.M. Hall, *Pap. Geol. Surv. Can.*, 87–1A (1987) 458.
104 J. Ching, B. Schmidt and A. Mojica, *Precious Met.*, 17 (1993) 131.
105 B.J. Perry, R.R. Barefoot, J.C. Van Loo, A.J. Naldren and D.V. Speller, *Spec. Publ. Chem. Soc.*, 124 (1993) 91.
106 R.A. Romero, J.E. Tahan and J.A. Navarro, *Anal. Sci.*, 7 (1991) 829.
107 F. Luo, C. Lang, M. Wang and L. Jianyan, *Huaxue Fence*, 27 (1991) 230.
108 M.A.E. Wandt and M.A.B. Pougnet, *Analyst*, 111 (1986) 1249.
109 T.T. Gorsuch, *The Destruction of Organic Matter*, Pergamon, New York (1970).
110 P.R.M. Correia, E. Oliveira and P.V. Oliveira, *Anal. Chim. Acta*, 405 (2000) 205.
111 N.N. Meeravali and S.J. Kumar, *Anal. Chim. Acta*, 404 (2000) 295.
112 H. Polkowska-Motrenko, B. Danko, R. Dybczy'nski, A. Koster-Ammerlaan and P. Bode, *Anal. Chim. Acta*, 408 (2000) 89.
113 S.A. Bhandari and D. Amarasiriwardena, *Microchem. J.*, 64 (2000) 73.
114 D. Pozebon, V.L. Dressler and A.J. Curtius, *Talanta*, 51 (2000) 903.
115 C.C. Oliveira, E. Ayres, G. Zagatto, A.N. Araújo and J.L.F. Costa Lima, *Anal. Chim. Acta*, 371 (1998) 57.
116 G. Esslemont, W. Maher, P. Ford and F. Krikowa, *At. Spectrosc.*, 21 (2000) 42.
117 V.E. Negretti de Bratter, P. Bratter, A. Reinicke, G. Schulze, W.O.L. Álvarez and N. Álvarez, *J. Anal. At. Spectrom.*, 10 (1995) 487.
118 M.J. Campbell, C. Demesmay and M. Olle, *J. Anal. At. Spectrom.*, 9 (1994) 1379.
119 L. Duneman and M. Meinerling, *Fresenius J. Anal. Chem.* 342, (1992) 714.
120 E.J. Gawalko, T.W. Nowicki, J. Babb and R. Tkachuk, *J. AOAC Int.*, 80 (1997) 379.
121 A. Krushevska, R.M. Barnes and C. Amarasiriwaradena, *Analyst*, 118 (1993) 1175.
122 J. Liu, R.E. Sturgeon and S.N. Willie, *Analyst*, 120 (1995) 1905.
123 P. Quevauviller, J.L. Imbert and M. Olle, *Mikrochim. Acta*, 112 (1993) 147.
124 G. Schnitzer, A. Soubelet, C. Testu and C. Chafey, *Mikrochim. Acta*, 119 (1995) 199.
125 G.S. Banuelos and S. Akohoue, *Commun. Soil Sci. Plant Anal.*, 25 (1994) 1655.
126 S. Fridlund, S. Littlefield and J. Rivers, *Commun, Soil Sci. Plant Anal.*, 25 (1994) 933.
127 T.J. Gluodenis and J.F. Tyson, *J. Anal. At. Spectrom.*, 7 (1992) 301.
128 S.J. Haswell and D. Barclay, *Analyst*, 117 (1992) 117.
129 L.H.J. Lajunen and J. Pispanen, *Atom. Spectrosc.*, 13 (1992) 127.
130 B.S. Sheppard, D.T. Heitkmper and C.M. Gaston, *Analyst*, 119 (1994) 1683.
131 J.E. Tahan, V.A. Granadillo, J.M. Sánchez, H.S. Cubillan and R.A. Romero, *J. Anal. At. Spectrom.*, 8 (1993) 1005.
132 M.D. Mingorance, M.L. Pérez-Vázquez and M. Lachica, *J. Anal. At. Spectrom.*, 8 (1993) 853.
133 R. Mizushima, M. Yonezawa, A. Ejima, H. Koyana and H. Satoh, *Tohoku J. Exp. Med.*, 178 (1996) 75.
134 Q. Yang, W. Penninckx and J. Smeyersverbeke, *J. Agric. Food Chem.*, 42 (1994) 1948.
135 S. Wu, Y.H. Zhao, X. Fen and A. Wittmeier, *J. Anal. At. Spectrom.*, 12 (1997) 797.
136 D.C. Baxter, R. Nichol and D. Littlejohn, *Spectrochim. Acta*, 47 (1992) 1155.
137 M.J. Campbell, G. Vermeir, R. Dams and P. Quevauviller, *J. Anal. At. Spectrom.*, 7 (1992) 617.
138 M. Navarro, M. López, M.C. López and M. Sánchez, *Anal. Chim. Acta*, 257 (1992) 155.
139 J.C. Schaumloffel and W.F. Siems, *Rev. Sci. Instrum.*, 67 (1996) 4321.
140 R. Schlenz and E. Zeiller, *Fresenius J. Anal. Chem.*, 345 (1993) 68.

141 H. Vanhoe, *J. Trace Elem. Electrol. Health Dis.*, 7 (1993) 131.
142 N. Xu, V. Majidi, W.D. Ehman and W.R. Markesbery, *J. Anal. At. Spectrom.*, 7 (1992) 749.
143 A. Chatterjee, *J. Anal. At. Spectrom.*, 15 (2000) 753.
144 I. Rodushkin, F. Ödman, R. Olofsson and M.D. Axelsson, *J. Anal. At. Spectrom.*, 15 (2000) 937.
145 R. Sur, J. Begerow and L. Dunemann, *Fresenius J. Anal. Chem.*, 363 (1999) 526.
146 J.L. Burguera, M. Burguera, C. Rivas, C. Rondon, P. Carrero and M. Gallignani, *Talanta*, 48 (1999) 885.
147 A. Morales-Rubio, A. Salvador and M. de la Guardia, *Fresenius J. Anal. Chem.*, 342 (1992) 452.
148 A.G. Coedo, M.T. Dorado, I. Padilla and F.J. Alguacil, *J. Anal. At. Spectrom.*, 13 (1998) 1193.
149 I. Rodushkin, T. Ruth and A. Huhtasaari, *Anal. Chim. Acta*, 378 (1999) 191.
150 H.M. Kingston and L.B. Jassie, *J. Res. Natl. Bur. Stand.*, 93 (1988) 269.
151 R. Demura, S. Tsukada and I. Yamamoto, *Eisei Kagaku*, 31 (1985) 405.
152 A.A. Craveiro, F.J.A. Matos, J.W. Alencar and M.M. Plumel, *Flavour Fragance J.*, 4 (1989) 43.
153 G.J. Collin, D. Lord, J. Allaire and D. Gagnon, *Parfums Cosmet. Aromes*, 97 (1991) 105.
154 E. Vereda Alonso, A. García de Torres and J.M. Cano Pavón, *Analyst*, 117 (1992) 1157.
155 P. Hocquellet and M.P. Candillier, *Analyst*, 116 (1991) 505.
156 M. Krachler, H. Radner and K.J. Irgolic, *Fresenius J. Anal. Chem.*, 355 (1996) 120.
157 N.N. Meeravali and S.J. Kumar, *Fresenius J. Anal. Chem.*, 366 (2000) 313.
158 V. Ducros, D. Ruffieux, N. Belin and A. Favier, *Analyst*, 119 (1994) 1715.
159 C.C. Huang, M.H. Yang and T.S. Shih, *Anal. Chem.*, 69 (1997) 3930.
160 A. Carlosena, M. Gallego and M. Valcárcel, *J. Anal. At. Spectrom.*, 12 (1997) 479.
161 K.J. Lamble and S.J. Hill, *Anal. Chim. Acta*, 334 (1996) 261.
162 K.I. Mahan, T.A. Foferaro, T.L. Garza, R.M. Martínez, G.A. Maroney, M.R. Trivisonno and E.M. Willging, *Anal. Chem.*, 59 (1987) 983.
163 F.I. Onuska and K.A. Terry, *Chromatographia*, 36 (1993) 191.
164 M. Betinelli and U. Baroni, *J. Environ. Anal. Chem.* 43, (1991) 33.
165 R.M. Sterrit and J.N. Lester, *Sci. Total Environ.*, 16 (1980) 55.
166 I. de Gregori, H. Pinochet, E. Fuentes and M. Potin-Gautier, *J. Anal. At. Spectrom.*, 16 (2001) 172.
167 N. Elwaer and N. Belzile, *Int. J. Environ. Anal. Chem.*, 61 (1995) 189.
168 A.D. Hewitt and C.M. Reynolds, *At. Spectrosc.*, 11 (1990) 187.
169 C.J. Warren, B. Xing and M.J. Dudas, *J. Can. Soil Sci.*, 70 (1990) 617.
170 B. Kratochvil and S. Namba, *Can. J. Chem.*, 68 (1990) 360.
171 L. Xu and W. Shen, *Fresenius J. Anal. Chem.*, 333 (1989) 108.
172 J. Huang, D. Goltz and F. Smith, *Talanta*, 35 (1988) 907.
173 C.M. Reynolds, *Gov. Rep. Announce. Index*, 93 (1993) 37.
174 W.R. Kammin and M.J. Brandt, *Spectroscopy*, 4 (1989) 49.
175 G.S.R. Krishnamurti, P.M. Huang, K.C.J. Van Rees, L.M. Kozak and H.P.W. Rostad, *Commun. Soil. Sci. Plant Anal.*, 25 (1994) 615.
176 H.M. Kingston and P.J. Walter, *Spectroscopy*, 7 (1992) 20.
177 A.D. Wewitt and J.H. Cragin, *Environ. Sci. Technol.*, 25 (1991) 985.
178 J.R. Wakakuwa and D.E. Krimbrough, *Environ. Sci. Technol.*, 26 (1992) 1849.
179 A.D. Wewitt and J.H. Cragin, *Environ. Sci. Technol.*, 26 (1992) 1848.
180 J. Nieuwenhuize, C.H. Poley-Vos, A.H. Van den Akker and W. Van Delft, *Analyst*, 166 (1991) 347.
181 L. Xu, W. Shen and J. Zhu, *J. Environ. Sci.*, 1 (1989) 69.
182 D.A. Binstock, P.M. Grohse, A. Gaskill Jr, K.K. Luk, P.L. Swift and H.M. Kingston, C. Sellers, *ASTM Spec. Tech. Publ.*, 1062 (1990) 259.
183 W. Van Delft and G. Vos, *Anal. Chim. Acta*, 209 (1988) 147.
184 C.G. Millward and P.D. Klickner, *J. Anal. At. Spectrom.*, 4 (1989) 709.
185 E. Boffelli, *Inquinamento*, 30 (1988) 96.
186 A.G. Coedo, I. Padilla, T. Dorado and F.J. Alguacil, *Anal. Chim. Acta*, 389 (1999) 247.
187 W. Lautenschlaeger and G. Kopp, *LaborPraxis*, 24 (2000) 76.

188 M.T. Dorado López, M.A. Palaios Vida and A. Gómez Coedo, *Congr. Nac. Cienc. Technol. Metal, 7th*, 3 (1990) 205.
189 P. Halmos, E. Gegus and J. Borszeki, *Magy. Kem. Foly.*, 99 (1993) 420.
190 P.A. Vozzella and D.A. Condit, *Anal. Chem.*, 60 (1988) 2497.
191 I. Hlavacek and I. Hlavackova, *J. Anal. At. Spectrom.*, 6 (1991) 535.
192 B.D. Zehr and M.A. Fedorchak, *Am. Lab.*, 23 (1991) 40.
193 T.Y. Chang, F.J. Isieh and C. Lee, *MRL Bull. Res. Dev.*, 4 (1990) 5.
194 L.A. Fernando, W.D. Heavner and C.C. Gabrielli, *Anal. Chem.*, 58 (1986) 511.
195 B. Berglund and C. Wichardt, *Anal. Chim. Acta*, 236 (1990) 399.
196 P.G. Riby, S.J. Haswell and R. Grzeskowiak, *J. Anal. At. Spectrom.*, 4 (1989) 181.
197 CEM Corporation, P.O. Box 200, Matthews, NC 28106, USA.
198 S. Mann, D. Geilenberg, J.A.C. Broekaert and M. Jansen, *J. Anal. At. Spectrom.*, 12 (1997) 975.
199 M.T. Larrea, I. Gómez-Pinilla and J.C. Fariñas, *J. Anal. At. Spectrom.*, 12 (1997) 1323.
200 S. Rauch, M. Lu and G.M. Morrison, *Environ. Sci. Technol.*, 35 (2001) 595.
201 L.B. Allen, P.H. Siitonen and H.C. Thompson Jr, *J. Anal. At. Spectrom.* 11 (1996) 529.
202 K. Swami, C.D. Judd, J. Orsini, K.X. Yang and L. Husain, *Fresenius J. Anal. Chem.*, 369 (2001) 63.
203 F.H. Ko, J.K. Lu, T.C. Chu, C.T. Chou, L.T. Hsiao and H.C. Lin, *J. Anal. At. Spectrom.*, 15 (2000) 715.
204 J. Diemer and K.G. Heumann, *Fresenius J. Anal. Chem.*, 364 (1999) 421.
205 F.H. Ko, M.Y. Wang and T.K. Wang, *Anal. Chem.*, 71 (1999) 5413.
206 H. Lachas, R. Richaud, K.E. Jarvis, A.A. Herod, D.R. Dugwell and R. Kandiyoti, *Analyst*, 124 (1999) 177.
207 H.A. McKenzie, *Trends Anal. Chem.*, 13 (1994) 138.
208 J. Alvarado, M. Márquez and L.E. León, *Anal. Lett.*, 21 (1988) 861.
209 M. Feinberg, C. Suard and J. Ireland-Ripert, *Chemom. Intell. Lab. Syst.*, 22 (1994) 37.
210 L.W. Collins, S.J. Chalk and H.M. Kingston, *Anal. Chem.*, 68 (1996) 2610.
211 C.J. Mason, G. Coe, M. Edwards and P.G. Riby, *Analyst*, 124 (1999) 1719.
212 J. Sherma, *Anal. Chem.*, 63 (1991) 118R.
213 J.P. Hsu, H.G. Wheeler Jr, D.E. Camann, H.J. Schattemberg, R.G. Lewis and A.E. Bond, *J. Chromatogr. Sci.*, 26 (1988) 181.
214 A. Pastor, E. Vázquez, R. Ciscar and M. de la Guardia, *Anal. Chim. Acta*, 344 (1997) 241.
215 C. Molins, E.A. Hogndoorn, E. Dijkman, H.A.G. Heusinkveld and R.A. Baumann, *J. Chromatogr.*, 869 (2000) 487.
216 I. Silgoner, R. Krska, E. Lombas, O. Gans, E. Rosenberg and M. Grasserbauer, *Fresenius J. Anal. Chem.*, 362 (1998) 120.
217 J. Fish and R. Revesz, *LC–GC*, 14 (1996) 231.
218 A. Bouaid, A. Martín Esteban, P. Fernández and C. Cámara, *Fresenius J. Anal. Chem.*, 367 (2000) 291.
219 G. Xiong, J. Liang, S. Zou, Z. Zhang, *Anal. Chim. Acta*, 371 (1998) 97.
220 G. Xiong, B. Tang, X. He, M. Zhao, Z. Zhang and Z. Zhang, *Talanta*, 48 (1999) 333.
221 C. Molins, E.A. Hogendoorn, H.A.G. Heusinkveld, D.C. Van Harten, P. Van Zoonen and R.A. Baumann, *Chromatographia*, 43 (1996) 27.
222 C. Molins, E.A. Hogendoorn, H.A.G. Heusinkveld, A.C. Van Beuzekom, P. Van Zoonen and R.A. Baumann, *Chromatographia*, 48 (1998) 450.
223 S.J. Stout, A.R. daCunha and D.G. Allardice, *Anal. Chem.*, 68 (1996) 653.
224 V. López-Ávila, R. Young, R. Kim and W. Beckert, *J. Chromatogr. Sci.*, 33 (1995) 481.
225 V. López-Ávila, R. Young, J. Benedicto and C. Charan, *Environ. Sci. Technol.*, 29 (1995) 2709.
226 V. López-Ávila, R. Young, J. Benedicto, P. Ho and R. Kim, *Anal. Chem.*, 67 (1995) 2096.
227 K. Hummert, W. Vetter and B. Luckas, *Chromatographia*, 42 (1996) 300.
228 W.H. Ho, S.J. Hsieh, *Anal. Chim. Acta*, 428 (2001) 111.
229 G.E. Berendsen and L. de Galán, *J. Chromatogr.*, 196 (1980) 21.
230 Y.M. Neff, *Polycyclic Aromatic Hydrocarbons in Aquatic Environment*, Applied Science, London (1979).

231 M.L. Lee, M.V. Novotny and K.D. Bartle, *Analytical Chemistry of Polycyclic Compounds*, Academic Press, New York (1992) 462.
232 G. Grimmer, *Environmental Carcinogens. Polycyclic Aromatic Hydrocarbons*, CRC Press, Boca Raton, FL (1983).
233 M. Tomaniova, J. Hajslova, J. Pavelka Jr, V. Kocourek, K. Holadova and T. Klimova, *J. Chromatogr.*, 827 (1998) 21.
234 P. Bruno, M. Caselli, G. de Gennaro, M. de Rienzo and A. Traini, *J. Environ. Monit.*, Jun (2000) 223.
235 L. Pensado, C. Casais, C. Mejuto and R. Cela, *J. Chromatogr.*, 869 (2000) 505.
236 I. Barnabas, R. Dean, I. Fowlis and S. Owen, *Analyst*, 120 (1995) 1897.
237 B. Vallejo-Pecharromán, L.E. García-Ayuso and M.D. Luque de Castro, *Chromatographia*, 53 (2001) 5.
238 A. Blanco Prevot, M. Gulmini, V. Zelano and E. Pramauro, *Anal. Chem.*, 73 (2001) 3790.
239 V. López-Ávila, R. Young, R. Kim and W. Beckert, *Anal. Chem.*, 66 (1994) 1097.
240 J.R.J. Paré and J.M.R. Bélanger, *Trends Anal. Chem.*, 13 (1994) 176.
241 F.I. Onuska and K.A. Terry, *Chromatographia*, 36 (1993) 191.
242 K. Ganzler, A. Salgo and K. Valco, *J. Chromatogr.*, 371 (1986) 299.
243 M.P. Llompart, R.A. Lorenzo, R. Cela, K. Li, J.M.R. Bélanger and J.R.J. Paré, *J. Chromatogr.*, 774 (1997) 243.
244 V. López-Ávila, R. Young and N. Tepltsky, *J. AOAC Int.*, 79 (1996) 142.
245 R. McMillin, L.C. Miner and L. Hurst, *Spectroscopy*, 13 (1997) 41.
246 R. Hoogerbrugge, C. Molins and R.A. Baumann, *Anal. Chim. Acta*, 348 (1997) 247.
247 M. Letellier, H. Budzinski, L. Charrier, S. Capes and A.M. Dorthe, *Fresenius J. Anal. Chem.*, 364 (1999) 228.
248 H. Budzinski, M. Letellier, S. Thompson, K. LeMenach and P. Garrigues, *Fresenius J. Anal. Chem.*, 367 (2000) 165.
249 H. Budzinski, P. Baumard, A. Papineau, S. Wise and P. Garrigues, *PAC*, 9 (1996) 225.
250 M. Letellier, H. Budzinski, P. Garrigues and S. Wise, *Spectroscopy*, 13 (1997) 71.
251 S. Jensen, L. Renberg and L. Reuterdardh, *Anal. Chem.*, 49 (1977) 316.
252 S.A. Wise, M.M. Schantz, B.A. Benner, M.J. Hays and S.B. Schiller, *Anal. Chem.*, 67 (1995) 1171.
253 F. Smedes and J. de Boer, *Trends Anal. Chem.*, 16 (1997) 503.
254 O. Zuloaga, N. Etxebarria, L.A. Fernández and J.M. Madariaga, *Trends Anal. Chem.*, 17 (1998) 642.
255 G. Xiong, X. He and Z. Zhang, *Anal. Chim. Acta*, 13 (2000) 49.
256 A. Eiguren Fernández, Z. Sosa Ferrera and J.J. Santana Rodríguez, *Anal. Chim. Acta*, 433 (2001) 237.
257 A.K. Djien Liem, *Trends Anal. Chem.*, 18 (1999) 499.
258 O. Zuloaga, N. Etxebarria, L.A. Fernández and J.M. Madariaga, *Talanta*, 50 (1999) 345.
259 S.P. Frot, J.R. Dean, K.P. Evans, K. Harradine, C. Cary and M.H.I. Comber, *Analyst*, 122 (1997) 895.
260 L. Croteau, M. Akhtar, J. Belanger and J. Pare, *J. Liq. Chromatogr.*, 17 (1994) 2971.
261 M. Abuín, A.M. Carro and R.A. Lorenzo, *J. Chromatogr.*, 889 (2000) 185.
262 R.A. Lorenzo, M.J. Vázquez, A.M. Carro and R. Cela, *Trends Anal. Chem.*, 18 (1999) 410.
263 M.J. Vázquez, A.M. Carro, R.A. Lorenzo and R. Cela, *Anal. Chem.*, 69 (1997) 221.
264 J. Szpunar, V.O. Schmitt, O.F.X. Donard and R. Lobinski, *Trends Anal. Chem.*, 15 (1996) 181.
265 I. Rodríguez Pereiro, A. Wasik and R. Lobinski, *Fresenius J. Anal. Chem.*, 363 (1999) 460.
266 O.F.X. Donard, B. Lalère, F. Martin and R. Lobinski, *Anal. Chem.*, 67 (1995) 4250.
267 E. Vázquez Blanco, P. López Mahía, S. Muniategui Lorenzo, D. Prada Rodríguez and E. Fernández Fernández, *Fresenius J. Anal. Chem.*, 366 (2000) 283.
268 H.J. Vandenburg, A.A. Clifford, K.D. Bartle, J. Carroll, I. Newton, L.M. Garden, J.R. Dean and C.T. Costley, *Analyst*, 122 (1997) 101R.
269 B. Marcato and M. Vianello, *J. Chromatogr.*, 869 (2000) 285.

270 G.A. Csiktusnádi Kiss, E. Forgács, T. Cserháti, T. Mota, H. Morais and A. Ramos, *J. Chromatogr.*, 889 (2000) 41.
271 L.E. García-Ayuso, J. Velasco, M.C. Dobarganes and M.D. Luque de Castro, *Chromatographia*, 52 (2000) 103.
272 L.E. García-Ayuso, J. Velasco, M.C. Dobarganes and M.D. Luque de Castro, *Int. Dairy J.*, 9 (1999) 667.
273 L.E. García-Ayuso, J. Velasco, M.C. Dobarganes and M.D. Luque de Castro, *J. Agric. Food Chem.*, 47 (1999) 2308.
274 A. Agazzi and C. Pirola, *Microchem. J.*, 67 (2000) 337.
275 G. Knapp, B. Maichin, P. Fecher, S. Hasse and P. Schramel, *Fresenius J. Anal. Chem.*, 362 (1998) 508.
276 D. Florian, R.M. Barnes and G. Knapp, *Fresenius J. Anal. Chem.*, 362 (1998) 558.
277 E. Campos, E. Barahona, M. Lachica and M.D. Mingorance, *Anal. Chim. Acta*, 369 (1998) 235.
278 J.L. Gómez-Ariza, D. Sánchez-Rodas, I. Giráldez and E. Morales, *Analyst*, 125 (2000) 401.
279 D.W. Bryce, A. Izquierdo and M.D. Luque de Castro, *Analyst*, 120 (1995) 2171.
280 B. Pérez Cid, A. Fernández Alborés, E. Fernández Gómez and E. Falqué López, *Anal. Chim. Acta*, 431 (2001) 209.
281 B. Pérez Cid, I. Lavilla and C. Bendicho, *Anal. Chim. Acta*, 378 (1999) 201.
282 A. Chaterjee, *Talanta*, 51 (2000) 303.
283 Questron Corporation, P.O. Box 2387, Princeton, NJ 08543, USA.
284 Floyd Associated Inc., Lake Wyle, SC, USA.
285 Parr Instrument Company, 211 Fifty-Third St., Moline, IL, USA.
286 Prolabo, 12 rue Peleé, 75011, Paris.
287 S.D. Chen and E.M. Chiu, *Food Sci. Agric. Chem.*, 1 (1999) 264.
288 H.D. Isengard and M. Walter, *Lebensm. Unters. Forsch.*, 207 (1998) 377.
289 C.C. Grimm, S.W. Lloyd, R. Batista and P.V. Zimba, *J. Chromatogr. Sci.*, 38 (2000) 289.
290 M. Zhu, F.J. Avilés, E.D. Conte, D.W. Miller and P.W. Perschbacher, *J. Chromatogr.*, 833 (1999) 223.
291 C.C. Grimm, S.W. Lloyd, P.V. Zimba and M. Palmer, *Am. Lab.*, 32 (2000) 40.
292 S.W. Lloyd and C.C. Grimm, *J. Agric. Food Chem.*, 47 (1999) 164.
293 L. Joergensen and H.N. Thestrup, *J. Chromatogr.*, 706 (1995) 421.
294 E. Marconi, G. Panfili, L. Bruschi, V. Vivanti and L. Pizzoferrato, *Amino Acids*, 8 (1995) 201.
295 H. Engelhardt, M. Krämer and H. Waldhoff, *Chromatographia*, 30 (1990) 523.
296 S.T. Chiou and K.T. Wang, *J. Chromatogr.*, 491 (1989) 424.
297 C. Woodward, L.B. Gilman and W.G. Engelhart, *Int. Lab.*, 20 (1990) 40.
298 A. Peter, G. Lans, D. Tourwe, E. Gerlo and G. Van Binst, *Peptide Res.*, 6 (1993) 48.
299 M. Reichelt, C. Hummert and B. Luckas, *Chromatographia*, 49 (1999) 671.
300 J. Thuery, *Microwaves: Industrial, Scientific and Medical Applications*, Artech House, Norwood, MA (1992).
301 L.P. Kok and M.E. Boon, *Microwave Cookbook for Microscopists: Art and Science of Visualization*, 3rd edn, Coulomb Press, Leyden, Netherlands (1992).
302 K.R. Foster and A.W. Guy, *Sci. Am.*, 255 (1986) 32.

Chapter 6

High-pressure, high-temperature solvent extraction

6.1. INTRODUCTION

Over the past few years, a technique based on the use of solvents at a high pressure and/or high temperature without reaching the critical point has gradually emerged as an efficient means for increasing automatability, shortening process times and reducing the amounts of solvent required to digest or leach solid samples. The use of this technique has grown dramatically in recent years, as shown by the more than 150 papers published in barely six years (from 1994 to 2000). Such features as the low consumption of organic solvents and the expeditiousness of determinations make this technique especially suitable for environmental analysis, the area where it has expanded to the greatest extent.

Extraction with an aqueous or organic solvent at a high pressure and/or temperature can be made by using the *static mode*, the *dynamic mode* (the solvent is circulated in a continuous manner through the sample) or a combination of both.

Ever since the inception of this technique, there has been much confusion regarding the name it should be given. In 1996, Richter *et al.*, of Dionex Corporation, published a paper in *Analytical Chemistry* called "Accelerated solvent extraction: a new tool for the environmental field", where they reported a new extraction methodology based on the use of organic solvents at a high pressure and temperature as a means for accelerating and increasing the efficiency of leaching processes [1]. The previous year, however, these authors had published two studies where they used the technique to extract PCBs and PAHs [2], and organophosphorus pesticides and various herbicides [3], from soil. Also, several 1994 internal documents of Dionex Corporation already referred to the extraction of BNAs (bases, neutrals and acids) [4], and of organophosphorus pesticides, chlorinated herbicides and PCBs [5], using a technique that the firm then called "accelerated solvent extraction" (ASE). Simultaneously to the previous studies, in September 1994 a research group at the North Dakota University headed by Professor S.B. Hawthorne published a paper in *Analytical Chemistry* [6] about the potential of using water at a high pressure (50 bar) and temperature (250°C) for extracting compounds of different polarity based on changes in the dielectric constant of water with variations in temperature; the authors referred to this technique as "subcritical water extraction". The essential differences between the method of Hawthorne's group and that under development by Dionex at the time were the use of water instead of an organic solvent as leaching agent, and the manner in which the extraction proper was conducted. While the prototype developed by Dionex — which was subsequently made commercially available — performed static extraction (i.e. with no circulation of the solvent across the sample), the equipment used by Hawthorne *et al.* included a modified supercritical fluid extractor and the extraction was conducted in a dynamic manner with a view to comparing subcritical and supercritical water extraction — the latter required the use of corrosion-resistant equipment, which was more expensive.

In principle, the commercial success of the Dionex proprietary model led many authors to adopt the term "accelerated solvent extraction" (ASE), which was also patented by the firm. So much so, that US EPA itself adopted it in Method 3545 [7], endorsed in July 1994, to refer to the extraction of volatile and semi-volatile compounds from soils, clays, sediments, sludges and waste solids by use of solvents at high pressures and temperatures. The virtually exclusive use of this term in the first few years was largely the result of the sole commercially available extractor for this purpose being that manufactured by Dionex. With time, however, alternative names such as *pressurized fluid extraction* (PFE), *pressurized hot solvent extraction* (PHSE), *high-pressure solvent extraction* (HPSE) and *subcritical solvent extraction* (SSE), among others, have gradually replaced ASE, a commercial designation that bears no relationship to the foundation of the technique. In fact, in revising Method 3545 in December 1996, the EPA substituted ASE by PFE, which was confirmed in the November 2000 update of the method.

The use of different names for the technique can lead to confusion as it may leave the impression that the names refer to different techniques rather than a single one involving extraction with solvent at a high pressure and temperature. However, authors tend to use different names for the static and dynamic modes, even though the names provide no clue as to which mode was used.

For consistency, we shall use *pressurized hot solvent extraction* (PHSE), which is the most universally accepted designation, throughout this chapter unless the specific mode involved is stated. Adopting PFE was discarded because it might have induced even greater confusion; in fact, using the word "fluid" rather than "solvent" extends the scope to supercritical fluid extraction, which is also conducted at a high pressure. One term that accurately reflects the foundation of the technique is *high-pressure, high-temperature solvent extraction*, which is the title of this chapter; however, the resulting acronym (HPHTSE) is too complicated and, therefore, impractical. Although static pressurized hot solvent extraction (SPHSE) is the most suitable designation, ASE is used here to refer to the static mode because it is the acronym most widely employed at present and will therefore be much more familiar to readers; in addition, the commercially available extractor that bears this name can only perform static extractions. The dynamic mode uses water as extractant in most cases, so authors tend to refer to it as "subcritical water extraction". However, because some applications use alternative solvents, we shall henceforward refer to it as *dynamic pressurized hot solvent extraction* (DPHSE) in order to clearly state that it constitutes the dynamic mode.

This chapter discusses the principal aspects of the technique in its two modes, the devices typically employed by each and their amenability to coupling with subsequent operations of the analytical process. Also, the main analytical applications of both modes in analytical chemistry are described, and their advantages and disadvantages with respect to alternative techniques such as Soxhlet, ultrasound-assisted, microwave-assisted or supercritical fluid extraction, discussed.

6.2. VARIABLES AFFECTING THE EXTRACTION PROCESS

The influence on the extraction process of the different parameters that affect performance depends on a number of steps that govern the transport of the analytes from the matrix to the bulk fluid. Although some details remain to be worked out, a number of steps can be envisaged to control the extraction. Thus, the desorption of analytes from solid samples can be described in the light of a model consisting of three steps, namely: (1) *desorption* from a solid particle; (2) *diffusion* through the solvent located inside a particle core; and (3) *solubilization* into the bulk of the flowing fluid (see Fig. 6.1). Each step depends on a number of factors including temperature, pressure, solvent type, solvent volume, solvent flow-rate, matrix composition, sample size and extraction time, the effects and potential manipulation of which should be examined on a case-by-case basis.

6.2.1. Temperature

Temperature is the single most important parameter influencing the kinetics of mass transfer in PHSE or, in other words, the most important factor contributing to increasing

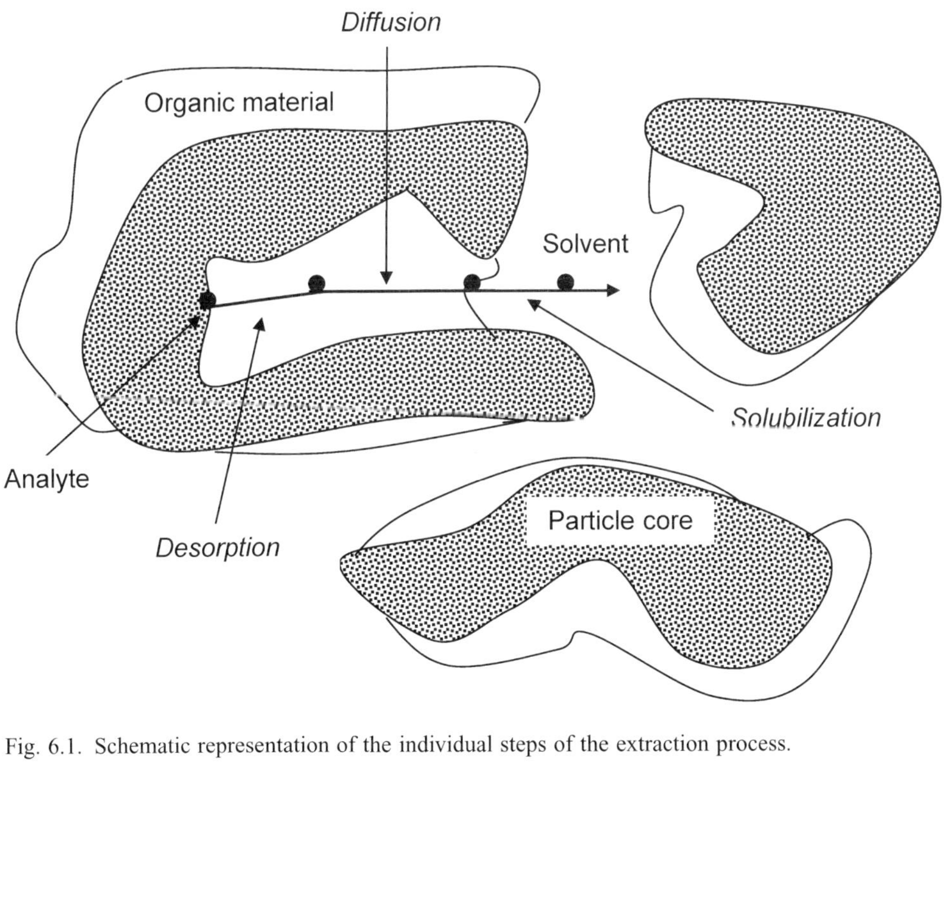

Fig. 6.1. Schematic representation of the individual steps of the extraction process.

the extraction efficiency in, not only PHSE, but also other extraction techniques such as SFE — where the dielectric constant of the leaching agent and hence its ability to determine some species changes considerably with temperature [8] — and microwave-assisted extraction (MAE) — where temperature is highly influential on reproducibility [9].

The use of high temperatures increases the *solubility* of solvents. Calculations of the temperature dependence of the solubility of an ideal solution allows the increase in solubility resulting from a given increase in temperature to be calculated. For example, the solubility of anthracene increases almost 13-fold as the temperature increases from 50 to 150°C. The solubility of hydrocarbons such as *n*-eicosane can increase several hundred-fold over the same temperature range [10]. The fact that the solubility of water in organic solvents increases with increasing temperature [11] is especially interesting in those cases where the extraction efficiency is decreased as a result of organic solvents at low temperatures and pressures being excluded from water-sealed pores containing the analytes. The increased solubility of water in organic solvents provided by high temperatures increases the availability of such sealed pores and of the analytes they contain as a result.

An increased temperature also has a favourable effect on *diffusion rates* and hence on the extraction efficiency. It is difficult to derive an exact relation for the effect of temperature on diffusion rates, especially with finite concentrations and a multicomponent system. In most systems, diffusion rates are estimated to increase by a factor of 2–10 on increasing the temperature from 25 to 150°C [12].

Temperature changes affect *surface equilibria*. In fact, they alter strong solute–matrix interactions caused by van der Waals forces, hydrogen bonding and dipole attractions of the solute molecules and active sites of the matrix, and can lead to disruption of the surface equilibrium in response to positive changes in temperature. Thermal energy can overcome cohesive (solute–solute) and adhesive (solute–matrix) interactions by decreasing the activation energy required for desorption. In addition, hydrogen bonds are weakened by increased temperatures, as noted by Pimentel and McClellan [13].

High temperatures also decrease the *viscosity* of liquid solvents, thus facilitating penetration of matrix particles and enhanced extraction. By way of example, the viscosity of 2-propanol decreases 9-fold as the temperature is raised from 25 to 200°C [12]. In addition to reducing viscosity, increased temperatures also decrease the *surface tension* of the solvent, solutes and matrix, thereby allowing the solvent to better "wet" the sample matrix. Both changes improve contact of the analytes with the solvent and hence the extraction efficiency. A decreased solvent surface tension also allows solvent cavities to form more easily [14], thus permitting the analytes to be more rapidly dissolved in the solvent [1].

Notwithstanding the foregoing, using a high temperature does not always result in increased extraction efficiency. Adverse effects are especially outstanding with thermolabile species; however, increased temperatures also reduce the *density* of the extractant, which can lead to a reduction of its extracting power. Thus, Schäfer *et al.* found the yield of extracted saturated and unsaturated fatty acids (FAs) from cereals to increase as the temperature was raised from 100 to 120°C; however, no further increase in efficiency was obtained by further raising the temperature. Rather, as the temperature increased to

150°C, FA extraction was slightly decreased. The decreased recoveries obtained were not a result of the thermal instability of the unsaturated FAs at 150°C as the yield of extracted saturated FAs decreased at least to the same extent as that of the unsaturated FAs. It seems more likely that the extraction efficiency started to decrease at temperatures above 120°C [15]. Whether thermolabile species are involved or an increased extractant density has an adverse effect on the extraction efficiency, intermediate temperatures should be used to ensure reliable extraction [16].

6.2.2. Pressure

A certain minimum pressure is required in PHSE to maintain the extractant in the liquid phase. Fortunately, the overpressure need not be too high. For example, 20 atm is sufficient to keep *n*-hexane (atmospheric boiling point 68.7°C) in the liquid state at 209°C. The minimum pressure required can be estimated from straightforward equations as described by Hass and Newton [17].

As shown in several studies, the pressure is usually a minor variable for the resulting efficiency in PHSE [18–20] provided the level used is high enough to maintain the solvent in the liquid state. In a study by Saim *et al.* [20], the total amount of PAHs extracted at different pressures (85 and 165 bar) with all other variables kept constant (120°C and 9 min static extraction) was similar (differences were within experimental error). In some cases, however, pressure can be a key to ensuring complete analyte removal. The use of high pressures facilitates extraction from samples where the analytes have been trapped in matrix pores. The pressure increment forces the solvent into areas of the matrices that would not normally be contacted by them under atmospheric conditions. For example, if analytes are trapped in pores, and water (or even an air bubble for small pores) has "sealed" pore entrances, then organic solvents may not be able to contact such analytes and extract them. The pressure increase (along with elevated temperatures and reduced solvent surface tensions) forces the solvent into the pore to contact the analytes.

The overpressure provides an advantage for samples in which the analytes are on the surface or in readily accessed pores, as seen in liquid chromatography. One good analogy is the comparison of standard column liquid chromatography to flash chromatography or even high-performance liquid chromatography. In those cases where the analytes occur mostly on the surface, the process of pressurized hot solvent extraction can be best described as adsorption chromatography. Pressurized flow aids in the solubilization of air bubbles, so the solvent comes in close contact with the entire sample matrix more rapidly [1].

Overpressure also has favourable effects in terms of the time needed to fill the extraction cell with solvent, especially for samples with small particle sizes [21]. However, the increased pressure can induce changes in the matrix surface resulting in a reduced extraction efficiency of analytes in some types of matrices [22–24]. Such is the case with the extraction of 4-nonylphenol from kaolin, which was quantitative at 100 atm at 100°C after 10 min of dynamic extraction, but required 30 min at the same temperature and 150 atm [25]. A similar effect was observed in the extraction of linear alkylbenzene-sulphonates from kaolin; as can be seen from Fig. 6.2, the time needed for quantitative

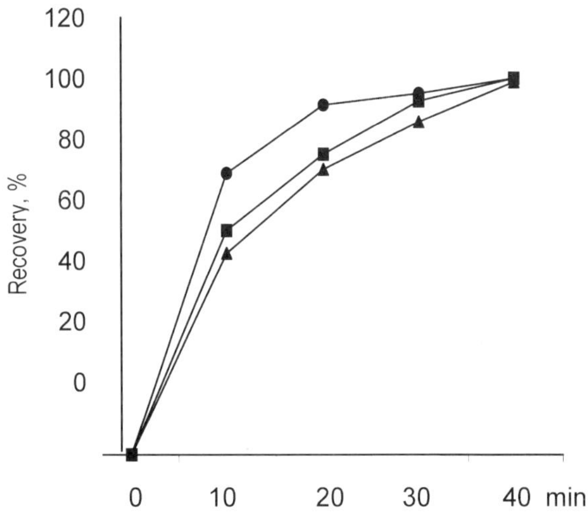

Fig. 6.2. Effect of pressure on the recovery of spiked linear alkylbenzenesulphonates, extracted from kaolin samples at a temperature of 100°C. (●) 100 atm, (■) 50 atm, (▲) 150 atm. (Reproduced with permission of Elsevier.)

recovery of the analytes decreased with increasing pressure from 50 to 100 atm (at a constant temperature of 100°C in both cases), but increased from 100 to 150 atm (to an extent that it surpassed the time needed for quantitative recovery at 50 atm) [25].

6.2.3. Solvent type

Optimal development of an extraction process entails an appropriate choice of solvent. The PHSE technique is compatible with a wide range of solvents others than those with an auto-ignition temperature of 40–200°C (e.g. carbon disulphide, diethyl ether, 1,4-dioxane) or a strong polar character (e.g. *n*-hexane) [20]. As a rule, strong bases and acids should also be avoided as solvents because they are corrosive — and their corrosion power increases with increasing pressure and temperature [26]. Transferring the extraction conditions used in conventional methods to PHSE usually does not require changing the solvent; however, the PHSE technique provides an opportunity for the use of a variety of solvents or solvent mixtures — even those that are not effective in conventional methods. This is done by adjusting the temperature and pressure during the process, which generally increases the solubilizing power of the solvent.

The static mode uses both organic solvents such as toluene [27], methanol [28] or acetone [29] and solvent mixtures (usually in a 1:1 ratio) including dichloromethane–acetone [20,28], acetone–hexane [30,31], heptane–acetone [31], acetone–isohexane [32] or methanol–water [33]. The use of mixed solvents as extractants provides improved extraction in terms of expeditiousness and recovery [20,28,30–35] as a result of the solubility parameter for a binary mixture being roughly proportional volumewise to the parameters of its components [36]. Thus, in the extraction of Irganox 1010 from polypropylene, the addition of 20% of cyclohexane to 2-propanol doubles the extraction

TABLE 6.1

MEAN RECOVERIES OF THIODIGLYCOL (TDG) FROM PORTON SOIL AS OBTAINED AFTER 1 DAY OF TREATMENT AT 150°C WITH A RANGE OF SOLVENTS FOLLOWING SPIKING OF THE SOIL WITH 10 μg TDG/g. STANDARD DEVIATIONS ARE GIVEN IN BRACKETS

Solvent	Recovery
Methanol–water (9:1)	8.89 (0.29) [(a)]
Acetone–dichloromethane (1:1)	5.45 (0.90) [(a)]
Acetonitrile	2.81 (1.15) [(a)]
Ethyl acetate	2.95 (0.58) [(a)]
Methanol	2.16 (0.51) [(b)]
Methanol–water (3:1)	4.97 (1.65) [(a)]
Water	1.10 (0.17) [(a)]

[(a)] $n = 5$ [(b)] $n = 4$

rate and enables quantitative extraction of the target analytes in less than 1 h as compared to 140 min with pure 2-propanol [34]. Also, in the extraction of atrazine and alachlor from aged contaminated soil, the use of a mixture of dichloromethane and acetone provides recoveries 3–30% higher for atrazine and up to 25% higher for alachlor than those obtained with pure hexane or methanol [28]. Finally, in the extraction of thiodiglycol from soil, the use of binary solvent mixtures increases the extraction efficiency by up to 65% (see Table 6.1). Water has rarely been used as extractant in the static mode [37,38]; however, it is the extractant most frequently used in the dynamic mode, with a variety of samples such as soils [39], foods [40] and plants [41], and analytes of widely variable polarity including acid herbicides [42], oils [43], PAHs [44], PCBs [6] or carbamates [45].

The use of modifiers occasionally improves the extraction process. Water as extractant can be modified with organic solvents such as methanol, acetone or acetonitrile in low proportions ($<$ 5%) in order to decrease its dielectric constant — and hence its polarity — without the need for a drastic temperature increase [37]. Also, an acid or base can be used to alter the pH in those cases where it significantly influences the extraction yield [29,46]. On the other hand, surfactants facilitate the extraction of non-polar compounds by formation of micelles [47]. Modifiers are less frequently used with extractants other than water. One example is the addition of sodium acetate to methanol to extract organotins (OTs); the additive increases the efficiency in two ways, namely: (a) acetate ion by complexing OTs and (b) sodium ion through cation exchange of OTs sorbed to the clay fraction of sediments [21].

6.2.4. Solvent volume

The amount of solvent required for extraction strongly depends on the extraction mode used. As noted earlier, PHSE can be conducted in the static mode, the dynamic mode or a combination of both. In static PHSE (or ASE), the sample is extracted with a minimum volume of solvent (< 15 ml) with no outflow [1]. Once extraction reaches equilibrium, the analytes are rapidly collected by flushing the extraction cell with solvent and an inert gas; as a result, the matrix contains a residual amount of the original analyte content that depends on its partition constant. When the partitioning equilibrium is slightly displaced to the solubilization of the analytes in the extractant, a larger volume of solvent is required — which, in the static mode, is limited by the volume of the extraction cell. When the solvent volume used in the static mode does not ensure quantitative extraction of the target analytes, several static extraction cycles or the dynamic mode must be used. In dynamic PHSE (DPHSE), the extractant continuously flows through the extraction cell, so the volume of solvent that comes into contact with the sample is a direct function of the flow-rate of the circulating extractant and the extraction time. Obviously, the dynamic mode uses larger volumes of solvent than the static mode, so it is less well-suited to trace analysis [18] — however, as shown below, the dilution effect can be minimized in various ways.

6.2.5. Solvent flow-rate

The flow-rate at which the extractant is circulated through the extraction cell holding the solid sample is a characteristic variable of the dynamic extraction mode. The effect of the flow-rate on the efficiency and expeditiousness of DPHSE can be used to determine whether the extraction is limited by analyte solubility in the extractant, diffusion in the particles and/or transfer from the particle surface to the extractant. If the extraction recoveries do not change when the flow-rate is increased, then the extraction is limited neither by solubility nor by the equilibrium of mass transfer between the matrix and extractant, so the rate-determining step of the process is diffusion inside the solid particles. In such a case, the extraction rate must be increased by raising the extraction temperature. If the limiting factor is analyte solubility, then doubling the flow-rate or the amount of fluid used doubles the amount of analyte that is extracted during the same time. On the other hand, if the analyte undergoes different re-adsorption/desorption steps during elution from the extraction chamber, doubling the flow-rate also doubles the extraction rate [48].

The flow-rates used in most reported PHSE methods range from 0.5 to 3.0 ml/min. The use of very low flow-rates (< 0.5 ml/min) is not recommended because it may easily lead to blockage of the restrictor used to maintain the pressure in the system; on the other hand, flow-rates above 3 ml/min increase dilution of the analytes in the extracts and this causes difficulty in subsequently determining the extracted species [18].

6.2.6. Matrix composition

The composition of the sample has a dramatic influence on extraction. Solid samples can differ significantly in their physico–chemical properties, the type of compounds they contain and their degree of granulation (particle size). These factors affect sorption and retention of analytes.

The influence of particle size on PHSE yields depends on the particular variable that governs the extraction efficiency. Obviously, if the PHSE rate is determined by diffusion of the analytes, it can be greatly increased by decreasing particle size [18]. However, the risk of losing volatile and reactive analytes during grinding is higher with samples consisting of particles less than 2 mm in size [49,50]. These two opposing effects must thus be taken into account in reducing the particle size of samples to be extracted.

Other variables with a direct effect on PHSE efficiency include the following:

(a) Aging of the sample. The influence of the solvent on the extraction efficiency has occasionally been found to change with the sample's aging time. Thus, in the extraction of pesticides from soil, no appreciable difference in terms of extraction efficiency for atrazine and alachlor was observed among dichloromethane–acetone, hexane and methanol two weeks after application to the soil, when the residues were still relatively fresh. However, following "aging" for 8 or 26 weeks, pesticide recovery was much higher with dichloromethane–acetone and methanol than with hexane, with no substantial difference between the former two solvents. For instance, after 26 weeks of incubation, the recovery of atrazine and alachlor with hexane was only 80% of that obtained with dichloromethane–acetone [28].

(b) The moisture content of the sample. It is advisable to use dry samples or place a desiccant together with the sample in the extraction cell in order to avoid any adverse effects of moisture on analyte recovery [26,51]. The effect can also be minimized by using a high temperature. For example, in a study on the extraction of hydrocarbons from dry and wet solids, dichloromethane–acetone, hexane–acetone and heptane–acetone provided essentially equivalent results at the different temperatures used for dry soils; with wet soils, however, dichloromethane–acetone and hexane–acetone performed worse than heptane–acetone at 125°C. In any case, quantitative recovery of all hydrocarbons was achieved with the three solvents at temperatures above 150°C [31].

(c) The addition of a dispersant, which can improve the extraction efficiency by decreasing sample compaction, thereby facilitating contact between sample and extractant. Thus, in the extraction of additives from polymers, mixing the ground polymer with sand resulted in faster extractions at higher temperatures and with stronger solvents when the polymer was softened to a greater extent. If no sand was used, particles partially coalesced and extraction rates were reduced [52].

6.2.7. Sample size

Sample size is one other variable to be considered in undertaking PHSE. Usually, sample sizes for PSE range from 0.5 to 10 g. Obviously, the amount of sample used must be large enough to ensure homogeneity and obtain adequate sensitivity for trace analyses. However, large samples require also large amounts of solvents for quantitative extraction and are much more likely to clog the system or its restrictor [18]. Usually, a small sample size (< 5 g) is preferred so as to avoid compaction of the sample in the extraction cell — particularly in the dynamic mode, where the solvent is continuously circulated, so the sample tends to build up at the cell outlet and can clog the system, thereby producing a high overpressure or even causing the cell to burst. On the other hand, the smaller the amount of sample used is, the greater is the cell volume that can be occupied by the extractant and hence the higher is the extractant/analyte ratio and the more markedly shifted is the partitioning equilibrium towards solubilization of the analytes — which facilitates their extraction. Sample size should therefore be as small as possible, provided sample homogeneity and sensitivity are still adequate.

6.2.8. Extraction time

Extraction times in PHSE are very short relative to those required by conventional solid–liquid extraction techniques such as Soxhlet extraction, and depend on which particular process determines the extraction rate. In the static mode, 5 min is often sufficient to achieve complete recovery (particularly with spiked samples); this is the recommended time in EPA Method 3545 for the extraction of volatile and semi-volatile organic compounds such as organophosphorus and organochlorine pesticides, chlorinated herbicides, PCBs, polychlorodibenzodioxins, polychlorodibenzofurans and diesel range organics (DROs) from soil, clay, sludge, sediment and waste water [7]. However, with complex matrices and/or very strongly retained analytes such as PAHs in bituminous fly ash, the extraction should be continued over 30 min in order to release high-molecular weight target analytes [53]; the need for extended extraction has been ascribed to the fact that such analytes are very strongly adsorbed on carbon-rich samples. The extraction of native (not added) PCBs from sediments and sewage sludge [54,55] requires two 5-min static extraction cycles. As much as 14% of some PCBs is extracted in the second cycle [54]. Also, optimum extraction of semi-volatile organics such as chlorobenzenes, hydrocarbons, pesticides, PCBs and PAHs from pine needles and moss requires three consecutive 10-min static extraction steps [56]. However, recovery in the static mode is not always quantitative, especially with a single step since, as noted earlier, the species to be extracted partition between the extractant and matrix, so, depending on the partition coefficient for the system in question, the process will be more or less quantitative. The dynamic mode, where the extractant is continuously circulated through the sample, held in the extraction chamber, at a preset flow-rate over a given period, is more efficient as the sample is continuously in contact with fresh extractant. Dynamic extraction times can be as long as 90 min in highly unfavourable cases such as the extraction of selenium, mercury and arsenic from coal [45]; by contrast, the extraction of PAHs and selenium from soil requires barely 15 min [6,57].

6.3. ACCELERATED SOLVENT EXTRACTION

Static pressurized hot solvent extraction (SPHSE), which shall henceforward be referred to as "accelerated solvent extraction" (ASE) for the reasons stated above, is the less flexible PHSE mode in terms of alteration or coupling to other techniques but is so far the more widely used — in fact, it accounts for over 65% of the PHSE publications reported since 1994. This is mainly the result of the sole commercially available extractor (the Dionex 200 model) implementing the static mode alone and also of the large number of studies conducted by different or even the same authors on the same analytes in the same matrices, which have therefore contributed little or nothing new in this area [58–63].

For example, from 1995 to 2000, more than 15 published studies on ASE involved the extraction of PAHs from soil or sediments in a virtually identical manner.

This section discusses the principal aspects of the ASE technique, which include the steps of the extraction process, the devices used, their amenability to coupling to other operations and the most salient applications reported so far (with special emphasis on those that have introduced some innovation). In the final sub-section, ASE is compared with alternative leaching techniques.

6.3.1. Steps involved in the ASE process

The ASE process normally involves the following six steps (Fig. 6.3):

(a) Loading the sample into the extraction cell.
(b) Filling and pressurizing the cell with solvent.
(c) Heating the cell at the selected temperature for equilibration at a constant pressure.
(d) Static extraction at a constant pressure and temperature over a predetermined period.
(e) Transferring the solvent to the collection vial and flushing the cell with fresh solvent to clean the sample or repeat the static extraction.
(f) Purging residual solvent from the sample to the collection vial using a suitable gas (usually nitrogen).

Prior to loading into the extraction cell, the sample is often pretreated in some way. For solid samples, the pretreatment usually involves drying: whether plain air-drying for 28–54 h [20,27,37,64] or freeze-drying [15,33,38,65]. Drying the sample is important since moisture in it may diminish the extraction efficiency. As in Soxhlet extraction and SFE, the addition of sodium sulphate [19,30,49] or an alternative desiccant such as Extrelut particles [51,66] is recommended in handling large amounts of water. Its efficiency in ASE, however, is yet to be established. The drying step is often followed by sieving [20,37] or grinding [15,38] of the sample.

Sample loading has been done in various ways, but some similarities between applications exist. Thus, in order to prevent clogging of the metal frit at the outlet of the extraction cell, the frit can be covered with a cellulose filter [28,67] or a small amount of celite [15]. Some authors recommend filling up the cell's dead volume with an inert matrix to ensure proper sample–solvent contact, to reduce solvent consumption and to prevent the cell from clogging. A number of inert matrices have been used for this purpose including diatomaceous earth [15], anhydrous sodium sulphate [19], glass fibre [31], high-density glass beads [33,65], sand [1,68] and hydromatrix [56,64]. The last also helps reduce the amount of residual moisture in the final extract [28]. Other materials such as Al_2O_3, both alone and in mixtures with Florisil, allow the extraction cell to be filled up and preliminary clean-up to be performed [27,69]. Recently, the possibility of in situ derivatization of the analytes during extraction has been demonstrated by placing the sample and a suitable derivatizing reagent together in the extraction cell [67,70–72].

Filling and pressurization of the extraction cell are done once the cell is loaded and tightened. Extractions can be performed either by preheating the cell before filling with

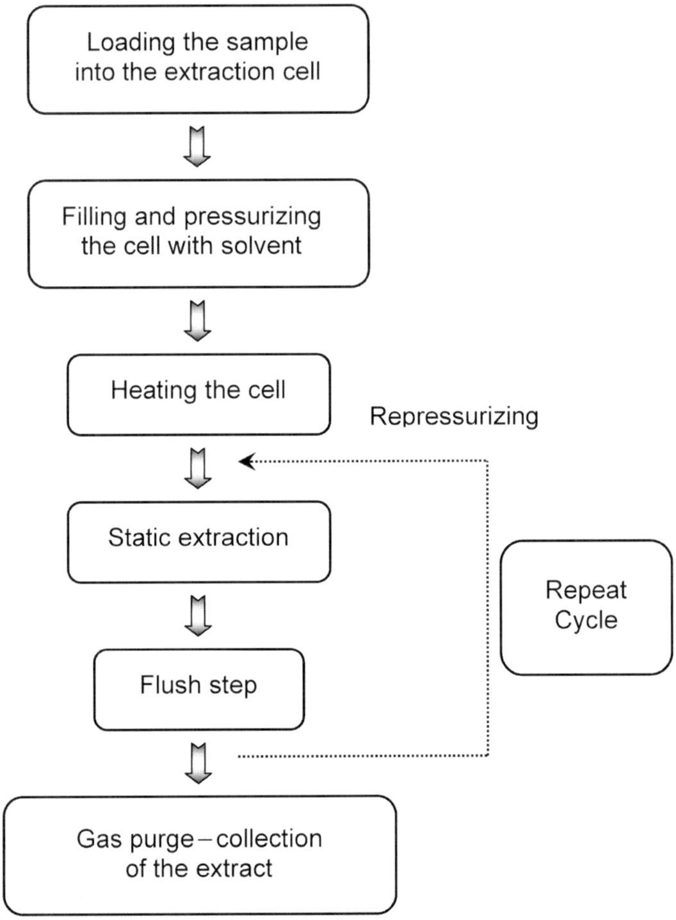

Fig. 6.3. Steps involved in an ordinary ASE procedure.

the solvent (preheat method) or by filling it with solvent before heating (prefill method). Only the prefill method is normally used, however.

Heating of the cell at the selected temperature for equilibration at a constant pressure is done in accordance with the temperature to be used in the extraction and with how rapidly the oven can reach it. Usually, 5 min suffices to equilibrate the system.

Static extraction is developed after filling the cell with solvent and reaching the preset temperature and pressure. During it, the analytes are released from the sample to the extractant.

The *transfer step* begins immediately after the static extraction step finishes. The pressure valve is opened and the solvent is allowed to flow to the collection vial. Cooling the vial is probably unnecessary as it appears to result in no difference in terms of recovery or precision. A filter can be placed in the bottom of the extraction cell to avoid collecting suspended powders in the vial [30,73].

Purging the solvent residue entails passing fresh extractant and an additional gas (after the last extraction cycle) through the cell to collect residual extract in the cell. A "rinse-solvent" volume corresponding to 60% of the empty extraction cell volume has been found to effectively avoid carry-over between extractions [74]. The extracts in the collection vials are subsequently processed in accordance with the particular determinative technique to be used.

The earliest applications of ASE involved static extraction over a preset period; however, as stated above, not always was quantitative extraction of the target species achieved with a single extraction cycle, which entailed using the dynamic mode [46,75] or performing several static extraction cycles under the same [21,33] or different conditions. For example, the use of extractants of variable dielectric constant allows the sequential extraction of species of variable polarity in different static extraction cycles [76].

6.3.2. ASE devices

The basic design of the devices used to develop an ASE process as reported in the first study using the Dionex® ASE 200 extractor [2,3], and also in some later investigations conducted using laboratory-built systems [1,18,77], is shown in Fig. 6.4. As can be seen, the equipment required to implement ASE consists of seven basic units, namely:

(a) A *reservoir* for storing the liquid extractant.
(b) A *high-pressure pump* for aspirating the extractant from the reservoir into the extraction cell.
(c) An *electrically heated oven* for placing the extraction cell.
(d) A stainless steel *extraction cell* for holding the sample.
(e) A *pressure valve* for producing and keeping an appropriate pressure in the system.

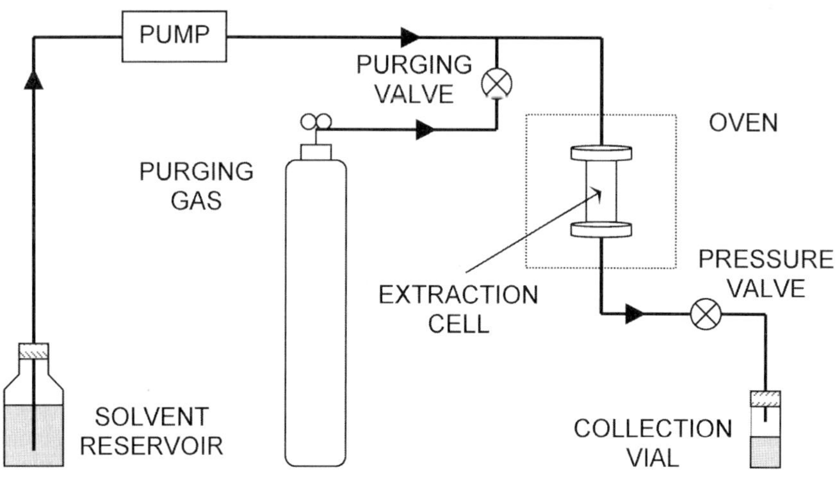

Fig. 6.4. Schematic diagram of an ASE system

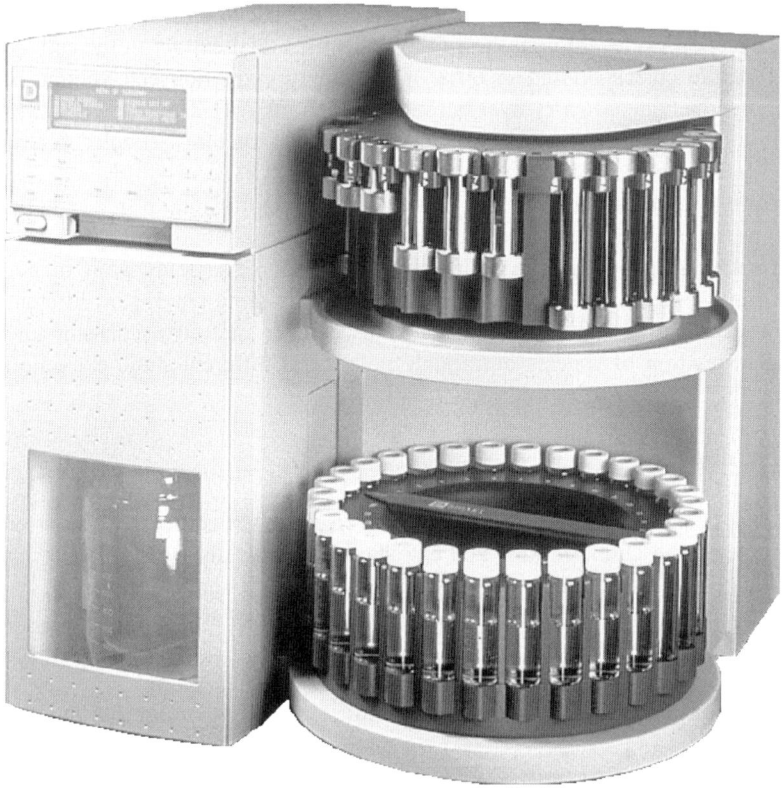

Fig. 6.5. Commercial extractor from Dionex Corporation (ASE 200 model). (Reproduced with permission of Dionex Corporation.)

(f) A *cylinder of gas* (usually nitrogen) for purging the system after extraction.
(g) A *collection vessel* (usually a vial).

Until recently, only two different ASE models were commercially available, namely: the Dionex® ASE 200 and ASE 300, both of which are based on the design by Richter *et al.* [2]. The main difference between these two extractors is the amount of sample they can process, which is in the region of 10 g in the former and as much as 100 g in the latter. The ASE 200 model (Fig. 6.5) was launched simultaneously with the inception of the technique as its developers were collaborators of the Dionex Corporation. The good results obtained and the wide scope of application to solid and semi-solid samples initially envisaged have been confirmed by the large number of papers published on the subject and of extractors currently in use in both research and, especially, routine laboratories.

Both extractors include a cell carousel holding up to 24 cells in the ASE 200 model and 12 in the ASE 300 model. Samples are sequentially processed in a fully automatic manner, using extraction cells of 1, 5, 11, 22 or 33 ml volume with the ASE 200, and of 34, 66 and up to 100 ml in the ASE 300 — depending on the amount of sample to be treated. Once prepared, samples are loaded into the fingertight, stainless steel extraction cells, which are placed on the cell carousel. The carousel rotates each sample cell into position for transfer to the oven chamber. The cell is then transferred to the oven and automatically sealed under pressure. Then, it is filled with solvent, pressurized and heated. After it reaches the set temperature, it is held in the oven, at a constant temperature and pressure, for a user-set time. The analytes and solvent are collected in the vial, and the cell is then flushed and purged with nitrogen gas. Finally, the cell is returned to the carousel and the next sample processed.

Recently, the firm Applied Separation (Allentown, Pennsylvania), in cooperation with the Institute of Analytical Chemistry of the Academy of Sciences of the Czech Republic, developed commercial equipment similar to that from Dionex; unlike this, however, it does not use a carousel to process samples in a sequential manner, but rather handles up to six samples simultaneously.

In addition to the previous three commercial extractors, some authors have developed custom models [78,79], adapted in most cases from a supercritical fluid extractor [25,58,80]. For example, Heemken *et al.* altered a Suprex SF extractor for use in ASE [81]; they disconnected the syringe pump from the CO_2 cylinder and filled it with a suitable ASE solvent. The restrictor was replaced with a stainless steel capillary tube leading into the trapping vial and an additional nitrogen pipe was installed at the inlet valve of the extraction vessel for purging after extraction. In the extraction of PAHs from soil [79], a custom extractor and commercially available equipment provided equivalent results; on the other hand, in the extraction of benzene and toluene from soil [78], the former provided even better results than the latter.

6.3.3. Combination of ASE with other operations of the analytical process

Accelerated solvent extraction as implemented in commercial equipment is basically discrete in nature, so it is rarely coupled to other operations of the analytical process. In fact, only in two reported applications was the static mode coupled on-line to other operations such as chromatographic separation, preconcentration and detection. Both used custom extractors as the compact design of the commercial models precluded their adaptation.

ASE–HPLC

Figure 6.6 depicts the experimental design used to couple a laboratory-made accelerated solvent extractor to a high performance liquid chromatograph with the aim of passing the extract through the analytical column in an automatic manner after static extraction at the preset temperature over the required time. This coupled assembly was used for the extraction and determination of benzaldehyde, benzene, toluene and ethyl benzene in soil, using water as both the extractant and the chromatographic mobile phase [82]. During the static extraction, water from P2 was continuously flowed through the injection valve (IV), the attached analytical column and, subsequently, the UV–Vis detector, in order to establish the baseline. After the extraction, the system was depressurized by opening the valves and the leachate was swept by water from pump P1 to the injection valve, in its

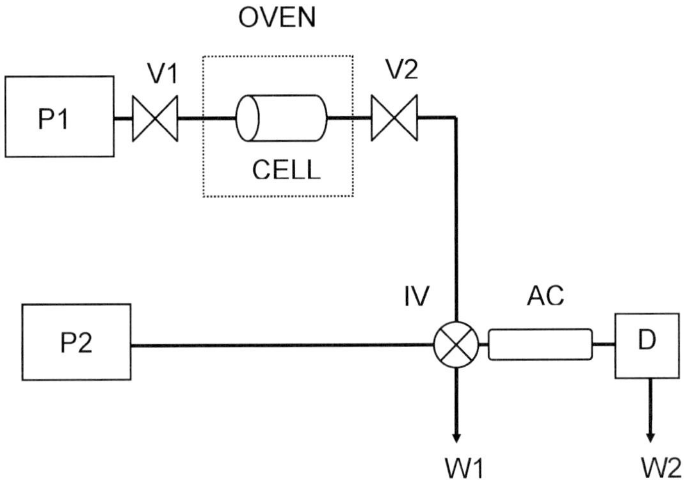

Fig. 6.6. Scheme of an ASE–HPLC set-up. P1 and P2 water pumps for the ASE and HPLC system, respectively; V1 and V2 needle valves; IV injection valve; AC analytical column; D UV–Vis detector; W waste. (Reproduced with permission of Elsevier.)

load position, so that a loop placed in the valve was filled with a fraction of the extract. After a preset time, the valve was switched to the inject position and part of the extract held in the loop of IV (10 μl) was inserted into the chromatographic column and swept by the mobile phase (water) from pump P2. In this way, the whole process was automated, so much so that the next sample in the sequence could be extracted simultaneously with the chromatographic analysis of that being processed at a given time. Obviously, because no preconcentration is done, the target species must be present at relatively high concentrations in the sample if this coupled technique is to be effective as a fraction of only a few microlitres is inserted into the chromatographic column.

ASE–Fluorimetry

This coupling was primarily used to develop an independent matrix extraction method allowing the extraction kinetics to be monitored with a view to determining the number of extraction cycles required to quantitatively extract the analytes irrespective of the kind of matrix concerned and of whether the species concentrations in it were known. The extract obtained after each cycle was driven to a graduated reservoir from which it was aspirated into a flow-injection manifold connected to a detector. This system was used for the extraction and monitoring of PAHs in soil, sediment and food, using a fluorimetric detector [83]. The intrinsic fluorescence of each analyte was monitored by injecting the extract into the carrier at a preset time, the emission signal thus obtained being proportional to the concentrations of the analytes extracted during the cycle. As stated above, this allows one to identify the time when extraction is complete, irrespective of the type of sample concerned or the species concentrations in it.

6.3.4. Applications of ASE

Accelerated solvent extraction has so far been exclusively applied to solid samples, especially in environmental analysis — which accounts for over 80% of its reported uses. In recent years, however, ASE has increasingly been used in the pharmaceutical and industrial fields (especially for the extraction of additives from polymers) as well.

This section discusses the principal applications of the ASE technique to different types of matrices and analytes.

Soil and sediments

Soil and sediments constitute the principal environmental target samples, and organic pollutants the main target analytes, for ASE, which, with a few exceptions that are described in the following section, provides results comparable to those obtained with other techniques such as Soxhlet, supercritical fluid and microwave-assisted extraction. Thus, ASE has been used for the extraction of PAHs [58–63] under the typical conditions required for quantitative recovery [viz. static extraction with 1:1 dichloromethane–acetone at 100°C for not more than 15 min]; that of PCBs [30,54,84] using acetone–hexane mixtures in variable proportions as extractants; those of hydrocarbons such as diesel range organics (DROs) [31,85], waste oil organics (WOOs) [31,85] and hexachlorocyclohexane isomers [73]; those of phenols [77] and chlorophenols [37] using superheated water at 125°C and 100 bar containing an organic modifier such as methanol, acetone or acetonitrile in a proportion of 3–5% — which proved more selective than extraction with water alone or a pure organic solvent; that of long-chain trialkylamines from sediments and sludges [86], where the concentrations found were an order of magnitude higher than those obtained with conventional methods (Soxhlet and ultrasound-assisted extraction) and 4–5 times the SFE values — as stated above, the efficiency with which thiodiglycol can be extracted [35] depends strongly on the particular solvent employed (see Table 6.1); and those of organochlorine pesticides and various herbicides [76,87–92].

The ASE technique has also been used for the extraction of inorganic compounds such as phosphorus in lake sediments [93], where 90 min of extraction provided results comparable to those obtained with conventional liquid–solid extraction for 3 days; that of heavy metals [92] with a view to assessing their mobility in solid wastes and soil; and that of organometals [21,94]. Especially prominent in relation to this last application is the extraction of organotin compounds, which provides valuable advantages over conventional solvent extraction methods (sonication and shaking) based on the use of acids and/or organic solvents of medium polarity; such advantages include reduced extraction times (from as long as 10 h when using acetic acid to a few minutes), preservation of the structural integrity of butyl- and phenyltin compounds — which are degraded by other techniques such as MAE — and quantitative extraction of compounds such as monophenyl- and diphenyltin, conventional solvent extraction of which would degrade them to inorganic tin.

In 1996, the US EPA adopted the ASE technique in producing Method 3545 [7] in view of the good results it provided at an early research stage. This official method is a procedure for extracting water-insoluble or slightly water-soluble semi-volatile organic compounds such as chlorinated herbicides, organophosphorus and organochlorine pesticides, PCBs, and polydibenzodioxins and polybenzofurans from soils, clays, sediments, sludges and waste solids. The method uses a temperature of 100°C and high pressures (1500–2000 psi) to achieve analyte recoveries equivalent to those from Soxhlet extraction but using less solvent and taking significantly less time. It has been validated for a number of analytes in different materials such as pesticides and PAHs in soils [28,49], and PCBs in both sediments [54,64] and sewage sludge [54], all with results similar to or even better than those provided by conventional methods such as Soxhlet or shaking extraction. However, the recommended conditions for implementation of this method (viz. 5 min of static extraction, 5 min of pre-heat equilibration and one static cycle) are not always sufficient to ensure quantitative extraction, particularly with natural or aged samples, which retain analytes more strongly than freshly spiked samples; this often requires altering the methods by using a static extraction time of 5–50 min [60,61,79], raising the extraction temperature to 150–200°C [78,84,88,95] or performing several extraction cycles — usually not more than three — [21,27,37,96,97]. In some cases, the extraction must be extended to as long as 90 min to ensure acceptable results [93]. Alternatively, one can perform sequential extractions with different solvents [93] or even combine ASE with SFE to accomplish the extraction of substances of rather different polarity [22,98]; such is the case with alkylphenols, alkylphenolethoxylates and linear alkylbenzenesulphonates from sediments, where ASE completely extracts linear alkylbenzenesulphonates but a second extraction step using CO_2 at 100°C at 450 atm is required for quantitative recovery of the other species [22]. Also, the extraction of hydrocarbons, phosphonates and phosphonic acids involves three steps, namely: one with supercritical CO_2 that extracts over 95% of the hydrocarbons; another with methanol-modified CO_2 that recovers the phosphonates; and a third with superheated water at 150°C at 100 bar that extracts polar phosphonic acids [98].

One effective alternative to ASE for the extraction of polar species is in situ derivatization by adding a suitable agent to the extraction cell or, directly, to the extraction solvent. This substantially improves the extraction of compounds such as 2-chlorovinylarsonous acid, which is derivatized with 1,2-ethanedithiol directly added to the extraction solvent [67] — by contrast, in SFE, the derivatizing reagent is added to the matrix before extraction, so the derivatization takes place in the supercritical fluid during a static step. Benzenediols [70] were subjected to in situ acylation at 120°C for 15 min using 2% v/v acetic anhydride in toluene in a single extraction cycle. Fatty acids [70] were converted into their methyl esters using boron trifluoride as derivatizing reagent; and acid herbicides [71,72] were derivatized to less polar esters by adding a solution of pentafluorobenzyl bromide directly to the extraction cell, thereby enabling not only the extraction of the target species, but also the subsequent determination of the derivatized acids in a direct manner using gas chromatography.

Recently, Hawthorne *et al.* coupled the ASE technique to solid-phase extraction using sorption discs that were placed in the extraction chamber together with the sample in order to simultaneously concentrate and extract polar compounds such as acid herbicides and their esters from soils, using strong anion-exchange (SAX) discs [99]. They also used this combined methodology to develop a convenient method of shipping extracts from field sites to the analytical laboratory [100]. In the former application, the esters of the acid herbicides were hydrolysed into their corresponding acids using water at 100–150°C and the extracted acids were trapped onto the SAX discs as the extraction cell was cooled. The use of discs in conjunction with the ASE technique allows one not only to concentrate or store analytes prior to analysis, but also to increase the extraction efficiency by displacing the analyte transfer equilibrium from the analytes to the solvent used as extractant by effect of extracted species being withdrawn from the solvent as they are retained on the discs.

Other environmental matrices

Accelerated solvent extraction has also been used to extract pollutants from matrices of environmental interest other than soil and sediments. Thus, it has been employed to extract dioxins, PAHs and chlorinated pesticides from fly ash [101,102], which requires using temperatures of up to 180°C to overcome their strong retention by the matrix; dioxins in urban dust [101,103] with results similar to those provided by Soxhlet extraction but in a shorter time (14 min versus 24 h) and using less solvent (15 ml versus about 100 ml); and PAHs in coal [104,105] with recoveries lower than those provided by SFE (particularly for low-molecular weight PAHs and fragrance materials in particulate matter [81,106]). The accelerated solvent extraction of PCBs from polyurethane foam air sampling cartridges [107,108] reduces solvent consumption by 80% and the extraction time by more than 13 h (30 min versus 14 h) compared to the Soxhlet method. Also, the ASE results obtained in the extraction of explosives such as 2,4-dinitrotoluene and diphenylamine from gunpowder were consistent with those of standard solid–liquid extraction techniques such as Soxhlet and ultrasound-assisted extraction [109].

Biological and pharmaceutical samples

The ASE technique has been used in the analysis of biological samples for two main purposes, namely: to determine contaminants in foods and animal tissues, and to extract target species from animal and plant tissues [110]. In the pharmaceutical field, ASE has also been used for two primary purposes, namely: to extract pharmacologically active substances from plants and, especially, for quality control of tablets and medical foods [111].

As regards food and animal tissue contaminants, ASE has been used in the determination of a number of organic and inorganic compounds in a wide variety of matrices including fish tissue, fruits and vegetables, and plant material. Thus, organic compounds such as *N*-methylcarbamates [51], organophosphorus [66] and organochlorine [112–114] pesticides, fungicides (extracted with methanol or acetonitrile at 100°C at 2000 psi) [115], dioxins [116] and PCBs [56,116,117] have been determined using ASE. The extraction of PAHs from real plant samples [56,118,119] required two steps with *n*-hexane and toluene at 40°C and 120°C, respectively, to obtain comparable or better extraction efficiency than that achieved with ultrasounds. Musk aroma compounds used in household chemicals and cosmetics [120]; and arsenicals such as inorganic species (arsenite and arsenate) and their methylated forms (monomethylarsonic acid and dimethylarsinic acid) [33,38,65,121] have also been extracted with ASE. As regards target species in foods, the ASE technique has been used to extract fat matter from cereals [15], meat [15,122], milk products [123], oilseeds [124] and fish tissue [125] using pure solvents such as hexane and petroleum ether or mixtures such as hexane–acetone and hexane–dichloromethane–methanol. In the extraction of flavanones and xanthones from root bark of the osage orange tree [126], similar results were obtained by using 45 or 35 min extraction with supercritical CO_2 or dichloromethane, respectively, at 80°C. On the other hand, in the extraction of flavours and off-flavours from fish, fruits, cheese and sausages, ASE provided results comparable to those of Soxhlet extraction but taking only 5% of the time and using only 10% of the amount of organic solvent, and with much less prominent matrix effects than in SFE [127].

The ASE technique has also proved an advantageous choice for the extraction of pharmacologically active compounds from plants, where it surpasses classical extraction methods such as those endorsed by pharmacopoeias (which are used as official standards for quality control of a large number of medicinal plants and rely on a broad range of techniques such as Soxhlet extraction, percolation, maceration, digestion, extraction under reflux and steam distillation) and others based on ultrasonication or turboextraction [76,128,129]. The ASE technique is a firm candidate for use in high-throughput screening programmes for natural product discovery, where large numbers of small-scale extractions have to be performed in an efficient, reproducible manner [76]. A number of ASE applications involving the extraction of pharmacologically active species from — mainly — plants [76,128,129] have been reported for substances such as hypericins, escin, silybin, curcuminoid and essential oils, and matrices such as St John's wort, horse chestnut seed, milk thistle fruit, tumeric rhizome and thyme. The amount of solvent used was reduced by a factor of 2 (e.g. horse chestnut seed) to 5 (e.g. milk thistle fruit), and time savings were significant (especially in those cases where the plant material required consecutive extractions with solvents of increasing polarity). Extraction of St John's wort, for example, was accomplished within less than 80 min instead of the 38 h required by pharmacopoeia methods.

Drugs [130] and vitamins [131] in medical foods; compounds such as felodipine [132] and diltiazem [133] in tablets; and corticosteroid residues in bovine liver [134] have also been extracted by ASE, with results comparable to those of conventional techniques such as Soxhlet, ultrasonic or supercritical fluid extraction.

Polymers

The additive content of polymers must be known for quality and regulatory reasons. The additives are usually extracted from the polymer prior to analysis. Traditional methods such as Soxhlet extraction, boiling under reflux or dissolution of the polymer followed by re-precipitation are often highly time- and solvent-consuming, and difficult to automate. Accelerated solvent extraction has been used to extract various additives from different types of polymers, all with good results and the intrinsic advantages of this technique, which include high throughput, solvent economy and the production of extracts that require no filtration [135,136]. Thus, ASE has enabled the efficient extraction of flame-retarded thermoplastic additives [135]; plasticizers such as dioctyl adipate, trioctyl phosphate, dioctyl phthalate and trioctyl trimellitate [137]; and antioxidants such as

Irganox and Irgafos [50,138,139] from various polymers including poly(vinyl chloride) (PVC), polyethylene and polypropylene.

One key aspect in using ASE to extract additives from polymers is choosing an appropriate solvent. Vandenburg *et al.* used Hildebrand solubility parameters as guides for solvent selection in ASE [140]. The Hildebrand parameter, δ, is defined as the square root of the internal energy of vaporization divided by the molar volume and is referred to as the "cohesive energy density". Broadly speaking, a polymer will be more soluble in a solvent if the solubility parameters for the two are similar. The larger the difference in solubility parameter between the solvent and polymer is, the higher is the temperature at which the polymer becomes soluble [141]. The Hildebrand parameters for a number of solvents and polymers are widely available from a variety of published sources [36,142].

In early studies [18,140], the extraction rate was found to increase with increasing temperature for each solvent until melting or dissolution of the polymer occurred. This was a result of increasing temperatures both increasing diffusion coefficients and causing swelling of the polymer. For extractions from polypropylene (PP), while the temperature was below the softening point, solvents with closer solubility parameters to the polymer gave faster extractions, presumably because they swelled the polymer to a greater extent [140]. However, the solvents with the closest solubility parameter to PP (viz. cyclohexane and chloroform) dissolved or fused the polymer, thereby causing maximum swelling, at a moderate temperature. The dissolution temperature, and hence the maximum degree of swelling, occurred at a higher temperature with a weaker solvent. Because the diffusion coefficient increased with increasing temperature, extraction from the polymer at a higher temperature was faster on equal degrees of swelling. For solvents with a solubility parameter widely different from that for PP (viz. acetonitrile), the rate was still fairly low at temperatures where the additive, Irganox 1010, was likely to have decomposed (150°C). The optimum extraction solvent (2-propanol) caused significant swelling at 150°C, without dissolving the bulk of the polymer. However, identification of the optimum solvent from theoretical considerations alone is not easy; in fact, although the extraction rate increases with increasing proximity of the solubility parameter to that for PP, the relationship is not linear [141].

6.3.5. Comparison of ASE with other solid–liquid extraction (leaching) techniques

Accelerated solvent extraction, like many other recent techniques, has frequently been compared with well-established, other solid–liquid extraction techniques in order to assess its potential in various fields and identify its advantages and disadvantages relative to existing alternatives.

It has for some time been a common practice to compare new extraction techniques with Soxhlet extraction, and ASE has been no exception in this respect. The ASE technique has many advantages over Soxhlet extraction, especially prominent among which are the following: reduced operating times and consumption of organic solvents, and partial automatability. In many cases, ASE provides results similar to (e.g. in the extraction of organophosphorus pesticides, organic matter, hydrocarbons and PAHs from soil

[2,3,31,59,85,87,91]) or even better than those obtained with Soxhlet extraction (e.g. in the extraction of trialkylamine and munitions residues from sludges and sediments [86,143]).

Reduced extraction time is one of the most salient advantages of ASE over the Soxhlet technique; the former typically takes 10–30 min compared to 8–24 h for the latter [73,107,109]. Solvent consumption is also much lower in ASE, which uses amounts up to 40% smaller than Soxhlet extraction [16]. Automatability is one other asset of ASE; in fact, commercially available ASE equipment allows every extraction step to be automated and pre-programmed extraction sequences to be developed. As noted earlier, ASE recoveries are usually similar to or better than Soxhlet recoveries. Worth special note in this context are the substantially improved recoveries of high-molecular weight PAHs from environmental reference materials [19] and of hydrocarbons (HCHs), pesticides (DDX), chlorobenzenes (Cl-Bz), PCBs and PAHs from real contaminated soil, where Hubert *et al.* obtained recoveries 10, 19, 5, 23 and up to 24 times higher with ASE than with Soxhlet for 24 h (see Table 6.2, where *f* represents the ratio of the ASE-extracted amount to the Soxhlet-extracted amount of each substance) [27]. In fact, PCBs went undetected with Soxhlet extraction. This study revealed that, even though Soxhlet extraction continues to be the reference method, it does not always provide quantitative recoveries — particularly with natural samples containing strongly retained analytes.

On the other hand, the use of accelerated solvent extraction as recommended in EPA Method 3545 (i.e. using a single, static extraction cycle) does not always provide better or even similar recoveries to those obtained with the Soxhlet method [144]. Such is the case with the extraction of DDT and its metabolites DDE and DDD, where, as can be seen from Table 6.3, ASE recoveries from aged contaminated soil were much lower than Soxhlet recoveries [144]. This was ascribed to the fact that, in Soxhlet extraction, a large volume of fresh organic solvent is continuously recycled through the sample, whereas, in a single ASE cycle, the sample never comes into contact with new portions of fresh solvent; as a result, if the partitioning equilibrium of the analyte between the matrix and solvent is not strongly displaced to the solvent, then a single extraction cycle does not suffice to completely remove the analyte. In order to determine whether solvent recycling was the reason for the apparent poorer ASE recoveries of DDT, DDD and DDE from aged contaminated soil, the influence of the number of static cycles performed was examined. Based on the results, three static cycles were required to quantitatively remove DDT, DDD and DDE from the samples; the results thus obtained were consistent with those provided by Soxhlet extraction (see Fig. 6.7). Therefore, the number of cycles to be used is one other variable to be studied in developing a new ASE method and comparing it with its Soxhlet counterpart.

The ASE technique has also been compared — to a lesser extent, however — with ultrasound-assisted extraction (USE) and found to provide better [49,59,81] or at least similar results [31,85,109] in most cases. The advantages of ASE over USE are similar to those of the former in relation to Soxhlet extraction, i.e. increased throughput, automatability and decreased solvent consumption. While ASE extraction times are also short relative to USE, this is not the most salient advantage as USE times are also fairly short. The reduced consumption of solvents is indeed a major advantage, particularly in cases such as the extraction of organic pollutants from marine particulate matter, where

TABLE 6.2

COMPARISON OF THE EXTRACTION EFFICIENCIES OF ASE AND SOXHLET EXTRACTION IN SOILS FROM SOUTHERN RUSSIA (CONCENTRATIONS ARE GIVEN IN ng/g DRY MASS)

Pollutant group	Technique	Soil 1	Soil 2	Soil 3	Soil 4	Soil 5	Soil 6	Soil 7	Soil 8	Soil 9
Σ HCHs	ASE	4.24	0.83	1.43	9.76	11.6	2.65	22.5	1.23	5.46
	Soxhlet	0.83	1.23	1.01	0.97	3.37	0.82	4.08	0.52	0.72
	f	5.1	0.7	1.4	10.1	3.4	3.2	5.5	2.4	7.6
Σ DDX	ASE	5.90	0.70	0.41	3.30	13.2	9.36	157	1.47	1.35
	Soxhlet	0.65	0.23	0.16	0.17	0.78	3.32	19.7	0.08	0.60
	f	9.1	3	2.6	19.4	17.0	2.8	9.9	18.4	2.3
Σ Cl-Bz	ASE	0.56	0.40	0.05	0.18	0.09	0.14	0.25	nd	nd
	Soxhlet	0.11	0.16	0.11	0.06	0.13	0.13	0.17	0.4	0.11
	f	5.1	2.5	0.5	3.0	0.7	1.1	1.5	—	—
Σ PCBs	ASE	0.69	0.87	0.34	77.5	0.69	0.32	0.92	0.08	0.71
	Soxhlet	nd	nd	nd	nd	0.03	nd	0.19	nd	0.20
	f	—	—	—	—	23.0	—	4.8	—	3.6
Σ PAHs	ASE	219	6.87	22.6	149	42.6	22.8	183	5.68	9.73
	Soxhlet	26.1	10.3	5.52	6.32	8.10	9.38	66.9	2.86	4.05
	f	8.4	0.7	4.1	23.6	5.3	2.4	2.7	2.1	2.4

f represents the ASE-to-Soxhlet extracted concentration ratio; nd = not detected

TABLE 6.3

COMPARISON OF SOXHLET EXTRACTION FOR 24 H WITH ASE (1 BATCH CYCLE) OF AGED CONTAMINATED SOIL USING DICHLOROMETHANE AS EXTRACTANT

	DDE		DDD		DDT	
Parameter	ASE	Soxhlet	ASE	Soxhlet	ASE	Soxhlet
Mean, μg/g	10.9	13.1	40.0	96	226	421
RSD, %	5.7	5.9	5.0	3.9	8.1	10.7

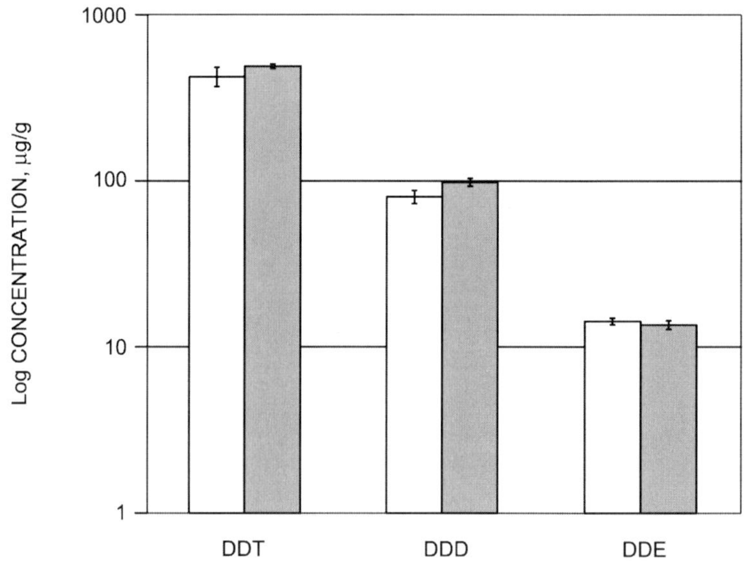

Fig. 6.7. Extraction of DDT and its metabolites from aged contaminated soil. Empty bars correspond to ASE (three static cycles) and shaded ones to Soxhlet extraction. (Reproduced with permission of Elsevier.)

the use of ASE saves 90% of the solvent volume used in ultrasonic extractions [81]. The reduced amount of solvent used results in shorter concentration steps or even none at all. As with Soxhlet extraction, ASE recoveries clearly surpass USE recoveries in some cases such as the extraction of semi-volatile organic compounds from the wax fraction of pine needles, where, as can be seen from Table 6.4, ASE provides recoveries up to 130 times as high [56].

The advantages of ASE over other techniques such as supercritical fluid extraction (SFE) and microwave-assisted extraction (MAE) are not so clear [16,60,62,116]. In fact, SFE features similar extraction times and uses little or no organic solvent. Also, SFE

TABLE 6.4

COMPARISON OF USE AND ASE FOR THE EXTRACTION OF SEMI-VOLATILE COMPOUNDS FROM THE WAX FRACTION OF PINE NEEDLES (CONCENTRATIONS ARE GIVEN IN ng/g DRY MASS)

Pollutant group	Technique	Soil 1	Soil 2	Soil 3	Soil 4
Σ HCHs	ASE	8.10	2.50	12.1	3.64
	Soxhlet	2.81	1.27	1.65	2.31
	f	2.88	1.97	7.33	1.58
Σ DDX	ASE	2.84	0.95	23.0	0.74
	Soxhlet	0.39	0.56	0.93	0.89
	f	7.28	1.70	24.73	0.83
Σ Cl-Bz	ASE	0.09	nd	nd	nd
	Soxhlet	nd	nd	nd	nd
	f	—	—	—	—
Σ PCBs	ASE	5.46	2.2	3.88	1.35
	Soxhlet	0.04	0.51	nd	0.66
	f	136.5	4.31	—	2.05
Σ PAHs	ASE	81.9	32.9	19.4	24.1
	Soxhlet	9.69	9.14	9.77	15.7
	f	8.45	3.60	—	1.54

f represents the ASE-to-Soxhlet extracted concentration ratio; nd = not detected

provides much higher preconcentration factors and is more readily coupled to other techniques [8]. On the other hand, ASE uses small volumes of organic solvents, so it is usually more economical as far as extractant consumption is concerned, and also less polluting — SFE uses high-purity CO_2, which is very expensive and a strong contaminant. In specific cases such as the extraction of polar compounds, where SFE is almost completely ineffective, ASE provides a valuable alternative [67,70–72]. Consequently, in comparing these two techniques with ASE, one should consider not only the previous aspects, but also those discussed below.

Of special importance in relative terms is the selectivity of an extraction method for target analytes in preference to bulk matrix organic compounds (e.g. humic matter from soil); in fact, the presence of co-extracted matrix organics frequently requires post-extraction clean-up steps before chromatographic analysis. In the extraction of PAHs from soil, the organic solvent extracts obtained by ASE or Soxhlet extraction were very dark and highly turbid (the colour of black coffee), while those obtained with SFE, which were collected in dichloromethane, were light yellow and clear, simply because pure CO_2 does not extract matrix organics as easily as organic solvents do [79]. Also, the organic solvent extracts yielded more artefact peaks in the GC–MS and GC–FID chromatograms

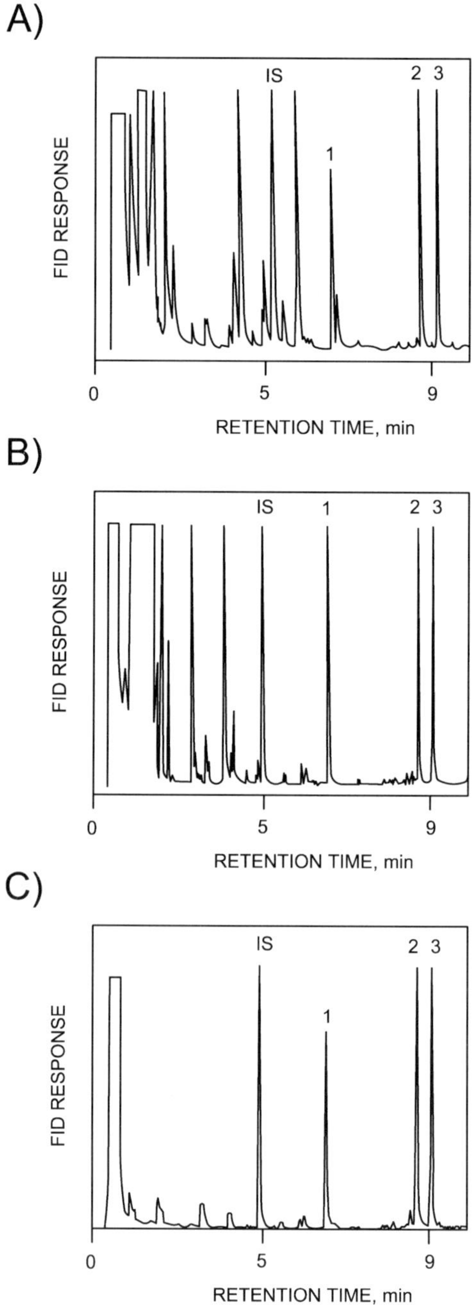

Fig. 6.8. GC–FID chromatograms for the early-eluting peaks for PAH-contaminated soil extracted using the (A) Soxhlet, (B) ASE and (C) SFE techniques. IS internal standard, 1 naphthalene, 2 2-methylnaphthalene, 3 1-methylnaphthalene. (Reproduced with permission of Elsevier.)

than did the supercritical CO_2 extracts (see Fig. 6.8) [79], which required clean-up of the organic solvent extracts and increased the risk of analyte losses [60].

Comparing microwave-assisted extraction (MAE) with ASE is as complicated as comparing the latter with SFE. In fact, the comparison cannot rely exclusively on recoveries as these are usually similar [30,62,73,114,116], so it must be made in other terms. For example, while ASE performs better than MAE and SFE in the extraction of hexaconazole from aged contaminated soils — the more aged the soil, the better — [68], MAE is to be preferred to ASE for the extraction of hexachlorocyclohexane isomers from soils when expeditiousness is crucial (ASE allows up to 12 samples to be processed simultaneously, which results in substantially decreased total analysis times) [73].

6.4. DYNAMIC PRESSURIZED HOT SOLVENT EXTRACTION

As can be seen from Fig. 6.9, dynamic pressurized hot solvent extraction (DPHSE) has evolved similarly to ASE; however, as noted earlier, DPHSE has been the subject of many fewer reports, primarily as a result of the lack of commercially available equipment for implementation. In any case, the relatively scant reported applications of DPHSE are of especial interest as regards automation of the analytical process; in fact, the dynamic nature of the system facilitates its coupling to other dynamic systems with a view to accomplishing preconcentration [39,42,45,145], filtration [42,45], chromatographic separation [145,146], derivatization [46,57] and detection [44,147], among others, and the partial or total automation of the analytical process.

This section discusses the more relevant aspects of this PHSE mode. The discussion begins with the equipment used to implement it, which is bound to provide an under-

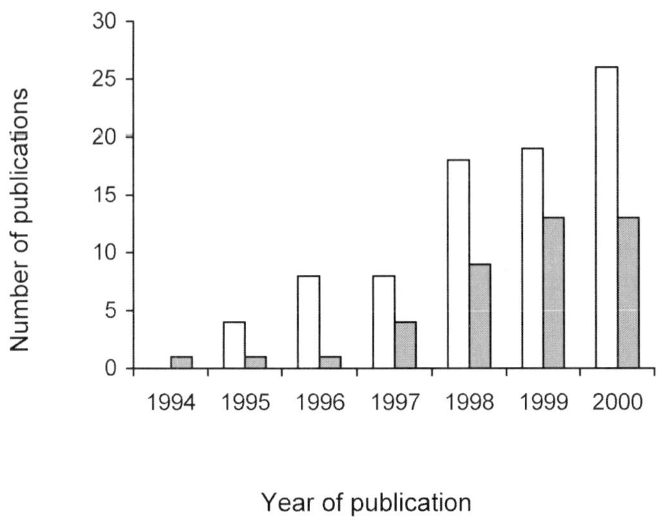

Fig. 6.9. Growth in the number of publications on ASE and DPHSE over the period 1995–2000. Empty bars correspond to ASE and shaded ones to DPHSE.

standing of the steps involved in the extraction process, and examines the principal advantages of using water — the solvent most widely employed in this context — as extractant. The main choices for correcting the dilution effect that arises from extraction under a dynamic regime — most of which rely on the ability to couple the extraction to subsequent operations of the analytical process — are also discussed. As with the static mode, the most salient reported applications of DPHSE are reviewed, and its advantages and disadvantages with respect to other leaching techniques are commented on.

6.4.1. DPHSE devices

Because of the above-stated lack of commercially available DPHSE equipment, applications of the dynamic extraction mode have all been developed using laboratory-built configurations that comprise the following basic elements:

(a) A *solvent reservoir* for storing the extractant;
(b) A *high-pressure pump* for propelling the extractant from the solvent reservoir to the extraction cell;
(c) An *electrically heated oven* for keeping the extraction cell at the desired temperature;
(d) A *pre-heater* located prior to the extraction cell for ensuring that the solvent is at the required temperature when coming into contact with the sample;
(e) An *extraction cell* for holding the sample;
(f) An *inlet valve* for performing static extractions and an *outlet valve* for both static extraction and producing the overpressure required at the beginning of extraction;
(g) A *cooling system* for cooling the extract as it emerges from the oven;
(h) A *restrictor* for keeping the pressure within the system at the preset level so that the solvent can be maintained in the liquid state at the operating temperature.

A schematic depiction of a typical dynamic extractor is provided in Fig. 6.10. The propulsion system can be a dual piston pump [43,57,148] or a syringe pump [18,82,149]. Dual piston pumps have the advantages over syringe pumps that they deliver a continuous supply of solvent (limited by the barrel size in syringe pumps) and allow easy solvent change-over. Unlike piston pumps, however, syringe pumps deliver a non-pulsating flow. All tubing and elements that constitute the system must be made of stainless steel in order to avoid corrosion by the solvents (particularly water) when operating at high temperatures. Special alloys such as hastelloid, which allows working temperatures as a high as 500°C to be used, have also been employed for this purpose [149].

The ovens used in DPHSE range from those for supercritical fluids [39,150,151] to gas chromatographic models [42,82] to specially designed units [75,146,148,152].

The extraction cell is preceded by a pre-heater intended to ensure that any solvent reaching the cell comes into contact with the sample at the oven temperature. Coils of variable length and empty extraction chambers have been used as pre-heaters. The presence of one is especially important with high extractant flow-rates: the length of the coil or the volume of the empty cell used should ensure that the solvent resides in them for long enough to reach the oven temperature before it enters the extraction cell.

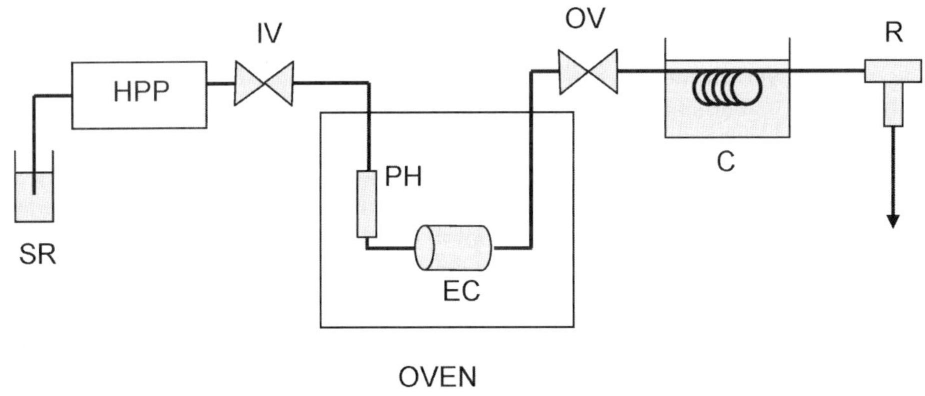

Fig. 6.10. Typical experimental set-up for dynamic pressurized hot solvent extraction (DPHSE). SR solvent reservoir, HPP high-pressure pump, IV inlet valve, OV outlet valve, PH pre-heater, EC extraction cell, C cooler, R restrictor.

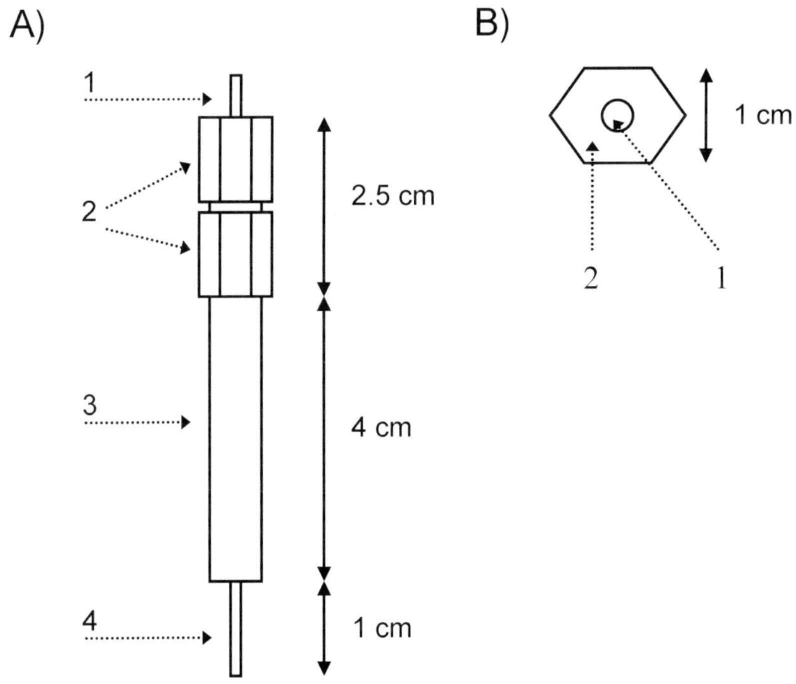

Fig. 6.11. Schematic diagram of a manual variable restrictor for coupling to an extraction system. (A) Side view. (B) Upper view. 1 1/16 in. OD stainless steel tubing, 2 adjustable parts, 3 restrictor body, 4 outlet part. (Reproduced with permission of Elsevier.)

The extraction cells used in DPHSE include those typically employed in SFE [109,151], empty HPLC columns [149] and specially designed stainless steel chambers [57,75,148]. The use of an SFE chamber as extraction cell is not recommended because the polyimide seals in it can fail at high temperatures [6]. On the other hand, the problem with empty HPLC columns is that they usually have small diameters and hence low volumes — this can result in excessive pressure build-up and bursting of the column. As a rule, DPHSE extraction cells consist of a stainless steel cylinder closed with screws at either end that permit the circulation of the leaching solvent through them. The screw caps also contain stainless steel filter plates to ensure that the sample remains in the extraction chamber.

The inlet valve need only be used if static extraction is to be performed; otherwise, only the outlet valve is required. Because the outlet valve is located outside the oven, some deposition of analytes in the valve may occur. However, the lack of detectable analytes in blank extractions and the quantitative collection of spiked analytes testify the absence of detectable analyte losses [6]. Normally, high-pressure needle valves [75,152] are used as inlet and outlet valves; six-way valves have been used for this purpose in some cases, however.

Extracts are usually cooled by using a water bath at ambient temperature [57,148]; however, when the risk of losing volatile species is high, the extract is normally cooled with ice [147,150].

A restrictor is placed at the end of the system to maintain the pressure inside the extraction cell during dynamic extraction. Fused-silica capillaries of variable inner diameter and length have been used as restrictors in DPHSE. However, some studies have shown that fused silica restrictors are subject to frequent plugging; this must be avoided by either keeping the restrictor at a high temperature (*ca.* 100°C) [39], which increases the risk of volatile losses, or by replacing it with stainless steel needle valves [149,150] or a custom restrictor [44,75], both of which provide good results. Figure 6.11 depicts a custom variable restrictor design. The inner diameter of the restrictor is hand-modified by using a pair of wrenches to screw or unscrew its upper part (adjustable pieces) in order to raise or lower the pressure in the system.

Recently, a laboratory-made miniaturized device for DPHSE was developed by Ramos *et al.* [153]. The most salient element of this device is a heatable 10 × 3.0 mm i.d. stainless steel holder that serves as the extraction cell (Fig. 6.12). Stainless steel frits are also used to prevent clogging of the exit tubing and valve by sample particles. The lower part is sealed by a laboratory-made, manually removed 5 μm stainless steel screen. Two PTFE rings are positioned at the top and bottom ends of the extraction cell that can be fitted to two adaptors for connection to standard nuts and stainless steel tubing. The two adaptors and the cell are pressed together to ensure leak tightness by tightening a large nut at the top of a cartridge. The extraction cell is surrounded by a stainless steel ring to which a resistive wire and a thermocouple are attached for heating and temperature control, respectively. Isolation is accomplished by using a ceramic ring around the stainless steel ring [154]. For direct control of the pressure inside the extraction cell, this is placed between a syringe pump (which is used to deliver the extractant) and an automated six-port valve. The tubing connected to the extraction cell is 0.13 mm i.d. and that connecting the valve port to the collection vial is 0.20 mm o.d. and 0.075 mm i.d. to improve heat dissipation before the solvent is collected.

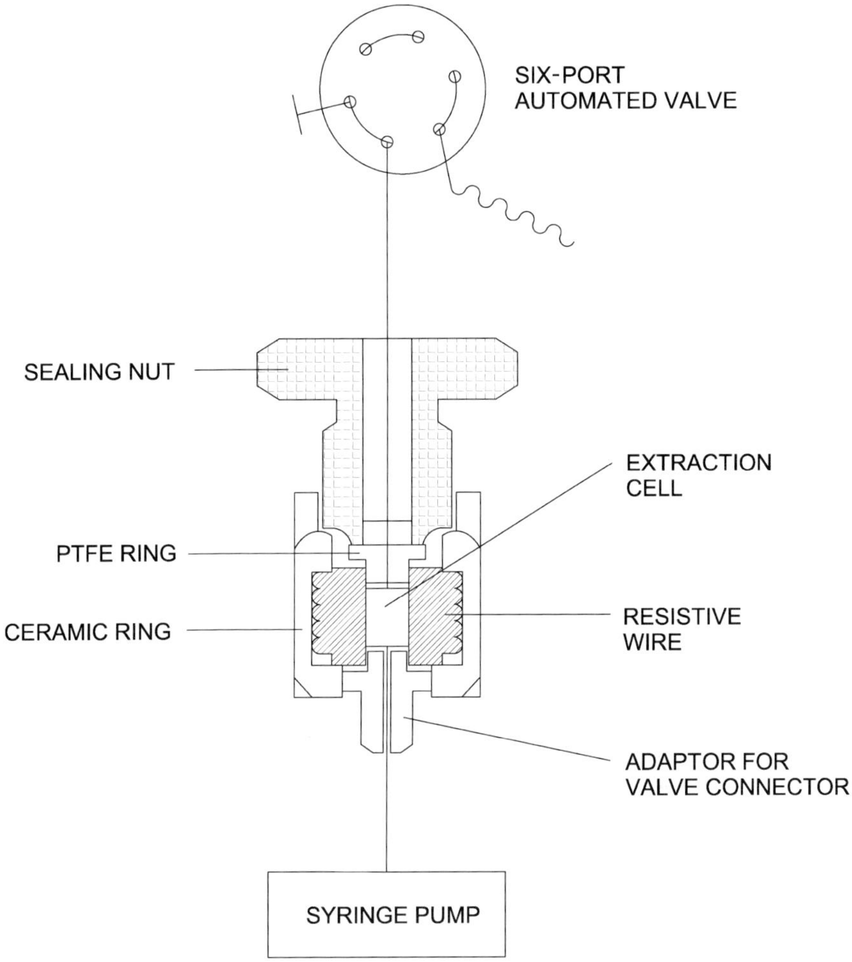

Fig. 6.12. Miniaturized device for DPHSE of solid and semi-solid samples. (Reproduced with permission of Elsevier.)

6.4.2. Steps of the DPHSE process

The steps involved in a dynamic pressurized hot solvent extraction process are similar to those of ASE, with only a few, slight differences — particularly at the last stages. Thus, the DPHSE process involves the following five steps (see Fig. 6.13):

(a) Loading of the sample into the extraction cell — jointly with a dispersant, if required;
(b) Filling and pressurizing of the extraction system with the leaching agent, propelled by a high-pressure pump;
(c) Pre-heating of the cell at a preset temperature and a constant pressure;

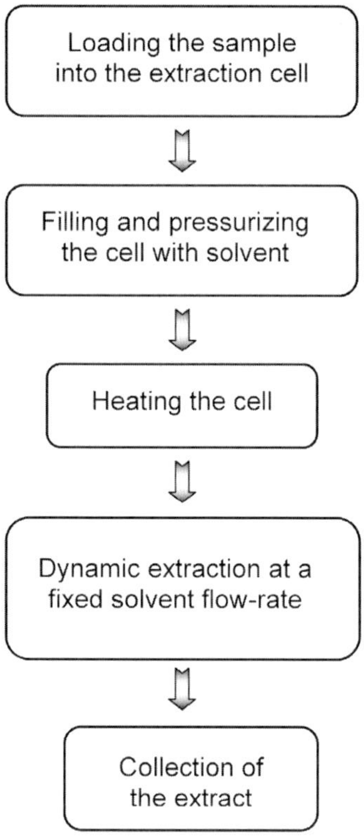

Fig. 6.13. Steps involved in a DPHSE procedure.

(d) Dynamic extraction at a constant temperature and pressure, using a fixed solvent flow-rate for a preset time;
(e) Collection of the extract as required for the subsequent step of the analytical process.

No purging of the system with a gas after extraction is required in this mode (Fig. 6.10) as the solvent is continuously circulated through the sample to sweep extracted analytes.

As in ASE, the sample can be subjected to various pretreatments prior to insertion into the extraction cell. Such pretreatments include air drying [39,47], oven drying [48], sieving [39,47] and grinding [18,75,155], which must be done immediately prior to extraction in order to avoid losses of volatile compounds, or specific procedures such as those involving the use of 1 M HCl for 2 h, which greatly improves the efficiency of dioxin extraction from fly ash [156], or freeze-grinding of polymers under liquid nitrogen and sorting of freeze-ground particles by sieving, which facilitates subsequent leaching of the sieved fractions separately in the extraction of additives from polymers [52].

The extraction cell is filled with sample, as in ASE. A fibreglass filter [34,43,157] can be placed between the sample and metal frits to avoid clogging of the latter. It should be noted that the solvent is continuously circulated in the same direction through the cell throughout the extraction, so a dispersant such as sand [34,149,158], diatomaceous earth [146] or glass beads [25,157] can help avoid compaction of the sample and facilitate penetration of the solvent through it. In loading the cell, the sample should be placed at the end where the solvent penetrates into the cell, thereby delaying compaction of the sample at the opposite end (that through which the leachate leaves the cell) for as long as possible.

Unlike ASE, because the leaching agent used in the dynamic mode is most often water, no desiccant need be added to the extraction cell together with the sample. As in ASE, however, additional materials such as surfactants [47] or sorption discs [39] can be placed in the cell to facilitate displacement of the analytes from the sample to the micellar medium or preconcentrate them in the sorbent, respectively.

Once the extraction cell is loaded, it is placed in the oven and the solvent is propelled through the system by a high-pressure pump. The extraction cell should be mounted vertically in the oven, with the solvent flowing from top to bottom so that any extracted analytes are immediately swept from the cell.

Before the oven temperature is raised, the system is pressurized by using an outlet valve in such a way that the flowing solvent, on reaching the valve — which is kept closed during this time, produces the overpressure required for the system to reach the oven temperature while maintaining the solvent in the liquid state. Once the preset temperature is reached, the outlet valve is opened and the solvent flows through the system for a preset time (dynamic extraction time). The effluent from the extractor is cooled — usually at ambient temperature — and collected as required for the subsequent step of the analytical process.

After extraction, the cell is replaced with a stainless steel connector that allows passage of the solvent to flush the system at a high flow-rate in order to avoid carry-over between samples.

In many cases, the extraction process includes an additional, static extraction step. To this end, the outlet valve (OV in Fig. 6.10) is supplemented by an inlet valve (IV in Fig. 6.10) between the high-pressure pump and the extractor. Once the system has been pressurized, the inlet valve is closed, the high-pressure pump stopped and the oven temperature raised. After the desired temperature is reached, the system is maintained under a static regime with both valves closed, and then the valves are opened and the pump restarted to allow the solvent to flow over the dynamic extraction period. Several studies [147,150,153] have shown that a combination of both extraction modes can result in substantially improved extraction and shorter extraction times. This is commented on in greater detail in discussing specific applications below.

6.4.3. Water as a leaching agent

The use of liquid water at a high temperature and pressure provides a promising alternative to other sample pretreatments as this is the ideal solvent for developing environmentally friendly methods. Water at a high pressure and temperature has traditionally been called "subcritical water"; however, the term "superheated water" is probably more accurate to refer to water heated above 100°C and subjected to the pressure required to keep it in the liquid state as subcritical water can be any water below the critical point

— whether or not it is heated above 100°C. We shall thus henceforward use "superheated water" to refer to water at a temperature above 100°C and "subcritical water" as a generic designation for water as an extractant — irrespective of its temperature.

The most salient feature of subcritical water as a leaching agent is the ease with which its dielectric constant, ε, can be altered. In fact, the dielectric constant of water can be changed over a wide range by changing the temperature under moderate pressure because it depends mainly on temperature and only slightly on pressure (the lower the pressure is, the lower is ε). At ambient pressure and temperature, water has a dielectric constant of *ca.* 80, which makes it an extremely polar solvent. This value, however, can be dramatically lowered by raising the temperature at a moderate pressure. For example, subcritical water at 250°C (and a pressure above 40 atm to maintain the liquid state) has $\varepsilon = 27$, which is similar to the values for ethanol ($\varepsilon = 24$) and methanol ($\varepsilon = 33$) at room temperature. This significant drop in dielectric constant allows hydrophobic compounds, which possess a limited water solubility at room temperature, to be more readily dissolved in water. An illustrative example of how effectively high-temperature water can increase solubility is provided by benzo[e]pyrene, the solubility of which increases from 4 ng/ml under ambient conditions [159] to approximately 10% (w/w) at 350°C and 100 bar (i.e. by a factor of 25 million) [160]. Also, because the dielectric constant of water decreases gradually with increasing temperature, and the dielectric constant is a measure of the solvating power of a solvent, considerable selectivity is gained through the use of water. This may be compared with supercritical CO_2 having a dielectric constant range of 1–1.6 [161]. One adverse feature of water is its increased corrosiveness at high temperatures and pressures, particularly in the supercritical state.

6.4.4. Automation and improvement of steps subsequent to DPHSE

The most analytically unfavourable feature of DPHSE is that analytes are diluted in the liquid extract, which requires concentration (usually by static liquid–liquid or solid-phase extraction). However, the increased flexibility of DPHSE relative to ASE has been used to develop various approaches to the partial or complete automation of methods based on the use of hyphenated techniques. Such approaches not only allow the dilution problem to be overcome but also enable automation and/or facilitate the development of other steps of the analytical process such as filtration, detection or chromatographic separation. The principal approaches to automation and improved implementation of analytical steps subsequent to DPHSE are discussed below.

DPHSE/liquid–liquid extraction

Two different approaches to the liquid–liquid extraction of leached analytes have been reported. One, which is discrete in nature, uses a vial containing the immiscible organic solvent, through which the aqueous leached phase is percolated. The other operates in a continuous manner, by percolating the effluent from a restrictor through a flow-cell located at the detector and filled with an organic phase that is lighter than water. The percolation of water emerging from the restrictor outlet through methylene chloride contained in a glass vial allows the partitioning of extracted analytes into the organic solvent. This method has been used for collection and concentration following subcritical water extraction of PCBs from soil and sediment [149]. The same system, with chloroform as collecting solvent, has been used for the subcritical water extraction of organic pollutants from environmental solid samples [6]. Continuous percolation of the solution emerging from the leaching chamber through a conventional photometric cell containing an immiscible solvent with a dissolved complexing agent (CCl_4–dithizone) was used as a collection–concentration system after DPHSE of iron from soil using superheated water [146], thus providing a means for indirect monitoring of the extraction kinetics.

DPHSE/solid-phase extraction

Solid-phase extraction (SPE) is currently the most widely used technique for preconcentrating analytes (particularly from aqueous solutions) on account of the high preconcentration levels it provides, its expeditiousness and its operational simplicity [162,163]. As stated above, the dynamic mode of PHSE (DPHSE) usually requires implementing a preconcentration step prior to the determinative step and, because the extracted analytes are dissolved in a liquid — usually aqueous — phase, SPE is a highly useful tool for avoiding the dilution effect. For this purpose, SPE cartridges [47] and columns packed with appropriate sorbents and coupled on-line to the extractor outlet [42,45] have been employed. A solid-phase trap fitted on-line with the extractor provides an excellent means for concentration prior to detection. The mini-column can be placed either in the transport tube or in the loop of an auxiliary injection valve of a flow-injection system, thus enabling elution in the opposite direction to retention. Analytes can be eluted manually, by disconnecting the column from the extractor [29], or automatically (either by passing a volume of organic solvent with the aid of a pump and collecting the eluate, which is made to volume in a vial [164], or by using a second injection valve to control the eluted volume to ensure reproducibility [42,45]). As shown in dealing with the DPHSE/SPE/HPLC coupling, the latter mode has been used for direct transfer of the eluate to the liquid chromatograph, also in an automatic manner. In both modes, the solid phase must be allowed to dry before the analytes are eluted; to this end, air or a nitrogen stream is circulated after the extract has been passed through the column.

Extraction discs placed directly over the exit frit of the extraction cell have been used for the preconcentration of highly water-soluble acidic compounds from soil [39]. A strong anion-exchange (SAX) sorption disc facilitated not only preconcentration but also derivatization of the analytes.

Miniaturized retention has been accomplished by using solid-phase microextraction (SPME) to collect PAHs from soil, air particulate matter and urban particulates [62,165], and pyrethrins from pyrethrum flower [165] in the fibre after quantitative extraction with superheated water [62,166].

DPHSE/filtration

One very common, undesirable occurrence in DPHSE is the presence of solid particles in the liquid extract due to oversaturation during cooling. This shortcoming can be circumvented by fitting a filtration unit in-line with the extractor; the unit can be placed either in the transport tube or in the loop of the injection valve — the latter choice allows the filter to be cleaned between samples by passing a rinsing solution at a high flow-rate in the opposite direction to filtration [42,45].

DPHSE/derivatization

When the extracted analytes provide no response on passing through a flow-cell located in a detector connected on-line to the continuous extractor via a dynamic manifold, the extractor outlet can also be connected to a manifold where the extract is merged with a stream of an appropriate reagent to derivatize the analytes as they are extracted, thereby enabling their subsequent on-line determination. This approach has been used to extract selenium from sand and sludge [57], and selenium, arsenic and mercury from coal. In the latter application, on-line derivatization with sodium tetrahydroborate (for selenium and arsenic) and tin chloride (for mercury) allowed the analytes to be determined in a direct manner using the atomic fluorescence technique [46].

DPHSE/SPE/HPLC

The DPHSE/SPE/HPLC coupling has been implemented with two different configurations. In one, the chromatographic mobile phase also acted as the eluent for retained analytes on the SPE column; in the other, elution was done by using an injection valve that allowed the eluent volume to be controlled in a reproducible manner [42,45]. In the latter configuration, the eluate was swept to the high-pressure injection valve in the chromatograph using air as carrier to avoid dilution. In both assemblies, the SPE column was placed in the loop of a six-way valve to allow continuous passage of the mobile phase through the chromatographic column and detector in order to establish the baseline; also, in the latter, elution was done in the opposite direction to retention to avoid changes in compaction of the sorbent material [165].

Integration of retention and determination

Sorbent materials can also be used in the flow-cell of a non-destructive detector for continuous monitoring of the retention step; this provides an indirect way of controlling the extraction kinetics. The kinetics can be monitored in a direct manner if the analytes possess appropriate intrinsic properties (e.g. in the extraction of pyrene from soils [44]) or following on-line derivatization (e.g. in the extraction of iron from soils [146] with photometric monitoring following complexation with SCN^-). In both cases, completion of extraction is signalled by the obtainment of a plateau (a constant signal) as a result of the constancy in the analyte content of the solid support. Once the signal levels off, the analytes can be eluted to waste — thus making the manifold ready for processing a new extract — or a portion of the extract can be inserted into a gas chromatograph for their individual determination as the sorbent is regenerated for a new extraction.

Other couplings

The potential of the DPHSE technique for coupling to subsequent operations of the analytical process is only limited by the analyst's ingenuity and material resources. The above-described systems can be combined by altering the sequence of steps and introducing appropriate modifications to develop fully automated systems for specific purposes. Thus, as many as four different steps have been coupled for the determination of pesticides in soil [42] and food [45] (DPHSE, filtration, SPE and HPLC) and that of Hg in soil [42] (DPHSE, SPE, derivatization and detection); in both, the analytical process was conducted in a completely automated manner.

6.4.5. Applications of DPHSE

Like ASE, pressurized hot solvent extraction has been virtually exclusively applied to solid samples. There is only a single reported use with liquid samples that involved altering the extractor and is dealt with separately at the end of this section on account of its innovative character. As in ASE, most applications are concerned with environmental samples; there are however, several interesting uses in the biological field. This section discusses the more interesting applications of DPHSE in terms of the matrix types and analytes involved.

Soil and sediments

As in ASE, more than one-half of existing publications on DPHSE focus on the extraction of organic pollutants from soil and sediments. Thus, pesticides such as herbicides [29,39,42,45,152,158] and insecticides [150] have been recovered using water as extractant and an extraction time shorter than 60 min in all instances. Because of the highly non-polar nature of PAHs, their quantitative extraction by DPHSE [6,47,147,151,167] requires using high temperatures (250°C), where the dielectric constant of water (ε = 27) is much lower than at ambient temperature (ε = 80) [6,147]; a micellar medium [47]; or an organic solvent such as methylene chloride [151]. The extraction of PCBs [145,149,168,169] also requires using water above 250°C since, although low-chlorinated PCBs such as mono-, tri- and tetrachlorobiphenyls are quantitatively extracted below 200°C, penta-, hexa- and heptachlorobiphenyls require temperatures from 250 to 300°C for complete recovery. Hydrocarbons such as benzene, toluene, ethylbenzene, naphthalene and xylenes [82,147], and brominated organic compounds [170], are extracted much more efficiently by DPHSE than by the Soxhlet technique. In fact, the latter could only extract (with recoveries lower than 36%) two of the five brominated compounds extracted by DPHSE; also, DPHSE provided cleaner extracts that required no subsequent clean-up. The dynamic technique also provided better results with surfactants such as alkylbenzenesulphonates and polyethoxy carboxylates [24,25,157,171].

The extraction of inorganic species such as Fe [146], and Cu, Pb, As, Se, Hg and Cd [172], from soils was investigated on a pilot plant scale with a view to developing an industrial-scale method for decreasing allowed levels of the metals used to manufacture cement. To this end, kinetic curves for the extraction rate were used that allowed the time needed to decrease the concentrations of these pollutants below the legal levels to be determined. The speciation of organic and inorganic selenium by continuous derivatization (hydride formation) and atomic fluorescence detection has also been reported [57].

In many of the previous applications, a combination of static and dynamic extraction was found to reduce the extraction time and provide better recoveries. Such is the case with the extraction of linear alkylbenzenesulphonates [24] and 4-nonylphenols [25] from sediments, and that of PAHs from soil [47], where a combination of 15 min static extraction and 10–20 min dynamic extraction ensured quantitative extraction of the analytes, thus minimizing the dilution effect arising from extended dynamic extraction.

Pure water at a high pressure and temperature was the solvent used as extractant in most applications. However, the addition of a modifier [157,173] or a co-extractant [47] can dramatically improve the extraction of some substances. Such is the case with the extraction of nonylphenol polyethoxy carboxylates from industrial and municipal sludges, where recovery was increased by more than 30% in the presence of 30% (v/v) ethanol in the water used as leaching agent [157]. Because of the hydrophobic nature of PAHs, the increased dielectric constant of water at a high temperature did not suffice to ensure quantitative extraction from soil. However, as can be seen from Fig. 6.14, the addition of a co-extractant (viz. sodium dodecyl sulphate, SDS, which forms charged micelles) dramatically improved the extraction of these hydrophobic compounds; also, it substantially reduced the extraction time and enabled the quantitative recovery of benzo(a)-acenaphthene [47].

Plant materials

The application of DPHSE to plant samples has revolved around the use of water as extractant for valuable essential oils from plants [41] such as fennel [43], marjoram [75], laurel [155], rosemary [174], oregano [175] and eucalyptus [176]. The isolation, concentration and purification of essential oils have been important processes for many years as a result of the widespread use of these compounds by the perfume, pharmaceutical, and human nutrition sectors. The use of water as DPHSE extractant allows oil composition to be altered by changing the extraction parameters. Thus, increasing the static extraction time increases the proportion of oxygenated compounds — the most odoriferous — in the extract, which are the main factors governing the quality of the oil; on the other hand, monoterpenes are extracted slowly and only very small amounts of sesquiterpenes, waxes and lipids are removed [155,174]. Compared to the traditional method based on hydrodistillation, the DPHSE method is more expeditious (1 h versus 4 h with hydrodistillation) and efficient (it provides higher yields for most compounds); also, it uses about twenty times less energy. This makes DPHSE an effective alternative to hydrodistillation on an industrial scale.

One of the few DPHSE applications using an extractant other than water is that involving the extraction of spice red pepper oil with subcritical propane for the determination of the carotenoid and tocopherol contents [177].

As regards foods, water has been used for the dynamic extraction of fungicides such as thiabendazole and carbendazim from fruits and vegetables, with recoveries of 80.9–100.5% at fortification levels of 0.14–100 ppm [40].

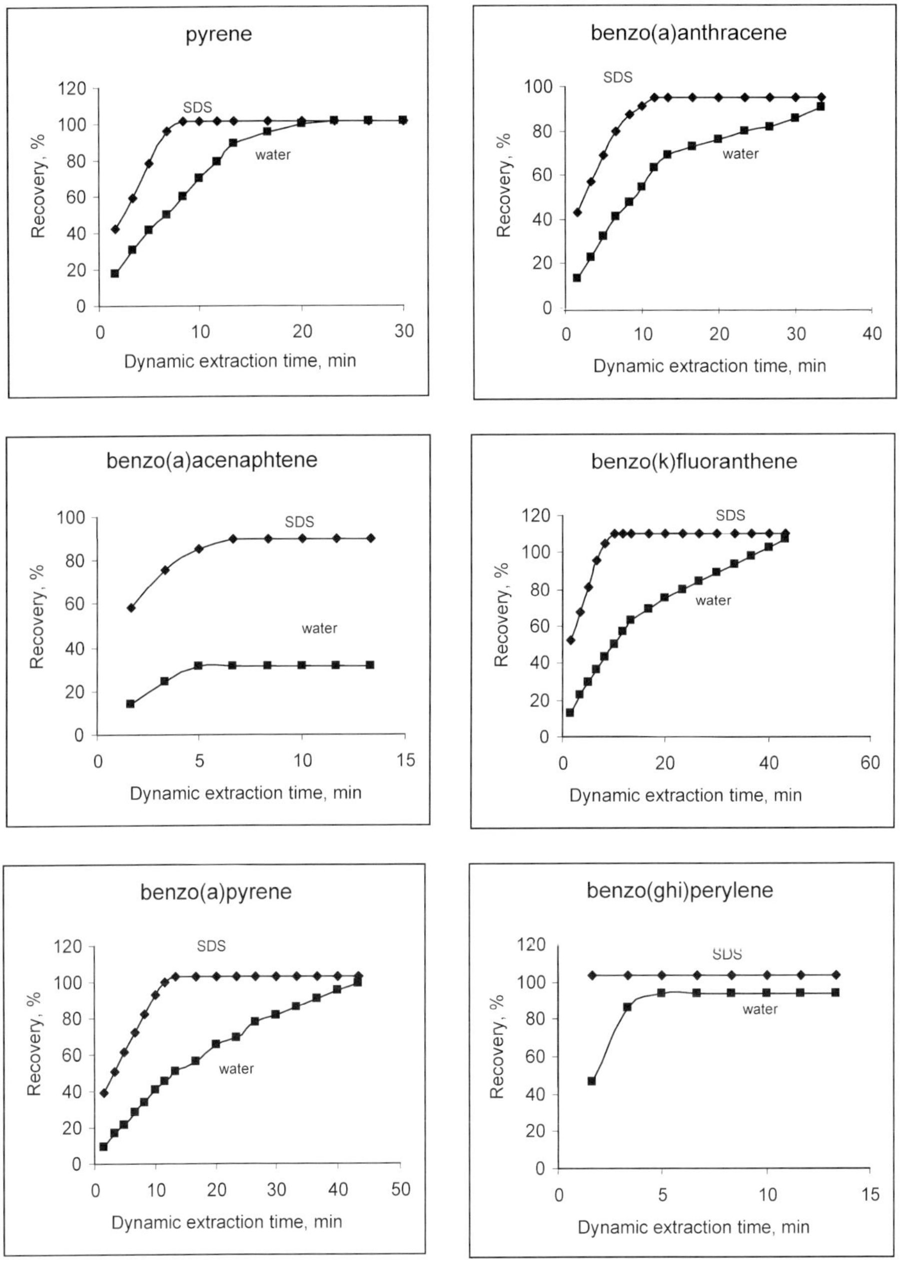

Fig. 6.14. Kinetic curves for six PAHs leached under optimum operating conditions by SDS–DPHSE and DPHSE. (Reproduced with permission of Elsevier.)

Polymers

Like ASE, DPHSE has been used to extract additives from various types of polymers including nylon [18,178], poly(1,4-butyleneterephthalate) [18] and polypropylene [34,52, 178], with good results in most instances. As in ASE, choosing an appropriate solvent and temperature is crucial with a view to ensuring quantitative extraction of the additives without dissolving or degrading the polymers. The solvents used to date for this purpose include hexane [18,178], propane [52] and 2-propanol [34], but not water.

Fly ash and coal

The DPHSE technique has also been used for the determination of organic pollutants and metals in fly ash and coal, respectively. The extraction of dioxins [48,179] and PAHs [180] from fly ash was accomplished with toluene [48,180] or a toluene–methanol mixture [179], with results as good as or even better than those provided by Soxhlet extraction for 24 h. On the other hand, the extraction of major ash-forming elements (Fe, Al, Ca, Mg, Na and K) [148] and minor inorganic pollutants (As, Se and Hg) [46] from coal was done with acidified water. In the latter case, a combination of static and dynamic extraction was found to provide quantitative recoveries within a shorter time and with less dilution of the extracts than dynamic extraction alone. Acidified water is more corrosive than pure water, so the high temperatures required for extraction (150–200°C) call for the use of an extractor made of a material more corrosion-resistant than steel: hastelloid. However, in proportions above 4%, nitric acid — the acidulant most frequently added to the water — has been found to result in clogging of the system and the restrictor, so the recommended acid concentration is much lower than that.

High-pressure liquid–liquid extraction

The use of subcritical water to extract compounds from liquid matrices has so far been the subject of a single application, namely: the demetallation of used industrial oils [181]. The experimental set-up employed is depicted in Fig. 6.15. The dynamic system consisted of a merging point beyond which alternate segments of the aqueous extractant (water acidified with 4% HNO_3) and the industrial oil were formed that circulated through the mixing coil — used in place of the extraction cell — where metals were transferred to the aqueous phase following oxidation by the nitric acid in it. The use of high temperatures was found to increase the oxidizing power of the acid. The foundation of this process is the same as that of a conventional continuous liquid–liquid extraction except that it is conducted at a high pressure and temperature. The used industrial oil and extractant are continuously delivered at the required pressure by a dosifier piston pump capable of propelling up to 10 litres of oil per hour and a high pressure pump, respectively. Metals (Cu, V, Pb, Ni, Cd and Cr) are oxidized at the water–oil interface and extracted into the aqueous phase. After passing through the coil in the oven, the water–oil system is cooled and collected in a vessel where both layers are spontaneously separated. A residence time in the mixing coil of 4–5 min provides oil samples demetallized to an extent of 75–98%.

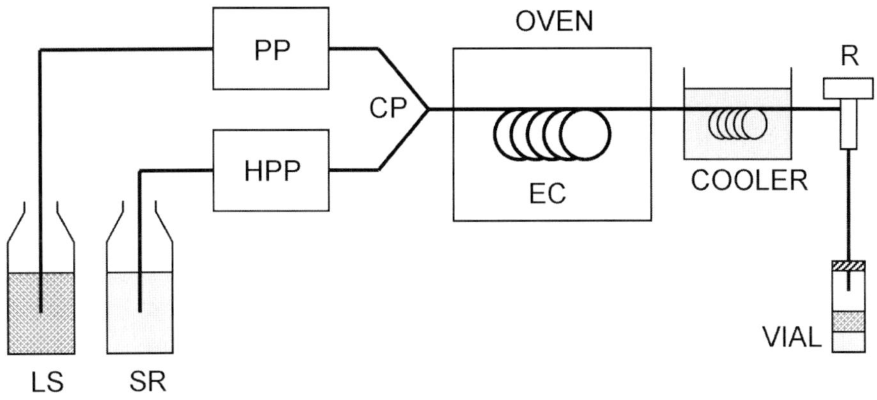

Fig. 6.15. Experimental set-up for high-pressure liquid–liquid extraction. LS liquid sample, SR solvent reservoir, PP piston pump, HPP high-pressure pump, CP confluence point, EC extraction coil, R restrictor. (Reproduced with permission of Springer-Verlag.)

6.4.6. Comparison of DPHSE with other leaching techniques

As in ASE, the efficiency of DPHSE for extracting compounds from soils has been compared with that of other techniques such as Soxhlet, microwave-assisted (MAE) and, especially, supercritical fluid extraction (SFE).

The advantages of DPHSE over Soxhlet extraction include not only decreased solvent consumption and operation times [25,158,179], but also the avoidance of organic solvents as extractants (water is the usual leaching agent) and the amenability to coupling to other operations of the analytical process, as described in the previous section.

As with ASE, the advantages and disadvantages of DPHSE with respect to MAE and SFE are not so clear and depend on the particular application.

In fact, DPHSE has rarely been compared with MAE. The former has been found to provide results similar to [167] or better than [52] those of the latter in the extraction of Irganox from polypropylene, where prolonged microwave exposure resulted in degradation of the polymer.

SFE has so far been the technique most frequently used to validate DPSE methods such as those for the extraction of dioxins from high- and low-carbon fly ash [48], triazolopyrimidine sulphonanilide herbicides, trichloropyridinol and PCBs from soil [150,152, 168], and carotenoids and tocopherol from spice red pepper [177]. As noted earlier, neither technique can be said to be better than the other; it depends on the characteristics of the analytes to be extracted (e.g. on their polarity and high-temperature stability). Thus, in the extraction of cloransulam-methyl from soil, while the use of subcritical water provided higher recoveries than SFE, the analyte was not hydrolytically stable above 150°C, which entailed using a lower temperature and hence an increased extraction time [152]. In the extraction of PAHs from bituminous coal fly ash [180], extraction with supercritical CO_2 yielded better recoveries than DPHSE using toluene and methylene

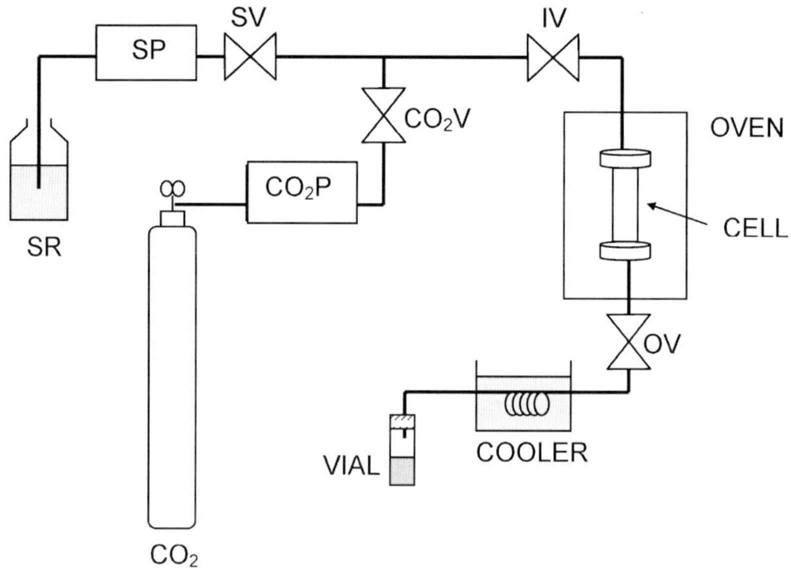

Fig. 6.16. Scheme of a DPHSE–SFE system. SR solvent reservoir, SP solvent pump, SV solvent valve, CO_2P CO_2 pump, CO_2V CO_2 valve, IV inlet valve, OV outlet valve. (Reproduced with permission of Elsevier.)

chloride as extractant, particularly for low-molecular weight PAHs, which cannot be quantitatively extracted by DPHSE. On the other hand, the use of water at 250°C and 200 bar in the extraction of trichloropyridinol from soil [150] provided better recoveries and precision, with decreased costs and in half the time as with supercritical CO_2.

One interesting possibility of DPHSE is that of performing sequential extractions (one with supercritical CO_2 and the other with superheated water) to recover compounds of differential polarity from the same matrix. The process is very simple: as shown in Fig. 6.16, DPHSE can be implemented with SFE equipment, simply by adding a pump to propel the solvent. In this way, PAHs of different molecular weights have been extracted from marine sediments using a first extraction with methylene chloride at 120°C followed by a second with supercritical CO_2; a whole extraction time of 30 min ensured quantitative recoveries [151]. Similarly, dacthal and its acid metabolites were extracted from soil using an extraction with CO_2 followed by another with water at 50°C and 200 bar [39]. With this approach, a single sample suffices to determine both polar and non-polar analytes; also, no sample clean-up is required and the method uses as little as 5 ml of non-chlorinated organic solvent.

References

1 B.E. Richter, B.A. Jones, J.L. Ezzell and N.L. Porter, *Anal. Chem.*, 68 (1996) 1033.
2 B.E. Richter, J.L. Ezzell, D. Felix, K.A. Roberts and D.W. Later, *Am. Lab.*, 27 (1995) 24.
3 J.L. Ezzell, B.E. Richter, W.D. Felix, S.R. Black and J.E. Meikle, *LC–GC*, 13 (1995) 390.

4 B.E. Richter, J.L. Ezzell and D. Felix, Singel Laboratory Method Validation Report: Extraction of TCL/PPL (Target Compound List/Priority Pollutant List) BNAs and Pesticides Using Accelerated Solvent Extraction (ASE) with Analytical Validation by GC/MS and GC/ECD, *Document 116064.A*, Dionex Corporation, 16 June (1994).
5 B.E. Richter, J.L. Ezzell and D. Felix, Singel Laboratory Method Validation Report: Extraction of Organophosphorus Pesticides, Chlorinated Herbicides, and Polychlorinated Biphenyls Using Accelerated Solvent Extraction (ASE) with Analytical Validation by GC/NPD and GC/ECD. *Document 101124*, Dionex Corporation, 2 December (1994).
6 S.B. Hawthorne, Y. Yang and D.J. Miller, *Anal. Chem.*, 66 (1994) 2912.
7 EPA Method 3545, Pressurised Fluid Extraction, Test Methods for Evaluating Solid Waste; EPA SW-846; US GPO, Washington, DC (1994).
8 M.D. Luque de Castro, M. Valcárcel M.T. Tena, *Analytical Supercritical Fluid Extraction*, Springer-Verlag, Heidelberg (1994).
9 H.M. Kingston and S.J. Haswell, *Microwave-Enhanced Chemistry: Fundamentals, Sample Preparation and Applications*, American Chemical Society, Washington (1997).
10 K.S. Pitzer and L. Brewer, *Thermodynamics*, 2nd edn, McGraw–Hill, New York (1961) Chapter 18.
11 T. Sekine and Y. Hasegawa, *Solvent Extraction Chemistry*, Marcel Dekker, New York (1977).
12 R.H. Perry, D.W. Green and J.O. Maloney, *Perry's Chemical Engineers' Handbook*, 6th edn, McGraw–Hill, New York (1984) Chapter 3.
13 G.C. Pimentel and A.L. McClellan, *The Hydrogen Bond*, Freeman, San Francisco (1960).
14 H.J. Möckel, G. Welter and H. Melzer, *J. Chromatogr.*, 388 (1987) 253.
15 K. Schäfer, *Anal. Chim. Acta*, 358 (1998) 69.
16 M.D. David and J.N. Seiber, *Anal. Chem.*, 68 (1996) 3038.
17 H.B. Hass and R.F. Newton, *Handbook of Chemistry and Physics*, 62nd edn, R.C. Weast and M.J. Astle, Eds, CRC Press, Inc., Boca Raton, FL (1982) 189.
18 X. Lou, H.G. Janssen and C.A. Cramers, *Anal. Chem.*, 69 (1997) 1598.
19 M.M. Schantz, J.J. Nichols and S.A. Wise, *Anal. Chem.*, 69 (1997) 4210.
20 N. Saim, J.R. Dean, Md.P. Abdullah and Z. Zakaria, *Anal. Chem.*, 70 (1998) 420.
21 C.G. Arnold, M. Berg, S.R. Müller, U. Domman and R.P. Schwarzenbach, *Anal. Chem.*, 70 (1998) 3094.
22 A. Kreisselmeier and H.W. Durbeck, *J. Chromatogr.*, 775 (1997) 187.
23 T.M. Fahmy, M.E. Paulaitis, D.M. Johnson and M.E.P. McNally, *Anal. Chem.*, 65 (1993) 1462.
24 W.H. Ding and J.C.H. Fann, *Anal. Chim. Acta*, 408 (2000) 291.
25 W.H. Ding and J.C.H. Fann, *J. Chromatogr.*, 866 (2000) 79.
26 ASE 200, Accelerated Solvent Extractor Operator's Manual, Dionex Corp., *Document No. 031149*, Revision 03, Sunnyvale, CA (1997).
27 A. Hubert, K.D. Wenzel, M. Manz, L. Weissflog, W. Engewald and G. Schüürmann, *Anal. Chem.*, 72 (2000) 1294.
28 J. Gan, S.K. Papiernik, W.C. Koskinen and S.R. Yates, *Environ. Sci. Technol.*, 33 (1999) 3249.
29 C. Crescenzi, G. D'Ascenzo, A. Di Corcia, M. Nazzari, S. Marchese and R. Samperi, *Anal. Chem.*, 71 (1999) 2157.
30 O. Zuloaga, N. Etxebarria, L.A. Fernández and J.M. Madariaga, *Trends Anal. Chem.*, 17 (1998) 642.
31 B.E. Richter, *J. Chromatogr.*, 874 (2000) 217.
32 O. Zuloaga, L.J. Fitzpatrick, N. Etxebarria and J.R. Dean, *J. Environ. Monit.*, 2 (2000) 634.
33 P.A. Gallagher, J.A. Shoemaker, X. Wei, C.A. Brockhoff-Schwegel and J.T. Creed, *Fresenius J. Anal. Chem.*, 369 (2001) 71.
34 H.J. Vandenburg, A.A. Clifford, K.D. Bartle and S.A. Zhu, *Anal. Chem.*, 70 (1998) 1943.
35 N.V. Beck, W.A. Carrick, D.B. Cooper and B. Muir, *J. Chromatogr.*, 907 (2001) 221.
36 A.F.M. Barton, *Handbook of Solubility Parameters and Other Cohesion Parameters*, CRC Press, Boca Raton, FL (1983).
37 L. Wennrich, P. Popp and M. Möder, *Anal. Chem.*, 72 (2000) 546.
38 A.C. Schmidt, W. Reisser, J. Mattusch, P. Popp and R. Wennrich, *J. Chromatogr.*, 889 (2000) 83.

39 J.A. Field, K. Monohan and R. Reed, *Anal. Chem.*, 70 (1998) 1956.
40 T.M. Pawloski and C.F. Poole, *J. Agric. Food Chem.*, 46 (1998) 3124.
41 M.D. Luque de Castro, M.M. Jiménez-Carmona and V. Fernández-Pérez, *Trends Anal. Chem.*, 18 (1999) 708.
42 J.L. Luque-García and M.D. Luque de Castro, *J. Chromatogr.*, in press (2002).
43 L. Gámiz-Gracia and M.D. Luque de Castro, *Talanta*, 51 (2000) 1179.
44 M.M. Jiménez-Carmona and M.D. Luque de Castro, *Anal. Chim. Acta*, 342 (1997) 215.
45 M.C. Herrera, R.C. Prados-Rosales, J.L. Luque-García and M.D. Luque de Castro, *Anal. Chim. Acta.*, submitted for publication (2002).
46 V. Fernández-Pérez, M.M. Jiménez-Carmona and M.D. Luque de Castro, *J. Anal. At. Spectrom.*, 14 (1999) 1761.
47 V. Fernández-Pérez and M.D. Luque de Castro, *J. Chromatogr.*, 902 (2000) 357.
48 I. Windal, D.J. Miller, E. De Pauw and S.B. Hawthorne, *Anal. Chem.*, 72 (2000) 3916.
49 J.A. Fisher, M.J. Scarlett and A.D. Stott, *Environ. Sci. Technol.*, 31 (1997) 1120.
50 M. Waldebäck, C. Jansson, F.J. Señoráns and K.E. Markides, *Analyst*, 123 (1998) 1205.
51 M. Okihashi, H. Obana and S. Hori, *Analyst*, 123 (1998) 711.
52 H.J. Vandenburg, A.A. Clifford, K.D. Bartle, J. Carroll and D. Newton, *Analyst*, 124 (1999) 397.
53 D.W. Kenny and S.V. Olesik, *J. Chromatogr. Sci.*, 36 (1998)
54 E. Björklund, S. Bowadt, T. Nilsson and L. Mathiasson, *J. Chromatogr.*, 836 (1999) 285.
55 P. Popp, P. Keil, M. Möder, A. Paschke and U. Thuss, *J. Chromatogr.*, 774 (1997) 203.
56 K.D. Wenzel, A. Hubert, M. Manz, L. Weissflog, W. Engewald and G. Schüürmann, *Anal. Chem.*, 70 (1998) 4827.
57 C.M. Rico-Varadé and M.D. Luque de Castro, *J. Anal. At. Spectrom.*, 13 (1998) 787.
58 K. Li, M. Landriault, M. Fingas and M. Llompart, *Analusis*, 26 (1998) 365.
59 V. Camel, *Spectra Anal.*, 29 (2000) 21.
60 J.D. Berset, M. Ejem, R. Holzer and P. Lischer, *Anal. Chim. Acta*, 383 (1999) 263.
61 K.J. Hageman, L. Mazeas, C.B. Grabanski, D.J. Miller and S.B. Hawthorne, *Anal. Chem.*, 68 (1996) 3892.
62 J.R. Dean, *Anal. Commun.*, 33 (1996) 191.
63 N. Saim, J.R. Dean, M.P. Abdullah and Z. Zakaria, *J. Chromatogr.*, 791 (1997) 361.
64 C. Bandh, E. Björklund, L. Mathiasson, C. Näf and Y. Zebühr, *Environ. Sci. Technol.*, 34 (2000) 4995.
65 J.W. McKiernan, J.T. Creed, C.A. Brockhoff, J.A. Caruso and R.M. Lorenzana, *J. Anal. At. Spectrom.*, 14 (1999) 607.
66 H. Obana, K. Kikuchi, M. Okihashi and S. Hori, *Analyst*, 122 (1997) 217.
67 X. Chaudot, A. Tambuté and M. Caude, *J. Chromatogr.*, 888 (2000) 327.
68 S.P. Frost, J.R. Dean, K.P. Evans, K. Harradine, C. Cary and M.H.I. Comber, *Analyst*, 122 (1997) 895.
69 F. Hoefler, *LaborPraxis*, 44 (1996) 53.
70 J. Pörschmann and J. Plugge, *Fresenius J. Anal. Chem.*, 364 (1999) 643.
71 S. Campbell and Q.X. Li, *Anal. Chim. Acta*, 434 (2001) 283.
72 M.D. David, S. Campbell and Q.X. Li, *Anal Chem.*, 71 (2000) 3665.
73 O. Zuloaga, N. Etxebarria, L.A. Fernández and J.M. Madariaga, *Fresenius, J. Anal. Chem.*, 367 (2000) 733.
74 E. Björklund, T. Nilsson and S. Bowadt, *Trends Anal. Chem.*, 19 (2000) 434.
75 M.M. Jiménez-Carmona, J.L. Ubera and M.D. Luque de Castro, *J. Chromatogr.*, 855 (1999) 625.
76 B. Benthin, H. Danz and M. Hamburger, *J. Chromatogr.*, 837 (1999) 211.
77 J.R. Dean, A. Santamaria-Rekondo and E. Ludkin, *Anal. Commun.*, 33 (1996) 413.
78 K.M. Meney, C.M. Davidson, D. Littlejohn, N.J. Cotton and B. Fields, *Anal. Commun.*, 35 (1998) 173.
79 S.B. Hawthorne, C.B. Grabanski, E. Martin and D.J. Miller, *J. Chromatogr.*, 892 (2000) 421.
80 C. Woolley, B. Ramsey, A. Fiorante and B. Martin-Dangler, *Supelco Rep.*, 14 (1995) 4.
81 O.P. Heemken, N. Theobald and B.W. Wenclawiak, *Anal. Chem.*, 69 (1997) 2171.

82 T.E. Young, S.T. Ecker, R.E. Synovec, N.T. Hawley, J.P. Lomber and C.M. Wai, *Talanta*, 45 (1998) 1189.
83 S. Morales-Muñoz, J.L. Luque-García, M.D. Luque de Castro, *Anal. Chem.*, submitted for publication (2002).
84 J.R. Donnelly, A.H. Grange, N.R. Herron, G.R. Nichol, J.L. Jeter, R.J. White, W.C. Brumley and J. Van-Emon, *J. AOAC Int.*, 79 (1996) 953.
85 Dionex Application Note 338 (2000) 4.
86 R. Alzaga, C. Maldonado and J.M. Bayona, *Int. J. Environ. Anal. Chem.*, 72 (1998) 99.
87 F. Hoefler, J. Ezzell and B. Richter, *LaborPraxis*, 19 (1995) 58.
88 J. Huau and A.M. Compiano, *Spectra Anal.*, 190 (1996) 34.
89 E. Conte, R. Milani, G. Morali and F. Abballe, *J. Chromatogr.*, 765 (1997) 121.
90 S.M. Pyle and A.B. Marcus, *J. Mass Spectrom.*, 32 (1997) 897.
91 L. Guzzella and F. Pozzoni, *Int. J. Environ. Anal. Chem.*, 74 (1999) 123.
92 S. Marchese, D. Perret, A. Gentili, R. Curini and A. Marino, *Rapid Commun. Mass Spectrom.*, 15 (2001) 393.
93 M. Waldeback, E. Rydin and K. Markides, *Int. J. Environ. Anal. Chem.*, 72 (1998) 257.
94 S. Chiron, S. Roy, R. Cottier and R. Jeannot, *J. Chromatogr.*, 879 (2000) 137.
95 D. Schwesig, A. Gottlein, L. Haumaier, R. Blasek and G. Ilgen, *Int. J. Environ. Anal. Chem.*, 73 (1999) 253.
96 M. Hauck, F. Hoefler and W. Schulz, *LaborPraxis*, 22 (1998) 44.
97 D.D. McCant, L.S. Inouye and V.A. McFarland, *Bull. Environ. Contam. Toxicol.*, 63 (1999) 282.
98 X. Chaudot, A. Tambuté and M. Caude, *J. Chromatogr.*, 866 (2000) 231.
99 X. Lou, D.J. Miller and S.B. Hawthorne, *Anal. Chem.*, 72 (2000) 481.
100 S.B. Hawthorne, S. Trembley, C.L. Moniot, C.B. Grabanski and D.J. Miller, *J. Chromatogr.*, 886 (2000) 237.
101 B.E. Richter, J.L. Ezzell, D.E. Knowles, F. Hoefler, A.K.R. Mattulat, M. Scheutwinkel, D.S. Waddell, T. Gobran and V. Khurana, *Chemosphere*, 34 (1997) 975.
102 P. Popp, P. Keil, M. Moeder, A. Paschke and U. Thuss, *J. Chromatogr.*, 774 (1997) 203.
103 B.E. Richter, J.L. Ezzell, D.E. Knowles, F. Hoefler and J. Huau, *Spectra Anal.*, 27 (1998) 21.
104 D.V. Kenny and S.V. Olesik, *J. Chromatogr. Sci.*, 36 (1998) 59.
105 L.M. Assis, J.S.S. Pinto and F.M. Lancas, *J. Microcolumn Sep.*, 12 (2000) 292.
106 S.L. Simonich, W.M. Begley, G. Debaere and W.S. Eckhoff, *Environ. Sci. Technol.*, 34 (2000) 959.
107 Dionex Application Note 333 (2000) 3.
108 S.G. Reddy, K.E. Keeler, M.W. Coyle and J.L. Ezzell, *Int. Lab.*, 29 (1999) 8.
109 A. Eisner, K. Kureckova and K. Ventura, *Chem. Listy*, 94 (2000) 235.
110 R.E. Majors, *LC–GC*, 13 (1995) 542.
111 K. Bartle, M. Cikalo, T. Clifford, A. Lewis, P. Myers, M. Robson and H.J. Vandenburg, *LC–GC*, 16 (1998) S32.
112 Anonymous, *LaborPraxis*, 24 (2000) 44.
113 Dionex Application Note 332 (2000) 3.
114 M. Weichbrodt, W. Vetter and B. Luckas, *J. AOAC Int.*, 83 (2000) 1334.
115 S. Kakimoto, H. Obana, M. Okihashi and S. Hori, *Shokuhin Eiseigaku Zasshi*, 38 (1997) 358.
116 A.K.D. Liem, *Trends Anal. Chem.*, 18 (1999) 499.
117 F. Schroeter, M. Anastassiades and E. Scherbaum, *CLB Chem. Labor Biotech.*, 50 (1999) 4.
118 K.D. Wenzel, A. Hubert, W. Engewald and G. Schueuermann, *GIT Labor Fachz.*, 19 (1999) 72.
119 G. Wang, A.S. Lee, M. Lewis, B. Kamath and R.K. Archer, *J. Agric. Food Chem.*, 47 (1999) 1062.
120 R. Draisci, C. Marchiafava, E. Ferretti, L. Palleschi, G. Catellani and A. Anastasio, *J. Chromatogr.*, 814 (1998) 187.
121 N.P. Vela, D.T. Heitkemper and K.R. Stewart, *Analyst*, 126 (2001) 1011.
122 Dionex Application Note 334 (2000) 5.
123 Dionex Application Note 340 (2000) 4.
124 B. Matthaeus and L. Bruehl, *J. Am. Oil Chem. Soc.*, 78 (2001) 95.
125 Dionex Application Note 337 (2000) 3.

126 C.T. da Costa, S.A. Margolis, B.A. Benner Jr and D. Horton, *J. Chromatogr.*, 831 (1999) 167.
127 J.G. Wilkes, E.D. Conte, Y. Kim, M. Holcomb, J.B. Sutherland and D.W. Miller, *J. Chromatogr.*, 880 (2000) 3.
128 S. Morf, B. Debrunner, B. Meier and H. Kurth, *LaborPraxis*, 22 (1998) 59.
129 Dionex Application Note 335 (2000) 3.
130 J.R. Williams, E.D. Morgan and B. Law, *Anal. Commun.*, 33 (1996) 15.
131 G.W. Chase Jr and B. Thompson, *J. AOAC Int.*, 83 (2000) 407.
132 E. Björklund, M. Jaremo, L. Mathiasson, L. Karlsson, J.T. Strode III, J. Eriksson and A. Torstensson, *J. Liq. Chromatogr. Relat. Technol.*, 21 (1998) 535.
133 F. Hoefler, R. Neufang, B. Raffelsberger and O. Stahler, *LaborPraxis*, 23 (1999) 52.
134 R. Draisci, C. Marchiafava, L. Palleschi, P. Cammarata and S. Cavalli, *J. Chromatogr.*, 753 (2001) 217.
135 Anonymous, *Lab. 2000*, 12 (1998) 54.
136 M. Riess, T. Ernst, G. Biermann and R. van Eldik, *GIT Labor Fachz.*, 42 (1998) 1008.
137 Dionex Application Note 336 (2000) 3.
138 F. Hoefler, *GLB Chem. Labor Biotech.*, 51 (2000) 56.
139 L.Y. Zhou, M. Ahraf-Khorassani and L.T. Taylor, *J. Chromatogr.*, 858 (1999) 209.
140 H.J. Vandenburg, A.A. Clifford, K.D. Bartle, S.A. Zhu, J. Carroll, I.D. Newton and L.M. Garden, *Anal. Chem.*, 70 (1998) 1943.
141 H.J. Vandenburg, A.A. Clifford, K.D. Bartle, R.E. Carlson, J. Carroll and I.D. Newton, *Analyst*, 124 (1999) 1707.
142 A.F.M. Barton, *Handbook of Polymer–Liquid Interaction Parameters and Solubility Parameters*, CRC Press, Boca Raton, FL (1990).
143 F.I. Onuska, A.H. El Shaarawi, K. Terry and E.M. Vieira, *J. Microcolumn Sep.*, 13 (2001) 54.
144 L.J. Fitzpatrick, J.R. Dean, M.H.I. Comber, K. Harradine and K.P. Evans, *J. Chromatogr.*, 874 (2000) 257.
145 B. Li, Y. Yang, Y. Gan, C.D. Eaton, P. He and A.D. Jones, *J. Chromatogr.*, 873 (2000) 175.
146 Y. Yang and B. Li, *Anal. Chem.*, 71 (1999) 1491.
147 M.M. Jiménez-Carmona, M.P. da Silva and M.D. Luque de Castro, *Talanta*, 45 (1998) 883.
148 M.M. Jiménez-Carmona, V. Fernández-Pérez, M.J. Gualda-Bueno, J.M. Cabañas-Espejo and M.D. Luque de Castro, *Anal. Chim. Acta*, 395 (1999) 113.
149 Y. Yang, S. Bowadt, S.B. Hawthorne and D.J. Miller, *Anal. Chem.*, 67 (1995) 4571.
150 M. Notar and H. Leskovsek, *Fresenius J. Anal. Chem.*, 366 (2000) 846.
151 M.S. Krieger, J.L. Wynn and R.N. Yoder, *J. Chromatogr.*, 897 (2000) 405.
152 M.M. Jiménez-Carmona, J.J. Manclús, A. Montoya and M.D. Luque de Castro, *J. Chromatogr.*, 785 (1997) 329.
153 L. Ramos, J.J. Vreuls and U.A. Th. Brinkman, *J. Chromatogr.*, 891 (2000) 275.
154 Th. Hankemeier, A.J.H. Louter, J. Dalluge, J.J. Vreuls and U.A.Th. Brinkman, *J. High Resolut. Chromatogr.*, 21 (1998) 450.
155 V. Fernández-Pérez, M.M. Jiménez-Carmona and M.D. Luque de Castro, *Analyst*, 125 (2000) 481.
156 R.M.M. Kooke, J.W.A. Lustemhouwer, K. Olie and O. Hutzinger, *Anal. Chem.*, 53 (1981) 461.
157 J.A. Field and R.L. Reed, *Environ. Sci. Technol.*, 33 (1999) 2782.
158 A. Di Corcia, A.B. Caracciolo, C. Crescenzi, G. Giuliano, S. Murtas and R. Samperi, *Environ. Sci. Technol.*, 33 (1999) 3271.
159 D. Mackay and W.Y. Shiu, *J. Chem. Eng. Data*, 22 (1977) 399.
160 N.D. Sanders, *Ind. Eng. Chem. Fundam.*, 25 (1986) 171.
161 B.D. Drakes and R.L. Smith Jr, *J. Supercrit. Fluids*, 3 (1990) 162.
162 J.S. Fritz and M. Macka, *J. Chromatogr.*, 902 (2000) 137.
163 V. Pichon, *J. Chromatogr.*, 885 (2000) 195.
164 T. Hyötyläinen, K. Hartonen, S. Säynäjoki and M.L. Riekkola, *Chromatographia*, 53 (2001) 301.
165 M. Krappe, S.B. Hawthorne and B.W. Wenclawiak, *Fresenius J. Anal. Chem.*, 364 (1999) 625.
166 H. Daimon and J. Pawliszyn, *Anal. Commun.*, 33 (1996) 421.

167 S. Dupeyron, P.M. Dudermel, D. Couturier, P. Guarini and J.M. Delattre, *Int. J. Environ. Anal. Chem.*, 73 (1999) 191.
168 S. Pross, W. Gau and B.W. Wenclawiak, *Fresenius J. Anal. Chem.*, 367 (2000) 89.
169 S.B. Hawthorne, C.B. Grabanski, K.J. Hageman and D.J. Miller, *J. Chromatogr.*, 814 (1998) 151.
170 M. Valcárcel and M.D. Luque de Castro, *Non-Chromatographic Continuous Separation Techniques*, Royal Society of Chemistry, Cambridge (1991).
171 M.S. Krieger, W.L. Cook and L.M. Kennard, *J. Agric. Food Chem.*, 48 (2000) 2178.
172 E. Priego-López and M.D. Luque de Castro, *Talanta*, submitted for publication (2001).
173 M.S.S. Curren and J.W. King, *Anal. Chem.*, 73 (2001) 740.
174 A. Basile, M.M. Jiménez-Carmona and A.A. Clifford, *J. Agric. Food Chem.*, 46 (1998) 5205.
175 R. Soto-Ayala and M.D. Luque de Castro, *Food Chemistry*, 75 (2001) 109.
176 M.M. Jiménez-Carmona and M.D. Luque de Castro, *Chromatographia*, 50 (1999) 578.
177 V. Illes, H.G. Dadood, P.A. Biacs, M.H. Gnayfeed and B. Meszaros, *J. Chromatogr. Sci.*, 37 (1999) 345.
178 H.J. Vandenburg, A.A. Clifford, K.D. Bartle, J. Carroll, I. Newton, L.M. Garden, J.R. Dean and C.T. Costley, *Analyst*, 122 (1997) 101R.
179 H. Bautz, J. Polzer and L. Stieglitz, *J. Chromatogr.*, 815 (1998) 231.
180 D.V. Kenny and S.V. Olesik, *J. Chromatogr. Sci.*, 36 (1998) 66.
181 V. Fernández-Pérez, M.M. Jiménez-Carmona and M.D. Luque de Castro, *Anal. Chim. Acta*, 433 (2001) 47.

Chapter 7

Analytical supercritical fluid extraction

7.1. INTRODUCTION

In 1822, Cagniard de la Tour showed the existence of a critical temperature for each individual substance above which such a substance can only occur as a fluid and not as either a liquid or a gas. This critical point is reached as one moves upward along the gas–liquid coexistence curve, where both temperature and pressure increase. The original liquid becomes less dense through thermal expansion and the gas becomes more dense as the pressure rises. At the critical point, the densities of the two phases are equal, the distinction between the gas and liquid vanishes and the coexistence curve comes to an end at the critical point, where the substance is described as a fluid. Supercritical fluids exhibit key features such as compressibility, homogeneity and a continuous change from gas-like to liquid-like properties.

Over the past twenty years, fluids above their critical temperatures and pressures have been used for a variety of analytical purposes: first as mobile phases in chromatography [1,2], thus giving rise to supercritical fluid chromatography, and then as solvents for the selective extraction of species — from solid samples in most cases [3–5].

7.2. PROPERTIES OF SUPERCRITICAL FLUIDS

In the supercritical state, a fluid possesses the typical density of a liquid (between 0.1 and 1.0 g/ml) and hence its characteristic dissolving power. As a result, a supercritical fluid can also be defined as a heavy gas with a controllable dissolving power or as a form of matter in which the liquid and gaseous state are indistinguishable.

Figure 7.1 shows a typical phase diagram for a pure substance. The regions where the substance occurs as a single phase [viz. solid (s), liquid (l) or gaseous (g)] are bounded by curves indicating the coexistence of two phases s-g, s-l and l-g. The three curves intersect at the so-called "triple point" (TP), where the three states coexist in equilibrium. At higher temperatures and pressures (i.e. following the l–g coexistence branch), the curve breaks at the so-called "critical point" (CP), the coordinates of which are designated by a critical temperature (T_c) and a critical pressure (P_c) above which no liquefaction will take place if the temperature is further raised and no gas will be formed if the pressure is increased. This region of pressures and temperatures above P_c and T_c, respectively, is called the "supercritical region".

The critical point is characteristic of each substance. Table 7.1 lists the critical temperature and pressure for various solvents. As can be seen, the critical pressure and

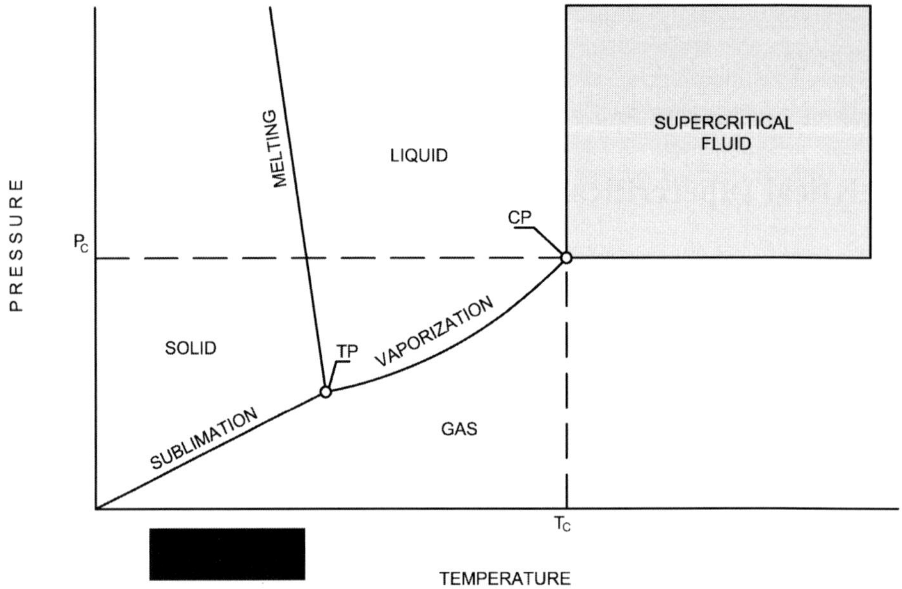

Fig. 7.1. Solid–liquid–gas–supercritical fluid phase diagram. TP triple point, CP critical point, P_c critical pressure, T_c critical temperature.

TABLE 7.1

SELECTED SUBSTANCES USED AS SUPERCRITICAL FLUIDS

Substance	T_c (°C)	P_c (bar)	ρ_c (g/ml)
Carbon dioxide	31	74	0.47
Water	374	221	0.23
Ethane	32	49	0.20
Ethene	9	50	0.23
Propane	97	43	0.22
Xenon	17	58	1.11
Ammonia	132	114	0.23
Nitrous oxide	37	72	0.45
Fluoroform	26	49	0.52

T_c critical temperature P_c critical pressure ρ_c critical density

temperature for carbon dioxide are among the lowest; on the other hand, those for water are the highest.

One other parameter of interest in a supercritical fluid is its critical density, ρ_c, which is also listed in Table 7.1 for the different fluids considered.

7.2.1. Properties of the supercritical region

Some properties of supercritical fluids endow them with improved extraction capabilities relative to liquids. Their individual influence is discussed below.

Density

The density of a supercritical fluid (SF) depends on the pressure and temperature to which it is subjected. Also, the dissolving power of a given fluid depends on its density. This is the origin of the good dissolving properties of supercritical fluids, where interactions between the fluid and solute molecules are quite strong. Density in the supercritical region increases sharply with increasing pressure at a constant temperature, and decreases with increasing temperature at a constant pressure. This property endows SFs with a high flexibility relative to liquid solvents, whose dissolving power is essentially independent of their pressure as the liquid density is not affected by this variable.

Diffusion coefficient

The diffusion coefficients of solutes in supercritical fluids are in between those they possess in liquids and gases. Because diffusion coefficients in SFs are higher than those in liquids, mass transfer is usually more favourable in the former. The diffusion coefficient can also be altered to advantage as diffusivity in a supercritical fluid decreases with increasing pressure and increases with increasing temperature, especially in the vicinity of the critical point.

Viscosity

Supercritical viscosity lies in between the viscosity of liquids and gases. For this reason, supercritical fluids exhibit more favourable hydrodynamic properties than do liquids. Also, the low surface tension of SFs allows them to readily penetrate porous solids and packed beds. The viscosity of SFs, like that of conventional fluids, is temperature-dependent; however, while pressure has little effect on the viscosity of fluids, it exerts a strong influence on that of SFs. As a result, increased pressures lead to increased supercritical viscosity and hence to diminished solute diffusivity and hindered transport phenomena, but also, most often, to increased solubility through decreased density.

Dielectric constant

The dielectric constant is one of the most relevant physico–chemical parameters with a view to defining solubility in fluids. Table 7.2 lists the constants for the solvents most

TABLE 7.2

DIELECTRIC CONSTANTS FOR SELECTED SUBSTANCES USED IN SFE

Substance	Dielectric constant (C^2/NM^2)	Substance	Dielectric constant (C^2/NM^2)
Carbon dioxide	1.60	Isopropyl alcohol	18.3
n-Hexane	1.89	Acetone	20.7
Benzene	2.28	Ethanol	24.3
Diethyl ether	4.34	Methanol	32.6
Supercritical water	≃5.0	Ethylene glycol	37.0
Ethyl acetate	6.05	Formic acid	58.0
Benzyl alcohol	13.1	Water at 25°C	78.5
Ammonia	16.90		

frequently used as supercritical extractants. As can be seen, the dielectric constant for water drops dramatically from 25°C to the supercritical state (viz. from 78.5 to *ca.* 5). Because of its low constant at high temperatures, water shields the electrostatic potential between ions weakly, thereby allowing dissolved ions to freely form ion-pairs. Under these conditions, supercritical water behaves as a non-polar solvent. These properties account in part for its ability to dissolve non-polar solutes. By contrast, the dielectric constant for CO_2 increases with increasing pressure, and so does its density [6]; thus, in a highly dense state (e.g. 200 bar at 40°C), the dielectric constant is 1.5, so carbon dioxide can be assimilated to a highly non-polar solvent capable of dissolving non-polar substances [7]. One other parameter of water that changes in the supercritical state is the dissociation constant, K_w, which increases from 10^{-14} to 10^{-8} or even greater values. An increase in the ion concentration by four orders of magnitude makes supercritical water highly ionic. In summary, water is less polar but more ionic under supercritical conditions than in its natural state.

Polarity

Polarity is one of the properties most markedly influencing solubility and also one that can be altered in order to modify the selectivity of an extraction process. A given molecule is polar when the centre of its negative charge does not coincide with that of its positive charge. One such molecule acts as a dipole (viz. two charges of the same magnitude but opposite sign that are separated in space). Molecules such as those of CO_2 have a zero dipole moment and are said to be non-polar. Non-polar and scarcely polar solvents with moderate critical temperatures (e.g. N_2O, CO_2, ethane, propane, pentane, SF_6 and some freons) have limited dissolving power for solutes of a high polarity or molecular weight. On the other hand, such solutes are highly soluble in polar fluids (e.g. NH_3), the use of which may be restricted by other factors such as a high toxicity and/or reactivity.

The effect of pressure on the dissolving power of various supercritical fluids has been examined spectroscopically, through comparisons on the π^* polarity and polarizability scale of Taft *et al.* [8]. Thus, the polarizability of CO_2 has been found to be lower than those of all hydrocarbons except methane. Also, for its polarizability per unit volume to be comparable to that of liquid cyclohexane at 45°C, CO_2 must be subjected to a pressure of 2700 bar [9].

Because unary polar solvents such as NH_3 pose major analytical problems, small portions of other solvents called "modifiers" are frequently added to scarcely polar solvents such as CO_2 in order to increase their dissolving power [10]. The ability to adjust the nature and degree of interaction of binary fluids provides separations and extractions with better dissolving features and modified selectivity.

The properties of SFs are frequently expressed in terms of reduced, rather than absolute, values. A reduced value is defined as the ratio of the actual absolute value to the critical point value (denoted by P_r and T_r for reduced pressure and temperature, respectively). Accordingly, if these two parameters are greater than unity, the substance in question will be in its supercritical state.

7.2.2. The solubility parameter

The solubility of a solute in a supercritical fluid can be quantitatively estimated using Giddings' theory, which relies on differences between the Hildebrand solubility parameters for the SF and solute concerned. Solubility in a supercritical fluid can be understood by examining the Gibbs–Helmholtz equation:

$$\Delta G = \Delta H - \Delta S \tag{7.1}$$

According to Hildebrand [11], the heat of mixing can be defined in mathematical terms as

$$\Delta H = \nu_1 \nu_2 (\delta_1 - \delta_2)^2 \tag{7.2}$$

where ΔH is the energy change resulting from the formation and cleavage of intermolecular bonds; ν_1 and ν_2 are the partial volumes of the solvent and solute, respectively; and δ_1 and δ_2 are the solubility parameters for the solvent and solute, respectively. Solubility therefore depends on the differences between the solubility parameters of the fluid and solute: the smaller the difference between such parameters is, the higher is the solubility.

The solubility of a solute can be defined in terms of its solubility parameters:

$$\delta = \sqrt{\frac{\Delta E_v}{n}} = \sqrt{\frac{\rho(\Delta H_v - RT)}{M}} \tag{7.3}$$

where ΔE_v, ν, ρ, ΔH_v and M are the vaporization energy, molar volume, density, heat of vaporization and molecular weight of the solute, respectively; R is the gas constant; and T is temperature.

On the other hand, the solvent power of a supercritical fluid is defined by the equation of Guiddings as applied by Li and Lee [12], where the solubility parameters of the fluid are given by

$$\delta_1 = 1.25 \frac{\sqrt{P_c}\,\rho_{r,SF}}{\rho_{r,L}} \tag{7.4}$$

P_c being the critical pressure, and $\rho_{r,\,SF}$ and $\rho_{r,\,L}$ the reduced density of the supercritical fluid and that of the fluid in a liquid state, respectively. This equation reflects the variation of the solvent power of the gas as a function of pressure and temperature, and is related to the fluid properties through P_c.

The solubility parameter for a supercritical fluid, δ_1, increases with increasing density, so it is markedly dependent on pressure. This parameter also depends strongly on temperature, particularly at high pressures. Consequently, the solubility parameter for a fluid can be adjusted through both pressure and temperature.

Although a number of substances are considered to be potentially useful for SFE, in practice, carbon dioxide is the most common choice on account of its convenient critical temperature, chemical stability, non-flammability, stability in radioactive applications, non-toxicity and moderate cost. Large amounts of accidentally released CO_2 can be hazardous owing to its tendency to blanket the ground, but appropriate detectors for this purpose are available. Carbon dioxide is an environmentally friendly substitute for organic solvents. Its polar character as a solvent is in between that for a truly non-polar solvent such as hexane and those for weakly polar solvents. Water has recently gained ground in analytical SFE applications. Both the high values of its critical parameters and its corrosivity under supercritical conditions have so far restricted more extensive use, however.

7.3. LABORATORY-BUILT AND COMMERCIAL SUPERCRITICAL FLUID EXTRACTORS

The essential purpose of supercritical fluid extraction is to use a fairly inexpensive, safe material for the extraction of, mainly, organic compounds from a matrix instead of a conventional solvent, thereby reducing manipulation, increasing automatability, decreasing the extraction time and avoiding the problems associated with the use and disposal of organic solvents.

7.3.1. Basic components of a supercritical fluid extractor

All SF extractors, whatever their complexity and cost, share the same essential components, which are schematically depicted in Fig. 7.2 and described below.

Source of the extraction fluid

This unit consists of a steel, aluminium or stainless steel cylinder (preferably furnished with a dip tube), a pressurized head space and a device allowing cooling for long enough to ensure equilibration.

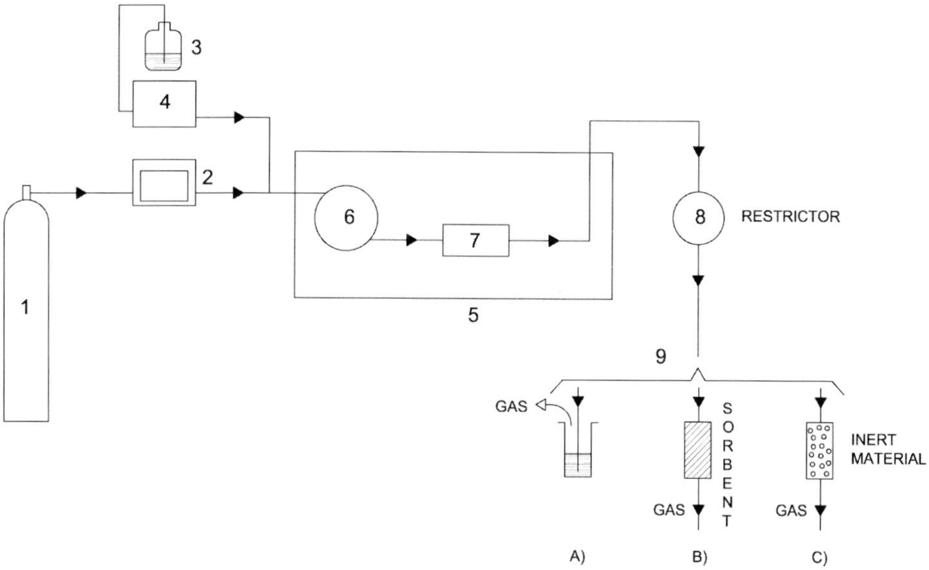

Fig. 7.2. Schematic diagram of a straightforward supercritical fluid extractor. 1 extracting fluid source, 2 extractant propulsion unit, 3 modifier reservoir, 4 modifier propulsion unit, 5 oven, 6 equilibration coil, 7 chamber containing the thimble or sample cell, 8 back-pressure regulator, 9 collection system (A bubbling, B sorption, C cryogenic trapping).

Propulsion system

The devices typically used to propel supercritical fluids consist of high-pressure pumps similar to those used in HPLC or SFC and are of two main types: syringe pumps and dual piston pumps. The former type of pump has the advantage that it provides a pulseless flow; however, the volume of fluid that can be used is limited by that of the syringe body. Only a single propulsion device need be used in SFE; however, an additional syringe or piston pump may be used to facilitate the addition of modifiers. These substances can also be inserted by adding the entrainer in the pump during loading, using a modifier chamber between the pump and extraction chamber, soaking the sample in the modifier or using a commercially available supercritical fluid already containing the modifier. Each option has specific advantages and shortcomings [3].

The interface between the cylinder and the propulsion system consists of a connecting line, which should be as short as possible, furnished with a series of safety devices including a bleed valve — which is especially useful to avoid the unwanted presence of gas in the pump when the connecting line is not short enough; a safety valve to avoid pressures above those the system and material used can reasonably withstand; a check valve to minimize contamination of the fluid stored in the cylinder; and filters of small pore size to remove any contaminant particles from the fluid.

Sample chamber and cell

The extractor cell or thimble, which should be accommodated in an appropriate chamber, must meet a series of requirements, namely:

(a) It should enable ready, convenient insertion of the sample;
(b) It should be easy to shut without the need for any tools;
(c) It should be amenable to sealing via software;
(d) It should be made of an inert, corrosion-resistant material such as Hastelloy-C or Inconel 702 if it is to hold supercritical water — supercritical CO_2 can be processed in stainless steel;
(e) It should lend itself to ready, rapid mounting and disassembling from the system;
(f) It should encompass a broad range of capacities (the usual range is 1–10 ml);
(g) It should be capable of withstanding pressures up to at least 700 bar.

As shown in Section 7.4, the sample size-to-cell volume ratio is one of the variables to be optimized in every SF extraction method. Its effect depends on the particular matrix–analyte couple and on the extraction mode (static or dynamic) to be used.

The sample cell can be single or multiple, depending on the specific goal of the extraction process. Thus, a single-cell extractor is usually more appropriate for research purposes, whereas 1–9 units are typically used in routine extractions.

An equilibration unit (viz. an adequately long coil) is placed in between the propulsion system and sample compartment in the oven to allow the reactant to reach the programmed pressure and temperature conditions before coming into contact with the sample.

Oven

The temperature required for proper development of the extraction process is usually provided by an oven allowing easy, precise setting and changing of temperatures between room level and 300°C — commercially available equipment encompasses broader ranges, however. A current trend exists to using water as the extractant, usually under subcritical conditions as shown in Chapter 6, but also in the supercritical state — despite the complications; this has promoted the development of commercial extractors operating over expanded temperature ranges. Many available extractors, however, are made of stainless steel, so they cannot be used with supercritical water.

The thermostat equipment must allow programming of the chamber, restrictor, restrictor outlet and collection system temperatures — and the propulsion unit to be cooled, if needed. Ovens from gas chromatographs have often been used to construct laboratory SF extractors. Rapid attainment of the preset temperature and cooling capabilities are desirable with a view to reducing the extraction time and the interval between successive extractions, respectively.

Back-pressure regulator or restrictor

The depressurization system acts as an interface between the supercritical conditions in the extraction cell and the atmospheric conditions to which the extract is eventually subjected when the extractor is not connected on-line to a chromatographic system for the individual separation of extracted species with a view to their subsequent detection. A wide variety of commercially available devices for this purpose exists that range from straightforward glass capillaries — the end of which can be readily cut off in the event of clogging — to hand-operated restrictors to computer-controlled units. This is one of the characteristic components of commercial SF extractors (one that can differ markedly among manufacturers).

Collection system

The unit intended to collect the effluent from the extractor following depressurization is only included in an SFE assembly when the extractor is not coupled on-line to another unit, whether a chromatograph or detector, and is located after the depressurization unit. There are three main ways of collecting solutes, namely: solvent bubbling, collection on a sorbent material and cryogenic trapping.

Solvent bubbling is the simplest way of collecting the effluent from an SF extractor as it only requires inserting the restrictor outlet through the septum of a vial containing a few millilitres of solvent. The efficiency of this collection system depends on (a) the solubility of the analyte in the collecting solvent, which is a function of the solvent polarity; (b) the solvent volume used, which may not be close to the saturation value; (c) the size of the bubbles formed, which dictates the exchange interface value; (d) the time of contact with the collector, which depends on the SF output flow-rate and the distance travelled by the bubbles; and (e) the solvent temperature, which has a decisive influence on solubility. The different types of collection system found in commercial SF extractors have been designed with a view to overcoming the adverse effects of the previous factors.

Collection on a sorbent involves the use of a solid material, either in the line or at the restrictor outlet. A number of materials have been [13] and continue to be investigated [14–16] with a view to optimizing collection of different types of analytes. This collection mode involves an additional step: desorbing the analytes from the sorbent by elution with a small volume of solvent for their subsequent determination or, alternatively, thermal desorption and sweeping by the eluent if an on-line coupled extraction–chromatographic system is being used.

With *cryogenic trapping*, the extraction mixture is cooled down until the supercritical fluid expands and the analytes deposit. The trapping temperature to be used depends on whether the analytes are to be isolated from the fluid or this is to be liquefied and the collection vessel sealed in order to avoid losses of analytes through partial crystallization or the formation of aerosols during cooling. When the temperature of the cryogenic trap is very low, the restrictor must be heated in order to avoid the formation of two phases. As with collection on a solid sorbent, an additional preparative step is required.

7.3.2. Operational modes

Depending on the way the sample and extractant are brought into contact, supercritical fluid extraction can be implemented in two different operational modes, viz. static and dynamic.

In the *static extraction* mode, the sample is soaked in the supercritical fluid and the fluid flow is halted over a preset interval after which it is propelled to the collection vessel, where the analytes are concentrated. This extraction mode allows penetration of the SF and is thus especially useful when the analytes cannot be expeditiously removed from the matrix.

In the *dynamic extraction* mode, the extractant is pumped through the sample into the collection system or interface when the extractor is connected on-line to a chromatograph or detector. In this way, the supercritical fluid is passed through the sample once before it is driven to the restrictor.

Which extraction mode is the better remains a controversial issue. While the static mode provides longer contact between the sample and solvent, swells the matrix and facilitates penetration of the extractant in its interstices — thereby increasing its efficiency — the dynamic mode allows the analyte to be continuously exposed to the pure (clean) solvent, thus favouring displacement of the analyte's partitioning equilibrium to the mobile phase. Most SFE methods use both modes: a static step is employed to ensure close contact between the sample and supercritical fluid without consuming much extractant that is followed by a dynamic step where the extracted analytes are driven to the restrictor and equilibrium is allowed to complete.

7.3.3. Commercial supercritical fluid extractors

Supercritical fluid extraction aroused the interest of commercial equipment manufacturers well before its potential was experimentally realized. However, the high expectations raised by this technique have not materialized — basically because of the strong dependence of the extraction efficiency on the properties of the sample matrix — as reflected in the declining interest of manufacturers in it. Thus, the number of manufacturers offering commercial supercritical fluid extractors fell by almost one half between the 1991 Pittsburgh Conference and 1997 — the year this number seemingly levelled off.

Commercial supercritical fluid extractors have evolved in parallel with the consolidation of specific types of applications of the SFE technique [17]. By way of example of a commercial extractor and its changes over time, let us consider the one manufactured by Hewlett-Packard. This extractor has evolved from the oldest model (HP 7860 A, Fig. 7.3A), which was launched in 1992, to the present one (HP 7680 T, Fig. 7.3B). Essentially, however, the extractor continues to consist of the following elements:

(a) A *dual high-pressure piston pump* that provides a continuous, consistent CO_2 flow.
(b) An *extraction module*, where the extraction step takes place. This module includes the *extraction chamber*, which accommodates the *thimble* containing the sample. The thimble ranges from 1.5 to 7 ml in size.

A)

B)

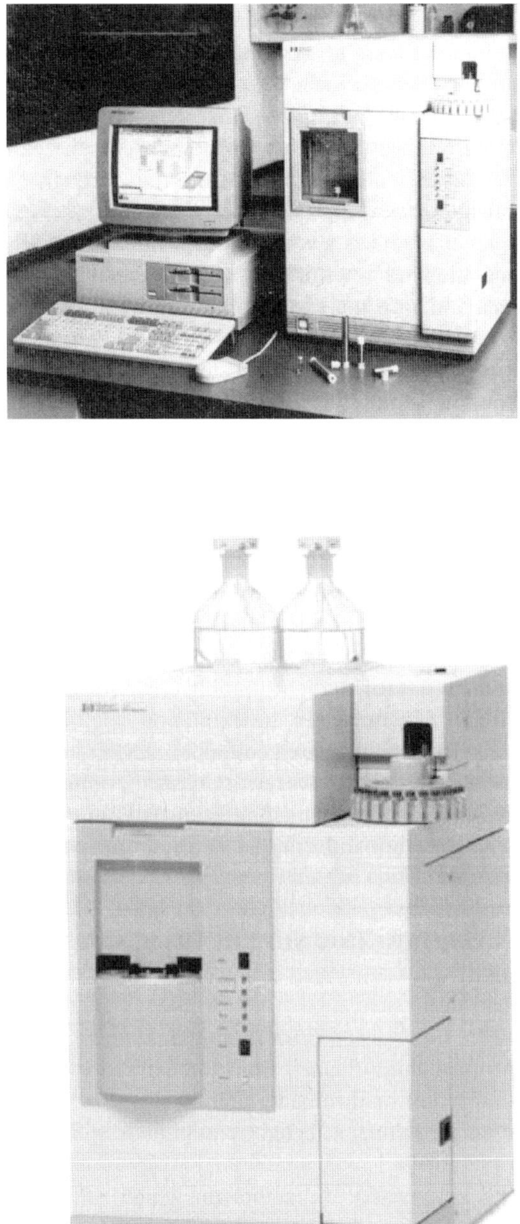

Fig. 7.3. Commercially available supercritical fluid extractors from Hewlett-Packard. (A) HP 7680 A model. (B) HP 7680 T model. (Reproduced with permission of Hewlett-Packard.)

(c) A microprocessor-controlled *restrictor* that resists plugging and allows independent control of the extracting CO_2 flow-rate and pressure.
(d) A *trap* where the extracted sample components are collected. The trap is located behind the cover on the front of the module and can use cryogenic trapping, solvent bubbling or sorption. Supercritical CO_2 containing the extracted components enters the trap through a nozzle. During the rinse or eluting step (depending on whether cryogenic trapping or sorption, respectively, is used), the extracted components are dissolved in the solvent and moved from the trap through an exit line to either an autosampler vial or waste.
(e) A Vectra *computer* that acts as a window allowing one to follow what happens inside the extraction module, and to tell it what steps to execute during an extraction process and when to start and stop the extraction through the options on the menu.
(f) A *printer* that provides three types of report, viz. the extraction method listing, which shows what happens during all extraction steps and rinse substeps, as well as information entered about the sample; the logfile, which lists what actually happened during the extraction; and the sample fraction report, which lists the extracted fraction outputs.

The different extractor models launched by Hewlett-Packard since 1992 have incorporated improvements based on both the manufacturer's research and clients' recommendations. The most salient among them are as follows: (a) finger-tightened thimble caps that allow faster, easier sample loading; (b) an interface between the HP 7680 T extractor and an HP chromatograph (via the automatic liquid sampler) that enables automatic extraction and analysis; (c) an eight-thimble turret for increased sample throughput; and (d) a modifier pump that increases the extraction selectivity of CO_2.

The high performance of sophisticated commercial SFE equipment makes it especially appealing for routine extractions. However, its typical compactness makes it difficult to alter, so simpler, laboratory-made extractors are usually preferred for research purposes as they lend themselves more readily to the incorporation of innovations.

7.4. VARIABLES INFLUENCING SUPERCRITICAL FLUID EXTRACTION

In a supercritical fluid extraction process, the analytes are transferred from their host matrix to the supercritical fluid, flushed from the extraction cell by the fluid and finally collected or sent to a detector or chromatograph for analysis. Unsurprisingly, the performance of the SFE technique is thus affected by a number of variables the most significant of which are listed in Table 7.3.

The large number and variety of factors on which SFE performance relies makes optimizing it rather a difficult task. Multivariate optimization approaches have been used from the beginning of this technique to both minimize the processing time and increase the extraction efficiency [18,19]. Simplex models [20] were the first to be used to examine the influence of interdependent variables. Two- and three-level orthogonal factor designs were developed to optimize up to nine extraction variables (viz. CO_2 flow-rate, fluid

TABLE 7.3

EXPERIMENTAL VARIABLES AFFECTING SFE EFFICIENCY

Properties of the fluid	*Dynamic and geometric factors*
• Nature (polarity)	• Extractant flow-rate
• Pressure (density)	• Extraction cell dimensions
• Temperature	• Extraction time
• Presence of a modifier	
• Modifier concentration	*Sample treatment*
	• Addition of liquids, solvents, derivatizing reagents, etc.
Properties of the solute	• Addition of solids
• Type of analyte (volatility, polarity, molecular weight)	
• Concentration	*Analyte collection mode*
	• Solvent bubbling
Properties of the solid	• Sorption
• Sample size	• Cryogenic trapping
• Particle size	• On-line coupling to a detector or chromatograph
• Nature of matrix	
• Presence of other extractable substances	
• Sample conditions	

density, extraction cell temperature, static extraction time, nozzle and trap temperatures, amount of derivatizing reagent, pyridine concentration and time of contact between the derivatizing reagent and sample prior to extraction) in a method for the supercritical fluid extraction of phenol in soil, which was derivatized by acetylation [21]. Composite designs of this type were also used to evaluate interactions between selected factors such as pressure, temperature, and the nature and proportion of the polar modifier in the SFE of cocaine from coca leaves [22]. A five-way ANOVA was conducted to identify both the most significant factors in the extraction of carbamate insecticides and their most statistically significant interactions [23].

Supercritical fluid extraction variables can be altered with a view to improving the extraction efficiency and/or the selectivity.

Variables can be adjusted in order to increase the extraction rate and/or the maximum amount of analyte that can be extracted. While increasing the extraction rate expedites the process, it is probably more interesting to improve the overall analyte recovery. In some cases, the analyte can only be recovered to a given extent, however much the extraction is prolonged. The variables on which increased extracted levels rely are the most significant as regards optimization when quantitative extraction is sought.

The variables of an SFE process can be optimized by determining the amount of analyte extracted over a preset interval at different values of the variable concerned or by running the whole cumulative extraction versus time curve for each variable value considered. The former choice is more expeditious but, as shown in Fig. 7.4, may lead to spurious conclusions. Thus, if measurements are made at a preset time t_1 shorter than that required for maximum efficiency (t_2 or t_{max}), then the process will involve comparing extraction rates, so no information on the extraction efficiency will be obtained. Conversely, if measurements are made at a time $t > t_{max}$, then one will obtain information on the extraction efficiency but not about the extraction rate. In the situations of Figs 7.4A and 7.4B, using t_1 or t_2 only would lead to errors of opposite sign [20]. On the other hand, under the conditions of Fig. 7.4C, any conclusions drawn from t_1 would be basically correct. Therefore, in studying the effect of a given variable, one should always determine the recovery at both a relatively short time and a long enough one.

7.4.1. Properties of the supercritical fluid

Although the first step in optimizing an SFE method should be selecting the extractant, in practice, CO_2 is almost invariably the choice as alternative extractants are only occasionally employed.

Experimentally, the performance of an SFE process relies on two essential properties of the supercritical fluid, namely: its density, which is dictated by its pressure and temperature; and its chemical nature (polarity). As shown below, however, other factors such as the presence of a chemical modifier and the volume of extracting fluid are also influential.

Influence of the fluid density

The solvent power of a supercritical fluid is a function of its density. No other separation technique allows it to be altered in such a simple manner as by changing the physical conditions. By way of example, Fig. 7.5A shows the density–pressure isotherms for carbon dioxide at a variable temperature and Fig. 7.5B illustrates the influence of the CO_2 density on the extraction efficiency for *trans*-β-carotene when all experimental variables except pressure are kept constant [24]. As can be seen, the recovery changed virtually linearly with the fluid density throughout the studied range. The effect is similar with other types of analytes such as fatty acids in red seaweed [25].

At a constant temperature, the extraction of non-polar analytes is usually favoured by a low pressure, whereas that of polar analytes requires an increased one. This behaviour is used for class-selective extractions, an example of which is the sequential extraction of air particulates with supercritical CO_2. Alkanes are extracted at 75 bar (45°C), whereas PAHs remain unextracted until the pressure is raised to 300 bar [26]. In some cases, an increase in pressure increases not only the solubility of the analyte but also its diffusivity. Such is the case with polymers, which absorb large amounts of CO_2 under supercritical conditions, thereby swelling and facilitating diffusion of the solute.

Influence of the fluid polarity

However influential it may be, the solubility of the analyte in the supercritical fluid is not the sole variable to be considered. In fact, it is necessary to overcome matrix–analyte interactions, which are very strong in some samples. The fluid polarity is a key factor under these conditions as diffusion in the solid matrix does not seem to be the extraction rate-determining step; otherwise, there should be no change in the recovery rate on

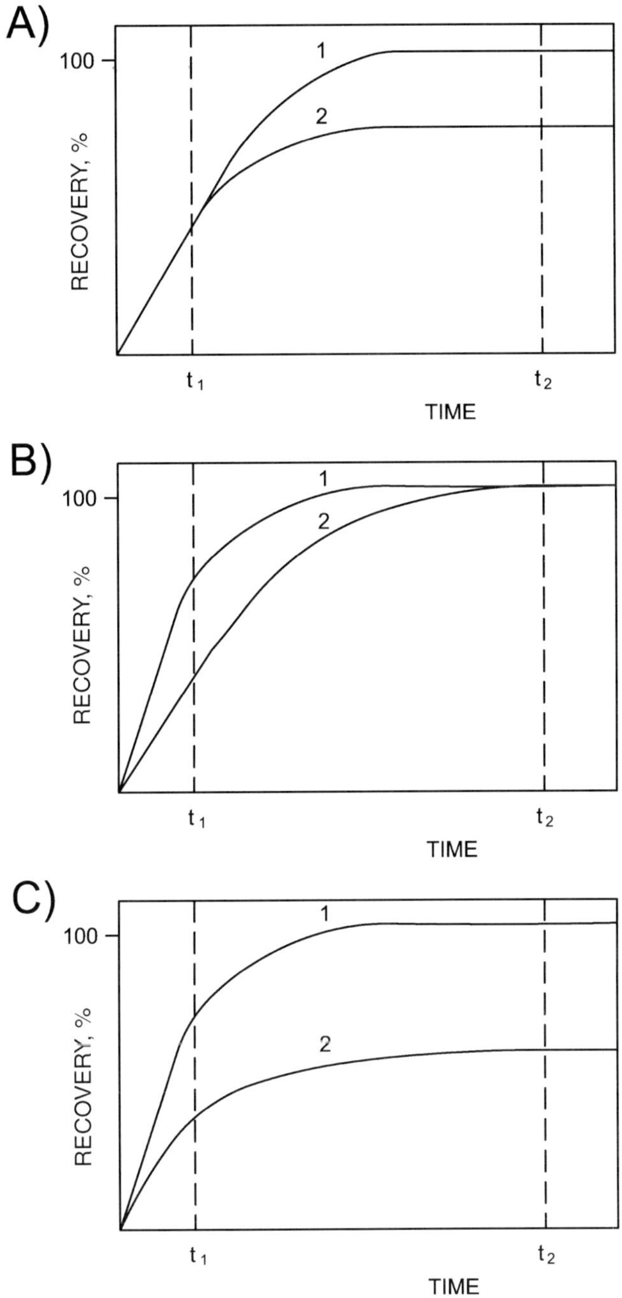

Fig. 7.4. Optimization of an SFE process. Study of the influence of an operating variable at two different values (1 and 2). A to C are graphs of the extraction kinetics in three different situations. The importance of using appropriate values for time t_1 and/or t_2 is apparent. (Reproduced with permission of Springer-Verlag.)

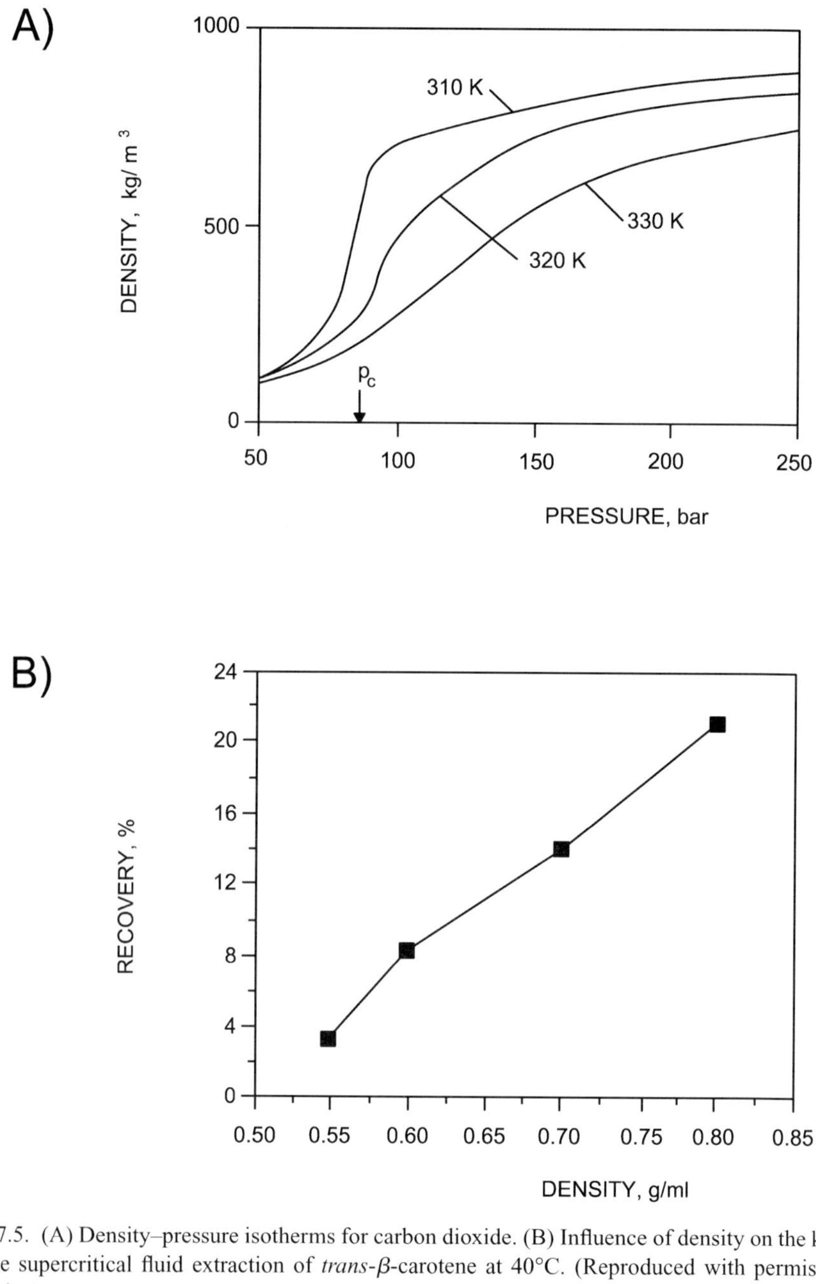

Fig. 7.5. (A) Density–pressure isotherms for carbon dioxide. (B) Influence of density on the kinetics of the supercritical fluid extraction of *trans*-β-carotene at 40°C. (Reproduced with permission of Elsevier.)

changing the nature of the supercritical fluid — unless the fluid induces a deep physical change in the matrix. The increased suitability of polar fluids is a result of their ability to interact with the matrix and efficiently compete with the analyte for its active sites.

Choosing a supercritical fluid on the basis of its Hildebrand solubility parameter is pointless when different fluids are compared; rather, it seems more appropriate to rely on the dipole moments of the fluids. Thus, based on its dipole moment, $CHClF_2$ should be more suitable than CO_2 and N_2O for extracting polar analytes — the opposite conclusion is reached if their solubility parameters are compared, however.

Influence of the fluid temperature

Temperature is of paramount importance in the SFE technique as it affects all extraction steps (desorption, diffusion and dissolution). The relative significance of each step depends on the particular matrix and analyte. Supercritical fluid extraction must be done at a temperature at least 10°C higher than the critical temperature of the fluid. The maximum usable working temperature may be limited by thermolabile components of the sample. The solute volatility and diffusivity increase with increasing temperature; however, the density of a supercritical fluid at a given pressure — and hence its solvent power — decrease with increasing temperature. In addition, the temperature plays a central role in the desorption kinetics, which may determine the extraction rate of adsorbed analytes. The solubility parameter for CO_2 and solutes decreases as the temperature is raised. At high temperatures, a decrease in solubility parameter for the solute can be more significant than one in that for the solvent. All these factors have led to most supercritical CO_2 extractions being done at mild temperatures (e.g. 50°C) and a pressure that is raised as the polarity of the target analytes increases in order to avoid the risks inherent in thermolabile samples. However, Langenfeld *et al.* [27] have shown the temperature to strongly influence the extraction efficiency for polar and strongly retained compounds in environmental matrices such as sediments. The effect of temperature can be related to a reduction in the kinetic constraints associated to overcoming the energy barrier of the desorption step, or to thermal alterations of the matrix.

The rate of diffusion of the solute through the sample matrix and within CO_2 can also be increased by raising the temperature. In addition to increased solute diffusivity, an increase in CO_2 diffusivity can facilitate absorption of the fluid in a polymer and hence diffusion of solutes across it.

Influence of the presence of a modifier

The solvent characteristics of a supercritical fluid can be altered by adding a modifier (also known as an "entrainer" or "cosolvent"). The mechanism of action of the modifier depends on both the type of matrix concerned and the form in which the analytes occur in it. A modifier can have four different effects, namely: (a) increase the analyte solubility by interacting with the solute in the fluid phase; (b) facilitate solute desorption by interacting with bound solutes, the matrix active sites or both; (c) favour diffusion of the solute within the matrix; and (d) hinder diffusion of the solute within the matrix through contraction, which will result in decreased recovery.

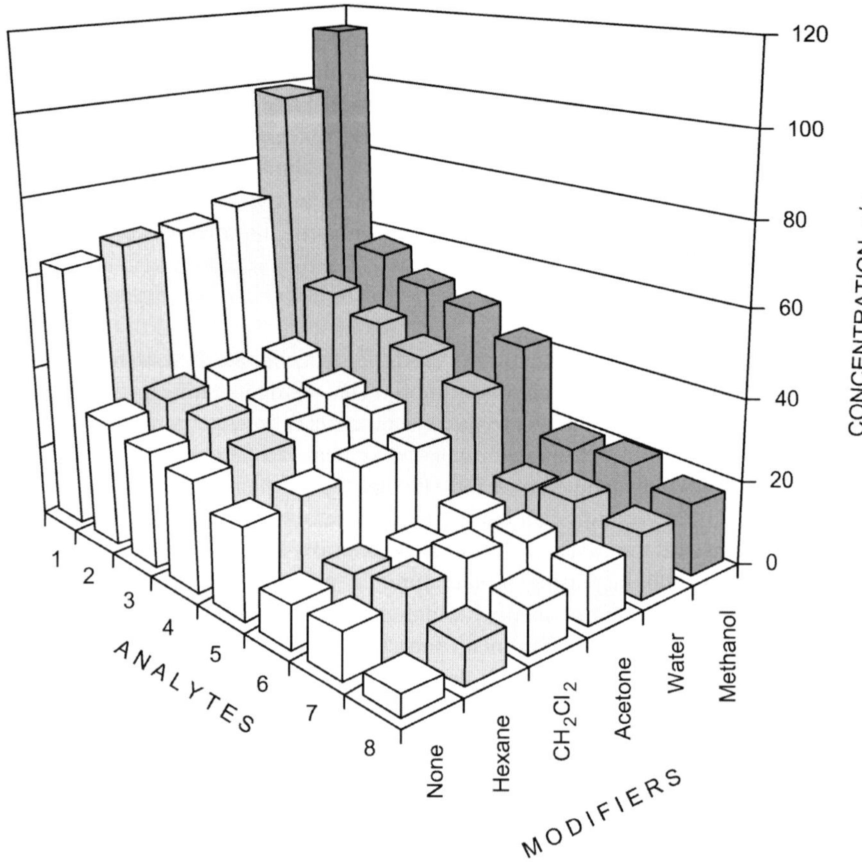

Fig. 7.6. Influence of various modifiers on the extraction efficiency for PAHs. 1 fluoranthene, 2 pyrene, 3 benzo(a)anthracene, 4 chrysene, 5 benzo(b)fluoranthene, 6 benzo(k)fluoranthene, 7 benzo(a)pyrene, 8 indeno(1,2,3-cd)pyrene. (Reproduced with permission of Pergamon Press.)

Modifiers very often accompany carbon dioxide as supercritical extractant. Because CO_2 is non-polar, it is classified as a non-polar solvent; however, because of its large molecular quadrupole, it exhibits some affinity for polar solutes.

Although modifiers are added to supercritical fluids to increase their polarity, they can also impart decreased polarity, aromaticity, chirality and the ability to further complex organometallic compounds. Just as carbon dioxide is the most popular substance for use as a supercritical fluid, it is also that to which modifiers are most frequently added. This is so because modifiers are seen as the means for enabling the use of CO_2 in situations where it may not be the best solvent. For example, methanol is added to supercritical CO_2 to increase its polarity, aliphatic hydrocarbons to decrease it, toluene to impart aromaticity, [*R*]-2-butanol to add chirality and tributyl phosphate to enhance the solvation of metal complexes. The amount of modifier to be added depends on the properties of the extractant and those of the analyte and matrix; usually, it ranges from a few

TABLE 7.4

CRITICAL TEMPERATURE AND PRESSURE FOR SELECTED SUBSTANCES USED AS MODIFIERS OF CARBON DIOXIDE

Substance	T_c (K)	P_c (bar)
Methanol	513	81
Ethanol	514	61
1-Propanol	537	51
2-Propanol	508	48
3-Propanol	536	42
Acetone	508	47
Acetonitrile	546	48
Acetic acid	593	58
Diethyl ether	467	36
Dichloromethane	510	63
Chloroform	536	54
Hexane	508	30
Benzene	562	49
Toluene	592	41
Tributyl phosphate	742	24

microlitres to 5–10% of the main extractant — Cleland *et al.* [28], however, added as much as 20% of methanol to CO_2 to recover arsenic from dogfish muscle. Figure 7.6 illustrates the effect of an identical amount (200 μl) of various modifiers added to supercritical CO_2 for the extraction of PAHs. The volume of methanol added to the sample can dramatically alter the extraction efficiency [29]. Thus, the addition of methanol resulted in drastically reduced recoveries in the extraction of vitamins D_2 and D_3 from pharmaceuticals; also, toluene had no significant effect, whereas 0.25 ml of diethyl ether increased the recovery by more than 10% [30]. The influence of modifiers on SFE performance has been [31] and continues to be documented [32–34]. The organic solvents most frequently used as modifiers for supercritical CO_2 are shown in Table 7.4.

In addition to its nature and concentration, the way the modifier is added (and the imbibition and equilibration times) must be considered in optimizing its usage. The manner in which the modifier is supplied, whether static or dynamic, is dictated by its mechanism of action, the available equipment and fluid costs.

In the static addition mode, a preset volume of modifier highly dependent on sample size is directly added to the sample, held in the cell, prior to extraction. The principal disadvantage of this mode is that, as the supercritical fluid begins to circulate through the sample, the modifier is swept from the extraction cell, so the matrix is brought out of contact with the modified supercritical fluid. Although dynamic addition is usually more effective than static addition — the modifier is continuously passed through the sample in the former — Ashraf and Taylor [35,36] found the static addition of methanol to CO_2 to be more effective in the extraction of PCBs from river sediments. Dynamic addition can be done by using pre-mixed fluid reservoirs or by mixing the pure components in the propulsion system, whether by placing the liquid modifier in the cylinder of a syringe pump or by using two pumps (one for the fluid and the other for dispensing the required amount of modifier).

Most SFE applications use methanol as modifier. In some, however, alternative cosolvents such as hexane, aniline, toluene and diethylamine have proved more effective.

Influence of the extractant volume

The influence of the volume of extracting fluid used is exerted mainly through the solubility of the target analytes in the supercritical fluid. Such a volume must be optimized in terms of the flow-rate, which will allow one to determine the influence of both in addition to that of the analyte solubilities and extraction kinetics.

7.4.2. Properties of the analyte

The solubility of a solute in a supercritical fluid is dictated by its polarity, volatility and molecular weight. Thus, the less volatile and more polar a given analyte is, the less readily soluble in a non-polar extractant such as supercritical CO_2 it will be.

Analytes of a high molecular weight containing polar groups call for high fluid densities and modified supercritical CO_2. On similar polarity, the extraction efficiency decreases with increase in analyte molecular weight. Figure 7.7 illustrates the influence of the molecular weight of PAHs on their extraction at two different efficiency levels [20]. An identical trend is observed in both cases that is consistent with the solubility of PAHs in

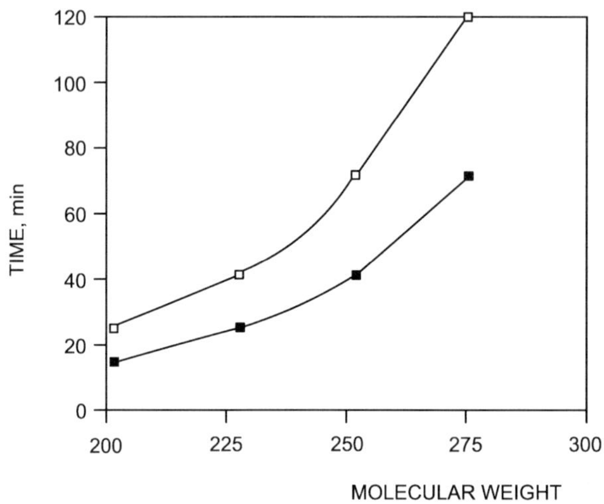

Fig. 7.7. Time required to attain 50% (■) and 80% recovery (□) of PAHs as a function of their molecular weight. (Reproduced with permission of John Wiley & Sons.)

supercritical CO_2 (this decreases with increasing molecular weight of the solute). By way of example of the differential behaviour of extraction depending on both the nature of the solutes and that of the extractant, the SFE efficiency achieved with $CHClF_2$ does not depend on the molecular weight of PAHs; by contrast, the efficiency with which PCBs are extracted by supercritical CO_2 and N_2O usually increases with increasing molecular weight of the solute [37].

The highest possible solubility is achieved when the solubility parameter for the extracting SF is similar to that for the solute. The extraction conditions must be chosen in such a way that the solute solubility in the fluid will be maximal whenever a large amount of solute (a major constituent) is to be extracted. The influence on the extraction efficiency of the analyte concentration in the target sample increases as the analyte solubility decreases.

7.4.3. Properties of the sample

The most influential characteristics of the sample in relation to SFE are the nature of the matrix and its porosity, moisture content, surface-to-volume ratio and size.

Influence of sample size

Increased amounts of sample call for proportionally increased volumes of supercritical fluid. The sample volume and extraction time can have strong effects in those cases where the analyte concentration in the sample is quite high. Figure 7.8A illustrates the recovery of PCBs from river sediment achieved with pure supercritical CO_2. As can be seen, recovery of the target analytes was almost quantitative after 50 min of dynamic extraction of 100 mg of sample; on the other hand, quantitative extraction from 1 g of sample was impossible even after 120 min, possibly because the extractant was saturated with PCBs [38].

Influence of particle size

The SFE rate increases with decreasing particle size and increasing surface area-to-weight ratio. Figure 7.8B illustrates the influence of the former variable on the SF extraction of cocaine from coca leaves [22]. As can be seen, quantitative extraction of cocaine from samples with particle sizes over the range 220–470 μm was impossible even after 60 min; on the other hand, samples with particle sizes of 150–170 μm were quantitatively extracted within 15 min.

Influence of the nature of the matrix

The extraction efficiency depends strongly on the nature of the matrix; as a result, the best supercritical fluid in terms of analyte solubility may not also be the most efficient. Differences in extraction efficiency for a given analyte under identical conditions arise from the matrix. The presence of functional groups on the surface of the matrix or in its components and their ability to bind to the analytes, the organic matter and moisture

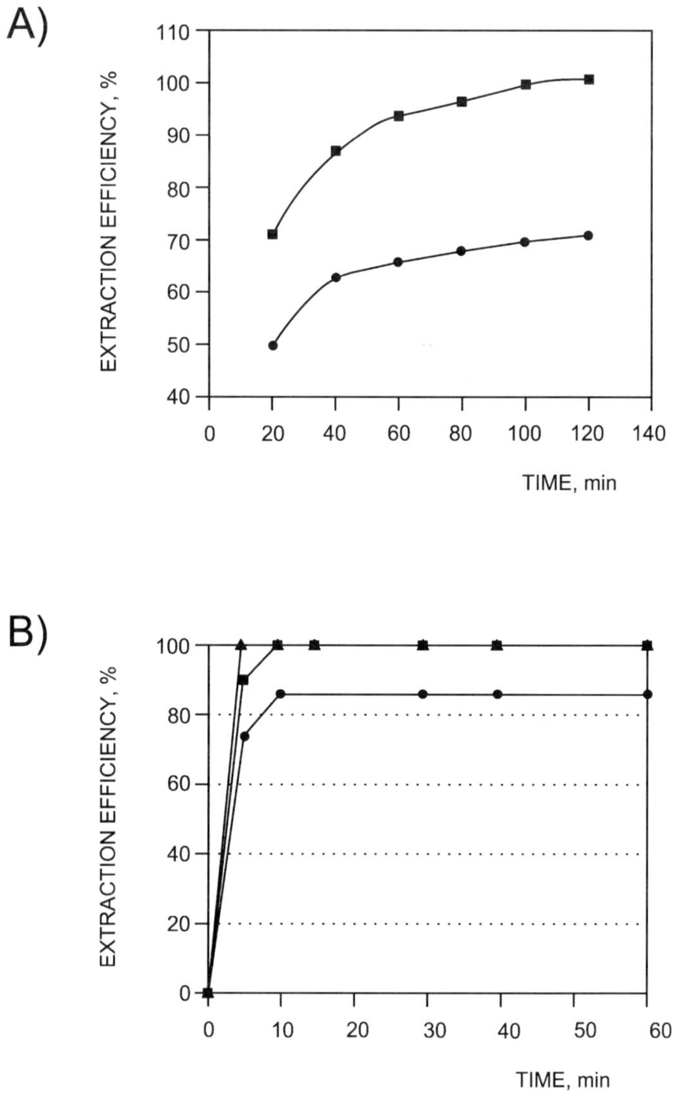

Fig. 7.8. Influence of sample size on the recovery of (A) PCB-52 from river sediment (■ 100 mg, ● 1 g) and (B) cocaine from coca leaves (● 220–470 μg, ■ 150–170 μg, ▲ less than 150 μg). (Reproduced with permission of Springer International and Elsevier, respectively.)

contents, and encapsulation phenomena, are among the matrix-related factors that dictate the ease with which analytes can be extracted.

The ability of the sample to covalently bind the analytes is a function of its (bio)chemical activity. Thus, 2,4-dichlorophenol can be readily removed from straw by SFE since, when spiked, the analyte can hardly bind to the matrix because straw consists almost solely of inactive tissue (lignin, which acts as a weakly absorbing surface for the analytes).

In field-treated matrices, however, some pesticide bound to lignin in the plant may remain in the matrix [39].

Influence of sample moisture

The effects of this variable depend on both the matrix and the analytes. Thus, hydrophobic matrices facilitate penetration of supercritical CO_2; also, because the fluid is water-immiscible, the presence of moisture can make the analytes inaccessible to it. Figure 7.9A illustrates the influence of this variable on the kinetics of extraction of pyrene from natural sludge, both as collected (45% moisture) and after air-drying (2% moisture) [40]. However, water added to the sample can facilitate extraction (e.g. that of caffeine from coffee beans, Fig. 7.9B). Natural samples with a high moisture content can plug restrictors as a result of the water they contain freezing at restrictor tips. This problem can be overcome by raising the restrictor temperature, at the expense of losses in the more volatile analytes.

Removal of fat in the sample together with the analytes can hinder collection of the analytes and require clean-up of the extract prior to analysis.

7.4.4. Dynamic and geometric variables

Dynamic factors are among the key variables to be optimized in an SFE process. In addition to extracting the analytes, the primary function of the supercritical fluid is to transport the solutes to the collecting vessel or to an on-line coupled chromatograph or detector. Ensuring efficient transportation of the analytes following separation from the matrix entails optimizing three mutually related variables, namely: the flow-rate of the supercritical fluid, the characteristics of the extraction cell and the extraction time. These factors must be carefully combined in order to allow the flow-cell to be vented as many times as required.

Influence of the flow-rate

In some cases, the flow-rate is the single factor most strongly influencing the extraction efficiency: the higher it is, the greater the amount of analyte that is extracted over a given interval [38]. In other cases such as the extraction of fluoranthene from soil, the flow-rate scarcely affects the results above a 0.3 ml/min level [41]. The optimum conditions for a particular extraction cannot be established on the basis of the number of times the extraction cell is vented because the extraction of most analytes is kinetically limited. Occasionally (e.g. in the extraction of cocaine from coca leaves), the process is governed primarily by the solubility of the analyte rather than by its desorption kinetics (see Fig. 7.10) [22].

Influence of the extraction cell

The size, geometry and void volume of the extraction cell, in addition to its amenability to decreasing compaction by stirring or ultrasonication, all influence the SFE process.

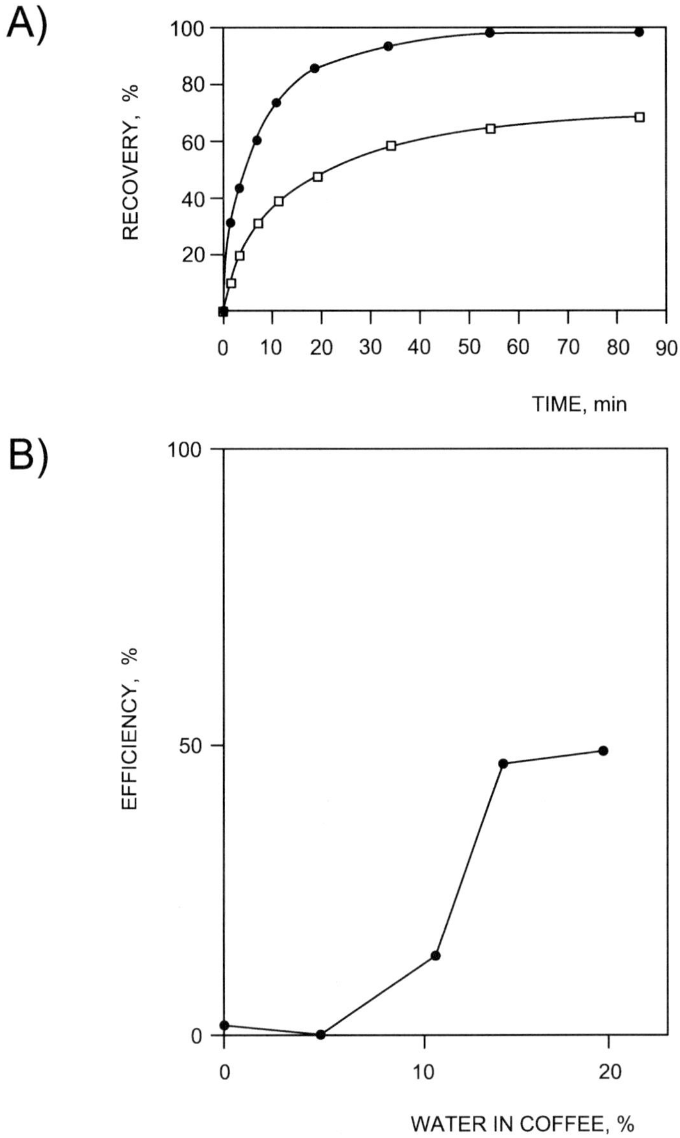

Fig. 7.9. Influence of moisture content on the efficiency of the SFE with CO_2. (A) Kinetic curves for the extraction of pyrene from sewage sludge (● 2% moisture, □ 45% moisture). (B) Extraction efficiency for caffeine in coffee with variable moisture content. (Reproduced with permission of the American Chemical Society and Elsevier, respectively.)

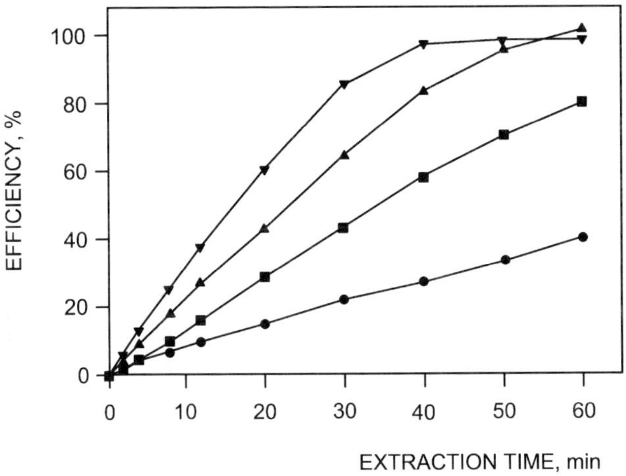

Fig. 7.10. Influence of the flow-rate on the extraction of fat from potato chips using supercritical CO_2 at 340 bar at 60°C. (●) 0.7 ml/min, (■) 1.4 ml/min, (▲) 1.9 ml/min, (▼) 2.6 ml/min. (Reproduced with permission of the American Chemical Society.)

As a rule, the volume of supercritical fluid passed through the cell in one extraction should be at least ten times higher than that of the cell; in this way, if the inner volume of the cell is reduced (e.g. by inserting a piece of an inert material), the extractant passing through it will be renewed more frequently. The effect of the cell void volume varies dramatically depending on whether the extraction is dynamic or static in nature [42]: while a high void volume favours static extraction (it allows more fluid to be held in the cell and the analytes to be more extensively solvated), a high cell volume can lead to incomplete extraction in the dynamic mode as a result of the flow velocity being inadequate to sweep the analytes when the sample deposits on the cell walls. The linear velocity of the supercritical fluid is probably the most influential factor related to cell size. Figure 7.11 illustrates the influence of this parameter on the extraction of two dioxins [43]. As can be seen, the smaller cell provided the better results in both instances.

The influence of cell geometry has been found to vary widely depending on the form (natural or spiked) in which the analytes are present in the matrix. At a constant cell inner volume, the effect of cell geometry on the extraction efficiency for natural samples such as lemon or soil is negligible provided elongated cells of small inner diameters or wide, short cells are used [31]. Conversely, recovery from spiked samples decreases with decreasing inner diameter-to-length ratio [44,45]. While the orientations of the flow and cell have minimal effects on the extraction rate for PAHs, the cell dimensions strongly influence their recovery.

Influence of the extraction time

As a rule, natural samples require longer times for quantitative extraction of analytes than do spiked samples. Thus, complete extraction of cholesterol from egg yolk takes

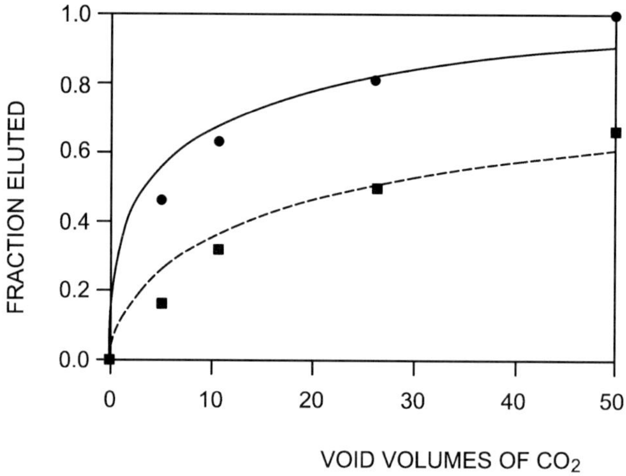

Fig. 7.11. Influence of the extraction vessel size on the SFE of dioxins from Florisil. (●) Volume 0.5 ml, length 6 cm. (■) Volume 5 ml, length 12 cm. (Reproduced with permission of the American Chemical Society.)

60 min compared to only 30 min for the quantitative extraction of this substance added to glass wool [18]. Even with natural samples, the time required to extract a given analyte varies markedly depending on the type of matrix that contains it. One case in point is the extraction of PAHs from certified reference materials (CRMs) with a matrix consisting exclusively of soil [46].

7.4.5. Analyte collection modes

The effectiveness of the different ways in which the depressurized mixture emerging from the restrictor can be collected depends on the characteristics of the particular extractant and analytes. When the extractant is a gas under atmospheric conditions (e.g. CO_2), the three collection modes described in Section 7.3 are applicable.

Solvent bubbling

The temperature of the collecting solvent is one of the most influential variables in this context as it dictates to what extent the solvent will evaporate or be swept by evolved gas bubbles. The temperature of the bubbling system is altered by effect of the restrictor end — which is usually kept at a high temperature to avoid plugging through saturation — being immersed in the solvent, and of the cooling resulting from expansion of the supercritical fluid. The collection efficiency can be boosted by altering several experimental variables (e.g. by using a vial with a small inner diameter to increase the distance travelled by the bubbles in a given collector volume or by placing a filter at the restrictor outlet in order to reduce bubble size).

Sorption

This collection mode entails the use of an appropriate solid, either in the line or at the restrictor outlet, whether packed in a column or as a bed. The material used for this purpose should ensure proper retention with minimal or no analyte losses, and enable the use of the best possible type and volume of eluent to desorb the target analytes. When the extracted analytes differ in polarity, a mixture of sorbents must be used to ensure adequate collection efficiency.

Comparative studies of potential sorbents for SFE have been conducted, using the same or different fluids as extractants. Thus, the extraction of sulfonamides from chicken liver using supercritical trifluoromethane or CO_2, both containing methanol as modifier, was followed by off-line collection on neutral alumina or C_{18} and a 1:1 phosphate/methanol rinse with small differences in performance among sorbents [47]. Subsequent experiments using variably sized grains of silanized glass, synthetic silica beads, C_{18}-modified silica, fused silica beads, cyano-modified silica and mixtures thereof, as well as cryogenic trapping on stainless steel balls, for the collection of compounds such as acetophenone, *N,N*-dimethylaniline, naphthalene, 2-naphthol and *n*-tetracosane following SFE with modified CO_2 revealed the best collecting material to be a C_{18}-silica bead mixed phase [48].

Supercritical fluid modifiers alter the behaviour of the sorbent materials, and hence their performance, by changing or saturating their active sites. The effect can be checked experimentally. Thus, the extraction of dinitroglycerin and trinitroglycerin retained in granulated active coal with pure CO_2 and the same fluid containing 2% methanol, followed by collection in a C_{18} sorbent, was found to be much more efficient with the modified supercritical fluid: 9.9% versus 2.9% for dinitroglycerin, and 97.3% versus 89.6% for trinitroglycerin [49]. The improved efficiency, however, could not be unequivocally ascribed to the presence of the modifier in the extraction step or the collection step. Mulcahey and Taylor studied the influence of the trap temperature on the collecting efficiency for a series of organic compounds. Figure 7.12 exposes the adverse effect of the modifier (4% and 8% methanol added to supercritical CO_2) on the collection of acetophenone and *N,N*-dimethylaniline on ODS [50].

Cryogenic trapping

This collection mode involves deposition on stainless steel by the sole effect of cooling the restrictor and pointing its outlet at metal balls. As a result, its performance is markedly temperature-dependent. Experiments with open and sealed collection systems have revealed the former to be subject to much more significant losses [51]. Obviously, the volatility of the target analytes plays a crucial role also in this step and, as such, must be carefully optimized.

7.5. APPROACHES TO IMPROVING SFE PERFORMANCE

Although supercritical fluids are widely regarded as "supersolvents", this designation is utterly unrealistic, as one can easily see by comparing their solvent powers with

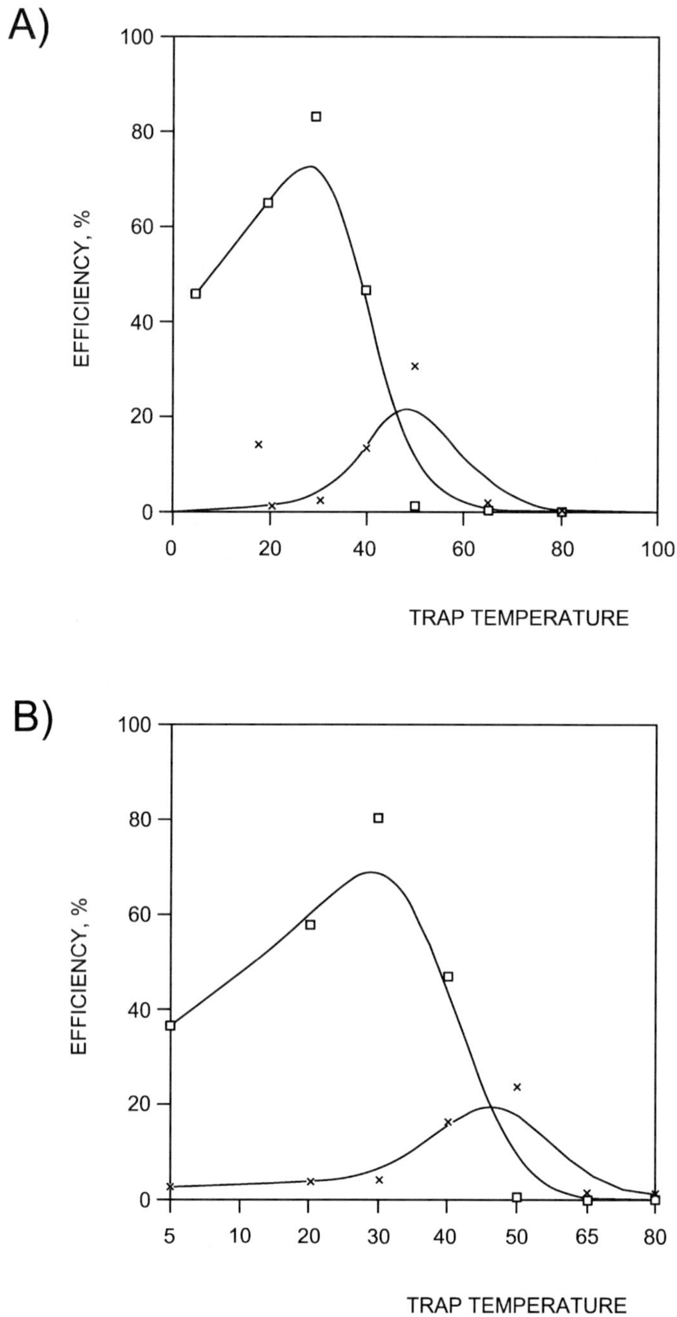

Fig. 7.12. Variation with trap temperature of the recovery of acetone (A) and *N*,*N*-dimethylaniline (B) on an ODS trap using supercritical CO_2 modified with (□) 4% and (×) 8% methanol. (Reproduced with permission of the American Chemical Society.)

TABLE 7.5

APPROACHES TO IMPROVING SFE PERFORMANCE

• Altering the properties of the supercritical fluid	• Esterification and related reactions
• Using a polarity modifier	• Formation of organometals
• Using an alternative supercritical fluid	• Chelation
• Ion-pairing	• Micellization

those of conventional liquids. In any case, such features as increased solute diffusivity, decreased viscosity and increased solvent strength, all of which are easier to control than in liquid solvents, have promoted the use of supercritical fluids for treating solid samples. Most analytical supercritical fluid extractions have focused on CO_2 as fluid on account of its high inertness and purity, low toxicity and moderate cost. Although CO_2 is an excellent solvent for non-polar organics, its polarity is occasionally too low to ensure efficient extraction, either because the analytes are poorly soluble or because the extractant is unable to displace them from the active matrix sites. In any case, the potential of CO_2 for analytical SFE has been exaggerated, mainly because of its excellent performance in the extraction of analytes spiked to inert supports. The far from natural behaviour of these solid samples has led some to consider supercritical CO_2 the ideal SFE leacher. However, matrix–analyte interactions dramatically reduce its extraction efficiency, the effect increasing with increasing polarity of the analytes.

The performance of SFE has been improved since its inception by circumventing the shortcomings that hinder quantitative leaching of polar and ionic species. The ensuing modifications have ranged from the mere alteration of extractant characteristics such as pressure and temperature to the use of complex derivatization sequences and include the manipulations shown in Table 7.5, which are used to raise the polarity of the extractant and lower that of the target species. The effects of changing the physical properties of the supercritical fluid and using a modifier are discussed in Section 7.4, so only the other approaches are dealt with here.

7.5.1. Using an alternative supercritical fluid

One way of improving SFE efficiency is by using a more suitable SF to extract the target analyte. Unfortunately, the choice of fluids other than CO_2 is restricted by the desire to have reasonable critical parameter values and costs, chemical inertness, low toxicity and little environmental impact. The use of supercritical N_2O has proved to increase the extraction efficiency for high-molecular weight PAHs and chlorinated dibenzo-*p*-dioxins from fly ash and sediment [52]. This extractant, however, does not always improve the extraction efficiency [53]; also, it can be explosive in the presence of reactive organics. Other polar fluids such as $CHClF_2$ (Freon-22) have exhibited increased efficiency in the extraction of nitrated and non-nitrated PAHs from particulate matter in diesel exhaust [54]. Freon-22 has also been found to allow significantly fast and effective extraction of steroids relative to CO_2 [55]. A comparison of Freon-22, N_2O and CO_2 as supercritical fluids for the extraction of native pollutants including PCBs from railroad bed soil showed the first to be the most efficient — most likely as a result of its high dipole moment. Trifluoromethane has also been tested as an extraction solvent for sulphonylurea herbicides [56]. Unfortunately, the high environmental impact of these alternative extractants significantly detracts from their use. Supercritical methanol is one other excellent solvent but is less attractive to use because it is liquid under ambient conditions, which complicates sample concentration following extraction [57]. Supercritical ammonia may also be an attractive solvent as regards strength [58] but is difficult to propel because it tends to dissolve pump seals, is chemically reactive and probably too hazardous for routine use.

One of the most promising alternatives to CO_2 is supercritical water. Like CO_2, water is an environmentally acceptable solvent, but has not yet received attention as an analytical extractant for environmental solids because, in contrast to CO_2, water is too polar to efficiently leach most non-ionic organics associated to environmental solids. As can be seen in Table 7.2, water has a dielectric constant of *ca.* 80 at ambient temperature and pressure; fortunately, however, the constant can be substantially lowered to accommodate much milder and potentially useful conditions simply by raising the temperature at a moderate pressure (see Fig. 7.13, which shows that water can be made even less polar than CCl_4). Thus, simply by changing its temperature, the solvent polarity of water can be expanded over a very broad range. By contrast, the dielectric constant of supercritical CO_2 ranges from 1 to 1.6 only. As shown by Hawthorne *et al.* [59], water can be

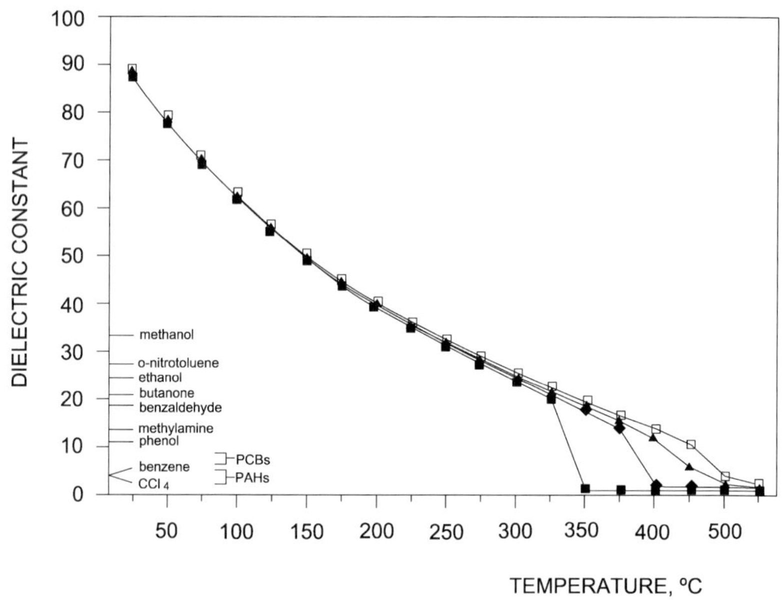

Fig. 7.13. Influence of temperature and pressure on the dielectric constant of water in relation to various typical organic solvents. (■) 100 bar, (◆) 200 bar, (▲) 300 bar, (□) 400 bar. (Reproduced with permission of the American Chemical Society).

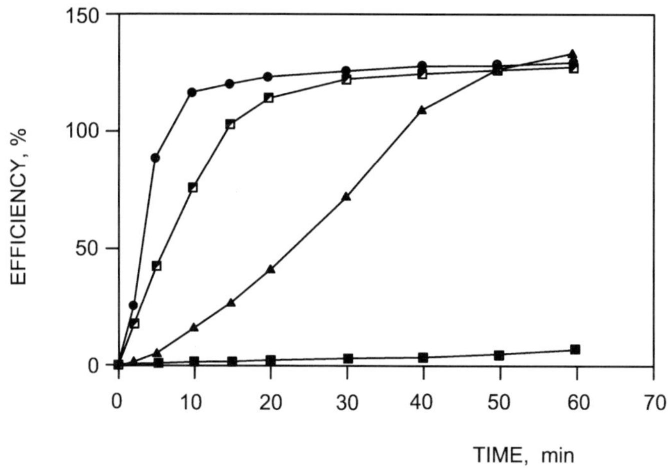

Fig. 7.14. Influence of the extraction temperature on the extraction rate for benzo(a)pyrene in highly polluted soil using water at 350 bar. (●) 300°C, (◪) 250°C, (▲) 200°C, (■) 120°C. (Reproduced with permission of the American Chemical Society.)

used to selectively extract a wide range of ionic to non-polar analytes by raising the temperature from subcritical to supercritical values. Figure 7.14 shows the effect of altering the temperature on the extraction of benzo(a)pyrene, which was recovered by only 5% after 1 h at 120°C but quantitatively leached in less than 5 min at 300°C [59].

The behaviour of water under supercritical conditions has been used to separate ionic species by precipitation as salts combined with the oxidation of organic species [60].

7.5.2. Ion-pairing

Neutralizing charged species by formation of ion-pairs is one way of lowering the polarity of ionic compounds and increasing their solubility in low-polar extractants such as supercritical CO_2 when the addition of a polar cosolvent is ineffective. Thus, secondary alkanesulphonate and linear alkylbenzenesulphonate surfactants were quantitatively determined in sewage sludge extraction with supercritical CO_2 in combination with ion-pairing using tetrabutylammonium hydrogen sulphate, followed by injection-port derivatization GC [61]. Another quaternary ammonium cation, trimethylphenylammonium, was used as a counterion for the quantitative recovery of sulphonamides with CO_2 at 40°C at 281 bar as extractant [62]. Clenbuterol was also extracted from both diatomaceous earth and food matrices (viz. feedstuff, freeze-dried milk and liver), following ion-pairing with 10-camphorsulphonate; supercritical CO_2 was used in the dynamic extraction mode at 383 bar at 40°C after the addition of the ion-pair forming reagent to the extraction cell containing the sample. The ammonium salt of the reagent was found to provide better results than its acid form [63].

7.5.3. Esterification and related reactions

In situ chemical derivatization under supercritical conditions is one other way of increasing the extraction efficiency for polar organics as it promotes the conversion of polar (hydroxyl and carboxyl) groups in the analytes to less polar functions (ether, ester and silyl derivatives) that facilitate dissolution in supercritical CO_2 according to Field [64]. Also, it conditions the analytes for their subsequent chromatographic determination. On-line derivatization with catalysts such as solid alumina [65] or immobilized enzymes [66] to transesterify triglycerides to their methyl esters cannot be included among the recovery improving approaches discussed in this section as it requires the prior extraction of the analyte.

Hawthorne *et al.* [67] succeeded in derivatizing, and extracting under supercritical conditions, the acid herbicides 2,4-dichlorophenoxyacetic acid and Dicamba from soil and sediment, microbial phospholipid fatty acids (as methyl ethers) from whole cells, and wastewater phenolics (as methyl ethers) from water and from C_{18} sorbent discs, using reagents including trimethylphenylammonium hydroxide and boron trifluoride in ethanol. Phenoxycarboxylic acid was also extracted in CO_2 following esterification with pentafluorobenzyl bromide in acetone and K_2CO_3 [68]. Methyl iodide was also used to methylate the same herbicide. Two different procedures for insertion of the derivatizing reagent were compared and pipetting it onto prepared samples was found to be less demanding of reagents and equipment, and also to provide more repeatable results than continuously supplying the derivatizing reagent to the extraction cell [69]. Pentafluorobenzyl bromide was also used in the in situ esterification and supercritical fluid extraction of a non-polar hydrophobic pesticide (Dacthal) and its mono- and dicarboxylic metabolites from soil and ground water, following solid-phase extraction on strong anion-exchange discs [70]. The in situ derivatization and extraction of volatile fatty acids entrapped on an ion exchange resin from aqueous solutions and urine as a test matrix using pentafluorobenzyl bromide in supercritical CO_2 has also been reported [71]. Catalysis by lipase, which converted various lipids from meat and oilseed to their methyl esters, was also used with an SFE cell where the presence of a small amount of water was found to result in improved efficiency [72].

In addition to increasing the solubility of polar compounds in supercritical CO_2, derivatization reagents may compete with the analytes and displace them from the active sites of the matrix. In this way, the extraction of non-reactive substances can also be improved when the active sites of the matrix are targets of the derivatizing reagent, as shown by the direct addition of a commercially available silylation reagent (hexamethyldisilane–tetramethylchlorosilane) to the sample matrix prior to the SFE of roasted coffee beans, roasted Japanese tea and marine sediment. The addition of the reagent improved the extraction yield for both derivatized (trimethylsilyl derivatives) and underivatized species [73]. The same reagent was also found to improve the extraction efficiency for PAHs, which lack reactive groups [74].

7.5.4. Formation of organometals

Supercritical fluid extraction has proved suitable for separating organometallic compounds from solid matrices, as shown for methylmercuric chloride, which was quantitatively extracted from a cellulose-based matrix by neat CO_2 supplied with a small amount of water as matrix modifier [75]; and also for organotin compounds, which were successfully extracted from environmental samples using SFE–GC–AES [76] and SFE–SFC–FID [77,78].

The behaviour of the metal towards supercritical fluid extraction depends on the number of organic molecules constituting the compound concerned and has led to the development of new approaches to lowering the polarity of analytes by formation of organometals and hence to enabling speciation. One particular method allows the simultaneous determination of tributyltin, diphenyltin and triphenyltin in sediment by in situ derivatization with ethylmagnesium bromide in the extraction cell [79,80]. Also, organic and inorganic arsenic from sand supports was speciated by GC following SFE with in situ derivatization with thioglycolic acid methyl ester [81]. Most reported SFE methods involving the formation of organometals were developed by using very simple matrices that were spiked with the target analytes, so they may not be as effective with natural samples.

7.5.5. Chelation

Ever since the first supercritical fluid extraction of metals from aqueous solutions and silica using in situ chelation was reported [82], a number of approaches based on the use of various types of ligands to extract metals and, usually, subsequent ion-pairing, have been developed. The principal types of reagents used for this purpose are described below.

Dithiocarbamate derivatives

A study of the influence of the type of tetraalkylammonium dialkyldithiocarbamate used in the recovery of metals from aqueous media confirmed the hypothesis that the non-polar character of the ion-pair formed depends primarily on the chain length of the alkyl substituent on the carbamate nitrogen and, to a lesser extent, on the chain of the alkyl substituent on the ammonium counterion [83]. This inference has been used to characterize cadmium, copper and zinc bound to metallothioneins previously isolated from rabbit liver by using supercritical CO_2 in conjunction with tetrabutylammonium and dibutyldithiocarbamate derivatizing reagents [84]. This approach has shown great promise for the speciation of protein-bound heavy metals. One other salient contribution in this field is the off-line complexation–supercritical fluid extraction of 13 organotin compounds from soil and sediments using diethylammonium diethyldithiocarbamate. The extracted material was subsequently treated with pentylmagnesium bromide to convert the ionic organotin compounds into their neutral derivatives, which are amenable to gas chromatography with atomic emission detection [85]. Especially worth noting among the results were the improved behaviour of modified CO_2 versus neat CO_2 and of a high pressure (450 bar) versus a low one (250 bar); also, the predicted higher recovery of tetrasubstituted compounds was found, rather, to decrease when the number of organic substituents was reduced [86].

Fluorinated dithiocarbamate derivatives yield metal complexes the solubilities of which are two to three orders of magnitude greater than those of the ligands containing no fluorine, as shown in the extraction of inorganic mercury [75], and in that of Zn, Cd, Co and Cu, from cellulose matrices [87]. Recently, fluorinated and non-fluorinated dithiocarbamates have been used to chelate mercury in matrices such as sludge, fly ash, soil, filter paper and sand; the extraction efficiency for some derivatives such as $Hg(CH_3)_2$ was found to be independent of the particular matrix, whereas that for others (e.g. $HgCH_3Cl$) was strongly matrix-dependent [88.]

Fluorinated β-diketones

Compounds such as fluorinated β-ketones have been used mainly for the extraction of lanthanides and actinides, as more effective ligands than fluorinated dithiocarbamates for complexation with f-block elements. In addition to complex formation, some analytes (e.g. tervalent lanthanides) require the presence of a small amount of water containing 5% methanol — which act as matrix and solvent modifier, respectively — for quantitative extraction [89]. Also, a synergistic effect on the SFE of actinides [90,91] and of lanthanides from cellulose [92] and acid solutions [93] was observed when using a mixture of tributylphosphate (TBT) and a fluorinated β-diketone. The effect was ascribed to competition of TBT with the matrix for the unoccupied coordination sites of lanthanides and actinides. Thus, the formation of adducts with the complexes of these analytes with fluorinated β-ketones in supercritical CO_2 facilitates their removal from the solid matrix.

Ionizable crown ethers

The use of ionizable crown ethers to improve SFE applications has so far been restricted to the extraction of mercury ions spiked to uncomplicated matrices such as sand, cellulose filter paper and liquids; in any case, the results obtained under mild SFE conditions (viz. 200 bar and 60°C) have been quite acceptable. In addition, the extraction is made highly selective: such ions as Cd(II), Co(II), Mn(II), Ni(II), Pb(II) and Zn(II) are virtually unextractable under these conditions [94].

7.5.6. Micellization

A more recent approach to decreasing the polarity of analytes with a view to facilitating their extraction with supercritical CO_2 involves the formation of reversed micelles. As the surfactant can provide the extractant with its non-polar or low-polar region for contact, micelles can be more easily accepted by CO_2 than a polar analyte if an appropriate micelle former is used. One typical example of this behaviour is the assay of non-ionic (Triton X-100), anionic (sodium dodecyl sulphate, SDS) and cationic (cetyltrimethylammonium chloride, CTAB) surfactants to improve the extraction of cholesterol. These surfactants were individually added to samples containing cholesterol, which, after drying at 70°C for 10 min, were extracted with supercritical CO_2. Table 7.6 shows the results obtained for diatomaceous earth spiked with the analyte. The addition of Triton X-100 to the sample and dynamic extraction at 383 bar at 40°C for 20–40 min ensured quantitative extraction of the target analyte from low-cholesterol and high-cholesterol foods as demonstrated with a certified reference material and the use of various derivatization reactions depending on

TABLE 7.6

EFFECT OF VARIOUS REVERSED MICELLE-FORMING REAGENTS ON THE SFE OF CHOLESTEROL[a]

Drying time,[c] min	Percent recovery of cholesterol ± SD [b] Triton X-100 microemulsion	SDS microemulsion	CTAB microemulsion
0	39.9 ± 1.4	17.5 ± 3.0	17.0 ± 2.6
10	65.5 ± 2.8	26.3 ± 2.0	57.2 ± 1.7

[a] Experimental conditions: P = 189 bar, T = 40°C, q_{CO2} = 1.0 ml/min, $t_{extract}$ = 10 min
[b] $n = 3$ [c] T = 70°C

the cholesterol content (viz. the Liebermann–Burchard reaction with high-cholesterol samples [95] and an enzymatic reaction with low-cholesterol ones [96]) prior to photometric detection. Based on the results, the use of organized media constitutes a simple, convenient way of expanding the scope of application of CO_2 as a supercritical extractant.

The performance of some of the previous approaches (e.g. in situ derivatization, chelation and ion-pairing) has only been assessed with spiked inert supports or samples involving no binding phenomena (e.g. drugs covalently bound to proteins or carbohydrates) or preferential reactivity of the reagent towards other matrix components. In some cases, recoveries were either not reported or reduced by more than 30% when a natural sample or a certified material was processed [70]. Even such straightforward systems as spiked supports may lead one to overestimate the potential of the SFE technique. In this respect, the complementary use of subcritical and supercritical water, or even mixtures thereof, together with the in situ application of focused microwaves, may enable the development of sample preparation methods encompassing the entire range of polarity of the target analytes.

7.6. HYPHENATED TECHNIQUES

Because SFE is a sample extraction–separation technique, it must precede other steps of the analytical process if the type and content of the species of interest in the extract are to be accurately determined. The analytical equipment required to develop the steps following extraction can either be coupled on-line to it or performed off-line. The way the extract should be treated with a view to identifying or quantifying the target analytes depends on its complexity and the type of information required. Thus, the analytes may require clean-up, individual separation or some other treatment prior to reaching the detector.

The off-line mode is to be preferred when a deep knowledge of the features of the extraction process concerned is required as this will allow such experimental variables as pressure, temperature, SF polarity and flow-rate, extractor volume and dimensions, extraction time and sample size to be optimized

The simplest step following supercritical fluid extraction is clean-up, which is frequently done off-line in optimizing a method (e.g. in the SFE of organochlorine pesticide residues from garlic [97]) despite the simplicity of using a cartridge coupled on-line to either the restrictor outlet or one of the usual collection systems (bubbling, sorption, cryogenic trapping). The on-line coupling of SFE to capillary zone electrophoresis (CZE) is technically more complicated; for this reason, the determination of carbamate residues in tobacco uses an off-line assembly (mainly because of the need to concentrate to dryness under a nitrogen atmosphere and to clean up the analytes by solid-phase extraction prior to insertion into the capillary) [98].

On-line coupling can be used to integrate the extraction step with others of varying complexity ranging from column clean-up [99] to HPLC–GC–MS [100]. On-line methods are especially sensitive as a result of the whole extracted material being transferred to the coupled equipment, which also minimizes potential errors arising from avoidance of some manipulation step. In addition to increased sensitivity and automatability, direct coupling of SFE to other techniques (particularly chromatographies) provides additional benefits derived from the fact that the target species are not exposed to atmospheric oxygen, light or high temperatures during analysis.

7.6.1. Supercritical fluid extraction–chromatography

All three types of chromatograph (gas, supercritical fluid and liquid) have been used in combination with SF extractors. The effectiveness of these combined assemblies rests on the availability of an appropriate interface between the two coupled techniques; this role can be assigned to an external accumulator or to a direct connection to the chromatographic column or the retention interface.

The function of external accumulator can be served by a six-way valve, the extract being trapped in its loop as the supercritical fluid is depressurized — the trap can optionally be packed with a sorbent. After trapping, the eluent sweeps the retained compounds to an HPLC instrument or the loop is heated to deliver the extract to a gas chromatographic column. In the latter case, the system can be used with volatile substances.

When the extracted analytes are to be retained directly on the chromatographic column or at the retention interface, their insertion can be accomplished in various ways, namely: (a) by injection into the column, whether directly (SFC, GC) or with the aid of a cooling system (GC, HPLC); (b) by split–splitless injection (SFC, GC); (c) by using a programmed temperature vaporizer (GC); or (d) by injection into a cold trap and subsequent thermal desorption (GC) or elution (HPLC).

SFE–GC

Although gas chromatography can only be applied to thermally stable compounds, it has been widely used in conjunction with SFE. Using these two techniques jointly, however, entails considering the following facts if proper performance is to be expected:

(a) Because the supercritical solvent is usually a liquefied gas, its return to the gaseous state on depressurization before reaching the interface causes it to expand by a factor around 1000.

(b) The solvent power of the most common supercritical fluid, CO_2, is restricted to non-polar and scarcely polar compounds, so substances dissolved in it are usually eligible for GC analysis. This is not the case, however, with highly polar compounds, which tend to decompose at the working temperatures typically used in GC.

(c) The amount of solute that is transferred to the column can be diminished by formation of a frozen CO_2 plug in its head. Any water present in the sample may also plug the column restrictor with ice deposits. Both shortcomings can be circumvented by using a hot, split or splitless injector as interface or a cooled thermal desorption injector.

(d) The presence of modifiers is occasionally incompatible with an on-line coupled SFE–GC system. Such is the case with the recovery of pyrene vapour deposited on silica gel after extraction with methanol- or toluene-modified supercritical CO_2, which results in the loss of all pyrene extracted even if a retaining device is used between the injection port and the column. By contrast, using neat CO_2 results in no significant loss of extracted pyrene [101].

(e) The restrictor diameter is a key variable in SFE–GC applications (not only with direct injection into the column, the diameter of which is dictated by the flow-rate of gas resulting from the depressurized supercritical fluid).

Both packed and capillary columns are used with the SFE–GC tandem. The detector is usually of the mass spectrometry, flame ionization or electron capture type. The most suitable interface for each application will be that allowing the extractant to be removed prior to the column — and hence to the detector.

The reproducibility of the SFE–GC hyphenated technique is comparable to that of off-line systems [102,103].

SFE–SFC

One of the most immediate advantages of coupling supercritical fluid extraction and chromatography is that the former constitutes a highly suitable means for delivering samples to the latter. Because the solvent used to inject the sample is the same as the mobile phase, the principal requisite for effectively coupling two techniques (viz. compatibility between the output of the first system and input of the second) is met. One additional advantage of this hyphenated technique over SFE–GC and SFE–HPLC is the low likelihood of sample components insoluble in the mobile phase reaching the chromatographic column.

The potential of the SFE–SFC couple has been assessed by a number of authors as this combined technique solves one of the major problems encountered in using SFC for bio-analysis. In fact, many pretreatments for biological samples use polar solvents, which are highly detrimental to the phase systems employed in this technique. This problem is readily solved by the SFE–SFC combination, proper performance of which, however, relies on the following conditions:

(a) The extraction chamber volume should be suited to the sample size afforded by the SF chromatograph;

(b) The pressure drop in the supercritical fluid during transfer of the extract to the chromatograph should be as low as possible;
(c) The chromatograph should be pressurized and equilibrated at the pressure to be used in the chromatographic determination before it receives the extract.

All three types of ordinary SFC columns (viz. open-ended, packed capillaries and normal-bore models) have been used with the SFE–SFC couple.

Many SFE–SFC applications use a second pump to pressurize the extraction chamber, the first pump being solely employed to effect the chromatographic separation. These systems pose few technical problems and are highly flexible by virtue of the extractor and chromatograph operating independently (in a static or dynamic manner).

Supercritical fluid extraction coupled to SFC has been used for the extraction, separation and identification of PAHs from coal. The supercritical extract was expanded with the aid of a frit restrictor accommodated in the sample cavity of a cooled micro-injector, the analytes being deposited by condensation while CO_2 was sent to waste through a vent valve. Subsequently, the loop contents were connected on-line to the mobile phase of the capillary chromatograph. The extracted analytes were detected by off-line FTIR spectroscopy following collection on a KBr disc and evaporation of the solvent [104].

Edible fat components were isolated and quantified by using open tubular columns in an SFE–SFC assembly consisting of three serially arranged ovens, namely: a household oven at 50°C for the extraction cell; a deactivated fused-silica capillary (1.8 m × 100 μm i.d.) as trapping tubing in the second oven; and a third accommodating an HP-5790 0.25 μm SB-cyanopropyl-50 column and an FID. The SFE–SFC interface was a fused-silica capillary (4 cm × 320 μm i.d.) glued to the trapping tubing via a coupled glass adaptor. During dynamic extraction, CO_2 was vented to the atmosphere via a short piece of fused silica tubing. Trimyristin, tripalmitin and cholesterol were determined in this way [104].

Commercially available SFE–SFC equipment such as that from Computer Chemical Systems, which comprises an extractor, an accumulation module and a chromatograph, has been used to monitor coating fibre manufacturing, either to determine all fibre components or the raw materials, for which no reliable quality control methods existed. Coating fibres typically consist of oils, waxes and oil mixtures (viz. solvents, vegetable oil triglycerides and alkylethoxy surfactants), the determination of which by GC and HPLC is hindered by the formation of aqueous emulsions and the absence of chromophoric groups, respectively. Figure 7.15 shows an assembly including two flame ionization detectors located at the accumulator outlet and chromatographic column for the joint and individual determination, respectively, of the extracted analytes [105].

SFE–HPLC

The SFE–HPLC combination is a logical extension of the above-described hyphenated techniques aimed at addressing the analysis of extracts inaccessible to GC and SFC owing to the high polarity, molecular weight or thermal lability of the analytes. Interfacing a supercritical fluid extractor to a liquid chromatograph is the most complex operation involved in SFE hyphenated techniques as a result of the difficulty of coupling a

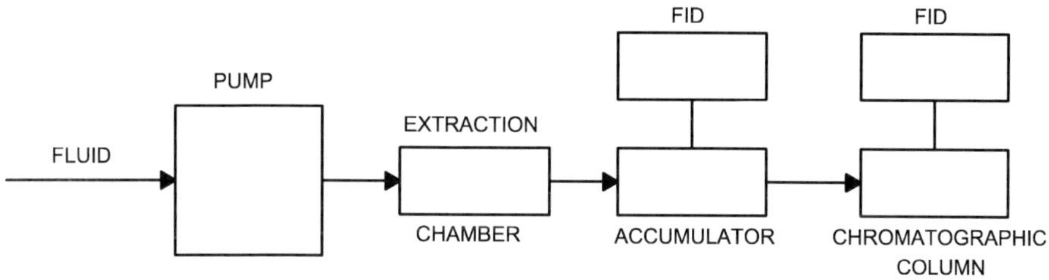

Fig. 7.15. Coupled SFE–SFC system with dual detector for the total and individual analysis of a coating fibre extract. (Reproduced with permission of Springer-Verlag.)

preparative technique that produces a gas at the interface with a chromatographic technique that handles a liquid mobile phase.

Figure 7.16A depicts a flexible SFE–HPLC coupled assembly developed by Ischi and Haerdi [106] that consists of three main parts [viz. the SFE system (A1), the interface (A2) and the HPLC system (A3)] each furnished with appropriate valves operating as shown in Fig. 7.16B. Thus, valve 5 in Fig. 7.16A is used to provide extraction with or without a modifier, via a tee connector; on the other hand, valve 10 allows switching between static and dynamic extraction. The former is done by having the valve close the outlet of the extraction cell after the desired temperature is reached. By switching the valve back, the dynamic state is restored. Valve 13 enables trapping of the extracted analytes, either on a C_{18} silica column placed in an oven for on-line preconcentration and insertion of non-polar or low-polar analytes into the chromatograph after elution or into a liquid phase to implement an off-line operation. When polar ionic analytes are to be preconcentrated, the eluent from the extractor is diverted to valve 18 and retained on the ion-exchange material packed in the column. Preconcentration of both non-polar, low-polar and polar ionic analytes can be accomplished by using both valves (13 and 18) [106].

Metal species such as rhodium and palladium spiked to sand or humic acids have been extracted as β-diketonates and on-line separated and determined by HPLC. The solid phase interface used for this purpose was a LiChrosphere 100 RP-18 cartridge [107].

The SFE–HPLC couple has also been used for sample preparation prior to GC analyses for the determination of a group of highly toxic PCBs usually occurring at very low concentrations relative to the bulk of PCB congeners. Following extraction and HPLC clean-up, the eluted PCBs were individually isolated off-line and determined by either GC–ECD or GC–MS [100].

Selecting the most suitable type of chromatography for coupling to SFE

Because all three types of chromatography can be coupled to SFE, the choice in each case should be dictated by the characteristics of the analytes to be isolated and determined. If GC can be used, it is to be preferred on account of its consolidated status and the high sensitivity and flexibility of the detectors with which it is compatible. Analysing supercritical fluid extracts by HPLC provides one special advantage over GC as most

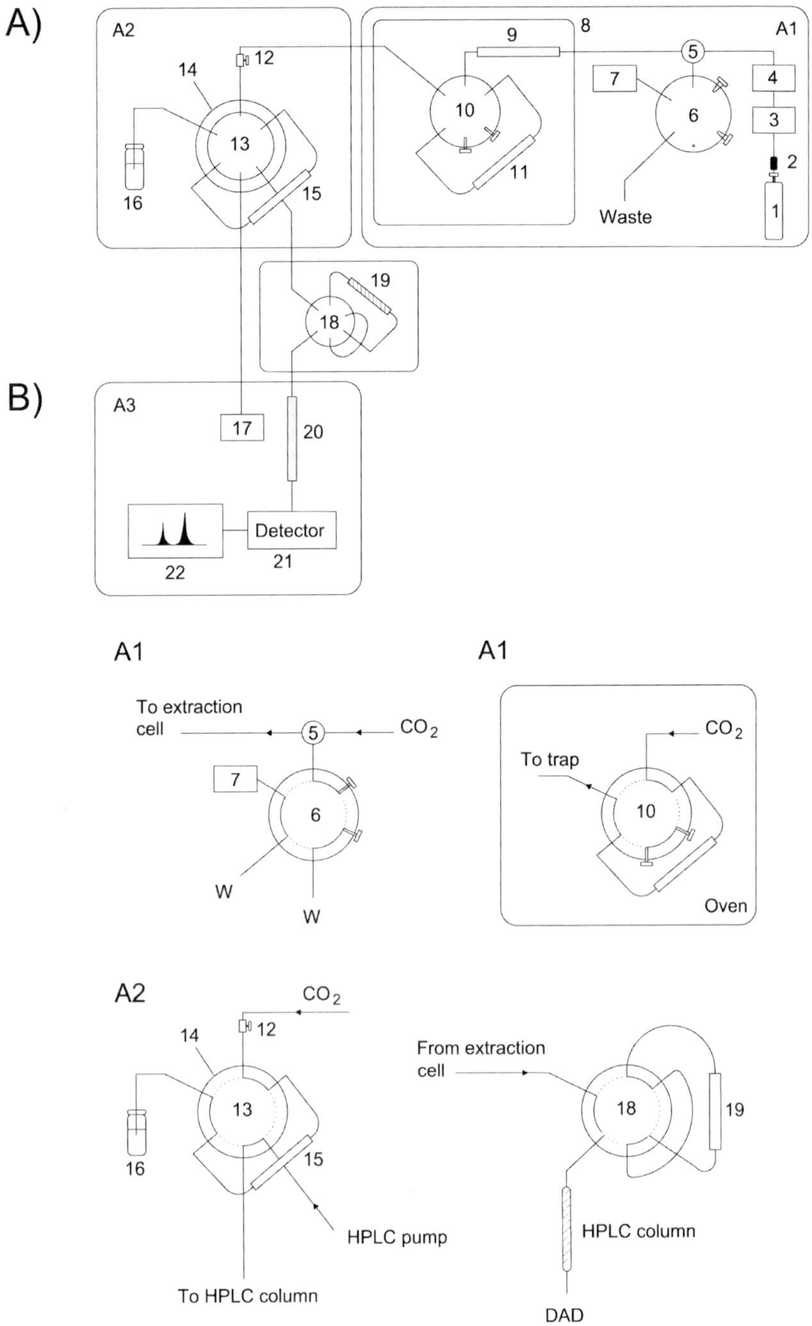
A)
A2
12
14
13
16
15
9
8
A1
5
4
7
10
6
3
11
2
Waste
1
19
18
B)
A3
17
20
Detector
21
22
A1
To extraction cell
CO_2
5
7
6
W
W
A1
CO_2
To trap
10
Oven
A2
CO_2
12
14
13
16
15
HPLC pump
To HPLC column
From extraction cell
18
19
HPLC column
DAD

coextracted species contain no chromophoric groups. On the other hand, most organic substances can be sensed with an ordinary GC detector (e.g. an FID).

7.6.2. Miscellaneous combinations

When the target analytes elicit a specific or highly selective response from a given detector, the detector can be coupled on-line to the extractor in order to enhance the determination with increased ease of automation of the overall process and the ability to monitor the extraction kinetics.

The SFE–detector interface

As in SFE–chromatography combinations, properly designing the interface between the two techniques is one of the most crucial steps in coupling SFE to any other type of technique. These alternative combinations have usually involved detectors and been aimed at expediting measurements of the extracted species or at monitoring the extraction kinetics. In coupling a supercritical fluid extractor to a detector (whether destructive or non-destructive) one should consider the drastic working conditions of the extraction process (i.e. high pressure, temperature and solvent strength, or corrosivity in the supercritical fluid) and the requirements of the particular detection technique. Table 7.7 classifies the prototypes used to this end according to the pressure level required.

Windowed *high-pressure flow-cells* constitute one of the most common types of interface used in the on-line detection with molecular spectroscopic techniques. Special attention should be given here to the characteristics of the window material [108], as well as to proper sealing of the window to the metal cell structure [109]. A special high-pressure flow-cell was recently developed for voltammetric measurements in supercritical CO_2 using a water-in-CO_2 microemulsion in which the water was intended to increase the conductivity of the fluid phase (Fig. 7.17A). The cell was machined from 316 stainless steel 3-in round stock and had an inner volume of 38 ml. Its body was tapped with tapered pipe threads to provide three feedthroughs two of which were used for the working and reference electrodes — which were positioned 0.3 cm apart — the third being used to insert and evacuate CO_2, reactants and products. The cell was kept electrically isolated to prevent contact with earth ground in order to minimize currents that would detract from the accuracy of the potential control [110]. However, leakage was found to be a common shortcoming, as was the difficulty of replacing damaged windows. In order to overcome these problems, various fibre optic-based flow-cells with a stainless steel cross structure, path lengths of 1–20 mm and inner volumes lower than 100 µl have been

---

Fig. 7.16. (A) Schematic diagram of SFE–HPLC coupled systems. (A1) SFE system: 1 CO_2 cylinder, 2 filter, 3 cooler, 4 SFE pump, 5 tee connector, 6 six-port valve, 7 modifier pump, 8 oven for extraction chamber, 9 mixing column, 10 six-port valve, 11 extraction cell. (A2) Interface: 12 restrictor, 13 six-port valve, 14 oven for the six-port valve, 15 cell trap, 16 liquid trap. (A3) HPLC system: 17 HPLC pump for H_2O or HCl/MeOH, 18 six-port valve, 19 anion-exchange column, 20 HPLC column, 21 diode array detector, 22 computer. (B) Exploded view of the valves. (Reproduced with permission of Pergamon Press.)

TABLE 7.7

TYPES OF INTERFACE BETWEEN AN SF EXTRACTOR AND A MOLECULAR SPECTROSCOPIC DETECTOR

High-pressure	Low-pressure flow injection
• Windowed flow-cell	• To a photometric detector
• Fibre optic-based flow-cell	• To a flow-through potentiometric sensor
	• To a piezoelectric detector

proposed [111,112]. Figure 7.17B depicts a typical flow-cell for fluorescence monitoring assisted by fibre optics. Input and output of the fibre are arranged normal to each other, whereas the inlet and outlet of the supercritical CO_2 used as extractant are aligned with the same axis [110]. Fibre optic materials such as silica and chalcogenides have been used in spectral zones from the ultraviolet to the near-infrared region with special transmission requirements [113–117].

Low-pressure flow injection interfaces have been used as links between the extractor and either a photometric detector [118], a flow-through potentiometric sensor [119] or a piezoelectric sensor [120] in dynamic flow injection (FI) systems. Figure 7.18 depicts these unusual types of interface. In the first (Fig. 7.18A), a membrane phase separator (total fluid volume 50 μl) was used to remove CO_2 from the extract. In this way, interferences were suppressed while ensuring quantitative transfer of the solutes (viz. chloramphenicol and penicillin G) to the hydrodynamic system.

The interface between the SF extractor and the photometric flow-through sensor for sulfaquinoxaline (Fig. 7.18B) is more complicated since, following expansion of CO_2 emerging from the detector, the analyte must be transferred to a continuously renewed 0.1 M HCl solution in a collector. This carrier solution is passed through a silica clean-up column that retains insoluble extracted compounds so as to avoid turbidity in the circulating stream. A debubbler is used to remove the remaining CO_2 prior to derivatizing the analyte by the Griess reaction, the coloured product formed being retained in the flow-cell (packed with C_{18} bonded silica) to enable photometric monitoring of the extraction kinetics.

The on-line piezoelectric system of Fig. 7.18C was used for the quantitative gravimetric determination of total fat in foods. The extract obtained by rinsing the cryogenic trap with *n*-hexane was aspirated to the low-pressure injection valve of the FI system, which injected the extract into an *n*-hexane carrier that drove the injected plug to the flow-cell for monitoring with the aid of an At-Cut 10 MHz piezoelectric quartz crystal coated with gold plated electrodes. The throughput and relative standard deviation thus achieved were 6 samples/h and 2.3%, respectively [120].

Atomic absorption interfaces have been used to couple SFE to an AAS detector. A study involving various silica T-tube interfaces for the on-line coupling of SFE to atomic

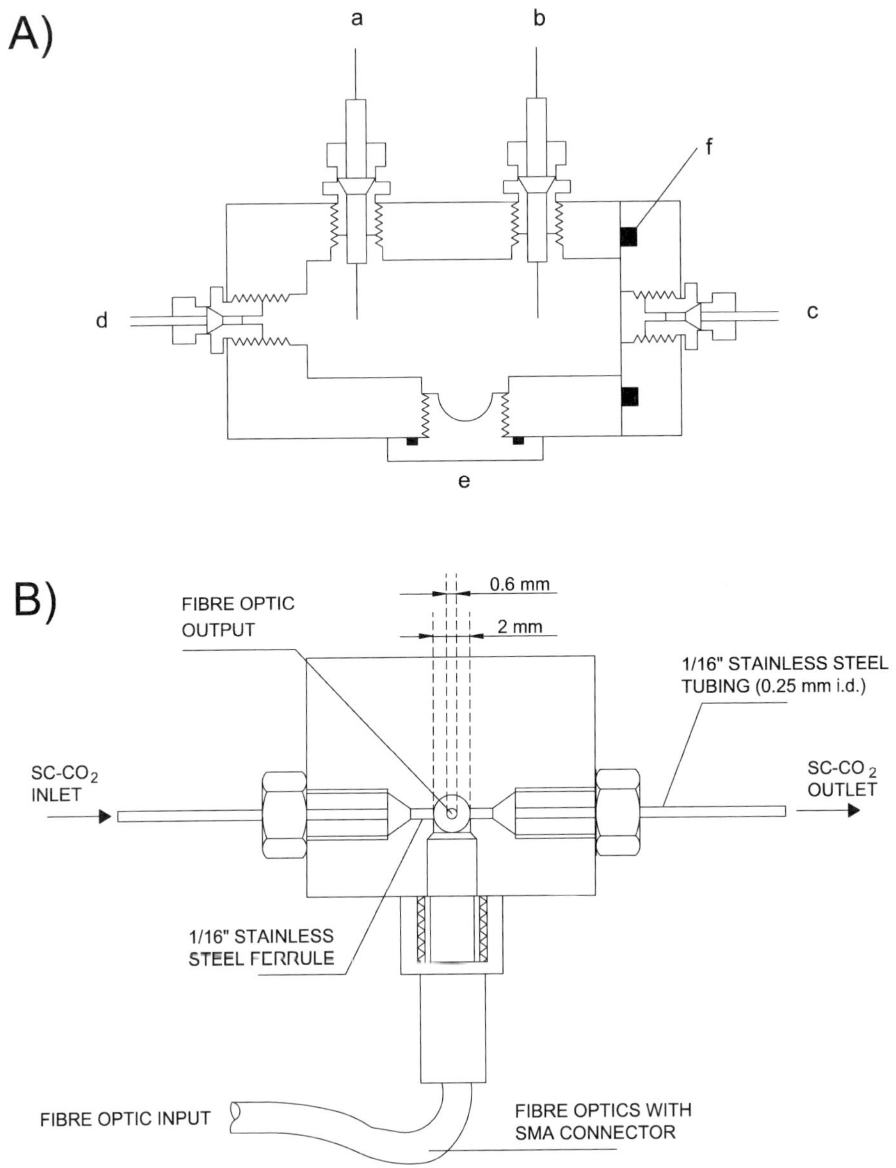

Fig. 7.17. High-pressure flow-cells. (A) Cell for voltammetric measurements in water-in-CO_2 microemulsions: a working electrode, b reference electrode, c supercritical fluid inlet, d supercritical fluid outlet, e sample port, f Teflon O-ring. (B) Fibre optic-assisted cell for fluorimetric measurements. (Reproduced with permission of the American Chemical Society.)

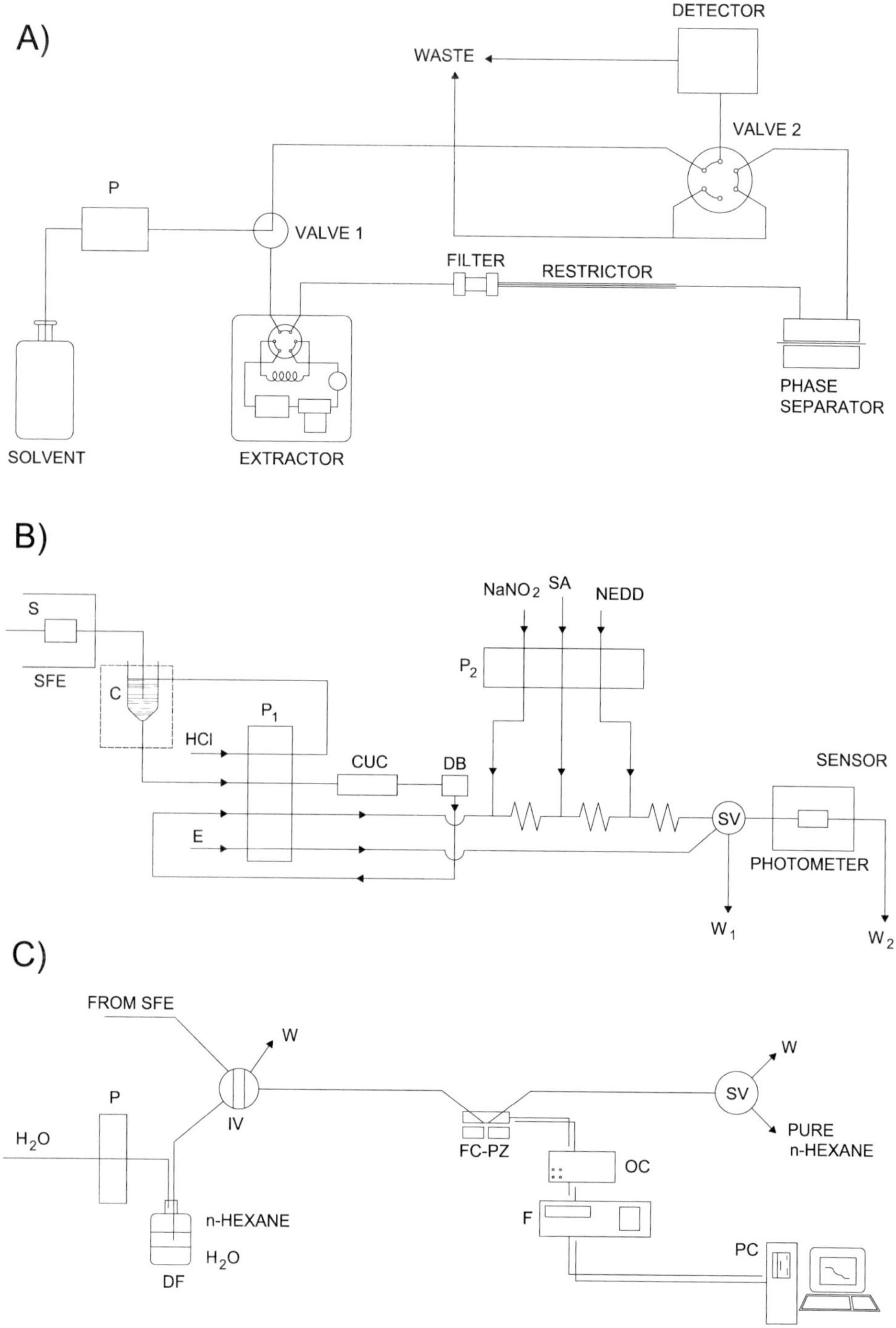
A)
DETECTOR
WASTE
VALVE 2
P
VALVE 1
FILTER
RESTRICTOR
SOLVENT
EXTRACTOR
PHASE
SEPARATOR
B)
S
SFE
C
$NaNO_2$
SA
NEDD
P_2
P_1
HCl
CUC
DB
SENSOR
E
SV
PHOTOMETER
W_1
W_2
C)
FROM SFE
W
P
IV
H_2O
SV
PURE
n-HEXANE
FC-PZ
OC
n-HEXANE
H_2O
DF
F
PC

absorption spectrometry showed the optimum model to consist of the following elements: (a) an optical tube mounted within the spectrometer optical beam; (b) a flame tube fitted with O_2 and H_2 gas entry ports and flow controllers; and (c) a sample insertion tube. Both the optical and the sample insertion tube must be heated with high-resistance heating wires. The extraction phase contained in a silica transfer line was superheated and sprayed into the diffused flame maintained within the upper region of the flame tube and in the optical tube [121,122].

Various other types of interfaces have been used to couple SFE to high-resolution detectors. Thus, direct fluid injection (DFI) has proved suitable for sample insertion into mass spectrometric detectors [123]. Capillary Pt-Ir tubing of 0.3 mm i.d. furnished with electrical and fluid circulation heating was used to connect the supercritical module to a mass detector.

A special pressure-proof probe head with a cell volume of 120 μl capable of withstanding pressures up to 400 bar and temperatures up to 100°C was specially developed for an on-line SFE–NMR prototype [124].

7.6.3. Types of detectors used in combination with supercritical fluid extractors

Both molecular and atomic detectors have been used in combination with SF extractors for monitoring purposes.

Infrared spectroscopy is one of the techniques most frequently coupled to SFE — it accounts for more than 30% of reported SFE hyphenated methods — which is unsurprising as it is one of the most powerful tools available for the elucidation of molecular structures. The specificity of IR spectral information is highly useful for the real-time monitoring of SFE processes with qualitative and quantitative purposes [125]. However, the partial transparency of supercritical CO_2 in the IR region — it exhibits strong absorption bands at 3800–3500, 2500–2150 and below 900 cm^{-1} — calls for careful background correction.

One of the key applications of SFE–FTIR is the extraction and on-line determination of hydrocarbons. Worth special note in this context are the determinations of *n*-tetracosane in Celite (limit of detection 74 ng) [126] and petroleum hydrocarbons (TPH) in soils (limit of detection 1.6 ppm), as well as the characterization of edible oils in foods by correlation of the fat vinylic C–H band intensity with the iodine number [108].

Fig. 7.18. Low-pressure interfaces to detectors based on flow injection. (A) Interface to a photometric detector across a membrane. (Reproduced with permission of the American Chemical Society.) (B) Interface to a flow-through photometric sensor with prior derivatization by the modified Griess reaction. (Reproduced with permission of the American Chemical Society.) (C) Interface to a piezoelectric detector. P peristaltic pump, C collector, CUC clean-up column, DB debubbler, SA sulfamic acid, NEDD *N*-(1-naphthyl)ethylenediamine dihydrochloride, SV switching valve, W waste, DF displacement flask, IV injection valve, FC–PZ flow-cell–piezoelectric crystal, OC oscillator circuitry, F frequency counter, PC personal computer. (Reproduced with permission of Elsevier.)

On an industrial scale, the SFE–FTIR couple has been successfully used to determine aramid, polyamide and polyurethane-based fibres, which require a different finish for textile processing [127]. By using CO_2 at 350 atm, 75°C and a flow-rate of 1.2 ml/min, the polyurethane finish [poly(dimethylsiloxane) oil] was thoroughly extracted within the first 10 min; by contrast, less than 35% of sorbitol derivatives present in the aramid finish could be separated after 20 min. The separation efficiency for polyamide finish components varied between 74 and 100%, with relative standard deviations for the most extractable finishes — which were recovered by more than 98% — ranging from 5.7 to 21%.

Spectrophotometry, which is one of the most useful tools for the automated determination of organic and inorganic compounds, possesses acceptable sensitivity at a low cost, and uses simple equipment, has scarcely been coupled to SFE owing to its low selectivity.

Spectrophotometric techniques can be readily coupled to SFE via a flow injection interface, as shown in Fig 7.18A for the determination of chloramphenicol and penicillin G [118], and in Fig. 7.18B for that of sulfaquinoxaline [119]. Also, a fibre optic interface with the detector using a high-pressure flow-cell similar to that depicted in Fig. 7.17B but with the fibre inlet and outlet aligned with the same axis was used for the extraction and determination of caffeine in roasted and green coffee. The absorbance–time curve at 275 nm was found to be related to the extraction yield; also, first-derivative spectra acquired in the region 260–320 nm allowed the selectivity of the method to be improved by minimizing the effect of other organic compounds co-extracted with the target analyte [115].

Fluorescence spectrometry is a powerful analytical tool by virtue of its high sensitivity; however, the usual presence of emission and excitation spectral interferences calls for additional sample pretreatments. This technique has been used in combination with SFE mainly in the determination of PAHs, which possess well-known luminescence properties. Screening these compounds from soil entailed the use of a high-pressure flow-cell such as that of Fig. 7.17B and allowed the extraction and determination of five individual PAHs in soil with relative standard deviations less than 5%; also, it provided qualitative and semi-quantitative information from the screening of natural soil samples [113].

Thermal lens spectrometry (TLS), a highly sensitive absorptiometric technique based on optical measurements of the thermal gradient produced by a compound following absorption of laser light, has limited routine application owing to the lack of affordable, commercially available TL spectrometers. The potential of the SFE–TLS combination has been demonstrated by using an Nd:YAG laser either to pump a dye laser (454 nm), using its second harmonic (355 nm) or, working either at its fundamental or first harmonic wavelength (1064 or 532 nm, respectively), to monitor compounds absorbing in the near infrared [128] or visible region [109]. The main advantage of this detection technique is its high sensitivity (the limit of detection corresponds to 5×10^{-6} absorbance units). On the other hand, its principal disadvantage arises from the influence of the physico–chemical properties of the supercritical fluid on the TLS signal.

Atomic absorption and *atomic emission* spectroscopies are praised for their high selectivity in the determination of a wide variety of metal and non-metal elements. On the other hand, extraction with supercritical CO_2 is used mainly to separate non-polar organic compounds, for which on-line detection with atomic techniques provides inadequate analytical information in most situations. In any case, incompatibilities in SFE–atomic spectroscopy combinations have been overcome by using various approaches (see Section 7.5). Thus, the use of tetraphenylborate as chelating agent has enabled the extraction and AAS determination of mercury and methylmercury [129], and that of alkylammonium compounds, the quantitation of various metals and anions [121,122]. Also, the SFE/ICP–AES tandem was successfully used to determine organic pollutants in soil by analysing for C, S, P and Si [130]. The two techniques were interfaced via a restrictor positioned concentrically inside the sample injection capillary of the plasma torch. A small amount of drying agent was added to the supercritical fluid in order to prevent plasma blow-out. The results were consistent with those obtained by off-line CO_2 SFE and Ar SFE.

Mass spectrometry has proved highly suitable for the elucidation of chemical structures and sensitivity. It is thus an invaluable tool for resolving extremely complex natural mixtures. Thus, on-line coupled SFE–MS constitutes a rapid, reliable choice for the determination of mycotoxins in wheat, which possess a high toxicological interest. For example, an SF extractor coupled to an MS chemical ionization source provided limits of detection in the picogram range for diacetoxyscirpenol and toxin T-2 [122]. Used in tandem with chemical ionization–collision induced dissociation MS–MS, SFE has also enabled the rapid detection and identification of species without the need for complete extraction [131].

Nuclear magnetic resonance spectroscopy is no doubt the most powerful choice for the elucidation of organic structures (isomers included). Although the proportionality between peak areas and the number of contributing nuclei has been exploited for quantitative purposes, the low sensitivity of this technique has hindered more general usage. Among others, the SFE–NMRS couple has been used to analyse food samples and to determine piperine and caffeine in pepper and coffee, respectively. The spectral shifts observed with linear pressure gradients degraded NMR resolution at the beginning of the extraction process — near the critical point — so stopped-flow measurements were required in order to obtain reliable spectral information [124].

Voltammetric measurements in supercritical CO_2 have been made by using water-in-CO_2 microemulsions to ensure adequate conductivity in the fluid phase. This allowed well-defined voltammograms for the redox reactions of ferrocene and tetramethyl-*p*-phenylenediamine to be obtained [112].

Piezoelectric measurements have also been made on-line with supercritical extraction by using an FI interface such as that of Fig. 7.18C, which allowed the determination of fat in foods using the procedure described in Section 7.6.2 [120].

The fact that SFE can be used in combination with a variety of techniques for on-line monitoring purposes may help overcome one of its most serious drawbacks — one that has caused it to fall short of the expectations aroused in the beginning — namely: that the extraction time varies among matrices, so no specific time can be adopted for routine analyses. Establishing the extraction time in terms of the signal provided by the on-line detector used may be a plausible solution and help expand the scope of application of this technique.

7.7. APPLICATIONS OF SUPERCRITICAL FLUID EXTRACTION

Based on the overwhelming prevalence of CO_2 as SF extractant, all applications described in this section refer to this fluid (whether neat or modified by a cosolvent) unless otherwise stated. Because of its special features, the uses of supercritical water are dealt with in another section.

Based on the number of reported applications, the unequivocal establishment of its scope, advantages and disadvantages from the results of such applications, and the inception of automated coupled systems [132,133], supercritical fluid extraction has already come of age.

The 1980s saw an avalanche of publications on SFE many of which reported contradictory results that deterred some potential users. This period has given way to one where the amount of SFE literature published each year has levelled off and controversy over its performance has virtually vanished. At present, newcomers to SFE can resort to a wide variety of experimental design approaches [18–23,134] in order to decide which variables are to be monitored, and in which situations, in order to ensure effective extractions. Research into the mechanisms that govern mass transfer during extraction [135] and the ensuing mathematic models [136], which have been thoroughly checked on a variety of matrix–analyte couples [137,138], have obviously helped the SFE technique to consolidate. The results of studies of SFE robustness [139], and interlaboratory evaluations [140], further support the establishment of accurate limits for SFE use. Comprehensive studies involving both spiked and natural samples [141–143] have led to the discouraging — but necessary — conclusion that the type of sample matrix and matrix–analyte interaction involved dictate to a great extent the duration of the process, which somehow restricts the use of SFE for routine purposes. The experiments of Benner [46] in relation to the time required for PAHs to be extracted from four natural standard reference materials (SRMs) two of which consisted of urban dust organics, one of organics in marine sediment and the fourth of diesel particulate matter, examined the effects of various extraction fluids (viz. carbon dioxide, chlorodifluoromethane and 1,1,1,2-tetrafluoroethane), cosolvents (dichloromethane and aniline), temperatures (60, 150 and 200°C) and amounts of added water. The results of this comprehensive study showed once more, and irrefutably, that the greatest shortcoming of SFE is that the extraction efficiency for specific compounds depends on the bulk composition of the host sample (i.e. on matrix effects). Thus, even if analyte recoveries are adequate for the intended purpose, the analyst should always weigh the benefits of SFE (viz. expeditiousness, a high selectivity and reduced waste solvent generation) against the potential restrictions imposed by matrix effects before deciding whether to adopt this technique as the primary extraction choice. If complete extraction of the target species is desired, the analyst should routinely compare the results from a conventional extraction method on the specific sample matrix with those of the SFE method in order to ensure that the latter meets the previous criterion; this is a labour-intensive, time-consuming task and hence impractical at routine analytical laboratories.

7.7.1. Selectivity in supercritical fluid extraction

The selectivity of supercritical fluid extraction is one of its most flexible properties relative to other separation techniques including liquid–liquid extraction and, especially, those starting with a solid sample. Virtually all internal and external parameters of the extraction process can be altered to enhance the selectivity of the overall analytical process. Thus:

(a) The type and composition of the supercritical fluid may be chosen in such a way to enable or prevent the extraction of specific substances in terms of molecular weight. Also, the need for a modifier is imposed by the polar or non-polar character of the species to be extracted.
(b) Analyte solubility can be adjusted via density. The latter in turn can be changed via pressure and temperature, and need not be taken to extreme values.
(c) The equilibration and extraction times are central to selectivity as they dictate which species are effectively removed from the matrix.
(d) In situ derivatization may have a strong impact on the extraction selectivity. In fact, the analyte properties can be altered by using an appropriate (bio)chemical reaction to establish significant differences from other sample components.

The external factors that can be adjusted to increase the selectivity of the overall analytical process are smaller in number, though not in significance — probably because their effects are better known. Essentially, there are two such factors, namely: (a) implementing a separation step after extraction by using an on-line or off-line coupled apparatus or instrument (usually a column chromatograph); and (b) using a detector to obtain a high enough selectivity to offset any deficiencies of the preliminary steps in this respect.

As in other techniques, SF extracts are not always free of unwanted matrix components. A number of clean-up methods have been used with and after the extraction step to overcome this problem. Sample clean-up is mandatory with fat-soluble analytes. On the other hand, neat supercritical CO_2 extracts from environmental samples are often clean enough for direct injection into a gas chromatograph [17].

Specific problems such as those posed by the presence of organic matter or sulphur in sediment extracts have elicited a wide range of solutions [144–147] depending on the type of chromatograph, detector and connection (on-line or off) between the SF extractor and the equipment used in the subsequent steps of the analytical process.

7.7.2. Scope of application

Supercritical fluid extraction has so far been applied mostly to solid samples and also, occasionally, to liquid ones (whether as such or following passage through a sorbent that was subsequently placed in the extraction chamber, a procedure also applicable to gaseous samples).

Solid samples

Changes in the focus of SFE can be easily followed through its reported applications. Thus, in 1993 [3], environmental applications prevailed (45.9% versus 21.9% devoted to foods and 11.6% to industrial analyses). By 1996, however, SFE applications to food analysis had risen to 38%, environmental uses fallen to 41% and industrial analyses levelled off at 11% [48]. More recently [17], the extraction of food components (particularly fat) has become one of the major applications of SFE, so much so that the current boom in SF extractor sales has been ascribed to it. The book by Luque de Castro *et al.* [3] contains comprehensive tables of SFE applications in various fields. Also, one review of SFE in food analysis [148] includes four tables with applications involving the extraction of fat from various types of sample (viz. meat and animal products, fish, cereal, seed and animal feed, plants and vegetables). On a more specific level, Eller and King reviewed determinations of the fat content in foods [149]. Finally, the *Analytical Chemistry* issues devoted to reviewing techniques provide periodic updates on SFE and SFC [150].

Liquid and gaseous samples

The few uses of SFE with liquid samples involve — usually aqueous — liquids, whether as such or following absorption into a solid sorbent to retain the target species. The former require that the extraction cell be altered in order to ensure prolonged sample–extractant contact times. The alteration usually involves inserting a capillary at each cell end so that the supercritical fluid can enter it at the top, reach the bottom, bubble through the sample and leave it through the lower capillary, from the end of which the fluid, containing the extracted analytes, emerges. In this way, the analytes can be extracted from the liquid sample while the matrix is held in the cell [151–154]. In these applications, one should consider the potential solubility of the liquid matrix in the supercritical fluid since, for example, water is slightly soluble (0.3%) in supercritical CO_2, so it may freeze at the restrictor outlet.

When the analytes are to be retained in a sorbent, the sample (which can be solid, semi-solid, liquid or gaseous) is inserted in the solid state into the extraction cell. Samples in the latter three forms are supported on an appropriate material in order to ensure effective attack by the supercritical fluid. Solid supports are not used for liquid, gaseous and semi-solid samples only, however. Some research work conducted so far on solid samples has involved not natural samples but synthetic ones prepared from a selected sorbent (a natural matrix where the presence of the analytes of interest was previously excluded or a synthetic support such as polyurethane foam or glass wool) with which a solution containing the analytes was homogenized. Quantitative evaporation of the analyte solvent is mandatory as any residual solvent may alter the polarity of the supercritical fluid and hence its action to an extent dependent on the particular fluid and solvent properties, and also on the amount of solvent retained.

Available procedures for removing aqueous matrices, which entails concentrating the analytes to be isolated, are quite varied in nature; the best choice in each case depends on the properties of the analytes and interferents. Such procedures include freeze-drying [155], the use of sorbents (e.g. anhydrous sodium sulphate [156] or diatomaceous earth

[157]) and mixing the sample with an appropriate sorbent (sand [158], paper strips, Celite [159]) and subsequently evaporating the solvent as described above for the preparation of spiked solid samples. In any case, the most flexible approach is that involving the use of discs of a sorbent material [160,161] or a pre-column packed with a material providing not only preconcentration but also enhanced selectivity in the separation. Small columns packed with a hydrophobic sorbent have proved highly efficient with biological (e.g. plasma) [162,163] and environmental samples [164]. The sample is pumped through the pre-column, which is subsequently eluted with water to remove hydrophilic compounds and dried in a nitrogen stream to evaporate the polar solvent used. Finally, the pre-column itself, accommodated in the sample chamber, acts as the extraction cell in the selected SFE mode.

Environmental analyses frequently involve gaseous samples, which is unsurprising taking into account the large number and variety of toxic substances that are currently discharged into the atmosphere. The chief problem posed by volatile pollutants is that conventional collection and desorption methods result in analyte losses. Most studies on air pollutants have focused on the less volatile organics, which are trapped with solid sorbents or as particulates on filters. Sorbents such as XAD, Tenax, Chromosorb polymers and polyurethane foam are routinely used to collect substances including PAHs; hydrocarbons; PCBs; nitrogen-, sulphur- and halide-containing compounds; and pesticides. Such substances are subsequently recovered by SFE in a manner similar to those of traditional Soxhlet extraction and sonication.

7.7.3. Comparison with other analyte removal techniques

SFE versus separation techniques in general

One mandatory step in developing and facilitating the consolidation of a new technique involves comparing it with those it may compete with in order to clearly establish its advantages and disadvantages and to provide potential users with clear decision-making arguments. Supercritical fluid extraction is no exception to this rule as its performance has been thoroughly compared with both traditional and modern solid-matrix extraction techniques for a wide variety of analytes (particularly in the environmental field). Thus, the performance of SFE, high-pressure solvent extraction (also called "accelerated solvent extraction", ASE) and Soxhlet extraction (SOX) in the recovery of organophosphorus hydraulic fluids from soil [165] was compared and ASE found to provide the best results with both spiked (Table 7.8) and native samples (Table 7.9). On the other hand, SOX, microwave-assisted extraction (MAE), SFE and ASE were found to provide similar results in the extraction of hexaconazole from variably aged weathered soils [166]. The results of multiple extractions from soil containing 1.5% or organic matter using ASE, SFE and MAE compared well with those provided by SOX. Based on relative standard deviations (RSD), however, ASE generally produced far superior results, both with samples aged for 1 and for 52 weeks (viz. 4.5% versus up to 14.6% with MAE and 8.4% with SFE); in any case, SOX invariably provided the best results, with RSD values in the region of 0.1%. The results for soils with higher organic matter contents (5.7%) were poorer. Thus, SFE provided recoveries about 50% lower than those achieved with SOX, but with acceptable RSD

TABLE 7.8

RECOVERIES (ppm) OF ORGANOPHOSPHORUS HYDRAULIC FLUIDS (OPS) FROM SPIKED SOIL (RSD, $n = 3$)

OP soil contaminant	Soxhlet extraction	SFE	SFE relative recovery,[a] %	ASE	ASE relative recovery,[a] %
TPP	61.7 (0.6)	63.3 (3.9)	102	74.8 (4.7)	121
ToCP	26.4 (1.0)	23.6 (3.6)	90	27.4 (5.9)	104
TmCP	20.0 (1.3)	20.1 (2.9)	100	22.2 (4.6)	111
Bm, pCP	33.9 (1.1)	30.2 (3.3)	89	35.5 (5.1)	105
Bp, mCP	18.5 (1.0)	14.8 (2.7)	80	19.6 (6.9)	106
Total TCP	72.4 (1.1)	65.1 (3.0)	90	77.3 (5.4)	107

[a] Relative recoveries as compared to those obtained with Soxhlet extraction
Reproduced with permission of the American Chemical Society

TABLE 7.9

RECOVERIES (ppm) OF ORGANOPHOSPHORUS HYDRAULIC FLUIDS (OPS) FROM NATIVE SOIL SAMPLES (RSD, $n = 3$)

OP soil contaminant	Soxhlet extraction	SFE	SFE relative recovery,[a] %	ASE	ASE relative recovery,[a] %
TmCP	14.4 (2.4)	13.0 (6.7)	90	15.6 (7.1)	108
Bm, pCP	28.9 (1.5)	25.6 (6.7)	89	28.3 (7.1)	98
Bp, mCP	16.2 (0.9)	14.5 (7.1)	90	14.8 (8.3)	91
Total TCP	59.4 (0.4)	53.1 (6.7)	89	58.7 (7.4)	99

[a] Relative recoveries as compared to those obtained with Soxhlet extraction
Reproduced with permission of the American Chemical Society

values (5–7%). Similar results were obtained with MAE — the RSD for the sample aged for 52 weeks, however, was 23.0%. SFE gave recoveries consistent with the ASE ones at all levels of soil weathering, RSD values ranging from 1 to 4%. In terms of the actual extract yielded by each technique, ASE provided the cleanest chromatographic samples, with no interfering co-eluting peaks. By contrast, MAE provided far dirtier extracts that required pre-chromatographic clean-up — and so did SFE with samples containing abundant organic matter — in order to allow the analyte signal — which consisted of many elution peaks — to be properly resolved.

The results obtained with easily removed analytes such as pyrethrins in flowers and allethrin on paper strips were comparable to those provided by less drastic techniques including ultrasonic extraction (USE) and SOX in terms of recovery and precision [167].

The extraction of methylmercury compounds from both biological samples [168] and marine sediments [169] by manual, supercritical fluid and microwave-assisted extraction

TABLE 7.10

COMPARISON OF SFE AND MAE FOR METHYLMERCURY

CRM		Manual extraction (n = 6)		SFE		MAE	
Sample	Certified concentration	Concentration	Recovery, %	Concentration	Recovery, %	Concentration	Recovery, %
DORM-1	0.785 ± 0.060	0.731 ± 0.040	93 ± 5	0.760 ± 0.020	96 ± 3	0.746 ± 0.010	97 ± 1
CRM 463	3.040 ± 0.160	2 830 ± 0.280	93 ± 4	2.640 ± 0.090	87 ± 3	2.940 ± 0.080	97 ± 3
CRM 464	5.500 ± 0.170	5.200 ± 0.130	95 ± 5	4.530 ± 0.060	82 ± 1	5.160 ± 0.100	94 ± 2
IAEA-085	24.600 ± 1.400	24.700 ± 2.200	107 ± 9			23.510 ± 1.900	96 ± 8

revealed the last choice to be the most efficient with a view to removing the analytes from both types of sample; in fact, MAE used less organic solvents than the manual method and exhibited shorter extraction times and better efficiency than SFE (see Table 7.10). Because MAE allowed the simultaneous extraction of up to 12 samples, it was deemed the technique of choice for routine analyses. However, as can clearly be seen from Fig. 7.19, following clean-up the extract provided by MAE was considerably dirtier than that obtained with SFE.

One special extraction mode uses liquids of enhanced fluidity. These are prepared by combining a high proportion of an associated organic solvent with a low-viscosity liquid such as CO_2. A comparison of this approach with SFE and SOX in terms of performance in the extraction of phenol pollutants from house dust revealed extraction with enhanced-fluidity liquids to provide results similar to those of SFE and better than those of SOX [170].

On the other hand, the results obtained in extractions of essential oils from various plants and fruits are inconclusive as they appear to depend on the particular type of sample. Thus, the results provided by SFE and steam distillation (SD) in the isolation of Coriander oil from fruits were similar; with a commercial oil, however, SFE exhibited higher recoveries for the oxygenated compounds of interest [171]. Also, the SF extract from Galbanum latex was found to contain fewer compounds than the SD extract from the same sample type [172]. On the other hand, modified SFE exhibited lipid recoveries from *Sargassum hemiphyllum* that surpassed those of MAE and were comparable to those of SOX — the material extracted by SFE contained greater amounts of ω-3 fatty acids, however. Increasing the pressure in SFE — and decreasing the solvent density as a result — gave rise to extracts with increased contents in polyunsaturated fatty acids and decreased contents in saturated ones. This ability to adjust the extract composition by altering pressure is one special feature of SFE [173]. The extraction of volatile secondary metabolites from *Spilantes americana* by steam distillation–solvent extraction (SD–SE) and the use of a supercritical fluid provided extracts that contained 46–50 compounds at concentrations above 100 ppb in the former case, and 51–67 in the latter. The SFE technique was found to extract sesquiterpenes, heavy hydrocarbons and nitrogen-containing compounds (especially amides) quite selectively; by contrast, SD–SE exhibited higher yields in monoterpenes and oxygen-containing compounds [174].

Based on the foregoing, efficiency in the extraction of essential oils from plants appears to depend strongly on the particular type of oil and also, possibly, on the sample matrix, so whether SFE surpasses SD-SE in this respect, or vice versa, cannot be stated unambiguously. The SFE technique does surpass ultrasonic extraction (USE) for the removal of organic, low-polar compounds [175–177]. On the other hand, it compares unfavourably with MAE in this respect [177–179]. Thus, López-Ávila *et al.* used MAE, SFE, SOX and USE to extract 94 compounds on the list of EPA method 8250 and found the first technique to provide recoveries above 80% for 51 such compounds, over the range

Fig. 7.19. Chromatograms obtained in the separation of methylmercury from DORM-1 CRM extracts using (A) manual extraction, (B) supercritical fluid extraction and (C) microwave-assisted extraction. (Reproduced with permission of the American Chemical Society.)

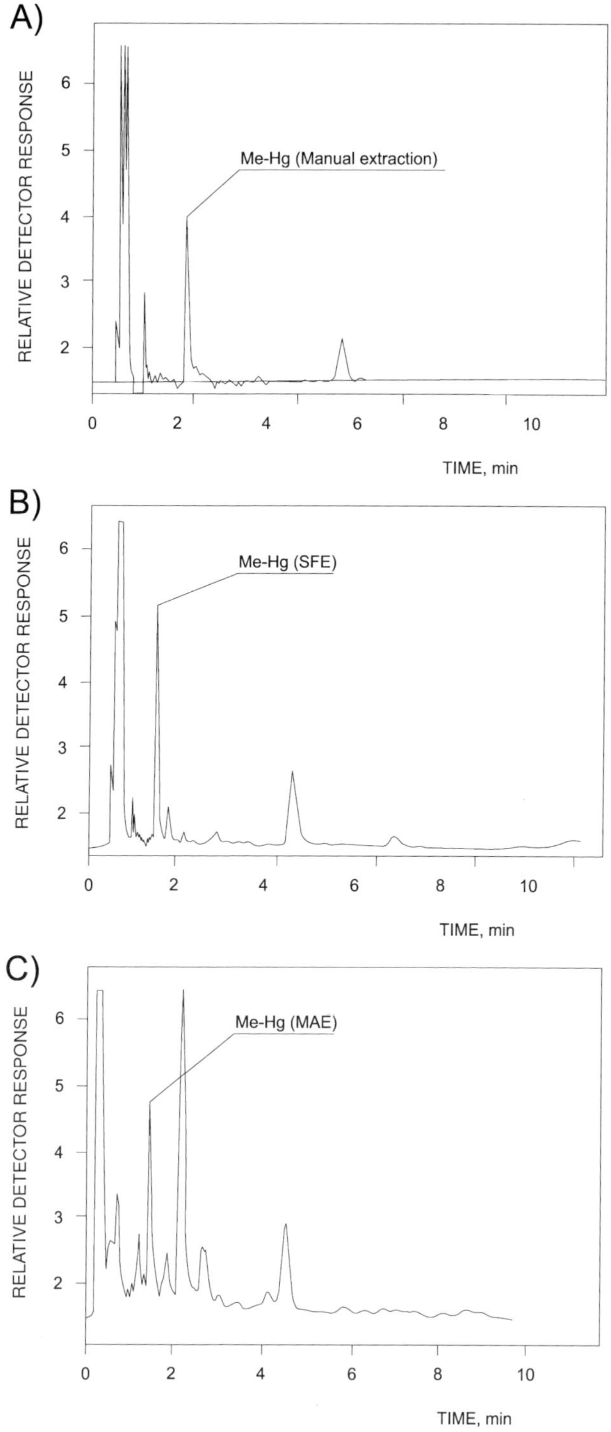
A)
RELATIVE DETECTOR RESPONSE
Me-Hg (Manual extraction)
TIME, min
B)
RELATIVE DETECTOR RESPONSE
Me-Hg (SFE)
TIME, min
C)
RELATIVE DETECTOR RESPONSE
Me-Hg (MAE)
TIME, min

50–79% for another 33, between 20% and 49% for a further 8, and less than 19% for the remaining 2. All three techniques provided unacceptable results for 15 polar compounds, 10 of which, however, were recovered by more than 70% using MAE with acetonitrile at 115°C. Again, these results lead to the conclusion, largely demonstrated when more drastic extraction techniques are compared, that the energy used by SOX (viz. that of the condensed solvent exclusively) is inadequate to break strong analyte–matrix bonds; this makes SOX unsuitable as a model technique.

One special application of SFE involves its joint use with subcritical water to extract Dacthal and its acid metabolites from soil; the herbicide is converted to monoacid and diacid derivatives that are more readily soluble in water than the parent compound. Thus, following extraction of Dacthal with supercritical CO_2 at 400 bar at 150°C for 15 min, its monoacid and diacid metabolites were extracted from soil in 10 min, using subcritical water conditions (viz. 200 bar and 50°C). The metabolites were trapped in situ on a strong anion-exchange disc placed over the exit frit in the extraction cell and subsequently combined with Dacthal by placing the disc into the GC autosampler vial containing the SFE extract to allow their simultaneous elution from the disc and derivatization to their ethyl esters. Only a single sample was analysed; also, because the disc-catalysed alkylation reaction did not transesterify Dacthal, its speciation was maintained [180].

Comparison of PAH extraction methods

The extraction of PAHs is worth separate discussion here on account of the vast number of applications it has elicited — over 130 literature references, most involving soil samples — relative to other species. A comparison of SFE and SOX for the removal of 11 PAHs from soil using neat CO_2 revealed the latter technique to require eight hours to match the efficiency achieved in one with the former [181]. One application involving the analysis of polluted land samples using both neat and modified supercritical CO_2 revealed the presence of a modifier and the extraction time to be the two operating variables most strongly influencing the recoveries of 16 PAHs. A subsequent comparison of the use of SFE for 1 h, MAE for 20 min and SOX for 6 h showed them to provide extraction efficiencies of 458.0 (RSD 3.1%), 422.9 (RSD 2.4%) and 297.4 mg/kg (RSD 10.0%), respectively. The precision of SFE was found to be decreased by the presence of sulphur in the soil [182]. In one other comparison of SFE, SOX and MAE involving the extraction of 17 PAHs from a native polluted soil, the first was also found to provide better results (viz. a PAH content of 458.0 mg/kg in the extract after 5 min of static extraction plus 1 h of dynamic extraction with supercritical CO_2 containing 20% methanol) than MAE with 100 ml of dichloromethane for 20 min (297.4 mg/kg) and SOX with acetone or dichloromethane for 6 h (422.9 mg/kg). The shortest time needed with MAE was partly offset by the time the vessel had to be left unopened until the control vessel reached atmospheric pressure (about 30 min) [183].

The effect of the presence of a modifier, which is critical in the supercritical fluid extraction of PAHs, has been thoroughly studied and compared with that on Soxhlet extraction. The modifier effectively displaces the analytes from the active sites in the matrix, thereby dramatically shortening the extraction time. A series of binary modifiers

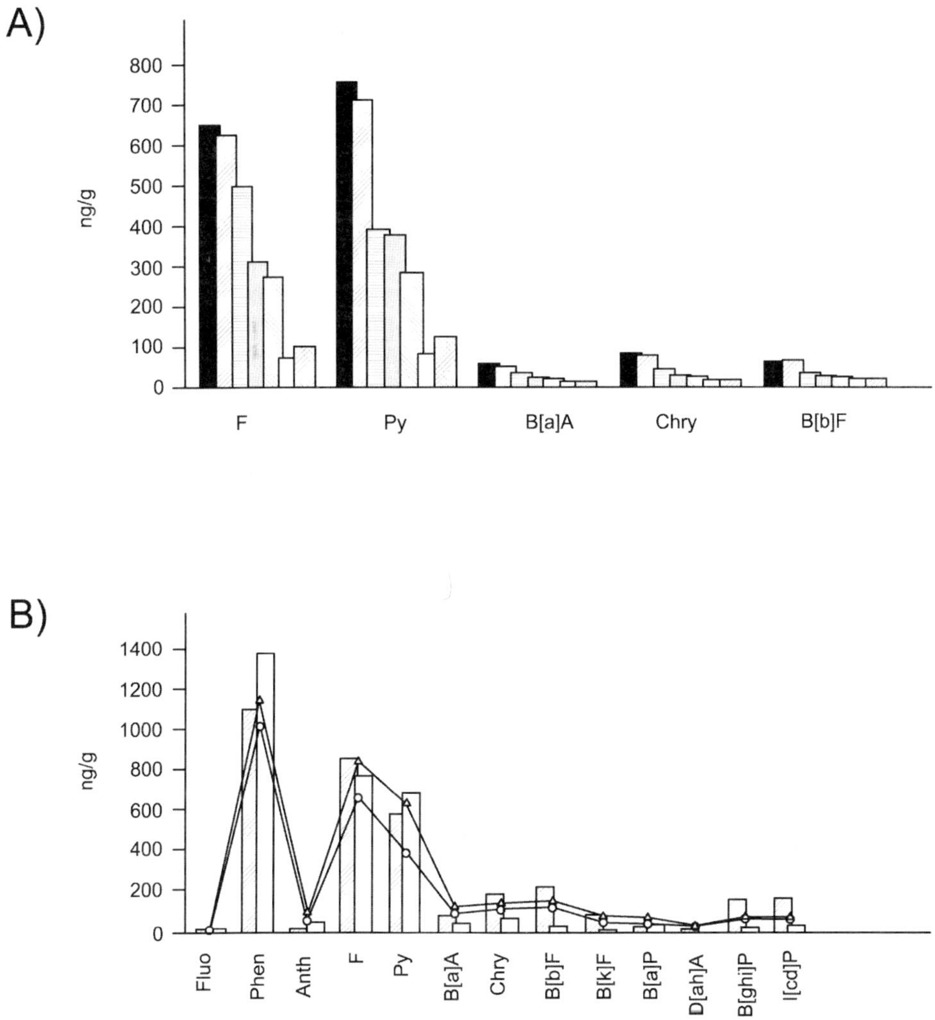

Fig. 7.20. Extraction efficiency achieved in the SFE of PAHs from fly ash. (A) By adding various binary modifiers to CO_2 [bars represent the following combinations from left to right: trifluoroacetic acid (TFA) in toluene, triethylamine (TEA) in toluene, isopropylamine in toluene, pure toluene, tetrabutylammonium hydroxide in methanolt-toluene, citric acid in a methanol–toluene mixture and the absence of modifier]. (B) By successive application of a liquid modifier (four times) compared to Soxhlet extraction (shaded, left bars) and conventional SFE with CO_2/pure toluene (right bars); CO_2/TEA in toluene (-○-) and CO_2/TFA in toluene (-△-). Fluo fluorene, phen phenanthrene, Anth anthracene, F fluoranthene, Py pyrene, B(a)A benz(a)anthracene, Chry chrysene, B(b)F benzo(b)fluoranthene, B(k)F benzo(k)fluoranthene, B(a)P benzo(a)pyrene, D(ah)A dibenz(a,h)anthracene, B(ghi)P benzo(g,h,i)-perylene, I(cd)P indeno(1,2,3-c,d)pyrene. (Reproduced with permission of Springer-Verlag.)

were tested for the extraction of fluoranthene (F), pyrene (Py), benz(a)anthracene [B(a)A], chrysene (Chry) and benzo(b)fluoranthene [B(b)F]; the results are shown in Fig. 7.20A. Binary modifiers were found to provide better results. A comparison with the Soxhlet extraction of a variety of PAHs in fly ash, sediments and sewage sludge was made. As can be seen from Fig. 7.20B, SFE required successive modifier applications to the extraction chamber, each followed by 10 min equilibration at 90°C and 30 min SFE with neat CO_2, which made the process rather labour-intensive and time-consuming [184]. The use of methanol as modifier and BF_3 as derivatizing reagent was found to increase the extraction efficiency [185]. The extraction of PAHs with a supercritical fluid similar to that used for the Soxhlet extraction of these compounds (viz. CO_2 or dimethyl ether, either pure or modified with toluene) revealed toluene-modified CO_2 to provide an overall extraction yield of 43 mg/kg compared to 50 mg/kg with toluene-modified dimethyl ether and 32 mg/kg with SOX [186]. The ASE and SFE techniques were compared mutually and with SOX, USE and methanolic saponification extraction (MSE) in terms of efficiency in the removal of PAHs from marine sediments. The best results, which were provided by SFE, MSE and SOX, are shown in Figs 7.21A, 7.21B and 7.21C, respectively; such results correspond to different types of sediment, which again confirms the strong influence of the nature of the matrix on the extraction efficiency [187] and is consistent with more recent findings of Benner [46]. All PAH extracts from complex matrices required clean-up prior to the isolation and separation of their components.

Solid-phase extraction (SPE) is nearly always required prior to the supercritical fluid extraction of PAHs from liquid samples; alternatively, samples can be spiked to a solid material such as diatomaceous earth. A comparison of SPE, SFE, SPE–SFE and liquid–liquid extraction (LLE) for the removal of PAHs from drinking water revealed the SPE–SFE combination to provide the best results and LLE the poorest [188].

Recently, the supercritical fluid extraction of PAHs was used to correlate the contents thus obtained with the bioremediation behaviour of the target analytes in a field treatment plot for 1 year; based on the results, mild SFE was deemed an expeditious, useful tool for predicting the bioavailability of PAHs in polluted soil [189].

Use of supercritical water as extractant

The special features of the uses of supercritical water justify dealing with them in a separate section. No commercially available or laboratory-made extractor can operate in a continuous manner under the conditions required by supercritical water. In fact, most of the few reported applications allegedly involving supercritical water extraction actually used subcritical water conditions with not more than two extractions in the supercritical state; this can readily be inferred from the equipment used — stainless steel SF extractors can hardly withstand the drastic conditions needed for work with supercritical water. As a

---

Fig. 7.21. Comparison of various techniques for the extraction of PAHs, using linear regression and the following CRMs: HS-6 (A), SPM I (B) and SPM II (C). (□) SFE, (○) MSE, (+) ASE, (×) SOX and (△) USE in the three figures [by exception, crosses correspond to SOX in C]. (Reproduced with permission of the American Chemical Society.)

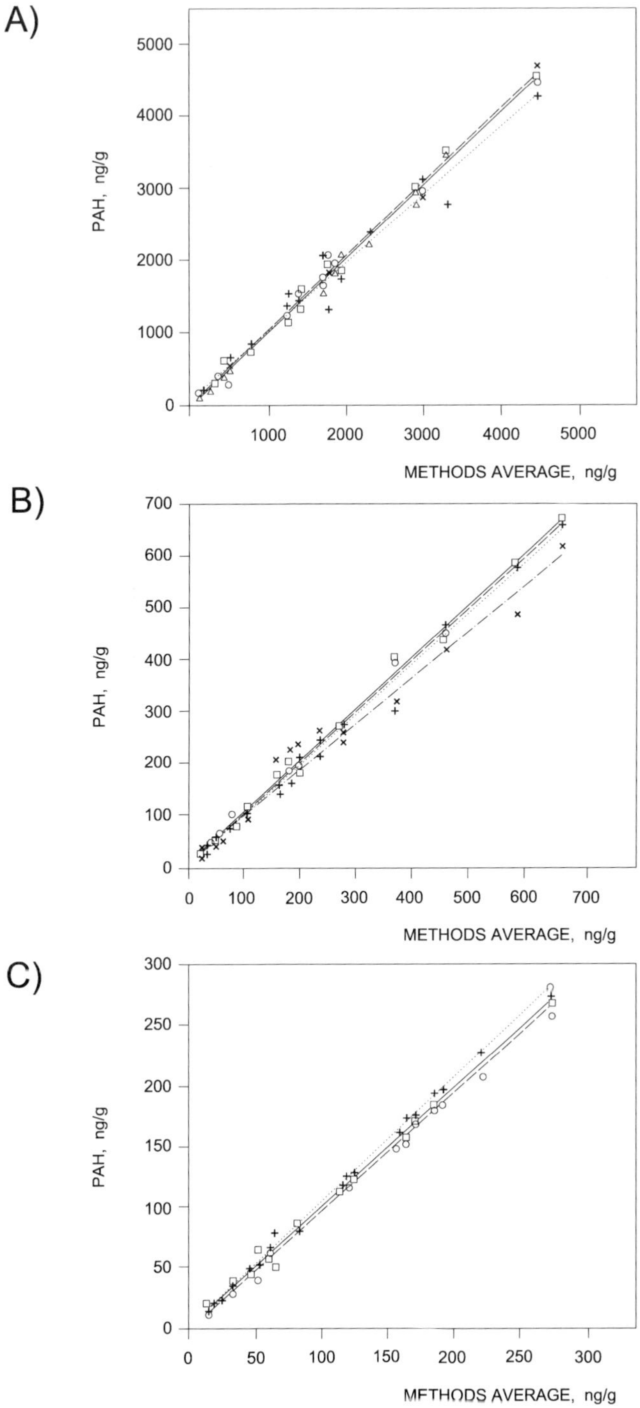
A)
PAH, ng/g
0
1000
2000
3000
4000
5000
1000
2000
3000
4000
5000
METHODS AVERAGE, ng/g
B)
PAH, ng/g
0
100
200
300
400
500
600
700
0
100
200
300
400
500
600
700
METHODS AVERAGE, ng/g
C)
PAH, ng/g
0
50
100
150
200
250
300
0
50
100
150
200
250
300
METHODS AVERAGE, ng/g

result, the performance of supercritical H_2O has not been examined as thoroughly as that of supercritical CO_2.

The Research Centre of Karlsruhe conducted tests with a view to finding materials capable of withstanding the oxidizing, corrosive conditions involved in the use of supercritical water. The materials examined included various alloys such as austenite, AISI type 316 SS austenitic CrNi stainless steel and several Ni-based alloys (viz. Inconel 625, Hastelloy C-276, Nicrofer 5923, Nicrofer 6025 and Haynes alloy 214) that were used to fabricate pressure tubes for exposure to supercritical water containing 0.5–5 mol O_2/kg and/or 0.05–0.5 mol HCl/kg over a period of 150 h (a preliminary step on this line) [190].

Most research work on the use of supercritical water has been conducted batchwise and involved non-analytical determinative applications. Thus, supercritical water oxidation (SCWO) was proposed as an alternative treatment for hazardous waste disposal [191] and also as a commercial tool for decomposing trichloroethylene, dimethyl sulphoxide and isopropyl alcohol on a pilot plant scale [192]. Current commercially available equipment (the aqua Critox® system) is usable with industrial and municipal sludge, mixed (radioactive and organic, liquid and solid) waste and military waste. This commercially available treatment has a number of advantages, namely: (a) because it uses an on-site treatment method, it avoids the need to transport hazardous materials; (b) it ensures complete destruction of organic wastes and allows reuse of the effluent as process water with results that meet the regulations for drinking water; and (c) no licence for effluent or air emissions is needed.

Organic reactions in supercritical water are very fast. The dehydrogenation of lactic acid to acrylic acid in a Hastelloy C-276 annular reactor at 310 bar, 320–400°C and residence times of 25–110 s in the presence of various catalysts was used to examine three main pathways [193]. Alkylation of isobutane with light olefins for the production of gasoline with a high octane number [194], the hydrolysis of vegetable oils [195] and organic synthesis [196] are among the reported uses of supercritical water. However, supercritical water does not seem to be an effective extractant as, only in special situations, are the analytes so strongly retained by the matrix to require supercritical water conditions — under which the sample is highly likely to be degraded anyway.

7.8. PRESENT AND FUTURE OF SUPERCRITICAL FLUID EXTRACTION

After the initial euphoria of the late 1980s and 1990s, SFE has consolidated as a powerful tool for the analysis of environmental, pharmaceutical, polymer and, especially, food samples (particularly fat analyses, which have boosted SF extractor sales recently) [17]. Notwithstanding its major restrictions (especially in relation to the extraction of polar analytes and the treatment of natural samples), SFE can undoubtedly expedite the pretreatment of solid samples with the aid of special approaches to the extraction of medium-polar and polar analytes. Foreseeing the direction in which SFE applications will move in the future entails considering the nature of its latest uses.

Coupled supercritical CO_2–subcritical water systems may facilitate the class-selective extraction of non-polar substances and families of compounds of different polarity by adjusting the temperature of the water.

The development of new analyte collection methods using decreased amounts of solvent — and hence providing more concentrated extracts — is one other focus of current SFE research [43].

One other potential trend for SFE aims at field work. So far, only a single SFE method for the field extraction of PCBs and PAHs from soil appears to have been reported; the method is used in combination with GC–ECD or FID and requires no clean-up following extraction [197]. A portable system developed by the US Department of Defense is currently at the testing stage. The ensuing method uses dry ice — thus avoiding the need for compressed CO_2 tanks — and has proved effective in the extraction of non-polar analytes; however, it requires the presence of modifiers to extract polar analytes. Such techniques as ASE, MAE and microwave-assisted SOX have lately experienced major breakthroughs [198–200]; consequently, SFE should only be considered the technique of choice for extraction purposes in those cases where its advantages over the previous alternatives (viz. cleanliness, expeditiousness and simplicity) are bound to offset the potential shortcomings arising from its inability to efficiently process specific matrices or analytes.

References

1 T.A. Berger, *Packed Column SFC*, Royal Society of Chemistry, Cambridge (1995).
2 K. Anton and C. Berger, *Supercritical Fluid Chromatography with Packed Columns. Techniques and Applications*, Marcel Dekker, New York (1997).
3 M.D. Luque de Castro, M. Valcárcel M.T. Tena, *Analytical Supercritical Fluid Extraction*, Springer-Verlag, Heidelberg (1994).
4 L.T. Taylor, *Supercritical-Fluid Extraction*, John Wiley & Sons, Chichester (1996).
5 J.R. Williams and A.A. Clifford, Eds, *Supercritical Fluids. Methods and Protocols*, Humana Press, New Jersey (2000).
6 B.A. Charpentier and M.R. Sevenants, Eds, *Supercritical Fluid Extraction and Chromatography. Techniques and Applications*, ACS Symposium Series (1988) 130.
7 G.M. Schneider, E. Stahl and G. Wilke, Eds, *Extraction with Supercritical Gases*, Verlag Chemie, Weinheim (1992) 73.
8 M.J. Kamlet, J.L. Abboud and R.W. Taft, *J. Ame. Chem. Soc.*, 99 (1977) 6027.
9 S. Kim and K.P. Johnston, *Effects of Supercritical Solvents on the Rates of Homogeneous Chemical Reactions*, ACS Symposium Series, 329 (1986) 42.
10 R.S. Smith, B.W. Wright and C.R. Yonker, *Anal. Chem.*, 60 (1988) 1323[a].
11 M. Ashraf-Khorassani, L.T. Taylor and P. Zimmermann, *Anal. Chem.*, 62 (1990) 1177.
12 S.F.Y. Li, C.P. Lee and H.K. Lee, *J. Chromatogr.*, 31 Aug. (1990) 515509.
13 L.J. Mulcahey and L.T. Taylor, *Anal.Chem.*, 64 (1992) 2352.
14 A.L. Howard and L.T. Taylor, *PittCon'92*, New Orleans (1992) 308.
15 B.W. Wenclawiak, O.P. Heemken, D. Sterzenbach, J. Schipke, N. Theobald and V. Weighlt, *Anal. Chem.*, 67 (1995) 4577.
16 G. Madras, C. Erkey and A. Akgermann, *Environmental Progress*, 13 (1994) 45.
17 B. Erickson, *Anal. Chem.*, 70 (1998) 333 A.
18 Z.K. Otero, Optimization of SFE by statistical methods, *Pittcon'92*, New Orleans (1992).
19 V. López-Ávila, N.S. Dodhiwala and W.F. Beckert, *J. Chromatogr. Sci.*, 28 (1990) 468.
20 M.T. Tena, M.D. Luque de Castro and M. Valcárcel, *Lab. Rob. Autom.*, 5 (1993) 255.
21 M.P. Llompart, R.A. Lorenzo and R. Cela, *J. Chromatogr. Sci.*, 34 (1996) 43.
22 A. Braachet, P. Christe, J.Y. Gauvrit, R. Longeray, P. Lantéri and J.L. Veuthey, *J. Biochem. Biophys. Methods*, 43 (2000) 353.
23 I.A. Stuart, R.O. Ansell, J. Maclachlan, P.A. Bather and W.P. Gardiner, *Analyst*, 122 (1997) 303.

24 J. Amador-Hernández, J.M. Fernández-Romero, G. Ramis-Ramos and M.D. Luque de Castro, *Appl. Spectrosc.*, 52 (1998) 1465.
25 P.C.K. Cheung, *Food Chem.*, 65 (1999) 399.
26 S.B. Hawthorne and D.J. Miller, *J. Chromatogr. Sci.*, 24 (1986) 258.
27 J.J. Langenfeld, S.B. Hawthorne, D.J. Miller and J. Pawliszyn, *Anal. Chem.*, 65 (1993) 338.
28 S.L. Cleland, L.K. Olson, J.A. Caruso and J.M. Carey, *J. Anal. At. Spectrom.*, 9 (1994) 975.
29 M.T. Tena, M.D. Luque de Castro and M. Valcárcel, *Chromatographia*, 38 (1994) 431.
30 L. Gámiz-Gracia, M.M. Jiménez-Carmona and M.D. Luque de Castro, *Chromatographia*, 51 (2000) 428.
31 J.J. Langenfeld, S.B. Hawthorne, D.J. Miller and J. Pawliszyn, *Anal. Chem.*, 66 (1994) 909.
32 S.R. Sargenti and F.M. Lancas, *J. Microcolum. Sep.*, 10 (1998) 213.
33 D.J. Pyo, K.Y. Park, H. Sin and M.H. Moon, *Chromatographia*, 49 (1999) 539.
34 F. Guo, Q.X. Li and J.P. Alcántara-Licudine, *Anal. Chem.*, 71 (1999) 1309.
35 M. Ashraf-Khorassani and L.T. Taylor, *Am. Lab.*, 27 (1995) 23.
36 M. Ashraf-Khorassani and L.T. Taylor, *Int. Lab.*, 26 (1996) 16.
37 S.B. Hawthorne, J.J. Langenfeld, D.J. Miller and M.D. Burford, *Anal. Chem.*, 64 (1992) 1614.
38 F. David, M. Verschuere and P. Sandra, *Fresenius J. Anal. Chem.*, 344 (1992) 479.
39 C.A.Thomson and D.J. Chesney, *Anal. Chem.*, 64 (1992) 848.
40 L.A. Dolata, J.M. Ley, A.C. Rosselli and R.M. Ravey, Making SFE work for environmental applications, *Pittcon'92*, New Orleans (1992).
41 S.A. Westwood, Ed., *Supercritical Fluid Extraction and Its Use in Chromatographic Sample Preparation*, Blackie Academic & Professional, London (1993).
42 Hewlett-Packard Application, No. 228–145.
43 J.J. Langenfeld, S.B. Hawthorne, D.J. Miller and J. Pawliszyn, *Anal. Chem.*, 67 (1995) 1727.
44 K.G. Furton and Q. Lin, *Chromatographia*, 34 (1992) 185.
45 K.G. Furton and J. Rein, *Chromatographia*, 31 (1991) 297.
46 B.A. Benner, *Anal. Chem.*, 70 (1998) 4594.
47 M. Ashraf-Khorassani, L.T. Taylor and F.K. Schweighardt, *J. AOAC Int.*, 79 (1996) 1043.
48 J.M. Levy, R.M. Ravey and R. Panella, *LC–GC*, 14 (1998) 570.
49 G.C. Slack, H.M. McNair, S.B. Hawthorne and D.J. Miller, *J. High Resolut. Chromatogr.*, 16 (1993) 473.
50 L.J. Mulcahey and L.T. Taylor, *Anal. Chem.*, 64 (1992) 981.
51 B.W. Wright, C.W. Wright, R.W. Gale and R.D. Smith, *Anal. Chem.*, 59 (1987) 38.
52 N. Alexandrou, P. Pawliszyn and M. Lawrence, *Anal. Chem.*, 64 (1992) 301.
53 S.B. Hawthorne, J.J. Langenfeld, D.J. Miller and M.D. Burford, *Anal. Chem.*, 64 (1992) 1614.
54 T. Paschke, S.B. Hawthorne, D.J. Miller and B. Wenclawiak, *J. Chromatogr.*, 609 (1992) 333.
55 S.F.Y. Li, C.P. Ong, M.L. Lee and H.K. Lee, *J. Chromatogr.*, 31 Aug. (1990) 515515.
56 A.L. Howard and L.T. Taylor, *Pittcon'93*, New Orleans, MA (1993) Abstract 1181.
57 P. Capriel, A.I. Haisch and S.U. Khan, *J. Agric. Food, Chem.*, 34 (1986) 70.
58 J.S. McPartland and R.G. Bautista, *Sep. Sci. Technol.*, 25 (1990) 2045.
59 S.B. Hawthorne, Y. Yang and D.J. Miller, *Anal. Chem.*, 66 (1994) 2912.
60 L.X. Li and E.F. Gloyna, *Sep. Sci. Technol.*, 34 (1999) 1463.
61 J.A. Field, D.J. Miller, T.M. Field, S.B. Hawthorne and W. Ginger, *Anal. Chem.*, 64 (1992) 3161.
62 M.T. Tena, M.D. Luque de Castro and M. Valcárcel, *Chromatographia*, 40 (1995) 197.
63 M.M. Jiménez-Carmona, M.T. Tena and M.D. Luque de Castro, *J. Chromatogr.*, 711 (1995) 269.
64 J.A. Field, *J. Chromatogr.*, 785 (1997) 239.
65 J.W. King, J.E. France and J.M. Snyder, *Fresenius, J. Anal. Chem.*, 344 (1992) 474.
66 B.E. Berg, E.M. Hansen, S. Gjorven and T. Greibrokk, *J. High Resolut. Chromatogr.*, 16 (1993) 358.
67 S.B. Hawthorne, D.J. Miller, D.E. Nievens and D.C. White, *Anal. Chem.*, 64 (1992) 405.
68 R. Hillmann and K. Bachmann, *J. High Resolut. Chromatogr.*, 17 (1994) 350.
69 M.Y. Croft, E.J. Murby and R.J. Wells, *Anal. Chem.*, 66 (1994) 4459.
70 J.A. Field and J. Monohan, *Pittcon'95*, New Orleans (1995) Abstract 370.
71 M.T. Cummins and R.J. Wells, *J. Chromatogr.*, 694 (1997) 11.

72 J.M. Snyder, J.W. King and M.A. Jackson, *M. Am. Oil Chem. Soc.*, 74 (1997) 585.
73 J.W. Hills, H.H. Hill Jr and T. Maeda, *Anal. Chem.*, 63 (1991) 2152.
74 J.W. Hills and J.J. Hill, *J. Chromatogr. Sci.*, 31 (1993) 6.
75 C.M. Wai, Y. Lin, R.D. Brauer, S. Wang and W.F. Beckert, *Talanta*, 40 (1993) 1325.
76 Y. Liu, V. López-Ávila, M. Alcaraz and W.E. Beckert, *J. High Resolut. Chromatogr.*, 16 (1993) 106.
77 J.W. Oudsema and C.F. Poole, *Fresenius J. Anal. Chem.*, 344 (1992) 426.
78 J.W. Oudsema and C.F. Poole, *J. High Resolut. Chromatogr.*, 16 (1993) 199.
79 Y. Cai, R. Alzaga and J.M. Bayona, *Anal. Chem.*, 66 (1994) 1161.
80 R. Alzaga and J.M. Bayona, *J. Chromatogr.*, 655 (1993) 51.
81 B.W. Wenclawiak and M. Krah, *Fresenius J. Anal. Chem.*, 351 (1995) 134.
82 K.E. Laintz, C.M. Wai, C.R. Yonker and R.D.Smith, *Anal. Chem.*, 64 (1992) 2875.
83 J. Wang and W.D. Marshall, *Anal. Chem.*, 66 (1994) 1658.
84 J. Wang and W.D. Marshall, *Analyst*, 120 (1995) 623.
85 Y. Liu, V. López-Ávila and M. Alcaraz, *Anal. Chem.*, 66 (1994) 3788.
86 Y.K. Chau, F. Yang and M. Brown, *Anal. Chim. Acta*, 304 (1995) 85.
87 V. López-Ávila, M. Alcaraz, W.F. Beckert and E.M. Heithmar, *J. Chromatogr. Sci.*, 31 (1993) 310.
88 M. Ashraf-Khorassani and L.T.Taylor, *Anal. Chim. Acta*, 379 (1999) 1.
89 Y. Lin, R.D. Brauer, K.E. Laintz and C.M. Wai, *Anal. Chem.*, 65 (1993) 2549.
90 Y. Lin, C.M. Wai, F.M. Jean and R.D. Brauer, *Environ. Sci. Technol.*, 28 (1994) 1190.
91 K.G. Furton, L. Chen and R. Jaffé, *Anal. Chim. Acta*, 304 (1995) 203.
92 Y. Lin and C.M. Wai, *Anal. Chem.*, 66 (1994) 1971.
93 K.E. Laintz and E. Tachikawa, *Anal. Chem.*, 66 (1994) 2190.
94 S. Wang, S. Elshani and C.M. Wai, *Anal. Chem.*, 67 (1995) 919.
95 M.M. Jiménez-Carmona and M.D. Luque de Castro, *Anal. Chim. Acta*, 358 (1998) 1.
96 M.M. Jiménez-Carmona and M.D. Luque de Castro, *Anal. Chem.*, 70 (1998) 2100.
97 J.H. Wang, Q. Xu and K. Jiao, *J. Chromatogr.*, 818 (1998) 138.
98 F.M. Lancas, S.R. Rissato and M.S. Galhiane, *Chromatographia*, 42 (1996) 323.
99 W.J. Yoo and L.T. Taylor, *J. AOAC Int.*, 80 (1997) 1336.
100 H.R. Johansen, G. Becher and T. G. Reibrokk, *Anal. Chem.*, 66 (1994) 4068.
101 R.F. Mauldin, J.M. Vienneau, E.L. Wehry and G. Mamanton, *Talanta*, 37 (1990) 1031.
102 G.P. Blanch, G. Reglero and M. Herraiz, *J. Agric. Food Chem.*, 43 (1995) 1251.
103 W.K. Modey, D.A. Mulholland, H. Mahomed and M.W. Raynor, *J. Microcolumn Sep.*, 8 (1996) 67.
104 B.E. Berg, H.S. Lund and T. Greibrokk, *Chromatographia*, 44 (1997) 399.
105 S.G. Yocklovich, J.C. Watkins and E.J. Levy, *J. Chromatogr.*, 505 (1990) 273.
106 F. Ischi and W. Haerdi, *Chromatographia*, 41 (1995) 238.
107 B.W. Wenclawiak, T. Hees, C.E. Zoeller and H.P. Kabus, *Fresenius, J. Anal. Chem.*, 358 (1997) 471.
108 P.B. Liescheski, *J. Agric. Food. Chem.*, 44 (1996) 823.
109 J. Amador-Hernández, J.M. Fernández-Romero, G. Ramis-Ramos and M.D. Luque de Castro, *Anal. Chim. Acta*, 390 (1999) 163.
110 H. Ohde, F. Hunt, S. Kihara and C.M. Way, *Anal. Chem.*, 72 (2000) 4738.
111 J. Zagrobelny, M. Li, R. Wang, T.A. Betts and E.V. Bright, *Appl. Spectrosc.*, 46 (1992) 1895.
112 J.K. Rice, R.A. Dunbar and R.V. Bright, *Appl. Spectrosc.*, 48 (1994) 1030.
113 M.T. Tena, M.D. Luque de Castro and M. Valcárcel, *Anal. Chem.*, 68 (1996) 2386.
114 M.T. Tena, M.D. Luque de Castro and M. Valcárcel, *Analyst*, 119 (1994) 1625.
115 M.T. Tena and M. Valcárcel, *J. Chromatogr.*, 753 (1996) 299.
116 R.W. Current and D.C. Tilotta, *J. Chromatogr.*, 785 (1997) 269.
117 S.G. Kazarian, *Appl. Spectrosc.*, 32 (1997) 301.
118 J.D. Brewster, R.J. Maxwell and J.W. Hampson, *Anal. Chem.*, 65 (1993) 2137.
119 M.T. Tena, M.D. Luque de Castro and M. Valcárcel, *Anal. Chem.*, 67 (1995) 1054.
120 L. Manganiello, A. Ríos and M. Valcárcel, *J. Chromatogr.*, 874 (2000) 265.

121 J. Wang and W.D. Marshall, *Anal. Chem.*, 66 (1994) 3900.
122 J. Wang and W.D. Marshall, *Analyst*, 121 (1996) 817.
123 R.D. Smith and H.R. Udseth, *Anal. Chem.*, 55 (1983) 2266.
124 U. Braumann, H. Handel, K. Albert, R. Ecker and M. Spraul, *Anal. Chem.*, 67 (1995) 930.
125 D.C. Tilotta, D.L. Heglung and S.B. Hawthorne, *Am. Lab.*, 28 (1996) 36.
126 C.H. Kirschner and L.T. Taylor, *Anal. Chem.*, 65 (1993) 78.
127 C.H. Kirschner, S.L. Jordan and L.T. Taylor, *Anal. Chem.*, 66 (1994) 882.
128 J. Amador-Hernández, J.M. Fernández-Romero and M.D. Luque de Castro, *Talanta*, 49 (1999) 813.
129 Y.C. Sun, J. Mierzwa, Y.T. Chung and M.H. Yang, *Anal. Commun.*, 34 (1997) 333.
130 S. Liang and D.C. Tilotta, *Anal. Chem.*, 70 (1998) 4487.
131 H.T. Kalinoski, H.R. Udseth and B.W. Wright, *Anal. Chem.*, 58 (1986) 2421.
132 R.J. Trigg, G.A. Keenan, D. McMullan and A.I. Coner, *Int. J. Environ. Anal. Chem.*, 70 (1998) 47.
133 M. Ashraf-Khorassani, S. Hinman and L.T. Taylor, *J. High Resolut. Chromatogr.*, 22 (1999) 271.
134 M. Kane, J.R. Dean and S.M. Hitchen, *Anal. Chim. Acta*, 271 (1993) 83.
135 E. Björklund, M. Jaremo, L. Mathiasson, J.A. Jonsson and L. Karlsson, *Anal. Chim. Acta*, 368 (1998) 117.
136 K.D. Bartle, T. Boddington, A.A. Clifford and N.J. Cotton, *Anal. Chem.*, 63 (1991) 2371.
137 M. Fischer and T.M. Jefferies, *J. Agric. Food. Chem.*, 44 (1996) 1258.
138 S. Ashraf, K.D. Bartle, A.A. Clifford, R. Moulder, M.W. Raynor and G.F. Shilstone, *Analyst*, 117 (1992) 1697.
139 R.C. Hale, M.O. Gaylor, J.F. Thames, C.K. Smith and R.F. Mothershead, *Int. J. Environ. Anal. Chem.*, 64 (1996) 11.
140 V. López-Ávila, Y. Liu and W.F. Beckert, *J. Chromatogr.*, 785 (1997) 279.
141 T.M. Young and J.W. Walter, *Anal. Chem.*, 69 (1997) 1612.
142 M. Palma and L.T. Taylor, *Anal. Chim. Acta*, 391 (1999) 321.
143 M.D. Burford, S.B. Hawthorne and D.J. Miller, *Anal. Chem.*, 65 (1993) 1497.
144 S. Bowadt and B. Johansson, *Anal. Chem.*, 66 (1964) 667.
145 J.M. Brooks, M.C. Kennicutt, T.L. Wade, A.D. Hart, G.J. Denoux and J. McDonald, *J. Environ. Sci. Technol.*, 24 (1990) 1079.
146 R. Bossi, B. Larsen and G. Pramazzi, *Sci. Total Environ.*, 121 (1992) 77.
147 P. Tong and T. Imagawa, *Anal. Chim. Acta*, 310 (1995) 93.
148 M. Valcárcel and M.T. Tena, *Fresenius J. Anal. Chem.*, 358 (1997) 561.
149 F.J. Eller and J.W. King, *Seminars Food Analysis*, 1 (1996) 145.
150 T.L. Chester and J.D. Pinkston, *Anal. Chem.*, 72 (2000) 129 R.
151 J. Hedrick and L.T. Taylor, *Anal. Chem.*, 61 (1989) 1986.
152 J. Hedrick and L.T. Taylor, *J. High Resolut. Chromatogr.*, 13 (1990) 312.
153 C.P. Ong, H.M. Ong, S.F. Li Yau and H.K. Lee, *J. Microcol. Sep.*, 2 (1990) 69.
154 R. Aranda and P. Kruus, *Int. J. Environ. Anal. Chem.*, 68 (1997) 59.
155 K.R. Jahn and B. Wenclawiak, *Chromatographia*, 26 (1988) 345.
156 K.S. Nam, S. Kapila, A.F. Yanders and R.K. Puri, *Chemosphere*, 20 (1990) 873.
157 J. Damian, L. Myer, P. Liesdheski and J. Tehrani, Supercritical fluid extraction of organic analytes from aqueous media and wet matrices, *Pittcon'92*, New Orleans (1992).
158 R.S. Michael, L.D. White and D.R. Hill, Optimization of supercritical fluid extraction of additives from polyolefins, *Pittcon'92*, New Orleans (1992).
159 J.L. Ezzell and B.D. Richter, Supercritical fluid extraction and chromatography following the solid phase extraction of analytes from water, *Pittcon'92*, New Orleans (1992).
160 M. Di Maso, B. L'Archeveque, A. Vas and S.A. McClintock, The use of supercritical fluid extraction for samples of pharmaceutical importance, *Pittcon'92*, New Orleans (1992).
161 L.A. Howard and L.T. Taylor, Supercritical fluid extraction of chlorosulfuron and sufemeturon methyl from aqueous matrices, *Pittcon'92*, New Orleans (1992).
162 H. Liu, L.M. Cooper, D.E.Raynie, J.D. Pinkston and K.R. Wehmeyer, *Anal. Chem.*, 64 (1992) 802.

163 W.M.A. Niessen, P.J.M. Tjaden and J. van der Greef, *J. Chromatogr.*, 454 (1988) 243.
164 J. You, Y.X. Chen and G.J. Wang, *Fenxi Huaxue*, 27 (1999) 337.
165 M.D. David and J.N. Seiber, *Anal. Chem.*, 68 (1996) 3044.
166 S.P. Frost, J.R. Dean, K.P. Evans, K. Harradine, C. Cary and M.H.I. Comber, *Analyst*, 122 (1997) 895.
167 A. Otterbach and B.W. Wenclawiak, *Fresenius, J. Anal. Chem.*, 365 (1999) 472.
168 A.M. Carro, R.A. Lorenzo, M.J. Vázquez, M. Abuin and R. Cela, *Intern. Lab.*, November (1998) 23.
169 R.A. Lorenzo, M.J. Vázquez, A.M. Carro and R. Cela, *Trends Anal. Chem.*, 18 (1999) 410.
170 T.S. Reighard and V. Olesik, *Anal. Chem.*, 68 (1996) 3612.
171 G. Anitescu, C. Doneanu and V. Radulescu, *Flavour Fragrance J.*, 12 (1997) 173.
172 A. Ghassempour and M.R. Arshadi, *Sepu*, November (1998) 463.
173 P.C.K. Cheung, A.Y.H. Leung and P.O. Ang, *J. Agric. Food Chem.*, 46 (1998) 4228.
174 E.E. Stashenko, M.A. Puertas and M.Y. Combariza, *J. Chromatogr.*, 752 (1996) 223.
175 X. Chaudot, A. Tambute and M. Caude, *J. High Resolut. Chromatogr.*, 21 (1998) 457.
176 M.G.R. Vale, L.P. Luz, A.F. Martins, E.B. Caramao, C. Dariva and J.V. Oliveira, *J. Microcol. Sep.*, 10 (1998) 259.
177 M.P. Llompart, R.A. Lorenzo, R. Cela, K. Li, J.M.R. Belanger and J.R.J. Paré, *J. Chromatogr.*, 774 (1997) 243.
178 A. Kreisselmeier and H.W. Durbeck, *J. Chromatogr.*, 775 (1997) 187.
179 V. López-Ávila, R. Young and N. Teplitsky, *J. AOAC Int.*, 79 (1996) 142.
180 J.A. Field, K. Monohan and R. Reed, *Anal. Chem.*, 70 (1998) 1956.
181 B. Wenclawiak, C. Rathmann and A. Teuber, *Fresenius J. Anal. Chem.*, 344 (1992) 497.
182 I.J. Barnabas, J.R. Dean, W.R. Tomlinson and S.P. Owen, *Anal. Chem.*, 67 (1995) 2064.
183 J.R. Dean, I.J. Barnabas and O.A. Fowlis, *Anal. Proc.*, 32 (1995) 305.
184 C. Friedrich, K. Cammann and W. Kleiböhmer, *Fresenius J. Anal. Chem.*, 352 (1995) 730.
185 C. Lutermann, W. Dott and J. Hollender, *J. Chromatogr.*, 811 (1998) 151.
186 B. Wenclawiak, T. Paschke and M. Krappe, *Fresenius J. Anal. Chem.*, 357 (1997) 1128.
187 O.P. Heemken, N. Theobald and B. Wenclawiak, *Anal. Chem.*, 69 (1997) 2171.
188 S.R. Sargenti and H.M. McNair, *J. Microcol. Sep.*, 10 (1998) 125.
189 S.B. Hawthorne and C.B. Grabanski, *Environ. Sci. Technol.*, 34 (2000) 4103.
190 N. Boukis, R. Landvatter, W. Habicht, G. Franz, S. Leistikow, R. Karft and O. Jacobi, *private communication*.
191 H.J. Bleyl, J. Abeln, N. Boukis, H. Goldacker, M. Kluth, A. Kruse, G. Petrich, H. Schmieder and G. Wiegand, *J. Separation Sci. Technol.*, (in press).
192 A. Suzuki, T. Oe, N. Anjo, H. Suzugaki and T. Nakamura, *4th International Symposium on Supercritical Fluids*, Sendai, Japan (1997) 895.
193 C.T. Lira and P.J. McCrackin, *Ind. Eng. Chem. Res.*, 32 (1993) 2608.
194 E.J.A. Schweitzer, J. de Graauw, O.S.L. Bruinsma and P.F. van den Oosterkamp, *4th International Symposium on Supercritical Fluids*, Sendai, Japan (1997) 523.
195 R.L. Holliday, J.W. King and G.R. List, *Ind. Enf. Chem. Res.*, 36 (1997) 932.
196 R.L. Holliday, B.Y.M. Jong and J.W. Kolis, *J. Supercrit. Fluids*, 12 (1998) 255.
197 S. Bowadt, L. Mazeas, D.J. Miller and S.B. Hawthorne, *J. Chromatogr.*, 785 (1997) 205.
198 L.E. García-Ayuso, M. Sánchez, A. Fernández de Alba and M.D. Luque de Castro, *Anal. Chem.*, 70 (1998) 2426.
199 L.E. García-Ayuso, J.L. Luque García and M.D. Luque de Castro, *Anal. Chem.*, 72 (2000) 3627.
200 L.E. García-Ayuso and M.D. Luque de Castro, *Trends Anal. Chem.*, 20 (2000) 28.

Chapter 8

Devices for solid sample treatment prior to introduction into atomic spectrometers: electrothermal devices and glow-discharge sources

8.1. INTRODUCTION

Atomic techniques have proved to be highly powerful tools for determining metals in a wide variety of matrices, particularly since the inception of special devices that have substantially improved on the atomization efficiency of flames. The use of electrothermal [1] and non-electrothermal devices to perform the steps involved in converting the target elements into the required atomic states has endowed absorption, emission and fluorescence atomic techniques with their present roles in analytical chemistry. Although mass spectrometry is not an atomic technique, it is also widely used for the determination of metals. For this reason, it is also dealt with in this chapter alongside the atomic techniques typically employed following atomization of a sample. This chapter is thus concerned with the sampling devices used prior to introduction into absorption, fluorescence emission and mass spectrometers. The former two always require that the target analyte be in the atomic state before the absorption or excitation source comes into play and hence the use of an electrothermal atomizer (ETA); the latter two only require vaporization of the sample (by means of an electrothermal vaporizer, ETV) if the atomic or ionized state can be reached by using a flame or plasma, vaporization–atomization–excitation (using a glow-discharge source) in emission spectroscopy and vaporization–atomization–ionization in mass spectrometry. None of these detection techniques is dealt with as such here, although they appear in some schemes illustrating their role in specific applications. As in previous chapters, readers are referred to monographs on these well-established techniques for further details [2–11].

In addition to its well-known advantages over sample dissolution (viz. reduced sample pretreatment and thus increased expeditiousness in the overall analytical procedure; low contamination risks, which is an essential requirement in determining trace metals; decreased analyte losses through sample pretreatment or retention by insoluble residues; and avoidance of corrosive or hazardous chemicals), solid sampling overcomes some of the shortcomings arising from the way liquid samples are introduced into flame or plasma spectrometers. Thus, liquid sampling requires relatively large volumes of solutions (at least 5 ml for analysis under typical operating conditions). This frequently calls for dilution of the sample. Pneumatic nebulizers have the additional drawback that usually less than 5% of the solution actually reaches the flame or plasma. Ultrasonic nebulizers are more efficient but are expensive and highly sensitive towards high acid or salt concentrations.

Also, the insertion of solvents — particularly into plasmas — causes a number of spectral interferences and, at dissolved solid concentrations above *ca.* 1%, signal suppression and salt build-up on the nebulizer, the interface or the injection tube of the torch.

Solid sampling devices have been used to either atomize samples into absorption and fluorescence spectrometers or to vaporize them into flames and plasmas prior to emission or mass detection. The most usual types of devices used in this context are as follows: (a) electrothermal vaporizers (ETV) and atomizers (ETA); (b) glow-discharge (GD) sources; (c) laser ablation devices; (d) laser-induced breakdown devices; (e) direct sample insertion (DSI) devices; and (f) arc nebulizers. Because these devices have received dissimilar attention by users, they are discussed here to an extent proportional to the interest they have aroused. Thus, electrothermal atomizers and glow-discharge devices are the main subjects of this chapter, which also provides a brief description of direct sample insertion devices and arc nebulizers. Such a rapidly growing field as laser sampling warrants separate coverage and is thus the subject matter of Chapter 9.

Atomizers and glow-discharge devices feature a key difference: the sample is introduced in the former and is part of the sampling device in the latter.

In essence, glow-discharge sampling is more closely related to the techniques described in Chapter 9 than to electrothermal devices; however, it is discussed here both to facilitate comparisons and to balance the contents of the two chapters.

8.2. ELECTROTHERMAL ATOMIZERS AND VAPORIZERS

This section is devoted to the types of devices most frequently used for both liquid and solid sampling prior to introduction into atomic spectrometers [12–14]. Atomic techniques and mass spectrometry make massive use of electrothermal devices, the maturity of which has been endorsed by IUPAC, which has included it in its "Nomenclature, Symbols, Units and their Usage in Spectrochemical Analysis. XII. Terms Related to Electrothermal Atomization", published in 1992 and subsequently reprinted in *Spectrochimica Acta* [1].

8.2.1. Fundamentals of electrothermal vaporizers and atomizers

Carbon atomizers

Some of the physical and chemical constraints on the flame atomization process — which usually precluded application to solid samples — were overcome with the advent of "flameless" atomization, initially accomplished with the pyrolytic coated graphite tube (or carbon rod-type) furnace atomizer. The graphite tube is a confined furnace chamber where pulsed vaporization and subsequent atomization of the sample is achieved by raising the temperature with a programmed sequence of electrical power. A dense population of ground state atoms is produced as a result for an extended interval in relation to the low atom density and short residence time of the flame. The earliest use of "furnace" devices in analytical atomic spectroscopy is credited to a simultaneous development by L'vov [15] and Massmann [16]; however, the first application of one such device to a solid sample was reported by L'vov in 1959 [17] and followed 12 years later by the first instance of solid sampling with a commercial electrothermal atomizer [18] — one year after the first commercial graphite-furnace atomizers (ETAs) were marketed by firms such as Varian and Perkin–Elmer for use in atomic emission spectrometers. Development of ETA-based solid sampling applications started in the late 1970s and early 1980s.

Basically, a graphite-furnace atomizer (Fig. 8.1A) consists of a graphite tube, electrical contacts, an enclosed water-cooled housing and inert purge gas controls.

The graphite tube is the usual heating element of the graphite furnace. The cylindrical tube is aligned horizontally with the spectrometer's path in atomic absorption and fluorescence techniques, and serves as the spectrometer sampling cell in AAS and AFS. The sample is placed at the centre of the tube wall onto the inner tube wall — which can be cup-shaped — or on a graphite platform. The tube is held in place between two graphite contact cylinders that provide electrical connection. An electrical potential applied to the contact causes current to flow through the tube, the effect being heating of the tube and the sample. Quartz windows at each end of the housing allow light to pass through the tube. The heated graphite is protected from oxidation by atmospheric air by the end windows and two argon streams. An external gas flow surrounds the outside of the tube and an independently controlled internal gas flow purges its inside. The system should regulate the internal gas flow so that the internal flow is reduced or, preferably, completely stopped during atomization. This helps maximize the sample residence time in the tube and boosts measured signals. The graphite tube is heated by passing electrical current from the graphite contacts at the ends of the tube throughout its length. This type of furnace, which is longitudinally heated, is similar to the original design of Massmann [16] and is the basis for most currently available commercial graphite furnaces. It has a major liability: as the electrical contacts at each end of the tube must be cooled, there must always be a temperature gradient along the length of the tube, which results in the tube end adjacent to the electrical contacts being cooler than the central portion. This temperature gradient can cause vaporized atoms and molecules to condense as they diffuse to the cooler tube ends. This may in turn result in interferences, the most common type being incomplete removal of the analyte or matrix from the tube. Incomplete removal of the matrix during pyrolysis can increase background absorption during atomization; incomplete removal of the analyte is more serious as it results in carry-over or memory effects, whereby part of the analyte in the sample being processed remains in the tube and contributes to the analytical signal for the next. In order to reduce carry-over, most longitudinally heated furnace heating programmes use one or more clean-out steps after atomization or both increased pyrolysis temperatures and an in situ digestion procedure [19], which often entails altering the inside surface of the graphite tube (e.g. by coating with Ir sputtered in a low-pressure Ar discharge). A clean-out step involves the application for several seconds of full internal gas flow and a temperature equal to or greater than that used for atomization to remove residual sample components. While this additional step works well with the more readily atomized analytes, it does not always succeed with those that require higher atomization temperatures. The use of a high-temperature clean-out step may also reduce tube lifetime.

The transversely heated furnace (THGA, Fig. 8.1B) avoids many of the problems associated with the longitudinally heated furnace. The graphite tube of a transversely heated

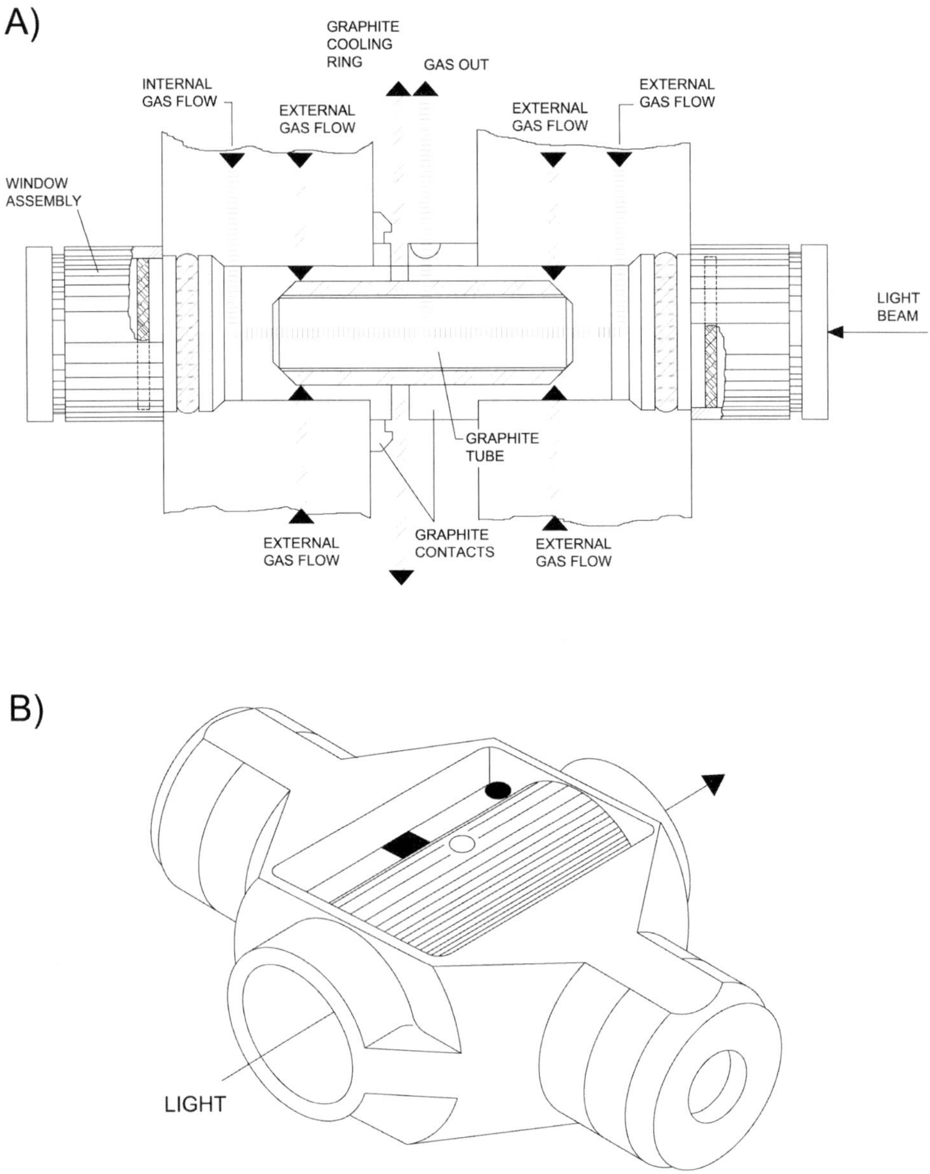

Fig. 8.1. Typical graphite furnaces: (A) longitudinally heated model, (B) transversely heated model. For details, see text. (Reproduced with permission of Perkin–Elmer.)

furnace includes integral tabs that protrude from each side. The tabs are inserted into the electrical contacts and, when the power is applied, the tube is heated across its circumference (i.e. transversely). In this way, the tube is heated evenly over its entire length, thus eliminating or significantly reducing the sample condensation problems typical of longitudinally heated furnaces. One additional advantage of the transversely heated furnace is that it enables longitudinal Zeeman-effect background correction without the need to use a polarizer in the optical system.

The comparison of a transversely heated graphite atomizer having longitudinal Zeeman-effect background correction with a Massmann-type furnace having continuum-source background correction in the determination of aluminium in nickel-based alloys by ETA–AAS revealed the THGA to be less sensitive — owing to its shorter path length — but also to provide better limits of detection by a factor of 2 as a result of its background correction system [20].

The Frech two-step furnace, with separate control of the vaporization and atomization functions, represents a substantial improvement on commercial Massmann-type and THGA furnaces for interference-free analyses by ETA–AAS. However, it has the disadvantage that it relies on diffusion and convection to transport sample vapours from the cup vaporizer to the tube atomizer. Transport by purging is one solution to this shortcoming. For this purpose, the Massmann-type atomizer is heated to a steady-state atomization temperature and the THGA vaporizer is pulse-heated to have the purge gas drive the analyte from the vaporizer to the atomizer [21].

Platforms for carbon atomizers

One source of interference with carbon atomizers arises from the analyte appearing in the furnace atmosphere before the furnace has reached the actual temperature set for the atomization step in the furnace programme. The problem results from the temperature of the tube wall rising rapidly at the beginning of the atomization step. The sample, which is in direct contact with the tube wall, immediately experiences the increase in wall temperature. When the tube wall reaches the temperature at which analyte atoms are produced, atoms are released from the hot tube surface into the inert gas atmosphere, the temperature of which lags behind that of the tube walls and is cool relative to the surface. This sudden cooling inhibits atomization of the vaporized molecular species and results in non-spectral interferences. Elimination of this delay favours the formation of free atoms and maximizes the sensitivity, which remains constant irrespective of the sample matrix. One of the earliest devices (usually platforms) used to avoid the delay was the L'vov platform, which is a flat piece made of solid pyrolytic or pyrolytically coated graphite that is placed in the bottom of the graphite tube as shown in Fig. 8.2A. When the sample is located in the shallow depression of the platform, the sample experiences the temperature of the platform rather than that on the tube wall. Since the platform is, by design, in poor physical contact with the round tube surface, the platform is heated primarily by radiant rather than conductive energy. This means that the platform will heat more slowly than the graphite tube. During the rapid heating in the atomization step, when the platform does finally reach the atomization temperature for the analyte, atoms are released into a furnace environment the temperature of which is similar to that of the

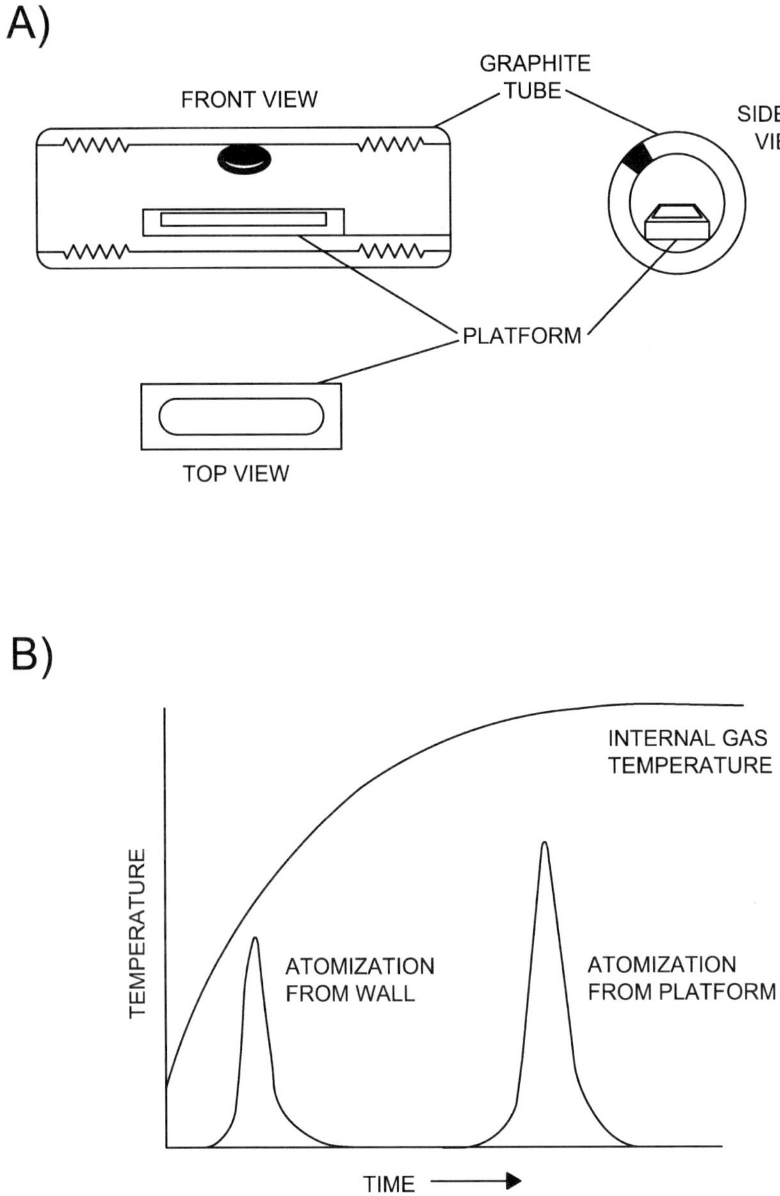

Fig. 8.2. Use of the L'vov platform for electrothermal atomization. (A) Scheme of the device. (B) Behaviour of the analyte as compared with the absence of the device in the atomizer. (Reproduced with permission of Perkin–Elmer.)

platform (see Fig. 8.2B). Under these conditions, dissociation is favoured and recombination minimal. The increased sensitivity exhibited by the signals of Fig. 8.2B arises from the fact that, when the furnace tube reaches a temperature at which the analyte will vaporize, the metal is driven from the surface into the gas phase. Both the temperature at which the analyte first starts to volatilize and the rate at which it does will depend on the specific nature of the matrix constituents and their amounts. Since the platform is heated by radiation from the tube walls, heating of the sample on the platform is delayed relative to the tube walls and hence to the vapour within the tube. Instead of volatilizing the analyte as the temperature changes, appropriate conditions can be found to volatilize it after the tube walls and gas phase have reached a more stable or steady-state condition. The effect of the platform is thus to delay the volatilization of the analyte until the atomizer has reached a stable (high) temperature. A number of users have attempted to improve the features of commercially available platforms as regards both shape and material with little or no success [22].

A series of additional alterations of the graphite tube atomizer intended to improve performance (e.g. the cup and cup cuvette atomizer, boat and microboat platforms, carbon rods, graphite probes, ring chambers and surface atomizers) have also been reported [23]. Interested readers are referred to specific sources [2–7] for more details.

Coatings for carbon atomizers

Although graphite — particularly pyrolytic graphite — remains a unique material with many desirable properties for an atomizer, a number of metal coatings on the graphite substrate can be used to increase the sensitivity for some elements. Almost thirty years ago, the use of high-melting Ta and Nb carbides was proposed to raise the sensitivity of ETA–AAS for boron by using the L'vov cuvette as atomizer. Since then, interest in this idea has not declined, as shown by the more than four hundred papers on research into carbide-modified graphite tubes. Based on the structure of the carbide layer used for this purpose, modified graphite tubes can be classified into the following three categories:

(a) PVD-modified tubes, where the solid layer of carbide on the surface prevents the interaction of analytes and matrices with the graphite. As a rule, these tubes have very long lifetimes and their surface activity is lower than that of the original tubes.
(b) Tubes treated under atmospheric pressure with solutions of modifying elements, which distribute over the graphite surface as "islands" of the corresponding carbides. In this way, the analysed sample is kept in contact with the graphite.
(c) Tubes modified under a low or high pressure with solutions or pure liquids containing the modifying elements. These tubes are placed between the previous two. After several treatments under optimal conditions, a solid carbide layer eventually forms.

The tubes can also be classified as follows according to the properties of the specific carbides used in the coatings:

(a) Tubes modified with metal-like carbides such as TiC, ZrC, HfC, VC, MoC and WC, which are more chemically active than pyrolytic graphite;

(b) Tubes modified with metal-like carbides such as TaC and NbC, which are more stable to oxidation by acids and possess roughly the same reactivity to oxygen as pyrolytic graphite;
(c) Tubes modified with covalent B_4C or SiC, which exhibit a high chemical inertia;
(d) Tubes modified with compounds of the elements in the third group (La, Th and Y), the surface activity of which is scarcely dependent on the particular modifying method used.

An excellent review of the methods used to modify graphite tubes; investigations of the modified surface; the physical, chemical and catalytic properties of the modified tubes; and the main processes occurring in them was published in 1998 by Volynsky [24].

Non-carbon atomizers

Although graphite furnaces are the most widely used in atomizers, alternative materials have also been employed for this purpose.

Tungsten-coil atomizers have to some extent been used in connection with atomic techniques in general and ETA–AAS in particular, and their popularity continues to rise. Their high simplicity and low cost make them attractive alternatives to graphite furnaces for many applications. However, they are much more interference-prone than their graphite counterparts. The interferences experienced by W-coil atomizers have been overcome in various ways, however. Thus, Barbosa *et al.* [25] avoid the interference from a salt matrix and implement a preconcentration step by electrochemically reducing Pb onto the coil surface. They use a flow injection system to deliver the sample through an anode inserted in the tip of the autosampler and the W-coil itself as cathode. The use of a W-coil as a platform inside a Massmann-type furnace provides no appreciable improvement over wall atomization [26].

Metal tube atomizers, though not available commercially, are occasionally preferred to graphite furnaces, especially for the determination of less volatile "carbide-forming" metals, mainly in liquid samples but also in solids [27].

The most salient features of laboratory-built atomizers for direct solid sampling are as follows:

(a) They are usually larger than commercial atomizers, which facilitates sample insertion, prevents blocking of the radiation beam and avoids the high background signals typically yielded by solid matrices.
(b) The analyte is volatilized and atomized separately.
(c) They use large amounts of sample to avoid errors arising from inhomogeneous distribution of the analyte.
(d) They use a low heating rate during atomization in order to avoid a high background.

Atomizers for solid sampling can be heated inductively or resistively. Although resistively heated furnaces are more common, some induction furnaces can readily accommodate solid samples.

Interfaces

The relationship established between electrothermal atomizers and detectors depends on the nature of the detector to be used for identification and/or quantitation. Thus, an absorption spectrometer uses the atomizer as its detection cell: the beam from the atomic radiation source passes through the atomizer and is absorbed to an extent dependent on the analyte concentration. Fluorescence detectors require no special adjustments for excitation of the analyte in the atomizer other than those derived from the need of collecting the emission at a right angle to the excitation source. Up to four different commercial furnaces and one tungsten furnace were used for the comparative determination of titanium by laser-excited (LE) AFS. The limit of detection for the analyte was found to be similar with all types of furnace: around 1 pg for furnaces not exposed to substantial amounts of Ti and a few picograms for those exposed to more than a few hundred picograms of this element. A commercial transversely heated graphite atomizer was found to provide the most reproducible atomization of Ti. While the W furnace exhibited no memory effects for the analyte, it gave signals that were about 250 times weaker than those of the graphite furnaces owing to incomplete atomization [28]. A special "envelope" of a commercial atomizer was designed in the form of a laboratory-built cylindrical vacuum chamber of stainless steel intended to lower the working pressure inside the graphite furnace to 7 Pa with a view to reducing matrix interferences; however, the sensitivity was dramatically affected by the high diffusion rates observed under these conditions [29]. The single 90° off-axis ellipsoidal mirror fragment used as dispersive detection system in order to enhance the collection efficiency of atomic fluorescence from ETAs may be viewed as an interface between these devices and the detector [30].

The needs arising in coupling an electrothermal vaporizer (ETV) to an ICP, ICP–MS or MS instrument are rather different. In the three cases, the vaporized or atomized sample must be efficiently transported to the plasma source, for excitation, or directly to the detector. In all instances, the performance of the ensuing methods is strongly dependent on the transport efficiency of the volatilized analytes from the ETV to the ICP torch or MS instrument. In other words, the vaporizer must be converted into an efficient flow-through device. Figure 8.3A shows a commercial graphite furnace modified for minimizing transport losses of eight selected analytes; its transport efficiency was determined by monitoring a radiotracer. Figure 8.3B depicts selected interfaces used for this purpose [31].

8.2.2. Solid sampling modes in electrothermal vaporizers and atomizers

Solid samples can be introduced into electrothermal atomizers either as such or as slurries. Each insertion mode has its own features, advantages and disadvantages.

Direct sampling of the solid

With Massmann-type electrothermal atomizers, the solid sample can be introduced into the furnace either through a central hole in it or from the ends of the graphite tube. The use of this type of furnace in this sampling mode poses problems such as cumbersome

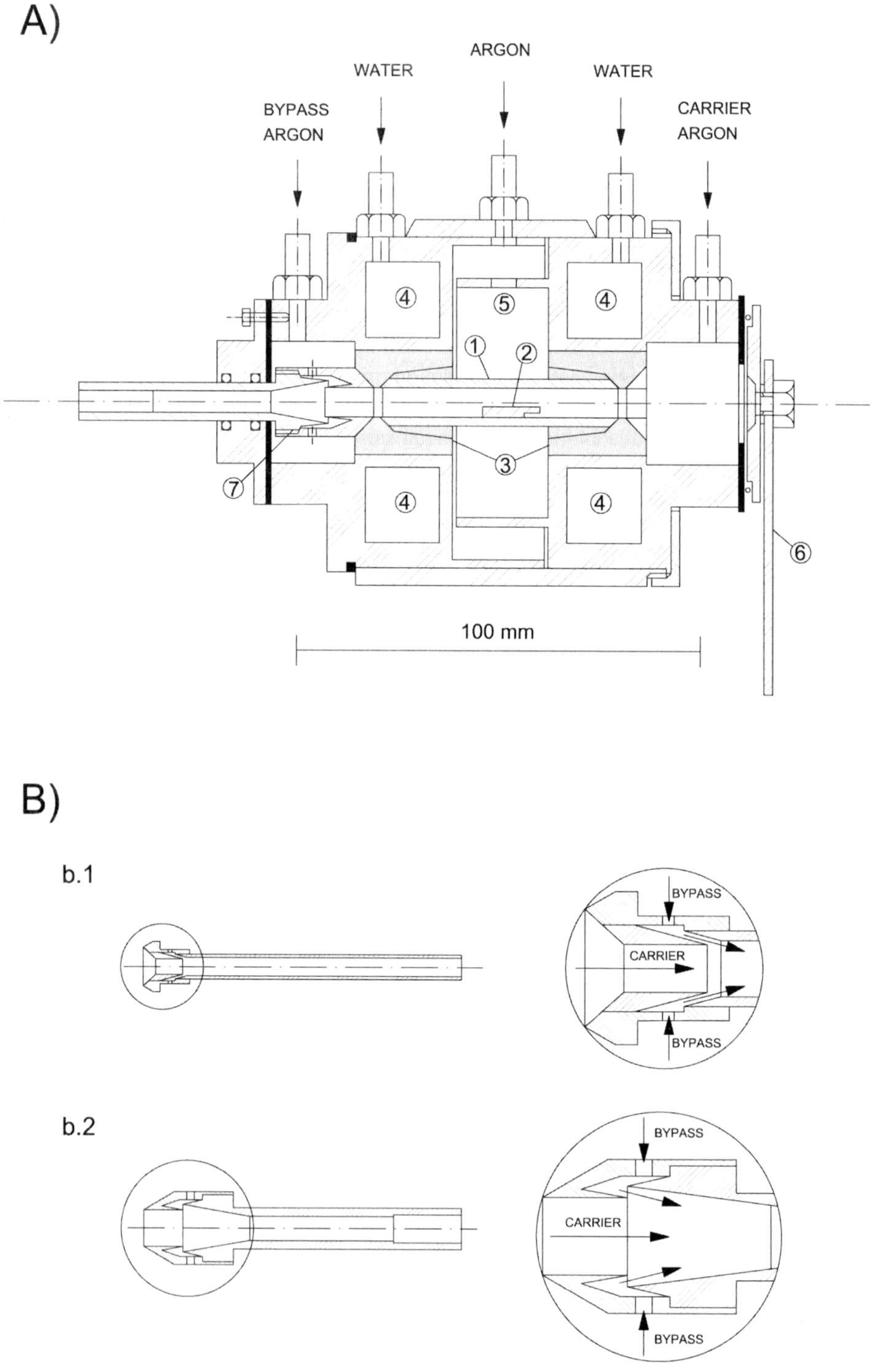
A)
ARGON
WATER
WATER
BYPASS
ARGON
CARRIER
ARGON
④
⑤
④
①
②
③
⑦
④
④
⑥
100 mm
B)
b.1
BYPASS
CARRIER
BYPASS
b.2
BYPASS
CARRIER
BYPASS

sample insertion, strong spectral interferences and irreproducible deposition of the sample on the atomizer walls. Accurate positioning of the sample in these devices is important in order to avoid irreproducible vaporization resulting from the temperature gradient along the tube.

Samples can be inserted in four different ways in the direct sampling mode. One involves introducing a support on which the sample is weighed; once the sample is inside the atomizer, the support is re-weighed. This procedure is used mostly with tube-type atomizers. The weighing and transfer operations require special care in order to minimize contamination and sample losses. The tantalum solid spoon commercially available from Perkin–Elmer is one device of this type. The sample can be inserted through the dosing hole of the graphite tube provided the hole is previously enlarged. Solid samples consisting of ion-exchange resins for the preconcentration of metals from solutions can be introduced into the tube by using a set of plastic tweezers and a small funnel [32].

A more advantageous approach uses a graphite support (a platform, boat, microboat, cup or probe) for both weighing and atomization. The use of inert supports made of ashless paper and cellulose acetate has also been tested [33]. However, inert supports tend to give high blank values; also, the residue remaining after atomization can degrade accuracy and reduce the atomizer lifetime.

One other procedure involves introducing the sample with a special injector. In order to facilitate insertion through the sampling hole in the graphite tube, solid sample injectors operated similarly to conventional autosamplers have been used that work like a microsyringe and introduce amounts of sample ranging from 0.5–5 mg with reproducibility comparable to that of manual solid sample insertion [34].

The fourth procedure for sample introduction in the direct sampling mode uses special devices to collect the sample (by filtration through porous graphite or by electrostatic accumulation). Such devices are especially useful for solid materials suspended in gases. Samples are collected on filters made of various materials including cellulose acetate and nitrate, and PVC.

Slurry sampling

Slurries for insertion into electrothermal vaporizers and atomizers are prepared by adding a liquid diluent to the solid material, previously ground, sieved — when necessary — weighed and placed in a container ensuring stability of the slurry during the time required for sampling. The amount of solid material to be weighed depends on the concentration of the analyte and the dilution in the final volume of slurry. For less homogeneous samples, the precision is improved when the slurry is made from increased amounts of sample. The slurry sub-sample is inserted into the atomizer either by hand, using an

---

Fig. 8.3. (A) Scheme of a graphite-ETV modified for ICP–AES work: (1) graphite tube, (2) graphite platform, (3) electrical contacts, (4) cooling water chambers, (5) protecting gas chambers, (6) automatic shutter, (7) ETV–ICP connecting interface. (B) Two different ETV–ICP connecting interfaces: (b.1) Nozzle-type aluminium-made interface with cooling bypass gas admittance; (b.2) same as b.1 but with a larger inner diameter prior to the bypass gas entrance. (Reproduced with permission of the American Chemical Society.)

autosampler or with the aid of a flow injection system. When autosampler cups are used to weigh the solid material, the maximum volume of slurry is limited to the cup volume (about 2 ml), which in turn limits the maximum amount of material that can be suspended. Once the slurry is prepared, the solid sample must be evenly distributed in the volume of liquid.

Slurry sampling gained increasing acceptance over the 1990s as a means for dealing with various types of samples with practical advantages over alternative procedures including the ability to use conventional liquid autosamplers, perform several replicates in a single aliquot, handle increased amounts of sample and alter the slurry concentration. Additional assets include improved analytical performance as a result of the combined benefits of solid and liquid sampling, and increased repeatability and reproducibility. On the other hand, the large number of variables to be considered complicates optimization of analytical processes involving slurry sampling.

8.2.3. Variables of solid sampling with electrothermal vaporizers and atomizers

The variables that influence the solid sampling process depend on the particular sample insertion mode used.

Variables affecting direct sampling of the solid

Sample preparation Treatment of samples for solid sampling should be kept to a minimum in order to avoid contamination or losses — grinding, sieving, homogenization and drying, among other operations, are often required, however. Because particle size influences accuracy and precision, samples are ground to obtain a powder or reduce the size of particles. Samples can be ground by using an agate mortar or a vibrational pulverizer. Grinding is suitable for brittle materials such as rocks, minerals, glass and soil. Also, biological materials can be converted into solids (e.g. by drying, dry ashing, plasma ashing or freeze-drying) for subsequent grinding. On the other hand, grinding metallic samples is more complicated and can introduce contamination.

Particle size Most materials require grinding in order to reduce particle size to an appropriate extent (e.g. to 5 μm). Grinding is also recommended for solid samples that are originally in powder form when the analyte is unevenly distributed. This operation improves contact between the atomizer surface and the sample. However, it is not always easy to perform (e.g. with metallic samples) and requires great care in order to avoid contamination. The effects of grinding solid samples are somewhat controversial. In some types of sample (e.g. biological materials), particle size has no influence on accuracy but only on precision; in others (e.g. refractory materials such as rocks and minerals), vaporization of the analyte is frequently influenced by particle size. Integrated absorbance measurements for repeated atomization of the sample until thorough vaporization of the analyte have revealed that increasing the amount of sample used does not always result in a proportional increase in the atomic absorption signal. Precision is poor when the sample is inserted into the atomizer as a single particle, as found in the determination of Cu following preconcentration on an ion-exchange column and use of wall atomization [35].

Sample homogeneity The effect of inhomogeneity is more critical when the amount of sample is small (sample masses are usually in the range 0.1–1.0 mg). The relatively large minimum amount of sample that can be effectively handled reflects the difficulty of weighing the sample and inserting it into the atomizer. Studies of trace distributions have shown that the influence of inhomogeneity on precision is negligible with some types of sample (e.g. metals), even with amounts as small as 2 mg [36]; on the other hand, homogeneity is unimportant with samples such as rocks and glasses in amounts down to about 1 mg [37].

Sample mass The amount of sample to be introduced into the atomizer is dictated by the analyte concentration in it. The range of suitable amounts is determined first by the availability of a microbalance to handle sufficiently small masses and the ability of the atomizer to accept larger amounts. Larger graphite tubes than those commercially available have occasionally been used to introduce up to 200 mg of sample. The use of larger amounts of sample results mainly in improved precision, particularly with inhomogeneous samples. However, the range of sample masses that can be introduced into the atomizer is limited and, although amounts as small as 1 μg have been used, masses below 0.1 mg are difficult to handle (viz. to weigh and transfer). When the concentration of the target analyte is too high for the analytically useful working range of the calibration graph, one can use less sensitive lines, dilute the sample with graphite or employ an increased flow of Ar. These procedures, however, are error-prone. Vapour phase temperatures are lower when an internal gas flow is used, so matrix interferences are enhanced as a result.

Analyte location in the sample The type of bond between the analyte and the solid matrix, in addition to the location of the analyte in the solid particles (viz. on the surface or occluded within) can influence absorbance measurements through differences in vaporization kinetics. One example is the determination of several metals in steel. Signals with double peaks were observed as a result of the analyte being present in two different locations of the steel grains [38]. The formation of several peaks in the atomic absorption signals was also observed in the determination of lead in plastics using a cup-in-tube device [39] and ascribed to the manner in which the plastic decomposed as the temperature was raised. Also, several peaks were observed when lead was atomized at a low pressure from copper alloys, which was also ascribed to the presence of various chemical forms of Pb or changes in its location [40].

Variables affecting slurry sampling

The main variables influencing this sampling mode are slurry stability and concentration, and particle size.

Slurry stability Slurries can be maintained stable for insertion into ETAs and ETVs using various approaches, namely:

(a) Stabilizing agents. Slurries prepared in aqueous solutions are rarely suitable because most powdered materials undergo rapid sedimentation. This occurs immediately after the slurry is mixed and can be quantified via Stoke's law. The sedimentation rate depends on the densities of the diluent and solid material, the viscosity of the diluent medium and the radius of the sample particles. The slurry can be stabilized using a highly viscous liquid medium such as Viscalex, glycerol, a non-ionic surfactant, an organic solvent or a thixotropic thickening agent, among others. The stabilizing capability of these agents depends largely on their concentration, as well as on the characteristics and particle size of the sample. On the other hand, the maximum slurry concentration that can be used depends on the concentration of the stabilizing agent. The time interval between complete mixing of the slurry and removal of an aliquot for analysis can be extended by using a highly viscous medium similar in density to the sample particles. Problems have been noted regarding the use of viscous media to prepare slurries, however, such as the sample aliquot being inefficiently pipetted into the atomizer when a high concentration of stabilizing agent is used. Also, the sample can remain around the dosing hole, which can degrade precision with both micropipettes and autosamplers. The use of the latter additionally requires flushing the capillary between successive determinations. It is also necessary to include an additional pyrolysis step in order to remove excess stabilizing agent. Agents are usually inefficient when the slurry contains particles of a high density. Apart from stabilizing agents, the addition of wetting and antifoaming substances such as Triton X-100 can improve slurry dispersion. Stabilizing agents are the preferred choice when samples are to be inserted via an autosampler as the slurry can be allowed to stand in the sampling cups without further homogenization.

(b) Magnetic stirring and vortex mixing. These were the two procedures most widely used in early work on slurry sampling. Slurries are prepared in a beaker and stirred in order to ensure homogeneous distribution of the solid material (usually for 3–5 min). The stirring action is then stopped and a sample aliquot is withdrawn for insertion into the atomizer. The sample aliquot can also be withdrawn under continuous stirring, however. The effectiveness of magnetic stirring and vortex mixing depends basically on the sedimentation rate of the suspended material and the error associated with sedimentation is a function of both particle size and the presence of particles having different compositions. As a result, large errors can be expected when the analyte is predominantly distributed in particles of a high density, which will undergo rapid sedimentation. After magnetic stirring or vortex mixing, slurries are usually introduced with a micropipette because of the difficulty of incorporating these systems into an autosampler. Some miniature stirring devices have been used in combination with an autosampler for this purpose [41].

(c) Ultrasonic agitation. This is by far the most effective method for homogenizing slurries for insertion into ETAs. Ultrasonic agitation can be used in combination with both manual and automated introduction of the slurry. One advantage of this method relative to magnetic stirring and vortex mixing is that the target analyte is partially extracted into the liquid phase under the ultrasonic action when slurries are prepared in an acid medium. Ultrasonic agitation is more effective than other agitation systems (see Chapter 3). Agitation with a small ultrasonic probe also allows slurries to be prepared directly in the sample cups of an autosampler, but special care is required to avoid contamination when the titanium probe touches the cup wall. A device of this type is commercially available from Perkin–Elmer. The precision thus achieved depends on both the extraction yield of the analyte and the degree of homogeneity of the sample. Precision of 2–10% as RSD was obtained in the determination of Se, Pb and Tl in coal and coal fly ash, with an analysis time of less than 1 min by using ultrasonic agitation of the slurry but no pyrolysis step [42]. Commercial ultrasonic probes for slurry homogenization provide additional advantages such as increased throughput, automatability of slurry preparation and insertion into the furnace, improved reproducibility of analyte extraction, and increased representativeness of the sample aliquot.

(d) Gas mixing of the slurry. Slurries can be efficiently homogenized by passing an argon stream through a narrow capillary tube inserted in the slurry medium. This allows samples to be directly prepared in autosampler cups. Homogenization requires a "bubbling time" with the Ar stream of only 30 s and provides a precision of about 6% as RSD in the determination of Cu, Co, Cr, Mn, Fe and Ni in glass [43].

(e) Predigestion of the slurry. This procedure can facilitate extraction of the target analyte into the liquid phase. In contrast to the extensive sample pretreatment inherent in the traditional fusion and digestion procedures, predigestion of a slurry only requires partial decomposition of the sample, so it is not time-consuming. For example, partial wet oxidation of biological and plant materials can be accomplished by using concentrated sulphuric acid with subsequent analysis of the carbonaceous slurry formed, all with increased expeditiousness and reliability relative to dry ashing and extensive wet oxidation. The difficulties of stabilizing slurries containing large particles in the analysis of sediments and particulate matter can be overcome by predigestion (viz. by adding a small amount of concentrated nitric acid to the sampling cups containing the slurry), which results in partial extraction of the analytes into the liquid portion of the slurry. Metals in glass can be predigested in the slurry by adding a low concentration (3%) of hydrofluoric acid and homogenizing by gas mixing. The predigestion step can be implemented either before or after loading the cup with sample.

(f) Electric dispersion in a condensed medium. The electric dispersion of metallic samples in a condensed medium produces a colloidal dispersion that is suitable for slurry preparation as judged by the characteristics of the sols formed upon spark ablation. The suspension contains metallic particles of size less than 1 μm and is highly stable. In this way, slurries of metallic materials can be prepared and the need to machine them for powdering avoided, which reduces contamination risks. The spark ablation can be done in both aqueous and organic solvents.

(g) Laser ablation, which is discussed in Chapter 9.

Particle size The particle size of the solid material used to prepare the slurry can influence its stability, deposition and atomization efficiency which, in turn, can affect accuracy and precision. Various types of equipment for grinding solids have been used to prepare slurries for ETA–AAS including mixer mills, other ball mills, and bottle and bead mills. The appropriate particle size varies among materials depending on various parameters such as the homogeneity of the original material with respect to the analyte, the amount of sample used for analysis and the particular diluent employed. Particle size can also influence the homogeneity of the slurry through the rate of particle sedimentation and the amount of analyte extracted into the liquid phase. When large particles are present in the slurry, problems may also occur due to discrimination of the larger particles by the autosampler and irreproducible injection of the sample into the atomizer.

Notwithstanding the decisive importance of particle size in the materials used to prepare slurries, information about particle size distribution is often scant or unreported in publications dealing with slurry sampling. Also, when particle size is reported, the sample has usually been ground to pass through a given mesh dimension and only the maximum particle size of the material is stated. Wibetoe *et al.* [44] studied the potential of using a Coulter particle analyser with a view to developing a method for measuring the effect of grinding of solid materials prepared for slurry sampling. This method would allow particle size distributions to be plotted and/or quantified in various ways including number–size and volume–size plots, size being expressed as a volume, surface area or (most often) diameter. Plotting volumes of particles against size is advantageous as detailed information about the larger particles is lost when the distribution is reported in terms of number-percentage. Plant material (viz. flowers, leaves, the stem and roots from the sample plant) was used to demonstrate the approach. Precision was found to increase with decreasing particle size, particularly in the determination of Ni and Cu; also, accuracy increased with increasing grinding time (see Table 8.1). Rather low concentrations were obtained with 1 min grinding for all plant parts and the three metals. Inaccuracy was probably the result not only of the poor precision achieved, but also of all the systematic errors probably present (all results were too low). From the data in Table 8.1 it follows that grinding for as long as 60 min was not always necessary for this type of material as the precision and accuracy obtained after 5 min of grinding was more than adequate in many cases.

Slurry concentration One other important factor in preparing a slurry is its concentration. Samples with a high analyte content are easier to analyse with the slurry technique than by direct solid sampling as the slurry can be readily diluted in the former. However, the slurry can only be diluted to a certain extent: precision degrades with a high dilution as a result of only a small number of particles remaining in the slurry. On the other hand, when the analyte content in the original sample is very low, the concentration of the slurry can be increased accordingly — the pipetting efficiency, however, can deteriorate with more concentrated slurries. One other factor to be taken into account is that increasing the slurry concentration can enhance matrix effects. In food analyses, the RSD is typically less than 4% when using particles less than 50 μm in diameter — provided the slurry concentration is below 10% m/v. Accuracy deteriorates when the slurry concentration exceeds 5% as a result of the matrix effects. Slurry concentrations above 5% result in inefficient deposition of the slurry aliquot [41]. The errors associated with slurry sampling have been characterized and found to arise from uncertainty in the sample volume and the number of particles it contains, as well as from variations in the mass of individual particles. These errors can be minimized by using small particles, concentrated slurries and narrow particle size distributions [45].

Some authors [46] claim that using slurry nebulization is only advisable when no efficient alternative digestion procedure is available. The dilution factor of the sample in slurry nebulization is at least as high as in normal solution nebulization. The concentrations

TABLE 8.1

RESULTS OF THE ULTRASONIC SLURRY SAMPLING ETA–AAS DETERMINATION OF Co, Ni AND Cu IN PLANT PARTS AND OLEA LEAVES CRM, WITH INCREASING GRINDING TIMES; $N = 5$ (DIFFERENT SLURRY PREPARATIONS). THE CERTIFIED OR INDICATIVE VALUE FOR THE CRM IS GIVEN IN BRACKETS

	Slurry sampling grinding time						Decomposed samples[a]	
	1 min		5 min		60 min			
	Mean, µg/g	RSD, %	Mean, µg/g	RSD, %	Mean, µg/g	RSD, %	Mean, µg/g	RSD, %
Co								
Flowers	3.5	18.1	6.5	13.1	6.7	3.6	6.0	7.0
Leaves	5.8	19.7	9.2	13.2	11	7.7	10.0	11.5
Stem	2.2	21.0	5.7	15.1	4.8	3.0	5.3	11.5
Roots	4.4	20.6	8.3	17.5	8.5	3.5	8.7	2.6
Olea Leaves CRM	0.17 (0.2[c])	22.2	0.18	19.1	0.19	16.7	0.17	11.8
Ni								
Flowers	43	18.8	63	8.7	65	6.0	60	7.2
Leaves	50	15.4	77	13.2	75	10.3	73	7.3
Stem	12	15.2	20	7.0	21	4.8	21	13.3
Roots	37	15.2	63	11.7	64	2.3	65	10.6
Olea Leaves CRM	3.8 (8[c])	19.3	7.3	10.2	8.1	6.3	7.5	16
Cu								
Flowers	52	15.0	115	12.0	120	8.7	102	9.0
Leaves	112	13.3	188	11.4	190	7.5	195	5.1
Stem	43	17.4	125	10.6	130	6.3	140	10.7
Roots	265	13.4	401	10.4	395	4.9	408	1.3
Olea Leaves CRM	44 (46.6±1.8[d])	14.1	45	12.3	46	6.7	46	4.6

[a] $n = 6$; [b] 45 min grinding for CRM; [c] value given on CRM certificate (not certified value); [d] certified value (mean ± 95% confidence interval of the mean)

of dissolved solids in analytical solutions are even higher as the dispersing agents alone account for 0.5 to 1%. In addition, contamination due to impurities in the grinding agents may result in analytical errors. Polyatomic ions formed from the components of the grinding agents or the dispersants can also cause interferences.

8.2.4. Steps of an electrothermal solid sampling process

In addition to sample-related aspects, spectrometer control functions must be programmed to occur at specified times within the graphite furnace programme. While the number of steps within each programme is variable, 6 steps constitute the typical ETA programme for slurry samples (similar to that for liquid samples), which reduces to 5 steps when the solid sample is introduced as such and the drying step is thus omitted. The operational sequence thus involves (1) drying, (2) pyrolysis or ashing, (3) cooling down (optional), (4) atomization, (5) clean-out and (6) cooling down. Figure 8.4 shows a typical temperature programme, the steps of which are discussed below.

Drying

Once the slurry has been driven to the furnace using one of the procedures described in the previous section, it must be dried at a low enough temperature to avoid sample splattering, which would result in degraded analytical precision. The temperature to be reached depends on the nature of the liquid phase in which the solid is dispersed. Temperatures around 100–120°C are common for aqueous solutions, but a value just above the boiling point of the dispersant is advisable. Use of a temperature "ramp" provides a variable time over which the temperature is raised. An extended ramp time provides a slower, more gentle increase in heating. When a platform is used, the temperature lag of the platform relative to the tube walls provides a natural ramping effect. Longer times are used when the sample is to be atomized from the tube walls. After the temperature ramp, the furnace is held at the selected drying temperature until drying is complete. Since the volume of slurry used is typically of only a few microlitres, the drying hold time is usually less than 1 min (usually 1–3 seconds per microlitre of dispersant). The pyrolytic platform invariably requires the use of somewhat higher temperatures than those needed for sampling from the tube walls. During the drying process, the internal gas flow is normally kept at its default maximum value (250–300 ml/min) to purge the vaporized solvent from the tube.

Pyrolysis or ashing

The purpose of this step is to volatilize inorganic and organic matrix components selectively from the sample, leaving the analyte element in a less complex matrix for analysis; as a result, the duration of this step is a function of the complexity of the sample and can vary between 10 and 30 s. During it, the temperature is increased as rapidly as possible to volatilize matrix components but is kept below the level at which analyte losses might occur. The temperature to be used in the pyrolysis step will depend on the particular analyte and matrix. Recommended temperatures are usually provided in the

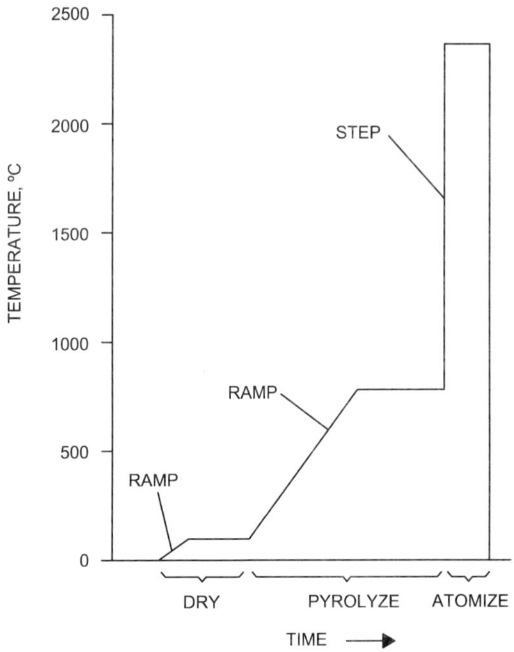

Fig. 8.4. Typical temperature programme for an electrothermal atomizer (ETA).

documentation accompanying each commercially available graphite furnace. The internal gas flow is again kept at 250–300 ml/min in order to allow volatilized matrix materials to be driven off. With some samples, it is advantageous to change the internal gas (e.g. to air or oxygen) during the pyrolysis step to facilitate sample decomposition.

Pre-atomization cool-down

This optional step is used mainly with longitudinally heated furnaces in order to cool the furnace prior to atomization as the heating rate is a function of the temperature range to be encompassed. The rate of heating increases as the temperature range is broadened. The use of a cool-down step prior to atomization maximizes the heating rate and extends the isothermal zone within the tube immediately after heating. The extended isothermal zone has been shown to improve sensitivity and reduce peak tailing for a number of elements including those which are typically difficult to atomize in the graphite furnace. This step is not required with transversally heated furnaces as the isothermal zone extends the length of the tube in this type of system.

Atomization

In order to produce an atomic vapour of the analyte elements, the temperature of this step is increased to the point where dissociation of volatilized molecular species occurs.

The atomization temperature is a function of the analyte element and must be increased as quickly as possible. Therefore, ramp times will normally be set to a minimum value resulting in the highest possible heating rate. It is also desirable to reduce or, preferably, completely stop the internal gas flow during atomization in order to extend the residence time of the atomic vapour in the furnace, thus maximizing sensitivity and reducing some interferences. It is worth noting that the analytical signal does not increase linearly with increase in the atomization temperature because the rate of loss of atomic vapour from the atomizer also increases with increasing temperature.

Clean-out and cool-down

After atomization, the furnace may be heated to still higher temperatures to burn off any sample residue remaining in it. An optional cool-down step then allows the furnace to return to a near-ambient temperature before the next sample is introduced.

8.2.5. Instrument parameters affecting solid sampling with electrothermal atomizers and vaporizers

As stated above, the sample dispensed into the furnace is subjected to a multi-step temperature programme. When the temperature is increased to the point where sample atomization or vaporization occurs, measurements are made directly or after transporting the vapour to the ICP torch or MS instrument. The variables under operator control in this situation are as follows:

(1) Temperature (viz. the final level);
(2) Ramp time (viz. the time for temperature rise);
(3) Hold time (viz. the time during which the final temperature is maintained);
(4) Internal gas (viz. its type and flow-rate).

The value of each of these instrument parameters to be used in each step depends on the instrument itself, the sample matrix and the analyte.

8.2.6. Use of modifiers in electrothermal solid sampling

The concept of *chemical modification* (CM) is extremely popular in modern electrothermal-assisted atomic techniques. As per IUPAC's recommendations [47], "in order to influence processes taking place in the atomizer in the desired way, reagents called 'chemical modifiers' may be added. These can help to retain the analyte to higher temperatures during pyrolysis, remove unwanted concomitants or improve atomization in other ways". However, there is a tendency towards broadening the scope of this term, starting from the classical and still used term "matrix modifier", "matrix/analyte modifier" or "instrumental matrix modification" to indicate the useful effects of the type, pressure and flow-rate of protective gas or gas mixtures; "internal matrix modifier" for matrix constituent(s) with favourable effects on processes in the atomizer, either by themselves (e.g. refractory components) or upon addition of suitable promoters; "permanent modifiers" for graphite surfaces, which are coated with high-boiling point noble metals or carbides. A simplistic current classification based on the duration of their effect divides modifiers into two categories: conventional and permanent.

Conventional modifiers

Conventional modifiers must be added to the atomizer prior to every atomization heating cycle because they are expelled during the atomization and/or clean-out steps. Some research into transient modifiers is currently being devoted to their mechanisms of action; such is the case with the studies on the film produced by the decomposition of ascorbic acid on the platform surface, which absorbs Au atoms preferentially [48], or the use of synchrotron X-ray absorption spectroscopy to further elucidate the mechanism of Pd modification of selenium [49]. However, a number of publications have been devoted to the applications of conventional modifiers over the past two decades, but particularly during the 1990s. Specially prominent in this context are the studies of Burguera *et al.* [50] on the addition of modifiers to Massmann-type furnaces for the determination of Cr in urine, which led to increased background absorbance signals — which is generally not the case with THGAs, however. Also worth noting is the study of the effectiveness of Pd and phosphate modifiers in a tungsten furnace, where both modifiers were found to increase the maximum permissible pyrolysis temperature for Cd but not to the extent achieved in a graphite furnace [51].

Permanent modifiers

Permanent modifiers are a result of the search for a solution to the high blanks caused by repeatedly adding transient or conventional modifiers as even the highest purity salts and compounds of the modifiers may contain measurable concentrations of metals. Unsurprisingly, interest in the use of permanent modifiers has risen in recent years. A permanent modifier interacts strongly enough with the graphite furnace to be retained through multiple firing cycles, and any trace metal impurities are usually removed during its initial thermal conditioning. Permanent modifiers can also increase the lifetime of the graphite furnace.

Existing chemical modifiers span all aggregation states and a wide variety of chemical families, namely:

(a) Inorganic salts — the most popular group — the modifiers being both cations such as Pd(II), Ni(II) or NH_4^+, and anions such as nitrate, phosphate, vanadate or tungstate. Nitrate and ammonium ions, respectively, have proved the best counterions for these salts.
(b) Organoelements and complexed forms of modifiers, which are useful when samples or analytes are transferred into organic phases such as extracts, leachates, oils, fats, petroleum products, organometal [lead acetate or acetylacetonate] solutions, etc.
(c) Inorganic acids (HNO_3, H_3PO_4) and bases (aqueous ammonia).
(d) Organic acids (acetic, tartaric) and bases (e.g. solvents for biological tissues including TMAH, TAAH and thyamine hydroxide).

(e) Complexing agents such as EDTA, citric acid or citrate ion for keeping the analyte, modifier and/or matrix components in solution.
(f) Reductants for conditioning noble metal modifiers (e.g. ascorbic acid, citric acid, NH_2OH/HCl, H_2–Ar).
(g) Oxidants [HNO_3, $Mg(NO_3)_2$, O_2, air alternate gases] for promoting in situ ashing of organic matrices.
(h) Reactive alternate gases (H_2, O_2, freons).
(i) Miscellaneous organic additives with complexing action (e.g. organic acids and solvents, surfactants).

Obviously, some of these single-component modifiers may play more than one role. In addition, the efficiency of chemical modifiers can be improved in various ways, namely:

(a) Palladium and other noble metals often perform better in the presence of reductants, resulting in a "reduced Pd modifier". Common reductants include ascorbic, oxalic and citric acids, hydroxylamine hydrochloride and hydrazine sulphate. Thermal conditioning of noble metals by injection and pretreatment at 1000°C, followed by cool-down and then sample insertion are also widely used. Pre-reduction of noble metals (Ga, In and Tl mainly) has proved especially important with some difficult analytes forming strong bonds with chloride and other halides.
(b) Metals other than Pd (Ir, Pt, Rd, Ru) may perform better in special analytical situations by virtue of their lower volatility.
(c) Mixed noble metal modifiers may be preferred on account of their lower volatility and higher vaporization temperature.
(d) Mixed modifiers containing noble metals and oxide- or carbide-forming elements are frequently not only more efficient but also more universally applicable to larger numbers of analytes.
(e) (Noble) metal modifiers often exhibit improved efficiency when applied on carbide-coated graphite tubes or platforms, an effect similar to, yet lower in magnitude, than that observed in the presence of their corresponding mixed modifiers.
(f) Permanent modifiers such as carbide coatings (e.g. NbC, TaC, WC, ZrC), less volatile noble metals (Ir, Rh) or noble metals on carbide coatings are becoming increasingly popular.

The most salient effects sought in modifiers are as follows:

(a) Efficient thermal stabilization of low to moderate volatile analytes (over 30 elements including such important analytes as Bi, As, Cd, P, Pb, Sb, Se, Tl and Sn), particularly in matrices of volatility comparable to that of the analyte.
(b) Isoformation of various analyte species such as arsenite, arsenate, monomethyl arsonate, arsenobetaine, arsenocholine, etc., with a view to levelling off their integrated absorbance by transformation into less volatile chemical forms.
(c) Increasing the volatility of unwanted contaminants, which facilitates their removal during the pyrolysis step (e.g. elimination of the bulk chloride matrix in analyses of seawater or urine by adding NH_4NO_3, dilute HNO_3, ammonium salts of organic additives or H_2–Ar as alternate gas during pyrolysis.

(d) Facilitating in situ ashing rather than charring of organic matrices using air or O_2 alternate gas during the (low-temperature pyrolysis step).
(e) Thermal stabilization of some interfering species such as P_2 or PO in phosphate-rich samples (e.g. bone, biological fluids) upon addition of noble metals or Ni so as to delay their vaporization during the atomization step (as P atoms rather than molecules).
(f) Transformation of some interfering compounds into less harmful species (e.g. binding sulphate by mixed modifiers containing salts of alkaline-earth metals such as Pd–Mg–Ba or Pd–Sr).
(g) Modifier action as volatilizers (i.e. by facilitating low-temperature atomization of the analyte before the bulk matrix).
(h) Ensuring better contact between the sample and atomization surface by adding wetting agents and/or organic reagents for smooth drying and/or pyrolysis.
(i) Enabling in situ speciation analysis, as in the fraction vaporization of Se(IV) alone as a volatile piazselenol complex during the pyrolysis step, Se(VI) remaining in the atomizer for selective quantitation.
(j) Increasing sensitivity. While very attractive, this aim and (eventually) result of chemical modification may just reflect the situation that conditions for efficient atomization cannot be obtained in the absence of modifiers in many analytical cases (e.g. in the determination of high-volatile organotins, where premature losses can be avoided or greatly reduced in the presence of an organopalladium modifier).
(k) Trapping volatile hydrides in hydride generation prior to electrothermal atomization by using graphite surfaces modified with carbides, noble metals or both.
(l) Improved resistance of some Ta, W and Nb carbide-coated surfaces to attack and permeation by acid digests, organic solvents and extracts, corrosive matrices and modifiers, among others.
(m) Providing better conditions for simultaneous multi-element assays through thermal stabilization of volatile analytes and employing more uniform programmes for groups of several analytes (e.g. As and Se; Cd, Cr, Pb and Ni).

The following are a few selected examples of the use of both conventional and permanent chemical modifiers of diverse nature and causing different effects.

(a) The use of a reactive gas such as Freon 1,2 as a halogenating reagent to improve the release of analytes such as Al, Ag, As, Bi, Ca, Co, Cr, Cu, Fe, Ga, K, Li, Mg, Na, Ni and Pb in graphite and silicon carbide for their determination by ICP–AES [52]. The chlorinated compounds of these elements are much more volatile than the elemental species themselves. By this means, the vaporization temperature can be considerably reduced, the release of analytes from the samples improved and, in some instances, the reaction of carbide-forming elements with the furnace diminished. This should lead to improved transport efficiency and hence to increased sensitivity. Figure 8.5 shows the emission signals obtained for Li with and without the addition of Freon in processing an aqueous standard solution and a graphite sample. Polytetrafluoroethylene (PTFE) has also been used in slurry introduction sampling (SIS) for fluorination (and efficient volatilization) of cadmium from solid biological materials [53,54], and of Co and Ni from silicon dioxide powder [55].

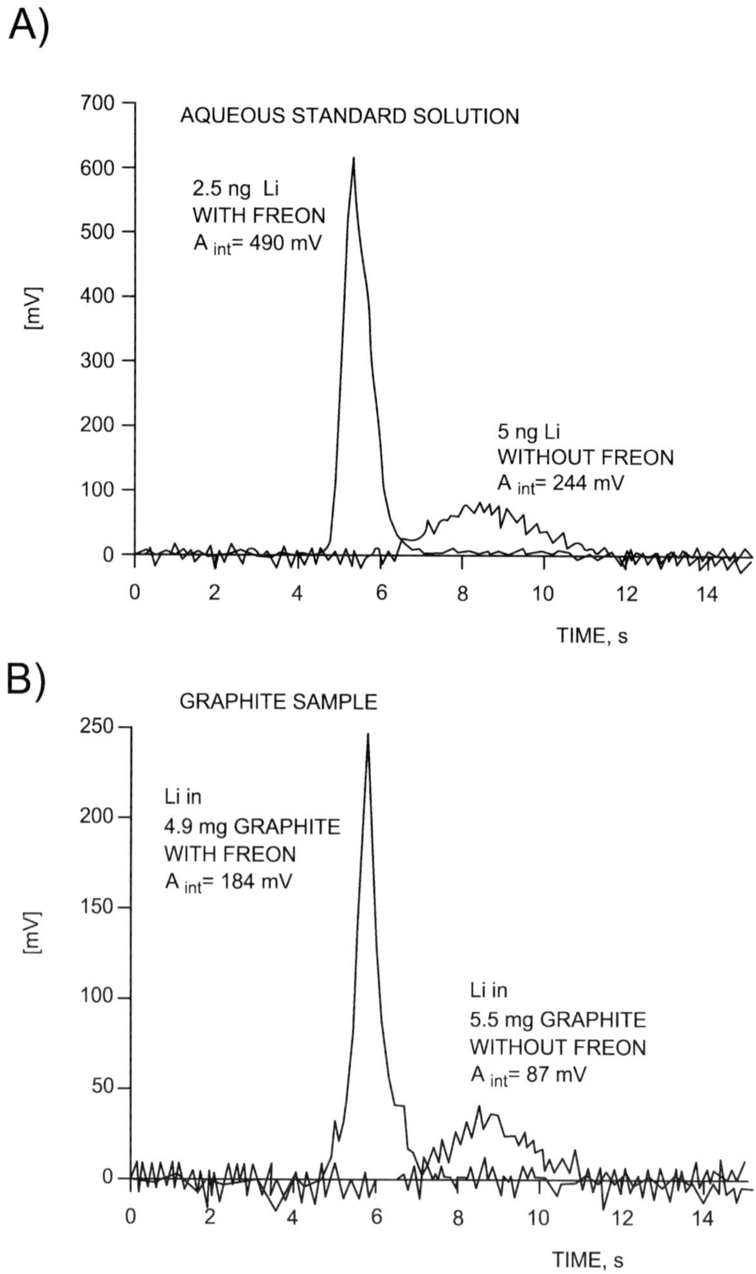

Fig. 8.5. Emission signals for Li as obtained with and without the use of Freon as additive. (A) Aqueous standard solution. (B) Graphite sample. (Reproduced with permission of the American Chemical Society.)

TABLE 8.2

DETERMINATION OF ARSENIC IN SEDIMENT AND SOIL SAMPLES BY ETA–AAS (n = 5). PERMANENT MODIFIER, 250 μg OF W + 200 μg OF Rh; CONVENTIONAL MODIFIER, 5 μg OF Pd + 3 μg OF $Mg(NO_3)_2$. CONFIDENCE INTERVAL AT 95% CONFIDENCE LEVEL ($t_{Student}$ = 2.776)

Sample	As, μg/g			
		Slurry		
	Certified value	W + Rh	Pd + Mg	Digested Pd + Mg
Buffalo River sediment SRM 2704	23.4 ± 0.8	23.3 ± 0.3	23.6 ± 0.4	22.8 ± 0.5
Pond sediment NIES 2	12.0 ± 2	12.1 ± 0.4	11.8 ± 0.5	13.1 ± 0.8
Lake sediment SL-1	27.5 ± 2.9	26.9 ± 1.0	28.7 ± 1.2	26.4 ± 1.6
River sediment CRM 320	76.7 ± 3.4	78.2 ± 1.3	75.2 ± 1.8	79.1 ± 2.2
Soil-5 IAEA	93.9 ± 7.5	90.4 ± 2.5	92.7 ± 3.6	89.4 ± 4.4
Soil-7 IAEA[(a)]	13.3	13.7 ± 0.5	13.1 ± 0.7	13.8 ± 1.4
Sediment I	—	65.9 ± 2.4	66.8 ± 3.0	67.4 ± 4.0
Sediment II	—	44.2 ± 1.2	46.1 ± 2.2	45.4 ± 3.8

[(a)] Confidence interval: 12.5–14.2 μg/g
Reproduced with permission of the Royal Society of Chemistry

(b) A recent study on the use of W–Rh permanent modifiers in ETA–AAS for the determination of lead in slurries of biological materials revealed that the modifier remained stable over approximately 300 analytical measurements when a volume of 20 μl of slurry containing up to 1.5% m/v was delivered to the atomizer; the permanent modifier increased the tube lifetime by approximately 100% relative to untreated integrated platforms. Also, the sensitivity decrease during the atomizer lifetime was less marked than with conventional modifiers, which resulted in increased sample throughput. The atomizer lifetime was restricted to the THGA wall durability; in fact, the W–Rh treated platform remained intact after more than 650 analytical firings. The results thus obtained in the determination of lead were consistent with those provided by digested solutions and the use of Pd/$Mg(NO_3)_2$ [56].

(c) A comparison of the use of a W–Rh permanent modifier and a Pd/$Mg(NO_3)_2$ conventional modifier for the determination of As after slurry introduction sampling of sediments into ETA–AAS equipment revealed the results of both approaches to be consistent (see Table 8.2) [57].

(d) Lead as a modifier has been used to improve the determination of various analytes in highly different matrices and also using different detectors. Thus, the presence of palladium and ammonium nitrate in the determination of lead in fish by slurry sampling and ETA–AAS substantially improved the signal by increasing the atomization temperature and reducing differences in analyte transport from the sample and calibration solutions

[58]. The use of Pd and NaCl was found to have a similar effect on the determination of As, Mn, Pb and Se in coal by slurry introduction sampling (SIS)–ETA–ICP–MS [59].

Despite their advantages, modifiers are subject to a number of side effects and restrictions that are increasingly better known. While some of their problems can be reduced, others can only be avoided by using an alternative modifier. Their most serious shortcomings are as follows:

(a) Impaired effectiveness with real matrices owing to competitive side reactions with, or inhibition by, matrix and/or solvent constituents. Pyrolysis temperatures often happen to be lower than those established with simple aqueous solutions.
(b) The need for longer temperature programmes, which arises from the unfavourable kinetics of the solid-phase reactions by which analytes are stabilized and matrix interferences avoided.
(c) High reagent blanks produced by large amounts of modifier, which can lead to the need for unacceptably cumbersome blank corrections. This shortcoming can be circumvented by using high-purity — and thus expensive — reagents or additional purification steps involving ion-exchange or liquid extraction.
(d) Overstabilizing effects resulting from the use of large amounts of modifier, extended pyrolysis times at high pyrolysis temperatures, gradual accumulation of non-volatile modifiers or migration of the modifier to the cooler ends of a longitudinally heated atomizer.
(e) Corrosive action on graphite and pyrocoatings, which has a strong effect on the atomizer lifetime.
(f) Irreversible contamination of graphite parts and their surrounding areas by Cr, Mg, Mo, Ni, Pd or V modifiers, which precludes their subsequent determination as analytes.
(g) High background absorption from the modifier itself, which requires effective correction.
(h) High toxicity or other harmful effects.
(i) Compatibility problems (some modifiers must be stabilized with high acid concentrations in order to avoid hydrolysis and their disappearance from the solution).
(j) Restrictions derived from the lack of non-empirical information and markedly discrepant results.

Permanent modifiers are subject to specific additional shortcomings such as the following:

(a) Partial, non-uniform coating of platform and tube surfaces, which produces "islands" that act as atomization sites;
(b) Relatively short lifetimes;
(c) Restrictions upon temperature programmes;
(d) Limited applicability to fairly complex solutions;
(e) Overstabilization of some carbide-forming and less volatile analytes;
(f) A lack of interest among equipment manufacturers.

Modifiers are essential for many ET–AAS applications. The new generation, permanent modifiers, is receiving increasing attention by researchers, which will help to overcome their present drawbacks.

8.2.7. Shortcomings of electrothermal sampling

The direct introduction of solids into electrothermal atomizers and vaporizers poses a series of problems especially prominent among which are the following:

(a) Inserting a solid sample into an electrothermal device is not as easy as introducing a liquid sample; in fact, direct insertion of a solid sample requires extreme care and training. Some loss or contamination can easily occur during weighing and/or transfer from the balance to the atomizer. This problem is less critical with slurry sampling.
(b) In contrast to liquid samples, which are inserted in a direct manner, a separate weighing operation is required for each replicate. This can make the analytical procedure time-consuming to an unacceptable extent for routine determinations. Slurries allow several determinations to be performed with a single sample preparation. The automation of the solid sample insertion operation has resulted in some improvement in routine determinations.
(c) Standardization poses problems not encountered with liquid samples. Matrix dependence of peak shape and incomplete release of the analyte from the solid sample are two common occurrences. Calibration with aqueous standards is not always possible and the method of standard additions fails when the added analyte is not uniformly affected by the matrix. Calibration with certified reference materials with matrices highly similar to those of the samples is the most reliable approach in this context.
(d) Chemical modifiers are less effective for the direct atomization of solids as they cannot interact in an effective manner with the analyte, which is occluded within the sample particles. Various chemical modifiers have been tested with a view to improving the interaction.
(c) Solid samples are more difficult to dilute when the analyte is highly concentrated as this entails lowering the instrument sensitivity or previously diluting the sample with an appropriate solid buffer. The slurry sampling technique affords dilution but requires extreme care to avoid errors (e.g. when only a few solid particles remain in the slurry).
(f) Precision is worse than with liquid samples. Relative standard deviations are typically in the region of 10% but can be improved by either decreasing particle size or increasing the amount of sample used for the determination when sample inhomogeneity is the most influential factor. With many certified reference materials (CRMs), sample inhomogeneity must be considered for amounts of sample below 200 mg, which is much larger than the 0.1–10 mg typically used in solid sampling.
(g) Parts of the matrix can remain in the furnace after atomization and damage the atomizer. Residual matrix build-up can also block the light beam emitted by the radiation source and affect analytical performance. This shortcoming can be circumvented by using a clean-out step after each series of measurements.

(h) Spectral interferences are more pronounced when the solid sample is atomized in a direct manner. High and structured background signals can cause large errors. The use of a powerful background correction system is thus required if accurate results are to be obtained. Oxygen ashing can also help to remove non-specific absorbance interferences.
(i) Slurry introduction is subject to specific problems arising from the need to maintain a stable, homogeneous slurry.
(j) The determination of refractory metals is hampered by incomplete release during atomization and by occlusion in the solid matrix. High atomization temperatures relative to liquid samples are required under these conditions to release the analyte within a reasonable time.
(k) Preparing blanks can be cumbersome, particularly with ground samples.

Many of the previous problems have been partly solved with the implementation of new instrument developments and improvements in analytical methodology over the past two decades. Progress in three different lines (viz. the design of graphite atomizers specially adapted for handling solid samples, the automation of solid sample insertion and the use of stable-temperature platform furnaces) has helped overcome some of the drawbacks of solid sampling with electrothermal atomizers and vaporizers.

8.2.8. Calibration in methods involving electrothermal sampling of solid samples and slurries

Notwithstanding their selectivity, atomic techniques are largely used for quantitative rather than qualitative analysis. Some theoretical evaluations have been made with a view to using solid sampling ETA–AAS for screening purposes by simulating more than 18 000 results corresponding to different possible situations and analysing them using information theory. In the light of the results, the median can be used to better effect than the mean value as the former is less markedly affected by the potential presence of outliers. Using the median and typical working conditions, the minimum number of measurements to be made in order to ensure a recall of over 0.95 in a screening method ranges from 5 to 20, which involves between 15 and 60 minutes' work [60].

The solid matrix can strongly influence the shape of the signal provided by the atomic spectrometer. Such influence hinders the use of the peak height for the target analyte. The mechanisms governing the processes occurring in atomizers have been modelled with a view to minimizing or suppressing the effect of the solid matrix [61–64]. The fast transient signal calls for a fast instrumental response, which should preferably be integrated. Linearization algorithms allow results to be improved and the linear range to be extended [65]. The use of external standards (e.g. Ar^{2+} in ICP–MS) to compensate for variations in mass sampled and analysed is a usual practice for correcting matrix-induced signal suppression effects [66]. Calibration can be done using various approaches involving CRMs, synthetic solid standards, the method of standard additions or aqueous standards.

The procedure involving CRMs requires that a reference material containing an appropriate concentration of the target analyte be available. This is not a general calibration method owing to the lack of reference materials with certified analyte concentrations spanning the range of interest (e.g. in environmental samples). In any case, calibration with CRMs has been used with metal-containing solid samples [67], sewage sludge [68], biological samples [69] and plant material [70]. One substantial disadvantage of reference materials used as standards is that the certified concentration usually has an uncertainty that is high compared to the uncertainty of the method. Consequently, the use of CRMs as calibrants introduces high, frequently ignored uncertainty in the final value. In some instances, when CRMs are not available, a synthetic solid sample matching the composition of the original sample can be used instead.

Obtaining synthetic solid standards is a difficult, time-consuming process that is only justified when both the number of samples and that of analyses is high (e.g. in the steel industry).

The method of standard additions has been used to suppress the influence of the bulk composition of solid samples. It requires that the analyte contained in the solid sample and the added analyte be identically affected by the matrix. This method can be used in three different ways in connection with solid samples. One involves adding increasing amounts of analyte to fixed amounts of sample. In the second, a single amount of sample is subjected to a single addition of analyte from a standard solution, which can lead to inaccuracy in the final result. In the third approach, both the amount of solid sample used and that of standard added are varied; the instrumental signal is a function of two independent variables and can be extracted by multiple linear regression.

The use of aqueous standards entails assuming that the instrumental signal will be independent of the bulk matrix and dependent on the analyte concentration alone. This requirement is usually not met because of both spectral and chemical interferences from the matrix. For calibration with aqueous standards to be successful, a series of instrumental requirements including expeditious signal processing, the use of integrated measurements, appropriate background correction and atomization under isothermal conditions must be met. With solid samples, absolute calibration of the autosampler or pipette used to dispense the standard solutions is also important because the accuracy of the determinations is directly dependent on the accuracy of the pipette. A comparison of the use of CRMs and multi-element solution standards revealed that the results of both approaches were similar, with not much influence of the residual organic matrix on the transport efficiency. Adding dilute nitric acid to the solid sample inside the graphite boat of the furnace allowed removal of most of the organic matrix during the pyrolysis step. Consequently, calibration against aqueous standard solutions provided analytical results that were as good as those obtained with the method of standard additions [71].

A comparison of three calibration methods based on the use of CRMs, generalized simplified standard additions and multiple standard additions revealed the last to be the most practical and reliable [72].

The use of solid standards introduces a different concept of calibration. Traditional calibration is done by changing the analyte content. With micro-weighings, accurately replicating the same sample mass is difficult. Therefore, calibration with solid standards for ETA is done with powdered reference materials by varying the sample mass. Conventional two-dimensional calibration focuses on the analyte mass without considering the corresponding matrix mass. The analyte mass is an analytical quantity derived

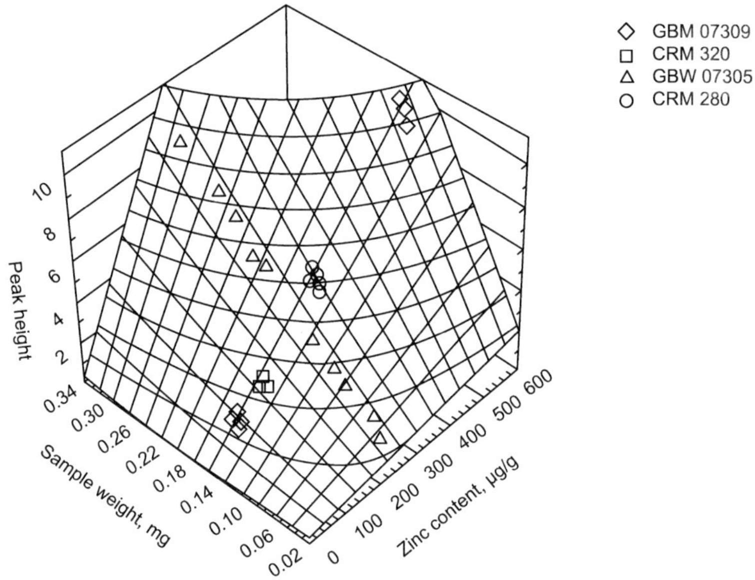

Fig. 8.6. Three-dimensional plot for Zn illustrating different calibration methods. Traditional calibration by variation of the analyte content is shown using four different samples. Calibration by variation of the sample weight is shown using the certified reference material GBW 07305. (Reproduced with permission of Springer-Verlag.)

TABLE 8.3

RESULTS OF THE DETERMINATION OF Pb USING SOLID SAMPLING GF–AAS

CRM/RM	Certified value, µg/g	Analysed value[(a)] 3D, µg/g	Recovery %	Analysed value[(a)] 2D, µg/g	Recovery %
W-1	7.5 ± 1.5	7.1 ± 1.6	95	6.8 ± 1.8	91
BM	13 ± 2.0	15.0 ± 1.9	106	14.8 ± 1.9	114
Mica-Mg	13 ± 2.0	8.7 ± 1.6	97	8.5 ± 1.8	94
Mica-Fe	9 ± 3	11.1 ± 0.6	85	10.6 ± 1.3	82
TB	13 ± 3	6.4 ± 0.7	80	6.3 ± 0.5	79
GBW 083802	8 ± 1.4	13.5 ± 1.1	95	12.9 ± 1.3	91
	14.2 ± 2.7				

[(a)] ± standard deviation ($n = 5$)

Reproduced with permission of Springer-Verlag

from the sample weight and analyte content. The disadvantage is that, in materials with a strong matrix influence, the matrix mass is not taken into account. In order to include the sample weight and analyte content in the calibration, the transition to three-dimensional calibration is required. Figure 8.6 shows a 3D plot for Zn that illustrates the different approaches to calibration in this context, namely: traditional calibration by changing the analyte content or by varying the amount of sample. Table 8.3 summarizes the results provided by the 2D and 3D calibration methods. As can be seen, recovery and precision are better with 3D calibration. Confidence intervals for the 3D calibration method have recently been modelled [73].

8.2.9. Applications of electrothermal solid sampling prior to introduction into an AAS, ICP–AES, AFS or ICP–MS instrument

The number of applications of atomic techniques based on solid or slurry sampling is so large that only a comparatively minute fraction is discussed in this section. Interested readers are referred to the biannual reviews of *Analytical Chemistry* and the atomic spectroscopy update in the *Journal of Analytical Atomic Spectrometry*, among other sources, for more extensive information. A specific review of the uses of graphite atomizers modified with high-melting carbides has been published by Volynsky that includes virtually all metals determined in this manner [74].

As with liquids, the four atomic techniques have complementary uses with solid samples. When axial inductively coupled plasma atomic emission spectroscopy (ICP–AES) hit the market in the early 1990s, some manufacturers of graphite furnace atomic absorption (GF–AAS) systems feared it was time to start looking for a new job. Despite the increased pressure from ICP-based and other competing techniques, however, GF–AAS continues to hold a place in the elemental analysis market [75]. Considerable efforts have been made over the last decade to combine solid sample insertion methods with ICP–MS. Many systems which had stood the test with ICP–AES were adopted for use with ICP–MS. Based on its excellent multi-element determination capabilities, its high sensitivity and its ability to handle transient signals, ICP–MS is regarded as a more suitable combination with solid sampling devices than is ICP–AES; however, there has been a marked decline in work concerned with the direct introduction of solids into ICP instruments in the last few years.

Below are described selected examples of the use of the four techniques following electrothermal pretreatment.

Atomic absorption spectrometry was the first in its time and perhaps the most widely used with solid samples. There are very few references to the use of flame as atomizer in handling solid samples in combination with an air–acetylene flame [76] or slurry sampling [77,78] prior to AAS. When the concentration of the target analytes is high enough (e.g. in the determination of Cd, Pb, Hg, Sb and As in soil, rock and sediment samples), both the sensitivity and the accuracy can be improved by dilution with graphite powder [79]. The determination of other metals in high-purity titanium using a calibration graph obtained with standard solutions provides limits of detection ranging from 0.02 (Mg) to 30 ng/g (Sn) [80]; those obtained in high-purity tungsten range from 0.01 (Mg) to 1.7 ng/g

(Zn), so they are far better than the limits typically provided by digested samples [81]. In dealing with slurries, very long shaking times (1–4 h) with ultrasonic vibrators are most often required [82,83]. Some prefer to use ultrasonic assistance to leach solid samples in the autosampler cups (using a small ultrasonic probe) [84–86]. Other methods such as that for the determination of Ag are special; following preconcentration on a chelate resin, the solid is suspended in 1 ml of dilute HNO_3 and inserted into the graphite furnace [87]. The determination of total mercury in natural gas liquid and condensate by carbon adsorption and slurry sampling [88] is an unusual application of GF–AAS. The speciation of mercury in human hair requires the use of both solid and liquid samples [89]. Table 8.4 compares the direct analysis of pieces of high-purity titanium for a series of metals such as Al, As, Ca, Co, Cr, Cu, Fe, K, Mg, Mn, Na, Ni, Pb, Sn and Zn using ETA–AAS following sample dissolution with the results provided by other, well-established techniques such as neutron activation analysis (both RNAA and INAA). As can be seen, the need for no sample dilution in solid sampling allows the determination of many components that cannot be determined after diluting the sample in the dissolution step. The limits of detection achieved are compared with those provided by RNAA, INAA and isotope dilution mass spectrometry (IDMS) in Table 8.5; as can be seen, the ETA–AAS technique yielded better results than all others. Figure 8.7 depicts the procedure used to treat the solid sample prior to ETA–AAS, as well as other steps of the overall analytical process [80]. Table 8.6 compares solid sampling (SoS) and slurry sampling (SlS) in the determination of trace impurities in niobium pentoxide; contents are given as mean values and standard deviations based on 6 and 5 separate determinations for SoS–AAS and the other two methods, respectively. The relative standard deviations (RSDs) of SoS–AAS, which include the inhomogeneity of the material at the milligram level, are between 4 and 10% — by exception, those for Co, Mn and Na are 17–20%. Taking into account the low contents determined (between 5 ng/g for Mn and 160 n/g for Fe), the RSDs can be considered excellent and indicative of high homogeneity in the material. The RSDs for the solid sampling method are similar to those for slurry sampling; however, they are significantly lower (by a factor up to more than 2) than those for solution ETA–AAS. This can largely be ascribed to the effect of the relatively large scatter of the blank in the determination of such low analyte concentrations. The limits of detection of the methods are compared in Table 8.7 with those provided by arc-AES and ICP–AES; as can be seen, SoS–AAS performs better than the other methods. The usefulness of 3D calibration for SoS–GF–AAS was recently demonstrated [90].

The earliest uses of *inductively coupled plasma atomic emission spectrometry* (ICP–AAS) for the excitation and monitoring of the signal from solid samples provided unacceptable results (recoveries between 60% and 136%) [91] which have since been improved on [92–93], particularly with slurry sampling — a currently well-established choice for solid sampling analysis. Applications span a wide range of sample types from plant to refractory materials [94–103]. Table 8.8 compares the results of ETV–ICP–AES and ETA–AAS in the direct analysis of high-purity tantalum powders for other metals; as can be seen, the combination of ICP–AES with solid sampling and ETV is a powerful tool for multi-element characterization at trace levels, which is also well suited to routine analyses when provided with automated platform handling. As shown in Table 8.9, this technique can compete with ETA–AAS plus sample digestion as regards limits of detection [104].

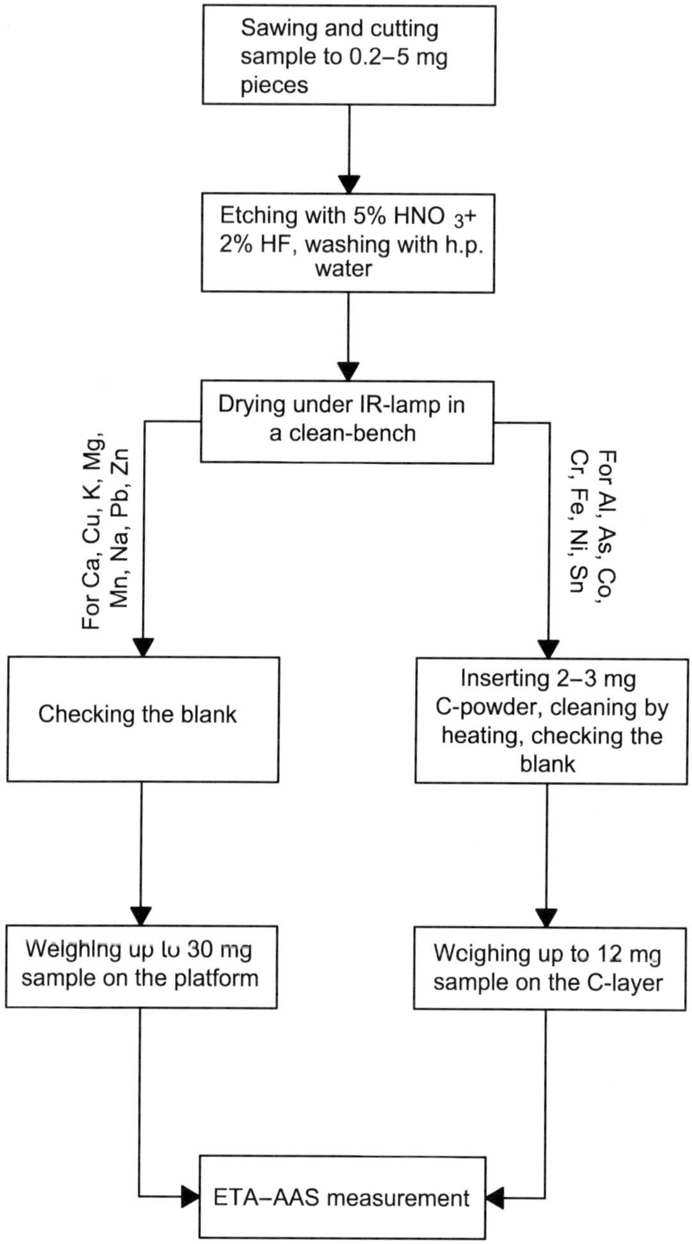

Fig. 8.7. Procedure for the treatment of high-purity titanium samples for the determination of various metal impurities.

TABLE 8.4

COMPARISON OF THE RESULTS OBTAINED IN ANALYSES OF TWO TITANIUM SAMPLES BY SOLID SAMPLING ETA–AAS WITH THOSE OF SOLUTION ETA–AAS, INAA AND RNAA

		Content, ng/g			
Element	Sample	SoS-ETA–AAS	Solution-ETA–AAS	INAA	RNAA
Al	Ti-1	19 100 ± 200	15 800 ± 600		
	Ti-2	15 900 ± 1900	12 700 ± 600		
As	Ti-1	9400 ± 1000		7500 ± 500	
	Ti-2	42 ± 6			
Ca	Ti-1	0.95 ± 0.15			
	Ti-2	0.31 ± 0.04			
Co	Ti-1	380 ± 30		420 ± 10	430 ± 10
	Ti-2	34 ± 3			
Cr	Ti-2	2700 ± 200	2900 ± 100		
Cu	Ti-1	6200 ± 300			5600 ± 1800
	Ti-2	29 ± 3	30 ± 5		
Fe	Ti-2	4500 ± 600	4700 ± 200		
K	Ti-1	0.13 ± 0.01		< 110	< 30
	Ti-2	< 0.1			
Mg	Ti-1	23.8 ± 2.4			
	Ti-2	0.4 ± 0.1			
Mn	Ti-1	8000 ± 300		7200 ± 1000	
	Ti-2	1.9 ± 0.1			
Na	Ti-1	0.8 ± 0.2		< 3	< 3
	Ti-2	1.8 ± 0.4			
Ni	Ti-2	1200 ± 100	1300 ± 200		
Pb	Ti-1	46 ± 3			
	Ti-2	< 3			
Sn	Ti-1	83 000 ± 7000		79 000 ± 2000	69 000 ± 4000
	Ti-2	450 ± 50			
Zn	Ti-1	0.46 ± 0.08		< 180	< 6
	Ti-2	0.19 ± 0.04			

Atomic fluorescence spectrometry may be the most sensitive of the four techniques — particularly with laser assistance; it has rarely been used with solid or slurry sampling and largely for determinations of metals in biological fluids, urine [105–107] and blood [106–110]. Typical examples of solid sampling with this technique include the determination of Li in lithium oxalate [111], Ti in electrothermal atomizers [112], Pb and Tl in nickel-based alloys [113], and Co in high-purity tin [114].

TABLE 8.5

LIMITS OF DETECTION (LoD) OBTAINED IN DIRECT SOLID SAMPLING ETA–AAS AND COMPARISON WITH THOSE REPORTED FOR OTHER METHODS USED FOR THE ANALYSIS OF TITANIUM

	LOD, ng/g			
Element	SoS-ETA–AAS	INAA	RNAA	IDMS
Al	20			
As	6	4		
Ca	0.08			
Co	2	3	2	
Cr	2	130		
Cu	1		1	1
Fe	2	1800	30	35
K	0.1	80	30	
Mg	0.02			
Mn	0.3	5		
Na	0.04	0.2	0.05	
Ni	6	1500	9	4
Pb	3			
Sn	30	1700	250	
Zn	0.05	180	1	

For Al, As, Co, Cr, Fe, Ni and Sn, based on an applied sample amount of 12 mg; for the other elements, on 30 mg
Reproduced with permission of the American Chemical Society

Over the past decade, *inductively coupled plasma–mass spectrometry* (ICP–MS) has evolved from a delicate research tool for the well-trained scientist into a well-established, robust analytical tool for trace and ultra-trace element determinations. This technique owes its popularity to its extremely low limits of detection, its multi-element capabilities and wide linear dynamic range, its high sample throughput, and the ability to survey spectra and to derive isotopic information on the analyte elements. Most authors reporting on the analysis of solid samples using ETV–ICP–MS prefer to analyse slurries over dry solid samples [115–121]. Wang *et al.* [122] explored the feasibility of "real" solid sampling ETV–ICP–MS and concluded that, in order to ensure adequate accuracy, external calibration based on a solid CRM of composition as similar as possible to that of the sample should be used. However, Vanhaecke *et al.* showed that, with appropriate precautions, (a) single-standard addition [123,124], (b) isotope dilution [125] and, to a lesser extent, (c) external calibration with either an aqueous standard solution [123] or a solid CRM with a matrix composition comparable to that of the sample [126], was possible. One unusual method with characteristics similar to both solid and liquid sampling uses in situ digestion of the sample in the graphite furnace [127]. It was developed in order to overcome the difficulty arising in forgeries of antique silver objects to

TABLE 8.6

CONTENTS OF TRACE IMPURITIES AS DETERMINED IN NIOBIUM PENTOXIDE BY DIRECT SOLID SAMPLING ETA–AAS AND COMPARISON WITH THE RESULTS OBTAINED BY SLURRY SAMPLING ETA–AAS (SIS-ETA–AAS) AND SOLUTION ETA–AAS (Sol-ETA–AAS)

Element	Content, ng/g		
	SoS-ETA–AAS ($n = 6$)	SIS-ETA–AAS ($n = 5$)	Sol-ETA–AAS ($n = 5$)
Al	52 ± 3		
Co	6 ± 1	< 28	< 26
Cr	34 ± 2	42 ± 7	27 ± 4
Cu	< 1	< 11	< 25
Fe	159 ± 13	108 ± 8	< 655
K	48 ± 5	50 ± 4	< 58
Mg	55 ± 3	72 ± 6	60 ± 4
Mn	5 ± 1	< 2	< 15
Na	94 ± 19	122 ± 22	110 ± 32
Ni	52 ± 2		
Zn	157 ± 16	195 ± 16	201 ± 72

Reproduced with permission of the Royal Society of Chemistry

TABLE 8.7

LIMITS OF DETECTION (LoD) ACHIEVED IN THE ANALYSIS OF NIOBIUM PENTOXIDE USING SoS-ETA–AAS AND COMPARISON WITH THOSE REPORTED FOR OTHER METHODS

Elements	LOD, ng/g				
	SoS-ETA–AAS	SIS–ETA–AAS	Sol-ETA–AAS	Arc-AES	ICP–AES
Al	2			3000	5000
Co	2	28	26		7000
Cr	2	3	15		2000
Cu	1	11	25		46 000
Fe	2	19	660	3000	1000
K	1	2	58		20 000
Mg	0.5	0.5	8	3000	1000
Mn	0.5	2	15	3000	3000
Na	0.5	6	49		10 000
Ni	2				
Zn	0.5	5	16		

For Al, Co, Cr, Cu, Fe, Mn and Ni, based on a sample amount of 15 mg, Na and Zn, on 6 mg
Reproduced with permission of Springer-Verlag

TABLE 8.8

ELEMENT CONTENTS DETERMINED IN TWO TANTALUM SAMPLES USING SOLID SAMPLING ETV–ICP–AES IN COMPARISON WITH THE RESULTS OBTAINED BY SOLID SAMPLING ETA–AAS

		Content, µg/g	
Element	Sample	ETV–ICP–AES	ETA–AAS
Cu	Ta-1	0.065 ± 0.007	0.068 ± 0.012
	Ta-2	0.053 ± 0.008	0.046 ± 0.004
Fe	Ta-1	8.7 ± 0.4	9.1 ± 0.4
	Ta-2	6.5 ± 1.3	6.2 ± 0.7
K	Ta-1	< 0.5	0.17 ± 0.01
	Ta-2	1.6 ± 0.3	2.0 ± 0.15
Mg	Ta-1	6.6 ± 0.5	
	Ta-2	6.9 ± 2.4	6.4 ± 0.6
Na	Ta-1	0.12 ± 0.02	0.088 ± 0.014
	Ta-2	0.25 ± 0.02	0.35 ± 0.03
Al	Ta-1	0.16 ± 0.08	
	Ta-2	4.6 ± 0.6	
As	Ta-2	1.6 ± 0.3	
Ca	Ta-1	0.20 ± 0.1	
	Ta-2	0.23 ± 0.03	

Reproduced with permission of Springer-Verlag

be identified by their fingerprints of impurities; dissolution of the sample caused severe problems since Au and Ag had to be dissolved simultaneously. In situ digestion with nitric acid in the graphite tube allows the determination of Au, Bi, Cd, Pb, Sb, Sn and Zn over the range 10 µg/g–1% with an accuracy of 10–50% in less than 1 mg of sample. Table 8.10 compares the performance of ETV–ICP–MS and ETA–AAS in the determination of ruthenium in photographic materials, using both liquid and solid sampling with the former and pneumatic nebulization with the latter. As can be seen, consistency between the results of both approaches was very good; in fact, the differences between the average ETV–ICP–MS and the reference values (obtained by ETA–AAS or CP–MS with pneumatic nebulization) were all less than 10%. In addition, the precision was quite satisfactory (average RSD < 10%, $n = 5$). The absolute limit of detection achieved with solid sampling was *ca.* 1 pg and the relative value with liquid sampling was 0.1 µg/ml, using a typical volume of 10 µl (for solid sampling this is equivalent to 1 ng/g, taking 1 mg as a typical sample mass) [128]. A recent contribution to slurry sampling electrothermal vaporization ICP–MS involves the use of organic acids as modifiers for the determination of metals in soil with limits of detection of 0.3 and 9 ng/g for Tl and Pb, respectively [129].

TABLE 8.9

LIMITS OF DETECTION OF AN ETV–ICP–AES METHOD FOR THE ANALYSIS OF TANTALUM POWDERS AS COMPARED TO SOLUTION ICP–AES USING NEBULIZATION (Sol-ICP–AES) AND SOLUTION ETA–AAS (Sol-ETA–AAS)

	Limit of detection, ng/g		
Element	ETV–ICP–AES[a]	Sol-ICP–AES[b]	Sol-ETA–AAS
Ag	5	—	—
Al	25	500	—
As	180	—	—
Ca	30	20	—
Cd	100	200	—
Co	70	200	—
Cu	5	100	9
Fe	20	100	15
Ga	60	—	—
K	250	400	15
Li	40	—	—
Mg	30	20	13
Na	40	—	—
Pb	250	10 000	—

[a] Based on 3σ of the procedure blank ($n = 15$), calculated for 40 mg Ta-sample per measurement
[b] Involving analyte–matrix separation
Reproduced with permission of Springer-Verlag

TABLE 8.10

COMPARISON OF ETV-ICP-MS AND ETA-AAS FOR THE DETERMINATION OF RU IN PHOTOGRAPHIC EMULSIONS [Results, in ppm (μg Ru per g Ag). Standard deviations are given in brackets[a]]

	Emulsion I	Emulsion II	Film I	Film II
Ru added	1 ppm	10 ppm	10 ppm	50 ppm
Liquid sampling ETV–ICP–MS	—	11.0 (0.1, $n = 5$)	—	—
Solid sampling ETV–ICP–MS	1.36 (0.12, $n = 5$)	11.0 (0.8, $n = 5$)	13.3 (0.9, $n = 5$)	52.1 (1.6, $n = 5$)
ETA–AAS	1.06 (0.09, $n = 3$)	11.1 (0.2, $n = 3$)	12.6, 12.6	45, 49.1
Pneumatic nebulization ICP–MS	1.34 (0.03, $n = 4$)	11.4 (0.7, $n = 4$)	—	—

[a] The Ag concentrations for emulsions I and II were 88.8 and 88.6 mg/g, respectively. Films I and II contained 1.94 and 2.07 g/m^2, respectively
Reproduced with permission of the Royal Society of Chemistry

8.3. GLOW-DISCHARGE SAMPLING

8.3.1. Introduction

The increased interest in glow-discharge techniques, initially in the academic world and later among instrument manufacturers, can be ascribed largely to their performance in solid sample analysis.

The glow-discharge (GD) is an old source that found many uses in analytical spectroscopy in the past two decades. Its operational simplicity and high flexibility aroused increasing commercial interest. From an obscure analytical approach focusing primarily on metal analysis, glow-discharge has developed into a sophisticated technique suitable for the analysis of non-metals, thin films, semiconductors, insulators and organic materials. Efforts have been devoted in the past two decades, and continue into the present, to better understand the many chemical and physical phenomena that influence the effectiveness of the GD source.

The glow-discharge is a low-pressure (0.1–1.0 torr) plasma composed of two electrodes immersed in a partially ionized noble gas [130]. The name comes from the relatively bright central glow originating from excited gas atoms emitting their characteristic optical radiation. A complex combination of atomic ionic species from both the discharge gas and the cathode sample is available for analytical use. Compared with other plasma sources such as the inductively coupled plasma, the GD is a compact, low-volume source that operates at a low power. It is inexpensive to construct and maintain, and plasma gases are consumed at quite modest rates (in the millilitre per minute region). The deceptively fragile appearance of the GD may account for the fact that analysts have failed to take it seriously for many years. In actual practice, the atomic collision processes that occur in the GD are robust enough to break many tenaciously bonded species. Growing interest and acceptance have led to the commercial availability of a variety of instruments that has made GD sources more competitive with other, well-established solid sampling alternatives in the last decade.

Figure 8.8 illustrates the primary analytical techniques that currently incorporate GD in their experimental schemes. Although most GD measurements are designed to use only one of these approaches, simultaneous application of more than one technique is possible. As the discharge ablates the cathodic sample material into the plasma, elemental species are available for a number of different measurements. The large atom population in the discharge makes sensitive absorption (AA) measurements possible; and, because a fraction of these atoms is excited, the subsequent decay produces atomic emission (AE). In addition, atoms may be stimulated by external sources for atomic fluorescence (AF). Some atoms can even be sufficiently excited to lose an electron; the resultant ions can be sampled by mass spectrometry (MS). For two of these techniques (viz. GD–AES and GD–MS), the glow-discharge is self-sufficient (i.e. atomic excitation and ionization result without any external assistance). For GD–AA and GD–AF, an external stimulation probe is required to measure the atomic absorption or to pump sputtered atoms to an energy level suitable for atomic fluorescence.

One major strength of GD is the separation or sampling step (sputter atomization) from the subsequent analytical steps of excitation or ionization. The GD process can be

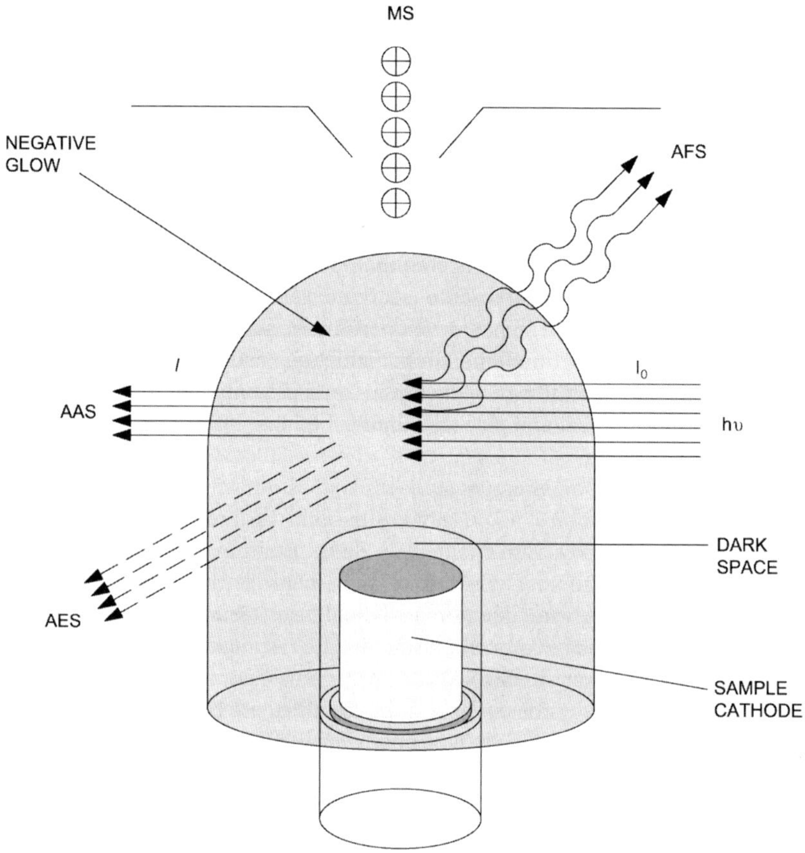

Fig. 8.8. Analytical techniques most frequently coupled to glow-discharge sampling. I_0 incident radiation, I attenuated radiation. (Reproduced with permission of the American Chemical Society.)

thought of as an atom generator delivering a steady-state population of sample atoms to analyte measurement sites separated in space and time from the sample. The atoms have no "memory" of their original chemical environment, so matrix effects from the sample are minimized.

8.3.2. Principles of the glow-discharge

The key to the success of GD is the ease with which it can create an atom reservoir of the sample material directly from the solid state. There are many types of glow-discharge [130], including the common neon light.

Direct-current glow-discharge

As a rule, direct-current (dc) glow-discharge equipment is a simple two-electrode device filled with a rare gas to about 0.1–10.0 torr. A few hundred volts applied across the electrodes causes breakdown of the gas and formation of the ions, electrons and other species that endow the glow-discharge with its analytical usefulness. Figure 8.9 is a simplified sketch showing the basic components and discharge regions. The sample to be analysed serves as the cathode — the anode material is not particularly critical.

Only two plasma zones need be considered here [131], viz. the cathode dark space and the negative glow. The former is a thin layer that exhibits relatively little light emission from electron atom collisions because of the high energy of the accelerated electrons. By contrast, the negative glow is a region of bright emission arising from excitation of atoms by low-energy electrons that have been slowed by inelastic collisions.

The process develops as follows: the sample is atomized into the discharge by so-called "sputtering" [132]. As shown in Fig. 8.10, positive ions and fast neutrals strike the cathode sample, penetrating a few atomic distances before losing their momentum through a sort of three-dimensional billiard cascade of atomic lattice collisions. Some atoms at the surface receive sufficient energy through these collisions to overcome their binding energy and are normally ejected as neutral atoms; a small portion, however, may be released as ions. Positive ions are returned to the surface by the cathode dark space field, but the neutral atoms diffuse into the negative glow region. Glow-discharge-based analytical methods exploit this large sputtered neutral population.

The glow-discharge is a collision-rich environment. At 1 torr, atomic mean free paths are in the 0.05–0.10 mm range, meaning that both incoming sputter agents and outgoing sputtered species undergo frequent collisions. For example, bombarding argon ions originating at the cathode dark space–negative glow interface (Fig. 8.10) may experience

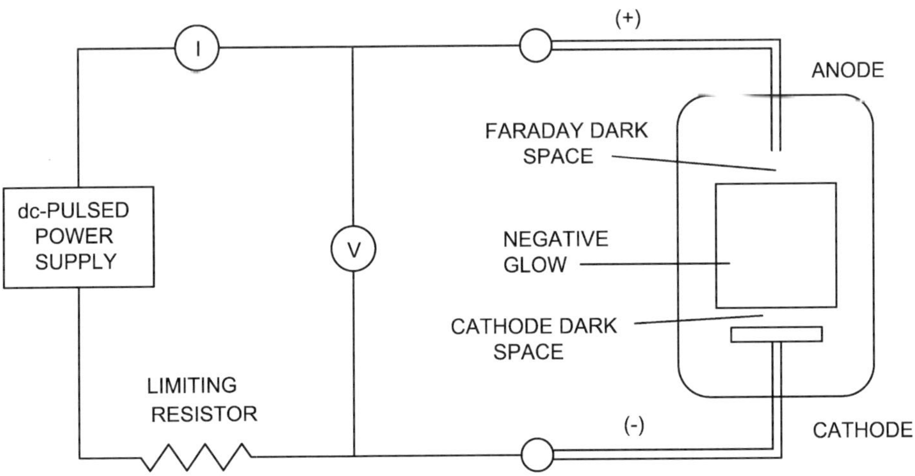

Fig. 8.9. Scheme of a dc glow-discharge source and its associated components. (Reproduced with permission of the American Chemical Society.)

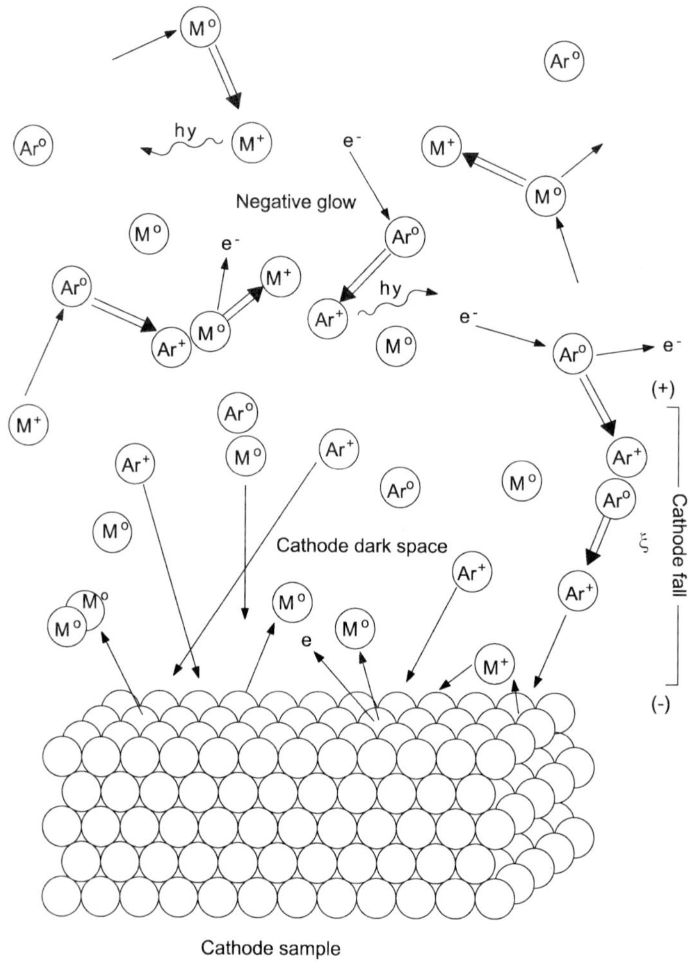

Fig. 8.10. Sputter atomization and ionization processes occurring in a glow-discharge. M cathode metal atom, ξ potential gradient of cathode fall. (Reproduced with permission of the American Chemical Society.)

one or more collisions before striking the cathode. Thus, for a 500 V discharge, relatively few 500 eV ions reach the cathode; the average energy might be closer to 100–200 eV; this, however, is sufficient to sputter-remove large amounts of sample. Similar collisional effects are experienced by sputtered particles that are deflected or even knocked back onto the sample surface for subsequent re-sputtering. Sputter yield depends on the mass and energy of the sputter ion — one or two ejected atoms per incoming particle are not unusual — but with glow-discharges one must consider the effective sputter yield, wherein the net weight loss per unit of current incorporates re-deposition effects.

Almost all the discharge potential is dropped across the cathode fall (dark space), a region very close to the cathode surface. Argon ions are accelerated across this dark space by the cathode fall potential and strike the sample surface, thereby creating a collision cascade with sufficient energy to dislodge one or more surface atoms for each incoming ion. Secondary ions are also released, but the cathode fall potential returns them to the surface. Secondary electrons are accelerated by the same field into the negative glow and help maintain the discharge. The negative glow region appears as a luminous cloud surrounding the dark space and extending outward towards the chamber wall, which normally serves as the anode. The glow results from the discharge gas colliding with electrons moving through the region, causing the atoms of the fill gas (usually argon) to be excited. Relaxation of these atoms produces emission at a wavelength characteristic of the discharge gas (e.g. blue for argon). Because the electric field is dissipated across the dark space, the negative glow is considered essentially a field-free region. Atoms, ions and electrons all possess relatively low energies.

At typical GD pressures, a short mean free path results, so most accelerated ions undergo many charge-exchange collisions before they reach the cathode, causing the average energy to be much smaller than the discharge voltage. As shown in Fig. 8.10, a fast argon ion, Ar_f^+, can collide with a slow argon atom, Ar_s^0, and yield a fast atom and a slow ion. Some of the fast atoms that result from the charge-exchange collisions bombard the cathode and contribute to the net sputtering rate. Sputtered atoms tend to diffuse away from the cathode; however, two-thirds or more are re-deposited on the target cathode by collisions with the dominant argon atoms. Those atoms that escape re-deposition diffuse into the negative glow, where they can be excited or ionized by collisions with metastable argon atoms or electrons. Thus, the sputtered sample exists in the discharge as both ground state atoms, excited atoms and ions. Although some poly-atomic species also exist in the glow-discharge, the sputter-plasma process is primarily atomic in nature. The reactions shown in Fig. 8.10, however, represent a gross over-simplification of the plasma chemistry involved. Many other species of transient stability are formed and may be observed, particularly by GD–MS. The glow-discharge chamber is essentially a small chemical reaction cell that can be controlled to the analyst's benefit. One aspect of GD that has been extensively studied for analytical purposes is the possibility of altering the plasma chemistry to obtain unique chemical environments. In addition, pulsed GD operation provides opportunities for time-resolved decoupling of plasma reactions and products.

Radio frequency glow-discharge

Atomization–excitation/ionization processes in radio frequency glow-discharge (rfGD) are also maintained through two electrodes — one of which is the analytical sample — also immersed in an inert gas such as argon. Unlike dc discharges, in which both electrodes must be electrically conducting in order to sustain the required flow current, one of the electrodes in an rfGD cell can be an electrical insulator. The fact that a plasma can be established at the surface of an insulating electrode/sample is intrinsic to rf operation and is also the key point of interest in the introduction of the rf powering scheme in elemental analysis.

The application of a high-voltage pulse (e.g. the rf voltage) to an insulating surface can be compared to the charging of a capacitor. When a high negative voltage ($-V_s$) is applied to the insulator, the surface potential initially drops to $-V_s$. If positively charged species produced by the ionization of the plasma gas are near the cathode, they are accelerated to the surface. At this point, the potential on the target increases with time to more positive potentials through ion neutralization reactions at the cathode surface. The characteristic discharge time of this capacitor (sample) is in the region of 1 μs in low-pressure environments and the application of voltage pulses at frequencies of the order of ⩾1 MHz results in a *pseudo*-continuous plasma.

The most important result of the application of a high-frequency ac potential is the self-biasing that sustains the discharge and reduces sputtering in the plasma. Consider the 2 kV peak-to-peak square wave potential V_a applied to an insulating surface (creating a GD plasma) and the resultant potential of the cathode (V_b) as shown in Figs 8.11A and 8.11B, respectively. In the initial half-cycle, the voltage on the sample cathode goes to −1 kV and then begins positive charging to about −0.7 kV. During the next half-cycle, the applied voltage is switched to 2 kV to the positive, which produces a surface potential of +1.3 kV. The positively charged surface during this fraction of the cycle causes the electrons in the plasma to accelerate towards the cathode (sample). The increased mobility of the plasma electrons relative to the much heavier positive ions results in fast neutralization of the surface charge during this second half-cycle such that electrode surface potential approaches zero much more rapidly than during the previous half-cycle; it reaches +0.7 kV as the next cycle starts. As the polarity of the voltage is switched, the surface voltage will reach −1.3 kV (instead of −1.0 kV), offsetting the voltage by −0.3 kV. As successive cycles proceed, the surface voltage on the insulator sample will be further offset in the negative direction until it reaches an equilibrium level at which the electrode virtually bears a dc-like negative voltage called the "dc bias".

The dc bias potential is generally one-half the applied peak-to-peak voltage and is a function of the general electrode design and discharge operating parameters. The insulating target material is alternately bombarded by high-energy ions and low-energy electrons; however, the discharge is, for all intents and purposes, continuous with respect to most analytical measurement systems. Most importantly, the sample serves as a time-averaged cathode and is subjected to the cathodic sputtering characteristic of GD devices. It is important to bear in mind that the electrode processes, by virtue of the differing mobilities of electrons and positively charged ions, possess a capacitive nature as this defines the dc bias, which in turn ultimately has a pronounced effect on the atomization, excitation and ionization characteristics.

The above-described capacitive response of an insulating target will obviously not be present in an electrically conducting cathode. In fact, the bias potential in a non-capacitive (conducting electrode) system is zero. As with atmospheric pressure, rf plasma, the rf potential in these devices is typically coupled to the insulating material via a matching (L–C) network, which serves to maximize the power delivered to the plasma by balancing the system impedance of the rf generator to that of the discharge. As such, the capacitance of this circuitry provides the capacitive response for conducting targets.

Because it is governed by the same principles as the dc discharge, this dc bias voltage induces the sputtering of the cathode (sample) material, causing the sample to atomize. Sputtered atoms (predominantly) are then available to diffuse into the negative glow region for subsequent excitation and ionization. Detection of analyte species by atomic

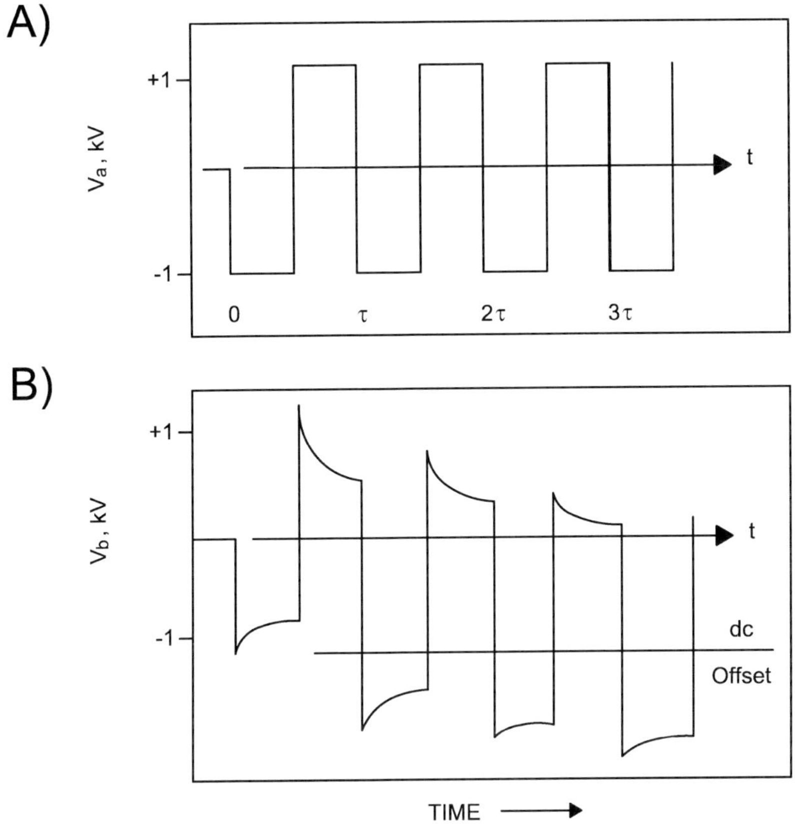

Fig. 8.11. (A) Establishment of a dc bias potential at an insulating surface under the influence of a high-frequency, square-wave potential. (B) The resulting potential on the cathode surface. (Reproduced with permission of the American Chemical Society.)

absorption and emission, MS, and a variety of laser-based methods, is possible. The exploration of rfGD sources for analytical use has so far focused primarily on the development of rfGD–AES and rfGD–MS applications.

The operating characteristics of any rfGD source (viz. its current, voltage and pressure) ultimately define the atomization–excitation–ionization characteristics. Briefly, discharge gas ions must be accelerated to the cathode surface and arrive with sufficient energy so that the collision cascade can produce a sputtering event (viz. atomization). The ion kinetic energy is thus affected by the discharge voltage (dc bias) and the source pressure, which affects the ion and atom mean free paths. Most of the discharge electrons are in fact created as secondary products in the sputtering process. Such electrons are accelerated from the cathode by the fall potential to the negative glow region. Within the negative flow, these electrons cause excitation of sputtered and discharge gas species as well as direct ionization of both. Long-lived (metastable) excited-state discharge gas atoms can in turn affect the ionization of sputtered species via a Penning-type process.

Therefore, the extent of atomic excitation and ionization in this region is also dictated by the discharge power. As the application of rf powered GDs evolves, fundamental studies relating changes in operating parameters to the density and energy of sputtered atoms, electrons and ions, and metastable atoms, will become increasingly important with a view to obtaining better insights into analytical source design and operation.

Comparison of the dc and rf operational modes in GC

Table 8.11 compares analytically relevant particle densities and energies in rf and dc GD devices. As can be seen, the values given for the two plasma types are fairly consistent. Therefore, one would expect that the analytical characteristics of rf and dc GD devices should be very different. A comparison of sputtering rates of metals for the rf atomic emission sources and conventional Grimm-type devices reveals no differences (0.1–1.5 mg/min). The key difference is of course the fact that the rf powered devices are capable of direct atomization of insulating materials. One other advantage, which has yet to be investigated, may result from the lower operating resources of rf devices producing a higher degree of depth resolution for thin film analyses.

In practice, there exists a number of substantial differences in plasma energetics between dc and rf powered discharges. Comparisons of the two operating modes have been made with the same source simply by changing the power supplies — in principle, any rf-designed discharge can operate in the dc mode, but the opposite is not necessarily true. Direct comparisons are shown in Table 8.11. The ion and electron densities, plasma potentials and electron energies were obtained with a Langmuir probe. The most salient differences are that rf plasmas are composed of electrons of much higher energy, whereas dc plasmas possess higher ion and electron number densities. Based on a simple Boltzmann analysis, these data suggest that rf plasma probably has more electrons of suitable energy for atomic excitation than does dc plasma and that high-lying excited states may be more extensively populated in the former. A number of mechanisms have been proposed to account for the enhanced electron energies in rfGDs [130], but the effect is most easily understood by picturing the 13.56 MHz. Additional energy is obtained by electrons when collisions with atoms give them a directionality in phase with the

TABLE 8.11

COMPARISON OF CHARGED PARTICLE CHARACTERISTICS IN THE SAME SOURCE

Parameter	rf	dc
Electron density, cm^{-3}	$2 \times 10^{10}-6 \times 10^{10}$	$6 \times 10^{10}-18 \times 10^{10}$
Average electron energy, eV	4–7	0.7–1.0
Excitation temperature, K	5000–8000	25 000–4000
Electron temperature, eV	1.5–2.5	0.2–0.6
Ion number density, cm^{-3}	$3 \times 10^{10}-2 \times 10^{10}$	$4 \times 10^{10}-20 \times 10^{10}$
Plasma potential, eV	9–16	2.4

Reproduced with permission of the Royal Society of Chemistry

instantaneous field change. Electrons in a dc plasma feel no such polarity changes in the effectively field-free negative flow.

While Wagatsuma and Suzuki [133] found no significant differences between the two operational modes of GD in 1997, Pan *et al.* found the following the next year: (a) an rf discharge usually exhibits higher analyte signal-to-background ratios than does a dc discharge, but the latter exhibits higher emission intensities; (b) an rf discharge becomes stable more quickly than a dc discharge and is more stable over an extended period, so the rf mode generally yields lower limits of detection; (c) the two modes provide the same between-sample reproducibility in terms of signal-to-background ratio, but the dc mode provides better such reproducibility in terms of emission intensity; and (d) a dc discharge sputters more rapidly than an rf discharge, but the latter exhibits much higher electron energies and hence higher excitation efficiencies [134].

8.3.3. Glow-discharge sources: geometry and improvements

The first concern in designing a source with an optimal power is to ensure the maximum possible atomization–excitation–ionization capabilities for the applications at hand.

Source geometry

The geometry of the discharge is critical in terms of isolating the bombarding species to the sample cathode and determining the energy of the bombarding ions as related to the dc bias in the case of rf sources. The first criterion is to maximize the ratio of the respective surface areas of the anode to the cathode. Only in highly asymmetric rf discharges is the dc bias located at the particular electrode. Specifically, ion bombardment is isolated to the smaller electrode (termed by analogy to dc sources as the cathode), with the negative bias approaching one-half of the rf peak-to-peak voltage in the ideal case.

The actual distribution of electrode potential as related to both discharge symmetry and power coupling was reported by Kohler [135]. As shown in Fig. 8.12, six basic cases of symmetry and coupling can be described that expose the effect of geometry and mode of rf coupling (direct or capacitive) on both the dc bias voltage, V_{dc} (broken line), and plasma potential, $V_p(t)$ (solid lines) — a sinusoidal waveform was assumed. As illustrated, increasing the anode-to-cathode ratio increases the dc bias voltage and decreases the plasma potential. The distribution voltage is due to differences in the capacitances of the dark space sheaths of the two electrodes. At frequencies above a few MHz, the sheaths can be assumed to behave capacitively because the rf frequency is below the resonance frequency for the electron and above that for the ion. This means that an electron can cross the respective sheaths in less than one rf cycle, while a more massive ion experiences a number of rf oscillations.

The glow-discharge can be adapted to many discharge types and ion source configurations. Occasionally, the sample type itself dictates a certain cathode geometry; other source models arise from more fundamental considerations. Most available glow discharge sources can be assigned to one of the five simplified models depicted in Fig. 8.13. In each

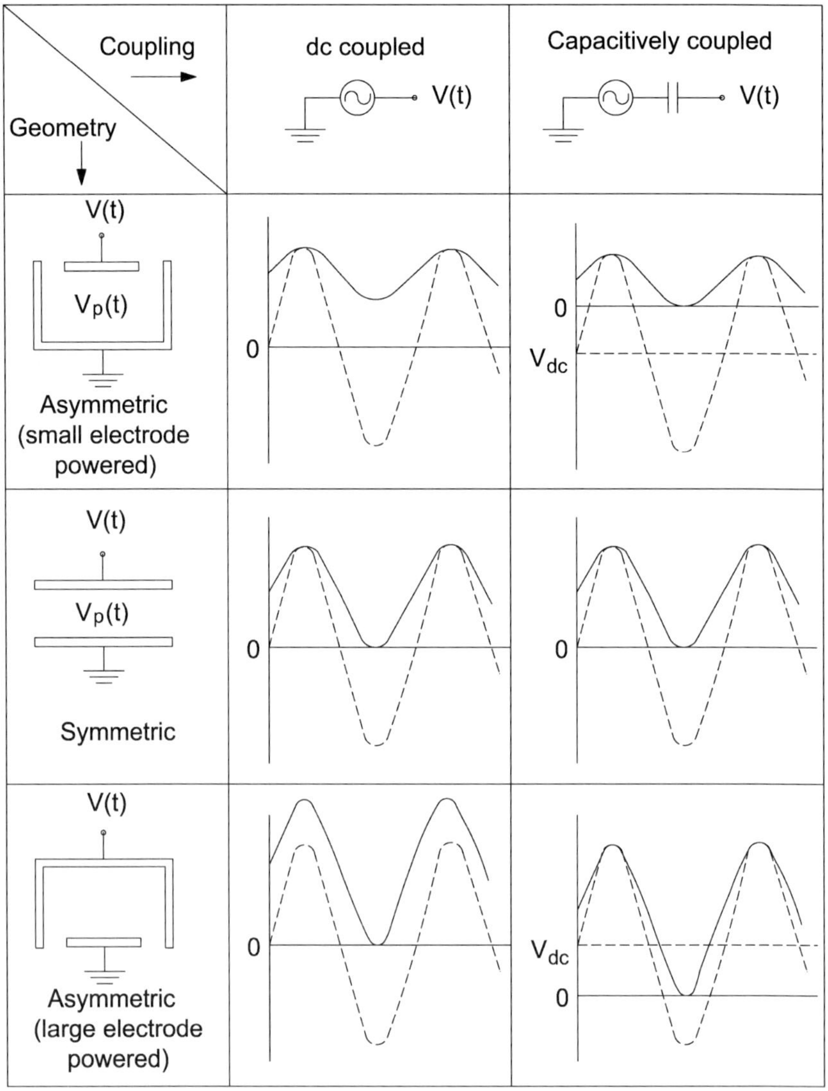

Fig. 8.12. Depictions of relative plasma potentials [$V_p(t)$ (solid lines)], excitation electrode voltages [$V(t)$, broken lines)] and dc bias voltages [V_{dc}] as a function of the discharge coupling and geometry. (Reproduced with permission of the American Chemical Society.)

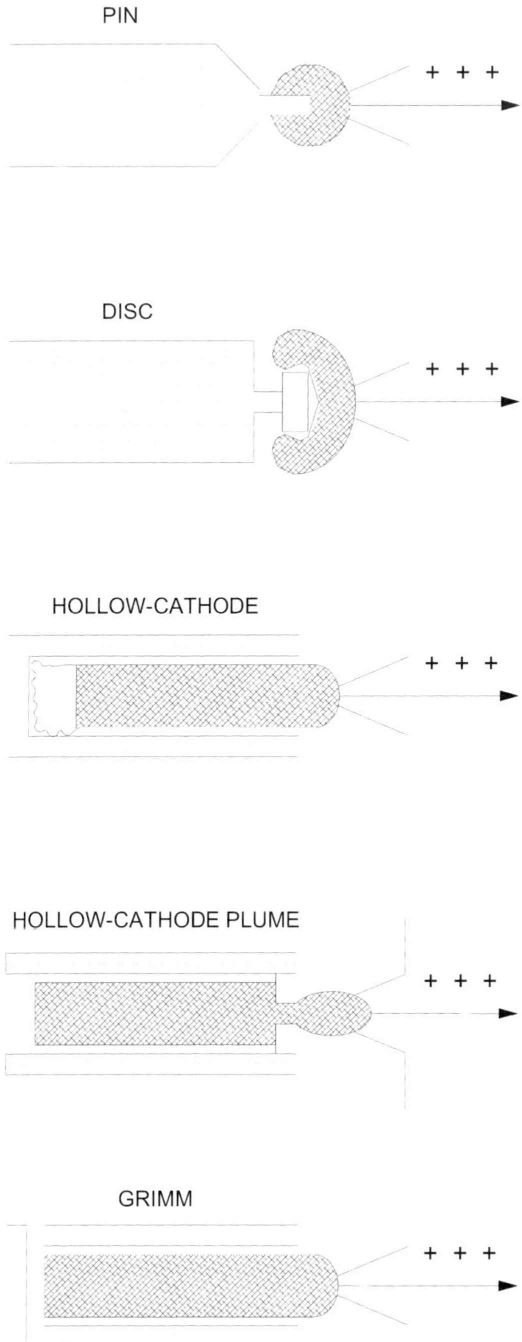

Fig. 8.13. Simplified representation of five common source configurations for glow-discharge. (Reproduced with permission of the American Chemical Society.)

case, ions are extracted from the negative glow of the plasma, which contains ions representative of the sample composition. A sampling orifice is located near the negative glow or a sampling cone is used to advantageously probe specific regions of the discharge.

The most flexible source configuration is probably that of the pin source, which uses a spring clip or pin mount adaptable to samples ranging from thin wires to rods. Placed near the ion exit orifice, a small sample cathode (e.g. a 2 × 10 mm pin) at a low discharge current (1–5 mA) produces sufficient ion signal for sensitive elemental analyses. For sample materials that can be more readily cut into discs, the disc ion source may be preferable. Compared with the pin source, a somewhat different shape for negative glow is observed; however, equivalent current densities yield comparable signals from the two types of source.

The hollow-cathode source provides one other form of glow-discharge [136]. It can be viewed as two planar cathodes brought close enough to each other to ensure that their negative glows will overlap and coalesce into a single common negative glow. This enhances cathodic sputtering and plasma ionization. Hollow-cathode ion sources provide excellent sensitivity; however, compared to pin or disc preparation, the machining of a hollow cathode from a bulk metal is occasionally more cumbersome. This configuration is particularly useful with samples in the form of metal foil or fragments that can be inserted into the cathode cavity for flow discharge.

The hollow-cathode plume (HCP) [137] makes rather different use of the hollow-cathode effect. The hollow-cathode base is a sample disc with a 1–2 mm orifice. By choosing appropriate gas pressures and discharge currents, the energetic plume is ejected through the orifice. A constricted hollow-cathode discharge localized in the narrow orifice creates intense sputter action on the surface of the disc and causes direct transport of the atoms into the plume for excitation — and ionization, if required.

The Grimm glow-discharge [138] is considered to be an obstructed discharge. By bringing the tubular anode within the cathode dark space, the discharge is obstructed or prevented from striking between the adjacent electrodes, thus restricting the discharge. This source can be used to analyse large samples without cutting or shaping of the material as the anode defines the sample area to be sputtered.

Improvements in GD sources

Glow-discharge sources, both dc and rf, have traditionally been operated in the continuous mode. The simplest original arrangement uses a low-cost supply with a discharge power often as low as 5–10 W; plasma species were thus formed with a steady-state stability of *ca.* 1–2%. Subsequently, advantages were gained by pulsing the discharge, boosting it using microwaves or magnetic fields, or employing secondary cathodes.

Pulsed sources Some commercially available power supplies are essentially power operational amplifiers that can be triggered by an external signal source. A pulse generator allows the power supply to be driven at selectable repetition rates and duty cycles. For a given average current, pulsed operation uses a higher voltage, higher current discharge during its "on" cycle than in dc operation. More energetic argon ions are produced, which enhances sputter yield and increases atom density in the plasma. Pulsed discharges also enable recording of time-resolved mass spectra by gating the detector at the desired interval along the pulse envelope. For example, more favourable ion signals from the sputtered species are obtained in the immediate post-pulse interval. A series of these pulsed sources have been reported that use various GD configurations; they were reviewed by Harrison [139], who subsequently conducted primary studies on a microsecond pulsed system based on a conventional hollow-cathode lamp [140] and on one based on the Grimm design [141]. In the latter, the authors investigated potential gains in signal-to-background ratio (S/B) and limits of detection compared to a conventional dc operating mode [142]. They found the high instantaneous pulsed power itself to increase S/B by a factor of about 2 and obtained gains in the range 5–10. For nitrogen, they found that the background from leakage of air could be reduced to approximately one-third compared to the dc mode. One other interesting observation on the use of the pulsed mode was that some non-conducting materials that could not be run in the dc mode were successfully run in the microsecond pulsed mode.

Boosted sources Discharge sources can be supplemented using various approaches that have been reviewed by Leis and Steers [143].

Magnetic enhancement of an ionization source for GD–MS was reported in the early 1980s [144], so its advantages and disadvantages have by now been clearly established. Results similar to those of a Grimm-type source [145]; much higher sputtering rates for insulating and conducting samples than an analogue source can provide without the field as a result of electrons being trapped in the discharge until they are deflected by collision [146]. Increased intensity in non-resonant emission lines for Cu(I) but decreased intensity in resonant ones are both the likely result of self-absorption [147]. Limits of detection 3–40 times lower in the presence of the magnetic field [148]; increased emission intensity of Al, Cu and Ni (by 46%, 26% and 77%, respectively) [149]; and improved limits of detection ranging from 10 to 100 ng/g with ICP–MS [150] are a few other improvements afforded by GD-assisted methods with magnetic enhancement, which, however, require longer warm-up times and large arrays of sample types containing the analyte at concentrations over a wide range for calibration [151].

From the very beginning, the microwave-boosted glow-discharge lamp reported by Leis *et al.* [152] exhibited an enhancement factor of about 5 relative to a GD lamp; also, it diminished sample erosion but provided less uniform craters. Subsequent research on this type of lamp led to improved net emission intensities and S/B ratios, as well as to improved precision in the emission intensity and reduced self-adsorption in the glow-discharge [153]. A comparison of a glow-discharge, microwave-plasma and microwave plasma-enhanced glow-discharge with a steel sample as the cathode revealed differential distributions of the Ar spectral lines [154]. Because the intensities of the Ar(I) lines and OH bands were largely independent of the dc discharge current — unlike the analytical lines, which depended strongly on it — discrimination was achieved by modulating the dc current and using lock-in amplification, thus substantially reducing the interferences with the Bi(I) and Al(I) lines [155]. One straightforward, flexible form of microwave-boosted glow-discharge source is that based on a slab-line microwave cavity, which has been used to investigate fundamental excitation processes in unboosted and microwave boosted GD sources [156]. The changes in spectral output can be described by the

TABLE 8.12

GENERAL EFFECTS OF THE SUPPLEMENTARY MICROWAVE DISCHARGE ON VARIOUS GROUPS OF LINES

Emitting species	Group of lines	Typical values of F
Inert gas atoms	All lines	2–5
Inert gas ions	All lines	0.2–0.5
Atoms sputtered from the cathode	Lines from low lying upper levels	20–50
	Series of upper levels with same electron configuration *n*	Falls with increasing *n*
	Upper levels close to ionization (above or below)	100–200
Ions of cathode material	Lines excited directly by charge transfer (CT)	0.1
	Lines excited by cascade from CT excited levels	0.4
	Lines not excited directly or indirectly by CT (upper level too high or too low, or CT prevented by selection rules)	3

Reproduced with permission of the Royal Society of Chemistry

enhancement factors, *F*, of the different lines, *F* being the ratio of the intensity with the supplementary discharge to that without it, at constant current and pressure. Table 8.12 shows typical values for *F*. In addition to their expanding analytical applications, boosted GD sources can be of assistance in characterizing the excitation processes that occur in the discharge by examining the spectral changes involved [143].

Dual cathodes Couples of cathodes of the same or different nature have been used for various purposes including the simultaneous comparison of a target sample and a reference standard — thus avoiding the need for relative sensitivity factors — [157]; correction of line interferences caused by the GD in AES (using double voltage modulation via the supplementary electrode) [158]; simultaneous AA and MS detection for comparison of atomic and ionic populations [159]; and, mainly, as secondary cathodes for the analysis of non-conducting samples using dcGD [160–163].

Other improvements Jet-boosted GD in various configurations has been developed to improve precision, linearity and limits of detection in combination with AES [164]. Also, a two-cascade GD ion source proved highly effective with double-focusing MS, with which it provided sensitivity comparable to that of the ICP technique [165]. Finally, new hardware for rfGD from a Grimm-type excitation source enabled the analysis of very thin layers of target samples [166].

8.3.4. Nature of the sample for glow-discharge sampling

Samples to be used in dc discharges must be electrically conducting. Metals, alloys and even semiconductors work quite well in this mode. Non-conducting samples must be mixed with a conducting matrix such as pure copper or silver and pressed into a suitable form. On the other hand, an rf discharge allows non-conducting samples to be handled directly, thereby avoiding the need for matrix modification.

Radio frequency-powered GD sources have been coupled with various analytical spectrometries for more than two decades [167]. Analyses are not limited to conducting samples; in fact, electrical insulators such as glass, ceramics and geological materials — some of the most difficult samples to dissolve for solution methods of elemental analysis — can be readily examined.

8.3.5. Sample state and preparation for glow-discharge sampling

Clearly, the principal use of GD is for solid analysis. At a time when many complex procedures are being examined for introducing solid samples into plasmas that are, by nature, better suited to solution injection, GD sputter sampling provides a simple procedure for conversion of a solid into an atomic vapour.

Samples for dcGD solid sampling were originally mixed and compacted in conducting (metal) powders to form conducting sample discs for analysis. This matrix modification methodology has been used for many years in arc and spark emission spectrometries, and also in MS. Although relatively straightforward in implementation, this approach still poses a number of challenges as regards sample preparation (viz. grinding, drying and mixing) and the introduction of concomitant residual gases that affect source stability and analytical precision. These matrix modification steps are not required by rfGD analysis methods. The principal, distinct advantage of rfGD over traditional spectrochemical sources is its inherent capability for direct solid analysis of both conducting and non-conducting samples.

Sample preparation for metals and alloys usually involves brief cleaning and/or polishing, although the sputter process itself serves as a self-cleaning mechanism for removing surface impurities. Powdered samples are compacted in a special die to form pins or discs. Special treatments (e.g. that for high-purity titanium, which is machined using an aluminium oxide cut-off wheel and double etched with a hydrogen peroxide-based acid etchant) reduce preparation times from 3 h to 20 min and pre-sputtering times from more than 1 h to 20 min [168]. Non-flat surface analysis requires special designs of both the anode and the cathode (where the sample is held) [169].

Samples from soil, rock or sediments are usually dried, homogenized and ground in a mill to a specific particle size (less than 50–100 μm) prior to analysis. They are thoroughly mixed with an amount of pure conducting matrix powder which is several times greater than the sample mass, and pressed into pellets, which produces dilution of the matrix. The conducting matrix usually consists of copper, but graphite, silver and tantalum have also been used for this purpose. The choice of copper as a binding material is particularly popular because the pressed pellet has good mechanical properties, excellent thermal and electrical conductivity and a high sputtering efficiency. However, this approach is invariably limited by signal drift effects caused by preferential sputtering if the sputtering rates of the conducting host matrix and the non-conducting analyte material differ significantly.

Samples obtained from animal or plant tissue must be processed prior to GD analysis. The sample may either be dried in an oven or ashed. The dry residue is homogenized and ground, after which it can be handled in the same manner as particulate matter.

Solution analysis is also possible, but requires special manipulations. Volumes in the microlitre range are placed on a host electrode; as the solvent is removed, a thin-film sample residue remains that can be sputtered like any other solid. Alternatively, one can use the flow injection technique to insert small volumes of liquid samples. In any case, reduction of the dissolved sample into a thin surface residue allows very low limits of detection to be obtained from the transient signal. Marcus and Davis [170] recently used GD–AES at atmospheric pressure for direct sampling of liquid media, using an electrolytic solution containing the analyte specimen as one of the discharge electrodes. The passage of electrical current (either electrons or positive ions) across the solution/gas interface caused local heating and the volatilization of the analyte species. Collisions in the discharge region immediately above the solution surface resulted in optical emission that was characteristic of the analyte elements. The analyte solution acted as either the cathode or the anode. Preliminary limits of detection for Na, Fe and Pb were found to lie in the range 11–14 ppm for 5 μl sample volumes.

Gases and volatilized samples can also be analysed with GD sources. Target analytes in a gas phase can be introduced in three different ways, namely: (a) by direct insertion of the gaseous sample; (b) by subjecting the sample to prior thermal vaporization; and (c) by using a chemical reaction (e.g. hydride generation or oxidation) to convert the analytes to volatile species prior to insertion into the GD source. In the last case, by-products of the gas generation step such as water vapour, carbon dioxide and molecular hydrogen must be removed before they reach the discharge. A preconcentration step involving freezing of gaseous or volatilized samples on a cooled trap is often used prior to inserting the sample with a view to significantly improving limits of detection. One of the most salient applications in this context is the use of gas chromatography for the separation of gaseous or volatilized samples prior to introduction into the glow-discharge.

8.3.6. Variables affecting glow-discharge sampling

The variables that influence performance in GD sampling can be classified according to whether they are shared by the source and detector or pertain exclusively to one or the other. The optimum value for each variable depends on the particular aim of the work at hand (e.g. quantitation in homogeneous samples, surface and depth profiling studies in layered samples).

The most influential parameters in dcGD are the *current intensity* that circulates between the electrodes, the *applied voltage* and the *pressure of the plasma gas*. In 1993, Payling and Jones devised a method for separating the contributions of the three key variables in these sources and proposed a quantitative approach in which the emission yield was assumed to depend on pressure alone [171]. This hypothesis was shown to hold for the rfGD mode as well [172].

The analytical performance of an rfGD for a specific sample is known to depend on three key operational parameters, namely: the *pressure in the discharge chamber*, the *voltage between the electrodes* and the *applied power*. However, owing to the interdependence between the three parameters, two of them can be fixed while the other is a free variable. The way the three discharge parameters affect the emission yield is not fully understood and is also a controversial issue. In the past, it was assumed that the emission yield depended on the discharge pressure only; today, however, it is accepted that the emission yield depends on the three key discharge parameters. Parker *et al.* [173] studied the role of the power and pressure parameters on the emission yield using rfGD and found slight changes in discharge pressure to affect it. On the other hand, Bengtson and Hänström [174] reported that the emission yield for a given spectral line reached similar values when the pressure was modified if the electrical parameters were kept constant. Recently, Sanz-Medel *et al.* studied the influence of the three operational modes (leaving the applied power, dc bias or Ar pressure as the free parameter) on the resulting qualitative and quantitative profiles giving information about the chemical composition of coatings and their thickness in a variety of samples. The results obtained in the three operating modes were consistent with those provided by dcGD–AES in a round-robin exercise conducted within the framework of a European Union project [175]; in any case, the worst relative standard deviations were obtained in the operating mode where the gas pressure was allowed to change [176].

Time-related variables are critical in GD work. One of the key features of any analytical method used in a production facility is the sample analysis time. Because GD is a kinetic-based system, the evolution of the plasma processes to, hopefully, an equilibrium situation, is of special concern. It is sufficient to say that any analysis (spectral acquisition) performed in a sequential mode requires that the plasma reach a steady state before analyses can be started — which is obviously arguable for simultaneous detection. Thus, for most analyses, the plasma stabilization time can be a substantial fraction of the overall sample analysis time.

Short- and long-term plasma stability

The stability of a plasma source can be considered within two main time frames. First, there is a nominal period (short term) that is required for most simultaneous or sequential analyses and is 0.5–15 min in length. Short-term stability is also important when evaluating limits of detection based on the stability of the spectral background (RDSB), which is commonly used in emission spectrometry. The second level of stability (long term) applies to those instances where long integration times are used across a broad spectral region, taking upwards of 30 min — which is typical of magnetic-sector MS.

Gas type, pressure and flow-rate

These factors clearly influence plasma formation, the requirements for which depend on whether the GD source is of the rf or dc type and also on the detection technique to which the GD is coupled. As early as 1976, Gough [177] reported improved direct solid atomic absorption analysis achieved by using a flowing gas instead of a static cell and

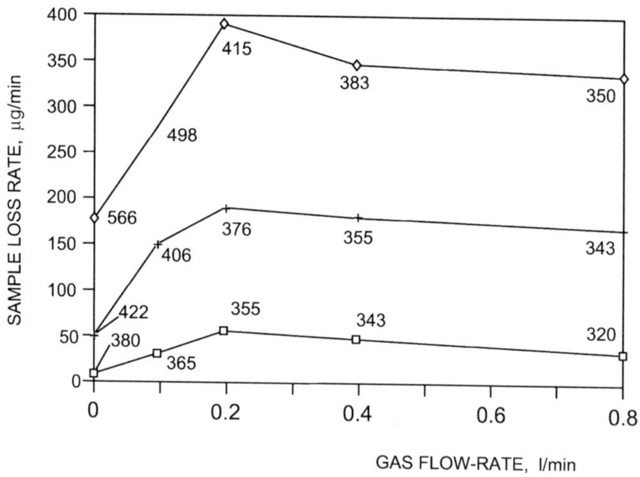

Fig. 8.14. Average sample loss rate for a 6-min sputtering time versus gas flow-rate at three different current densities. The number below each point is the cell voltage: (□) 30 mA/cm², (+) 40 mA/cm², (◇) 60 mA/cm². (Reproduced with permission of the American Chemical Society.)

correcting background absorption — which possibly resulted from aggregation of the sputtered atoms or particles with a dual-modulation amplifier. This contribution was greatly surpassed by the commercial six-jet Atomsource, which provided a higher sampling frequency in addition to proven enhancements in sensitivity and selectivity as shown by Kim and Piepmeier [178]. The influence of the gas flow-rate and current density on the sample loss rate — and hence on the amount of analyte available for adsorption — is illustrated in Fig. 8.14. Such influence — together with that of pressure, power and orifice diameter — has been confirmed with both conducting and non-conducting samples; in the case of rf sources, it is exerted not only on the sample loss rate, but also on the self-bias potential [179]. The direct effect of pressure on the atmospheric electrolyte-cathode GD on line intensity can be expressed using a mathematical model that also predicts the presence of an intensity peak [180].

The type of gas used to form the plasma is especially influential in some cases (particularly on line profiles and widths). Thus, operating a microwave-boosted GD source at 980 V at 30 mA with a microwave power of 40 W and Ne at 4.8 hPa as carrier gas prior to AES resulted in an erosion rate of 9 nm/s compared with 21 nm/s using similar conditions but Ar at 2.1 hPa as carrier gas [181]; the limits of detection for twelve metals in aluminium were similar with both, however. On the other hand, the type of gas employed dramatically affects the emission lines of halogens. Such is the case, for example, with the determination of fluorine using a Grimm-type GD source with Ne or Ar as the plasma gas. While detecting fluorine emission lines in an argon plasma is difficult, a neon plasma exposes not only the atomic lines but also the ionic lines for this element [182]. The excitation of fluorine may primarily be the result of collisional energy transfer from neon species having high internal energies. Conversely, use of the same source for sampling halogenated hydrocarbons in the determination of chlorine provided excellent performance when operated with helium [183]. In the case of alloys and a dc source coupled to AES, the use of pure Ar led to a strong dependence of the voltage–current characteristics of the discharge on the cathode material; on the other hand, a 2:23 Ar–He plasma mixture provided a discharge that was independent of the nature of the cathode [184]. Subsequent experiments with mixtures of these gases and an rf source also coupled to AES equipment revealed the measured electron and ion number densities, and the sputtering rate, to decrease with increase in the helium pressure but also to enhance the Al(I) and Cu(I) emission intensities, which was ascribed to measured increases in the average electron energies and electron temperature [185].

The spatial distribution of the analyte atoms is of paramount importance in atomic absorption techniques; in order to optimize atomic absorption signals, it is therefore desirable to find a region in the chamber where the concentration of sample atoms is maximal. In addition, the GD is an inhomogeneous medium, which makes knowledge of the spatial distribution of sputtered neutrals critical for AAS.

One additional aspect to be considered in coupling a GD to an MS detector is ionization. Thus, given that the presence of oxygen or water vapour was known to adversely affect performance in dcGD–MS [186–188], the relative influence of Penning ionization on the relative sensitivity factors (RSFs) in dcGD as a function of the plasma gas was studied by using Ar, Kr and Xe. The latter two provided higher RSFs than Ar for the majority of elements studied [189]. When Kr was used as plasma gas, the addition of a small amount of H_2 (0.2%) reduced matrix effects compared to using Kr alone in the case of alloys [190] in MS. With non-conducting materials, the effect of small amounts of H_2, He, N_2 or O_2 added to Ar increased the sputtering rate as a function of the concentration of the secondary gas, but with no effect on ion intensity [191].

One influential variable typical of rf sources is the aperture bias voltage, which has dramatic effects on the signal-to-noise ratio and has been extensively studied by Wagatsuma [192]; with a view to improving sensitivity, they controlled the voltage (either by introducing a new channel of dc current via a self-bias potential opened by using a low-pass filter circuit and a load resistor [193,194] or by applying a bias-current modulation at the frequency of the pulsed dc current) and achieved selective detection by using a lock-in amplifier at low noise levels [195].

One key parameter in the case of pulsed sources is the measurement gate, optimization of which dramatically improves the signal-to-noise ratio.

The influence of other, minor variables can also be adjusted in order to improve performance. Thus, in addition to the typical parameters [viz. rf power, power density, plasma gas flow-rate and pressure, ion sampling distance, anode orifice diameter, and sample size, thickness and shape (pin-like or planar)], the efficiency of GD sampling as regards analytical stability, reproducibility, and quantitation and detection limits, also depends on cell configuration — which has strong effects on radial diffusion processes.

Self-absorption, which has a critical influence on AES sensitivity, could not be accurately quantified by using an equation relating the emission intensity of a resonance line affected by this phenomenon to the sputtering rate of the analyte concentration in the matrix [196]. Subsequent studies revealed that, for Cu and V, ionic lines are more sensitive than atomic lines — the latter proved independent of self-absorption but were strongly influenced by the type of plasma used (Ar or Ne) [197].

8.3.7. Glow-discharge sampling as coupled to spectrometric detection

Glow-discharge–mass spectrometry

Glow-discharge–mass spectrometry (GD–MS) has received increasing attention over the last decade. In fact, it is possibly the oldest form of MS. Electric discharges were commonly used as "natural" ion sources in the 1920s and 1930s for the early mass spectrographs [198]. Much of the early analytical work in GD–MS was aimed at replacing the spark ion source (a vacuum discharge ionization source that yields broad energy spreads, an erratic ion beam and usually unreliable quantitative results).

The primary complicating factor in GD–MS source design is the need to accommodate the ion source volume within the main vacuum chamber. This allows cryogenic cooling of the discharge cell to remove trace impurities (which tend to decrease sputtering rates and result in the formation of molecular ions that increase spectral complexity) from the discharge gas. Thus, the sample must be transported to the cell volume through a vacuum interlock system by means of a direct insertion probe (DIP).

The DIP design used in these assemblies is similar to those used in organic MS. The rf-powered DIP must meet several design requirements including efficient coupling of the rf power to the sample, isolation of the sputtering to the sample surface and effective shielding of potential radiated power losses. The two designs in Fig. 8.15 are intended to afford great flexibility and allow the use of the same DIP for the actual sample insertion. Figure 8.15A is an expanded view of the pin-type sample holder implemented on the quadrupole instrument. This configuration allows direct mounting of the sample in the recessed region of the 8-mm diameter copper holder (mounted to the end of the rf feedthrough of the probe); the probe can thus be inserted into the discharge cell. Samples ranging from 1 to 5 mm in diameter can be analysed directly without excessive machining. This basic cell design has also been successfully implemented on the magnetic sector spectrometer. It affords easy sample changeover without breaking the source vacuum and provides for completely grounded coaxial protection of the powered rf lead.

With samples less than 5 mm in diameter, the flat sample holder depicted in Fig. 8.15B affords direct analyses provided the sample is nominally flat. The holder is mounted to the end of the DIP and comprises a PTFE clamp assembly, support rods, boron nitride spacers and either stainless steel or aluminium anode orifice plates. The DIP is placed through the clamp assembly and the rf feedthrough is placed directly behind the sample (cathode), which locks it in place. The spacers separate the anode plate and the cathode in the same manner as the O-ring seal in the rf-AES source design. The anode body in this case is the Ta ion volume enclosed by the anode plate and the ion exit orifice mounted to it; the entire cell assembly is affixed to the commercial source cryo-cooling ring.

The GD ion source gained popularity in the past two decades on account of its stability, sensitivity and operational simplicity. At least three commercial instruments were already available in the early 1990s.

Elemental analysis by GD–MS provides substantial advantages over competing techniques in that it responds to metals and non-metals, exhibits a high sensitivity, suffers from minimal matrix effects and provides isotopic information. Unsurprisingly, it has developed into the most prominent GD coupling.

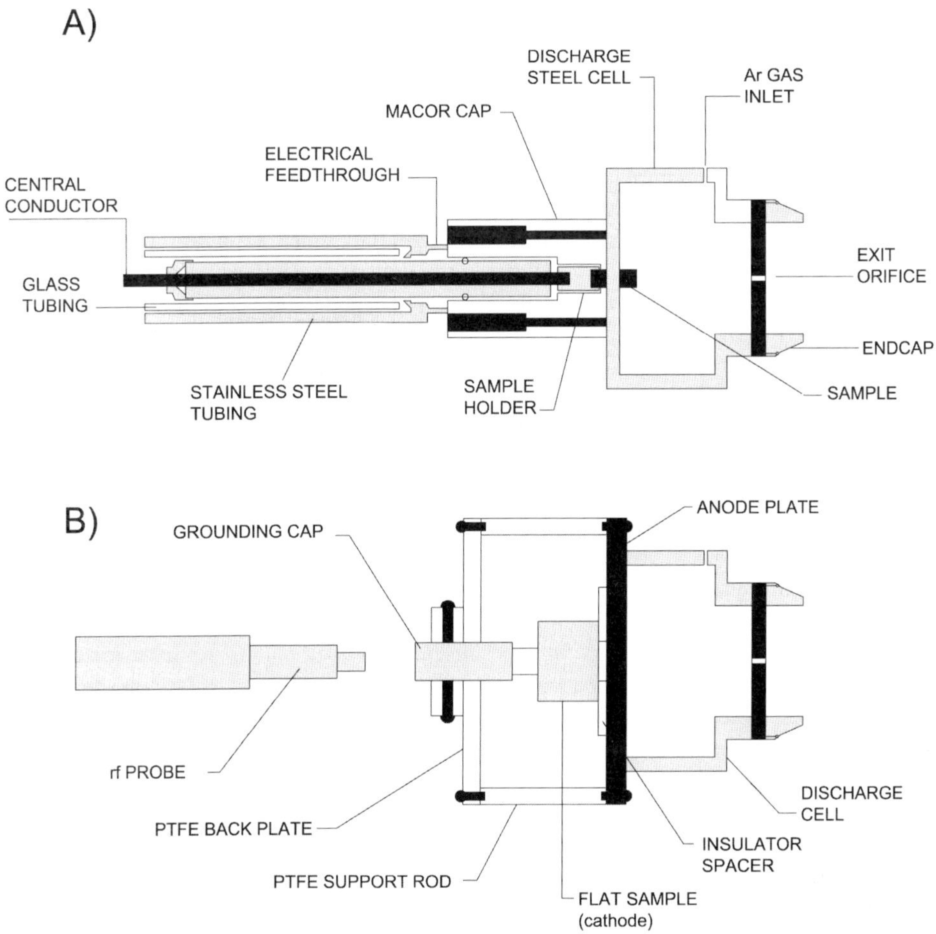

Fig. 8.15. Ion source designs for (A) pin and (B) flat sample types in rfGD–MS. (Reproduced with permission of the American Chemical Society.)

One key difference between MS and optical detection techniques is that the sample material must be physically transferred out of the source. Analyte ions are extracted by electrostatic lenses and transferred through a mass analyser for eventual detection. Although ions are formed throughout the source cell, only those created very close to the exit orifice can survive to the high-collision environment and depart in the charged state. Both magnetic-sector and quadrupole-based instruments have been used in GD–MS, and commercial versions of each are available. The typical discharge operation conditions for GD–MS are 1–5 mA, 800–1500 V and 0.2–2.0 torr.

Three-dimensional mathematical models of the behaviour of the different species in a dcGD in argon were used to study the influence of the dimensions of flat-cathode/hollow-anode cylindrical cells on the calculated plasma quantities. The cell dimensions were found not to affect the plasma quantities qualitatively but only as regards their

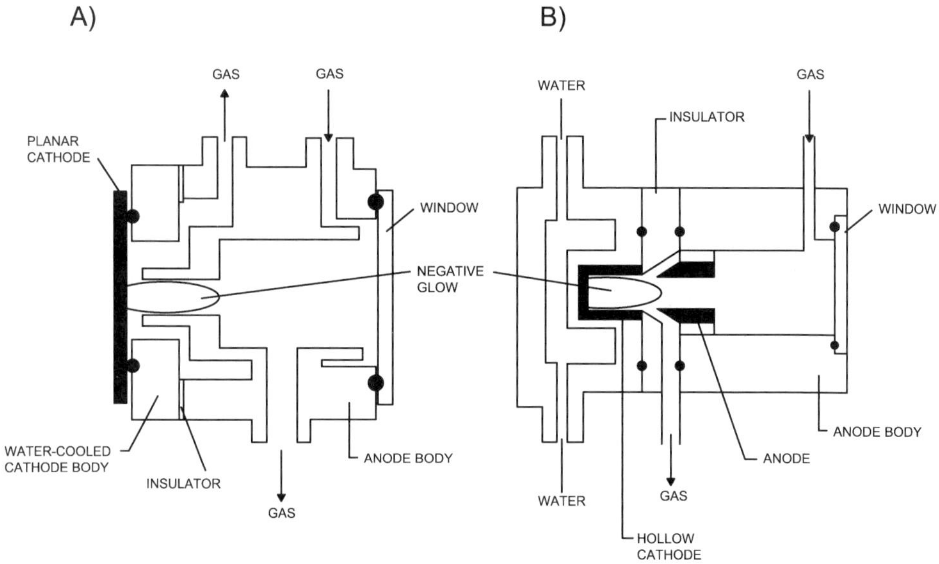

Fig. 8.16. Sources for GD–AES. (A) Grimm-type (planar cathode). (B) Hollow-cathode type. (Reproduced with permission of the American Chemical Society.)

absolute values. Thus, for a discharge of 1 kV and 2 mA at 1 torr, the optimum cell for MS is one having 2 cm as both length and radius [199].

Glow-discharge–atomic emission

Glow-discharge–atomic emission techniques (GD–AE) for solid analysis are instrumentally simpler and more inexpensive than GD–MS [200]. Major trade-offs, however, include poorer limits of detection and reduced elemental coverage. Although compact commercial GD–AE instruments were popular in Europe during the 1980s, their use in the USA remained modest until the 1990s.

Figure 8.16 depicts two commonly used versions of GD emission sources. The planar cathode or Grimm source in Fig. 8.16A is more common [138]. Also adapted for GD–MS applications, the Grimm source is an obstructed discharge in which the anode extends into the cathode dark space and defines the sample area that is sputtered. The sample is placed outside the source for easy access and changeover. Such a configuration is particularly adaptable to the analysis of metal sheets, discs or any flat conducting sample that can be placed close to the source opening; powdered samples pressed into discs are also suitable for this purpose. The typical discharge operation conditions for a Grimm source are 500–1000 V, 25–100 mA and 1–5 torr. The planar cathode discharge is featured in commercial instruments designed for applications to trace elemental analyses in solids. Of special interest is its use for depth profiling of layered metal samples and for the analysis of geological materials such as granite, basalt, mica and feldspar without the thermal effects and self-absorption apparent in arc-spark operation.

The other, less commonly used source for GD–AE is the hollow-cathode discharge (HCD) shown in Fig. 8.16B, which features a cavity cathode rather than the planar cathode of the Grimm source [201]. The net effect of this difference at a given discharge voltage is concentration of the plasma within the constraining cavity, which enhances elemental emission intensities. Typical operating conditions are 200–500 V, 10–100 mA and 0.1–1.0 torr. For the analysis of metals and alloys, either the hollow cathode is machined from the bulk metal, or the sample (e.g. chips or powder) is placed in the hollow cavity of a pure material such as graphite. Solutions can be analysed by evaporating the sample to a residual film in the hollow cathode. Limits of detection usually range from 0.1 to 10 ppm. The coupling of HCD with microwave sources and the development of microcavities have helped improve the sensitivity of this type of source.

Table 8.13 illustrates the excellent temporal features of rfGD–AES. Both metallic and non-conducting samples reach stability levels above 5% as RSD within an operating time of less than 30 s. In addition, analyte signals maintain a high degree of stability over single-element (1 min) and multi-element (5 min) analysis times. These temporal features suggest that the use of simultaneous detection is not required for bulk analyses (e.g. in arc and spark emission analyses). Analysis times could be more like those in arc and spark analyses (< 1 min) if multi-channel instruments were used. Table 8.13 also illustrates the external precision obtained for analyte concentrations at the tens of ppm level with both conducting and non-conducting samples. These values could be further improved by using internal standardization methods.

Although GD–AES is generally considered to have good precision, the ability to obtain high levels of precision with the rf source when sputtering non-conducting samples is somewhat surprising. The effects of residual atmospheric oxygen and water vapour, which effectively quench the important metastable levels of the discharge gas atoms and inhibit the cathodic sputtering process, are well documented in the GD literature [202,203]. The sputtering of oxide samples such as glasses and ceramics most certainly introduces large amounts of oxygen into the plasma. Glasses and ceramics, which tend to be somewhat porous, also act as sources of water vapour that is released from the sample matrix in the sputtering process. Seemingly, however, the likely introduction of gaseous contaminants does not affect the basic operational stability or reproducibility of rfGD sources when operating in the AES mode. Although, based on Table 8.13, GD–AES possesses better analytical characteristics than GD–MS, in fairness, the development of analytical methods for MS is very easy compared to the labour-intensive selection of analytical transitions (wavelengths) required in AES. The identification of spectral interferences is also far easier in MS. Unfortunately, the number of instrumental plasma parameters that must be optimized results in a process that is much more complex than simply "making photons".

Glow-discharge–atomic absorption

The glow-discharge is particularly suitable for atomic absorption (AA) techniques as the overwhelming majority of sputtered species are neutral atoms. This ability to produce a steady-state population directly from a solid matrix has obvious advantages that led Russell and Walsh to suggest the use of GD in the early development of AA [204].

TABLE 8.13

COMPARISON OF BASIC ANALYTICAL CHARACTERISTICS FOR rfGD–AES AND rfGD–MS

Method	Sample type	Stabilization time to 5% RSD, min	Short-term[a] precision, % RSD	Long-term[b] precision, % RSD	External precision, % RSD	LOD
AES	Metals	0.5	0.05	1	1	10–100 ppb
AES	Nonconductors	0.5	0.8			< 1 ppm
MS	Pin-type metals	3	< 1			1–100 ppb
MS	Pin-type nonconductors	30	< 2			< 1 ppm
MS	Flat metals	3	< 2			10–500 ppb
MS	Flat nonconductors	30	< 3			< 1 ppm

[a] Defined as 1 min for AES and 15 min for MS [b] Defined as 5 min for AES and 45 min for MS
Reproduced with permission of the American Chemical Society

Notwithstanding this appealing incentive, little concerted effort was made to capitalize on the idea until the inception of commercial instruments. Figure 8.17 shows a sputter atomization cell for AA measurements; compared to GD–MS and GD–AE sources, a long pathlength is required to enhance absorbance, which is integrated across a broad volume of the cell. The atom density is maximal near the cathode, with a gradient extending to the cell walls. As in other GD sources, a flow-through gas system is used. The gas jet is directed onto the sample surface, which reduces redeposition by sweeping sputtered atoms into the negative glow and results in significant sensitivity enhancements. The typical operating conditions are 300–800 V, 60–200 mA and 4–10 torr.

Sample preparation is much the same as for other GD couplings to instruments. Metal or alloy discs, and compacted conducting samples, are the most frequently used. Solution samples have been examined by deposition onto graphite, aluminium and copper cathodes. Quantitative analyses are usually performed by using a calibration graph run from standard reference materials, a process that takes considerably more time than preparing a solution-based curve. Conversely, time is saved in the direct analysis of solids; also, the inert atmosphere of a GD cell reduces the spectral interferences frequently encountered in flame atomizers.

Glow-discharge–atomic fluorescence

Because the glow-discharge creates a cloud of sputtered atoms in a low-pressure, rare-gas environment that is an excellent quenching medium, it is an effective source for atomic fluorescence. In addition, the inert gas acting as diluent poses few chemical interferences and the elemental absorption lines are relatively narrow. It should be noted that GD–AF has not been used to the same extent as other GD techniques, however.

The use of a laser to generate fluorescence from the GD atom reservoir is a natural consideration in implementing GD–AF. The low background noise from the GD leads to projected limits of detection at the attogram level. The use of a pulsed laser as an excitation source in GD–AF makes a complementary pulsed glow-discharge advantageous in terms of efficient sample use and background noise reduction.

Figure 8.18 shows an unconventional GD–AF system that uses two GDs for complementary purposes. The analytical Grimm discharge produces atomic emission from the elements in the sample. This radiation is directed into a second discharge, a resonance detector with a cathode composed of the analyte element. Radiation from the sample strikes the resonance discharge, causing fluorescence of the analyte elements only. By rotating successive pure cathodes into the resonance discharge, multi-element AF is possible without dispersion optics. A miniature (1.5 ml inner volume) stainless steel GD with three quartz windows allowing passage of a laser beam and collection of the fluorescence produced normal to the beam direction was used for liquid samples. A volume of 0.4 ml of sample was deposited on a Ni cathode and dried under N_2, and a vacuum of 0.1 torr was applied after flushing with Ar. The limit of detection thus achieved for europium was 0.08 fg [205].

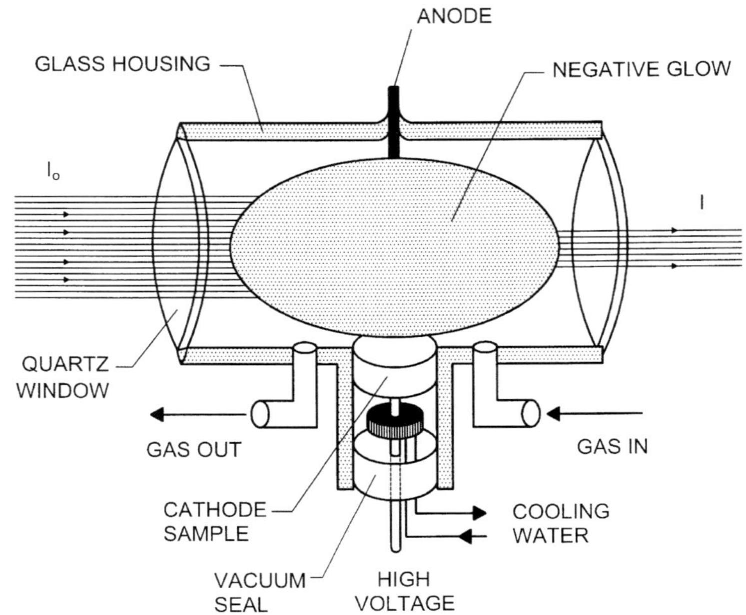

Fig. 8.17. Sputter atomization cell for GD–AAS. (Reproduced with permission of the American Chemical Society.)

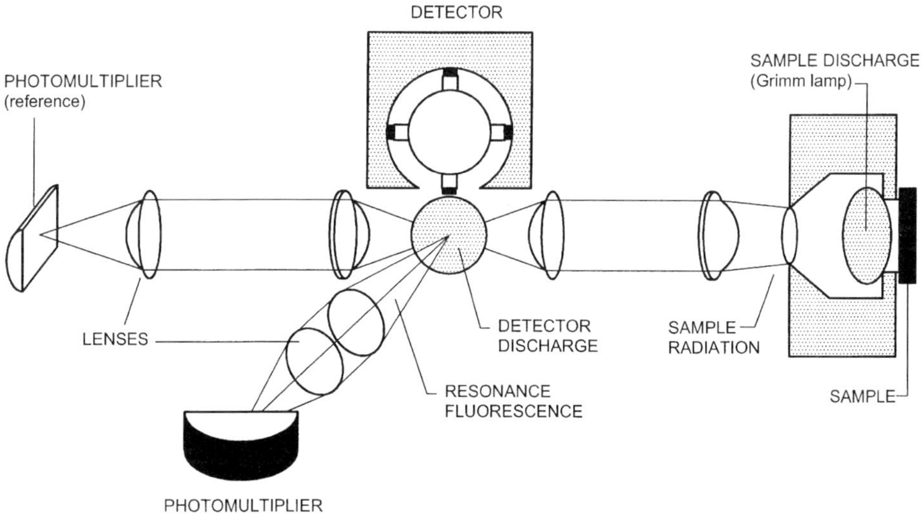

Fig. 8.18. Resonance detection instrument for GD–AFS. (Reproduced with permission of the American Chemical Society.)

Miscellaneous couplings

Furnace atomization non-thermal excitation Conventional uses of GD take advantage of the sputter atomization characteristics to convert a solid to vapour. The combination of an alternative atomization source, with the GD used as an excitation or ionization source, has drawn attention to furnace atomization non-thermal excitation spectrometry (FANES). Figure 8.19 shows a typical FANES source in which a graphite cylinder is used both for thermal atomization of solution residues and as the hollow cathode of a GD for excitation of the sample. A high current is applied to the graphite cylinder that produces a pulse of sample atoms and a subsequent pulsed emission signal as the atoms interact with the GD. The similarities to electrothermal atomization AA are obvious. With liquid samples, small volumes (5–50 μl) of sample are pipetted onto the cathode, dried and ashed at atmospheric pressure while the system is flushed with argon. Typical discharge conditions are 1–5 torr, 20–30 mA discharge current and up to 500 A atomization current, which produce temperatures near 3000°C. The transient emission signal persists for 0.3–1.0 s and limits of detection range from 0.04 pg for Ag to 800 pg for Se [206]. The complementary technique FINES (furnace ionization non-thermal excitation spectrometry) measures ionic emission and has also been used in combination with MS.

Laser-assisted GD Laser ablation has been used prior to glow-discharge for the efficient direct elemental analysis of ceramics by AES [207]. Effects of laser irradiation such as sensitivity enhancement and alleviation of matrix effects have been found, as shown in Chapter 9. The use of a diode laser after glow-discharge has also been proposed for isotopically selective excitation of gaseous uranium atoms produced by cathodic sputtering. Optogalvanic detection of the discharge atom population allowed identification of ^{235}U at

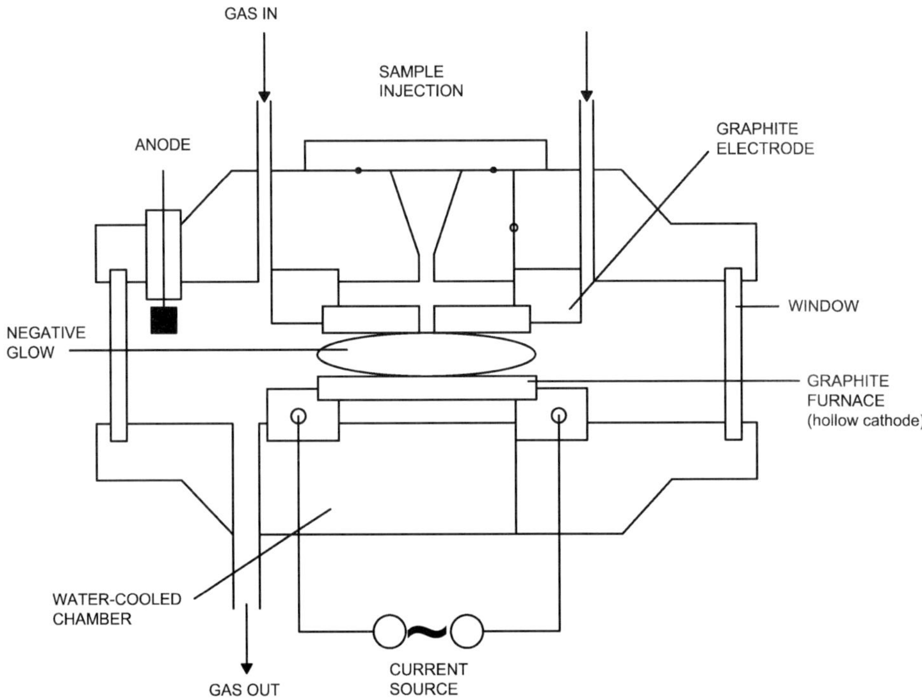

Fig. 8.19. Glow-discharge cell used as a furnace atomization non-thermal excitation source with electrothermal atomization. (Reproduced with permission of the American Chemical Society.)

depleted, natural and enriched abundances in uranium metal and uranium oxide samples based on the spectral signatures of two transitions [208]. When the detector used after laser-assisted GD with a tunable laser is a mass spectrometer, the sensitivity increases as a result of the unknown analyte being completely ionized, and so does the selectivity by effect of the laser being tuned at a frequency where only the analyte is ionized [209].

Microwave-induced plasma sources Microwave-induced plasma (MIP) has also been used in combination with GD. The glow-discharge source serves to provide free sample atoms produced by cathode sputtering and partial ionization, and the MIP source serves to enhance the ionization of the free atoms. Because the MIP is normally even farther from local thermal dynamic equilibrium than other plasma types, it seems to be more suitable for combination with a source having better atomization capabilities such as GD. The MIP–GD combined atomization source is usually coupled to a time-of-flight mass spectrometer. Despite the complexity of the experimental set-up, the enhancement factors achieved are of only 3–4 [210]. Use of a time-of-flight interface between glow-discharge sampling and mass spectrometry has, so far, rarely been successful [211,212].

X-ray tube-assisted GD A Grimm-type GD lamp was modified by incorporating a 15 mm thick PTFE support to insulate the cathode support from the anode body. With this modified-tube stable high-voltage discharge, plasma was created in the hollow-anode near the cathode at He gas pressures of 100–100 Pa and discharge currents of 0.1–10 mA. Electrons emitted from the cathode surface were accelerated in the cathode fall region and used to generate X-rays from suitable targets in either the end-window or side-window transmission mode. Experiments under different working conditions and with different samples led to the conclusion that the Grimm–GD X-ray tube may be useful as a portable X-ray tube [213].

8.3.8. Advantages and restrictions of glow-discharge sampling

Glow-discharge sampling provides a number of advantages that can be summarized as follows:

(a) It affords direct analysis of both conducting and non-conducting samples (the latter requires using an rf source or mixing with a conductive matrix if a dc source is employed);
(b) It is responsive to both metallic and non-metallic samples;
(c) Matrix effects are low and can be minimized using a pulsed source and gated monitoring;
(d) Limits of detection in the parts-per-billion range or even lower are typically obtained that can be improved with the aid of auxiliary means (e.g. laser or MIP assistance);
(e) Sensitivity is relatively uniform among elements;
(f) Reliable isotopic information can be easily obtained;
(g) Spectra are much simpler than their optical emission counterparts;
(h) The discharge is stable and has low power requirements;
(i) Glow-discharge sources operate in inert environments and are thus free from atmospheric contaminants;
(j) Both purchase and operational costs are moderate;
(k) Sampling permits rapid qualitative scanning of the periodic table;
(l) The stability of the source affords precise quantitative analyses.

On the other hand, glow-discharge sampling is subject to the following shortcomings:

(a) A differential pumping system must be used for gas loading;
(b) The discharge gas and gas-matrix molecules cause spectral interferences;
(c) Applicability is primarily restricted to solids — liquid and gaseous samples can also be processed, however, following easy preparation steps.

8.3.9. Models and calibration in methods involving glow-discharge sampling

Models

Hyperbolic, parabolic and exponential equations were developed at an early stage for evaluation in emission spectroscopy with GD lamps. Because of analogies between emission and X-ray spectrometry, the hyperbolic equation proved the most suitable for the types of curve obtained [214]. Two-dimensional mathematical models for describing electrons, Ar ions and fast Ar atoms [215], and a subsequent model for Ar metastable atoms, Cu atoms and Cu ions [216], both involving dcGD and Ar, provided an overall picture of the GD. The number of densities of different species, the relative contributions of different production and loss processes, the thermalization profile of sputtered Cu, the degree of ionization of this metal, different ionization mechanisms for Cu atoms, the Cu ion-to-Ar ion density ratio and the roles of Ar ions, and of fast atoms and Cu ions, in sputtering were thus discussed. The subsequent development of mathematical models including 2D and 3D Monte Carlo simulation models showed the 2D models to provide slightly better results, but the 3D models to provide additional details such as cathode crater profiles [217]. More recent mathematical models for dcGD used as ion sources in MS have been reported [218] and a 3D modelling network comprising seven different, previously reported models has been constructed. The electrical current as a function of pressure and voltage, the potential distribution throughout the discharge, density profiles and energies of the various plasma species, collision processes in the plasma, and cathode sputtering under the discharge conditions typically used in GD–AES have thus been modelled. The influence of pressure, voltage and current on the calculated results provided by the model were generally highly consistent with their experimental counterparts [219]. Additional contributions in this field include mathematical simulations and comparison with experiments involving Ar and Cu optical emission spectra [220]; the hybrid Monte Carlo fluid model for a microsecond pulse GD [221]; and the 3D modelling network for Ar GD operation in both the dc and rf modes [222]. Interested readers are referred to the original references for further details.

The influence of cathode geometry in a dcGD ion source for MS has also been modelled [223]. Flow charts of programmes for constructing databases of tables and of the programme to find emission lines under various search conditions have been provided by Wagatsuma [224].

The equation derived by Boumans in 1972 for the sputtering rate in a Grimm-type hollow-cathode lamp [225] has been revisited and the threshold voltage constant shown to be the combination of a minimum voltage for current flow and ionization, and the minimum voltage for sputtering [226]. Owing to energy losses that occur before the ions reach the surface, the minimum voltage for sputtering is roughly twice the sputtering threshold voltage. Based on this finding, a revised equation has been proposed and supported with experimental results [227].

Calibration

Calibration in glow-discharge sampling methods often involves the use of CRMs or a method of standard additions of well-known powders or liquid residues. Massive solid samples and pressed pellets behave differently when analysed using a glow-discharge. Therefore, solid CRMs usually cannot be used directly to calibrate analysis pellets. Accordingly, calibration standards should be prepared along the same processing route as the pellets containing the material to be analysed. For instance, for the GD analysis of an Antarctic sediment, four calibration standards were prepared from seven different high-purity powdered oxides [228]. The oxides were mixed in various concentrations, added at a mass ratio of 1:5 to copper powder and pressed into pellets to obtain the calibration standards. One other example is the analysis of non-conducting materials in a copper matrix, for which a set of calibration standards was prepared using a river sediment CRM, a bauxite CRM and several mixtures of high-purity Fe_2O_3 [229]. The highest sensitivity was obtained for discs with only a 5% non-conductor content; also, the results for the two CRMs revealed that the accuracy achieved was adequate for quantitative analysis. A set of new ZnAl reference materials was recently prepared with a view to supporting the depth profile analysis of zinc-based coatings by GD–AES. The reference materials (RMs) were subjected to a two-step certification procedure: in the first, they were analysed for composition using methods other than GD–AES (mostly wet chemical methods); in the second, sputter factors for the RMs were determined from GD–AES measurements. To this end, the standard model of matrix-independent emission yields was used to relate the GD–AES intensities of selected elements to the compositions of the RMs, found in the first step of the certification procedure. Based on the results, the model of matrix-independent emission yields was violated by the matrices studied [230].

Calibration standards for composition–depth profiles of non-stoichiometric titanium nitride coatings have been prepared by deposition of steel substrates using reactive magnetron sputtering at different N_2 partial pressures. The coatings were analysed by GD–AES, using a pure titanium probe and a stoichiometric TiN coating as standards. The coatings were found to contain 20–50% atomic nitrogen. They were removed from the substrate, digested with boiling H_2SO_4 and analysed for N and Ti using the Kjeldahl method and AAS, respectively. A comparison of the three measurements showed the AAS and Kjeldahl results to correlate well but to deviate systematically from the AES results. The deviation was ascribed to the use of poor calibration standards for the AES measurements, so the Kjeldahl method was proposed as a suitable calibration tool for AES standards [231]. The electrodeposition of Cu, Zn, Ni and Pb on a compressed Ag cathode for use as standard materials in GD–MS has also been investigated as the direct addition of aqueous standards to powdered Ag has proved unsuitable for standard preparation. The relative ion yields differed from those obtained from bulk metal powders, thus adversely affecting their use as standards [232].

Dynamic range extension in GD quadrupole/ion-trap MS based on selective ion-accumulation (e.g. by mass-selective instability, single-frequency resonance ejection, combined rf–dc and entrance end-cap dc methods) allows the selective accumulation of the analyte ions and enables the dynamic range to be increased by a factor of 105 [233]. The linearities and relative trapping efficiencies of the previous methods were assessed with respect to the injection time and the methods were used for the GD ion-trap MS determination of major and minor constituents in NIST SRM 1103 Free Cutting Brass.

Marcus *et al.* [234] used a glow-discharge source with direct injection of particles for special samples such as airborne particulates [234]. This novel concept was tested on NIST SRM 1648 Urban Particulate Matter (a composite of airborne particles collected in St Louis, Missouri) as well as on ground caffeine powder and various pure organic substances.

8.3.10. Applications of glow-discharge sampling in combination with spectrometries

Surface and depth profile analysis

This is one common application of GD sampling that uses both dc and rf sources for purposes including quantitative analysis comparisons [235,236]. A third type of source, the pulsed glow-discharge source, provides some potential analytical advantages over a steady-state dc source including enhanced control of the sputtering rate, increased emission intensities and possible temporal factors. Because the sputtering process can be altered by controlling the pulse frequency and width, extension of the pulsed technique to depth profiling can allow more flexibility in thin layer analysis. Layers of a few atomic distances can be removed over an extended time scale by using a longer duty cycle that creates a slower sputtering process.

Qualitative depth profiling depends on the sample material being removed uniformly, layer by layer, across the area exposed to sputtering by argon ions and fast atoms. The formation of crater curvature by non-uniform erosion tends to degrade measured time–intensity data, thereby decreasing the accuracy of quantitative measurements of film thickness. Because crater shape is a function of the operating parameters, abnormal shapes are often the result of poorly chosen parameters. It is therefore important to use optimal experimental conditions to sharpen the time–intensity profiles and enhance depth resolution. The influence of operating parameters on crater shape in dcGD [237–240], rfGD [241] and microsecond pulsed Grimm GD [242,243] has been examined and the last found to depart markedly from the characteristics of a dcGD and hence to result in more flexible operating conditions.

From the above it follows that, before sputtering rates can be determined, a relationship between the glow-discharge conditions and the rate at which the material is released from the surface must be determined. With a microsecond pulsed GD, high operating pulse voltages and pressures yield an increased sputtering rate; the opposite holds with low voltages and pressures. Moreover, concave crater shapes — which have an inherent dip in the middle of the profile — result from sputtering at low pulse voltages (e.g. lower than 1000 V). Conversely, convex crater shapes — which have a bulge in the middle of the profile — result from the use of high voltages (e.g. higher than 1500 V). One possible explanation for this trend is that crater shape results from the electric field focusing or defocusing the bombarding ions responsible for sputtering. Flat crater bottoms and good depth resolution are only obtained within a limited range of voltages and gas pressures. Pulse widths and pulse frequencies have little influence on crater shape provided the pulse voltage and gas pressure conditions remain constant. The principal advantage of these two pulse parameters is their ability to increase or decrease the sputtering rate without altering depth resolution. By reducing the duty cycle, less material can be removed per unit time. In this way, a wide range of layers from several nanometres to tens of microns can be analysed. One other parameter that warrants examination is the time allowed for the sample to sputter. In theory, time should have no effect on the sputtering rate for pure samples since sputtering will simply penetrate deeper into a sample with longer analysis times. However, if the increasing depth affects the net diffusion of atoms from the surface, then calculated sputter rate may be affected.

Sputtering rates and penetration rates In order not to elaborate excessively on these concepts, this section deals only with the theoretical aspects in relation to pulsed GD sources. Interested readers are referred to the original references for more details.

The sample sputtering rate, q (ng/pulse), occasionally referred to as the "net sputtering rate" — which recognizes the fact that the glow-discharge process involves redeposition effects — can be calculated from the following expression:

$$q = \frac{\Delta W}{tf} = \frac{\Delta W}{p_{\mathrm{N}}} \tag{8.1}$$

where ΔW is the amount of sample lost (mg) due to net sputtering, t the sputter time (s) and f the pulse frequency (Hz). If the sputtering time is measured, then the first part of eq. (1) is used and the number of pulses must be calculated. However, a pulse counter can be used to directly monitor the number of pulses, p_{N}, from the power supply, thus making the second part of eq. (1) applicable. Moreover, using a pulse counter is bound to yield more accurate results and precise sputtering rates by virtue of the constraints on sputter time measurements — some fractional time in each pulse is required for the plasma to stabilize, which makes it difficult to measure the "true" sputter time.

Obtaining a quantitative depth profile entails determining concentration as a function of depth. This can be accomplished by converting the sputtering rate of the sample to its penetration rate (w), which is the amount of sample layer eroded per unit time and is independent of the sample density. The penetration rate has units of distance per time (μm/s) for dc sources; however, the units are changed to distance per pulse (nm/pulse) to account for the use of a pulsed glow-discharge source. The penetration rate can be determined quantitatively from the following expression:

$$w = q/A\rho \tag{8.2}$$

where ρ is the density (g/cm^3) of the target material and A the sputtered area (πr^2, r being the radius of the Grimm anode, which is typically 2 mm). As can be seen from eq. (2), the penetration rate depends directly on the density of the material being analysed. Such a dependence poses no problem in determining the penetration rate for a pure element. However, the determination of the sputtering depth for a multi-layered sample meets with some difficulties: as one sample layer is sputtered away and the next is exposed, the interface involves a change in density of the pure element comprising the first layer to that of a second layer, along with changes in the sputtering and concentration of the two elements. If the transition from one layer to the next were instantaneous, this interference problem would not exist. However, sample material from both layers is removed at this interface, so the sample acquires multi-element characteristics rather than acting as a series of separate pure layers. This problem must thus be taken into account in using a quantitation method.

Calibration curves Depth profiling entails running calibration curves. A variety of bulk reference standards are used for this purpose as the analyses require wide concentration ranges in different matrices. For example, when analysing a layered sample, the element of interest must be calibrated over the whole concentration range (0–100%) because of the concentration range created during the transition. However, sputtering rates are unique due to matrix differences and produce a different curve for each family of materials. To correct for this, the sputtering rate for each standard can be used to offset inherent differences and allow the use of a single calibration curve. The variety of samples to be used for calibration should have different sputtering rates. By way of example, a procedure based on the SIMR (Swedish Institute for Metal Research) calibration model exists in which intensity correction relies on determining the sputtering rates for an iron alloy and using that for pure iron (which can be directly determined from weight loss measurements) for intensity correction. Provided the calibration of each element is normalized to the same reference sputtering rate, the reference sample (i.e. pure iron or an iron alloy) will not create a discrepancy. Intensity correction can be done by multiplying the measured intensities of each calibration sample by a factor equal to the ratio of the sputtering rate of the reference material to that of the sample. In mathematical form, this is expressed as

$$I_{\mathrm{norm}} = I\,(q_{\mathrm{ref}}/q_{\mathrm{s}}) \tag{8.3}$$

where I_{norm} is the normalized intensity, I is the "raw" intensity, and q_{ref} and q_{s} are the sputtering rates for the reference material and sample, respectively. Figure 8.20A is the non-linear copper calibration curve from which the sputtering rate-corrected calibration plot of Fig. 8.20B was obtained using the results provided by eq. (3). Use of this corrected calibration curve affords better estimations of unknown copper concentrations based on intensity measurements. One salient advantage of this calibration procedure is that bulk standard material can be used instead of expensive, less readily available layered samples. Also, the analytical procedure is simpler as it only requires one reference sputtering rate to be determined — rather than multiple rates for the reference material. The information from a linear calibration plot can be used in the subsequent quantitation procedure.

Quantitation in depth profiling The data typically obtained in depth analyses usually represent time–intensity profiles. However, the information desired is in the form of depth–concentration profiles, which allow more detailed information to be extracted. Thus, for a given sample, the concentration of a thin layer in the sample can be monitored at variable sample depths. The time-to-depth conversion process requires knowledge of the sputtering rate for each element and calibration of the instrument with a series of standard samples. This qualitative to quantitative conversion meets the requirements of the true depth profile better. Two major quantitation methods exist in this context, namely: the method of the Institute de la Recherche Siderurgie (IRSID) and that of the Swedish Institute for Metal Research (SIMR). The former, which was introduced by Pons-Courbeau in France [244,245], uses the emission yield concept as a basis for quantitation of depth profiles. It relies on the assumption that the integrated intensity from one element is proportional to the sputtered mass of that element, which implies that the emission yield (defined as the amount of emitted light per unit sputtered mass of an element) is independent of the sample matrix provided the excitation conditions remain at least nearly constant [9]. This assumption is widely accepted to hold — at least roughly. The SIMR

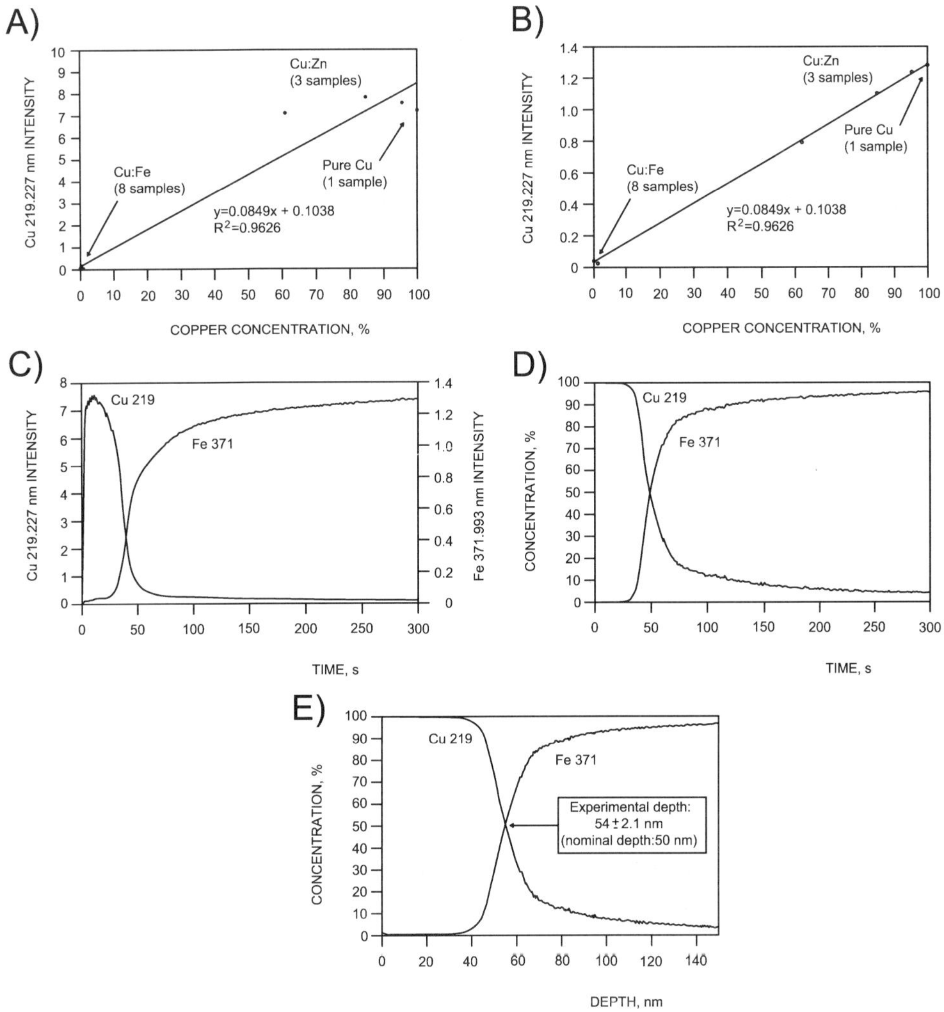

Fig. 8.20. Quantitative depth analysis. (A) Plot of uncorrected calibrated data for various copper samples (Cu 219.227 nm emission) in different matrices. (B) Plot of sputtering rate-corrected copper calibration data of (A). (C) Time–intensity plot for a 50 nm layer of copper deposited on a steel substrate (intensities were measured using the Cu 219.227 nm and Fe 381.993 nm emission lines). (D) Normalized concentration–time plot from (C) for a 50 nm copper layer on a steel base. (E) Concentration–depth plot from (D) for a 50 nm copper layer on a steel base. (Reproduced with permission of the Royal Society of Chemistry.)

method is not directly based on the emission yield concept. Rather, it relies on the proportionality between the sputtering rate and intensity. The essential difference from the IRSID method is that, in the SIMR method, the concentration rather than the sputtered mass of each element is used in the calibration function; this makes the SIMR method very similar to standard techniques for bulk analysis. Measurements are based on calibration at predetermined levels of voltage and current that are referred to as the "reference excitation conditions". One of the main goals of this method is to establish a calibration procedure resembling a standard bulk analytical method. Both quantitation methods are implemented with a steady-state dcGD. Pulsed GD introduces two additional parameters, viz. pulse width and pulse frequency, which provide further control over the sputtering process [243].

Quantitation principle The first step in the quantitation method, once the sputtering rate has been determined and the instrument calibrated, involves collecting a raw time-intensity plot for a 50 nm layer of copper deposited on a steel substrate. The plot of Fig. 8.20C shows the variation of the intensity of the Cu(II) 219.227 nm and Fe 371.993 nm atom lines. The data in this plot only show intensity as a function of time, so they allow no qualitative information such as depth to be extracted. The next step involves determining the concentrations of the selected elements. This is done by converting the intensity data of Fig. 8.20B into concentrations, using the sputtering rate-corrected calibration curve for bulk standards shown in Fig. 8.20A. These concentrations are normalized to 100%. The results of the conversion are shown in Fig. 8.20D. The third step of the method involves determining the composite sputtering rate and composite density as a function of time from the normalized concentrations obtained in the previous step. The composite sputtering rate is simply the summation of the percent contribution of each element to the sputtering rate of the reference material over all the elements in the sample. Similarly, the composite density is the summation of the percent contribution of each element to the density of the reference material over all the elements in the sample.

After the composite sputtering rates and densities have been calculated, they can be used in conjunction with the previously determined sputter area and sputter time to obtain the penetration depth (d). The sputtered depth increment, Δd, at a given point in time can be obtained by multiplying the penetration rate [eq. (2)] at such a point by the number of pulses per sampling point:

$$\Delta d_t = w_t p_{\mathrm{n}} \tag{8.4}$$

where Δd_t is the sputtered depth increment (nm), w_t the penetration rate (nm/pulse) at that point and p_{n} the number of pulses per sampling point. The overall penetration depth (d) can be obtained by summing all sputtered depth increments [eq. (4)] for each sampling point throughout the sampling range:

$$d = \sum_{t=0}^{n} \Delta d_t \tag{8.5}$$

Based on eqs (4) and (5), the time axis (*s*) of Fig. 8.20D can be converted to depth (nm) for the entire sampling depth (150 nm). Therefore, the final form of the data now reveals concentrations as a function of depth. Figure 8.20E shows the quantified result of the Cu layer on a steel substrate shown previously in Fig. 8.20C as a time–intensity plot. The experimental results obtained using the quantitative method predict a depth of 54 ± 2.1 nm (1 s at the 95% confidence interval), consistent with the nominal depth of Cu reported for the sample. In addition, this plot provides more valuable information than do the time–intensity plots as it allows the concentration of the multiple layers to be monitored as the sample is being sputtered.

Applications of GD–MS

Glow-discharge coupled to mass spectrometry has been widely used for the analysis of metal alloys and high-purity materials. This is possibly the only GD–detector coupling that combines the required sub-parts-per-billion limits of detection with the breadth of coverage required for survey analysis. Glow-discharge mass spectra are much simpler than line-rich optical spectra, which makes qualitative analyses rapid and straightforward. Elemental ratio intensities are used for quantitation, usually with the matrix element acting as an internal standard. Polyatomic mass spectral interferences resulting from chemical combination of the discharge gas and the sample matrix (e.g. $FeAr^+$) can create problems at low concentration levels. Given the capabilities of modern detection systems, however, GD–MS ion intensities are linear over a wide range of concentrations, from major components to trace constituents.

The advantages of GD–MS have turned it into a routine technique that provides reliable data on a daily basis in commercial laboratories. Quantitative analysis of such difficult elements as B, C, Si, and S in steels yields good results. The surface-active nature of GD ablation also enables the analysis of thin films and coatings. Although re-deposition phenomena somehow limit the sharpness of layer resolution, GD–MS can be effectively used for such samples. The routine uses of GD–MS are rarely reported in the literature, which has led to an understatement of its actual usefulness. In any case, the maturity of this technique is quite apparent from examples such as the use of rfGD as interface with a 3D quadrupole ion-trap MS/MS instrument with efficiencies approaching 100% for oxides and hydroxides [246]; the low mass resolution for in-depth analysis of technical surface layers [237]; the time-resolved experiments involving ionized sputtered atoms in pulsed rf-powered GD–MS [247]; the design and characterization of an rfGD for double-focusing mass spectrometers providing representative spectra for glass, soil and brass matrices [248]; the development of new software for data processing of mass spectra acquired by high-resolution double-focusing GD–MS [249]; and the adaptation of a GD–MS system to a glove-box for the analysis of nuclear materials [250].

A comparison of atomization and ionization processes in dc, rf and microsecond pulsed discharges revealed the greatest mass loss to occur with the last type of source (two orders of magnitude greater than with the other two) [251].

Quantitative determinations have recently focused on high-purity materials [252]; trace noble metals (Pd, Pt and Th) in automobile catalysts [253]; and non-conducting [254,255] and semi-conducting materials [256]. The determination of ^{40}Ca in the presence of ^{40}Ar

as an illustration of the usefulness of time-gated detection [257]; the characterization of polymers [258]; and the use of solid-phase microextraction as a sample insertion technique for the analysis of organometallic compounds [259] are additional salient applications. Finally, surface analysis continues to be a common use of GD–MS [260].

Other applications of the GD–MS technique worth mentioning here include the monitoring of depth distribution of B and Li in ZrO_2 layers, where Zr alloy was corroded to ZrO_2 in an autoclave in solutions containing Li and B, samples being collected at different reaction times for analysis; the GD–MS crater depth and sputtering rate were determined profilometrically under the discharge conditions used [261]. The mechanisms of corrosion due to diffusion of impurities were also studied [262]. A comparison of dcGD–MS and ICP–MS for the analysis of trace elements in nuclear samples [263]; the development and evaluation of a GD–MIP source for TOF–MS [264] and a study of practical and theoretical aspects of GD–MS interfaces [265] are also worth mentioning here.

Recent contributions have focused on coupling to gas chromatography. Such is the case with the dc gas sampling GD ionization source for the mass spectrometric analysis of halogenated hydrocarbons [266], where the continuous discharge is confined within the first vacuum stage of the differentially pumped spectrometer interface and can be either operated statically or rapidly switched between atomic and molecular ionization modes; both atomic and molecular spectra can be produced from a helium-supported plasma. Atomic and molecular limits of detection obtained with boxcar averager data collection were found to be comparable for static and switched operation (1–30 and 7–140 pg/s, respectively). The GC–rfGD–MS coupling has also been used in the analysis of organotin compounds [267].

Applications of GD–AES

Commercial GD–AES instruments designed for trace elemental analysis in solids use the planar cathode discharge [268]. Slightly different versions for correcting some deficiencies involving the use of a floating restrictor to decrease the curvature in the burning profile, Fourier transform (FT) for photon noise decrease, a boosting microwave discharge or a magnetic field are described in Section 8.3.3. Of special interest in this respect is the use for depth profiling of layered metals and the analysis of geological materials such as granite, basalt, mica and feldspar without the thermal effects and self-absorption apparent in arc-spark operation. For non-conducting samples such as airborne dust, an rfGD for the atomic emission determination of trace levels down to 2 μg/g of a number of elements in coal fly ash has been used [269]. One recent determinative application of Grimm-type GD–AES is the analysis for gold in precious alloys [270].

The hollow-cathode discharge, which relies on a cavity cathode rather than on the planar cathode of the Grimm source, is less frequently used. The net effect of this difference at a given discharge voltage is concentration of the plasma within the constraining cavity, which enhances elemental emission intensities. The performance of these lamps can be boosted with a view to lowering limits of detection and reducing interferences [271,272].

Most recent applications of GD–AES are concerned with surface analysis — mainly depth profiling — of anodic alumina films [273], painted coatings [274], titanium carbide and nitride coatings [275] and electroplated ZnNi [276]. Additional applications include studies on the influence of amplitude modulation in depth profiling thin surface layers, the use of microsecond pulsed GD in quantitative depth analysis [278] and the examination of the influence of surface roughness on depth resolution [279].

Cho *et al.* [280] compared dc and rf gas jet-boosted GD–AES for the analysis of steel in terms of the voltage–current relationship; they studied the effects of the gas flow-rate (0–800 ml/min) and pressure (3–5 torr) on the dc bias potential, sample weight loss and emission intensity. The use of a simultaneous spectrometer for the rf mode proved dispensable by virtue of the high stability (variations less than 0.3%) for both matrix and trace elements. Both the rf and the dc mode provided calibration graphs that were linear over two or three decades (or even more if the analytical signal was normalized to the signal of a matrix component); however, the limits of detection obtained in the rf mode for many trace elements in steel were at the level of tens of ppb, which was an order of magnitude better than in the dc mode.

GD–AES was found to surpass the energy dispersive X-ray (EDX) linescan technique in depth profiling capabilities; the former exhibited better turnaround time and spatial resolution, and required less sample preparation [281].

An additional, unusual application of rfGD–AES is the direct chemical analysis of bone [282].

GD–AES has also been used for GC detection in mercury speciation analyses [283]. Also, an atmospheric pressure GD atomic emission source was designed for direct sampling in liquid media; however, the preliminary limits of detection for elements such as Na, Fe and Pb were in the range 11–14 ppm (*ca.* 60 ng) for 5 μl sample volumes [284]. Specific applications to liquid and gaseous samples, and couplings to chromatographs for wet speciation analysis can be found in recent reviews such as that by Baude *et al.* [285].

Applications of GD–AAS

Various types of GD have proved highly suitable for use as atom reservoirs for AAS; however, this solid sampling approach has been less frequently used in atomic absorption than in atomic emission and mass spectrometry, possibly as a result of the wider commercial availability of electrothermal devices.

Studies of the effect of the discharge conditions [286] and those of the anode dimensions and location of the gas inlet port on the spatial distribution of the sputtered atoms [287] (both in an rfGD atomizer), and of the role of discharge parameters and the effect of limiting the orifice diameter [288], have been carried out. Additional applications of the GD–AAS couple include the evaluation of pulsed and transient atomization modes [289]; the use of a jet-enhanced sputtering device [290,291] and that of a thin-walled metallic hollow cathode [292]; the development of a new, two-component atomizer for the direct determination of medium and volatile elements in high-purity solid refractory metals [293]; and the use of an experimental design to optimize the analysis of non-conducting materials [294].

The glow-discharge also facilitates atomic absorption from non-ground state levels, which is often required in diode laser AAS.

Applications of GD–AFS

The glow-discharge is also highly suitable for use as an atom reservoir in fluorescence spectrometry as quenching of species through collisions is considerably reduced, which increases the populations of excited states in fluorescence work. Although conventional excitation sources such as an ordinary Xe arc lamp [295] — which performs best in the pulsed mode [296] — have also been used in this context, laser sources are far more commonplace here. For example, a miniature glow-discharge source assisted by laser in the excitation step was used by Winefordner *et al.* [205] for the determination of Pb with a limit of detection of 0.6 pg, and for those of europium, yttrium and thulium at the femtogram level [297]. Pulsed GDs have proved the most suitable for coupling to AF detectors [298,299] in most situations; however, a commercially available hollow-cathode lamp has also been used to approach single atom detection [300]. The GD–AFS coupling has also been used for in-depth resolution of microelectronic materials [301]. In any case, applications are still scant, especially if the high resolution and low limits of detection achieved so far are considered.

8.3.11. Trends in glow-discharge sampling

Compared to other plasma sources such as ICP, the glow-discharge has less stringent technical and financial demands as regards source construction and power supply. Most glow-discharges reported so far are operated at low pressures, so vacuum pumping is necessary — which is an obvious hindrance to the development of more affordable instrumentation. Two very recent developments in this context are the atmospheric electrolyte-as-cathode GD and the capillary GD, which overcome this shortcoming with liquid and gaseous samples, respectively.

In the recently developed atmospheric electrolyte-as-cathode GD sources, a liquid serves as the cathode for the discharge, which facilitates direct analysis of continuously flowing liquids [302]. The device is potentially applicable to the continuous analysis for metals in water and wastewater.

Using a different approach, Eijkel *et al.* [303] employed scaling theory to reduce electrode geometries by more than two orders of magnitude in comparison to the normal linear dimensions of a GD in order to generate a continuous glow-discharge in a capillary on a chip. They demonstrated the application as an emission detector for volatile organic compounds at either reduced or atmospheric pressures [304]. Although GD dimensions are extremely small, the microchip plasma can be successfully used for molecular emission detection; also, it exhibits limits of detection as low as 10^{-14} g/s for methane and calibration curves that are linear over two orders of magnitude. Simple instrumentation, a small detector size and a high sensitivity make the device highly suitable for integration into microanalysis systems for volatile compounds, which could lead to the development of either new GD detectors or portable spectrometers for screening and field sampling. In relation to this application, GDs may become competitive with other plasma sources operated on a chip such as the microwave discharge developed by Engel *et al.* [305].

One other potential analytical use of GDs is as tunable plasma sources. Tunable sources can be operated in two modes to detect either atoms or molecules of (mainly) organic compounds, depending on the conditions chosen for source operation. Details of various sources for this purpose are given in a special issue of the *Journal of Analytical Atomic Spectrometry* [306], which shows that they can be used for both emission and mass spectrometry. Normally, GDs are operated at a moderate or high power to ensure complete dissociation of molecules. When operated at a low power, the gas temperature is so low that soft ionization without molecular bond breakdown is possible. In emission spectroscopy, GDs can serve as detectors for elements such as C, N, O and H; in mass spectrometry, they can be used for the identification of molecules. They are thus theoretically suitable for direct (chemical) speciation analysis without the need for a prior separation. The most elegant feature of this approach is that, simply by changing the power, both operation modes can be realized. Glow-discharges can thus be used to solve analytical problems in areas of organic mass spectrometry where formerly only electron impact ionization, chemical ionization, fast atom bombardment, thermospray or electrospray ionization (ESI) could be used for soft ionization. Here, the borderline between inorganic and organic analysis is becoming increasingly blurred. Many innovative organic mass spectrometers equipped with ESI still have an additional atmospheric pressure ionization mode; they use a discharge that is highly similar to GD for enhanced detection of difficult to ionize organic compounds but they are mostly not operated in an atomic mode.

8.4. OTHER SOLID SAMPLING APPROACHES

A number of other devices are used for solid sampling, albeit to a much lesser extent than those discussed in the preceding sections. The most salient among them are described below.

8.4.1. Arc nebulization

Arc and spark sources for solid sampling are among the oldest described in the spectrometric literature. A relatively modern system of this type is the arc nebulization device developed by Jiang and Houk in 1986 [307] for the direct elemental and isotopic analysis of solid conducting materials (Fig. 8.21A). The sample was used as the cathode in an intermittent arc generated between a Cu anode and the sample cathode. The eroded sample material was transported through a 2 m long × 3 mm i.d. plastic tube into the torch of an ICP–MS instrument by an argon carrier gas flow. For the steel samples studied, the sample material was removed at a rate of *ca.* 1 mg/min. Standardization was done with steel SMRs the elemental composition of which was certified by NBS. Background spectra similar to those of LA–ICP–MS (see Section 9.2) were obtained that enabled the determination of non-metallic impurities. Limits of detection in the lower μg/g range were achieved for Al, Si, Ti, V, Cr, Co, Ni, As, Zr, Nb, Mo, Sn, Ta and W. The relative precision and accuracy were found to be *ca.* 5%. In a subsequent study, the arc nebulization system was used for the analysis of non-conducting ores and sediments [308], which were powdered, mixed with graphite and pressed into conducting pellets. Only small peaks for ArC^+, ArO^+ and Ar_2^+ were detected in the background spectrum above

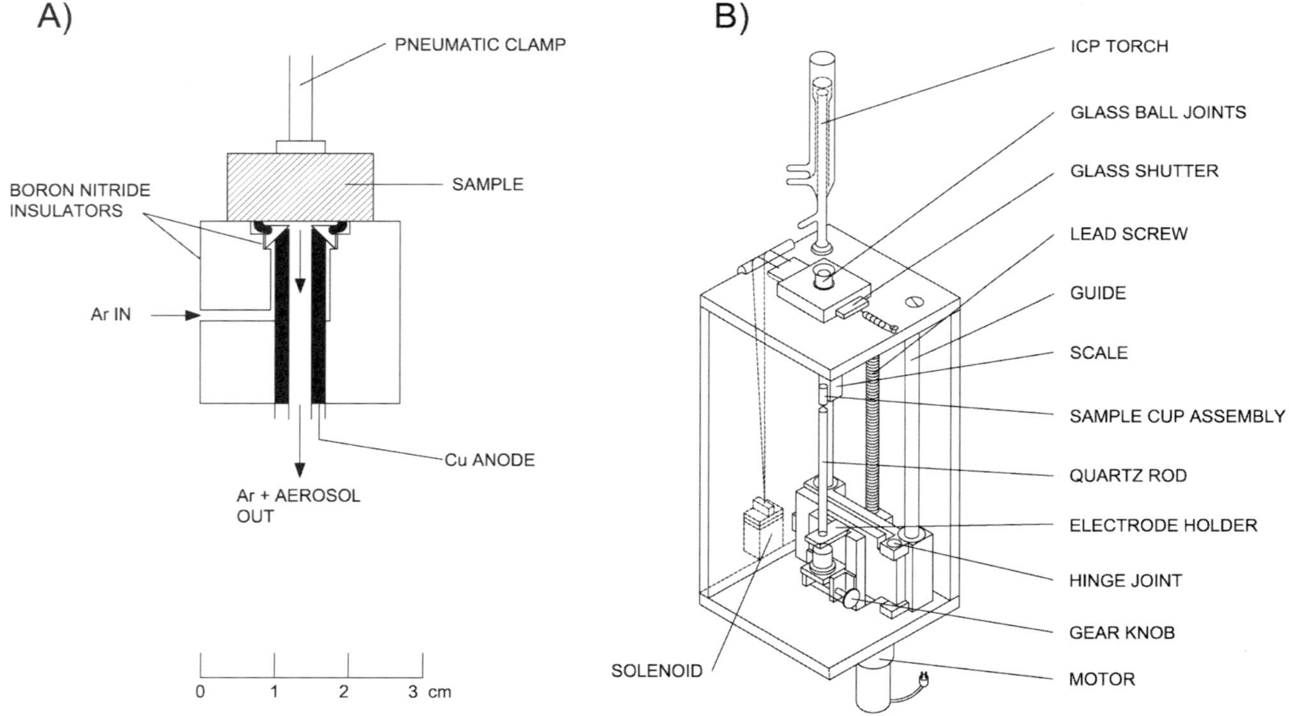

Fig. 8.21. (A) Scale diagram of an arc nebulizer. (B) Three-dimensional view of the DSI device of Horlick *et al.* (Reproduced with permission of the Royal Society of Chemistry.)

$m/z = 45$. The small O_2^+ peak observed at mass 32 enabled the determination of sulphur. The limits of detection for the above-mentioned elements were 30–50 μg/g. Isotope ratios were determined with a precision of *ca.* 1% as RSD for the major constituents.

8.4.2. Direct sampling insertion devices

At present, direct sampling insertion devices (DSID) exist for all atomic techniques. The earliest device of this type was developed by Horlick *et al.* for use with ICP–AES [309,310]. The system was subsequently refined to the final design shown in Fig. 8.21B. Thus, the Fassel-type torch was modified to accommodate the sample probe, which consisted either of laboratory-made graphite, tantalum or tungsten cups or of graphite sample cups used in classical dc arc emission spectroscopy. The cups sit on top of a quartz rod that is fixed on the sample cup holder. The bottom of the torch is sealed by a glass shutter — except during sample insertion — so as to maintain the gas pressure in the central tube of the torch and keep the plasma discharge at its normal position. During insertion, the glass shutter is opened. When insertion is complete, a Teflon stop at the end of the quartz rods seals the bottom of the central tube. The sample can be dried or ashed at a position 35 mm below the load coil. When fully inserted, the top of the cup is aligned with the top of the coil. In this position, the sample cup can reach temperatures of 2000 K, but even at such levels, non-volatile analytes such as carbide-forming elements do not vaporize completely. Sample size is of the order of a few milligrams (20 μl for liquid samples). This DSID is fully automated and features an overall time for full insertion of 5 s.

This type of device can also be used with ICP–MS [311], where it provides increased sample insertion efficiency and much simpler background spectra in relation to ETV. In addition, matrix interferences can be reduced by removing the matrix at lower temperatures before the sample probe is fully inserted into the plasma [312,313]. The very small peak widths obtained during vaporization of the sample into the plasma can cause serious problems when the ICP–MS software or electronics is too slow for acquiring a sufficiently large number of data points.

A recent, improved contribution is the use of the Rh–Cu couple for in-torch vaporization in direct elemental ICP-MS analyses in solid microsamples (sample mass *ca.* 130 μg), slurries and liquid samples (sample volumes in the microlitre range) [314].

References

1 T.A.M. Ure, L.R.P. Butler, B.V. L'vov, I. Rubeska and R. Sturgeon, *Spectrochim. Acta*, 52B (1997) 1559.
2 K.W. Jackson Ed., *Electrothermal Atomization of Analytical Atomic Spectrometry*, John Wiley, New York (1999).
3 B. Welz and M. Sperling, *Atomic Absorption Spectrometry*, 3rd rev. edn, Wiley–VCH, New York (1999).
4 J. Sneddon Ed., *Advances in Atomic Spectroscopy*, JAI Press, Stanford, CT. 1997, vols 3 & 4 (1997 & 1998, respectively).
5 D.J. Butcher and J. Sneddon, *Practical Guide to Graphite Furnace Atomic Absorption Spectrometry*, Wiley-Interscience, New York (1998).

6 L. Ebdon and E.H. Evans, *An Introduction to Analytical Atomic Spectrometry*, John Wiley, New York (1998).
7 J.R. Dean, *Atomic Absorption and Plasma Spectroscopy*, 2nd rev. edn, John Wiley, New York (1999).
8 R.K. Marcus Ed., *Glow Discharge Spectroscopies*, Plenum Press, New York (1993).
9 R. Payling, D.G. Jones and A. Bengtson Eds., *Glow Discharge Optical Emission Spectrometry*, John Wiley & Sons, New York (1997).
10 R.E. March and J.F.L. Todd, *Practical Aspects of Ion Trap Mass Spectrometry. Volume III: Chemical, Environmental and Biomedical Applications*, Springer-Verlag, Berlin (1995).
11 K. Tanaka Ed., *Glow-Discharge Hydrogenated Amorphous Silicon (Advances in Solid State Technology)*, Kluwer Academic Publishers (1990).
12 K.W. Jackson, *Anal. Chem.*, 72 (2000) 159 R.
13 V. Majidi, N. Xu and R.G. Smith, *Spectrochim. Acta*, 55 B (2000) 3.
14 V. Majidi, R.G. Smith, N. Xu, M.W. McMahon and R. Bossio, *Spectrochim. Acta*, 55 B (2000) 1787.
15 B.V. L'vov, *Atomic Absorption Spectrochemical Analysis*, Adam Hilger, London (1970).
16 H. Massmann, *Spectrochim. Acta*, 23 B (1968) 215.
17 B.V. L'vov, *Inzh. Fiz. Zh.*, 2 (1959) 22.
18 J.D. Kerber, *Absorpt. Newsl.*, 10 (1971) 104.
19 C.J. Rademeyer, B. Radziuk, N. Romanova and Y. Thomassen, *J. Anal. At. Spectrom.*, 12 (1997) 81.
20 S. Chang, H. Liu and S.J. Tsai, *J. Anal. At. Spectrom.*, 13 (1998) 1123.
21 I.L. Grinshtein, Y.A. Vilpan, L.A. Vasilieva and V.A. Kopeikin, *Spectrochim. Acta*, 5 B (1999) 745.
22 I.L. Shuttler and H. T. Delves, *J. Anal. At. Spectrom.*, 4 (1989) 137.
23 C. Bendicho and M.T.C. de Loos Vollebregt, *J. Anal. At. Spectrom.*, 6 (1991) 353.
24 A.B. Volynsky, *Spectrochim. Acta*, 53 B (1998) 509.
25 F. Barbosa, F.J. Krug and E.C. Lina, *Spectrochim. Acta*, 54 B (1999) 1155.
26 S.P. Quinaia and J. Nobrega, *Fresenius J. Anal. Chem.*, 364 (1999) 333.
27 K. Ohta, H. Uegomori, S. Kaneco and T. Mizuno, *Ann. Chim.*, 89 (1999) 435.
28 P. Ljung, E. Nyström, O. Axner and W. Frech, *Spectrochim. Acta*, 52 B (1997) 703.
29 R.F. Lonardo, A.I. Yuzefovsky, R.L. Irwing and R.G. Michel, *Anal. Chem.*, 68 (1996) 514.
30 A.I. Yuzefovsky, R.F. Lonardo and R.G. Michel, *Anal. Chem.*, 67 (1995) 2246.
31 U. Schäffer and V. Krivan, *Anal. Chem.*, 70 (1998) 482.
32 T. Takada and T. Koide, *Anal. Chim. Acta*, 198 (1987) 303.
33 W.J. Price, T.C. Dymott and P.J. Whiteside, *Spectrochim. Acta*, 35 B (1980) 3.
34 Z. Grobenski, R. Lehmann, R. Tamm and B. Welz, *Mikrochim. Acta*, 1 (1982) 115.
35 T. Takada and T. Koide, *Anal. Chim. Acta*, 198 (1987) 303.
36 E. Lundberg and W. Frech, *Anal. Chim. Acta*, 104 (1979) 67.
37 D.D. Siemer and H.Y. Wei, *Anal. Chem.*, 50 (1978) 147.
38 K. Takada and K. Hirokawa, *Talanta*, 29 (1982) 849.
39 G.E. Carnrick, B.K. Lumas and W.B. Barnett, *J. Anal. At. Spectrom.*, 1 (1986) 443.
40 D.C. Hassell, T.M. Rettberg, F.A. Fort and J.A. Holcombe, *Anal. Chem.*, 60 (1988) 2680.
41 S. Lynch and D. Littlejohn, *J. Anal. At. Spectrom.*, 4 (1989) 157.
42 D. Bradshaw and W. Slavin, *Spectrochim. Acta*, 44 B (1989) 1245.
43 C. Bendicho and M.T.C. de Loos-Vollebregt, *Spectrochim. Acta*, 45 B (1990) 679.
44 G. Wibetoe, D.T. Takuwa, W. Lund and G. Sawula, *Fresenius J. Anal. Chem.*, 363 (1999) 46.
45 J.A. Holcombe and V. Majidi, *J. Anal. At. Spectrom.*, 4 (1989) 423.
46 H. Baumann, *Fresenius J. Anal. Chem.*, 342 (1992) 907.
47 T.A.M. Ure, L.R.P. Butler, B.V. L'vov, I. Rubeska and R. Sturgeon, *Pure Appl. Chem.*, 64 (1992) 253.
48 E. Iwamoto, M. Itamoto, K. Nishioka, S. Imai, Y. Hayashi and T. Kumamaru, *J. Anal. At. Spectrom.*, 12 (1997) 1293.
49 M.M. Lamoureux, J.C. Hutton and D.L. Styris, *Spectrochim. Acta*, 53 B (1998) 993.

50 J.L. Burguera, M. Burguera, C. Rondon, L. Rodríguez, P. Carrero, Y. Petit de la Peña and E. Burguera, *J. Anal. At. Spectrom.*, 14 (1999) 821.
51 E. Krakosvka, G. Holeczyova and P. Pulis, *Proc. Semin. At. Spectrochem.*, 14th (1998) 299.
52 U. Schäeffer and V. Krivan, *Anal. Chem.*, 71 (1999) 849.
53 W. Fuyi and J. Zucheng, *Anal. Chim. Acta*, 391 (1999) 89.
54 S.Z. Chen, F. Li, Z.H. Liao, T.Y. Peng and Z.C. Jiang, *Fresenius, J. Anal. Chem.*, 364 (1999) 556.
55 F.Y. Wang, Z.C. Jiang and T.Y. Peng, *J. Anal. At. Spectrom.*, 14 (1999) 963.
56 E.C. Lima, F. Barbosa and F.J. Krug, *Fresenius J. Anal. Chem.*, 369 (2001) 496.
57 F. Barbosa, E.C. Lima and F.J. Krug, *Analyst*, 125 (2000) 2079.
58 S.J. Huang and S.J. Jiang, *Analyst*, 125 (2000) 1491.
59 S.M. Maia, J.B.B. da Silva, A.J. Curtius and B. Welz, *J. Anal. At. Spectrom.*, 15 (2000) 1081.
60 M.A. Belarra, M. Resano and J.R. Castillo, *J. Anal. At. Spectrom.*, 14 (1999) 547.
61 T.M. Rettberg and J.A. Holcombe, *Anal. Chem.*, 60 (1988) 600.
62 P. Mandzhukov, E. Vassileva and V. Simeonov, *Anal. Chem.*, 64 (1992) 2596.
63 M.W. Hinds, G.N. Brown and D.L. Styris, *J. Anal. At. Spectrom.*, 9 (1994) 1411.
64 G.N. Brown, D.L. Styris and M.W. Hinds, *J. Anal. At. Spectrom.*, 10 (1995) 527.
65 M. Berglud and D.C. Baxter, *Mikrochim. Acta*, 119 (1995) 311.
66 F. Ghent, G. Galbacs, S. Boonen, L. Moens and R. Dams, *J. Anal. At. Spectrom.*, 10 (1995) 1047.
67 J. Pauwels, F. De Angelis, F. Peetermans and C. Ingelbrecht, *Fresenius J. Anal. Chem.*, 337 (1990) 290.
68 U. Kurfürst, M. Kempeneer, M. Stoeppler and O. Schuierer, *Fresenius J. Anal. Chem.*, 337 (1990) 248.
69 I. Atsuya, K. Akatsuka and K. Itoh, *Fresenius J. Anal. Chem.*, 337 (1990) 294.
70 K.P. Schmidt and H. Falk, *Spectrochim. Acta*, 42 B (1987) 431.
71 J. Bernhardt, T. Buchkamp, G. Hermann and G. Lasnitschka, *Spectrochim. Acta*, 54B (1999) 1821.
72 G. Galbács, F. Vanhaecke, L. Moens and R. Dams, *Microchem. J.*, 54 (1996) 272.
73 W. Schrön, A. Liebmann and G. Nimmerfall, *Fresenius J. Anal. Chem.*, 366 (2000) 79.
74 A.B. Volynsky, *Spectrochim. Acta*, 53B (1998) 1607.
75 K.W. Jackson, *Anal. Chem.*, 72 (2000) 159 R.
76 R. Kanipayor, D.A. Naranjit, B.H. Radziuk and J.C. van Loon, *Anal. Chim. Acta*, 166 (1984) 39.
77 L.H. Liu, Q.K. Zhang and Y. Hu, *Guangpuxue Yu Guangpu Fenxi*, 19 (1999) 424.
78 L.H. Lui, S.B. Luan and Q.K. Zhang, *Guangpuxue Yu Guangpu Fenxi*, 19 (1999) 419.
79 A.M. de Kersabiec and M.F. Benedetti, *Fresenius J. Anal. Chem.*, 328 (1987) 342.
80 V. Krivan and H.M. Dong, *Anal. Chem.*, 70 (1998) 5312.
81 M. Hornung and V. Krivan, *Spectrochim. Acta*, 54 B (1999) 1177.
82 W. Fuyi and J. Zucheng, *Anal. Chim. Acta*, 391 (1999) 89.
83 X. Chen and W.D. Marshall, *J. Agric. Food Chem.*, 47 (1999) 3727.
84 I. Lavilla, J.L. Capelo and C. Bendicho, *Fresenius J. Anal. Chem.*, 363 (1999) 283.
85 L. Amoedo, J.L. Capelo, I. Lavilla and C. Bendicho, *J. Anal. At. Spectrom.*, 14 (1999) 1221.
86 J.L. Capelo, A.V. Filgueiras, I. Lavilla and C. Bendicho, *Talanta*, 50 (1999) 905.
87 Y. Liang, Y.W. Tang and C.Y. Wang, *Fenxi Shiyanshi*, 18 (1999) 41.
88 J. Shiowatana, A. Siripinyanond, W. Waiyawat and S. Nilmanee, *At. Spectrosc.*, 20 (1999) 224.
89 P. Bermejo-Barrera, E.M. Verdura-Constenla, A. Moreda-Pineiro and A. Bermejo-Barrera, *Anal. Chim. Acta*, 398 (1999) 263.
90 G. Nimmerfall and W. Schrön, *Fresenius J. Anal. Chem.*, 370 (2001) 760.
91 J.A.C. Broekaert and F. Leis, *Mikrochim. Acta*, II (1985) 261.
92 T.Y. Peng, Z.C. Jiang, B. Hu and Z.H. Liao, *Fresenius J. Anal. Chem.*, 364 (1999) 551.
93 B. Cai, B. Hu and Z.C. Jiang, *Fresenius J. Anal. Chem.*, 367 (2000) 259.
94 V.A.H. Van Borm, J.A.C. Broekaert and R. Klockenkämper, *Spectrochim. Acta*, 46 B (1991) 1033.

95 C.K. Manickum and A.A. Verbeek, *J. Anal. At. Spectrom.*, 9 (1994) 227.
96 L. Halicz, I.B. Brenner and O. Yoffe, *J. Anal. At. Spectrom.*, 8 (1993) 475.
97 C.J. Walker, D.E. Davies, K.E. Turner and I.C. Hamilton, *Fresenius J. Anal. Chem.*, 355 (1996) 801.
98 I. Varga, F. Csempesz and Gy. Záray, *Spectrochim. Acta*, 51 B (1996) 253.
99 T. Mochizuki, A. Sakashita, H. Iwata, Y. Ishibashi and N. Gunji, *Fresenius J. Anal. Chem.*, 339 (1991) 889.
100 G. Galbács, A. Szorcsik, Z. Galbács, N. Buzás and T. Haraszti, *Talanta*, 52 (2000) 1061.
101 D.L. McCurdy and R.C. Fry, *Anal. Chem.*, 58 (1986) 3126.
102 C.S. Saba, W.E. Rhine and K.J. Eisentraut, *Anal. Chem.*, 53 (1981) 1099.
103 B. Jirgensons, *Organic Coloids*, Elsevier, Amsterdam (1958).
104 K.C. Friese and V. Krivan, *Fresenius J. Anal. Chem.*, 364 (1999) 72.
105 R.Q. Aucelio, C.L.V. Johnson, B.W. Smith and J.D. Winefordner, *Anal. Chim. Acta*, 411 (2000) 57.
106 R.Q. Aucelio, B.W. Smith and J.D. Winefordner, *Appl. Spectrosc.*, 52 (1998) 1457.
107 R.Q. Aucelio, V.N. Rubin, B.W. Smith and J.D. Winefordner, *J. Anal. At. Spectrom.*, 13 (1998) 49.
108 D.J. Swartm and J.B. Simeonsson, *J. Anal. At. Spectrom.*, 14 (1999) 929.
109 R.Q. Aucelio, B.W. Smith and J.D. Winefordner, *Anal. Sci.*, 15 (1999) 321.
110 R.Q. Aucelio, V.N. Rubin, E. Becerra, B.W. Smith and J.D. Winefordner, *Anal. Chim. Acta*, 350 (1997) 231.
111 B.W. Smith, I.B. Gornushkin, L.A. King and J.D. Winefordner, *Spectrochim. Acta*, 53 B (1998) 1131.
112 P. Ljung, E. Nystrom, O. Axner and W. Frech, *Spectrochim. Acta*, 52 B (1997) 703.
113 R.L. Irwin, D.J. Butcher, J. Takahashi, G.T. Wei and R.G. Michel, *J. Anal. At. Spectrom.*, 5 (1990) 603.
114 M.A. Bolshov, A.V. Zybin, V.G. Koloshnikov and I.I. Smirenkina, *Spectrochim. Acta*, 43 B (1988) 519.
115 U. Voellkopf, M. Paul and E.R. Denoyer, *Fresenius J. Anal. Chem.*, 342 (1992) 917.
116 D.C. Grégoire, N.J. Miller-Ihli and R.E. Sturgeon, *J. Anal. At. Spectrom.*, 9 (1994) 605.
117 R.W. Fonseca and N.J. Miller-Ihli, *Spectrochim. Acta*, 51 B (1996) 1591.
118 M.J. Liaw and S.J. Jiang, *J. Anal. At. Spectrom.*, 11 (1996) 555.
119 S. Hauptkorn, W. Krivan, B. Gercken and J. Pavel, *J. Anal. At. Spectrom.*, 12 (1997) 421.
120 M.J. Liaw, S.J. Jiang and Y.C. Li, *Spectrochim. Acta*, 52 B (1997) 779.
121 Y.C. Li and S.J. Jiang, *Anal. Chim. Acta*, 359 (1998) 205.
122 J. Wang, J.M. Carey and J.A. Caruso, *Spectrochim. Acta*, 49 B (1994) 193.
123 F. Vanhaecke, S. Boonen, L. Moens and R. Dams, *J. Anal. At. Spectrom.*, 10 (1995) 81.
124 S. Boonen, F. Vanhaecke, L. Moens and R. Dams, *Spectrochim. Acta*, 51 B (1996) 271.
125 F. Vanhaecke, S. Boonen, L. Moens and R. Dams, *J. Anal. At. Spectrom.*, 12 (1997) 125.
126 F. Vanhaecke, I. Gelaude, L. Moens and R. Dams, *Anal. Chim. Acta*, 383 (1999) 253.
127 C. Moor, P. Boll and S. Wiget, *Fresenius J. Anal. Chem.*, 359 (2997) 404.
128 Y. Hu, F. Vanhaecke, L. Moens, R. Dams and I. Geuens, *J. Anal. At. Spectrom.*, 14 (1999) 589.
129 H.H. Lu and S.J. Jiang, *Anal. Chim. Acta*, 429 (2001) 247.
130 B.N. Chapman, *Glow Discharge Processes*, John Wiley & Sons, New York (1980).
131 J.D. Cobine, *Gaseous Conductors: Theory and Engineering Applications*, Dover, New York (1941).
132 W.D. Westwood, *Prog. Surf. Sci.*, 7 (1976) 71.
133 K. Wagatsuma and S. Suzuki, *Fresenius J. Anal. Chem.*, 358 (1997) 581.
134 X. Pan, B. Hu, Y. Ye and R.K. Marcus, *J. Anal. At. Spectrom.*, 13 (1998) 1159.
135 K. Kohler, J.W. Coburn, D.E. Horne and J.H. Keller, *J. Appl. Phys.*, 57 (1985) 59.
136 S. Caroli, *Prog. Anal. Atom. Spectrosc.*, 6 (1983) 253.
137 R.K. Marcus and W.W. Harrison, *Spectrochim. Acta*, 40 B (1985) 933.
138 W. Grimm, *Spectrochim. Acta*, 23 B (1968) 443.
139 W.W. Harrison, *J. Anal. At. Spectrom.*, 13 (1998) 1051.

140 H. Yan, W. Hang, B.W. Smith, J.D. Winefordner and W.W. Harrison, *J. Anal. At. Spectrom.*, 13 (1999) 1033.
141 C. Yang, M. Mohill and W.W. Harrison, *J. Anal. At. Spectrom.*, 15 (2000) 1255.
142 A. Bengtson, C. Yang and W.W. Harrison, *J. Anal. At. Spectrom.*, 15 (2000) 1279.
143 F. Leis and E.B.M. Steers, *Spectrochim. Acta*, 49 B (1994) 289.
144 B.L. Bents and W.W. Harrison, *Anal. Chem.*, 54 (1982) 1644.
145 M.J. Heintz, P.J. Galley and G.M. Hieftje, *Spectrochim. Acta*, 49 B (1994) 745.
146 M.J. Heintz and G.M. Hieftje, *Spectrochim. Acta*, 50 B (1995) 1109.
147 M.J. Heintz, K. Miffling, J.A.C. Broekaert and G.M. Hieftje, *Appl. Spectrosc.*, 49 (1995) 241.
148 A.R. Raghani, B.W. Smith and J.D. Winefordner, *Spectrochim. Acta*, 51 B (1996) 399.
149 M. Chen, J.S. Ren, H.B. Ma and G.S. Zhang, *Spectrochim. Acta*, 52 B (1997) 1161.
150 A.I. Saprykin, *J. Anal. Chem.*, 54 (1999) 671.
151 M.J. Heintz, J.A.C. Broekaert and G.M. Hieftje, *Spectrochim. Acta*, 52 B (1997) 579.
152 F. Leis, J.A.C. Broekaert and K. Laqua, *Spectrochim. Acta*, 42 B (1987) 1169.
153 Y.X. Su, P.Y. Yang, D.Y. Chen, Z.G. Zhang, Z. Zhou, X.R. Wang and B.L. Huang, *J. Anal. At. Spectrom.*, 12 (1997) 817.
154 Y.M. Li, Z.H. Du, Y.X. Duan, H.Q. Zhang, Q.H. Jin and H.S. Liu, *Guangpuxue Yu Guangpu Fenxi*, 17 (1997) 65.
155 E.B.M. Steers and F. Leis, *J. Anal. At. Spectrom.*, 12 (1997) 307.
156 M. Outred, M.H. Rümmeli and E.B.M. Steers, *J. Anal. At. Spectrom.*, 9 (1994) 381.
157 J.A. Klinger and W.W. Harrison, *Anal. Chem.*, 63 (1991) 2984.
158 M. Dogan and A. Ulgen, *Fresenius J. Anal. Chem.*, 355 (1996) 651.
159 T.J. Loving and W.W. Harrison, *Anal. Chem.*, 55 (1983) 1526.
160 W. Schelles, S. De Gendt, V. Muller and R. Van Grieken, *Appl. Spectrosc.*, 49 (1995) 939.
161 W. Schelles, S. De Gendt, K. Maes and R. Van Grieken, *Fresenius J. Anal. Chem.*, 355 (1996) 858.
162 W. Schelles and R.E. Van Grieken, *Anal. Chem.*, 68 (1996) 3570.
163 W. Schelles and R. Van Grieken, *J. Anal. At. Spectrom.*, 12 (1997) 49.
164 H.J. Kim, Y.S. Park, G.H. Lee, K.B. Lee, H. Kim and J.S. Kim, *Microchem. J.*, 59 (1998) 399.
165 G.G. Sikharulidze and A.E. Lezhnev, *J. Anal. At. Spectrom.*, 14 (1999) 45.
166 V. Hoffmann, H.J. Uhlemann, F. Praerssler, K. Wetzig and D. Birus, *Fresenius J. Anal. Chem.*, 355 (1996) 826.
167 D.C. Duckworth and R.K. Marcus, *Anal. Chem.*, 61 (1989) 1879.
168 D. Fang and P. Seegopaul, *J. Anal. At. Spectrom.*, 7 (1992) 959.
169 J.Ruste and F. Schwoehrer, *Fresenius J. Anal. Chem.*, 355 (1996) 861.
170 R.K. Marcus and W.C. Davis, *Anal. Chem.*, 73 (2001) 2903.
171 R. Payling and D.G. Jones, *Surf. Interface Anal.*, 20 (1993) 787.
172 R. Payling, D.G. Jones and S.A. Gower, *Surf. Interface Anal.*, 20 (1993) 959.
173 M. Parker, M.L. Hartenstein and R.K. Marcus, *Spectrochim. Acta*, 52 B (1997) 567.
174 A. Bengtson and S. Hänstrom, *J. Anal. At. Spectrom.*, 13 (1998) 437.
175 J. Pisonero, C. Pérez, R. Pereiro, N. Bordel and A. Sanz-Medel, *J. Anal. At. Spectrom.*, 16 (2001) 370.
176 C. Pérez, R. Pereiro, N. Bordel and A. Sanz-Medel, *J. Anal. At. Spectrom.*, 15 (2000) 67.
177 D.S. Gough, *Anal. Chem.*, 48 (1976) 1926.
178 H.J. Kim and E.H. Piepmeier, *Anal. Chem.*, 60 (1988) 2040.
179 S.K. Choi and H.J. Kim, *Fresenius J. Anal. Chem.*, 355 (1996) 308.
180 P. Mezei, T. Cserfalvi and M. Janossy, *J. Anal. At. Spectrom.*, 12 (1997) 1203.
181 F. Leis and E.B.M. Steers, *Fresenius J. Anal. Chem.*, 355 (1996) 873.
182 K. Wagatsuma, K. Hirokawa and N. Yamashita, *Anal. Chim. Acta*, 324 (1996) 147.
183 J.A.C. Broekaert, T.K. Starn, L.J. Wright and G.M. Hieftje, *Spectrochim. Acta*, 53 B (1998) 1723.
184 K. Wagatsuma and J. Hirokawa, *Surf. Interf Anal.*, 21 (1994) 631.
185 M. Belkin, J.A. Caruso, S.J. Christopher and R.K. Marcus, *Spectrochim. Acta*, 53 B (1998) 1197.
186 K. Wagatsuma and J. Hirokawa, *Anal. Chim. Acta*, 306 (1995) 193.

187 J.W. Teng, C.M. Barshick, D.C. Duckworth, S.J. Morton, D.H. Smith and F.L. King, *Appl. Spectrosc.*, 49 (1995) 1361.
188 P.H. Ratliff and W.W. Harrison, *Spectrochim. Acta*, 49 B (1994) 1747.
189 M. Saito, *Fresenius J. Anal. Chem.*, 351 (1995) 148.
190 M. Saito, *Bunseki Kagaku*, 48 (1999) 77.
191 T. Tanaka, M. Matsuno, J.C. Woo and H. Kawaguchi, *Anal. Sci.*, 12 (1996) 591.
192 K. Wagatsuma, *Bunseki Kagaku*, 48 (1999) 95.
193 H. Matsuta and K. Wagatsuma, *Anal. Sci.*, 15 (1999) 319.
194 H. Matsuta and K. Wagatsuma, *Spectrochim. Acta*, 54 B (1999) 527.
195 K. Wagatsuma and H. Matsuta, *Anal. Sci.*, 15 (1999) 517.
196 Z. Weiss, *J. Anal. At. Spectrom.*, 12 (1997) 159.
197 K. Wagatsuma, *Bunseki Kagaku*, 48 (1999) 457.
198 F.W. Aston, *Mass Spectra and Isotopes*, 2nd edn, Longmans, Green and Co., New York (1942).
199 A. Bogaerts and R. Gijbels, *J. Anal. At. Spectrom.*, 12 (1997) 751.
200 J.A.C. Broekaert, *J. Anal. At. Spectrom.*, 2 (1987) 537.
201 S. Caroli Ed., *Improved Hollow Cathode Lamps for Atomic Spectroscopy*, Ellis Horwood Ltd, Chichester, England (1985).
202 K.R. Hess and W.W. Harrison, *Anal. Chem.*, 60 (1988) 691.
203 P.L. Larkins, *Spectrochim. Acta*, 46 B (1991) 291.
204 B.J. Russell and A. Walsh, *Spectrochim. Acta*, 15 (1959) 883.
205 C.L. Davis, B.W. Smith, M.A. Bolshov and J.D. Winefordner, *Appl. Spectrosc.*, 49 (1995) 907.
206 H. Falk, E. Hoffman and Ch. Ludke, *Spectrochim. Acta*, 36 B (1981) 767.
207 T. Oguri, H. Inoue, S. Tsuge, K. Kitagawa and N. Arai, *J. Anal. At. Spectrom.*, 12 (1997) 823.
208 C.M. Barshick, R.W. Shaw, J.P. Young and J.M. Ramsey, *Anal. Chem.*, 66 (1994) 4154.
209 P.J. Savickas, K.R. Hess, R.K. Marcus and W.W. Harrison, *Anal. Chem.*, 56 (1984) 817.
210 Y. Duan, Y. Su and Z. Jin, *J. Anal. At. Spectrom.*, 15 (2000) 1289.
211 W. Hang, C. Baker, B.W. Smith and J.D. Winefordner, *J. Anal. At. Spectrom.*, 12 (1997) 143.
212 Y.X. Su, Z. Zhou, P.Y. Yang, W.R. Wang and B.L. Huang, *Spectrochim. Acta*, 52 B (1997) 633.
213 K. Tsuji, T. Sato and K. Wagatsuma, *Spectrochim. Acta*, 53 B (1998) 417.
214 R. Plesch, *GIT Fachz. Lab.*, 28 (1984) 596.
215 A. Bogaerts, R. Gijbels and W.J. Goedheer, *Anal. Chem.*, 68 (1996) 2296.
216 A. Bogaerts and R. Gijbels, *Anal. Chem.*, 68 (1996) 2676.
217 A. Bogaerts and R. Gijbels, *Fresenius J. Anal. Chem.*, 359 (1997) 331.
218 R. Gijbels and A. Bogaerts, *Spectrosc. Eur.*, 9 (1997) 8.
219 A. Bogaerts and R. Gijbels, *Spectrochim. Acta*, 53 B (1998) 437.
220 A. Bogaerts and R. Gijbels, *J. Anal. At. Spectrom.*, 13 (1998) 721.
221 A. Bogaerts and R. Gijbels, *J. Anal. At. Spectrom.*, 15 (2000) 895.
222 A. Bogaerts and R. Gijbels, *J. Anal. At. Spectrom.*, 15 (2000) 1191.
223 A. Fiala, L.C. Pitchford, J.P. Boeuf and S. Baude, *Spectrochim. Acta*, 52 B (1997) 531.
224 K. Wagatsuma, *Bunseki Kagaku*, 46 (1997) 229.
225 S. Boumans, *Anal. Chem.*, 44 (1972) 1219.
226 R. Payling, *Surf. Interf. Anal.*, 21 (1994) 785.
227 R. Payling, *Surf. Interf. Anal.*, 21 (1994) 791.
228 S. Caroli, O. Senofonte, S. Caimi and P. Karpati, *J. Anal. At. Spectrom.*, 11 (1996) 773.
229 S.J. O'Gram, J.R. Dean, W.R. Tomlinson and J. Marshall, *Anal. Chim. Acta*, 294 (1994) 95.
230 Z. Weiss and P. Smid, *J. Anal. At. Spectrom.*, 15 (2000) 1485.
231 H.R. Stock, F. Hoehl and P. Mayer, *Fresenius J. Anal. Chem.*, 349 (1994) 208.
232 D.A. Day, A.L. Zook, C.M. Marshick and K.R. Hess, *Microchem. J.*, 55 (1997) 208.
233 D.C. Duckworth, C.M. Marshick, D.H. Smith and S.A. McLuckey, *Anal. Chem.*, 66 (1994) 92.
234 R.K. Marcus, M.A.Dempster, T.E. Gibeau and E.M. Reynolds, *Anal. Chem.*, 71 (1999) 3061.
235 R. Payling, D. Jones and S. Gover, *Surf. Interf. Anal.*, 20 (1993) 959.
236 F. Prassler, V. Hoffmann, J. Schumann and K. Weitzing, *J. Anal. At. Spectrom.*, 10 (1995) 677.
237 N. Jakubowski and D. Stuewer, *J. Anal. At. Spectrom.*, 7 (1992) 951.
238 A Quentmeier, *J. Anal. At. Spectrom.*, 9 (1994) 355.

239 A. Raith, R.C. Hutton and J.C. Huneke, *J. Anal. At. Spectrom.*, 8 (1993) 867.
240 E. Rose and P. Mayr, *Mikrochim. Acta*, I (1989) 197.
241 M. Parker, M.L. Hartenstein and R.K. Marcus, *Anal. Chem.*, 68 (1996) 4213.
242 C.L. Yang, K. Ingeneri, M. Mohill and W.W. Harrison, *J. Anal. At. Spectrom.*, 15 (2000) 73.
243 E. Oxley, C. Yang and W.W. Harrison, *J. Anal. At. Spectrom.*, 15 (2000) 1241.
244 J. Takadoum, J.C. Pirrin, J. Pons-Corbeau, R. Berneron and J.C. Charbonnier, *Surf. Interf. Anal.*, 6 (1984) 174.
245 J. Pons-Corbeau, J.P. Cazet, J.P. Moreau, R. Berneron and J.C. Charbonnier, *Surf. Interf. Anal.*, 9 (1986) 21.
246 S.A. McLuckey, G.L. Glish, D.C. Duckworth and R.K. Marcus, *Anal. Chem.*, 64 (1992) 1606.
247 C. Pan and F.L. King, *Anal. Chem.*, 65 (1993) 3187.
248 D.C. Duckworth, D.L. Donohue, D.H. Smith, T.A. Lewis and R.K. Marcus, *Anal. Chem.*, 65 (1993) 2478.
249 J. Robben, D. Dufour and R. Gijbels, *Fresenius J. Anal. Chem.*, 370 (2001) 663.
250 M.Betti, G. Rasmussen, T. Hiernaut, L. Koch, D.M.P. Milton and R.C. Hutton, *J. Anal. At. Spectrom.*, 9 (1994) 385.
251 D. Pollmann, K. Ingeneri and W.W. Harrison, *J. Anal. At. Spectrom.*, 11 (1996) 849.
252 R. Shekhar, M.V. Balarama-Krishna, J. Arunachalam and S. Gangadharan, *At. Spectrosc.*, 20 (1999) 25.
253 D.M. Wayne, *J. Anal. At. Spectrom.*, 12 (1997) 1195.
254 R. Jaeger, J.S. Becker, H.J. Dietze and J.A.C. Broekaert, *Fresenius J. Anal. Chem.*, 358 (1997) 214.
255 M. Belkin and J.A. Caruso, *Can. J. Anal. Sci. Spectrosc.*, 43 (1998) 8.
256 R. Jaeger, A.I. Saprykin, J.S. Becker, H.J. Dietze and J.A.C. Broekaert, *Mikrochim. Acta*, 125 (1997) 41.
257 C.L. Lewis, E.S. Oxley, C.K. Pan and F.L. King, *Anal. Chem.*, 71 (1999) 230.
258 W. Schelles and R. Van-Grieken, *Anal. Chem.*, 69 (1997) 2931.
259 T. Goreck, M. Belkin, J. Caruso and J. Pawliszyn, *Anal. Comm.*, 34 (1997) 275.
260 D.M. Wayne, R.K. Schulze, C. Maggiore, D.W. Cooke and G.J. Havrilla, *Appl. Spectrosc.*, 53 (1999) 266.
261 L. Adalve de las Heras, O.L. Actis-Dato, M. Betti, E.H. Toscano, U. Tocci, R. Fuoco and S. Giannarelli, *Microchem. J.*, 67 (2000) 333.
262 L.O. Actis Dato, L. Aldave de las Heras, M. Betti, E.H-Toscano, F. Miserque and T. Gouder, *J. Anal. At. Spectrom.*, 15 (2000) 1479.
263 L. Aldave de las Heras, F. Bocci, M. Betti and L. Actis Dato, *Fresenius J. Anal. Chem.*, 368 (2000) 95.
264 Y.X. Su, Y.X. Duan and Z. Jin, *Anal. Chem.*, 72 (2000) 5600.
265 W. Hang, X.M. Yan, D.M. Wayne, J.A. Olivares, W.W. Harrison and V. Majidi, *Anal. Chem.*, 71 (1999) 3231
266 J.P. Guzowski and G.M. Hieftje, *Anal. Chem.*, 72 (2000) 3812.
267 M.A. Belkin, L.K. Olson and J.A. Caruso, *J. Anal. At. Spectrom.*, 12 (1997) 1255.
268 M. Dogan, *J. Serb. Chem. Soc.*, 55 (1990) 63.
269 C.K. Pan and F.L. King, *Appl. Spectrosc.*, 47 (1993) 2096.
270 O. Senofonte and S. Caroli, *J. Anal. At. Spectrom.*, 15 (2000) 869.
271 K. Kitegawa, S. Kanoh, K. Ohta and M. Yanagisawa, *Anal. Sci.*, 4 (1988) 153.
272 S. Caroli, O. Senofonte, N. Violante and N. Astrologo, *J. Anal. At. Spectrom.*, 3 (1988) 887.
273 K. Shimizu, G.M. Brown, H. Habazaki, K. Kobayashi, P. Skeldon, G.E. Thompson and G.C. Wood, *Surf. Interf. Anal.*, 27 (1999) 24.
274 M. Fernández, N. Bordel, R. Pereiro and A. Sanz-Medel, *J. Anal. At. Spectrom.*, 12 (1997) 1209.
275 F.L. Freire, L.F. Senna, C.A. Achete and T. Hirsch, *Nucl. Instrum. Methods Phys. Res.*, March (1998) B136.
276 C. Pérez, R. Pereiro, N. Bordel and A. Sanz-Medel, *J. Anal. At. Spectrom.*, 15 (2000) 1247.
277 K. Wagatsuma, *Surf. Interf. Anal.*, 27 (1999) 63.
278 E. Oxleyu, C. Yang and W.W. Harrison, *J. Anal. At. Spectrom.*, 15 (2000) 1241.

279 K. Shimizu, G.M. Brown, H. Habazaki, K. Kobayashi, P. Skeldon, G.E. Thompson and G.C. Wood, *Surf. Interf. Anal.*, 27 (1999) 153.
280 W.B. Cho, Y.A. Woo, H.J. Kim, I.J. Kim and W.K. Kang, *Appl. Spectrosc.*, 51 (1997) 1060.
281 M. Ives, D.B. Lewis and C. Lehmberg, *Surf. Interf. Anal.*, 25 (1997) 191.
282 R. Martínez, C. Pérez, N. Bordel, R. Pereiro, J.L. Fernández Martín, J.B. Cannata-Andía and A. Sanz-Medel, *J. Anal. At. Spectrom.*, 16 (2001) 250.
283 N.G. Orellana-Velado, R. Pereiro and A. Sanz-Medel, *J. Anal. At. Spectrom.*, 13 (1998) 905.
284 R.K. Marcus and W.C. Davis, *Anal. Chem.*, 73 (2001) 2903.
285 S. Baude, J.A.C. Broekaert, D. Delfosse, N. Jakubowski, L. Fuechtjohann, N.G. Orellana-Velado, R. Pereiro and A. Sanz-Medel, *J. Anal. At. Spectrom.*, 15 (2000) 1516.
286 G. Absalan, C.L. Chakrabarti, J.C. Hutton, M.H. Back, C. Lazik and R.K. Marcus, *J. Anal. At. Spectrom.*, 9 (1994) 45.
287 G. Absalan, C.L. Chakrabarti, K.L. Headrick, M. Parker and R.K. Marcus, *Anal. Chem.*, 70 (1998) 3434.
288 M. Parker and R.K. Marcus, *Appl. Spectrosc.*, 48 (1994) 623.
289 C.L. Chakrabarti, K.L. Eadrick, J.C. Hutton, B. Zang, P.C. Bertels and M.H. Back, *Anal. Chem.*, 62 (1990) 574.
290 S.J. O'Gram, J.R. Dean, W.R. Tomlinson and J. Marshall, *J. Anal. At. Spectrom.*, 7 (1992) 229.
291 S.K. Choi and H.J. Kim, *Fresenius J. Anal. Chem.*, 355 (1996) 308.
292 A.A. Ganeyev and S.E. Sholupov, *Spectrochim. Acta*, 53 B (1998) 471.
293 B. Docekal, *Spectrochim. Acta*, 53 B (1998) 427.
294 S.J. O'Gram, J.R. Dean, W.R. Tomlinson and J. Marshall, *Anal. Chim. Acta*, 294 (1994) 95.
295 H. Bubert, *Spectrochim. Acta*, 39 B (1984) 1377.
296 W.O. Walden, W.W. Harrison, B.W. Smith and J.D. Winefordner, *J. Anal. At. Spectrom.*, 9 (1994) 1039.
297 C.L. Davis, B.W. Smith, M.A. Bolshov and J.D. Winefordner, *Appl. Spectrosc.*, 49 (1995) 907.
298 B.W. Smith, N. Omenetto and J.D. Winefordner, *Spectrochim. Acta*, 39 B (1984) 1389.
299 J.B. Womack, E.M. Gessler and J.D. Winefordner, *Spectrochim. Acta*, 46 B (1991) 301.
300 B.W. Smith, J.B. Womack, N. Omenetto and J.D. Winefordner, *Appl. Spectrosc.*, 43 (1989) 873.
301 S. Grazhulene, V. Khvostikov and M. Sorokin, *Spectrochim. Acta*, 46 B (1991) 459.
302 Y.S. Park, S.H. Ku, S.H. Hong, H.J. Kim and E.H. Piepmeier, *Spectrochim. Acta*, 53 B (1998) 1167.
303 J.C.T. Eijkel, H. Stoeri and A. Manz, *J. Anal. At. Spectrom.*, 15 (2000) 297.
304 J.C.T. Eijkel, H. Stoeri and A. Manz, *Anal. Chem.*, 71 (2000) 2600.
305 U. Engel, A.M. Bilgic, O. Haase, E. Voges and J.A.C. Broekaert, *Anal. Chem.*, 72 (2000) 193.
306 Special Issue, *J. Anal. At. Spectrom.*, 15 (2000) 1–78.
307 S.J. Jiang and R.S. Houk, *Anal. Chem.*, 58 (1986) 1739.
308 S.J. Jiang and R.S. Houk, *Spectrochim. Acta*, 42 B (1987) 93.
309 E.D. Salin and G. Horlick, *Anal. Chem.*, 51 (1979) 2284.
310 E.D. Salin and R.L.A. Sing, *Anal. Chem.*, 56 (1984) 2596.
311 V. Karanassios and G. Horlick, *Spectrochim. Acta*, 44 B (1989) 1345.
312 V. Karanassios and G. Horlick, *Spectrochim. Acta*, 44 B (1989) 1363.
313 V. Karanassios and G. Horlick, *Spectrochim. Acta*, 44 B (1989) 1387.
314 H.R. Badiei and V. Karanassios, *J. Anal. At. Spectrom.*, 15 (2000) 1057.

Chapter 9

Laser-assisted solid sampling

9.1. INTRODUCTION

The time when physicists invented laser (about four decades ago, when some scientists referred to it as "a solution without a problem") seems very distant now. Today, lasers are everywhere around us (e.g. in bar code scanners, pointers, CD players, sophisticated lighting systems, etc.) in areas such as high-power plasmas for X-ray generation, pulsed laser deposition for thin-film coating, cutting and welding of workpieces with lasers, and tissue and bone surgery in laser medicine. Analytical chemistry has extensively exploited the advantages of laser by developing a wide variety of techniques that continue to grow in number day by day. In fact, the use of lasers in such a specific application as solid sample treatment is so wide, that this chapter has been limited in scope to direct analyses of the elemental composition of solids as a complement to Chapter 8. Hyphenated techniques involving time-of-flight (TOF) [1], matrix-assisted laser desorption ionization (MALDI) [2] or laser-enhanced Raman spectroscopy (SERS) [3] are not dealt with here because they are used mainly at the molecular level and, basically, in connection with liquid samples — despite the fact that the sample becomes solid after the typical pretreatment applied. Also, those techniques used for direct desorption from natural solid samples (e.g. single-step and two-step laser desorption–ionization aerosol mass spectrometry) with a view to facilitating surface analysis are only considered in those cases where the laser acts on atoms (whether neutral or ionized), even if the final goal is a molecular determination (e.g. when metal ions are used for the chemical ionization of organic compounds and biologically relevant molecules [4–7]). Only the parent technique is described in detail, its derivations being illustrated with a few selected examples.

This chapter thus deals with the use of lasers for removal of small portions of solid samples for subsequent elemental analysis. The physico–chemical form in which the material is removed, which dictates the nature of the subsequent steps of the analytical process, has given rise to different atomic techniques using previous laser sampling. Figure 9.1 shows the two principal groups of atomic techniques based on it. In one, known as "laser ablation" (LA), the sample is removed as an aerosol; no excitation of particles, atoms or ions is pursued in this step, but only the formation of sufficient small particles for transport to the atomizer, which can be a plasma source (particularly an inductively coupled plasma, ICP). In the ICP, the ablated material is heated at temperatures above 6000 K, which are high enough to cause breakdown and excitation and/or ionization of the atoms; subsequently, emission lines can be detected on deactivation of the target elements (LA–ICP–AES) or the species can be allowed to continue its path

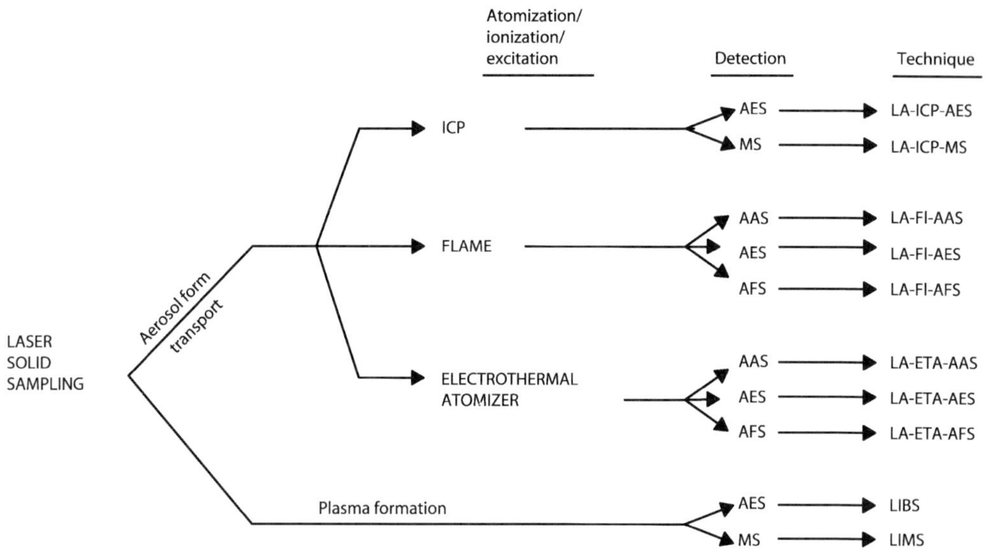

Fig. 9.1. Techniques based on laser-assisted solid sampling/atomic detection.

through the interface and detected using a mass spectrometer. The combination is highly sensitive and limits of detection at the parts-per-billion or even lower are available for almost all elements in the periodic table. Only a small number of them (e.g. H, He, O and noble gases) are difficult to ionize and hence difficult to determine. When a mass spectrometer is used as detector, the combined approach is designated LA–ICP–MS. A flame, with subsequent absorption, emission or fluorescence monitoring (LA–Fl–AAS, LA–Fl-AES and LA–Fl–AFS, respectively) or an electrothermal atomizer (LA–ETA–AAS, LA–ETA–AES, LA–ETA–AFS) can also be used for laser ablation. Glow-discharge subsequent to LA is so scantily used that it does not warrant inclusion in Fig. 9.1; however, a selected example of its use is described among the applications. The detector connected in-line with the atomizer constitutes the last part of the assembled equipment and also the last acronym of the hyphenated technique. These techniques have the advantage that they separate the excitation step from the sampling step. As a result, both steps can be controlled and optimized separately. Most excitation steps are developed at atmospheric pressure, so no vacuum is required for sampling.

The other group of atomic techniques based on laser sampling involves simpler couplings where only the detector is used after laser action — the laser energy is used not only for ablation, but also for atomization, ionization and excitation, and a plasma is formed — and comprises "laser-induced breakdown" (LIB) or "laser-induced plasma" (LIP) techniques. The emission lines for most elements are plentiful and, because their widths are in the region of several picometres, a large enough number of lines usually exists for identification and quantitative analysis of most of the elements in the periodic table. The detectors used to monitor these lines are typically either normal or intensified charge coupled devices (CCDs); conventional diode array detectors have also been employed for this purpose, however. The excited species (ionized or not) can be either deactivated in situ (in laser-induced breakdown spectrometry, LIBS) or transferred to a mass spectrometer (in laser-induced breakdown–mass spectrometry, LIB–MS or LIMS). Techniques involving the use of two lasers (one for ablation and the other for atomization, ionization and excitation) are also discussed in this chapter.

For a long time, both sampling approaches have been implemented by using a 1064 Nd:YAG laser providing 1 J pulses with a length of 1 ms. In addition to removing material, this laser was found to cause significant melting of samples, which was especially apparent from melt ejections in metals. As a result, the released material differed somewhat in composition from the sample: it contained more volatile than refractory elements by effect of fractionation. The use of UV lasers delivering high power densities (above 10^{10} W/cm^2), which are only available from pulsed lasers, suffices to cause non-thermal ablation and effectively restricts melting effects and fractionation.

9.2. LASER ABLATION

Laser ablation is the more simple and popular group of the laser sampling techniques for elemental analysis. Also, it can be considered the first step of an LIB process. The ablation process encompasses erosion, melting, vaporization and sublimation. An ideal laser-based solid sample insertion system for any of the above-described atomization–ionization–excitation sources should meet the following requirements: (a) it should require little surface preparation other than polishing and cleaning; (b) it should provide fine particles (no larger than a few microns in size); and (c) it should result in no selective vaporization and no matrix, memory or segregation effects. The extent to which a system fulfils these requirements depends on the properties of the laser (e.g. whether it is of the continuous wave or pulsed type, its wavelengths, its pulse duration and its repetition rate when pulsed). The type of analysis, either local or bulk, must also be considered. A local analysis can typically be accomplished by using a single-pulse or low-pulse rate laser. A transient signal will be obtained in both cases. Macro analyses require the use of a high pulse rate. Also, translation or rotation of the sample is needed to present a fresh surface to either each laser shot or set thereof. Finally, a steady state may be reached depending on the particular time constant of the transport configuration.

9.2.1. Features of ablation lasers

The performance of ablation lasers depends largely on two factors, namely: the nominal laser wavelength and intensity, and whether the continuous or pulsed mode of irradiation is used. Studies on the first two factors have yielded controversial results that can be ascribed to differences in sample nature or grain size. Thus, ruby (694 nm), Nd:YAG (1064, 532, 266 and, recently, 213 nm) and excimer (308, 248, 222, 193 and, recently, 157 nm) lasers have been used for laser ablation (LA). A comparison of the Nd:YAG laser operating at 1064 nm and 266 nm revealed better absorption of UV wavelength by most types of samples and also decreased fractionation. The particle sizes produced during ablation were smaller with 266 than with 1064 nm, which was possibly the reason for the reduced

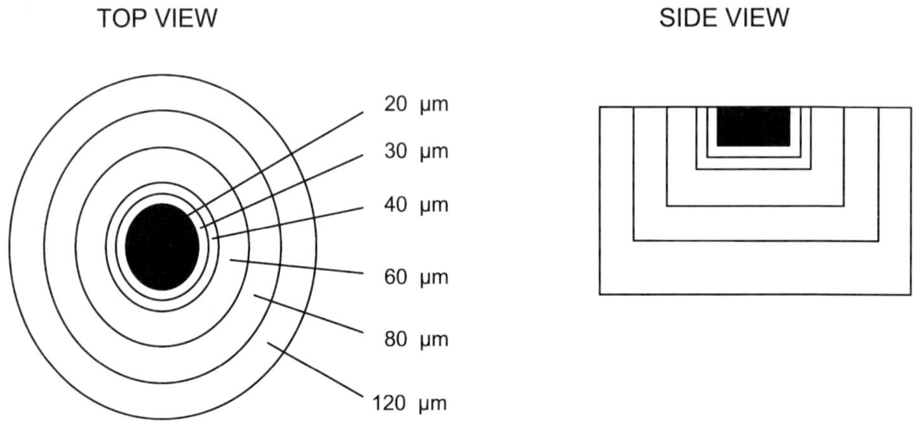

Fig. 9.2. Scheme of cascade ablation using crater sizes of 20 to 120 μm. (Reproduced with permission of Springer-Verlag.)

elemental fractionation observed at the former wavelength [8–10]. Experiments comparing the interaction of laser beams at 213 and 266 nm with solids [11] revealed improved ablation characteristics for highly transparent samples. In contrast to previous findings [12,13], non-sample related changes in signal intensities with time were found to be significantly smaller at the shorter wavelength. However, a recent comparison of laser beams at 266, 213 and 157 nm [13] revealed a similar extent of elemental fractionation with the three wavelengths. This suggests a decreased influence of wavelength relative to other factors such as energy density, focal point and the crater diameter to depth ratio. A comparison between a 266 nm Nd:YAG laser and a 193 nm excimer laser under closely matched conditions revealed improved ablation characteristics at 193 nm [13]. These results can be ascribed to differences in particle size distribution caused by the lasers. Little research, however, has been conducted at wavelengths below 266 nm.

ICP–TOF–MS experiments confirmed the reduced elemental fractionation provided by a 193 nm ArF laser equipped with a homogenizing optical array. If elements are assumed to separate between particles according to volatility, a significant change in the elemental ratios between fractionating and non-fractionating elements will occur within the first few seconds of ablation for a stepwise enlarged ablation spot. The resulting ablation pattern is shown in Fig. 9.2 and the variation of the Sr/Zr, Ce/La, Pb/U, Pb/Th and Pb/Pb elemental ratios as a function of crater diameter in Fig. 9.3. Notwithstanding the increased pit sizes (20–120 μm), the elemental ratios obtained were very stable and no significant indication for element-selective redeposition on the sample surface was obtained. The Sr/Zr and Ce/La elemental ratios should not change as these elements are assumed to be refractory, so they should be very similar in terms of ablation behaviour. It should be noted, however, that the depth-to-diameter ratio of the ablation crater never exceeded unity. Ablation depths exceeding this aspect ratio are indicative of elemental fractionation, as reported by Horn *et al.* [14]. A relationship between fractionation and ionic radius, and one between charge and melting temperature, have been proposed. Fractionation also

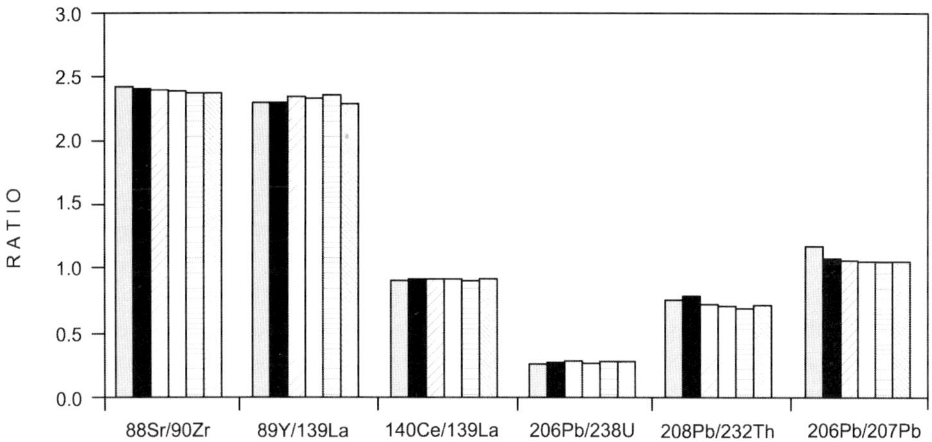

Fig. 9.3. Elemental ratios measured for various crater sizes on the same ablation spot. Bars correspond to the following values from left to right: 20, 30, 40, 60, 80 and 120 μm. (Reproduced with permission of Springer-Verlag.)

appears to correlate with the geochemical classification of elements (viz. lithophiles, siderophiles and chalcophiles) when using the 266 nm Nd:YAG laser to ablate geological samples [13].

Comparisons between laser wavelengths have been performed by using different samples under different output energies, with different optical arrangements, cell geometries and volumes. The results suggest improved ablation behaviour for highly transparent samples with lower (deep UV) wavelengths and less matrix-dependent ablation. In general, at any wavelength, the transparency of the sample influences the amount and extent by which it fractionates [14]. Recently, Günther *et al.* [15] conducted preliminary experiments using a 266 nm Nd:YAG laser ablation system with an optical assembly similar to that of the existing 193 nm unit for the purpose of examining the influence of wavelength (for equivalent optics) on the ablation characteristics. Experiments involving applying 3 mJ energy to the sample surface and the formation of a 100 μm crater resulted in the appearance of more transient signals than those observed at a low energy (0.5 mJ); however, the calculated fractionation index [16] for the Pb/Ca ratio was still significant. In short, laser wavelength is not the sole critical parameter influencing fractionation as this phenomenon can be minimized or enhanced in any sample and at any wavelength depending on laser irradiance. An irradiance value can be established at which fractionation is either not significant or completely eliminated. No unequivocal relationship between an increase in signal intensity (viz. in the quantity of mass ablated and transported to the atomizer) and laser wavelength has so far been established.

Lasers for solid sampling should always be operated in the pulsed mode. The continuous, CW mode, as often used for drilling and welding in engineering, would have serious drawbacks. The application of a CW laser provides much lower intensities than the peak intensities of pulsed laser. There would be continuous removal from the sample, which is closer to ordinary evaporation than the explosion-like ablation process of pulsed lasers.

The small area irradiated by a CW laser would continuously melt and the spatial composition in the liquid would change with time because of fractional evaporation and mixing in the liquid phase. Furthermore, the plasma above the sample would be continuously heated and, if the intensity is not low enough, it would shield the sample from the laser beam because of absorption by the plasma can be reduced by the choice of a laser with a shorter wavelength, a different kind of gas or a lower gas pressure.

The most popular lasers for ablation have nanosecond pulse lengths. Sufficiently high intensities for explosion-like ablation can be used and there is additional heating of the plasma gas. The intensities, the wavelength, the kind of gas and its pressure, must, however, also be chosen carefully in order to avoid plasma shielding of the sample.

The application of lasers with shorter pulses (e.g. in the picosecond or even femtosecond range) reduces heat dissipation by collisions of the high-energy electrons with the atoms in the solid, thereby minimizing melting of the laser crater region, as shown in Fig. 9.3, and fractional evaporation from the liquid phase. On the other hand, there is also less energy deposited in the gas plasma; this usually facilitates atomization of the material ablated from the sample.

9.2.2. Steps and thresholds in laser ablation

Steps of the laser ablation process

Laser ablation involves two main steps, namely: laser–solid interaction and removal of ablated material.

Laser–solid interaction In this step, a laser beam impinges on a solid surface. Raising the temperature of the sample to a high enough level entails using a high laser energy. This can be accomplished by using a pulsed laser and ensuring accurate focusing of the beam. Properly focused ordinary lasers with pulses of a few nanoseconds and peak energies of several tens of a millijoule can provide energy densities (irradiance values) around 10^6–10^9 W/cm^2. Under these conditions, the sample surface begins to absorb energy in the optical field, so its temperature increases gradually over the pulse duration.

Separation of the material (ablation) When the temperature reaches a high enough level, particles separate from the sample surface through vaporization, boiling or sublimation and form a vapour phase between the solid and the laser beam, all at moderate irradiance values (below about 10^6 W/cm^2). The vapour thus formed is quite light and essentially transparent; with higher irradiance values, however, the vapour tends to absorb strongly at the beginning and can easily reach supersaturation. A significant amount of material can thus condense and form submicroscopic drops capable of scattering the incident laser beam.

Thresholds in laser ablation

There are at least two borderline situations as regards the interaction of a laser pulse with a surface in an ablation process. These situations are usually described quantitatively as the lowest irradiance levels required to cause certain effects, namely: the damage threshold and the ablation threshold.

The *damage threshold* is defined as the pulse irradiance required to cause a surface change that can be detected under the optical microscope. Although this definition is somewhat vague, it provides a useful irradiance limit that usually results in a detectable change in the optical properties of the sample surface. As a rule, any formation on the sample surface or its vicinity capable of inducing or enhancing the release of mass by the laser energy can start the damage and hence influence the damage threshold. Typical surface formations include electronic defects, impurities and particle inclusions. Laser absorption can be enhanced by fractures, voids and holes, grains, steps and various other irregularities.

The *ablation threshold* is the irradiance value at which the material starts to separate from the sample. As a rule, the material separates through vaporization to a depth where the energy supplied by the laser beam equals the energy of vaporization. The material can separate in other forms including solid or liquid fragments and products of the photochemical decomposition of the sample. Because the energy deposited over the sample must exceed its latent heat of vaporization (H, J/kg), the threshold fluence (F_{min}, J/m^2) is given by

$$F_{min} = \frac{\rho H \sqrt{a}}{\sqrt{t_e}} \tag{9.1}$$

where ρ (kg/m^3) denotes sample density, a (m^2/s) thermal diffusivity and t_e (s) pulse width. Heat effects considered, the ablation threshold thus depends on both the laser (for a given material) and the focal conditions (fluence). The ablated mass can be calculated from the expression

$$M = \frac{E}{C(T_b - T_0) + H} \tag{9.2}$$

where E is the laser energy per pulse (J), C is the specific heat (J/kg K), and T_b and T_0 are the boiling temperature and initial ambient temperature (both in K), respectively.

The sudden increase in crater depth observed during high irradiance (> 10^{10} W/cm^2) laser ablation of silicon [17], which has been ascribed to phase explosion, can be used to establish a new threshold: the threshold irradiance for phase explosion. This threshold depends on two laser parameters, viz. beam spot size and wavelength. The larger the beam size and the longer the incident wavelength are, the higher is the laser irradiance required to cause phase explosion.

9.2.3. Craters and amount of ablated material

Based on existing results, the amount of material that is ablated from a sample ranges from 10 ng to 500 μg per pulse. The largest amounts (200–500 μg) are provided by free-running ruby lasers; ruby lasers in the Q-switched mode provide less ablated material (typically 1 μg/pulse). Nd:YAG lasers give amounts in the 1 ng to 1 μg per pulse range, depending on their operating mode. In any case, these amounts of material are inadequate to match the sensitivity obtained with aqueous solutions.

Ruby and Nd:YAG lasers produce craters ranging from 1 to 500 μm in size. The craters are usually wider than deeper and exhibit overlapping rough materials at the edge suggestive of thermal effects.

The rapid increase in crater depth above the threshold irradiance for phase explosion correlates with a significant increase in signal intensity. The ratio of crater volume to signal intensity, which represents the entrainment efficiency, remains the lowest at laser irradiances slightly above the phase explosion threshold. Such a ratio, however, increases at irradiances well above the threshold ($> 10^{11}$ W/cm^2).

Particles differ in diameter and shape. Spherical particles are formed from molten material. Angular fragments are ejected directly from the surface. Dimensions can be up to 10 μm, which is too large to ensure efficient transport and thorough volatilization in the subsequent plasma, flame or electrothermal atomizer.

The mass removal rate depends strongly on the laser power density, which is the main controllable parameter for a given material, and also on the laser wavelength when sampling solids. A wide variety of materials exhibit a decrease in the rate of change (roll-off) in the amount of mass removed with increasing incident laser power density. Roll-off results from a change in the efficiency of material removal by the laser beam and is primarily due to shielding of the target from the incident laser energy by a laser–vapour plume interaction. Studies on the amount of mass removed as a function of laser power density have revealed the change in ICP–AES intensity with laser power intensity to result from changes in mass removal efficiency. Roll-off in mass ablation is not due to a change in particle size distribution of the ablated species, fractionation of the sample or a change in transport efficiency to the ICP torch. Thus, accurate tracking of the laser ablation/ICP–AES process justifies the use of internal and external standards [18]. In the absence of matrix-matched standards, accuracy depends on the difference in degree of fractionation between the sample and standards.

9.2.4. Sample preparation for laser ablation

Laser sampling allows a large number of samples to be processed without the need for preparation. This is not the case with every type of sample, however. The extent to which a sample requires preparation for laser sampling depends on its nature and on the particular information required. Valuable experience has been gained and reported for preparing solid samples for other solid analysis techniques; many of the ensuing procedures have proved extremely useful and convenient for laser sampling as well. In general, these procedures tend to be less labour-intensive than those required for solution sample preparation.

With powdered samples, it can be of assistance to stabilize the sample, either by compacting it into a pellet (with or without a binding agent) or by fusing it to a borate glass. With non-homogeneous samples, it may be advisable to grind and mix the sample before stabilizing. A statistical study of the influence of sample inhomogeneity on the precision of analytical results was conducted by Scholze *et al.* [19], who applied a correlation function to the results obtained in the determination of selected elements in heterogeneous powdered soils. In order to obtain reliable information on the bulk composition of heterogeneous samples, a volume of about 20 mm^3 (equivalent to 50 craters *ca.*

1 mm deep each) was required for analysis. In this way, some analytical parameters such as the dependence of the expected relative standard deviation on the number of craters examined were determined. The minimum mass to be used was estimated from the number of craters to be obtained. The results were only semi-quantitative in most cases. In some special cases, quantitative analysis was possible provided the analyte was nearly uniformly distributed or the number of craters studied was increased in accordance.

In some cases, it helps to either polish or etch the surface of a solid sample — unless correlation between surface topography and composition is important. Small samples such as pins or crystals can be embedded in an epoxy cast and analysed either directly or after polishing the cast surface. Otherwise, small samples can simply be stuck into a piece of putty or onto tape or an adhesive pad. Because the sample cell is not evacuated, outgassing of the support material is not a problem with this approach.

9.2.5. Ablation cells

An efficient ablation cell must met several basic requirements, namely: (a) it should be easy to operate (particularly, it should allow easy sample changeover with some flexibility in sample size); (b) it should exhibit no memory effects; (c) it should allow the sample to be displaced in order to present a fresh surface to the laser action; and (d) it should possess a low dead volume.

A wide variety of sample cell designs exists [20–24]; most use a flow of argon tangential, normal or concentric to the vapour plume to sweep vaporized material out of an enclosed sampling cell through the transfer tube. Because sample transport efficiencies are lower than unity, some sample redeposition can occur in the sampling cell and transfer tube. Usually, this causes no memory effects unless the material deposits on the cell window through which the laser beam is passed. Subsequent laser pulses can revaporize material previously deposited on the window, thus interfering with the analysis. This problem is avoided by systems with removable windows, which can be either replaced or cleaned.

Ablation cell configurations can be classified into two broad categories depending on both the angle of laser incidence on the sample and the way the carrier gas is circulated through it, namely:

(a) With focusing of the laser beam normal to the sample surface and the lateral inlet and outlet of the carrier gas (Fig. 9.4A). The flat sample can thus be rotated. This design is widely used, even though it is subject to some drawbacks such as a turbulent flow and material deposits on cell walls. The cell can be made of glass or a metallic material (stainless steel). Usually, the window can be easily replaced. In an early design [25], the sample was continuously brought by a tape.
(b) With focusing of the laser beam either vertically downwards on the sample surface or with a given angle, up to a horizontal position (Fig. 9.4B) [26]. The carrier gas is injected into the cell. Usually, a tangential flow around the sample is used to provide a "sheathing" effect [21,26]. Material deposits onto walls are thus minimized. The carrier outlet is most often at the top of the cell [21,26]. One alternative design involves focusing of the laser beam on the lateral cylindrical surface of a

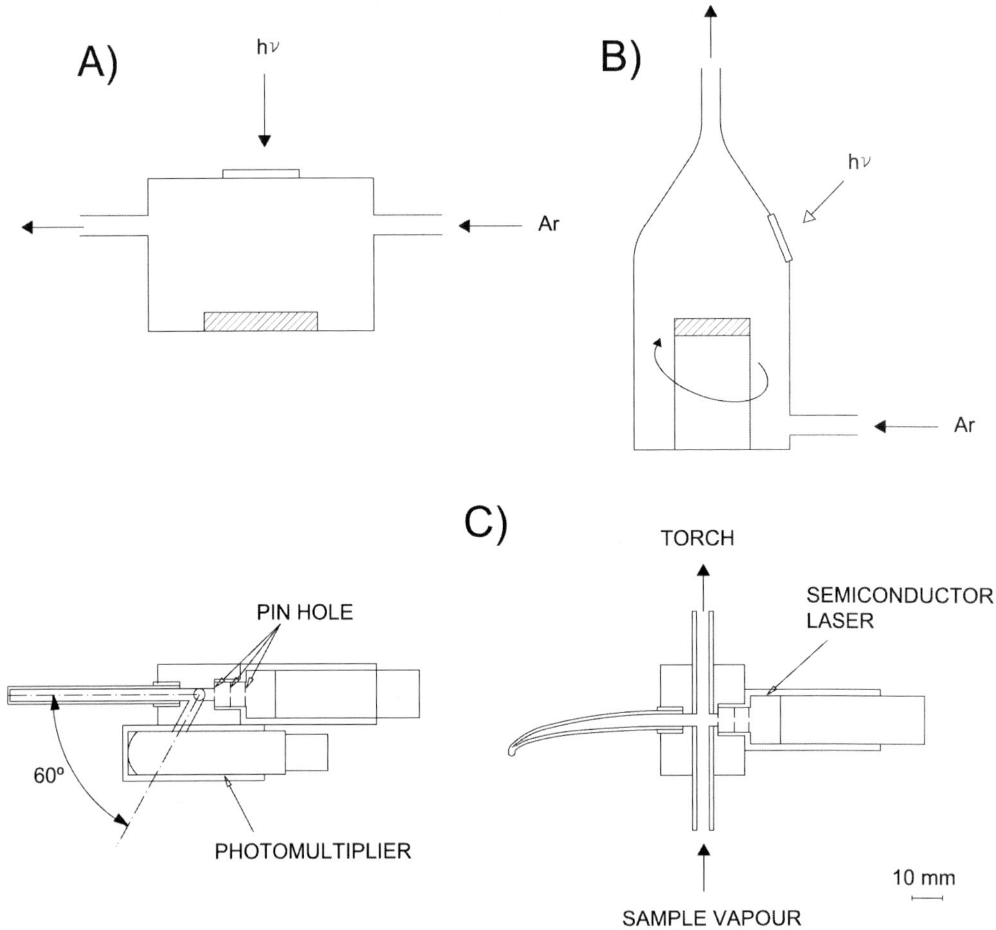

Fig. 9.4. Typical laser ablation cells with lateral inlet and outlet of the carrier gas (A) and tangential flow of the carrier gas (B). (C) Cross-sectional views of laser light-scattering cell.

rotating barrel with apparently inadequate ablation efficiency [27]. Other designs have been developed in order to improve the uptake rate [24], which constitutes a critical parameter.

Specially shaped cells are required for special uses. One case in point is the cell depicted in Fig. 9.4C, which was designed to correct the variation of aerosol density in LA–ICP–MS and was a laboratory-made laser light-scattering cell for insertion into the aerosol transport tube between the laser ablation system and the torch. In this way, the temporal variation of signals from the laser light-scattering cell was correlated with that of the ion signals from the mass spectrometer [28].

Special cells are also designed for safety reasons such as avoiding contact with radioactive substances. Thus, based on other, previously reported designs [26,29], Leloup *et al.* [30] developed a sophisticated cell used in a glove box to reduce the volume of radioactive waste in the ablation of uranium samples. The ablation cell consisted of two main parts, namely: a mechanical, mobile part for sample positioning and a glass part for ensuring that the ablated material was carried away to the plasma torch, also located in a glove box.

A more recently developed ablation cell (a jet-cell) directs the ablation gas flow of argon through a capillary onto the ablation side [31], which results in improved signals. However, the wide variety of potential target samples calls for different, specific ablation cells, so most laboratories using this technique employ two or more types of cell. A procedure for optimizing the geometry of these cells has been reported [32].

Most laser sampling systems use electronic stepper motors to move the stage on which the sample is mounted. Using a computer for remote control of the stage movement allows the sample to be moved precisely under the laser beam in various well-defined, reproducible translation programmes over a line or across a wide area, thus providing an alternative to single spot analysis. This approach is especially useful for analysing lateral compositional trends and for the bulk characterization of inhomogeneous materials.

9.2.6. Transport of ablated material

Using a transport interface is almost mandatory in connecting the laser ablated material with the atomizer. There are few references to a direct connection to the torch [33]. Most workers use a tube made of PVC, Tygon or polyethylene with an i.d. of 2–7 mm and a length up to 150 cm. This results in large differences among time constants. In addition, deposits along the tube are also frequently observed. Few authors have examined transport efficiency in connection with ICP [34] and values as low as 5% at the torch level have been reported. A cyclone chamber can be used to remove exceedingly large particles [35]. Depending on the time constant and laser pulse frequency [36], a *quasi*-steady state can be achieved that allows the use of conventional data acquisition and processing systems.

9.2.7. Mixed gas sample introduction

The ablation process has been studied under different gas environments to improve sample transport into the atomizer. Combinations of various gases (e.g. nitrogen, hydrogen and helium [37–41]) have been investigated. With ICP–MS, the use of nitrogen is leading to higher sensitivity in the high mass region, which is of especial importance for the determination of lead and uranium. Eggins *et al.* [39] achieved increased sensitivity (high-mass region) by adding helium to the main flow of argon used to flush the ablation cell. The improved sensitivity thus obtained was mainly the result of decreased material deposition around the ablation crater. However, using pure helium (which was subsequently mixed with argon) as an ablation cell carrier gas resulted in increased sensitivity and reduced background intensities, thereby improving limits of detection [37]. The degree of improvement was different for 266 and 193 nm laser systems. However, the source of this enhancement — which might be related to particle size distribution or a change in the ICP conditions — has not yet been quantified and requires further investigation. All mixed gas experiments (particularly those involving helium) alter gas dynamics at the interface, which makes it rather difficult to discard specific processes as the origin of the effects observed in alternative carrier gases for sample insertion.

9.2.8. Equipment

Most laser ablation researchers design their own cells and transport systems, i.e. the laser–atomizer interface. Some routine users, however, purchase commercial equipment the characteristics and complexity of which depend on both the atomizer and the detection system. Laser ablation equipment is manufactured by CETAC Technologies for coupling with mass and emission detectors, under the names CSX-200 and Soils-500, respectively; they are the successors to the 1994 versions from CETAC. As stated by the vendor, the new systems are endowed with a flat-top beam profile, a geological package with a binocular viewing microscope, enhanced Windows™ software, unique laser energy control through innovative beam attenuation, pulse-repetition rate controlled through the laser Q-switch, user-friendly integration and improved stability. Also, the systems are so compact that they require no resonator alignment.

9.2.9. Calibration techniques for laser ablation–atomization–ionization–excitation–detection

The existence of reference materials and appropriate calibration procedures are two essential issues to be considered in quantifying the components of a sample. Quantitation has been substantially improved by the commercial availability of an increasing number of certified solid reference materials, especially for low concentration levels. Recently [42], NIST archival leaf standards were used as matrix-matched standards for reliable quantitative elemental analysis of Spanish moss samples. LA–ICP–MS was used with mixing standards in order to produce at least three data points for each calibration curve; the results thus obtained were compared with those provided by microwave digestion ICP–MS/AES. Standard addition was also examined and found to be an effective method in the absence of matrix-matched standards.

An alternative calibration technique was developed and successfully applied to techniques hyphenated to LA [43–48]. In addition, an increasing number of matrix-independent calibration procedures using external reference materials and/or internal standardization procedures is being developed [49–51].

Although laser ablation usually dispenses with the need for sample preparation in the analysis of rocks, the use of a major element of the sample as internal standard requires the prior knowledge of its concentration; also, the addition of an internal standard element to a rock sample is difficult. Fusion of the sample with $Li_2B_4O_7$ allows the use of Li and B as internal standards. Linear calibration graphs (constructed from standard reference geological materials) have thus been obtained for Al, Ca, Fe, K, Mg, Na, Si and Ti, and relative standard deviations of 1–2% ($n = 12$) achieved.

Several methods other than those based on internal standards have been tested with a view to improving precision in analytical methods involving laser ablation. Attenuation of a helium–neon laser beam due to scattering by ablated particles [52] and the sound waves generated in the sampling cell during the LA process [53] have been used to correct variations in the transport of sample vapour. The precision was increased by a factor of about 2 by normalizing the analyte emission intensities with the measured amplitude of the sound wave. Also, a laser light-scattering cell for measuring aerosol density was inserted into the transport tube of the sample vapour from the LA sampling cell to the ICP torch and the ion signals from the detector were normalized to the scattered light signals. Even in the single-pulse mode, the precision of the ion intensities normalized with the laser light-scattering signals could be improved to the same level achieved with the internal standard method [28].

Based on the results of their experiments with liquid samples in LA–ICP–MS, Günther *et al.* [54] developed a new approach to calibrating microanalyses of solids by this hyphenated technique. They used a liquid standard covered with a thin plastic film through which a small hole was drilled with the laser. Later, the introduction of liquid standards into the ablation cell by formation of either pneumatic [55] or ultrasonic aerosols [56] has allowed full exploitation of these approaches, which are illustrated in Figs 9.5A and 9.5B, respectively. Pneumatic aerosols require the use of a matrix element as internal standard owing to the inherent differences in sample insertion rate with LA and pneumatic nebulization. This poses problems with elements having no isotope with a relative abundance less than 0.1% as the resulting ion signals are too high for direct measurement and lowering the ablation rate compromises the sensitivity for trace elements. This shortcoming can be circumvented by using a high-stability ICP unit and mass filter, which allows the tail of a mass peak for the matrix element to be used as internal standard [55]. In calibrating the matrix element by nebulization with liquid standards, the ion signal measured on the peak tail is directly proportional to the element concentration in the ICP, thus indicating that peak shape is not only stable, but also independent of peak height. The advantages of this method lie in the ease of preparing calibration standards for quantitative measurements with a laser ablation system and in access to homogeneous standard materials, which are difficult to homogenize in the solid state. Traces are calibrated relative to a fixed concentration of the matrix element. Calibrations thus performed for trace concentrations in high-purity copper provided good recoveries from high-purity reference standards.

The ultrasonic nebulization–LA approach (Fig. 9.5B) was designed for multi-element determinations in geological samples; in order to match the two techniques, standard solutions were nebulized with an ultrasonic nebulizer (USN) during solution calibration and a blank target (e.g. lithium borate) was simultaneously ablated with a focused laser beam. Homogeneous geological samples were measured using the same experimental set-up, where a 2% HNO_3 solution was nebulized simultaneously with the USN. Both homogeneous targets prepared from inhomogeneous geological samples by powdering, homogenizing and fusing with a lithium borate mixture in a muffle furnace at 1050°C, and homogeneous geological glasses, were examined, and relative standard deviations in the range 2–10% (n = 6) were obtained for trace elements.

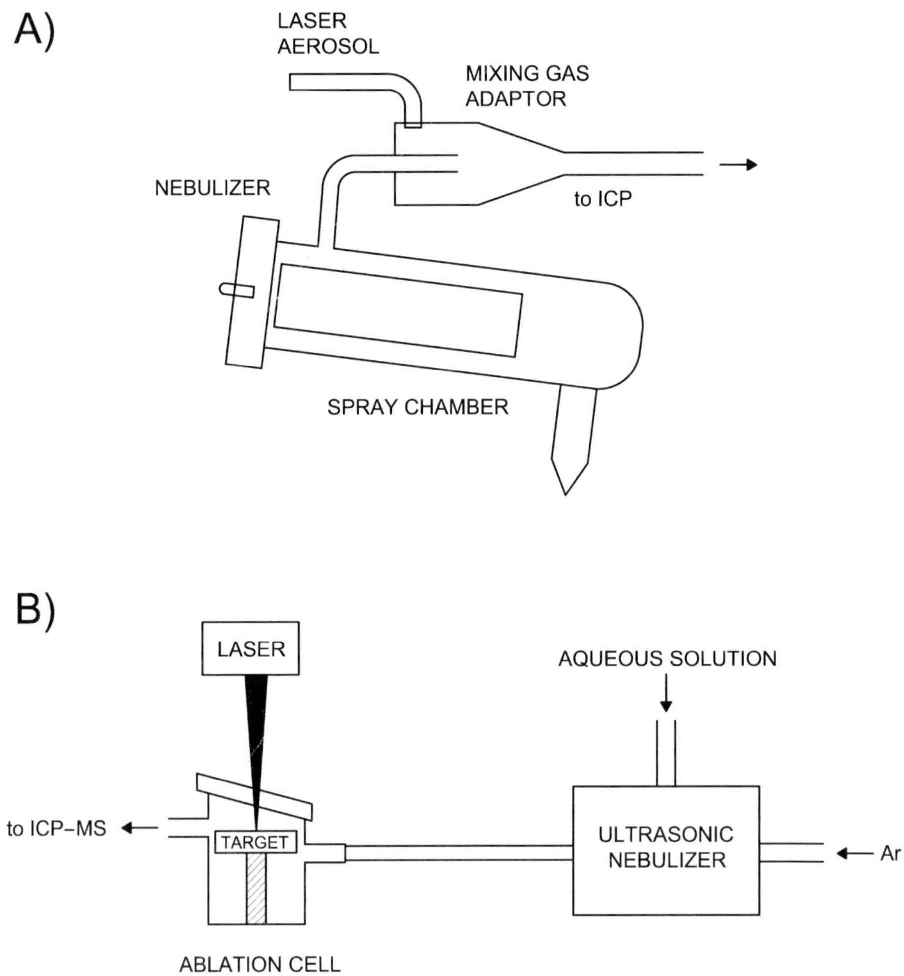

Fig. 9.5. (A) Schematic view of the combined use of laser ablation and pneumatic nebulization. (B) Experimental arrangement of LA–ICP–MS with solution calibration. (Reproduced with permission of Springer-Verlag.)

9.2.10. Shortcomings of laser ablation

Some of the problems encountered in using the LA technique arise from so-called *elemental fractionation*, a non-sample related change in the analyte response during the ablation process. This effect has been observed to some extent with all types of lasers, but other variables including energy density, focal point and the crater diameter-to-depth ratio also appear to exert a strong influence [9–13,39,51].

A number of specific approaches to minimizing elemental fractionation have been developed [11,14,54,57,58]. It remains impossible to pinpoint the source of fractionation:

ablation, transportation or excitation and ionization in the plasma, flame or electrothermal atomizer. Recent observations by Mank and Mason [12], and Russo [13], suggest that laser wavelength is not the dominant factor in elemental fractionation. The effects of the crater diameter-to-depth ratio appear to be more important. This has been confirmed by the results obtained with a 193 nm excimer laser ablation system for geochronological analysis [14].

Variations in laser irradiance constitute one other of the most serious shortcomings in LA. Fluctuations in the transversal mode structure and hot spots in the laser beam profile can give rise to significantly different ablation results. The beam profile should have at least Gaussian shape (TEM_{00}), which produces "bell-shaped" craters. "Flat-top" beams, which have good homogeneous intensity distributions, are advantageous in laser sampling. They can be generated from Gaussian laser beams, either by expansion of the beam and use of an aperture which selects the central part or by application of specially designed optical arrangements of lenses or mirrors. Poor shot-to-shot reproducibility is one of the hindrances to using single-shot laser ablation (viz. one laser pulse per analysis) for the direct elemental analysis of solid samples with spatial resolution — in addition to variations in the specimen being sampled. Proper data analysis appears to be one effective way of solving these problems. Two specific methods for improving single-shot measurement precision use the simultaneous full-spectrum acquisition capability of a time-of-flight mass spectrometer. In one, a normalization factor is computed from the combined signal yielded by all sample constituents, the reasoning being that the combined spectrum should be proportional to the total mass of sample ablated. This results in increased precision by a factor greater than 2, which is moderately better than is possible with a single internal standard. The increase in measurement precision is concentration-dependent, the greatest improvement (10–15 fold) being experienced by high-concentration elements. The other method correlates the attenuation of plasma matrix ions to analyte intensities. It provides no statistically significant improvement in precision, which is limited by the relatively low signal-to-noise ratio of the attenuated signals [59].

9.2.11. Selected applications of laser ablation sampling prior to atomization–ionization–excitation–detection

Laser ablation is used in two major fields, namely: (a) bulk analysis, with a low spatial resolution (80–350 μm crater diameters) [60–63]; and (b) local analysis, with a high spatial resolution (4–80 μm crater diameters) [64–70]. Because laser ablation removes very small amounts of sample, a high sensitivity is essential for trace and ultratrace analysis. This makes the combination of LA with ICP–MS the most attractive and wide-spread choice. Depending on the particular application, virtually all types of mass filters including the quadrupole, magnetic sector field (with single or double focusing, single or multi-collector detectors) and time-of-flight types can be used in conjunction with LA; they have been found to improve the sequential and simultaneous determination capabilities of LA–ICP–MS and to arouse increased interest in this direct solid sampling technique as a result [71]. In fact, this powerful tool has been used for solid micro sampling analyses in geological [11,31,49,72–77], biological [78,79], environmental

[80–82], nuclear [30,83,84] and metallurgical samples [85]. It is an expeditious technique of relatively moderate cost that combines the high-sensitivity multi-element capabilities of ICP–MS and the advantages of in situ micro solid sampling by LA. It is particularly attractive to scientists wishing to study dissolution-resistant solid materials or spatial distributions of trace elements in a micro-scale area on the sample surface. However, the complexity of data acquisition and reduction of transient signals has so far restricted application of this technique owing to the need for continuous intervention of an operator. Recently, an automated method for the in situ trace element analysis of solid materials was reported [86] the performance of which was demonstrated by determining 22 trace elements in certified reference materials.

The fact that laser ablation enables both bulk and local analysis has caused the demand for the flexibility of ICP mass spectrometers to rocket. The resolution of bulk analysis is typically 0.1–1 mm, whereas that of local analysis is usually between 5 and 50 μm; this results in a volume difference of 900 on the assumption of a constant ablation rate per pulse for a 10 μm crater relative to a 300 μm one. The linear dynamic range of the detector is therefore one of the most influential parameters for solid sampling. Major elements, usually employed as internal standards for quantitation, should ideally be measured together with minor and trace elements. Few commercially available systems are equipped with pulse counting and analogue mode amplifiers in the detection system resulting in 9–10 orders of magnitude in linear dynamic range.

Analyte sensitivity as reported for different instruments is mainly influenced by the experimental crater size, energy density and repetition rate, which influence the amount of ablated and transported material. Normalizing sensitivities obtained with a similar energy density to a crater diameter of 40 μm and a repetition rate of 10 Hz leads to rather equal sensitivities of *ca.* 1000–3000 cps/ng for most commercially available instruments. The ensuing limits of detection are thus influenced mainly by background noise.

One other critical parameter for multi-element analysis is the data acquisition rate, the effect of which has been extensively discussed by Longerich *et al.* [72]. Recent improvements in sector-field instruments — which have commonly been judged slow relative to magnet jump speed — have been implemented in this context. Thus, measurement efficiency has been improved by using a short, fast change in the accelerator voltage for a magnet jump in such a way that the differences between quadrupole-based instruments and sector-field instruments have been brought to within a factor of 2 [87]. The ideal detection for short transient signals in laser ablation would be an "all-collecting" sector-field instrument where a detector array would be located in the space and time focus collecting all isotopes fully simultaneously. The ICP–TOF–MS combination is near-ideal in this respect as all isotopes are sampled at once and detection is very fast [88,89]. The typical sampling time of an ICP–TOF–MS instrument is in the region of 5 μs; also, with a read-out time of 12.75 ms (255 averaged spectra) the problem of spectral skew in sequential mass analyses is avoided. This provides access to elemental composition even for ultra-short transient signals and precise isotope ratios. The high data acquisition rate is required for the fast transient signals encountered, for example, during ablation of fluid inclusions.

Environmental applications

There has been sustained interest in the development of analytical methods for real-time monitoring of environmental pollutants in recent years. Semi-volatile organic compounds tend to be adsorbed onto the surface of aerosol particles of respirable sizes and pose potential health hazards. Conventional methods used for the analysis of organic compounds that are present on the surface of aerosol particles are based on particle collection followed by extraction and chromatographic analysis of extracted species [90,91]. These methods require large amounts of sample and long analysis times.

Over the past decade, several research groups have reported real-time chemical analyses of aerosols [92–102]. Airborne particles are directly aspirated into a vacuum chamber where they are ablated and ionized by a laser pulse, followed by mass spectrometric detection of the ions. Several instruments are capable of providing size, number density and composition information about aerosols. The relatively high laser fluence required to ionize analyte molecules present on the surface of aerosol particles produces a high abundance of ions from the bulk material that can obscure weak signals from the surface-localized species. Charge-transfer processes specific to the matrix can also alter the analytical results [103]. Single-step laser desorption–ionization aerosol mass spectrometry is mainly suitable for the characterization of the chemical composition of the bulk material in the particles. Two-step laser desorption–ionization has been coupled with aerosol mass spectrometry to reduce fragmentation of organic species [104] or space charge defects in the ion source of the mass spectrometer [105], but none of the tests was specifically designed for analysing molecular adsorbates on particles. Recently, the surface analysis of particles by use of two-step laser desorption–ionization real-time aerosol mass spectrometry (L2RTA–MS) was accomplished [106]. The instrument used allowed real-time monitoring of polycyclic aromatic hydrocarbons (PAHs) adsorbed onto the surface of airborne particles without significant time delay between analysis and environmentally significant events. The one disadvantage of L2RTA–MS is that it requires two lasers, which complicates the equipment. This has led some authors to explore the use of real-time aerosol mass spectrometry (RTA–MS) for the detection of surface-adsorbed species as an alkali metal adduct, thus taking advantage of the presence of easily ionized alkali metals contained in environmental particles. During laser desorption of surface-adsorbed analytes, Na^+ and K^+ ions are produced by the interaction of the laser radiation with the particle's material. The alkali metal ions serve as in situ chemical ionization reagents for the neutral analyte molecules.

This new approach has been tested for monitoring tributyl phosphate (TBP) on the surface of environmental particles. The effect of laser fluence on the signal intensities of the potassium ion and cationized TBP was examined and the instrument found to perform best with laser fluences resulting in high abundances of K^+ but low abundances of ions from the particle's bulk material. The relatively low laser fluence needed to produce potassium ions avoided excessive fragmentation of the analyte. This approach allows real-time monitoring of sub-monolayer coverage of TBP on the surface of micro-sized particles [107]. Compared to detection by conventional L2–MS of aniline on silica surfaces as reported by Voumard *et al.* [108], the surface needed was at least 10^6 times greater than that required by RTA–MS with in situ ionization. The greatest disadvantages of the latter approach are fluctuations in the effective laser fluence — on which the ion signal exhibits a steep dependence, especially in the region close to the ion production threshold — and the unexplored capabilities of MS/MS in this context (the charge of dissociated ions is carried by alkali metals, so no information about the organic compound can be obtained). On the other hand, the formation of adduct ions has the advantage that it shifts the *m/z* ratio for the *pseudo*-molecular ion to higher levels, where the background intensity is lower and few ions from the particle bulk interfere with detection of the analyte.

Biological applications

The environmental monitoring of shellfish is a combined biological–environmental application of LA sampling; in fact, pollutants in seawater are incorporated into shells and can be used as indicators of environmental conditions. The hard parts of a shelly organism provide a time-based record of pollution. This additional information cannot be obtained with conventional soft tissue analyses. LA–ICP–MS as a direct solid sampling technique provides a spatial resolution of 150 μm with IR lasers, giving sub-parts-per-million limits of detection for carbonate materials [109]. Recent developments in laser technology can improve spatial resolution even further (to < 10 μm) by use of UV lasers, thereby enabling analyses between different growth bands of the shell and providing a way for establishing a timeline of pollution events. Such events may affect a whole population of shellfish, which can thus be used as a reference [110].

Determinations of metal components in biological materials provide responses to various problems, as shown by Cox *et al.* [111] in the trace element profiling of dental tissues. They used LA–ICP–MS to obtain elemental fingerprints for individual teeth, which were sectioned to produce a sample representing the time axis of tooth development. The ablated material was *ca.* 100 μm across, and the ablation process was reproducible and measured using phosphorus. The mercury and gold concentrations were found to decrease from the outer (mouth exposed) part of the tooth to the inner, newly deposited material. By contrast, the lead content was found to increase from the outer part to the inner part, thus reflecting the exogenous origin of the mercury and gold, and the endogenous origin of the lead from the blood supply. Determinations of the concentrations of other elements present in the teeth allowed the comparison of data between teeth through element-to-element ratios. The Ca/P ratio was recommended for this purpose. A more recent study based on a spatially resolved method enabled the determination of the migration of mercury from a filling into the tooth material and then into the human body, a process with obvious potential implications on human health. The interest in the use of other substances such as gold instead of amalgam for dental restoration was reinforced by the results [112].

The wide scope of LA–ICP–MS is reflected in its use for the characterization of annual growth rings in trees, which takes advantage of the fact that absorption at laser wavelength by a biological sample can differ from that of inorganic materials because the bonding energies of the molecules may be similar to photon energies and certain molecules can undergo photochemical changes. Therefore, in contrast to ablation of inorganic samples, organic material can show little or no evidence of melting or thermal damage. Calibration was done by using cellulose mixed with multi-element standard solutions and

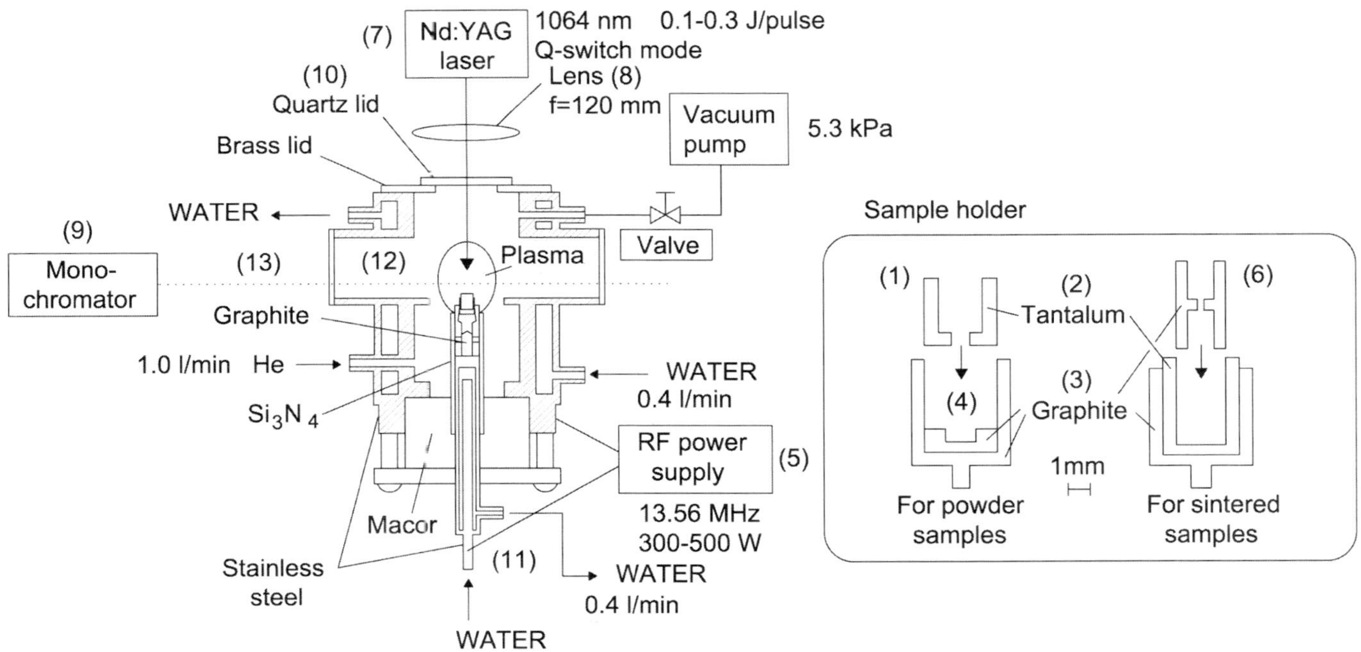

Fig. 9.6. Schematic diagram of the experimental system for laser ablation-assisted radio-frequency atomization excitation. (1) Sample holder, (2) tantalum lid, (3) graphite cup, (4) graphite disc, (5) rf power source, (6) different types of sample holder for sintered ceramics, (7) Nd:YAG laser, (8) laser beam-focusing lens, (9) spectrometer, (10) quartz window for laser irradiation, (11) central electrode, (12) discharge chamber, (13) quartz window for optical observation. (Reproduced with permission of the Royal Society of Chemistry.)

pressed into pellets. The elements studied (using a powerful Nd:YAG laser) included Mg, Al, Ca, Cr, Mn, Fe, Co, Ni, Cu, Zn, Cd, Ba, Tl, Pb, Bi and U. Although LA-based techniques provide only semi-quantitative results, the precision was high enough to allow the determination of elemental distribution and the identification of concentration trends [113]; this permitted the establishment of an autocorrelation function for the concentrations of elements in growth rings. The correlation time calculated from the autocorrelation function provided an indication of the transport of elements, thus allowing their migration from ring to ring to be studied [114].

One unusual application of LA–ICP–MS is for fingerprinting scene-of-crime evidence. The combination of this technique with ternary-plot pattern-recognition software was used to determine the origin of samples of steel (e.g. from safes, gun barrels, tools) and glass (e.g. floats, sheets, containers) [115].

A laser ablation-assisted glow-discharge excitation source with emission detection was used for the direct determination of elements in ceramics [116] using the experimental set-up depicted in Fig. 9.6. The sample was placed in a holder (1) that also served as hot electrode, and the atomizer consisted of a hollow Ta lid (2), a graphite cup (3) and a graphite bottom disc (4) with a cavity in which the powdered ceramic sample was loaded and irradiated with a Q-switch Nd:YAG laser (7) through a quartz window (10). Application of an rf power of 300–500 W at 13.56 MHz (5) to the central electrode (11) caused the formation of a stable He discharge plasma at a low pressure around the electrically conductive Ta hollow lid. Tantalum was also preferable for reaching higher temperatures because of lower radiative heat loss due to its low sensitivity relative to graphite. A different type of sample holder (6) was used for sintered ceramic samples in order to ensure a stable discharge, prevent the sample from melting with the Ta hollow lid and fix the sample location. Pulsed laser radiation of about 124 mJ per 10 ns was passed through the Ta hollow lid and successively hit the samples on the graphite disc at a rate of 10 pulses/s, resulting in increased atomization efficiency. The irradiation position of the sample surface was adjusted by focusing the lens (8). The atomic emission was detected by the spectrometer (9) and the analyte signal from the photomultiplier tube was acquired by a computer through an A/D interface. Either 5 μl of slurry or several milligrams of sintered silicon nitride was loaded in the cavity of the graphite bottom disc, the sample then being covered with the hollow lid and the sample cup fitted on top of the central electrode (11). The air in the discharge chamber was pumped out for 5 min to a pressure below 0.1 mm Hg and the helium was introduced into the chamber to attain a discharge pressure. Finally, the rf power was switched on and, concurrently, the computer started a programme to trigger the laser and acquire data in the boxcar integration mode. Effects of laser irradiation such as a sensitivity enhancement and alleviation of matrix effects were found in the direct atomization–excitation of ceramics. Figure 9.7 shows the marked enhancement of the emission signal for magnesium in the presence of laser sample irradiation and Table 9.1 shows the effect, in quantitative terms, of changes in the sample residue remaining in the sample holder, peak areas, relative standard deviations and normalized peak areas per ablated mass for Al. Peak area was increased by a factor about 4.9, which is greater than that obtained for a powder sample inserted into the system as a slurry. Also, the normalized peak area-to-ablated mass ratio for the ceramic material was increased by a factor of 2.0 and the reproducibility was improved.

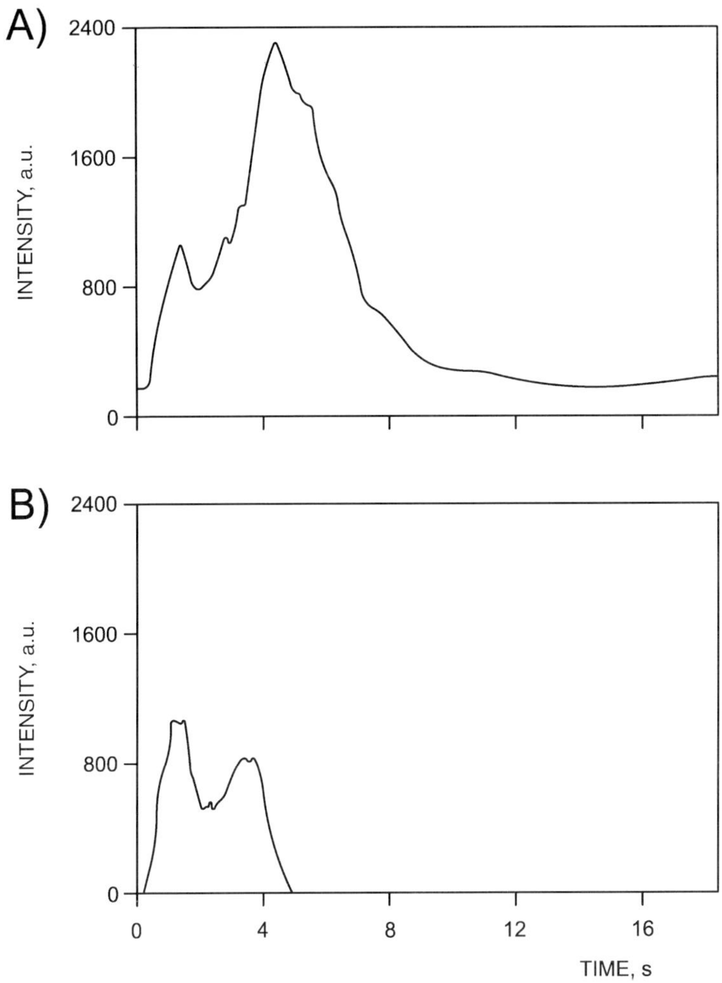

Fig. 9.7. Enhancement of the glow-discharge emission signal for magnesium in the presence of laser irradiation (A) as compared with no laser assistance (B). (Reproduced with permission of the Royal Society of Chemistry.)

Calibration using powder reference materials was possible for sintered ceramics. A very similar publication by the same authors, but in Japanese, was later reported [117].

Miscellaneous applications

Although LA is too expensive to be the method of choice for solution analysis, the use of synthetic solutions and dual sample insertion systems for laser-induced aerosols mixed with liquids (dried and wet) has offset the lack of solid reference materials for quantitation [14,45–48,55]. A specific sample preparation procedure exists [118] for

TABLE 9.1

EFFECT OF LASER IRRADIATION ON Al EMISSION PEAK IN DIRECT EXCITATION OF POWDER CERAMICS (n = 4)

Parameter	Without laser	With laser	Ratio
Average peak area (arbitrary units)	930 *	1600	2.6
RSD (%)	5.1	2.8	0.55

* Sample powder residue was occasionally found in the cup. This value is based on 100% introduction
Reproduced with permission of the Royal Society of Chemistry

modifying the absorption of aqueous solutions used as calibration standards in the laser ablation of solids. However, the ablation behaviour of solutions is wavelength-dependent, in contrast to IR light, which is transmitted to a high degree in dilute solutions and therefore requires additives or modifiers to perform direct ablation analyses [45]. In a study involving the ablation of high matrix containing solutions (0.1 g Sn/l), samples were filled into polyethylene microtubes, sealed and placed in the ablation cell to minimize evaporation and cross-contamination. The laser was imaged onto the tube, and an 80 μm hole (2 mJ pulse energy, 10 Hz repetition rate) was ablated. The greatest difference between solid and liquid ablation was signal stability, which was poorer for liquids owing to droplet splashing during ablation. Quantitation was done by using identically prepared synthetic solutions for calibration. The accuracy of the determination of 1 μg/g and the related limits of detection were at the sub-ng/g level for most of the elements. A total volume of 11 μl of solution was analysed three times from the same tube, which led to absolute limits of detection of 2 pg. The greatest assets of this ablation procedure are its very low sample uptake, and the greatly reduced spectral interferences and matrix effects. In addition, this approach might be useful with a view to investigating spectral interferences in laser ablation.

9.2.12. Comparison of laser-assisted sampling with other sampling techniques

Compared with many other techniques, sensitivity factors for LA–ICP hyphenated to either MS or AES are relatively uniform across the periodic table [21,22]. Also, the spectra for LA-assisted techniques are relatively simple and interferences between elements are more easily identified and corrected.

ICP–AES was recently used in combination with HRICP–MS for the determination of 70 elements in coal using various sample treatments. Four microwave-assisted digestion procedures involving different dissolution mixtures (viz. nitric and hydrofluoric acids, aqua regia, hydrogen peroxide), lithium metaborate fusion with and without previous sample ashing, and direct sampling by laser ablation, were tested. The use of high-purity reagents, prevention from sample contamination and various correction procedures for spectral interferences provided limits of detection in the low nanogram-per-millilitre range for most of the elements studied. The precision was assessed from replicate analyses (sample preparation included) of coal samples and found to average at 4–5% and 10–15% (as RSD) for procedures involving sample digestion and LA sampling, respectively. As regards dissolution mixtures, a combination of HNO_3/HF with H_2O_2 provided the best agreement with certified, recommended, literature-compiled and consensus values, even though quantitative recovery of Si, Cr, Hg, W, Zr and Y required the use of fusion. Overall, the results provided by LA fell within ± 20% of those obtained by digestion [119].

The practical implementation of various spectroscopic analysis methods involving ICP–AES or LA–ICP–MS and two rather different types of sample was assessed and the following conclusions drawn from the study:

(a) In the analysis of a magnetic material (NdFeB) by ICP–AES and LA–ICP–MS, accuracy in the determination of the major element (Fe) was found to depend on the particular dissolution method. Thus, Fe was preferably determined by classical titrimetry (see Table 9.2). However, the difference between the titrimetric and ICP results usually did not exceed 3%. When material other than Nd, Fe and B was to be examined, a survey analysis made use of a laser shot mandatory; the procedure revealed the presence of more than a dozen elements. Table 9.3 shows those that were subsequently determined by ICP–AES as well (following dissolution) to confirm the results. As can be seen from the table, the LA–ICP–MS results for the major elements were very poor but revealed at one glance the presence of substantial amounts of Co, Dy, Pr, Al, Si, Nb and Mo.

(b) Sintered substances such as $SrTiO_3$ posed dissolution problems. The question of whether a layer of $SrTiO_3$ deposited on a substrate of $SrTiO_3$ was possibly of lesser purity than the substrate itself was answered by using LA–ICP–MS (see Table 9.4). Subsequent neutron activation and ICP–AES analysis yielded the results also summarized in Table 9.4; prior dissolution of the powdered sample using an autoclave was required. These latter results confirmed the former. In fact, the initial LA–ICP–MS analysis provided all relevant information.

The principal conclusion from the previous studies was that ICP–AES was the most rugged and flexible approach, but also that LA–ICP–MS was uniquely suitable for direct solid local analyses [120].

A comprehensive comparison of LA–ICP–MS with other techniques for the analysis of ceramic layers of solid oxide fuel cells was conducted by Becker *et al.* [121]. They

TABLE 9.2

TYPICAL RESULTS OF ICP–AES ANALYSES OF THE MAGNETIC MATERIAL NdFeB

Analyte	Concentration (% w/w)	Analyte	Concentration (% w/w)
Nd	30.4	Pr	0.28
Fe [a]	66.0	O [b]	0.01
B	1.01	Other elements	0.71
Dy	1.49	Total	99.9

[(a)] Determined by titrimetry [(b)] Determined by gas analysis

TABLE 9.3

ANALYTE CONTENTS (% m/m) IN AN NdFeB MAGNET AS DETERMINED BY LA–ICP–MS AND ICP–AES

Analyte	LA–ICP–MS	ICP–AES
Nd	52	30.1
Fe	40	59.5
B	0.16	1.06
Co	2.0	3.15
Dy	3.4	2.00
Pr	1.3	0.70
Al	0.50	0.80
Si	0.28	—
Pb	0.50	0.12
Mo	0.092	0.095

Reproduced with permission of the Royal Society of Chemistry

TABLE 9.4

ANALYTE CONTENTS (μg/g) IN VARIOUS $SrTiO_3$ MATERIALS AS DETERMINED USING DIFFERENT METHODS

Analyte	Nd-doped $SrTiO_3$ target LA–ICP–MS	Nd-doped $SrTiO_3$ powder ICP–AES	Pure $SrTiO_3$ substrate	
			LA–ICP–MS	NAA
Ba	110	— [a]	25	17
Zr	15	— [a]	< 1	< 5
Nd	4100	— [a]	< 1	< 1
Dy	10	— [a]	< 1	< 3
Al	350	306	< 20	—
Cu	50	73	< 5	< 1
Fe	15	11	< 5	< 7

[a] Not determined
Reproduced with permission of Springer-Verlag

compared relative sensitivity coefficients (RSCs) in LA–ICP–MS and ICP–MS by using cathode layers consisting of mixtures of carbonates and oxides of La, Sr and Mn in the former case and a multi-element standard solution in the latter. As can be seen from Fig. 9.8, RSCs were similar for both mass spectrometric methods (they differed by a factor of only 0.1–3 for most of the elements studied). The RSCs thus determined were used for the analysis of $La_{0.6}Sr_{0.35}MnO_3$ perovskite layers and the target structure (a mixture of the initial oxides and carbonates or the perovskite) was found to have no effect on them.

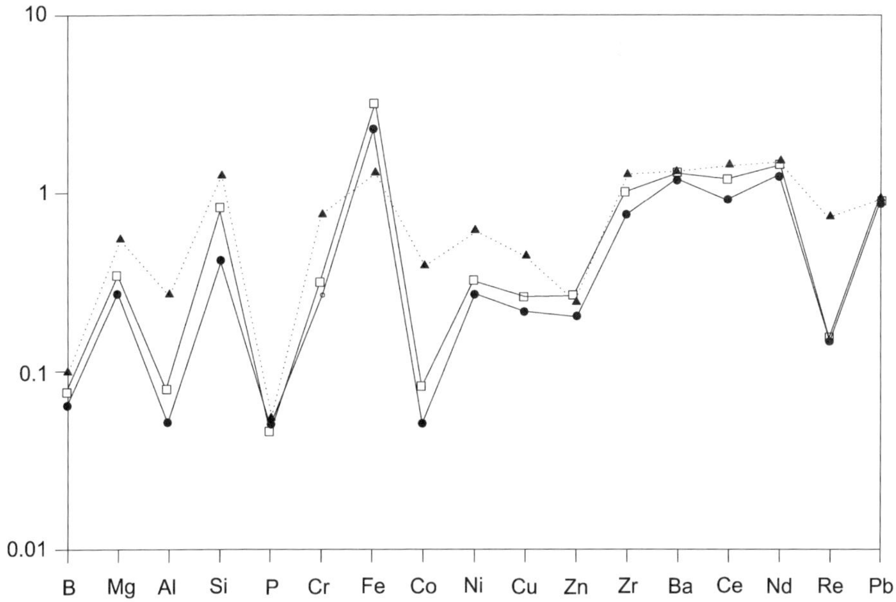

Fig. 9.8. Relative sensitivity coefficients in LA–ICP–MS [● $La_2O_3/SrCO_3/MnO_2$ stoichiometric mixture; □ perovskite ($La_{0.35}Sr_{0.65}MnO_3$)] compared to ICP–MS [▲ 0.7 ppm multi-element standard solution]. (Reproduced with permission of Springer-Verlag.)

The same perovskite was used to compare LA–ICP–MS and spark source mass spectrometry (SS–MS) for trace analysis [122], the results being quantified via RSCs in both cases. As shown in Table 9.5, the results of the SS–MS trace analysis were comparable to those provided by LA–ICP–MS. Limits of detection in the nanogram-per-gram range in SS–MS and in the low microgram-per-gram range in LA–ICP–MS were obtained. LA–ICP–MS was found to be subject to some problems — especially when a low-resolution mass spectrometer was used — arising from interferences of atomic ions of the analyte and molecular ions. Further inhomogeneity of the solid sample and possible memory or contamination effects could lead to spurious results. Such problems were exposed by comparing the results of LA–ICP–MS and ICP–MS for perovskite layers on yttria-stabilized zirconia. The Al, Co and Ni concentrations measured with both mass spectrometric approaches were quite consistent (see Table 9.6). The high concentration of Zr in the perovskite layers was ascribed to diffusion of this substrate element in the layers during sample annealing. Also, the porosity of the cathodic layer (layer I) seemingly led to apparent higher Zr values using LA–ICP–MS relative to ICP–MS because the substrate material (YSZ) was partly evaporated by the laser beam.

A comparison of laser ionization mass spectrometry (LI–MS) and LA–ICP–MS for the characterization of compact ceramics [123,124] provided limits of detection that were much lower (in the nanogram-per-gram range) with the former technique than with the latter. However, LI–MS is a time-consuming analytical technique with photoplate detection of mass-separated ions compared to LA–ICP–MS, which uses fast electrical ion registration.

TABLE 9.5

COMPARISON OF THE RESULTS (μg/g) OBTAINED IN THE ANALYSIS OF $La_{0.6}Sr_{0.35}MnO_3$ BY SS–MS AND LA–ICP–MS

Analyte	SS–MS	LA–ICP–MS
Mg	6.1	6.5
Al	2.0	7.1
Ca	17.0	—
P	4.9	6.0
V	1.5	—
Fe	2.5	—
Co	4.0	5.1
Ni	4.1	3.2
Cu	2.6	—
Zr	14.0	10.0
Mo	2.2	—
Ba	12.0	9.1
Ce	1.5	2.1
Pt	0.7	—
Pb	1.2	1.9

Reproduced with permission of Springer-Verlag

TABLE 9.6

RESULTS (μg/g) OF THE TRACE ANALYSIS OF $La_{0.6}Sr_{0.35}MnO_3$ LAYERS ON YSZ AS MEASURED BY LA–ICP–MS AND ICP–MS

	Layer I		Layer II	
Element	LA–ICP–MS	ICP–MS	LA–ICP–MS	ICP–MS
B	5.4	—	9.2	—
Al	83	80	92	80
P	17.1	—	79	—
Cr	16.8	< 10	32	65
Co	5.6	6.0	0.8[a]	5.0
Ni	29	30	32	40
Zn [a]	80	25	58	25
Zr	2173	760	1079	1200
Ba	18.6	12.0	19.3	5.0
Ce	7.2	< 1	6.1	1.5
Pb	0.9	—	0.5	—

[a] Possible inhomogeneity

Reproduced with permission of Springer-Verlag

9.3. LASER-INDUCED PLASMA SPECTROSCOPY

9.3.1. Introduction

The literature on the use of laser-induced plasma (LIP) or laser-induced breakdown (LIB) for chemical analysis has, since 1962, gone through three distinct periods of high interest. After the initial report by Brech at the X Colloquium Spectroscopicum Internationale [125], a period of enthusiastic, optimistic research ensued that peaked in about 1970. Most of this early work was carried out with commercially available equipment that used an auxiliary delayed electrical spark source for obtaining information of the spectrochemical plasma; however, the first spectrochemical use of the laser-induced plasma as such was reported in 1964 [126,127]. After about 1980, interest in the technique returned [128] and, since 1985, there has been a continual increase in research activity. At least 120 publications on the use of laser-induced plasmas for chemical analysis were reported over the period 1996–1997. Also, a number of acronyms have been proposed for this technique the most widely used of which is LIBS (for "laser-induced breakdown spectroscopy"). The renewed interest in LIBS stems from the availability of increasingly reliable, smaller, less expensive laser sources and, more importantly, from the development of sensitive imaging optical detectors such as the intensified charge coupled detector (ICCD). These detectors are time-gatable and hence capable of isolating optimum time intervals during evolution of the LIB plasma for spectrochemical measurement.

Laser-induced breakdown spectroscopy is a simple, rapid in situ analytical technique based on laser-induced plasma atomic emission. For analytical purposes, the plasma emission is spectrally resolved and atomic lines are subsequently analysed to determine elemental concentrations in the sample. The unique advantages of LIBS, which were recognized in the earliest work, remain compelling today: immediate in situ analyses can be carried out on any type of sample with no preparation. However, LIBS is subject to problems arising from a high continuum background, line broadening and self-absorption; these shortcomings have gradually decreased over the last few years.

9.3.2. Features of LIBS

Laser-induced breakdown spectroscopy (LIBS) features various assets and shortcomings worth considering in adopting it as the technique of choice for specific applications. The most salient advantages of LIBS for spectrochemical analysis are as follows:

(a) It requires little or no sample preparation, which results in increased throughput and reduced usage of labour-intensive, time-consuming preparation procedures — which, in addition, can introduce contamination;
(b) It is flexible enough for sampling all types of media and hence for analysing conducting and non-conducting materials, for example;
(c) It uses very small amounts (*ca.* 0.1 μg to 1 mg) of sample for vaporization, so it is often deemed a non-destructive technique;
(d) It enables the analysis of extremely hard materials that are difficult to digest or dissolve (e.g. ceramics and superconductors);

(e) Local analysis in microregions provides a spatial resolving power of *ca.* 1–100 μm;
(f) It possesses simultaneous multi-element determination capabilities;
(g) It can be used for direct detection of species in aerosols (in a solid or liquid particle present in a gaseous medium);
(h) It provides simple, expeditious analyses (ablation and excitation are done in a single step);
(i) It facilitates remote measurements.

By contrast, LIBS is subject to several shortcomings including the following:

(a) It uses equipment of high cost and complexity;
(b) Appropriate standards are difficult to obtain — which is why this technique is most frequently semi-quantitative in nature;
(c) It is subject to major interferences arising from the sample matrix or, in the case of LIBS for aerosols, and/or from particle size;
(d) It is less sensitive than conventional solution spectrometric techniques such as ICP–AES and GF–AAS;
(e) It exhibits poor precision (typically 5–10%) that depends on the sample matrix and degree of homogeneity, as well as on the excitation properties of the laser;
(f) The use of high-energy laser pulses can cause eye damage.

All these features except advantage (i) are shared by laser ablation. Current research in LIBS aims at minimizing these shortcomings while preserving or improving its performance.

9.3.3. Steps and thresholds in LIBS

Steps of LIBS

The LIBS technique basically involves four steps, namely: laser–solid interaction, material removal, plasma formation (also called "breakdown") and analysis of the photons emitted by the plasma formed. The conditions used in each step are usually optimized in relation to the particular application. Two of the previous steps (viz. laser–solid interaction and material removal) are also present in laser ablation (see Section 9.2.2).

Plasma formation (breakdown) The interaction of the laser beam with the removed material in the vicinity of the substrate surface causes strong thermalization and ionization of the vapour produced. As a result, the vapour is disrupted and a plasma is formed. Although some species can be vaporized directly as ionized fragments, the plasma forms via a number of processes involving both the solid and the vapour. Three types of process are specially effective in generating a plasma. In one, the electrons that migrate to the conducting band by effect of the heat supplied undergo an increase in kinetic energy through interaction with the radiation field in the form of reverse Bremsstrahlung processes. This causes additional ionization that gives rise to an avalanche effect. In the second type of process, ionization also results from collisions of high-energy ions with excited atoms and molecules. Finally, the use of a high enough beam energy causes the multi-photon ionization of neutral species in the ground state and the photoionization of excited atoms. The dynamic evolution of laser plasmas involves the supersonic expansion of photofragments, which produces a typical sound with each laser pulse. The expansion rate is proportional to the pulse irradiance. Plasma temperatures in the range 10^4–10^5 K and electron number densities in the region of 10^{15} to 10^{19} cm^{-3} have been measured at various stages of plasma evolution [129,130]. Eventually, the plasma decays through radiative, quenching and electron–ion recombination processes that lead to the formation of high-density neutral species in the post-plasma plume. The decay ends with the formation of clusters (dimers, trimers, etc.) via condensation and three-body collisions, and with the thermal and concentration diffusion of species into the ambient gas. This usually occurs within hundreds of microseconds (up to milliseconds) after the plasma has been ignited. This physical picture can be modified greatly in regimes of very short pulses (< 1 ns) [131–133], very high laser irradiances (> 1 GW/cm^2) and for unusual matrices (e.g. highly transparent media) [134,135].

Analysis of the photons emitted by the plasma This step is shared by other atomic emission spectrometries. However, the high electron temperature of laser-induced plasmas requires the use of temporal resolution to improve the detection sensitivity. The electron density of the plasma remains very high during the time the laser pulse interacts with the solid and the vapour. The emission spectrum obtained in this situation is a continuum due to ion–electron recombination. Ionic and atomic lines are superimposed on the emission background. Once the laser pulse has extinguished, the system rapidly loses energy and many atoms are excited. Observation of the plume at this time allows species to be much more easily detected.

Thresholds in laser-induced plasma formation

As with laser ablation, a number of borderline situations (defined via the so-called "damage threshold", "ablation threshold" and "plasma threshold") can be considered in describing the interaction of a laser pulse with a surface to induce the ablation and plasma formation associated to laser-induced breakdown. Only the plasma threshold is discussed here, however, as the other two are shared by laser ablation and are dealt with in Section 9.2.2.

The *plasma threshold* is the irradiance required to produce the optical breakdown of the vapour. It depends on the nature of the surface (particularly on various optical, thermophysical and thermodynamic properties of the material). The plasma threshold is usually higher than the ablation threshold. Observing optical emission entails applying more energy than that required to reach the plasma threshold.

Although phase explosion and the factors governing it [17] have not been studied in relation to LIBS, its threshold can be assumed to depend strongly on the laser beam spot size and wavelength.

Because neither ablation nor plasma formation require that the sample be of the conducting type, any type of surface can theoretically be examined. However, direct molecular analysis is unfeasible except in a few, specific cases. One other salient analytical feature of LIBS is that analyses can be performed without the need for special control of the sample environment at the times of ablation and plasma formation except in those cases where the emission lines for the target elements fall in the vacuum UV region.

9.3.4. Experimental set-up for LIBS

A typical assembly for development of an LIBS method consists of the different units depicted in Fig. 9.9, namely: (1) a laser device emitting at the appropriate wavelength, energy and repetition rate when pulsed; (2) beam steerers and lenses for proper guiding and focusing, respectively, of the laser beam; (3) a sample positioner with precise, reproducible *x-y-z* translation of the sample; (4) a fibre optics assembly comprising an optical fibre of an appropriate material dependent on the specific spectral working zone and lenses for collection and transmission of the radiation emitted by the plasma to a spectrograph of appropriate resolution for separation of the different atomic and ionic emission lines of the species in the plasma; (5) a detector (preferably an intensified coupled charge detector connected to a personal computer of adequate power running software for data processing). Proper performance of a pulsed laser requires the use of a delay-pulse generator to gate the detector, where the laser flash-lamp trigger signal is employed as the master control of the delay gate.

9.3.5. Basic aspects of LIBS

The limited number of quantitative applications of LIBS can be ascribed mainly to unsatisfactory figures of merit (accuracy, precision and limit of detection). This has largely been the result of an incomplete understanding of laser–material interaction processes and subsequent imprecise control of variables. Winefordner and coworkers developed a model for time-resolved measurement of the intensity variation of five lines along the height of a laser-induced breakdown plasma on a solid target under standard atmospheric conditions. Following background correction, peak stripping of the lines at 367 and 368 nm, and integration over the line profiles, the temporal and spatial development of the populations in the levels from which the fine line originated were calculated. The spatial and temporal population distribution suggested that the levels in question were not in equilibrium with one another in most measurements. Spatial and temporal excitation temperatures in the region of 14 000 K were obtained from Boltzmann plots, but the accuracy of these values was dubious in the light of non-equilibrium thermodynamic conditions, even with delay times as long as 14 μs and despite the fact that, based on theoretical models, the relaxation time needed for a transient plasma under such pressure and energy density conditions to reach thermal equilibrium was in the region of 1 μs.

One plausible explanation for the previous discrepancy is that the LIB plasma acts like an explosion in which the high-energy particles are mostly moving radially directed outward in space away from the target material. As a result, a longer relaxation time will be needed for sufficient collisional exchange to occur before energy equilibrium is reached compared to a similar source with random motion of energetic particles [136]. Two-dimensional images of the spatial and temporal distribution of chemical species associated

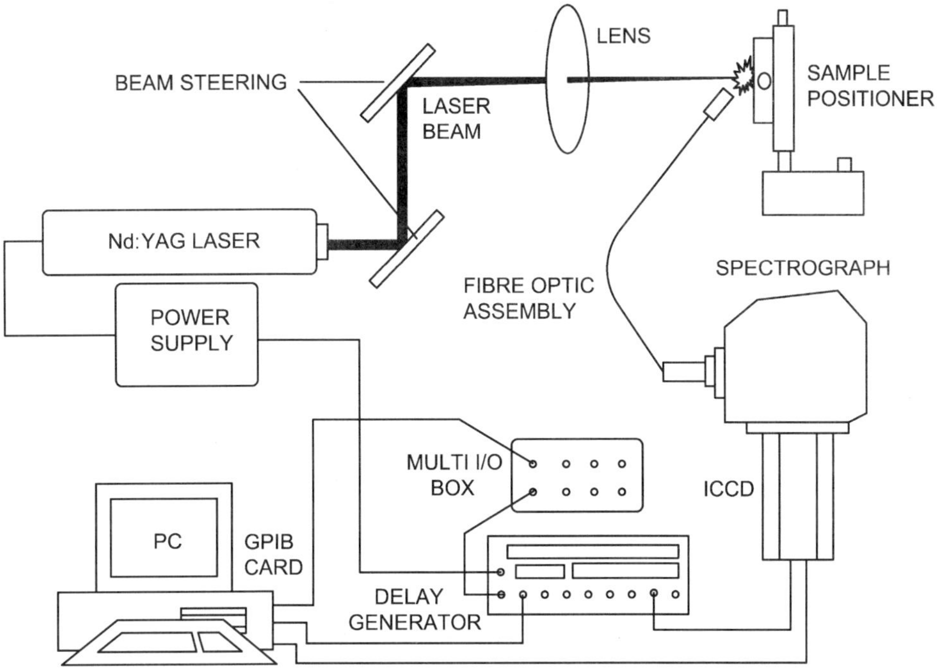

Fig. 9.9. Experimental arrangement for implementation of LIBS. (Reproduced with permission of the Royal Society of Chemistry.)

with the spectral emission were captured in a subsequent study [137]. The time-dependent spatial shape and size of the plasma, viewed simultaneously from two orthogonal directions, were measured in terms of the emission from a lead ionic line and a set of lead atomic lines. The main conclusion was that temporal and spatial behaviour were characteristically different for the different spectral lines: the ionic emission was confined to a smaller, more concentrated central part of the LIB plasma, whereas the atomic emission was more evenly dispersed over the entire plasma. These results revealed the need to consider two new variables (in addition to well-documented independent variables such as laser pulse duration and shape in time, laser wavelength and energy, the physical and chemical properties of the target material, and the composition and pressure of the surrounding environment), namely: the specific volume of plasma being observed and the line of sight of the detector [137].

The rotational, vibrational and excitation temperatures in plasmas formed by Nd:YAG and excimer lasers on graphite targets have been compared. Molecular emission from CN in the plasma was used to determine vibrational and rotational temperatures; a small amount of Fe or Pb was added to the graphite target and the emission from a series of these lines was observed to determine excitation temperatures. While the excitation and vibration temperatures were similar, the rotation temperatures determined in each case were almost one order of magnitude lower than either the vibrational or excitation values.

Also, the rotational populations were such that it was impossible to determine a single temperature, possibly because different mechanisms of CN formation resulted in this multi-temperature distribution. The vibrational and rotational temperatures were higher for the excimer laser (308 nm induced plasmas) than for the Nd:YAG laser (1064 nm plasmas), viz. 18 000 K versus 10 000 K.

9.3.6. Variables affecting LIBS performance

In developing an LIBS method, one should tune the equipment to be used in such a way as to obtain optimal values for those variables affecting performance — and hence the final results — to the greatest extent.

Laser features

Producing breakdown above the analysed surface entails using a high *laser power*. In theory, one can partly compensate for a low laser output energy by intense focusing up to a point where diffusion becomes dominant. However, a too strongly focused beam produces highly localized sparks that may result in local fluctuations in sensitivity. This is no restriction with homogeneous samples and even with inhomogeneous ones provided a proper sampling method is used to ensure accurate statistics. There is, however, a theoretical and a practical limit to the focusing ability that must be established in terms of the effect of the laser power on signal intensities and signal-to-noise ratios.

Aragón *et al.* [138] studied the spatial distribution of Cu, Mg and Fe in plasmas produced on copper, stainless steel and alumina matrices as a function of laser power. They found the logarithmic emission intensity for the alumina sample to increase linearly with increasing laser power density over the range 80–900 GW/cm^2. By contrast, emission intensity in the metallic samples remained constant up to 700 GW/cm^2 and then increased sharply. By observing emission and ratioing it to copper emission, the formation of a plasma in air was found to shield the metal targets at intermediate power values and the analysis of metallic samples in air to require power densities in excess of 700 GW/cm^2.

The *laser repetition rate* is a key variable in the LIBS technique. Laser breakdown produces a persistent mass of aerosol above the sample, the production rate increasing with increasing repetition rate and yielding a higher steady-state aerosol concentration above the sample.

Laser wavelength is also of paramount significance in LIB spectroscopy. Excitation temperatures examined as a function of laser wavelength revealed plasmas induced by UV wavelengths to provide higher temperatures than those induced by IR radiation, which was ascribed to the stronger shock caused by UV light.

Lens effect

Each lens produces a different energy density distribution in space that results in different plasma characteristics. A laser beam can be focused to a minimum size d given by the diffraction limit relation [139], $d = 2.44f\lambda/D$, where λ is the laser wavelength, *f* the focal length of the focusing lens and *D* the laser beam diameter before focusing. In practice, the focus size is larger than that predicted by the diffraction limit owing to irregularities in the focusing optics and to changes in laser-cavity transverse temporal modes. The use of high-quality optics and beams in the TEM_{00} mode provides focal diameters between 2*d* and 5*d*.

Too harsh focusing conditions (viz. too small a focal length) are commonly believed to provide poor results. Usually, relatively soft focusing (i.e. the use of a long focal length) is preferred, partly because, under such conditions, the exact location of the sample along the laser beam coordinate is not too important. The properties of laser plasmas are known to be highly dependent on the laser power density at the surface and vary substantially for a single lens. By using constant laser characteristics such as the power density and changing the focal length to a metal Cu target, Bulatov *et al.* [140] showed a lens of $f = 150$ mm to produce very small amounts of plasma owing to the high energy density localized at the focal point. This is the origin of the well-known poor analytical performance achieved under such conditions. On the other hand, a lens of $f = 1000$ mm is also unsuitable for analytical purposes owing to the long "tail" along the laser beam. At longer focal lengths, the high energy density required for the plasma to form is reached far ahead of the sample surface. Therefore, the plasma is created in the air, away from the sample. Much energy is lost ahead of the observation volume, thus considerably reducing analytical performance. The best conditions are obtained with $f = 400$ mm. The plasma "tail", which is clearly observed at long focal lengths but is also present under medium focusing conditions, originates from the laser pulses producing an aerosol above the sample. The next pulse is partially scattered and only too small a portion of the energy reaches the sample surface as a result. This reduces analytical performance. On the other hand, the same aerosol effect contributes to improving the results obtained under medium focusing conditions, where the energy density (at the same distance from the surface) is lower. In this situation, less light is scattered and the persistent aerosol is excited in the hot plasma to produce good analytical spectral lines. This aerosol consists of substrate material and produces the right spectral lines without the need to spend energy on surface ablation. This effect, which was previously observed and quantified [140,141], was explained by Bulatov *et al.* based on the imaging of the plasma. Generally, a long tail can also be established by using a short focal length if the power density is reduced, the results depending on the sample properties as well.

Crater shape

In a typical LIBS configuration, the laser beam (in the best instance following a near-Gaussian energy distribution) is focused on the sample surface by using a single-lens optical system. The sample area affected by the core of the incident pulse will be preferentially ablated, while the beam edges (irradiated with a less uniform, lower energy pulse) will be ablated to a lower extent. The resulting effect is the formation of an uneven crater profile with a typical conical shape. The depth information derived from such a crater tends to convolute the ablated material within the crater profile, with the deleterious effect of a loss of actual depth resolution. Beam masking is one way of filtering the beam-edge contribution and obtaining more homogeneous craters. This was checked by altering the spatial profile from an XeCl excimer laser by using a straightforward two-lens telescope to generate a flat energy profile beam that impinged on a layered sample (Zn-coated steel) without beam conditioning. The irradiance obtained (*ca.* 10^7 W/cm^2) was high enough to vaporize the target, cause plasma formation and allow the use of atomic emission spectrometry with acceptable signal-to-noise ratios. By modifying the beam energy distribution, flat ablated profiles and improved depth resolution (up to a few nanometres per pulse) were obtained [142].

The crater diameter and ablation depth per pulse depend primarily on laser irradiance but also on the transfer characteristics. Conducting materials exhibit large craters, while insulators have deeper craters as a result of poorer lateral heat conduction. Figure 9.10 shows a series of craters obtained at different laser energies. The aspect ratio (R) defines crater shape as a function of crater diameter (d) and depth (h): $R = d/h$. However, the ablation depth increases only up to an irradiance of 7.5×10^7 W/cm^2. Higher irradiances result in decreased ablation depths per pulse. This seemingly contradictory behaviour arises from shielding of the sample surface by the plasma, which decreases the effective amount of energy that is deposited on the surface. Clearly, because the laser is focused on the surface, the beam should be defocused along the propagation axis of the beam. The concomitant increase in beam area causes a decrease in laser irradiance. Consequently, the combined effect of shielding and the variation of irradiance along the sample depth makes it difficult to obtain an accurate description of the ablated depth per pulse. The measured ablation depth per pulse thus represents an average value over the studied depth. Based on the above, the best conditions for mapping are those of maximum focus and minimal energy flux. Under such conditions, performance is 70 μm for lateral resolution (minor axis of the crater) and 0.16 μm per laser shot for depth resolution.

Refining the LIBS approach entails improving the focusing optics and using lasers with a more uniform energy distribution across the beam. The influence of laser focus on crater shape is shown in Fig. 9.11. The parameter "working distance" (WD) is defined as the difference between the lens-to-sample distance (g) and the focal distance (f): WD $= g - f$. Advances in these areas will provide surface craters of controlled shape, which will in turn result in improved spatial resolution in LIBS.

The effect of ambient pressure on the propagation of plasma formed on an Al target in air has been studied [22]. The speed of propagation and the range of species travel speeds were found to increase with decreasing pressure.

Sample location

The location of the sample relative to the lens focus is usually optimized empirically; more comprehensive information, however, can be obtained from two-dimensional visualization [140]. The general knowledge is that placing the sample somewhat in front or behind the focus provides better results. The results obtained by using a length of the optimal focal distance and changing the sample location revealed that when the sample was placed too close to the lens, the plasma was very flat and relatively cold. Under these conditions, the plasma could not be produced in the air (it was induced solely by the surface material). As the sample was moved towards the focus, the plasma became more spherical and the results improved. The best conditions were obtained when the

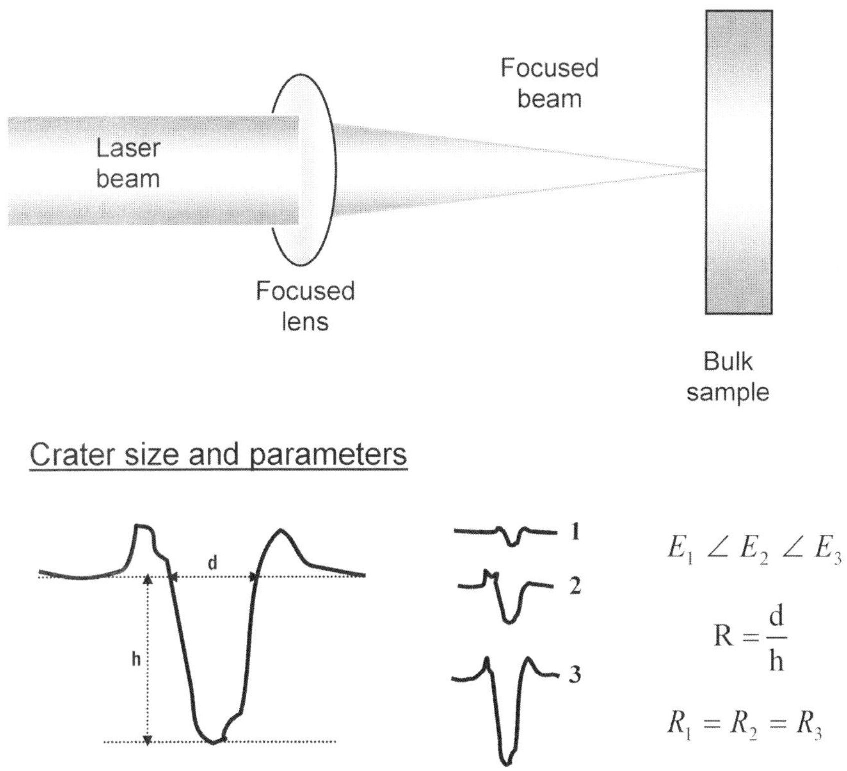

Fig. 9.10. Effect of beam power on crater shape and size. For details, see text.

plasma was induced in the air, right in front of the sample. When the sample was placed exactly in the focus, a hot ring (probably formed by scattered material) around the main plasma plume was observed that gave rise to impaired analytical performance.

Grain size

Size effects in chemical analysis are influences of sample size on the monitored concentrations. Either the analytical procedure is size-dependent or the distribution of the analysed compounds is. In other words, samples with the same overall (m/m) concentrations but different grain size may provide different readings. In LIBS, such effects stem from the laser's sampling process and from the nature of contamination in special cases. The laser pulse ablates a thin layer of the sample surface. Compared to sand, grains of soil are relatively uniform so no particle size effect is to be expected from them; on the other hand, sand grains are larger and more prone to having size and contaminating effects. Natural abundance elements that are evenly spread throughout have no size effects. However, the nature of some man-made contamination is such that contaminants

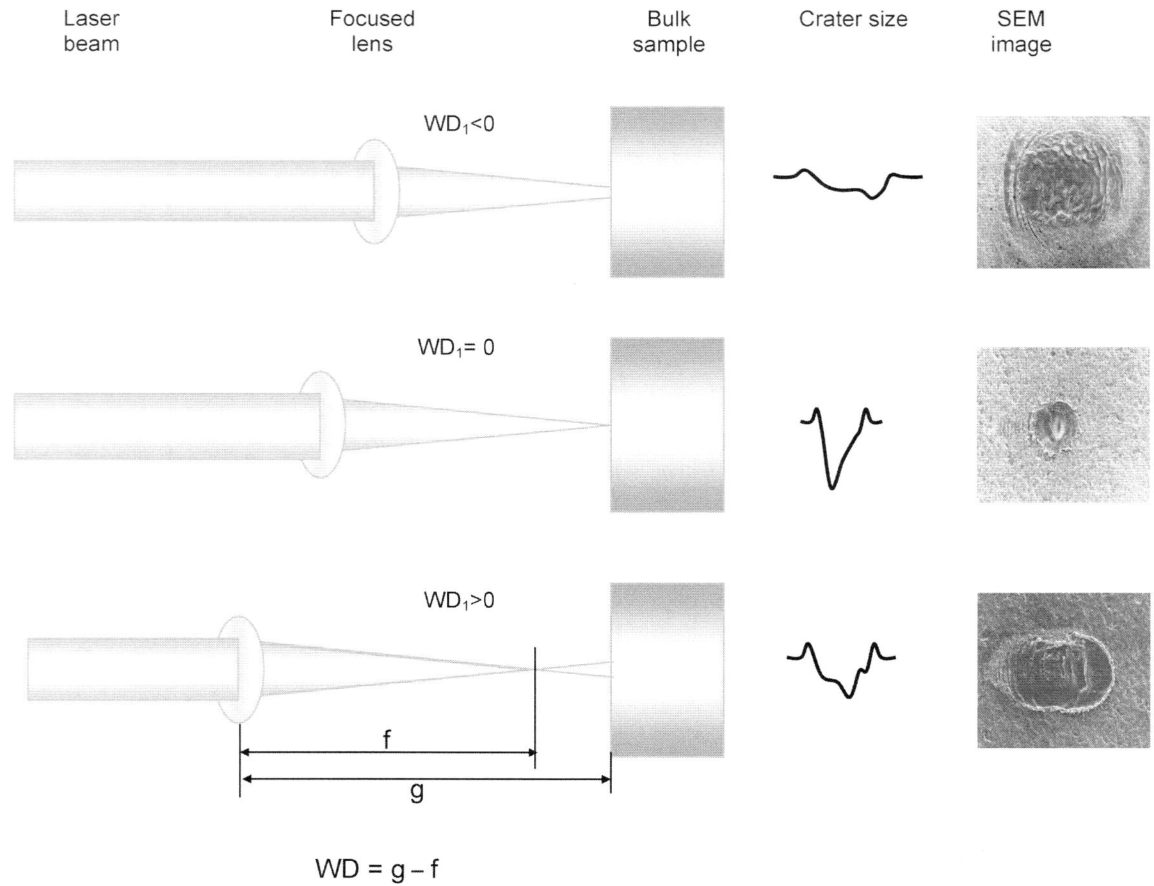

Fig. 9.11. Influence of the focusing conditions on crater shape and size. WD working distance, *g* lens-to-sample distance, *f* focal distance, SEM scanning electron microscopy.

are stuck mainly to the grain surface; such is the case when sand is contaminated by solutions containing heavy metals.

Sample moisture

The presence of water in the sample reduces the emission intensities from its plasma, mainly because some laser energy is wasted for water heating rather than for plasma formation. For example, about 25 mJ is required to evaporate a thin (10 μm) film of water over an area of 1 mm^2. Also, the aerosol production mechanism is suppressed by the water, which results in decreased signals at high repetition rates. On the microscopic level, water increases the energy requirement for ablation of the substrate. However, these effects act in different ways for different matrices. For a sand sample, the intensity is already reduced to one half at a 0.8% moisture content; by contrast, soil samples require 5% water for an equivalent reduction. The reason is that soil adsorbs water much more efficiently than does sand. The water does not penetrate into sand grains and, at the same moisture content (w/w), much more water is present at the surface compared to soil. The practical meaning of these results is that the sample has to be dried in order to obtain comparable results. However, only local, temporal drying is necessary. The laser samples a layer only a few microns thick and the plasma lasts only a few microseconds. Therefore, the drying device is required to deliver sufficient heat to win in the competition with capillary forces, which tend to rewet the site, just for the short duration of the measurement.

Experiments with $CaCO_3$ pellets exposed a linear decrease in emission intensity with increasing percent weight of water. The emission of the Ca ion lines at 393.4 nm was found to decrease by one order of magnitude over the water wt% range from 0 to 30. Temperatures derived from Boltzmann plots obtained by adding $PbCl_2$ to graphite and observing lead lines were *ca.* 10 000 K for all targets irrespective of moisture content. Plasma images showed the size of the plasma produced to be roughly independent of moisture content as well. This led to the conclusion that the decreased emission was largely the result of the amount of sample removed by the laser pulse decreasing with increasing sample moisture content.

A halogen lamp (100 W) suffices to dry samples with low moisture contents; however, a more powerful heating device is required for wet samples. Direct contact with a heating element provides good results, but introduces the risk of external contamination. Alternatives such as microwave heating can be effective for this purpose.

Atmospheric gas

The plasma spot is produced in the gas phase just above the sample surface. Therefore, the composition of the gas influences the plasma characteristics and the measurement results. This effect has been studied with and without a closed chamber.

When the gas flows over the sample in the absence of a chamber, the flow-rate must be low in order to avoid removal of particles and altering the surface shape. The gas improves the intensity of the atomic line of Zn at 481.1 nm and precision with respect to measurements in air. The signal is assumed to be enhanced by two types of factors.

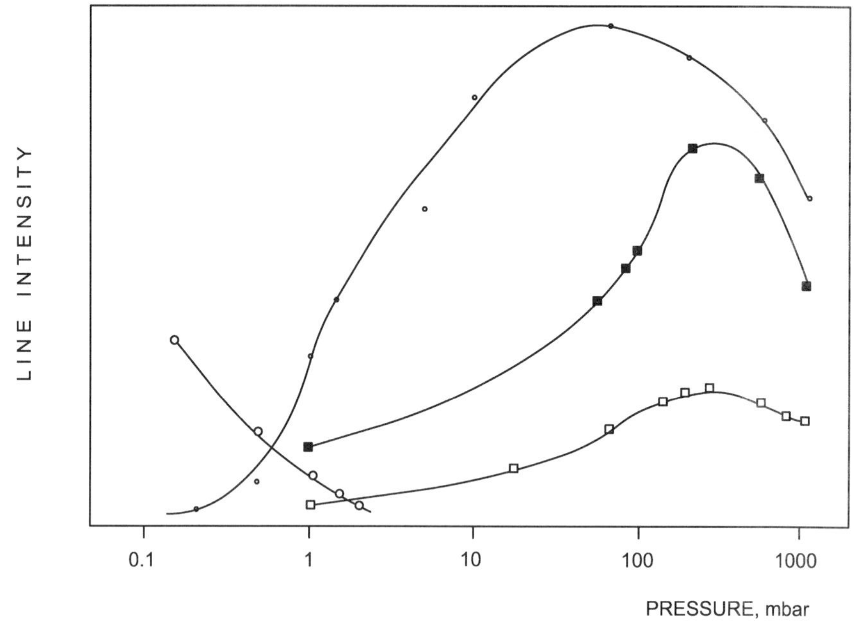

Fig. 9.12. Variation of the intensity of Cu lines as a function of the pressure of the surrounding air atmosphere (●) Cu I 578.21 nm, (■) Cu II 254.48 nm, (□) Cu II 236.98 nm, (○) Cu II 490.97 nm. (Reproduced with permission of Elsevier.)

One is the ionization potential of the gas: the lower it is, the easier the plasma is formed and the higher its temperature can be. The other effect is cooling of the excited analysed atoms in the plasma. As the surrounding masses are heavier, the transfer of collisional translational energy is less effective and the plasma lasts longer. As expected, the relative intensities increase as the mass of gas involved increases and the ionization potential decreases. Argon is an inexpensive, practical choice for increasing signals by a factor up to 1.8 with improved precision [140].

Studies on copper atomic and ionic lines over a pressure range for the surrounding gas of 0.1–100 mbar (Fig. 9.12) revealed the virtual absence of line emission at 1 mbar irrespective of line type. The line intensity of the ionic line with the highest energy became negligible above 10 mbar. Therefore, only Cu ionic lines with low excitation energies were observed in addition to the Cu atomic lines with an intensity lower by a factor less than 2 at atmospheric pressure. This suggests that the potential loss in line intensity due to atmospheric pressure was offset by a higher probing efficiency (i.e. a greater number of emitting atoms probed by the collimating system). Because no air plasma was observed, Cu was assumed to be the main plasma component. Thus, the species composition of the plasma was found to depend solely on the origin of the ablated material (i.e. on the ablation process), which was independent of the nature of the surrounding atmosphere. Also, the high intensity of the Cu continuum suggests that electron–ion recombination, which was largely pressure-dependent, played a major role in the ionic and atomic state populations.

Subsequent research by Millán *et al.* [143] using poly(methyl methacrylate) as a model to study spectral interferences by air in LIBS revealed that the emission spectrum from the plasma included the surrounding atmosphere (O, N and C lines mainly). In order to avoid either operating under vacuum conditions or the use of controlled atmospheres, these authors tested two different approaches, namely: (a) defocusing the sample to physically separate the target sample and air plasmas, which were efficiently resolved by using the spatial resolution capability of a CCD; and (b) using the different rates of decay of the sample and air spectra to selectively obtain the sample emission spectrum with a 1500 ns delay time, the effect being similar to that obtained with spatial resolution.

Observation height in the plasma

Proper selection of the observation height in the plasma is crucial with a view to obtaining the best operating conditions (particularly the highest possible signal-to-background ratio for analytical atomic lines in the absence of self-absorption and line broadening).

Emission line

The line of choice for monitoring the emission spectrum depends mainly on the following two factors:

(a) The concentration of the target analyte in the sample. For a minor component, a resonant line provides maximum sensitivity; for a major component (e.g. gold in jewels [144,145]), a non-resonant line is to be preferred that is selected with a view to minimizing the background contribution and potential interferences from other sample components.
(b) The sample matrix. The presence of emission from other sample components can severely restrict selection of the analyte emission line, so much so that a non-resonant line may provide better results than a resonant line. The analyst's expertise plays a decisive role in the choice.

9.3.7. Shortcomings of LIBS

Plasma analysis is known to be subject to spectroscopic disturbances including plasma line instability, self-absorption, line broadening, slit problems and grating imperfections.

Stability of laser-induced plasma emission

One of the shortcomings of LIBS, particularly in relation to quantitative elemental analysis, arises from the instability of the laser-induced plasma emission resulting from laser intensity fluctuations (1–5%); the amount of scattered light present depends on local matrix effects and on physical and chemical properties of the target material. The most common way of compensating for signal fluctuations in LIBS is by calculating the ratio of the spectral peak intensity to that of a reference intensity. However, this internal calibration method provides relative rather than absolute concentrations.

Shot-to-shot variations in laser power

One of the major drawbacks of the LIBS technique is the poor reproducibility of shot-to-shot spectra. This problem arises from various sources including laser instability; variations in lens-to-lens distance; optic coating or destruction; matrix effects (e.g. of the sample composition, degree of homogeneity, colour, moisture content) and the atmospheric conditions.

Sample inhomogeneity

The lack of homogeneity of the sample is a major source of variance. The laser samples an average area of 1 mm^2, which is not sampled uniformly as different parts are exposed to different temperatures. A routine measurement typically involves about 100 laser pulses, so the examined area is in the region of 1 cm^2. Inhomogeneities influence laser absorption, the mass ablated and the plasma conditions. On the other hand, there can also be substantial inhomogeneities in the standard reference samples that will pose problems with analyses. These problems become even more severe if the laser is focused more tightly in order to sample smaller amounts of material.

Optical instability

One other intrinsic problem of LIBS stems from optical instability as a result of the distance of the surface from the lens varying by effect of the usual roughness of the surface. As a result, the location of the spark relative to the collecting fibre also varies. In addition, the fraction of reflected light changes from pulse to pulse owing to differences in impact angle on the microscopic level. Optical instability causes fluctuations in the ablation process, the plasma profile and the inspected volume of plasma.

Self-reversal

Studies of profiles of Ba ion lines in plasmas formed by an Nd:YAG laser at 1064 nm on $YBa_2Cu_3O_7$ in air at atmospheric pressure as a function of laser power showed the line at 455.4 nm to be strongly self-reversed at all laser powers, and a third peak to evolve in the centre of the self-reversed profile at power densities above 16 GW/cm^2. This third peak was ascribed to fluorescence of Ba ions that absorbed emitted radiation, the absorption peaking 120 ns after the initial laser pulse.

The effect of atmospheric gas on the emission of an LIB plasma [146] exhibits self-absorption in He compared to Ar as a result of increased free atom populations in the outer regions of the plasma. Spectral line widths do not correlate well with atmospheric gas. This rules out Doppler effects as a major source of broadening in the laser-induced plasma. The use of a low pressure (*ca.* 1 torr) to examine the influence of this variable on the shock wave or secondary plasma revealed an increased emission intensity, which confirmed the assumption that the secondary plasma was excited by the shock wave.

The effect of Ar, He, N_2 and air on the self-reversal of Cu emission lines is illustrated in Fig. 9.13A, where spectra (recorded with a delay time 300 s) are normalized to the

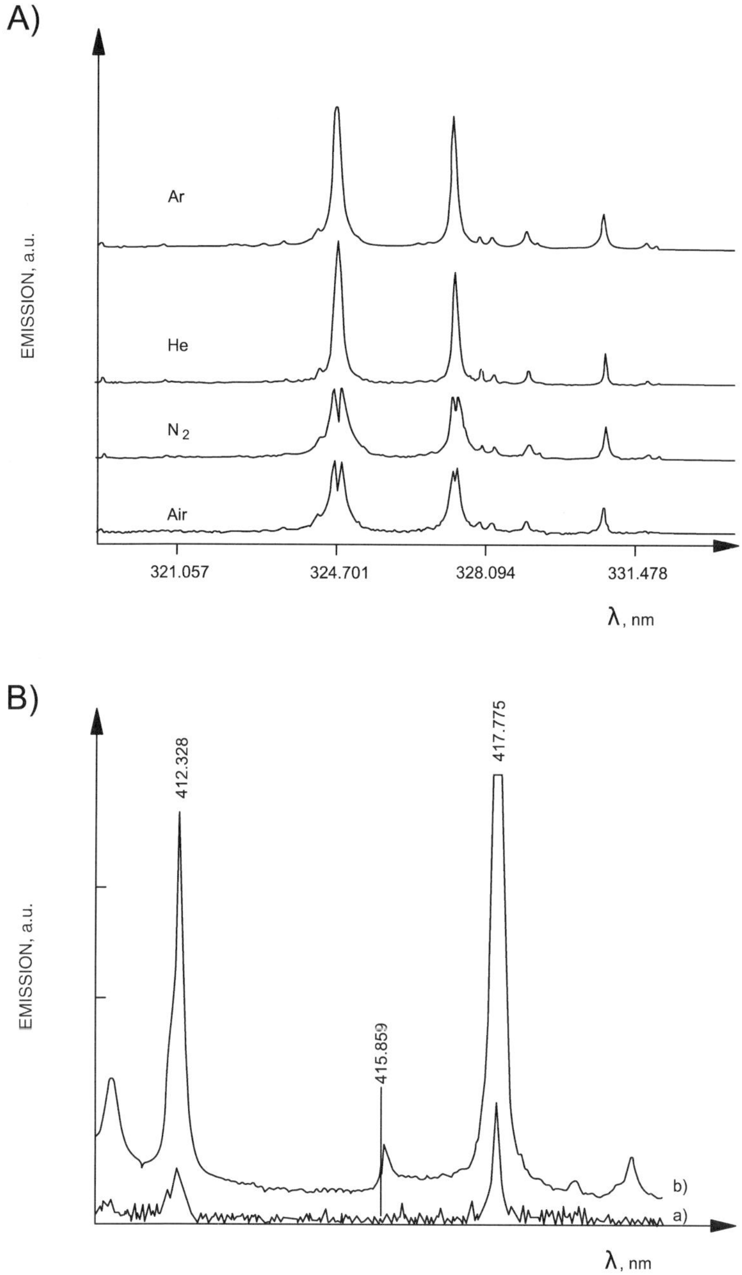

Fig. 9.13. (A) Influence of the nature of the surrounding atmosphere on Cu emission lines (the various spectra are normalized to the Cu 339.78 nm line). (B) Influence of the surrounding atmosphere on the intensity of the Cu I 412.328 nm line: (a) air, (b) Ar. Two Ar lines are present, namely: Ar I 415.859 nm and Ar I 417.775 nm. (Reproduced with permission of Elsevier.)

330.78 nm line, in the vicinity of the Cu I 324.7 and Cu I 327.4 nm resonant lines. The spectra suggest that the main difference between the air or nitrogen and He or Ar atmospheres is the strong self-reversal effect of the resonant lines. The higher sensitivity obtained for Ar is exemplified in Fig. 9.13B for the Cu I 412.328 nm line. As can be seen, some Ar lines (viz. those at 415.859 and 417.775 nm) are observed for the Ar surrounding atmosphere; by contrast, the air surrounding atmosphere gives no lines from air components.

Line broadening

This undesirable effect is usually observed at short delay times. Figure 9.14 shows a large width change in the Cu I 427.51 nm line for time delays from 0 to 1 μs, together with a wavelength shift of 0.21 nm linked to the line broadening. After 500 ns, the experimental linewidth is limited by the spectral bandpass of the monochromator. With shorter delay times, line broadening is more marked — partly as a result of self-absorption — but the different lines exhibit broadening to a different extent, which can only be ascribed to other factors such as pressure, and Doppler and Stark broadening, pressure being the most influential factor with times longer than 20 ns.

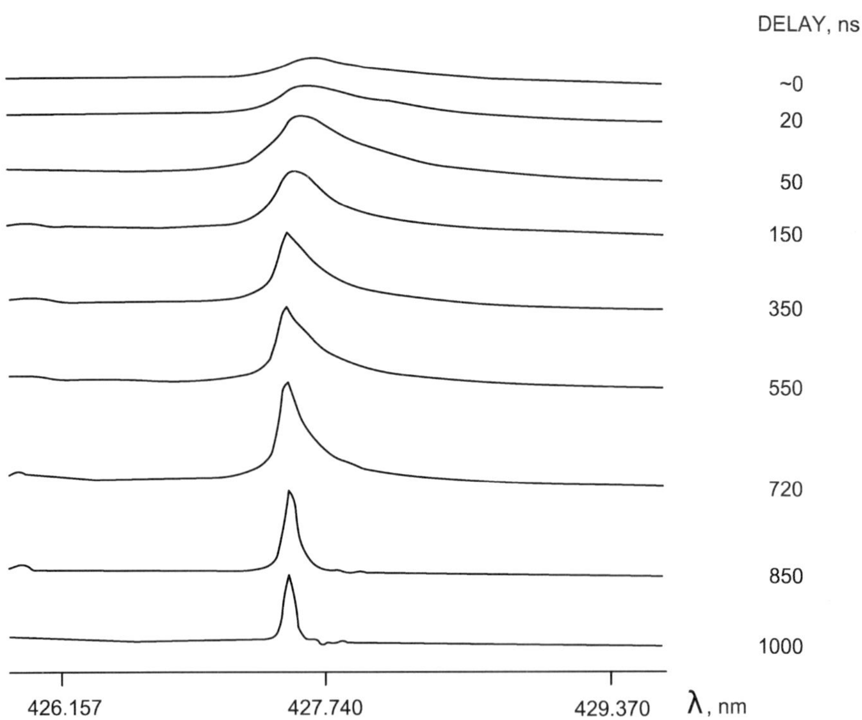

Fig. 9.14. Variation of the line width and shift of the Cu I 427.51 nm line with increasing time delays. (Reproduced with permission of Elsevier.)

9.3.8. Data collection: detector delay

The spectral emission of the plasma is time-dependent. Directly after the laser pulse, the emission is characterized by an abundance of ionic lines on top of a broad recombination background. The LIB spectrum is essentially featureless during the first 250 ns. At successively longer times, emission lines "emerge" from the background and increasingly dominate the spectrum. Also, lines are better resolved at longer times because their widths decrease as the plasma cools and the electron density decreases. The temperature is typically above 15 000 K and the broad spectra obtained are scarcely informative. The detector should thus be gated and light collected within the optimal time window. The characteristics of the emitted light depend on several factors, especially prominent among which is the plasma profile; this is dictated both by the laser (optical set-up included) and by the sample matrix. The influential factors include the individual response of each chemical element to the plasma conditions and even the differential behaviour of the spectral lines for the vapours of each element. Figure 9.15 shows the normalized average intensities as a function of time for three samples during the first 8 μs. In these measurements, the delay time was varied while the integration time was kept constant at 1 μs. The results reveal much shorter lifetimes for both sand and soil than for the metal surface. The soil matrix has the shortest lifetime, which means that its plasma cools down much faster. In metals, which can easily release electrons, the plasma is more easily produced

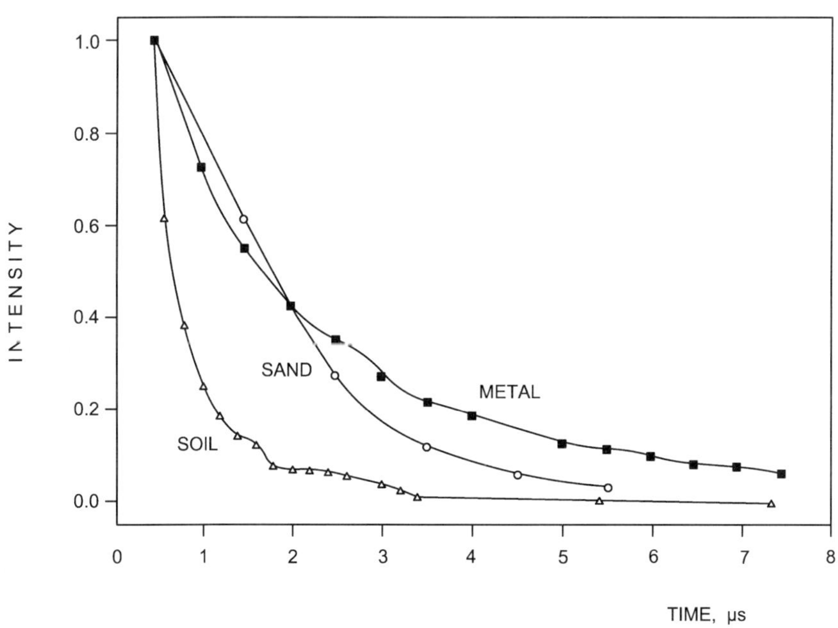

Fig. 9.15. Average intensities as a function of time after the laser pulse. Each sort of material is characterized by a different decay constant. Thus, each material has a different informative lifetime. (Reproduced with permission of the American Chemical Society.)

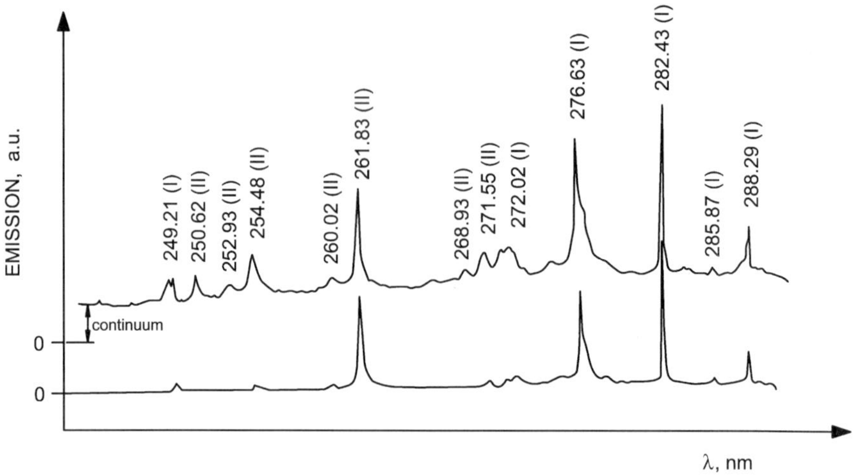

Fig. 9.16. Cu spectra obtained in air at atmospheric pressure. The upper spectrum was obtained with a time delay shorter than 200 ns after the laser pulse; the lower spectrum, with a time longer than 200 ns. (Reproduced with permission of Elsevier.)

and is heated to higher temperatures. Also, the plasma lasts and radiates for longer times. Plasma lifetime is also influenced by other effects that are responsible for differences between matrices. Such effects include the ablation of particulates (rather than vapours) into the hot plasma and heat transfer. There are thus two time parameters to be optimized, namely: the delay time and the integration time. Based on the results of various studies, the former requires much stricter control than the latter.

As can be seen from the Cu spectra of Fig. 9.16, which were obtained in air at atmospheric pressure over a short wavelength range using different delay times (shorter and longer than 200 ns), the continuum intensity was very high with short delay times; also, Cu atomic lines were abundant and Cu ionic lines scant. Above 200 ns, however, the continuum intensity was low and only Cu atomic lines were readily observed. No laser-induced breakdown of air was observed in the absence of a target at a laser energy and wavelength of 3 mJ at 337.1 nm. Also, although the discharge was formed in air, no lines or molecular bands from air components were observed, probably as a result of the fast expansion of the plasma and the absence of diffusion of air into the discharge. Breakdowns for both air and argon occur at a higher energy (e.g. 10 mJ at 1064 nm). Breakdown at 532 nm in He has also been observed in the determination of hydrides [147]. Experiments involving a more powerful XeCl excimer laser (up to 300 mJ) revealed no breakdown in air, which suggests that its absence was largely due to the use of a UV wavelength. Consequently, plasmas produced by a UV laser seem to differ in nature from those produced by a laser emitting a red line. The optimum delay time is element-dependent; thus, it is 2 μs for Cl, F, P, S and As, and 4 μs for Be and Na.

Constraints such as the presence of a continuum background, line broadening and self-absorption can be minimized by using time-resolved LIB spectroscopy (TRE-LIBS)

[148–152] and time-integrated spatially resolved LIBS [138,146,152–154] in a controlled atmosphere. In most LIBS work, TRE-LIBS has usually been employed to avoid the intense initial continuum emission and improve the line-to-background (L/B) ratio by gating off the early stages of plasma emission. A suitable choice of time delays in detecting the emission spectra allows selective assignment of the resolved line emission to different elements. However, a spatially resolved signal along the vertical path from the target surface has been shown to provide a spectrum similar to those obtained with time-delayed methods but without the expensive gated detection system required for TRE-LIBS.

9.3.9. Data analysis

In processing LIBS data, one must ensure that the effects described in Section 9.3.7 are minimized or avoided.

Formerly, the most common method used to compensate for fluctuations such as shot-to-shot variability was that based on an internal standard. However, the rationale for selecting an internal standard is not straightforward and, in many industrial applications, the use of such a standard is impractical or even impossible. This has fostered the use of alternative methods to compensate for shot-to-shot variations. One such method is post-measurement data manipulation or data analysis. Modern schelle spectrometers, which are equipped with array detectors, allow very large spectral regions to be captured for each individual laser pulse. The possibility therefore exists to obtain diagnostic information based on a very large amount of spectral data on a shot-to-shot basis. Such data, combined with the use of fundamental or empirical models, may allow individual spectra to be corrected for variations in plasma temperature or electron number density. Xu *et al.* [155] developed a new data acquisition approach followed by a data analysis routine for LIBS measurement by which the spectra resulting from individual laser shots are stored and evaluated as single measurements. A data analysis routine is then applied to the raw data that is based on the fact that part of the average background (the background signal averaged over a number of pixels) undergoes the same shot-to-shot variation as the analyte signals. In the ensuing model, the variations are assumed to be multiplicative in nature. The approach was validated with LIBS data and found to provide limits of detection one order of magnitude better than those obtained with the standard method.

Currently available software enables the accumulation of spectra from an arbitrary number of laser shots with an arbitrary number of successive repetitions. This mode of operation is specially suitable for depth profile analysis. By way of example, in order to determine two components in a depth profiling study, a series of pulses are accumulated by maintaining a constant laser fluence. The experimentally measured intensities of the lines selected for the two elements are normalized to their maximum values to account for the difference in oscillator strength of the lines. Such values are then normalized to the sum of the intensities of both elements. Normalization to the combined intensities is equivalent to normalization to the ablated mass. This procedure is unsuitable for the lower layers of a sandwich close to the substrate, for which normalization should also include the line intensity for the substrate element.

Pattern recognition techniques based on principal component analysis (PCA) and cluster analysis (CA) for treatment of the vast amount of data provided by each laser pulse applied to the in-depth characterization of screen-printed electrodes were recently used [156,157]. The experimental set-up used included an Nd:YAG laser operated at the fundamental wavelength (1064 nm). The enormous amount of data to be analysed can be readily inferred from the working plan: eight sampling zones were studied in order to characterize the different parts of the electronic device, viz. the inert support (I), insulating layer (II), working electrode (III), reference electrode (IV) and electrical contacts with (V, VI, VII) and without (VIII) protection of an insulating layer. Likewise, each sampling zone consisted of five sampling points used as replicates, separated by 250 μs, and ten laser shots were delivered at each sampling position. Visual Basic routines were used for computation of analytical signals as well as for the construction of three-dimensional data arrays subsequently processed using commercially available surface software. The excellent results obtained from these chemometric treatments testify to the suitability of these approaches for this type of data.

One other example of the use of chemometric approaches for solving a complex problem is provided by the determination of gold in jewellery pieces, which was started using Au–Ag–Cu alloys [144,145]. Laser-induced plasma emission spectra in the UV region were examined with a view to characterizing the spectral information from time-integrated and time-resolved data obtained with laser energies of 3 and 8 mJ, and 7 laser shots per sampling. Type 1 partial least-squares regression (PLS1) was used for data processing and its ability to overcome LIBS restrictions assessed. Spectroscopic information from three spectral ranges was used, namely: the entire laser-induced emission spectrum from 266 to 340 nm (which encompassed 990 experimental points); the spectral ranges comprising resonant line information for gold, silver and copper (266–269 and 326–340 nm, with 196 points); and the range including non-resonant emission lines for the three metals (269–313 nm, with 560 points). LIB spectra were analysed with and without background correction. First, a series of approaches including a cross-validation method, the prediction sum of squares (PRESS) and statistical criteria such as that of the first local minimum PRESS value [158] and the F-statistic comparison of PRESS values [159], as well as other chemometric tools included in the Pirouette software suite (e.g. statistical parameters, PLS loading factors) were tested with a view to avoiding overfitting and underfitting models through inclusion of non-representative factors. The internal cross-validation of calibration models was performed by using various statistical parameters, namely: R^2 (the squared correlation coefficient), SEC (the standard error of calibration) and REP (the relative error of prediction, expressed as a percentage). The predictive ability of the resulting calibration models was checked via validation tests. Recovery results, expressed as percentages, were estimated, as were REP (%) and SEP (the standard error of prediction) as external validation parameters. In this way, calibration models suited to the special types of sample involved were constructed.

9.3.10. Applications of LIBS

One of the most salient features of LIBS is that the multi-dimensional information it provides can be used to determine depth profiles and lateral distributions, as well as for

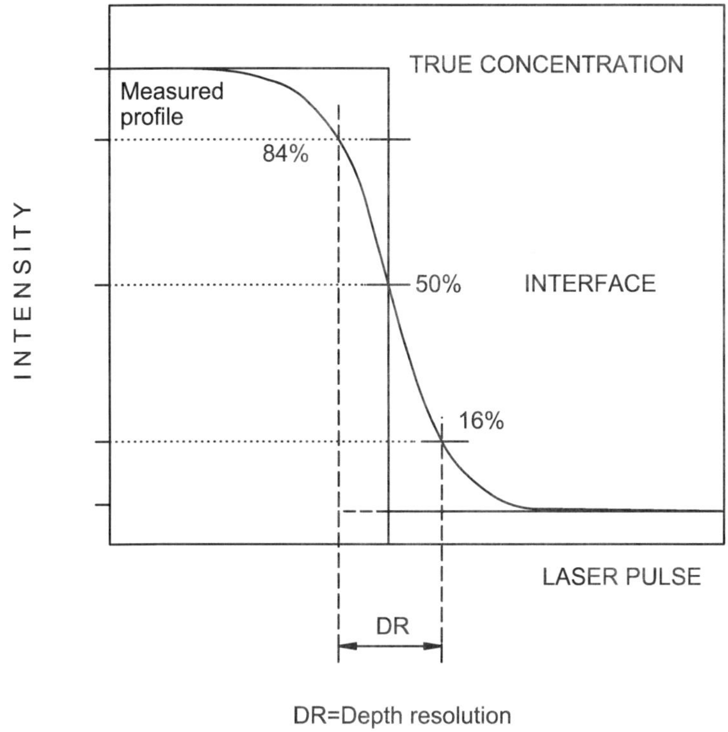

Fig. 9.17. In-depth profiles obtained by LIBS and definition of in-depth resolution (DR).

tomographic imaging studies. These applications require establishing the surface sensitivity and lateral resolution of the technique in terms of its operational parameters. As noted in Section 9.3.6, both parameters are dependent on the shape of the craters formed during ablation. The surface sensitivity of a particular technique can be defined as its capacity to provide data for a sample surface and is determined by the sampling depth (i.e. by the distance normal to the surface that is examined in the test). The sampling depth in LIBS is governed by the laser pulse energy (Fig. 9.10) and the focal conditions of the beam (Fig. 9.11). In both cases, an average ablation rate is calculated that is usually expressed as a mass per pulse. Because the ablation rate varies both as a function of the number of pulses used in the profiling measurement and along the profile (of known depth), the number of pulses to be used should be as small as possible. The depth profile for a simple interface between two layers of known thickness is of special interest. In this context, the depth resolution (DR) of the technique is the depth range (expressed as a number of pulses) across which the signal decreases (or increases) by a preset factor on crossing an abrupt interface between two different media. Usually, such a factor is chosen to be 16–84% of the maximum signal (Fig. 9.17).

Depth profiling

Depth profiling, which refers to the number of atomic layers that can be analysed simultaneously and from which compositional information is derived as a function of depth below the surface, is one of the most salient uses of LIBS. The irradiance to be used should be as low as possible to ensure that the information gathered belongs to a small number of atomic layers. The material is removed in a sequential manner by using short laser pulses on a preset sample position to obtain the emission spectrum for each plasma formed. The pulse axis is converted to a depth by topographic (profiling) analysis of the craters formed in response to a predetermined number of pulses. Alternatively, the depth produced by each pulse can be determined by measuring the number of pulses required to perforate a profile of known thickness. An average ablation rate is calculated in both instances. The depth resolution of the device used plays a key role and must be established on an individual basis.

The capability of LIBS to resolve depth profiles was demonstrated by Vadillo *et al.* [160] using electrolytically deposited brass samples where the emission corresponding to Cr (357.8 nm), Ni (341.4 nm), Cu (327.4 nm) and Zn (334.5 nm) was monitored. Since the nominal thickness of the layers was known, the ablated mass over the range 150–500 nm per pulse was estimated as a function of matrix type and laser irradiance. The amount of ablated mass was found to be strongly dependent on the laser pulse irradiance, which was varied by defocusing. Ablation depths of 6.5 ng/pulse were obtained that resulted in limits of detection in the femtogram-per-pulse range for elements present in the sample at concentrations of a few micrograms-per-gram. However, laser–matter interaction processes, which are poorly understood, restricted the applicability of the technique. Differences in ablation depth due to matrix effects were observed in the resolution of Cr–Ni and Cr–Fe interfaces using similar experimental conditions. Subsequent studies involving a compressed, collimated, filtered, non-focused XeCl excimer laser provided a flatter profile of the crater, which resulted in improved depth resolution (up to 8 nm/pulse) [161]. Until recently, only lasers with nanosecond pulse duration were used for depth profiling. The disadvantage of lasers with long pulses for this purpose is the predominantly thermal character of the ablation process. With metals, the irradiated spot is melted and a large portion of the material evaporates from the melt. Melting of the samples causes changes and mixing of different layers, followed by changes in phase composition during material evaporation (preferential volatilization) and bulk resolidification, which decreases the depth resolution of the LIBS technique. When a laser pulse of sub-picosecond duration is used, the deposition of laser energy onto the sample is so fast that the thermal diffusion length is dictated by the diffusion of hot electrons before the energy is transferred to the lattice. In this case, the diffusion length is of the order of less than 100 nm as opposed to approximately 1 μm in the case of nanosecond pulses. Less marked thermal diffusion provides better lateral and depth resolution, and is the basis for the successful applications of femtosecond pulses in material processing and microstructuring [162].

An in-depth study was conducted on screen-printed electrodes using wide, exhaustive sampling, which was required owing to the complexity of the sample (see the sampling and chemometric approaches described in Section 9.3.9). The vast amount of data produced required testing various chemometric approaches to select the most suitable one for the case under study. The spectroscopic data obtained for the overall components were autoscaled and analysed by PCA and CA. Carbon, gold, silver, platinum, palladium, titanium and aluminium were identified as major components from typical emission spectra such as those of Fig. 9.18. These major components were distributed as one-, two- or three-component monolayers in the different zones of the electronic device, the mixed components being the result of diffusion phenomena involved in the fusion process. Up to three distinct monolayers were distinguished in zones such as the electrical contacts, which testified to the suitability of LIBS pattern recognition techniques for in-depth characterization [156,157].

Mapping or spatial imaging

One other use of LIBS is the construction of distribution maps or chemical images. The term "multi-elemental chemical imaging" denotes a two-dimensional representation of the spatial distribution of single chemical species within a multi-component sample according to the spectral information provided by laser-induced breakdown spectrometry. The use of a multichannel detector such as a charge-coupled device enables the simultaneous acquisition of spectral information from multiple elements in a given target. Images are obtained by sequentially moving the sample using a step motor while the LIB spectrum is recorded at each new sample position. Chemometric processing of the previously stored spectra allows selective two-dimensional images to be reconstructed for each sample component. This application requires the use of multichannel detectors if information for several atomic species at each sample position is to be obtained.

Bulatov *et al.* [140] were the first to obtain two-dimensional images with full spectral resolution at each pixel using an electro-optical set-up consisting of imaging optics (zoom lenses) coupled to a step-scan Fourier transform visible spectrometer coordinated with a CCD camera. They used such spectral images to obtain information about optical and geometric effects on plasma formation. Similarity and classification maps of laser-induced plasma were obtained and used to allocate chemical components in the plasma. The scope of this approach was limited by the lack of temporal resolution, although it could be expanded by using a special temporal gating procedure.

Mapping of platinum group metals in automotive exhaust three-way catalysts was accomplished by Lucena *et al.* [163] using LIBS with a two-dimensional ICCD of spectral resolution 0.02 nm/pixel. Spectral detection, pulse energy and the beam focal conditions were optimized for the target material, thus providing quite good results for converters even with such a weak structure as intact washcoat. Because of the loose nature of the washcoat, lateral and depth resolution were both limited by the large craters formed by the focused laser beam. As a result, it was difficult to deal with typical problems arising from the deactivation of converters such as sintering or dispersion of elemental constituents on a submicron scale. Nevertheless, the ability of the method to work with large structures with minimal sample preparation is of great interest for monitoring the release of platinum group metals as a result of engine operation.

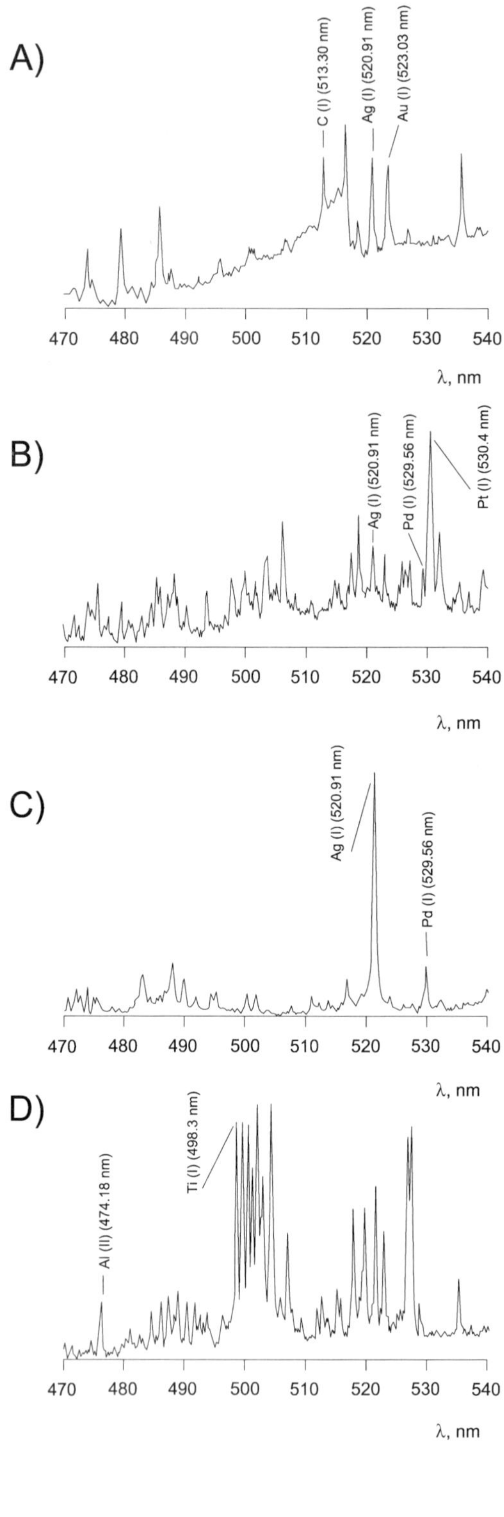
A)
C (I) (513.30 nm)
Ag (I) (520.91 nm)
Au (I) (523.03 nm)
470 480 490 500 510 520 530 540
λ, nm
B)
Ag (I) (520.91 nm)
Pd (I) (529.56 nm)
Pt (I) (530.4 nm)
470 480 490 500 510 520 530 540
λ, nm
C)
Ag (I) (520.91 nm)
Pd (I) (529.56 nm)
470 480 490 500 510 520 530 540
λ, nm
D)
Al (II) (474.18 nm)
Ti (I) (498.3 nm)
470 480 490 500 510 520 530 540
λ, nm

The pattern recognition techniques used for in-depth characterization of screen-printed electrodes (viz. PCA and CA) have also been employed for multi-elemental imaging in laser-induced breakdown spectrometry (MCI–LIBS) [157].

Spatially resolved chemical imaging has been achieved by combining a fibre-optic scanning probe microscope with LIBS in a single instrument called "TOPO–LIBS". The elemental composition of surfaces can be mapped and correlated with topographic data in a single experiment conducted in air with minimal sample preparation and in which the surface topography is analysed by scanning a sharp fibre-optic probe across the sample using shear force feedback [164].

Tomographic studies

In this context, the terms "tomographic analysis" and "lateral resolution" refer to the ability to discern close atomic neighbourhood. By coding the spatial chemical information alongside the depth dimension, the material under study can be described in tomographic terms. The above-described studies on depth resolution and mapping in screen-printed electrodes led to the final tomographic characterization of the target material. With this aim, 10 laser shots were systematically delivered on each of the 625 sampling positions in one of the two selected zones of the device (A) and 325 in the other (B), resulting in 6250 and 3250 plasma emission spectra, respectively. LIBS images were obtained by plotting the plasma emission intensities at the selected wavelengths as a function of locations on the sample surface. The spectral information thus obtained was represented for each element identified as a major component using a colour scale of ten levels where black corresponded to the absence of the element and white to the highest signal intensity (i.e. to the maximum amount detected), with grey levels in between. The results from the spatial distribution of the elements identified in zone A, which comprised the working electrode, the reference electrode and the electrical contacts covered by an insulating protection layer, are shown in Fig. 9.19 [157].

Analysis of interfaces

Interface analysis is one field of special interest for LIBS. The analytical aim is usually the description of a sample or zone thereof where an abrupt change in the identity or concentration of its components occurs. The suitability of LIBS for this type of application lies in the ability to associate spectral data to the spatial coordinates of the sample.

Quantitative analysis

The capabilities of LIBS for determinative applications are also well-documented. Notwithstanding its shortcomings, the LIBS technique can be made a fast tool — whether

---

Fig. 9.18. Typical plasma emission spectra obtained in different zones of a screen-printed electrode: (A) working electrode, (B) reference electrode, (C) electrical contacts and (D) polymeric coating used as insulator. Atomic emission lines used in the LIBS–PCA analysis are indicated. (Reproduced with permission of Elsevier.)

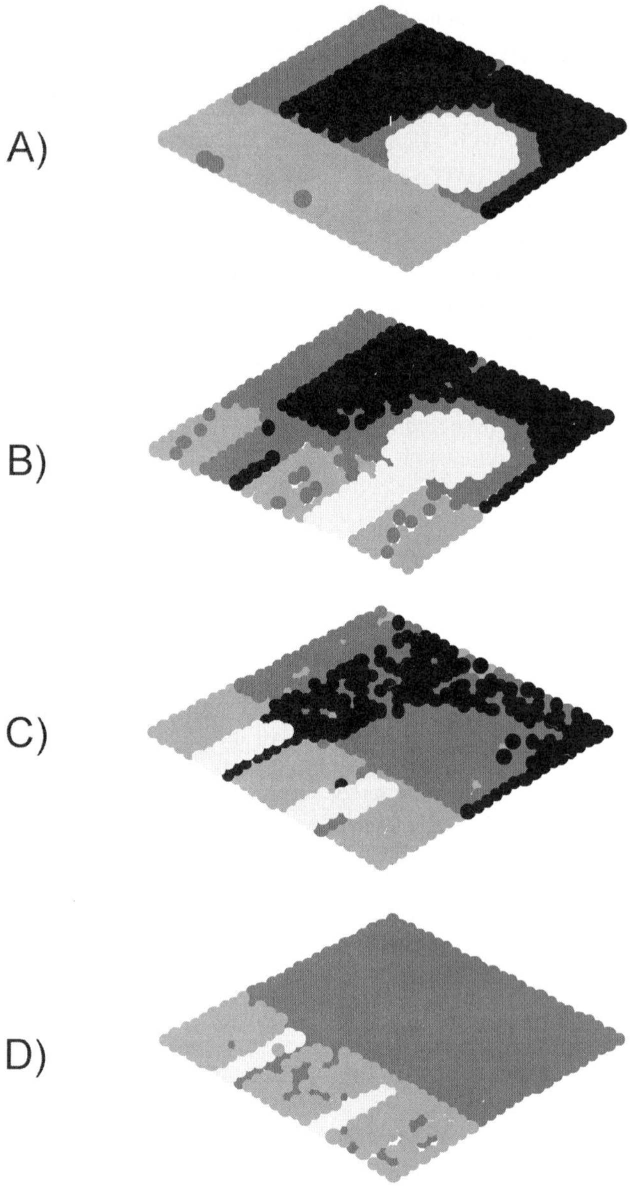

Fig. 9.19. Graphical representation of the results obtained by PCA–CA treatment of the spectroscopic information obtained after (A) the first, (B) the third, (C) the sixth and (D) the tenth laser shot in zone A of a target screen-printed electrode. (Reproduced with permission of Elsevier.)

or not the target material is conductive — by using more stable lasers, better gateable detectors and more resolutive chemometric approaches. Some of the promising results achieved so far in this area are commented on below.

Vanadium and titania supported silica catalysts were analysed by using an Nd:YAG laser operating at 532 nm (first harmonic) and collecting the resulting emission plasmas with a CCD [165]. The use of temporal resolution improved both linearity and signal reproducibility; on the other hand, the loss of sensitivity at delayed integration times was offset by accumulating 10 laser shots. Internal standardization was required to compensate for pulse-to-pulse variability and minimize matrix effects. Delayed integration decreased continuum emission but also affected the intensity of the lines for the internal standard. Thus, the intensity of the ionic Ti (I) line at 416.37 nm was significantly decreased, whereas that of the neutral Ti line at 407.85 nm was not altered. The latter atomic emission line was chosen for internal standardization as it met the regular requirements for an internal standard (viz. proximity to the analyte line and freedom from interferences with neighbouring spectral lines). Under such conditions, the limit of detection for V was estimated to be 38 μg/g in $2TiO_2/SiO_2$; also, the precision, expressed as RSD, was better than 6% over the concentration range 200–1000 μg/g.

The LIBS technique has been used as a fast, simple method for determining trace metal concentrations in such different matrices as starch-based flours [166] and pottery glaze [167]. Also, trace metals in stratum corneum of human skin were recently determined by Winefordner *et al.* [168] using an Nd:YAG laser operating at 1064 nm. By using glass slides, a poly(methyl methacrylate) solution to coat the slides and zinc as model analyte, standard solutions of the last were dispersed on the slides with the aid of a centrifuge with a precision of 6% (for 17 replicates), the slides being used for calibration after drying. This approach provided a calibration curve linear up to 1000 ng/cm^2 and a limit of detection of 0.3 ng/cm^2. Skin layers 2–3 μm thick each were removed using cyanoacrylate glue without causing any pain in the sampled individuals even though each zone was sampled six times. Each layer deposited on a glass slide was successfully analysed on an individual basis. This testifies to the usefulness of the technique for measuring metal concentrations in skin. The signal intensity provided by LIBS can be increased by a factor of 20, 20 and 11 for Al, Fe and Ti, respectively, using a pre-ablation laser pulse 2.5 μs before the ablation pulse [169].

In 1996, Nemet and Kozma showed the emission spectrometry of gold laser-produced plasma to be of interest for analytical purposes; a delay time of 800–1000 ns was found to ensure *quasi*-thermal equilibrium and thorough atomization in the plasma. The line profiles obtained under such conditions (both resonant and Stark-broadened) were fitted to a symmetric Lorentzian curve [170]. Recently, LIBS was used in combination with effective chemometric tools to develop a determination method for gold in homogeneous samples that allows the characterization of jewellery products. The results confirmed the LIBS technique as an effective alternative to the hallmark official methods [143,144,171].

Palanco and Laserna developed an instrument capable of reducing the quality assessment time in stainless steel factories by a factor of 25 [172]. The instrument performs LIBS-based quantitative analyses to check for the occurrence of accidental mixing during steel production, which may lead to incorrect shipping of the final products. Once again, multivariate calibration proved a valuable tool for correcting matrix effects and spectral interferences.

One other use of LIBS for quantitative analysis is that involving colloidal and particulate iron in water and the sequential use of two lasers. A custom coaxial sample flow apparatus was used to control the atmosphere of laser-induced plasma and thus determine iron concentrations as low as 16 ng/ml, both in an Fe(OH) colloidal suspension and in boiler water sampled from a thermal power plant. The ensuing method took only 100 s to apply, even with an accumulation of 2000 pulses [173].

Winefordner *et al.* [174] succeeded in identifying solid materials by correlation analysis in conjunction with LIBS. The primary goal was the instant identification of solid materials on the basis of computer-stored spectral libraries by using laser-induced plasma spectroscopy. With this aim, libraries were constructed prior to analysis that consisted of representative spectra from different groups of materials to be analysed. Special attention was paid to the identification of samples with very similar chemical composition such as some series of stainless steel and cast iron standards. Particulate material including iron and iron oxides was subjected to speciation analysis [175] using both linear and rank correlation methods. Rank correlation proved more reliable: it yielded near-unity probability of correct identification for almost all studied samples. Applications of this technique could be extended to the metallurgical, mineralogical, mining and semiconductor fields, and also to the medical, environmental and forensic sciences. Recently, the on-line analysis of ambient air aerosols by LIBS was reported [176].

Studies of art works both for pigment identification in paintings using LIBS and Raman microscopy [177] and for on-line optical diagnostics [178] have recently been reported.

Special determinative applications of LIBS involving liquid and gaseous samples

Although LIBS has so far been used largely in connection with solid samples, it has also been applied to liquid and gaseous samples in some instances.

Cremers *et al.* [179] reported the detection of elements of main groups I to III and the determination of uranium at the surface of aqueous solutions. Except for alkali metals and calcium, the limits of detection for all elements were above 1 μg/ml. Archontaki and Crouch [180] obtained improved limits of detection by generating the plasma not in a bulk solution, but on single droplets. Also, Cheung and Yeung [181] obtained better reproducibility with shorter measurement times by normalizing the emission intensities of photoacoustic signals, which are easy to measure thanks to the high amplitude of plasma-induced shock waves. The LIBS determination of cations such as Li^+, Na^+, Ca^{2+}, Ba^{2+}, Pb^{2+}, Cd^{2+}, Hg^{2+} and Er^{3+} has also been reported [182,183].

Laser-induced breakdown spectrometry was used for the analysis of gaseous samples containing elements such as F, Cl, S, P, As and Hg in air, and hydrides of column III and V elements (e.g. B_2H_6, PH_3) [184–189]. The aim was to measure trace amounts of analytes in hostile environments and gas impurities for hydride work. Mercury was detected at the parts-per-billion level in air using a photodiode array detector that recorded single-shot spectra over a range of 20 nm [186]. Cremers *et al.* [189] reported limits of detection of 8 and 38 μg/ml for chlorine and fluorine, respectively, the source of both elements being SF_6 and chlorinated fluorocarbons. A more recent research assessed the potential of the LIBS technique for the detection of candidate halon replacement compounds by focusing an Nd:YAG Q-switched pulsed laser into an air flow containing the analyte halocarbon compounds. The ultimate goal was to demonstrate the applicability of the technique in a severely hostile environment for in situ, real-time measurement of halons during use in full-scale fire suppression testing [190].

9.3.11. Comparison of LIBS with alternative techniques

The capabilities of LIBS, based on the multi-dimensional information it provides, have been compared with those of other techniques with similar competence. Thus, the use of LIBS for depth resolution of zinc-coated steel was compared with that of glow-discharge atomic emission spectroscopy (GD–AES). The depth resolution of LIBS was increased by collimating the laser beam. In principle, calibration with a set of reference materials of similar compositions to the unknown samples was not possible as the layers encountered in a depth profile represent widely different material types (matrices). Taking advantage of the number of GD–AES quantitative approaches based on the relationship between sputtered mass and emitted light (the so-called "emission yield"), the performance of LIBS and GD–AES was compared by converting the yield for the emission of a given element into sputtered mass, concentrations then being calculated by sum normalization — provided the experiment was conducted under constant conditions [117,118]. The results of the analysis of galvanized steel by GD–AES are shown in Fig. 9.20A; application of the algorithm enabled the thorough in-depth characterization of the material from a concentration versus depth plot. For comparison, the LIBS results were sum normalized as regards the Fe and Zn intensities to obtain a normalized peak area versus laser shot number plot. Figure 9.20B shows the results thus obtained, which were consistent with those provided by GD–AES [161]. Figure 9.21 provides another comparison between the two techniques as applied to the same sample; the emission intensity profiles for Zn and Fe were plotted against the depth, in microns, for GD–AES and the number of laser shots delivered at two different laser irradiances for LIBS. As can be seen, GD-AES delivered better resolution [160].

As a rule, LIBS performs relatively worse than other, well-established techniques. Thus, the ICP technique (which, like LIBS, uses plasma for excitation and the light it emits as the basis for analysis) provides better limits of detection (in the sub-nanogram-per-millilitre region). The greatest difference between both techniques is that inductively coupled plasma (ICP) is highly stable and influenced by non-linear ablation processes. One other major difference is that the ICP technique is usually applied to homogenized liquid samples whereas LIBS is used mainly for direct solid analysis.

The evaluation of LIBS and LIB–TOF–MS for in-depth analysis of layered materials has shown that both share advantages such as speed of analysis and negligible sample handling, but the absence of background contribution belongs to the latter only [191].

Based on the previous comparisons, the sensitivity of LIBS can be improved by overcoming the problems arising from plasma instability. Because this shortcoming is inherent in the technique, it can only be circumvented by proper characterization of the plasma.

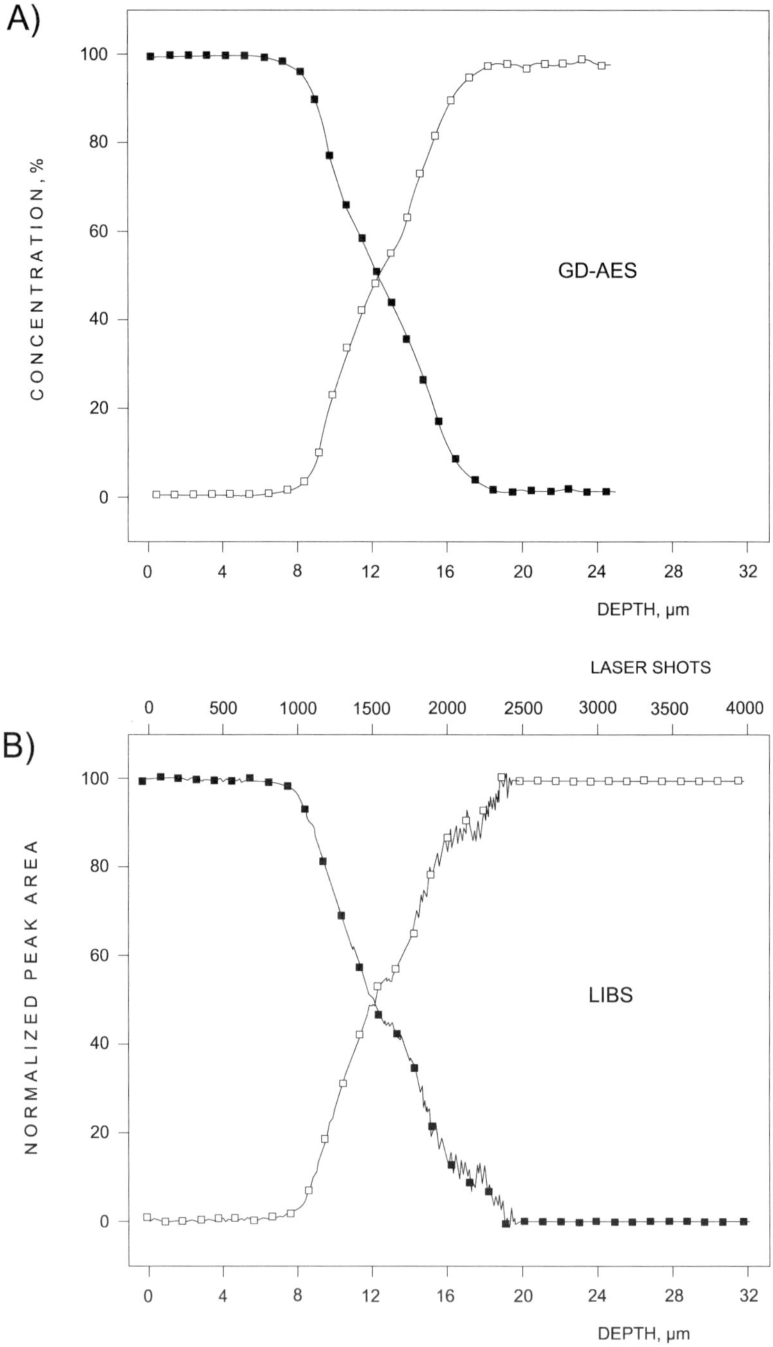

Fig. 9.20. (A) GD–AES and (B) LIBS emission profiles of Zn (■) and Fe (□). (Reproduced with permission of the Royal Society of Chemistry.)

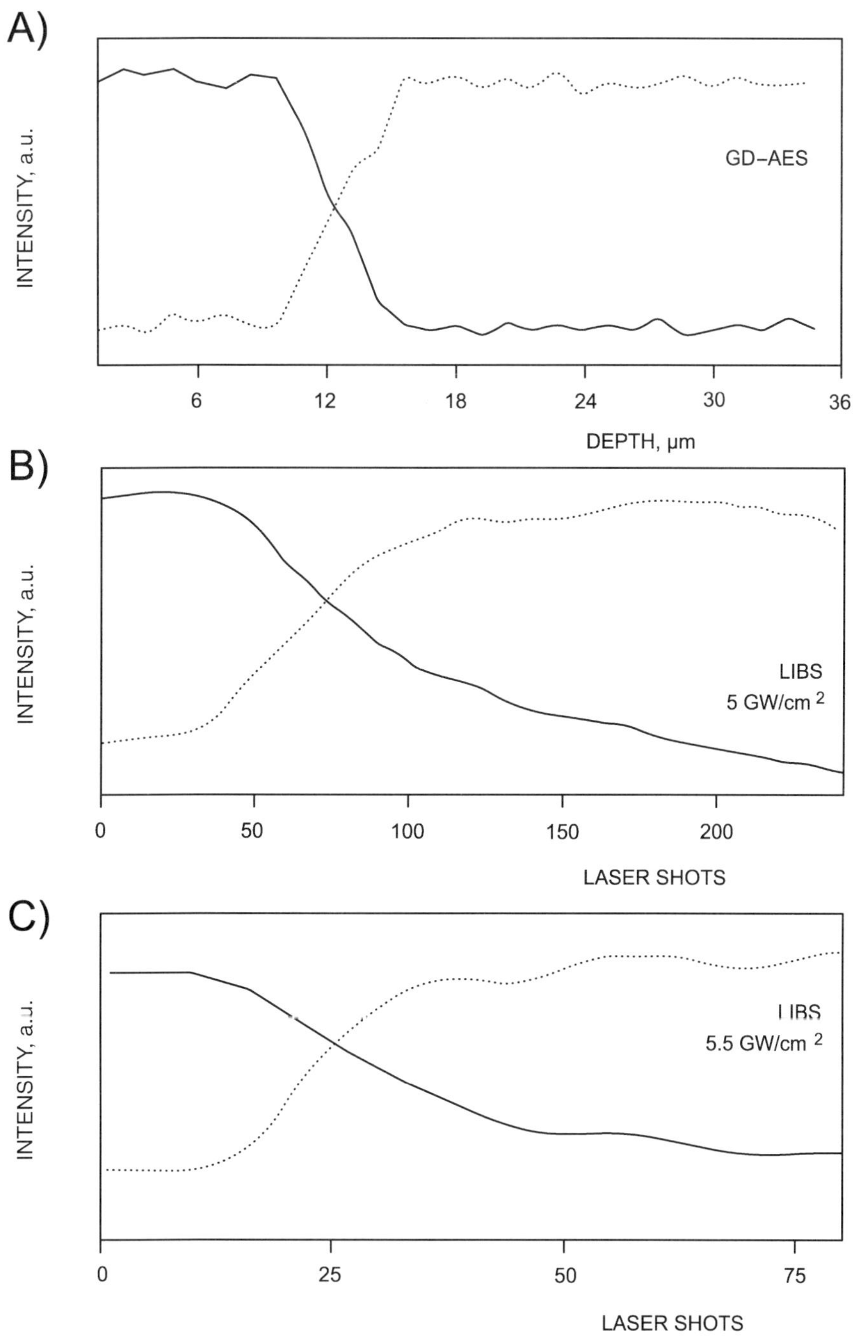

Fig. 9.21. Emission profile of Zn (solid line) and Fe (broken line) in a zinc-coated steel studied by LIBS at two different laser irradiances and also by GD–AES. (Reproduced with permission of the Royal Society of Chemistry.)

9.3.12. Laser-induced breakdown–mass spectrometry

The technique based on laser-induced breakdown coupled to mass detection, which should thus be designated LIB–MS, is better known as "laser plasma ionization mass spectrometry" (LI–MS). The earliest uses of the laser–mass spectrometry couple were reported in the late 1960s. Early work included the vaporization of graphite and coal for classifying coals, elemental analyses in metals, isotope ratio measurements and pyrolysis [192]. Later work extended these methods to biological samples, the development of the laser microprobe mass spectrometer, the formation of molecular ions from non-volatile organic salts and the many multi-photon techniques designed for (mainly) molecular analysis [192].

Figure 9.22 schematically depicts the experimental set-up used by Matus, Seufer and Jochum [193–198] for the direct analysis of rock surfaces [193], the microanalysis of mineral phases [194] and that of geological glasses [195] by LI–MS. These authors

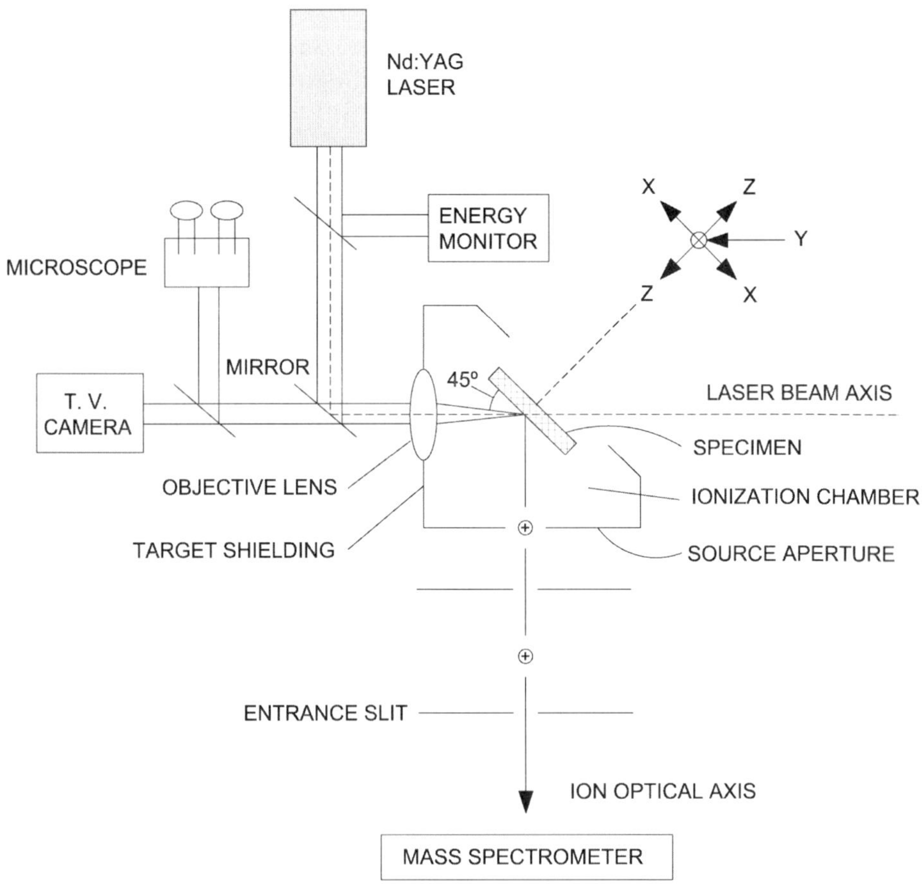

Fig. 9.22. Laser plasma ion source. (Reproduced with permission of Springer-Verlag.)

developed a procedure for the quantitative determination of element concentrations [196] where the concentration of an element x was calculated from

$$C_x = C_s H_{ik} \frac{b_k}{a_i} \frac{AW_x}{AW_s} \frac{1}{RSF_{x/s}} \tag{9.3}$$

where C_s is the concentration of an internal standard element; H_{ik} is the measured ratio of the ion intensities of isotope i of element k and of isotope k of element s; a_i and b_k are the abundances of isotopes i and k, respectively; AW_x and AW_s are the atomic weights of elements x and s; and $RSF_{x/s}$ is the sensitivity factor for element x relative to the internal standard element s [197].

Internal standardization in LI–MS results in improved precision and accuracy. An internal standard element corrects for potential interferences in ion yield and photoplate sensitivity. Geological glasses were analysed by using Ti as internal standard element, its concentration being determined by electron micropobe (EMP) and X-ray fluorescence (XRF) analysis. Relative sensitivity factors (RSFs) were used to correct for interferences in ion formation, transmission and detection of various elements. LI–MS exhibits only small differences in *RSF* because the high laser power density results in a uniform ionization efficiency [197,198]. Figure 9.23 shows the *RSF* values for 52 elements as obtained from analyses of certified reference materials. The experimentally determined $RSF_{x/\mathrm{Ti}}$ value ranged within 10% of that calculated from the following empirical formula:

$$RSF_{x/\mathrm{Ti}} = \left(\frac{I_x}{I_{\mathrm{Ti}}}\right)^{0.7} \left(\frac{R_x}{R_{\mathrm{Ti}}}\right)^{0.3} \left(\frac{T_x}{T_{\mathrm{Ti}}}\right)^{0.1} \tag{9.4}$$

where I_x is the ionization energy (eV), R_x the atomic radius and T_x the boiling point (°C) of element x.

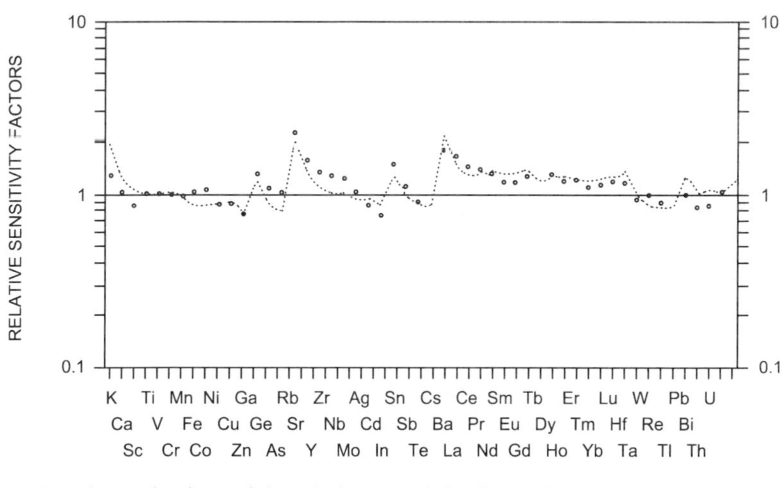

Fig. 9.23. Experimental values of the relative sensitivity factor for 52 elements from certified reference materials as compared with the theoretical values. (Reproduced with permission of Springer-Verlag.)

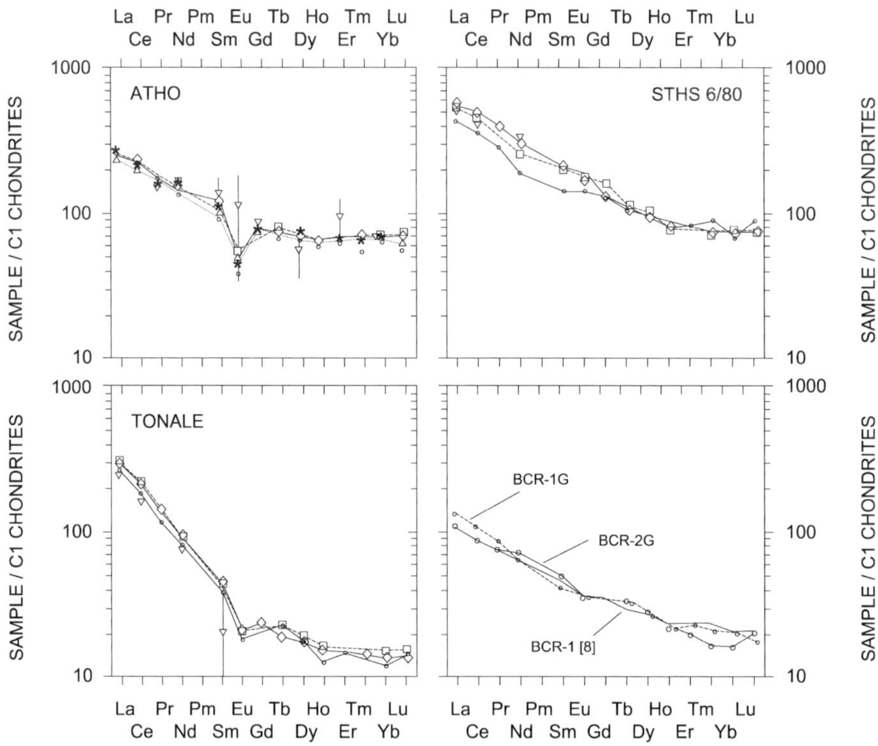

Fig. 9.24. C1 chondrite normalized rare earth element concentrations for geological glasses. The LI–MS data are compared with the results of other analytical techniques. (●) LI–MS; (◇) LA–ICP–MS; (△) HPLC; (▽) SY–XRF; (□) INAA; (×) TI–MS. (Reproduced with permission of Springer-Verlag.)

In order to demonstrate the analytical capabilities of LI–MS, the results it provided for glass samples were compared with those obtained using other multi-element and microanalytical techniques to determine 30 trace elements [195]. This comparison revealed the number of elements that can be determined by LI–MS to be similar to that of other multi-element techniques such as instrumental neutron activation analysis (INAA) or LA–ICP–MS. However, INAA was unable to determine some geochemically interesting trace elements such as Nb and Y, and LA–ICP–MS analyses were occasionally disturbed by the formation of argon clusters. In contrast to LA–ICP–MS, LI–MS can also measure single-isotope elements such as Nb, Y, Pr and Ho; also, it requires no wet chemistry.

In geochemistry, rare earth element (REE) patterns constitute a useful tool for interpreting geological processes. LI–MS produces reasonably good REE patterns, as shown in Fig. 9.24, which is a plot of CI normalized REE concentrations alongside the results obtained with INAA, HPLC, LA–ICP–MS, synchrotron X-ray fluorescence (SY–XRF) spectroscopy and secondary ion mass spectrometry (TI–MS). INAA and TI–MS surpass

LI–MS in precision but the last can determine this element group and other trace elements in areas as small as 0.1–1 mm^2 [197].

Regarding microanalytical capabilities, the LI–MS technique can be the best compared to LA–ICP–MS and SY–XRF. LI–MS has become a technique of choice for geochemical microanalysis (see Fig. 9.23). In contrast to LI–MS, the SY–XRF technique is non-destructive; however, LI–MS allows very small areas (less than 10 μm in size) to be analysed, so it causes negligible damage in most cases. In addition, the typical limits of detection of SY–XRF (5–10 μg/g) are often too poor to obtain reasonably good REE patterns. With elements such as Zn, Ni, Sr, Na, Zr and Nb in geological glasses, where their concentrations are often much higher, SY–XRF delivers acceptably accurate, precise results [198].

Notwithstanding the competitive results it provides, the potential of LI–MS has hardly been explored so far, and largely by the same workers.

References

1 R.J. Cotter Ed., *Time-of-Flight Mass Spectrometry: Instrumentation and Applications in Biological Reseach*, ACS Professional Reference Books,Washington, DC (1997).
2 G. Siuzdak Ed., *Mass Spectrometry for Biotechnology*, Academic Press, New York (1996).
3 M.J. Pelletier, *Analytical Applications of Raman Spectroscopy*, Blackwell Science, Ltd, Oxford (1999).
4 J.L. Beauchamp and R.H. Staley, *J. Am. Soc.*, 97 (1975) 5920–5921.
5 M.A. Posthumus, P.G. Kistemaker, H.L.C. Meuzelaar and M.C. TenNoever de Brauw, *Anal. Chem.*, 50 (1978) 985–991.
6 R. Stoll and F.W. Rollgen, *Org. Mass Spectrom.*, 14 (1979) 642–645.
7 L.M. Teesch and J. Adams, *Org. Mass Spectrom.*, 27 (1992) 931–943.
8 T.E. Jeffries, W.T. Perkins and N.J.G. Pearce, *Analyst*, 120 (1995) 1365.
9 M. Motelica-Heino, O.F.X. Donar and J.M. Mermet, *J. Anal. At. Spectrom.*, 14 (1999) 675.
10 M. Motelica-Heino, P. Le Costumer, J.H. Thomassin, A. Gauthier and O.F.X. Donard, *Talanta*, 46 (1998) 407.
11 T.E. Jeffries, S.E. Jackson and H.P. Longerich, *J. Anal. At. Spectrom.*, 13 (1998) 935.
12 A.J.G. Mank and P.R.D. Mason, *J. Anal. At. Spectrom.*, 14 (1999) 1143.
13 R. Russo, O. Borisov, X. Mao and H. Liu, *J. Anal. At. Spectrom.*, 15 (2000) 1115.
14 I. Horn, R.L. Rudnich and F.M. McDonough, *Chem. Geol.*, 164 (2999) 281.
15 D. Günther, I. Horn and B. Hattendorf, *Fresenius J. Anal. Chem.*, 368 (2000) 4
16 B.J. Fryer, S.E. Jackson and H.P. Longerich, *Can. Min. J.*, 33 (1995) 303.
17 J.H. Yoo, O.V. Borisov, X. Mao and R.E. Russo, *Anal. Chem.*, 73 (2001) 2288.
18 M.A. Shannon, X.L. Mao, A. Fernández, W.T. Chan and R.E. Russo, *Anal. Chem.*, 67 (1995) 4522.
19 H. Scholze, H. Stephanowitz, E. Hoffmann and J. Skolel, *Fresenius J. Anal. Chem.*, 363 (1995) 247.
20 A.L. Gray, *Analyst*, 110 (1985) 551.
21 P.A. Arrowsmith, *Anal. Chem.*, 59 (1987) 1437.
22 J.W. Hager, *Anal. Chem.*, 61 (1989) 1243.
23 P.A. Arrowsmith, in: *Lasers and Mass Spectrometry*, D.M. Lubman Ed., Oxford University Press, NY (1990).
24 P.A. Arrowsmith and S.K. Hugher, *Appl. Spectrosc.*, 42 (1988) 1231.
25 F.N. Abercrome, M.D. Silvester, A.D. Murray and A.R. Barringer, in: *Applications in Inductively Coupled Plasmas to Emission Spectroscopy*, R.M. Barnes Ed., Franklin Institute Press (1978) p. 121.

26 S. Guizhen and L. Soulin, *J. Anal. Atom. Spectrom.*, 3 (1988) 841.
27 M.E. Tremblay, B.W. Smith, M.B. Leong and J.D. Winefordner, *Spectrosc. Letters*, 20 (1987) 311.
28 T. Tanaka, K. Yamamoto, T. Monizu and H. Kawaguchi, *Anal. Sci.*, 11 (1995) 967.
29 F. Chartier, *PhD Thesis*, University Lauce Bernard, Lyon (1991).
30 C. Leloup, P. Marty, D. Dall'Ava and M. Perdereau, *J. Anal. At. Spectrom.*, 12 (1997) 945.
31 A. Moissette, T.J. Shepherd and S.R. Chenery, *J. Anal. At. Spectrom.*, 11 (1996) 177.
32 L. Moencke, M. Gäckle, D. Günther and J. Kammel, *R. Soc. Chem.*, Cambridge, Special Pub. 81 (1990) 1.
33 J.W. Carr and G. Horlick, *Spectrochim. Acta*, 37B (1982) 1.
34 S. Chenery, M. Thompson and K. Timmins, *Anal. Proc.*, 25 (1988) 841.
35 P.G. Mitchell, J. Sneddon and L.J. Radziemski, *Appl. Spectrosc.*, 40 (1986) 274.
36 F. Leis and L. Laqua, *Spectrochim. Acta*, 33B (1978) 727.
37 D. Günther and C.A. Heinrich, *J. Anal. At. Spectrom.*, 14 (1999) 429.
38 T. Hirata and R.W. Nesbitt, *Geochim. Cosmochim. Acta*, 59 (1995) 2491.
39 S.M. Eggins, L.P.J. Kinsley and J.M.G. Shelley, *Appl. Surf. Sci.*, 129 (1998) 278.
40 S.H. Jeong, O.V. Borisov, J.H. Yoo, X.L. Mao and A.E. Russo, *Anal. Chem.*, 71 (1999) 5133.
41 S. Durrant, *Analyst*, 117 (1992) 1585.
42 M. Bi, E. Austin, B.W. Smith and J.D. Winefordner, *Anal. Chim. Acta*, 435 (2001) 309.
43 V. Kanicky and J.M. Mermet, *Appl. Spectroc.*, 51 (1997) 332–336.
44 P. Richner, M.W. Borer, K.R. Bruskwyler and G.M. Hieftje, *Appl. Spectrosc.*, 44 (1990) 1290.
45 S.R. Chenery and J.M. Cook, *J. Anal. At. Spectrom.*, 8 (1993) 299.
46 E.F. Cromwell and P. Arrowsmith, *Anal. Chem.*, 67 (1995) 131.
47 D. Günther, R. Frischknecht, H.J. Müschenborn and C.A. Heinrich, *Fresenius J. Anal. Chem.*, 359 (1997) 390.
48 D. Günther, H. Cousin, B. Magyar and I. Leopold, *J. Anal. At. Spectrom.*, 12 (1997) 165.
49 S.E. Jackson, H.P. Longerich, G.R. Dunning and B.J. Fryer, *Can. Min.*, 30 (1992) 1049.
50 Z. Chen, W. Doherty and D.C. Grégoire, *J. Anal. At. Spectrom.*, 12 (1997) 653.
51 I. Horn, R.W. Hinton, S.E. Jackson and H.P. Longerich, *Geostand Newslet.*, 21 (1997) 191.
52 J.J. Leach, L.S. Allen, D.B. Aeschlimann and R.S. Houk, *Anal. Chem.*, 71 (1999) 440.
53 J.E. Reid, I. Horn, H.P. Longerich and L. Forsythe, *Geostand Newslet.*, 23 (1999) 149.
54 H. Pang, D.R. Wiederin, R.S. Houk and E.S. Yeung, *Anal. Chem.*, 63 (1991) 390.
55 H.F. Falk, B. Hattendorf, K.K. Rothensee, N. Wieberneit and S.L. Dannen, *Fresenius J. Anal. Chem.*, 362 (1998) 468.
56 C. Pickhardt, J.S. Becker and H.J. Dietze, *Fresenius J. Anal. Chem.*, 368 (2000) 173.
57 T. Hirata, *J. Anal. At. Spectrom.*, 12 (1997) 1337.
58 D. Günther and C.A. Heinrich, *J. Anal. At. Spectrom.*, 14 (1999) 1363.
59 A.M. Leach and G.M. Hieftje, *J. Anal. At. Spectrom.*, 15 (2000) 1121.
60 M. Odegard and M. Hamester, *Geostand Newslet.*, 21 (1997) 245.
61 L.P. Bédard, D.R. Baker and N. Machado, *Chem. Geol.*, 138 (1997) 1.
62 M. Odegard, *Geostand Newslet.*, 23 (1999) 173.
63 J.S. Becker and H.J. Dietze, *Fresenius J. Anal. Chem.*, 365 (1999) 429.
64 T.J. Shepherd and S.R. Chenery, *Geochim. Cosmochim. Acta*, 59 (1995) 3997.
65 A. Audetat, D. Günther and C.A. Heinrich, *Science*, 279 (1998) 2091.
66 D. Günther, A. Audetat, R. Frischknecht and C.A. Heinrich, *J. Anal. At. Spectrom.*, 13 (1998) 263.
67 A.M. Ghazi, T.E. McCandless, D.A. Vanko and J. Ruiz, *J. Anal. At. Spectrom.*, 11 (1996) 667.
68 T. Ulrich, D. Günther and C.A. Heinrich, *Nature*, 399 (1999) 676.
69 B. Schäfer, D. Günther, R. Frischknecht and D.B. Dingwell, *Euro. J. Mineral*, 11 (1999) 415.
70 A. Audetat and D. Günther, *Contrib. Min. Pet.*, 137 (1999) 1.
71 S. Durrant, *J. Anal. At. Spectrom.*, 14 (1999) 1385.
72 H.P. Longerich, S.E. Jackson and D. Günther, *J. Anal. At. Spectrom.*, 11 (1996) 899.
73 D. Günther, A. Audetat, R. Frischknecht and C. Eninrich, *J. Anal. At. Spectrom.*, 13 (1998) 263.
74 O.W.O. Hoskin and R. Wysoczanski, *J. Anal. At. Spectrom.*, 13 (1998) 597.

75 E.K. Shibuya, J.E.S. Sarkis, J. Enzweiler, A.P.S. Jorge and A.M. Figueiredo, *J. Anal. At. Spectrom.*, 13 (1998) 941.
76 M. Norman, W.L. Griffin, N.J. Pearson, M.O. García and S.Y. O'Reilly, *J. Anal. At. Spectrom.*, 13 (1998) 477.
77 Z. Chen, *J. Anal. At. Spectrom.*, 14 (1999) 1823.
78 S. Wang, R. Brown and D.J. Gray, *Appl. Spectrosc.*, 48 (1994) 1321.
79 R.D. Evans, P.M. Outridge and P. Richner, *J. Anal. At. Spectrom.*, 9 (1994) 985.
80 G.D. Price and N.J. Pierce, *Marine Pollution Bulletin*, 34 (1997) 1025.
81 C.D. Garbe-Schoberg, C. Reimann and V.A. Pavlov, *Environ. Geol.*, 32 (1997) 9.
82 S. Tanaka, N. Yasushi, N. Sato, T. Fukasawa, S.J. Santosa, K. Yamanaka and T. Ootoshi, *J. Anal. At. Spectrom.*, 13 (1998) 135.
83 J.S. Crain and D.L. Gallimore, *J. Anal. At. Spectrom.*, 7 (1992) 605.
84 M. Gastel, J.S. Becker, G. Kuppers and H.J. Dietze, *Spectrochim. Acta B*, 52 (1997) 2051.
85 A. Raith, R.C. Hutton, I.D. Abell and J. Crighton, *J. Anal. At. Spectrom.*, 10 (1995) 591.
86 Z. Chen, D. Canil and H.P. Longerich, *Fresenius J. Anal. Chem.*, 368 (2000) 73.
87 R. Pesch, M. Hamester and J. Wills, *Wint. Conf. Plasma Spectrochem.*, Ft Lauderdale, (2000) 274.
88 P.P. Mahoney, G.Q. Li and G.M. Hieftje, *J. Anal. At. Spectrom.*, 11 (1996) 401.
89 A. Bleiner, A. Plotnikov, C. Bogt, K. Wetzig and D. Günther, *Fresenius J. Anal. Chem.*, 368 (2000) 221.
90 G.R. Cass, *Trends Anal. Chem.*, 17 (1998) 356.
91 D.L. Poster, M.M. Schantz, S.A. Wise and M.G. Vangel, *Fresenius J. Anal. Chem.*, 363 (1999) 380.
92 P.J. McKeown, M.V. Johnston and D. Murphy, *Anal. Chem.*, 63 (1991) 2069.
93 O. Kievit, J.C.M. Marijnissen, P.J.T. Verheijen and B. Scarlett, *J. Aerosol Sci.*, 23 (1992) S301.
94 K.A. Prather, T. Nordmeyer and K. Salt, *Anal. Chem.*, 66 (1994) 1403.
95 K.P. Hinz, R. Kaufmann and B. Spengler, *Anal. Chem.*, 66 (1994) 2071.
96 B.A. Mansoori, M.V. Johnston and A.S. Wexler, *Anal. Chem.*, 66 (1994) 3681.
97 D.M. Murphy and D.S. Thompson, *Aerosol Sci. Technol.*, 22 (1995) 237.
98 W.D. Reents, S.W. Downey, A.B. Emerson and A.M. Mujsce, *Aerosol Sci. Technol.*, 23 (1995) 263.
99 M. Yang, P.T.A. Reilly, K.B. Boraas, W.B. Whitten and J.M. Ramsey, *Rapid. Commun. Mass Spectrom.*, 10 (1996) 347.
100 P.T.A. Reilly, R.A. Gieray, M. Yang, W.B. Whitten and J.M. Ramsey, *Anal. Chem.*, 69 (1997) 36.
101 R.A. Gieray, P.T.A. Reilly, M. Yang, W.B. Whitten and J.M. Ramsey, *J. Microbiol. Methods*, 29 (1997) 191.
102 P.T.A. Reilly, A.C. Lazar, R. Gieray, W.B. Whitten and J.M. Ramsey, *Aerosol Sci. Technol.*, in press.
104 B.D. Morrical, D.P. Fergenson and K.A. Prather, *A. Soc. Mass Spectrom.*, 9 (1998) 1068.
105 A. Zelenyuk, J. Cabalo, T. Baer and R.E. Miller, *Anal. Chem.*, 71 (1999) 1802.
106 A.C. Lazar, P.T.A. Reilly, W.B. Whitten and J.M. Ramsey, *Environ. Sci. Technol.*, 33 (1999) 3993.
107 A.C. Lazar, P.T.A. Reilly, W.B. Whitten and J.M. Ramsey, *Anal. Chem.*, 72 (2000) 2142.
108 P. Voumard, Q. Zhan and R. Zenobi, *Chem. Phys. Lett.*, 239 (1995) 89.
109 W.T. Perkins, R. Fuge and N.J.G. Pearce, *J. Anal. At. Spectrom.*, 6 (1991) 445.
110 A. Raith, W.T. Perkins, N.J.G. Pearce and T.E. Jeffries, *Fresenius J. Anal. Chem.*, 357 (1996) 789.
111 A. Cox, F. Keenan, M. Cooke and J. Appleton, *Fresenius J. Anal. Chem.*, 354 (1996) 254.
112 E. Hoffmann, H. Stephanowitz, E. Ulrich, J. Skole, C. Lüdke and B. Hoffmann, *J. Anal. At. Spectrom.*, 15 (2000) 663.
113 E. Hoffmann, C. Lüdke, H. Scholze and H. Stephanowitz, *Fresenius J. Anal. Chem.*, (1994) 350, 253–259.
114 E. Hoffmann, H. Stephanowitz and J. Skole, *Fresenius J. Anal. Chem.*, 355 (1996) 690.

115 R.J. Watling, B.F. Lynch and D. Herring, *JAAS* 12(12) (1997) 195–203.
116 T. Oguri, H. Inoue, S. Tsuge, K. Kitagawa and N. Arai, *J. Anal. At. Spectrom.*, 12 (1997) 823.
117 H. Matsuta, K. Wagatsuma and K. Kitagawa, *Bunseki Kagaku*, 49 (2000) 849.
118 F. Buoe-Bigne, B.J. Masters, J.S. Crighton and B.L. Sharp, *J. Anal. At. Spectrom.*, 14 (1999) 1665.
119 I. Rodushkin, M.D. Axelsson and E. Burman, *Talanta*, 51 (2000) 743–759.
120 P. Rommers and P. Boumans, *Fresenius J. Anal. Chem.*, 356 (1996) 763.
121 J.S. Becker, U. Breuer, J. Westheide, A.I. Saprykin, H. Holzbrecher, H. Nickel and H.J. Dietze, *Fresenius J. Anal. Chem.*, 355 (1996) 626.
122 H.J. Dietze, *Analytikertaschenbuch*, 10 (1991) 249.
123 H.J. Dietze and J.S. Becker, in: *Laser Ionization Mass Analysis*, A. Vertes, R. Gijbels and F. Adams Eds, Chemical Analysis Series, vol. 124, Wiley, New York, (1993), pp. 453–504.
124 J.S. Becker and H.J. Dietze, *Fresenius J. Anal. Chem.*, 344 (1992) 69.
125 F. Brech, *Appl. Spectrosc.*, 16 (1962) 59.
126 E.F. Runge, R.W. Minck and F.R. Bryan, *Spectrochim. Acta*, 20 (1964) 733.
127 E.F. Runge, S. Bonfiglio and F.R. Bryan, *Spectrochim. Acta*, 22 (1965) 1678.
128 W. Zapka and A.C. Tam, *J. Opt. Cos. Am.*, 71 (1981) 1585.
129 A. Lenk, Th. Witke and G. Granse, *Appl. Surf. Sci.*, 96–98 (1996) 195.
130 F.N. Beg, A.R. Bell, A.E. Dangor, C.N. Danson, A.P. Fews, M.E. Glinsky, B.A. Hammel, P. Lee, P.A. Norreys and M. Tatarakis, *Phys. Plasmas*, 4 (1997) 447.
131 D.W. Fradin, N. Bloembergen and J.P. Letellier, *Appl. Phys. Lett.*, 22 (1973) 635.
132 B.C. Stuart, M.D. Feit, S. Herman, A.M. Rubenchik, B.W. Shore and M.D. Perry, *Phys. Rev. B.*, 53 (1996) 1749.
133 P.K. Kennedy, *IEEE J. Quant. Elec.*, 31 (1995) 2241.
134 A.J. Glass and A.H. Günther, *Appl. Opt.*, 12 (1973) 637.
135 B.C. Castle, K. Visser, B.W. Smith and J.D. Winefordner, *Spectrochim. Acta*, 52B (1997) 1995.
136 B.C. Castle, K. Visser, B.W. Smith and J.D. Winefordner, *Appl. Spectrosc.*, 51 (1997) 1017.
137 F. Rocchi, V.G. Molinari, D. Mostacci and M. Sumini, *Spectrochim. Acta*, 56B (2001) 599.
138 C. Aragón and J.A. Aguilera, *Appl. Spectrosc.*, 51 (1997) 1632.
139 E.H. Piepmeier, *Analytical Applications of Lasers*, John Wiley & Sons, New York (1986).
140 V. Bulatov, L. Xu and I. Schechter, *Anal. Chem.*, 68 (1996) 2966.
141 R. Wisbrun, I. Schechter, R. Niessner, H. Schröder and K.L. Kompa, *Anal. Chem.*, 66 (1994) 2964.
142 J.M. Vadillo, C.C. García, S. Palanco and J.J. Laserna, *J. Anal. At. Spectrom.*, 13 (1998) 793.
143 M. Millán, J.M.Vadillo and J.J. Laserna, *J. Anal. At. Spectrom.*, 12 (1997) 441.
144 J. Amador-Hernández, L.E. García-Ayuso, J.M. Fernández-Romero and M.D. Luque de Castro, *J. Anal. Atom. Spectrom.*, 15 (2000) 587.
145 L.E. García-Ayuso, J. Amador-Hernández, J.M. Fernández-Romero and M.D. Luque de Castro, *Anal. Chim. Acta*, in press.
146 Y.I. Lee, K. Song, H.K. Cha, J.M. Lee, M.C. Park, G.H. Lee and J. Sneddon, *Appl. Spectrosc.*, 51 (1997) 959.
147 E.A.P. Cheng, R.D. Fraser and J.G. Eden, *Appl. Spectrosc.*, 45 (1991) 949.
148 A. Ciucci, V. Palleschi, S. Rastelli, R. Barbini, R. Fantoni, A. Palucci, F. Colao, S. Ribezzo and H.J.L. van der Steen, *Appl. Phys. B*, 63 (1995) 185.
149 R. Barbini, F. Colao, R. Fantoni, A. Palucci, S. Ribezzo, H.J.L. van der Steen and M. Angelone, *Appl. Phys. B*, 65 (1997) 101.
150 B.C. Castle, K. Visser, B.W. Smith and J.D. Winefordner, *Appl. Spectrosc.*, 51 (1997) 1017.
151 K. Song, H. Cha, J. Lee, J.S. Choi and Y.I. Lee, *J. Kor. Phys. Soc.*, 30 (1997) 463.
152 L. St-Onge, M. Sabsabi and P. Cielo, *J. Anal. At. Spectrom.*, 12 (1997) 997.
153 Y.I. Lee, T.L. Thiem, G.H. Kim, Y.Y. Teng and J. Sneddon, *Appl. Spectrosc.*, 46 (1992) 1597.
154 Y.I. Lee and J. Sneddon, *Analyst*, 119 (1994) 1441.
155 L. Xu, V. Bulatov, V.V. Gridin and I. Schechter, *Anal. Chem.*, 69 (1997) 2103.
156 J. Amador-Hernández, J.M. Fernández-Romero and M.D. Luque de Castro, *Surface Interface Anal.*, 31 (2001) 313.

157 J. Amador-Hernández, J.M. Fernández-Romero and M.D. Luque de Castro, *Anal. Chim. Acta*, 435 (2001) 227.
158 D.M. Haaland and E.V. Thomas, *Anal. Chem.*, 60 (1988) 1193.
159 J. Amador-Hernández, A. Cladera, J.M. Estela, P.L. López-de-Alba and V. Cerdá, *Analyst*, 123 (1998) 2235.
160 J.M. Vadillo and J.J. Laserna, *J. Anal. At. Spectrom.*, 12 (1997) 859.
161 J.M. Vadillo, C.C. García, S. Palanco and J.J. Laserna, *J. Anal. At. Spectrom.*, 13 (1998) 793.
162 V. Margetic, M. Bolshov, A. Stockhaus, K. Niemax and R. Hergenröder, *J. Anal. At. Spectrom.*, 16 (2001) 616.
163 P. Lucena, J.M. Vadillo and J.J. Laserna, *Anal. Chem.*, 71 (1999) 4391.
164 D. Kossakovski and J.L. Beauchamp, *Anal. Chem.*, 72 (2000) 4731.
165 H.H. Cho, Y.J. Kim, Y.S. Jo, K. Kitagawa, N. Arai and Y.I. Lee, *J. Anal. At. Spectrom.*, 16 (2001) 622.
166 Y. Yoon, T. Kim, M. Yang, K. Lee and G. Lee, *Michrochem. J.*, 68 (2000) 251.
167 D.N. Stratis, K.L. Eland and S.M. Ángel, *Appl. Spectroc.*, 54 (2000) 1719.
168 Q. Sun, M. Tran, B.W. Smith and J.D. Winefordner, *Talanta*, 52 (2000) 293.
169 I.B. Gornushkin, A. Ruiz-Medina, J.M. Anzano, B.W. Smith and J.D. Winefordner, *J. Anal. At. Spectrom.*, 15 (2000) 581.
170 B. Nemet and L. Kozma, *Fresenius J. Anal. Chem.*, 355 (1996) 904.
171 A. Jurado López and M.D. Luque de Castro, *Anal./ Bional. Chem.*, 372 (2002) 109.
172 S. Palanco and J.J. Laserna, *J. Anal. At. Spectrom.*, 15 (2000) 1321.
173 S. Nakamura, Y. Ito, K. Sone, H. Hiraga and K.I. Kaneko, *Anal. Chem.*, 68 (1996) 2981.
174 L.A. King, I.B. Gornuschkin, D. Pappas, B.W. Smith and J.D. Winefordner, *Spectrochim. Acta*, 54 B (1999) 1771.
175 I.B. Gornushkin, B.W. Smith, H. Nasajpour and J.D. Winefordner, *Anal. Chem.*, 71 (1999) 5157.
176 J.E. Carranza, B.T. Fisher, G.D. Yoder and D.W. Hahn, *Spectrochim. Acta*, 56 B (2001) 851.
177 L. Burgio, K. Melessanaki, M. Doulgeridis, R.J.H. Clark and D. Anglos, *Spectrochim. Acta*, 56 B (2001) 905.
178 R. Salimbeni, R. Pini and S. Siano, *Spectrochim. Acta*, 56 B (2001) 877.
179 J.R. Wachter and D.A. Cremers, *Appl. Specrosc.*, 41 (1987) 1042.
180 H.A. Archontaki and S.R. Crouch, *Appl. Spectrosc.*, 42 (1989) 741.
181 N. Cheung and E.S. Yeung, *Appl. Spectrosc.*, 47 (1993) 882.
182 R. Knopp, F.J. Scherbaum and J.I. Kim, *Fresenius J. Anal. Chem.*, 355 (1996) 16.
183 E.A.P. Cheng, R.D. Fraser and J.G. Eden, *Appl. Spectrosc.*, 45 (1991) 949.
184 M. Casini, M.A. Harith, V. Palleschi, A. Salvetti, D.P. Singh and M. Vaselli, *Laser Part. Beams*, 9 (1991) 633
185 C. Lazzari, M. De Rosa, S. Rastelli, A. Ciucii, V. Palleschi and A. Salvetti, *Laser Part. Beams*, 12 (1994) 525.
186 H.J. Hakkanen and J.E.I. Korppi-Tommola, *Appl. Spectrosc.*, 49 (1995) 1721.
187 J. Belliveau, L. Cadwell, K. Coleman, L. Huwel and H. Griffin, *Appl. Spectrosc.*, 39 (1985) 727.
188 D.A. Cremers, L.J. Radziemski and T.R. Loree, *Appl. Spectrosc.*, 38 (1984) 721.
189 D.A. Cremers and L.J. Radziemski, *Anal. Chem.*, 55 (1983) 1252.
190 C.K. Williamson, R.G. Daniel, K.L. McNesby and A.W. Miziolek, *Anal. Chem.*, 70 (1998) 1186.
191 C.C. García, J.M.Vadillo, S. Palanco, J. Ruiz and J.J. Laserna, *Spectrochim. Acta*, 56B (2001) 923.
192 R.J. Cotter, *Anal. Chem.*, 56 (1984) 485 A.
193 K.P. Jochum, L. Matus and H.M. Seufert, *Fresenius J. Anal. Chem.*, 331 (1988) 136.
194 L. Matus, H.M. Seufert and K.P. Jochum, *Int. J. Mass Spectr. Ion Proc.*, 84 (1988) 101.
195 H.M. Seufert and K.P. Jochum, *Fresenius J. Anal. Chem.*, 359 (1997) 454.
196 L. Matus, H.M. Seufert and K.P. Jochum, *Fresenius J. Anal. Chem.*, 350 (1994) 330.
197 K.P. Jochum and H.M. Seufert, *GIT Fachz Lab.*, 40 (1996) 517.
198 K.P. Jochum, A.W. Hofmann, M. Haller, M. Radtke, A. Knöchel, L. Vincze, K. Janssens, HASYLAB, 11 (1995) 1003.

Chapter 10

Robotic solid sample pretreatment

10.1. INTRODUCTION

Even though pick-and-place robots had been in use for some time previously, commercial laboratory robots were first introduced at the 1982 Pittsburgh Conference by the pioneering firm Zymark Corporation. The dramatic developments experienced by this type of robot over the following three years were again showcased at the Pittsburgh Conference, in 1985, which was considered the "year of the robot" by the journal *Analytical Chemistry* [1]. Although laboratory automation efforts had been directed in various ways for many years, it was the introduction and implementation of robotics in the chemical laboratory that focused efforts on a common target and provided a starting reference point.

Laboratory robotics is not an outgrowth of classical industrial robotics (manufacturing robotics). Developed independently, it focuses on the chemical process rather than on robotic hardware development. However, much of the technology that was previously developed and tested for industrial automation has found uses in laboratory robotics. Also, some classical terms are routinely used in connection with laboratory robotics and laboratory automation. By 1994, robotics had seemingly reached maturity, so a specific nomenclature for laboratory robotics and automation was issued by IUPAC [2,3]. Some of IUPAC's recommended terms are general and require the word "robot" or "robotics" for specific use (e.g. in "controlled-path robots", "corrosion-resistant robots", "feedback in robotics", "accuracy in robotics"); others are characteristic of robotic technology (e.g. "arm", "articulate structure", "flexible automation", "manipulator").

A robot (the name comes from the Czech word "robota", which means work or servant [2]) can be defined as "an automatically controlled, reprogrammable, multi-purpose, manipulative machine with several degrees of freedom, which may be either fixed in place or mobile for use in automation applications" [4] or, more loosely, as "a multipurpose machine which, like a human, can perform a variety of different tasks under conditions that may be unknown a priori" [5]. The different types of robots currently available can be classified according to physical features such as hardware construction, degrees of freedom, coordinate system or level of sophistication and technology [6].

Obviously, robotic techniques are more complex than other automated laboratory analytical techniques. Laboratory robots were originally marketed as do-it-yourself technology. Some of the first practitioners were chemists who were not skilled in the engineering and software aspects, so highly trained, dedicated personnel were needed. In many applications, the overall lead times to implementation became excessive and chemists were often sidelined for months awaiting the tools they needed to do their job.

Although robotic technology is claimed to provide flexible automation, in fact it possesses limited flexibility and applicability.

One other reason for the slow acceptance of robotics has been its presentation doing whimsical or trivial tasks, which caused potential users to disregard serious applications.

Fortunately, robot-based drug candidate screening systems represent a success story; for time-to-market and labour- and time-saving reasons, these systems are making a comeback in this specific pharmaceutical field.

The proprietary nature of robotic communications and software have also restricted their acceptance by laboratories. The lack of standardization among robot suppliers made it difficult to connect the robots to analytical instruments and other peripherals in a straightforward manner.

The initially envisaged uses of a robot in the laboratory — which led several renowned manufacturers to fail in the past — have changed with time. The early robotic units tended to mimic human manipulations, which made interfacing cumbersome and resulted in haphazard implementation of many applications. In general, robots were not really optimized for task-oriented work; their most general use was as sample-transport devices that moved samples or surrogates among a series of dedicated workstations which were optimized to do specific tasks such as solid-phase extraction or filtration. At present, robots are devoted to things that users could hardly figure out how to do easily in any other way. These unique needs range from massive robotic systems for handling thousands of samples produced in high-throughput drug-discovery and biochemical study programmes to stand-alone devices that drone away at repetitive tasks.

Depending on both the number of tasks to be performed and their complexity, users can currently choose between robotic stations and workstations. The former automate the entire processes; they often include facilities for sample insertion into analytical instruments and use the typical arm and an array of modules and peripherals to make tasks easier for the robot. The original idea of laboratory robotics revolved around a human-like arm capable of handling many analytical tasks. Today, manufacturers design arms to function as "movers of stuff" and operations such as transferring or filtering solutions are handled in modules around them. Workstations represent a newer vision of robotics. They are custom automation tools that assist workers in performing their job better and more efficiently. These stand-alone automated units are used for specific purposes such as handling liquids or preparing solid samples; they are growing in popularity as they can perform one to three tasks with reduced costs and sophistication. Some workstations can be interconnected or operated with robotic arms, while others use their own, dedicated arms (particularly to handle solid samples). One of the main differences between robotic arm systems and workstations is cost. Thus, a system built around a robotic arm costs more than € 100 000, whereas workstations can run from € 10 000 to 100 000, depending on their level of sophistication. However, it is the intended use as much as cost that often dictates the choice of design.

Because robotic technology continues to have some magical connotations in relation to laboratory automation, a number of manufacturers and users still use the words "robot" and "robotic" indifferently to refer to both robotic stations and workstations. In addition, any instance of automation is also indiscriminately associated with robotics by many. One case in point is the Internet page http://www.lab-robotics.org/manufact.htm, where about 100 companies summarize the highlights of their products, many of which bear little or no relation to robotics.

10.2. WORKSTATIONS, ROBOTS, MODULES AND PERIPHERALS

The most salient difference between robotic stations and workstations is that, whereas a workstation can only be used for the tasks (all or some) for which it was constructed, robotic stations can be modified by changing their software, modules or peripherals as required to undertake one or more specific tasks, or even a whole analytical process. As a result, describing a workstation is as simple as listing its intended functions, whereas characterizing a robotic station includes stating the type of arm it uses and the equipment that helps the arm perform its tasks.

Figure 10.1 depicts the scope of application of each option in terms of task complexity and throughput.

10.2.1. Workstations

The most simple and common workstations are those for dilution and/or reagent addition to a number of samples in a simultaneous manner, either to all samples in a rack or to a line with a slide *z*-axis (as in the Biomex® 2000 model from Beckman). Most workstations are designed to operate with liquid samples; such is the case with those from Cyberlab, Gilson, Zymark, SciLog, Sagian, Beckman and Hamilton, which manufacture specific equipment for liquid handling, solid-phase extraction and preparation of liquid

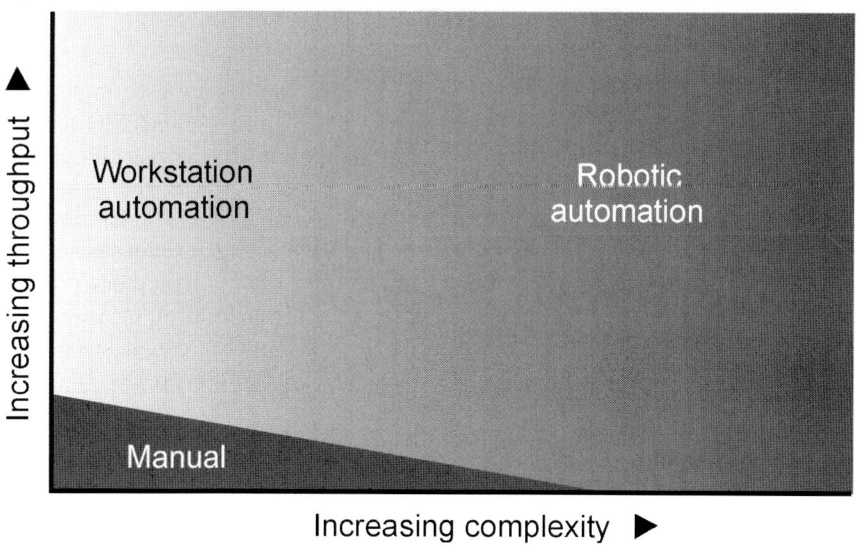

Fig. 10.1. Fields of application of workstations and robotic stations according to task complexity and throughput.

samples for insertion into chromatographs. Some firms (e.g. Zymark, Bohdan Automation, Source for Automation) also manufacture workstations for automatic dissolution of, mainly, pharmaceutical products; however, they can also be used to weigh and dissolve (or leach) other, specific types of solids.

The diversity of equipment with which manufacturers have flooded the market is exemplified by Zymark Corporation. In the last few years, this firm has launched six different types of workstations for handling liquid samples, namely:

(a) The Presto™ Liquid Handler, which uses a single-channel arm and a second, multichannel one that controls eight fixed cannulae for rapid, precise reagent addition, transfer or dilution to microplates with up to 384 wells.
(b) The Twister Universal Microplate Handler, which can be interfaced to microplate instruments such as washers, readers and dispensers to minimize the need for manual microplate handling.
(c) The Presto™ Microplate Labeler, a fast, reliable workstation for printing and applying adhesive bar code labels to microplates which is available as either a standalone workstation, integrated into a robotic station or loaded using the Twister™ Universal Microplate Handler.
(d) The Prelude™ Workstation, which is capable of automating solid sample treatments and includes options such as weighing, mixing, filtration and solid-phase extraction of samples for automatic insertion into HPLC systems, transfer to UV–Vis spectrophotometers and gathering in an EasyFill Sample Collection Module.
(e) The TurboVap™ Concentration Workstation series, which includes models II, LV, 500 and 96, all of which are equipped with shearing technology and differ mainly in the number of samples they can handle simultaneously.
(f) The RapidTrace™ SPE Workstation, which features an entirely modular configuration and can process up to 100 samples/h.

Most of the Zymark workstations designed for solid samples are used in pharmaceutical dissolution testing. In 1998, Zymark launched its MultiDose and MultiDose Plus workstations, which expanded the capabilities of similar previous devices by integrating a side-car workstation that attached and detached baskets or retrieved spent capsule sinkers. The following year, the Tablet Processing Workstation II (TPW II) improved on its predecessor by including an innovative wet grinding homogenization technique that provides reliable, reproducible results. Extracted samples can be filtered and diluted, the final sample solution being directly injected onto an HPLC instrument, measured via on-line UV–Vis spectroscopy or stored in test tubes or vials. Two electronic balances monitor solvent dispensing, liquid handling and sample handling operations. The TPW II uses Version 2.0 of Zymark's dedicated software for Windows™ NT v. 4.0, which enables the user a wide range of options including event handling and data archival, connection with on-line UV or LC data systems and the added flexibility of connecting with other common laboratory software packages. The TPW software can support up to three optional software add-in products, whether already available or under development. Such add-ins include a direct interface to the Hewlett-Packard ChemStation; an Advanced Data Channels Organizer that allows re-directioning of selected TPW II data to any network device in ASCII, Excel or ODBC format; and the Water Millennium Chromatography Data Manager. Supporting validation documents are included with the purchase of the upgrade.

Combinatorial chemistry workstations experienced a fresh impetus when Zymark acquired Scitec Automation Holdings, a Swiss-based laboratory automation solutions provider, in August 1999, thus introducing the new Scitec™ Combinatorial Chemistry Workstation (CCW). This handles solution and solid-phase synthesis with ease, can operate over a broad temperature range and is a reliable, flexible means for expediting the drug discovery process by enabling high-throughput parallel syntheses; also, it can be adapted to conveniently perform a wide variety of chemical conversions including multi-step reactions on solid supports.

Figures 10.2A and 10.2B depict the Prelude and the Tablet Processing Zymark workstations, both of which were designed for treating solid (general and specific) samples.

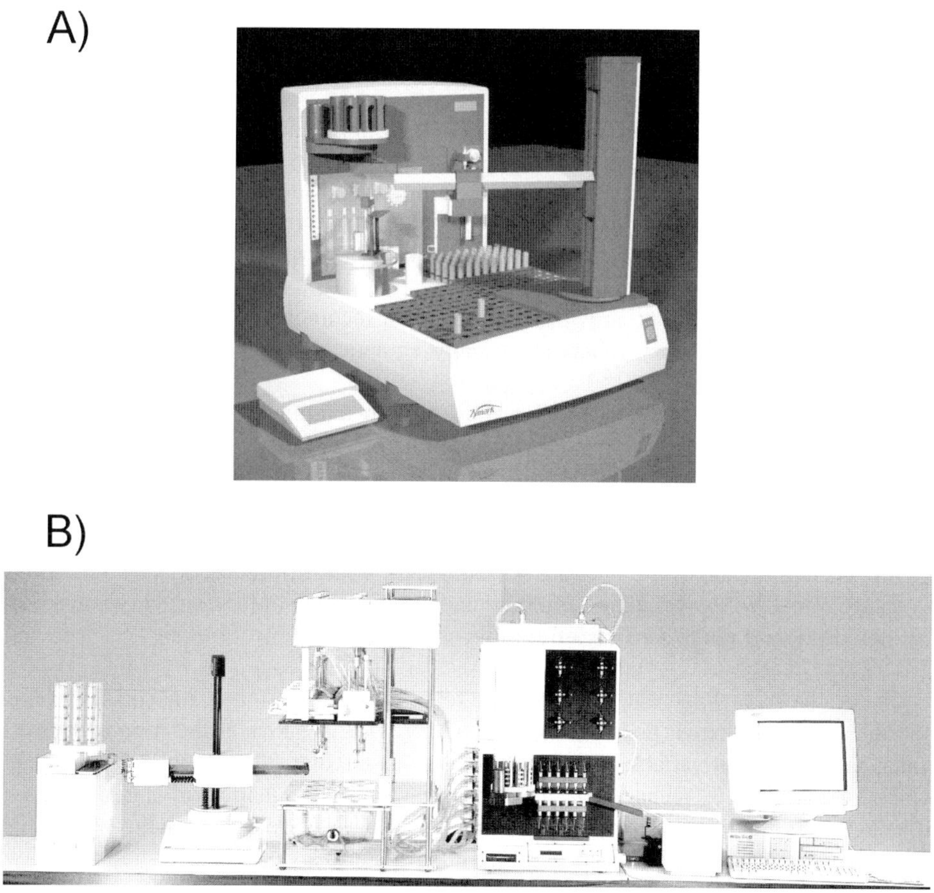

Fig. 10.2. Zymark workstations for solid sample treatment. (A) Prelude™ model. (B) Tablet Processing II model. (Reproduced with permission of Zymark Corporation.)

10.2.2. Robots

A laboratory robot constitutes the most flexible tool for automation as it can reproduce, with minimal adaptation, almost every task performed by an analyst.

A laboratory robot is essentially a machine consisting of various parts, namely:

(a) The *manipulator*, which is a mechanism usually comprising various segments, whether jointed or sliding relative to one another, for the purpose of grasping and/or moving objects, usually in several degrees of freedom. The essential parts of the manipulator are the body, the arm and a hand or end-effector. Robots can usually be furnished with various types of hand in order to handle objects of different size or shape. Based on the coordinate system they use, manipulators can be classified into *Cartesian* or *gantry* (defined by the x, y, z three-dimensional system, with translation motion but no radial motion), *cylindrical* (defined by the x, z two-dimensional system and the ϕ angle, with a radial motion and two translation motions), *spherical* [defined by the x one-dimensional system and two (ϕ and φ) angles, with one translation and two radial motions] and *revoluting* or *anthropomorphic* (with all motions of the radial type).

(b) The *controller*, an information processing device whose inputs are both desired and measured position, velocity and other pertinent variables of the process. The robot can be controlled by the user, in a point-to-point manner in the case of demonstration robots and by a computer otherwise. Computer-controlled robots are much more common and benefit from continuous improvement derived from the development of new software programming languages.

(c) The *power supply*, which provides the energy for locomotion of the different manipulator parts. Laboratory robots are usually equipped with electrical or pneumatic energy sources.

(d) *Sensors*, which are used to boost performance in intelligent robots. These devices can be of various types (e.g. stereoceptive sensors) and differ in their physical (e.g. the way variables are transduced into analogue or digital signals) and computational foundation (e.g. the way sensory data are processed to obtain sensory information). Sensors can be fixed or mobile and of the contact or non-contact type, depending on whether the target variables are sensed with or without physical contact with some body contained in the world. Force and tactile sensors are of the contact type; vision, proximity and range sensors, of the non-contact type.

One of the most important features of a robot, which determines its work envelope (i.e. the maximum extent and reach of the robot) is its position in the robotic station, which can be fixed or variable. The former is used in circular robotic stations, which are typical of Zymark's Py technology. In this configuration, the robotic arm is in the centre of a circle and the different peripherals are included in a work envelope radius as removable pieces of a pie. The work envelope of the arm in this case is 360° and the radius equal to, or shorter than, the extent of the arm (see Fig. 10.3A).

A mobile arm in a robotic station is supported in a track (Fig. 10.3B), which allows displacement of the arm in a length that varies, depending on the particular station, from

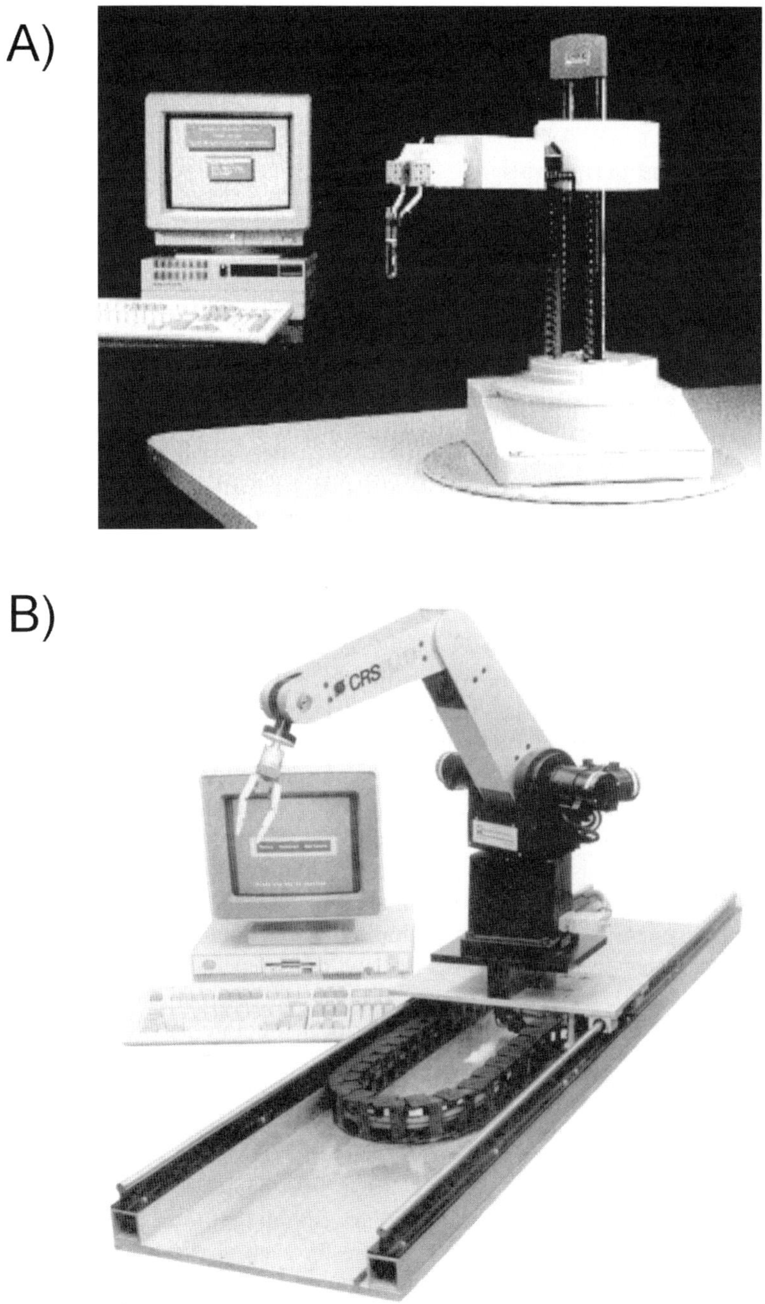

Fig. 10.3. Robotic stations with fixed and mobile arms. (A) Circular, PyTechnology, from Zymark. (B) Linear track, from Hudson. (Reproduced with permission of Zymark Corporation and Hudson Control Group, respectively.)

1 to 2 metres. The capacity of the linear robotic station for locating peripherals in the work envelope of the arm can be expanded by using a more complex arm capable of operating on both sides of the track. Linear robotic stations surpass circular ones in throughput but the two are similarly accurate.

10.2.3. Modules and peripherals

The modules of a robotic station are the devices (apparatus, instruments, racks) used by the arm to perform its tasks. In circular configurations, the modules are referred to as "peripherals".

The number of modules present in the work envelope of a robot arm varies with the number of tasks it is to perform in a given process. Some modules can be as complex as self-contained workstations. Although most of the modules required for the different steps of the process can be provided by either the arm's, or an alternative, manufacturer, some users design and construct their own modules, either because of the specificity of the task or with a view to reducing costs. A detailed discussion of such custom modules is obviously beyond the scope of this chapter. What follows is thus a brief description in relation to solid sample handling.

The *balance* is the most important module of a robotic station in dealing with solid samples. The only way of automating weighing is by using a robot or workstation. In addition to being connected to the Power and Event Controller (PEC), or central computer, the balance must be accessible to the robot arm, which should be furnished with a special end-effector or hand (a gripper hand) and a powder pouring device (usually vibration-based). Segregation of sample particles in terms of size can be caused by the programmed vibration of the source tube during the transfer and sample weighing step, which causes a bias in subsequent data. In this case, the pouring routine can be altered by rewriting the software subroutine to eliminate vibration and transfer the sample material in a single step [7]. Weight-based measurements of liquids are frequently used in robotic stations as they provide more accurate data than do volume measurements. Zymark's New Automated Filter Weighing System is an excellent example of the use of the balance with gaseous environmental samples. It provides automated and exact repetition of the analytical steps for compliance with EPA PM2.5 methods and uses no external air source in order to avoid introducing contaminants; also, filters are located before and after each weighing, and also ion-discharged before weighing.

A *dissolve and dilute* module is a dual peripheral that includes a vortex mixer to facilitate dissolution and the obtainment of a homogeneous solution, respectively.

A *master laboratory station* (MLS) is a module consisting of three syringes for dispensing liquids in conjunction with the dissolve and dilute module. However, the MLS can be used for additional purposes such as aspirating the phases involved in a liquid–liquid extraction (using a different syringe for each phase) [8].

A *centrifuge* is a key module when solid samples are to be leached or a precipitate forms during the process. The use of sensors to ensure correct positioning of the tubes with respect to the robot before and after the centrifugation step is mandatory here. Computer vision and neural networks have been used to detect errors in a robot system performing the automatic loading and unloading of a centrifuge [9].

All-purpose hands and *syringe hands*, available in a variety of designs, are also required elements of a robotic station. Differently sized objects (e.g. sample flasks, test tubes, probes, hold and press push-buttons) call for different types of hand. A syringe hand facilitates the withdrawal of liquids from vessels. Hand design has benefited from innovations devised by academic research groups [10].

Liquid–liquid, solid–liquid and *liquid–solid* extraction modules are also required devices for implementation of a number of methods.

Racks can hold a variable number of test tubes, the positions of which are numbered so that the robot can distinguish them. In addition to holding the tubes, racks can accommodate devices such as the aspirator assay probe or even the sample probe if one is used [8].

Additional modules not always required in a robotic station include a *capping–uncapping* module, used to remove and replace screw caps; a *bar-code reader*, which is usually a laser bar-code scanner combined with a turntable assembly capable of reading a label positioned anywhere around the circumference of a vial; and an *ultrasonic bath*, which is required for sonic mixing or cleaning, but also, occasionally, to facilitate dissolution or leaching.

The performance of robots in some processes can be improved by using various approaches to expedite one or more steps thereof. One such approach involves adapting non-automated devices for automated use in a robotic station. Such is the case with microwave digesters, which substantially accelerate solid sample pretreatment. For example, an ordinary focused-microwave extractor can be loaded and unloaded by impulsion and aspiration, respectively, using a peristaltic pump [8,11]; subsequently, specially constructed fingers can be used to handle the digestion vessels [12]. Also, the dawn of genome analysis protocols relied on automated laboratory work including the pipetting of reaction mixtures and reagents, loading of samples onto gels and grinding of clones onto filters [13]. Some researchers have succeeded in adapting robotic-based equipment to their specific needs [14,15]. Such is the case with the modular robotic pre-analytical workcell for coagulation analysis designed for timing studies intended to quantify the efficiency of the manual process and to identify areas in the processing of coagulation specimens where bottlenecks and long waiting periods have been encountered. The modular robotic system was developed to eliminate the bottlenecks and to allow a choice of specimen introduction by hand, by conveyor or by mobile robot [16].

Robot control is one other current research topic for various university researchers. This area encompasses new hardware concepts for robot control [17], software environments for optimizing the productivity of robotic laboratories [18] and dedicated software for use in environmental laboratories [19]. Finally, simulation and graphical animation of robots have proved highly useful with a view to both optimizing robot work and avoiding catastrophes on the workbench [20]. Incorporating robotic stations into large-scale laboratories entails establishing appropriate links with a laboratory information management system (LIMS) to ensure precise control of the laboratory as a whole [21].

Zymark's Clara 2000 software package exemplifies the incorporation of advances in computer science into commercially available robotic laboratory equipment. Its release followed the acquisition of Scitec Automation Holdings by Zymark. Based on an upgraded version of the Scitec Clara LT open architecture software, Clara 2000 provides substantial advantages in modularity, stability, Visual Basic customization capability and upgradeability for the future. Contributions from academic research groups have also been reported. One example is the practical approach to real-time scheduling and multitasking at the computer level developed by Entzeroth [22], which covers scheduling software, handling of errors and unexpected situations, and data management. A diagram of work and data flow during the screening process, and figures which show the scheduling or robotic movements of a receptor binding assay; an error handler for an exception-specific situation; and one for a general system error have also been reported.

10.2.4. Detectors

One of the essential units for completion of the overall analytical process by an automated system is the detector, which cannot be considered a module or peripheral as it preserves its identity as such and is only connected to the Power and Event Controller (PEC) for switching on, instrumental variable programming and data collection; the robotic arm acts as a bridge between the other steps of the analytical process and detection, either by inserting the treated sample into the detector or by connecting the treated sample with the detector via fibre optics (occasionally furnished with a sensor) or electrical wire.

Photometric detectors are the most common choice for monitoring the end product of a robotic station. The easiest way to detect products here is by using an optical fibre to drive the beam from the light source to the solution under study, with passage through it and then back to the detector. This approach has been widely used to monitor pharmaceutical dissolution tests, and also for on-line monitoring, using a diode array in both cases [23]. One other common way of photometrically monitoring the final solution of a robotic procedure is by using an LC sipping–injection module. For this purpose, the module is modified to act as a switching valve rather than as an injection valve. The solutions to be passed through the flow-cell can be aspirated in various ways, namely:

(a) By connecting one of the ports to a peristaltic pump. Following passage of a blank solution to zero the detector, the target solution is aspirated and the blank solution is circulated again both to sweep the target solution to waste and to zero the detector [24].
(b) By connecting two syringes of an MLS, assisted by a 3-way valve each, to the LC sipping–injection valve through the photometric cuvette, one syringe being used to aspirate the target solution into the cuvette and drain it after measurement, and the other to rinse the cuvette with distilled water [25]. When the target solution is in a test tube, the syringes of an MLS can perform the task without the aid of the LC sipping–injection valve [26]. When a derivatization reaction is required by a treated solid sample, a continuous system is to be preferred as it also drives the product to the flow-cell unassisted [8,27].

Luminescence molecular detectors have also been used for on-line monitoring of dissolution tests and the characterization of toxic residues using bioluminescence assays [28].

Atomic (AAS, DCP–AES) detectors have been coupled to robotic stations either through a continuous system acting as interface [11,29] or by direct aspiration into an ICP–AES instrument from a sample vial following treatment by the robot [30]. Mass spectrometric and NMR detectors used in this context are also based on direct aspiration.

Electrochemical detectors (particularly probe sensors) are also widely used in robotic stations. One salient application is the determination of soil pH [31] by inserting a glass electrode into a test tube containing the soil leachate. A straightforward washing–rinsing station allows cleaning of the electrode, which is fouled in the measuring operation. One other prominent application is the screening of polychlorinated biphenyls in used minerals by inserting a chloride ion-selective electrode connected to the LC2000/PCB Analyzer into a test tube containing the aqueous–alcoholic phase resulting from dechlorination with sodium, which is performed by the robot [32].

Finally, a conductivity detector has been used to determine soil salinity [33] and a balance to develop a gravimetric method for the determination of the oxidative stability of olive oil [34].

10.2.5. Miscellaneous considerations on workstations and robotic stations

Two essential facts to be considered when purchasing a robotic station are that no single firm builds everything one is bound to need to construct a robotic system tailored to their needs and that not all commercially available equipment operates by the same set of instructions or plays by the same set of rules. Compatibility between modules from different sources should thus be carefully checked prior to purchasing.

Reliability, which is one other key feature, depends on the particular components of the robotic system and its design. Because systems are linked to operate serially, even small losses of reliability at each device can be magnified at the end.

There is one other consideration in setting up robotic equipment that is shared by fully automated systems in general, namely: the surprising number of disposable items such as pipette tips, autosampler vials, filters and containers they go through. Even the microlitre amounts of solvents consumed can add up to large volumes in high-throughput operations. Users must bring these ancillary materials into the laboratory and then dispose of them; as a result, they may create an own small superfund site. Also, not all pipettes or autosampler vials are usable; thus, unlike a human, a robot cannot use a pipette with a crooked tip which means that disposables should be of a high quality to ensure reliability. Procurement of supplies and waste disposal are therefore two important issues in setting up and maintaining a robotic laboratory.

Because workstations can be designed and dedicated to a single, specific task, they are normally simpler mechanically and usually more reliable than are robotic stations. In addition, they can provide a data trail for regulatory compliance and are typically designed to be operated by non-experts. Commercial workstations vary in their level of sophistication, which allows the automation of even the most simple procedures. Economically, automation can benefit anybody running more than 150 samples a day and can justify purchasing a moderately priced workstation. More sophisticated workstations, however, are only profitable with a heavier workload. As a rule, a workstation will replace a € 40 000-a-year technician and have a one- to two-year payback period.

Workstations operate best when there are one to three functions being automated; more operations or less defined analytical steps require the sophistication of robotic arm systems, which are more flexible and work best early on in research, when the methodology is not well-defined. The combination of arms and modules with workstations has real advantages. A comparison of the efficiency of a tracked robot handling compound dissolution in collaboration with a liquid-handling workstation versus the robot alone revealed the robot–workstation team to increase productivity by 130% [5]. Where these types of system pay off the most is in high-throughput discovery and screening programmes. The automated systems developed to handle these chores can cost millions of euros to set up, require full-time experts in automation to design and maintain, and demand a huge financial commitment to operate. In any case, robotics is the only way to go for these types of applications.

Choosing between a robot plus peripherals and a workstation is a difficult task. From the beginning, automation held the promise of freeing analysts from cumbersome, time-consuming, repetitive tasks. This is especially true with the quality control (QC) laboratory, which must routinely test products such as pharmaceuticals or foods prior to release, often with a well-defined analytical procedure dictated by regulatory requirements. In these laboratories, workstations are typically the best solution as they are often more "hardwired" and are better in QC laboratories, where the analytical steps are well-understood and this equipment does save laboratories time and money. The best solution for implementing the complex treatments required by some solid samples is the sequential use of two workstations; when this is impossible, a robotic station is the next-best choice in most instances.

10.3. THE ROLE OF ROBOTS IN THE ANALYTICAL PROCESS

Automation of the analytical process by use of robotic equipment (robotic stations and workstations included) can reach from a single step to the whole analytical sequence. The number of steps that are robotized should be dictated by the user's experience and judgement, always as a function of the target process, costs, number of samples to be processed, etc. Straightforward single-task uses of robots, robotic sample preparation procedures and fully robotized methods are discussed below, as are more rational uses in combination with other techniques intended to ensure optimum development of each step of the analytical process.

10.3.1. Single-task robots and simple uses of robotics

One of the most important reasons of failure at the beginning of the laboratory robotics era, when no workstations were available, was the use of large robots to perform a single, repetitive task such as weighing, diluting or solid-phase extraction. One of the most common single tasks assigned to robots is weighing, which is the analytical step most difficult to implement using alternative automated approaches. Automating laboratory weighing may not increase the speed of weighing, but can free the valuable time of scientists and technicians. It has been claimed that, where a balance is in constant use for

4 h or more per day, the investment in a weighing automation system will be paid back [35].

Simple, inexpensive workstations designed for specific tasks include the TurboVap™ Concentration Workstations and the RapidTrace™ Workstation, which have found widespread use in clinical analysis. Non-commercial systems for single tasks include a microwave digestion system for dissolution of Ti(IV) oxide [36,37] and adaptations of workstations for special tasks such as the robotic–chromatographic method for the determination of glycosylated haemoglobin using a reprogrammed Hamilton Microlab 2200 pipetting Cartesian robot [38]. A robotic system was also used for the accurate, precise preparation of calibration standards, and automated, unattended multi-species preparation for both anion and cation analytical channels to support continuous on-line ion chromatography operations. Two robot Py-sections of a Zymark robot were designed, assembled and programmed in EASYLAB control language in command modules, the system resulting in substantial workforce savings [39]. Also, a robot was interfaced to an expert system for development of a standard-addition method. In this way, preliminary input of information by an operator and subsequent control of the robot table set-up, the concentration of cation addition solution, the monitoring wavelength, sample pH adjustments and the method of reagent addition were enabled. Three standard additions were run by the robot and a calibration line was constructed and tested for fit to a linear model, the system finally determining whether more additions were required [40].

10.3.2. Sample preparation

Sample preparation is the most general application of both workstations and robotic stations as the tasks involved in this step of the analytical process are the most time-consuming, error-prone and difficult to develop by unskilled operators; in addition, safety restrictions apply when toxic materials are to be handled. The use of a specific approach depends on the number of steps involved and their complexity. Table 10.1 summarizes the features of selected general and specific sample pretreatment procedures used in the environmental and clinical fields [41–77]. As can be seen, whereas most environmental samples subjected to a robotic treatment are solid, those dealt with by clinical laboratories are typically liquid and highly complex in nature. The type of analyte to be isolated from a given sample and the main operations to be performed by either the workstation or the robot to prepare the sample for subsequent steps, as well as the main steps following the treatment, are also listed in Table 10.1. The operations carried out after preliminary operations dictate the nature of the latter. When the step following treatment of the sample is insertion into a highly discriminating instrument such as an NMR or MS–MS instrument, the robotic station is aimed at a 24-h working day of the high-price instrument in an unattended manner, thus avoiding either the high maintenance personnel costs of continuous work or the purchase of another instrument. This task even justifies the use of a simple robotic station to insert samples with non-complex matrices into the instrument.

Although workstations and even robotic stations commonly perform liquid–liquid and liquid–solid separations, these steps can be implemented in a continuous manner by using a more inexpensive set-up and with shorter development times when the number of

TABLE 10.1

USE OF ROBOTIC STATIONS FOR SAMPLE PREPARATION IN ENVIRONMENTAL AND CLINICAL ANALYSIS

Analyte(s)	Sample	Main steps	Subsequent steps	Reference
Aflatoxin M1	Milk	SPE	HPLC, Fl. det.	45
Aflatoxins	Peanut butter	Filt., Evap.	HPLC, Fl. det.	40
Aldehydes	Surfactant	SLE, Filt.	HPLC	41
Atrazine, Alachlor	Soil	SLE, Evap.	GC, N-P det.	42
Cations	Lichen	Leaching	CE	53
Characterization	Toxic residues	Mix., washing	pHmetric determination	28
Diene value	Fuels	LLE	Titration	55
Isocyanates	Adhesive	SLE, Filt.	SEC	43
Metals	Used oils		ICP–AES	30
Metals	Soil	SLE	f.i. manifold	11
Micro-pollutants	Water	SPE	GC–ECD	48
Nutrients	Environm. samples	Hot block digestion	Kjeldahl method	46
PCBs	Mineral oil	Mix., LLE	Pot. det.	31
Pesticides	Vegetables	SPE	GC–ECD	49
Phosphorus-31	Water	Mix.	NMR	47
Siloxanes	Hair	LLE	ICP–AES	54
Sulphur (mercaptans)	Fuels	LLE	Titration	56
TCP	Soil and fruit	SLE	ELISA	50
TCP	Urine	Hydrolysis	LLE, GC–MS	52
Trace element	Titanium dioxide	MD	—	44
BOD	Waste water	Filt.	DOM	59
Alosetron	Plasma and serum	SPE	HPLC, Fl. det.	68
Amino-acids	Pharmaceutical samples	SPE	SEC	67
Cocaine	Urine	LLE	GC–MS	57
Cocaine	Blood	SPE	GC–MS	60
Diclofenac	Human plasma	LLE	HPLC, Phot. det.	63
Drugs	Biological fluids	Centr., Evap.	HPLC	64
Drugs	Animal feed	SLE, Centr.	HPLC	65
Drugs	Plasma	Precipitation	LC–MS/MS	77
Felbamate	Human plasma	Mix., Centr., Evap.	HPLC, Phot. det.	70
Frenolicin B	Poultry feed	SLE	LC	76
Heroin	Heroin	Mix.	GC–FID	71
Hydrochlorothiazide	Human plasma	LLE, Evap.	HPLC	72
Mycophenolic acid	Human plasma	SPE	HPLC	74
Nefazodone	Human plasma	LLE	HPLC, Phot. det.	62
Porphyrins	Faecal samples	LLE	Reversed-phase HPLC	73
Ranitidine	Serum	SPE	HPLC, Phot. det.	69
Sulphadimidine	Serum	Imm.	Phot. det	58
Terfenadine	Human plasma	SPE	HPLC, Fl. det.	61
Theophylline	Human plasma	LLE, Centr.	HPLC, Phot. det.	66
Vitamin C	Foods	Centr. Filt. SPE	f.i. manifold	51
Vitamins A, B	Dairy products	LLE	HPLC	75

Filt. Filtration; Evap. Evaporation; Fl. det. Fluorimetric detection; SLE Solid–liquid extraction; SEC Size-exclusion chromatography; MD Microwave digestion; LLE Liquid–liquid extraction; Imm. Immunoassay; Phot. det. Photometric detection; Fl. det. fluorimetric detection; N-P det. Nitrogen-phosphorous detector; BOD Biological oxygen demand; DOM Dissolved oxygen measurement; SPE Solid-phase extraction; Centr. Centrifugation; Mix. Vortex mixing; PCBs Polychlorinated biphenyls; Pot. det. Potentiometric detection; TCP 3,5,6-trichloro-2-pyridinol

samples is not very high and the results must be available within a short time. The unit operations that can be performed failure-free by a robot include weighing, centrifugation, filtration and special instances of solid and liquid transfer. On the other hand, reagent addition, dilution, heating, homogenization, derivatization and insertion into a measuring instrument are generally more rapidly, economically and efficiently performed by other automated alternatives, whether continuous or discrete.

10.3.3. Robotic development of the whole analytical process

Most frequently, using a robotic station to develop an entire analytical process is unwarranted. In the mid-1980s, when robotic technology first reached the analytical laboratory, robots were more of a novelty than a useful tool. At that time, conventional manual titrations and similarly easy tasks were entrusted to robotic stations. At present, however, well-established criteria exist to ensure correct use of the potential of robotic technology.

One well-justified use of a robotic station is in the gravimetric method for the determination of the oxidative stability of olive oil, where the station monitors the amount of oxygen absorbed by a sample heated under controlled conditions. The method was intended to relieve the typically heavy workload of olive oil analysis laboratories in winter and involves weighing the samples in test tubes that are heated for a few hours, allowed to cool and weighed, the cycle being repeated as many times as required, depending on the stability of the oil, until the weight gain of all samples reaches a pre-set value. The results obtained with this method are quite consistent with those provided by the well-established Rancimat^R method, which it clearly surpasses in throughput (150 versus only 6 samples/batch) [34].

An also effective gravimetric method for testing of environmental samples has been reported [78]. This type of sample has been handled by robotic systems for the determination of pH, BOD and suspended solids, using a sophisticated interface that allows the operator to start and stop the robot at any time during the analysis without the need for recovery software [79].

The use of automated titrators, whether photometric or potentiometric, to complete the work performed by the robot is not an instance of hyphenated batch methodologies but rather one of the functioning of the titrator as a module of the robotic station. Such is the case with the determination of fuel parameters including the diene value [55] and mercaptan sulphur [56]. In the former application, the robot weighs the sample, refluxes it in the presence of maleic anhydride and extracts the analytes into an aqueous phase which is poured into the titration vessel, where the robot inserts the photometric or potentiometric probe. For the determination of mercaptan sulphur, the robot also weighs the sample and removes sulphide by precipitation with a $CdSO_4$ solution and liquid–liquid extraction. Once the fuel is sulphide-free, which is checked by using a photometric probe, it is poured into the titration vessel by the robot arm, which also plunges an Ag electrode prior to starting addition of the titrant (an $AgNO_3$ solution). In both cases, the automated titrator acts as a module of the robotic station and is operated by the robotic arm.

10.3.4. Combining robotic and continuous systems for more reliable development of the whole analytical process

The "bridge" spanning the raw sample and reading of the analytical signal, which is a function of the analyte concentration, has traditionally been built around single batch, continuous or robotic methodologies for driving the sample to the detector. Each approach has its own advantages and disadvantages, and no single currently available choice for automation is the panacea. Combinations of a robotic station or a workstation with a liquid [41,44,45,61–66,68–70,71–76] or gas chromatograph [42,48,49,52,57,70,71], or a capillary electrophoresis system [53], have been reported in which an autosampler was used as an interface, the robot arm either placing the vials in it or loading them (following positioning on the autosampler) with the treated samples.

The Zymark–Waters partnership for the development, among other things, of an intelligent software interface between the Zymark TPW™ platform and the Water Millennium Chromatography Data Manager is an example of the need to make robotic equipment and chromatographs compatible. Users experienced in flow injection (FI) methodology proposed this continuous approach [80–82] as the most easy-going friend of robotics in order to take advantage of both. Thus, the expeditiousness and simplicity of FI systems are offset by their inability to automate certain operations (some as essential as weighing); on the other hand, robots, which are the most powerful systems as regards the variety of processes they can develop, have an inherent relative slowness compared to flow-injection systems in addition to a high purchase cost. Figure 10.4 illustrates the different degrees of difficulty involved in inserting samples in different aggregation states into an FI manifold: whereas liquid samples can be directly introduced, solid samples require one or more previous operations. The pretreatment steps can be shortened by using a more complex hydrodynamic system [83]. The figure also illustrates the need to use robotics to treat solid and/or complex samples. Thus, while the use of a single automated approach to circumvent the problems encountered in automating laboratory processes is desirable, the conservative policies of equipment manufacturers and the usual inertia of research groups to drastically change investigation guidelines have so far delayed advances in the combined use of several automated approaches. The complementary features of the FI technique and robotics warrant their joint use with a view to exploiting their respective advantages while avoiding their shortcomings in isolation. Surprisingly, however, the uses of such a promising couple have scarcely been explored so far.

Combined FI–robotic systems can in principle be used in two ways, namely: (a) by having the two individual systems operate independently of each other, the user acting as an active interface between both; and (b) by setting up an integrated system where the two sub-systems will interrelate to eliminate the need for human intervention. In the former approach, the robot is devoted to automating the complex preliminary operations the FI system cannot handle (e.g. placing partially treated samples in a rack for subsequent transfer to the FI system by the user); the FI system acts as an automated device for inserting samples into the instrument, with or without a prior separation and/or derivatization step. In the integrated approach, the FI manifold is a module of the robotic station that is used to insert the sample into the detector (also, with or without a prior separation and/or reaction). This latter approach involves no human intervention, which is an obvious advantage

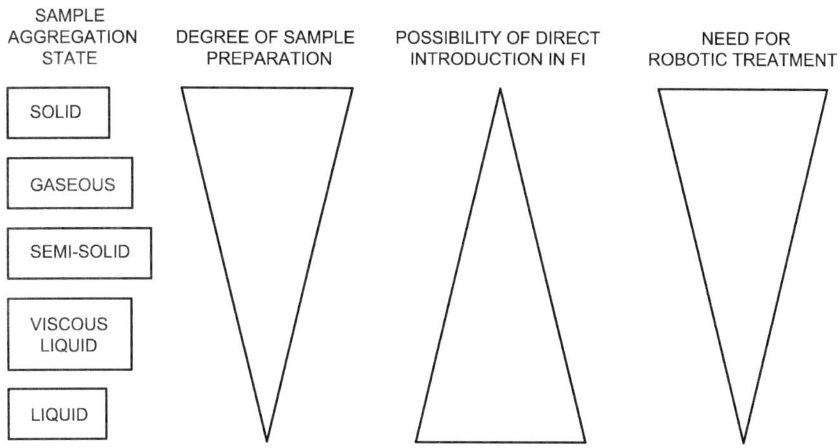

Fig. 10.4. Gradation in difficulty of sample manipulation by FI–robotic combined systems according to state of aggregation. (Reproduced with permission of Elsevier Science.)

with respect to the former. Below are briefly discussed several FI system–robot combinations reported so far.

Off-line coupling of a robot to an FI system

There are few reported instances of the combined use of an FI manifold and a robotic station operating independently. One is a method for the automatic analysis of used oils by ICP–AES [30]. The robot is used to weigh the oil after the sample is heated, to add the volume of xylene required to obtain a 1:9 weight ratio and to mix the two. The sample thus obtained is used to fill a vial fitted in the autosampler of the FI analyser, which acts merely as a device for automatic insertion of the sample into the ICP spectrophotometer. The link between the two sub-systems [viz. the transfer of both samples and information produced in the robotic operation (dilution of each sample)] is established manually. This approach does not completely dispense with human intervention, although it has been estimated to cut manpower requirements by two-thirds.

One other application is the automation of a method for the determination of total vitamin C in foods [51]. Here, the robotic station is used for homogenization of the sample, weighing, addition of an extractant, centrifugation, filtration and clean-up through a C_{18} column. After this treatment, the sample is manually transferred to the FI autosampler. A derivatizing reaction is implemented along the FI manifold to obtain a fluorescent product prior to insertion into the spectrofluorimeter. Although not specifically stated, the information produced is also transferred manually between both systems.

The previous two examples illustrate as many ways of transmitting information in off-line coupled FI–robot systems, namely: automatic processing by means of an on-line microcomputer or manual acquisition and treatment.

Integration of robotic stations with FI manifolds

When the FI manifold acts as a module of the robotic station (or rather, as an interface between the robotic station and detector), the different operations involved in the analytical process are distributed between modules in order to optimize the performance of the overall system while avoiding the need for human intervention in communicating the two systems physically and logically. This approach was developed by the authors' research group.

One typical application of this approach is the method for the determination of total polyphenols in virgin olive oil [84]. The sequence of steps involved in the standard method is as follows: (1) weighing the oil, (2) diluting it with *n*-hexane, (3) separating the analytes by liquid–liquid extraction, (4) derivatizing an aliquot of the extract with Folin–Ciocalteau reagent, (5) filtering the precipitate formed during the colour-forming reaction and (6) measuring the absorbance of the product. Apparently, this sequence provides little opportunity for FI systems owing to the high complexity of each step. Insertion into the detector is seemingly the single operation that the FI system could perform with ease; however, this would be of little help. The FI manifold can be made much more effective if its job starts when the analytes have been isolated. Thus, the robot directly introduces the extract into the manifold, which performs the derivatization reaction and transfers its product to the detector. In this way, the time required for the last derivatization step is dramatically shortened (1 min versus 1 h) and the filtration step is made redundant as no precipitate forms during the short interval between injection and detection. The borderline between the FI module and the robotic station in this approach is ill-defined as the robot, through

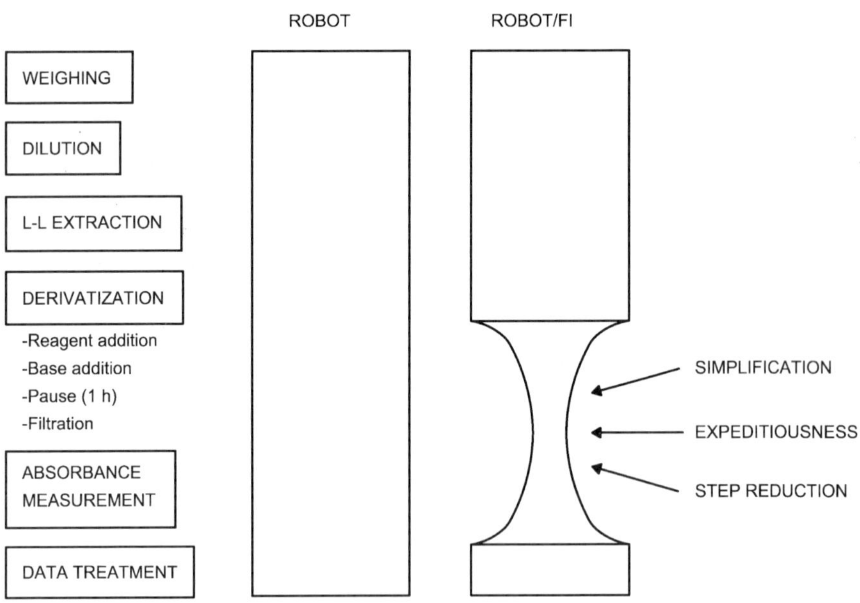

Fig. 10.5. Simplification of the determination of total polyphenols in virgin olive oil by use of a FI–robot integrated approach. (Reproduced with permission of Elsevier Science.)

its injection valve, inserts the extract from the test tube into the FI system, the peristaltic pump of which is controlled by the robot. Moreover, the prior calibration of the FI system is performed by the robot, which prepares and injects the different standard solutions from a stock solution of a polyphenol standard and runs the calibration graph, which is later used to quantify the analyte. The advantages of this combined approach are illustrated in Fig. 10.5. The information produced is processed by the controller of the robotic station, thus completely avoiding human intervention.

Even when the tasks to be conducted in the FI manifold are simple, coupling it to a robot provides some advantages. One example is the fully automated method for the determination of metals in soil. The use of a two-channel FI manifold to introduce robot-treated (weighed, microwave-digested, centrifuged) samples into the measuring instrument (e.g. an atomic absorption spectrometer) enables real and *pseudo*-dilutions with a dramatically broadened calibration range. In this way, the need for a prior estimation of the analyte concentration is avoided. In addition, such an expensive peripheral as a syringe hand is replaced with an inexpensive manifold, which reduces purchase costs [11]. The overall operational set-up for solid sample pretreatment plus determination without the FI interface is shown in Fig. 10.6.

One other FI–robotic approach clearly exemplifying the advantages of the joint use of both techniques is that to the determination of starch in foods [8]. The sequence of

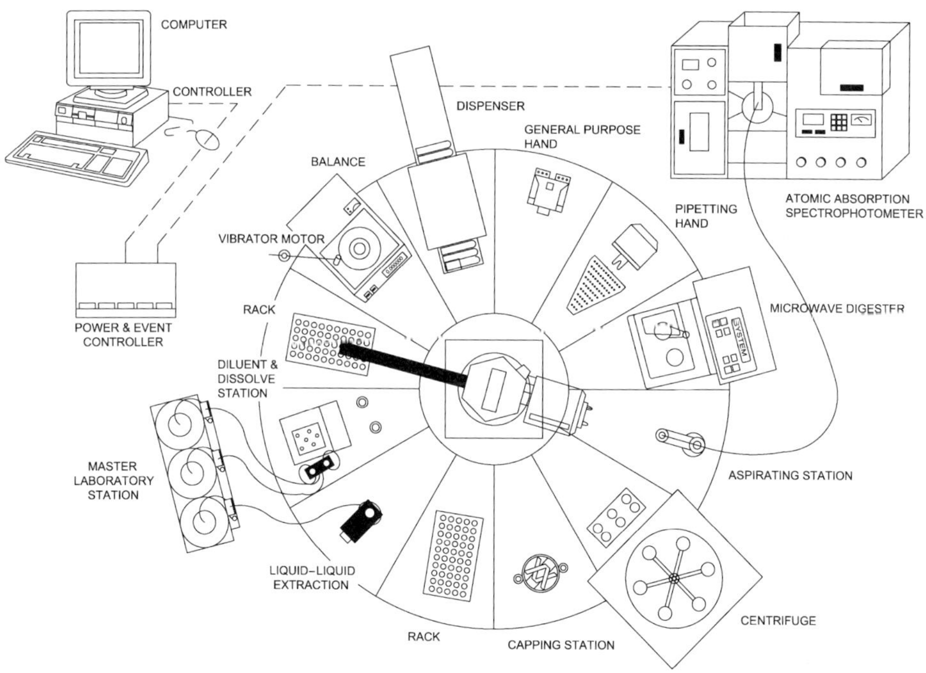

Fig. 10.6. Robotic station for the fully automated determination of metals in soil. (---) Passive interface. (—) Tubing. Active interfaces of the computer to the robot and its peripherals are not shown. (Reproduced with permission of Elsevier Science.)

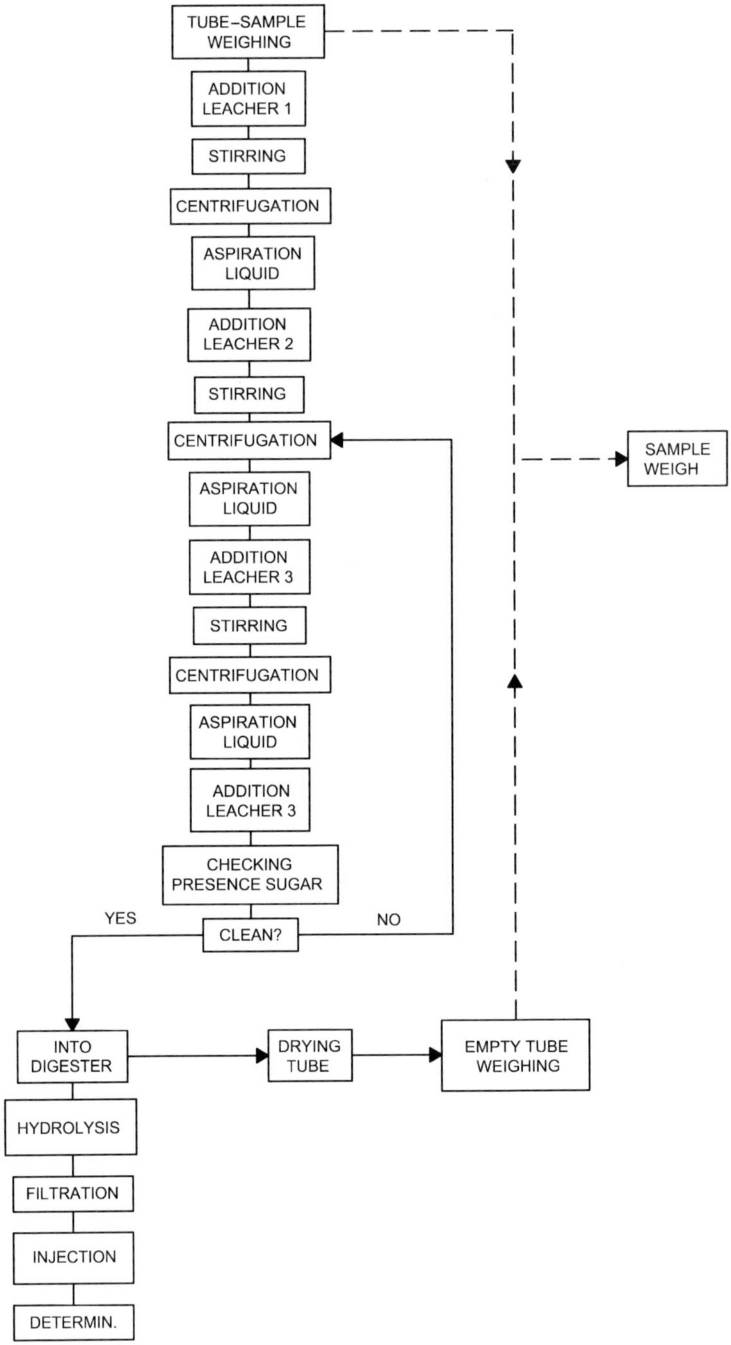

Fig. 10.7. Flow-chart of the robotic steps and flow-injection determination of starch in foods. (Reproduced with permission of Elsevier Science.)

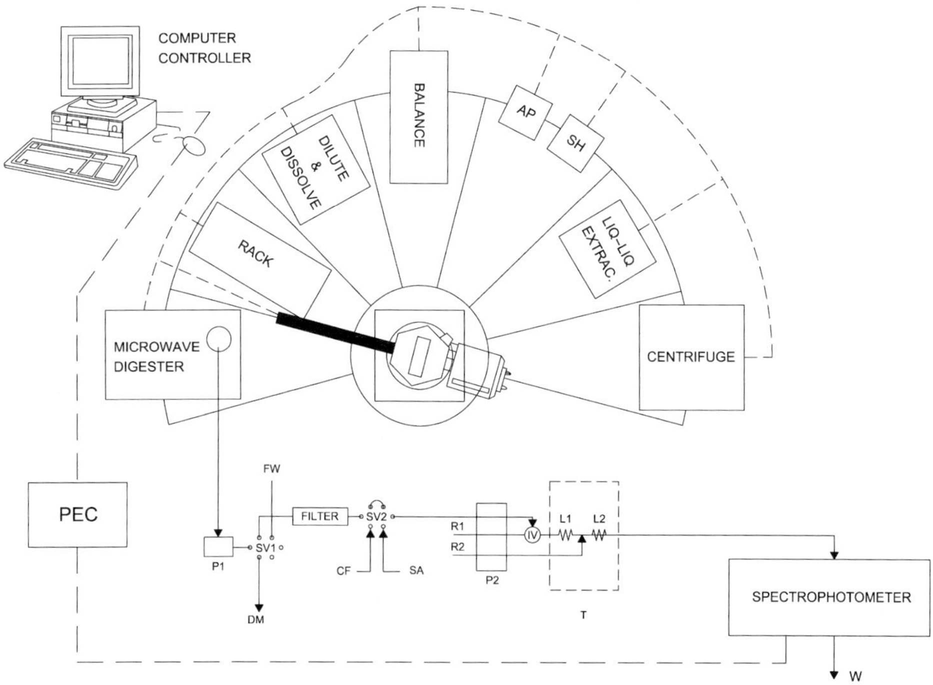

Fig. 10.8. Robotics–FI approach to the determination of starch in foodstuff. (– – –) Passive interface. (---) Active interfaces. PEC power and event controller, AP all-purpose hand, SH syringe hand, P peristaltic pump, FW filter waste, DW digester waste, CD clean-up filter solution, SA solution for assaying sugars, SV switching valve, R1 neocuproine channel, R2 NaOH channel, IV injection valve, L reactor, T thermostat, W waste. (Reproduced with permission of Elsevier Science.)

the time-consuming procedure for removal of sugars from the solid and the introduction of the sugar-free sample into the microwave-assisted digester are performed by the robot as shown in the chart of Fig. 10.7. Then, after hydrolysis, the liquid phase is aspirated into the injection valve of the continuous manifold through a filter. Upon injection of the solution from the digester into a carrier stream of neocuproine, the injected plug merges with a basic stream for development of the coloured product, which is monitored on passage through the flow-cell. Figure 10.8 depicts the overall combined system. The FI–robotic combined approach has also been used for the determination of nitrite, nitrate and chloride in fresh and cured meat [27].

10.4. ANALYTICAL SCOPE OF APPLICATION OF ROBOTICS

Although robotics can be used in virtually all analytical areas, its major applications encompass the clinical, pharmaceutical and biotechnological fields.

The large number of samples typically handled for analysis daily by a *clinical laboratory* promoted the manufacturing of the well-known air-segmented continuous autoanalysers, mainly from Technicon [85]. The original robot arms gradually replaced auto-analysers and were eventually superseded by workstations for specific tasks such as pipetting, diluting, reagent addition (i.e. sample-handling tasks) [86]. Dedicated clinical laboratory workstations are used to receive samples in capped, bar-coded vials in an indexing holding rack that the robot decaps in a decapper and introduces into the analyser(s) as a function of the specific bar code [87]. The use of robots in the health care sector is not limited to the clinical laboratory; in fact, it also encompasses tasks such as the assistance to surgical procedures, the elimination of mundane chores and the reduction of exposure of personnel to communicable diseases. Former problems arising from a lack of electronic communication, software, hardware and analytical standards have by now been overcome.

The *pharmaceutical field* has exploited robotic stations for the main purposes of dissolution testing and drug design. In fact, most of the procedures described in robotic manuals (e.g. Zymark's monographs and manuals [88]) are concerned with the determination of drugs in pharmaceutical preparations. Robotic stations as such were initially also programmed for such a simple use as dissolution testing. Later, workstations for dissolution testing have flooded the market with specific designs in variable degrees of sophistication (e.g. for continuous monitoring through fibre optics, insertion into a chromatograph). In any case, robotic stations continue to be used to develop innovations in the pharmaceutical field. The significance of the laboratory to product/process development, process improvement, market surveillance and general pharmaceutical business has been well understood by company managers [89].

One recent contribution to the clinical–pharmaceutical field took the form of an automated blood sampler for pharmacokinetic analysis [90]: the Culex ABS system, which was developed from the Raturn System for awake animals (equipment that allows painless blood sampling without impeding normal movements and feeding). The robot is instructed by the user on the method to be implemented, which includes the start time and the frequency of sampling, the sample volume (10–250 μl), the extent of dilution of the sample with heparinized saline and the return of any unused sample (following dilution with the saline to make up to volume of blood removed). The saline also acts as a carrier to transfer the diluted sample through tubing. The blood is collected in sealed refrigerated phials that are transferred to 96-well plates for analysis. The software in the system is controlled from a PC operating under Windows 95/98 and the method can be individually controlled for up to four rats. Blood can be taken from a single rat for up to several days. Urine and faeces are collected separately. The total floor space required by the system is less than 0.75 m^2.

However, the most important pharmaceutical use of robotics at present is in combinatorial chemistry in the search for new drugs. The impact of robotic high-throughput screening on drug discovery is a consequence of its ability to screen up to 100 000 samples per day per instrument; this in turn results from the demand for new compounds — which is being met by combinatorial chemistry [91]. By way of example, a system based on a Zymark XP Zymate robot and a minimal modification of the chemical method of reaction assembly, execution and isolation of the products, allowed the synthesis of more than 8000 new different products utilizing over 30 different types of chemistry [92].

Biotechnology is no doubt the most rapidly expanding field of application of robots. After years of hard work, the human genome was eventually deciphered in 2001. In fact, genomics continues to be one of the most exciting fields of research, and has been joined in this respect by proteomics, the word in the mouth of all attendants to the 2001 Pittsburgh Conference.

Most robotic biotechnological research work is conducted by workstations. Thus, accelerated recombinant protein purification processes have been developed by using a Vision Workstation equipped with a fully integrated sample-holding robot that allows preparative cation-exchange chromatographic runs to be completed and the collected fractions analysed by HPLC [14]. One other way to purify combinatorial proteins containing unreacted amines and other water-soluble byproducts or impurities is by liquid–liquid extraction (LLE). The method was automated by using 96-well plates and a robotic liquid-handling workstation. Crude combinatorial library samples were dissolved in a water-immiscible organic solvent and added to each plate well, packed with an inert support material coated with hydrochloric acid. Separation relied on the partitioning of the compounds between the two liquid phases and the product recovery, purity and amine removal efficiency were determined by HPLC, with and without pre-column derivatization. The clean-up procedure removed more than 98% of the amines, yielding an average product purity of 90% [93]. A robotic system was also used to image and excise 288 protein blots from an amido black stained polyvinylidene difluoride blot. After automated enzymatic digestion and MALDI-TOF/MS monitoring, data was analysed using novel automated peptide mass fingerprinting database interrogation software and 95 proteins were identified on the basis of peptide mass, isoelectric point and molecular weight in less than 10 days [94].

A dedicated workstation for protein isolation and identification that integrated 2D PAGE, gel imaging, sample handling/gel cutting robotics, mass spectrometry and bioinformatics is commercially available under the name ProteomeWorks System [95]. This station uses software such as PDQest for 2D gel spot excision and digestion, and ProteinLynx for the assignment of protein identity. It is claimed that the ProteinProbe of the system can be optimized for implementation of secure Intranet and matching each MS–MS spectrum in less than 1 s. One other piece of equipment, the Qiagen BioRobot 3000 workstation, allows the magnetocapture of cell lysates by means of magnetic beads [96].

Elucidating protein structures by X-ray crystallography obviously requires the use of crystallized proteins. However, these typically huge, loopy molecules frequently resist assembling into such ordered structures and scientists can spend months or even years figuring out what combinations of reagents, temperature and other conditions will cause a particular protein to crystallize. The high-throughput style of robotic combinatorial chemistry could be ideal for protein crystallization [97] and provide new avenues for exploring the use of robots in proteomics.

Various other fields benefit from the use of robotics. For example, the unstoppable advances in robotic technology have led to the development of a microrobotic arm constructed by using lithographic techniques that can be mobilized in an aqueous environment to pick up, lift and reposition a 100-μm glass bead. The 670 μm-long robot arm, attached to an oxide-coated silicon wafer, has doped polypyrrole/gold bilayer segments that serve as elbow, wrist and finger joints. By applying a voltage to the joints, cations are caused to flow into or out of the polymer, thereby making the material swell or shrink. This causes the specific joint to bend in a controllable manner. The microrobot may be an excellent tool for handling single cells, bacteria and other microstructures in aqueous media [98].

A comparison of the sequential injection (SI) technique and robotics for the automation of the rhFXIII fluorimetric activity assay [99] led to the adoption of the latter on account of its wider commercial availability; SI technology, however, is bound to match the current potential of robotics as soon as reliable SI systems reach the market [100].

Agricultural laboratories are increasingly taking advantage of the use of robotics as seasonal overloads call for a 24 h working day without involvement of temporary personnel. Thus, tasks such as weighing, leaching, filtering and measurement are frequently performed by robots in the determination of parameters such as pH [31], organic matter [26], conductivity [33], lime [101] and phosphorus in soil [25], or bitterness in virgin olive oil [24].

10.5. PRESENT AND FUTURE OF ROBOTICS

Robotics as implemented in workstations and robotic stations has proved to be one of the most accurate and autonomous automated facilities, a fact that directly influences both the productivity and the quality of the analytical results it provides. However, and despite the continuous advances, further expansion of robotic technology is meeting with several serious hindrances, namely:

(a) The scarcity or even lack of standards for checking and/or validating unit operations. No robotic method can be officially endorsed in the absence of an appropriate standard. This has promoted the search for new ones by public and private research groups.

(b) The high purchase cost of robotic stations, which makes them inaccessible to small laboratories — workstations are usually more affordable.

(c) The growing complexity of both hardware and, especially, software, which raises the need for a well-coordinated professional team to keep robotic equipment operational. This is particularly true of robotic stations, but altering the operational mode of a workstation also requires professional assistance, either from the manufacturer or from dedicated specialists.

(d) The scarcity of literature on robotics, which is a consequence of the limited use of robotic equipment for research purposes outside companies. In fact, the results of private research teams are rarely published in scientific journals and the information provided by manuals and propaganda publications hardly explains the underlying principle or working programmes. Obtaining detailed information for easy access to the functioning of robotic equipment is always a tough task for the user. On the other hand, the magnified advantages of robotics as expressed in manuals and, particularly, propaganda on this technology, can hardly help one acquire a true criterion about the benefits derived from its usage.

Current R&D efforts in robotics are being aimed at overcoming these weaknesses and provide clues for a number of strategies to be adopted in the future, namely:

(a) Cooperative alliances between robotics firms and other manufacturing companies are to be expected in the wake of the joint venture of Zymark Corporation and Waters for the development of a software interface between their respective equipment (e.g. the Zymark PTW™ platform and the Waters Millennium Chromatography Data Manager).

(b) The current drivers for new robotic technology are the high-throughput screening and combinatorial chemistry markets. This trend should continue and the next logical big step should be the automation of candidate drug screening procedures (e.g. by running in vitro tests on a massive scale). Cooperation between manufacturers is also bound to raise the quality of the results obtained in these fields and open new markets. This policy is exemplified by Zymark, whose alliance with Eppendorf-5 Primer Inc. (a leader in the development of high-performance reagents) has led to the integration of Zymark SciCloneTH based automated robotic and workstation technology validated with the PERFECTprep-96 VAC Plasmid DNA Purification System. This will allow customers to obtain better results in a high-throughput genomics environment.

(c) Experts also expect automation to become more sophisticated through adoption of new technical achievements in both this area and computer science, which are being incorporated day by day to new equipment. One salient example is the direct interface of the Tablet Processing Workstation II (TPW II) to the Hewlett-Packard ChemStation; an advance data channels organizer that permits the re-direction of selected TPW II data to any network device in ASCII, EXCEL or ODBC formats; and the Waters Millennium Chromatography Data Manager.

(d) The simultaneous treatment of several samples in different analysis steps will be increasingly fostered in the future. The scheduling modus operandi of the Clara open architecture software from Scitec is a clear example of this trend to increasing sample throughput by optimizing the temporal distribution of the unit operations performed by the robotic equipment.

(e) Robots will be driven to work in smaller dimensions (e.g. those of a 384-well plate) as they are still too large to deal with the increasingly smaller samples involved in clinical analyses. A saving of reagents will be one immediate result of this trend. Robotic devices themselves will also be reduced in size to free up valuable benchtop space.

(f) Internal and/or external sensors will be mandatory in order to construct robotic stations capable of performing new tasks (e.g. establishing when a solid sample has been completely dissolved, detecting the presence or appearance of a precipitate) and develop mobile robots. Optical, tactile, thermal and mass sensors are the targets of many research centres; these can be either used in isolation or incorporated into robots or robotic stations in order to ensure traceability in the results by use of well-established bar-coding systems. The latter are based mainly on two-dimensional codes that overcome the restrictions of linear symbols and enable whole new classes of applications.

(g) The true revolution in analytical robotics will probably come from the development of artificial intelligence. Current workstations and robotic stations are no more than sophisticated apparatus with simple feedback systems and, as such, have very limited capacity to take decisions in the different steps of the analytical process. Future robots will be managed by expert systems in such a way that they will be able to find solutions to analytical problems by using both the initial stored information and that obtained from the process it has developed previously. The "assimilation" to new information and its inclusion in the algorithm of the expert system (i.e. the apprenticeship from experience) is one of the target subjects of artificial intelligence research. The isolated modus operandi of current expert systems will change in the near future and will be incorporated into analytical equipment, thus endowing instruments with intelligence to take maximal advantage of their features. Robotic systems will be integrated into an intelligent surrounding and will act as both "clients" and "servers" of the analytical laboratory network. In short, the use of artificial intelligence and neural networks to integrate laboratory automation will be one of the most exciting areas in future robotics research.

Solid sample treatments will foreseeably benefit from the development of dedicated workstations capable of performing a series of steps, including weighing, required by a number of analytes and matrices; by using appropriate software, the selectivity of the operation will be tuned to the specific requirements of the sample. In this way, small laboratories will be able to programme their workstations to process different types of samples requiring different treatments during 24 h working days. It should be noted that the only way of completely automating solid sample pretreatment is by using robotic equipment — no other automated methodology allows samples to be weighed with absolutely no human intervention.

References

1 Year of the Robot, *Anal. Chem.*, 57 (1985) 651 A.
2 H.M. Kingston and M.L. Kingston, *J. Autom. Chem.*, 2 (1994) 43.
3 H.M. Kingston and M.L. Kingston, *Lab. Rob. Autom.*, 7 (1995) 3.
4 *Manipulating Industrial Robots: Vocabulary*, International Organization for Standardization, ISO/TR 8373 (1988).
5 D. Nitzan, "Development of Intelligent Robots: Achievements and Issues", *IEEE J. Rob. Autom.*, RA.1 (1985) 3.
6 H. Waldman, *Dictionary of Robotics*, MacMillan Pub. Co, New York (1985).
7 R.C. Messaros and S. Scypinski, *Lab. Rob. Autom.*, 7 (1995) 35.
8 A. Velasco-Arjona and M.D. Luque de Castro, *Anal. Chim. Acta*, 333 (1996) 205.
9 J. Gaba, M. Russo and A. Guez, *Lab. Rob. Autom.*, 10 (1998) 273.
10 H.R. Choi, Y.T. Lee, J.H. Kim, W.K. Chung and Y. Youm, *Lab. Rob. Autom.*, 6 (1994) 301.
11 P. Torres, E. Ballesteros and M.D. Luque de Castro, *Anal. Chim. Acta*, 308 (1995) 371.
12 B. Ciommer and N. Muller, *GIT Fachz. Lab.*, 40 (1996) 389.
13 C. Shumate, E. Mardis, L. Weinstock and D. Bruce, *Lab. Rob. Autom.*, 7 (1995) 73.
14 T. Londo, P. Lynch, T. Kehoe, M. Meys and N. Gordong, *J. Chromatogr.*, 798 (1998) 73.
15 S. Dunkerley and M.J. Adams, *Lab. Autom. Inf. Manage.*, 33 (1997) 93.
16 S. Graves, B. Holaman and R.A. Felder, *Clin. Chem.*, 46 (2000) 772.
17 J. Steinle, T.H. Connolly and F. Pteiffer, *Lab. Rob. Autom.*, 6 (1994) 313.
18 R. Buhlmann and J. Gil, *J. Chromatogr. Sci.*, 32 (1994) 243.
19 "Automated Environmental Lab Combines Robotics", *Chem. & Eng. News*, 14 Aug. (1996).
20 R. Schafer, C. Schulz and O. Stuckert, *4th International conference on automation, robotics and artificial intelligence applied to analytical chemistry and laboratory medicine*, Montreux, Feb. (1995).

21 N. Beasley, *Lab. Equip. Dig.*, 33 (1995) 19.
22 M. Entzeroth, *Lab. Autom. Inf. Manage.*, 33 (1997) 87.
23 P. Rogers, P.A. Hailey, G.A. Johnson, V.A. Dight, C. Read, A. Shingler, P. Savage, T. Roche and J. Mondry, *Lab. Rob. Autom.*, 12 (2000) 12.
24 J.A. García-Mesa, M.D. Luque de Castro and M. Valcárcel, *Lab. Rob. Autom.*, 5 (1993) 29.
25 P. Torres, J.A. García-Mesa and M.D. Luque de Castro, *Lab. Rob. Autom.*, 6 (1994) 233.
26 P. Torres, J.A. García-Mesa and M.D. Luque de Castro, *J. Autom. Chem.*, 16 (1994) 183.
27 A. Velasco-Arjona, J.A. García-Garrido, R. Quiles-Zafra and M.D. Luque de Castro, *Talanta*, 46 (1998) 969.
28 A. Velasco-Arjona and M.D. Luque de Castro, *Analyst*, 122 (1997) 123.
29 A. Velasco-Arjona, M.D. Luque de Castro, E. Ivanova and F. Adams, *Lab. Rob. Autom.*, 10 (1998) 293.
30 M.P. Granchi, J.A. Biggerstaff and P. Grey, *Spectrochim. Acta*, 42 (1987) 169.
31 P. Torres, J.A. García-Mesa and M.D. Luque de Castro, *Fresenius J. Anal. Chem.*, 346 (1993) 704.
32 A. Velasco-Arjona, A. Izquierdo and M.D. Luque de Castro, *Analyst*, 123 (1998) 509.
33 P.Torres, J.A. García-Mesa, M.D. Luque de Castro and M. Valcárcel, *Fresenius J. Anal. Chem.*, 346 (1993) 704.
34 J.A. García-Mesa, M.D. Luque de Castro and M. Valcárcel, *Talanta*, 40 (1993) 1595.
35 J. Marklew, *Lab. Update*, July–Aug. (2000) 16.
36 J. Norris and L. Ross, *Analytical Proc.*, 30 (1993) 164.
37 J.D. Norris, B. Preston and L.M. Ross, *Analyst*, 117 (1992) 3.
38 C.D. Herold, K. Andree and R.A. Felder, *Clin. Chem.*, 39 (1993) 143.
39 J.L. Chadwick, *Lab. Rob. Autom.*, 9 (1997) 21.
40 S.E. Eckert-Tilotta, T.L. Isenhour and J.C. Marshall, *Anal. Chim. Acta*, 254 (1991) 215.
41 L. Pieta, *Adv. Lab. Autom. Rob.*, 7 (1991) 303.
42 W.C. Koskinen, L.J. Jarvis and D.D. Buhler, *Soil Sci. Soc. Am. J.*, 55 (1991) 551.
43 A. Bruns, H. Waldhoff and W. Winkle, *J. Chromatogr.*, 592 (1992) 249.
44 M.D. Luque de Castro and P. Torres, *Trends Anal. Chem.*, 14 (1995) 492.
45 L.A. Gifford, C. Wright and J. Gilbert, *Food Addit. Contam.*, 7 (1990) 829.
46 H. Agemian, J.S. Ford and N.K. Madsen, *Chemom. Intell. Lab. Syst.*, 17 (1992) 145.
47 J.K. Gard, J.C. Burquin, W.B. Wise and B.S. Herman, *Spectroscopy*, 7 (1992) 28.
48 A. Millier and G. Wallet, *Chemom. Intell. Lab. Syst.*, 17 (1992) 153.
49 P. Torres and M.D. Luque de Castro, *Instrum. Sci. Technol.*, 24 (1996) 57.
50 A. Velasco-Arjona, J.J. Manclús, A. Montoya and M.D. Luque de Castro, *Talanta*, 45 (1997) 371.
51 J.T. Vanderslice and D.J. Hinggs, *J. Micronutr. Anal.*, 6 (1989) 109.
52 J.R. Ormand, D.A. McNett and M.J. Barttels, *J. Anal. Toxicol.*, 23 (1999) 35.
53 M. Arhoun, J.L.Martín-Herrera, E. Barreno and G. Ramis-Ramos, *Lab. Rob. Autom.*, 11 (1999) 121.
54 K. Ida and M. Karita, *Lab. Rob. Autom.*, 11 (1999) 25.
55 A. Velasco-Arjona and M.D. Luque de Castro, *Analyst*, 123 (1998) 1867.
56 A. Velasco-Arjona and M.D. Luque de Castro, *Fresenius J. Anal. Chem.*, 363 (1999) 311.
57 R.W. Taylor and S.D. Le, *J. Anal. Toxicol.*, 15 (1991) 276.
58 B.P. Ram, P. Singh and D. Allison, *J. Assoc. Off. Anal. Chem.*, 74 (1991) 43.
59 W.A. Michalik, R.M. Zerkel and J.B. Maynard, *Adv. Lab. Autom. Rob.*, 7 (1991) 325.
60 B. Houlihan, *Adv. Lab. Autom. Rob.*, 7 (1991) 583.
61 J.E. Coutant, P.A. Westmark and R.A. Okerholm, *J. Chromatogr.*, 108 (1991) 139.
62 J.E. Franc and K.A. Pittman, *J. Chromatogr.*, 108 (1991) 129.
63 L.A. Brunner and R.C. Luders, *J. Chromatogr. Sci.*, 29 (1991) 287.
64 K. Banno and R. Takahashi, *Anal. Sci.*, 7 (1991) 571.
65 C.N.C. Meredith and T.J. White, *Anal. Proc.*, 27 (1990) 182.
66 J. Hempenius, J. Wieling and D.A. Doornbos, *J. Pharm. Biomed. Anal.*, 8 (1990) 313.
67 D. Crahan, *Anal. Proc.*, 27 (1990) 23.

68 T.L. Lloyd, S.K. Gupta and J.R. Alianti, *J. Chromatogr.*, 678 (1996) 261.
69 T.L. Lloyd, T.B. Perschy and J.J. Tomlinson, *Biomed. Chromatogr.*, 6 (1992) 311.
70 J. Hempenius, G. Hendriks and C.C. Lin, *J. Pharm. Biomed. Anal.*, 12 (1994) 1443.
71 K. Pakulniewicz and J. Town, *LC–GC*, 13 (1995) 116.
72 J.Y.K. Hsieh, C. Lin and M.R. Dobrinska, *J. Pharm. Biomed. Anal.*, 12 (1994) 1555.
73 E.I. Minder and D.J. Vonderschmitt, *Chemom. Intell. Lab. Syst.*, 17 (1992) 119.
74 Y. Tsina and B. Wong, *J. Chromatogr.*, 675 (1996) 119.
75 L. Gámiz-Gracia, A. Velasco-Arjona and M.D. Luque de Castro, *Analyst*, 124 (1999) 801.
76 M.M. Walstead, E. Yapchanyk and Y. Haroon, *J. AOAC Int.*, 81 (1998) 931.
77 A.P. Watt, D. Morrison, K.L. Locker and D.C. Evans, *Anal. Chem.*, 72 (2000) 979.
78 L. Lindquist and F. Días, *Adv. Lab. Autom. Rob.*, 7 (1991) 341.
79 W. Anderson, *Adv. Lab. Autom. Rob.*, 7 (1991) 351.
80 M. Valcárcel and M.D. Luque de Castro, *Flow-Injection Analysis: Principles and Applications*, Ellis Horwood, Chichester (1987).
81 J. Ruzicka, E.H. Hansen, *Flow Injection Analysis*, John Wiley & Sons, New York (1988).
82 J.A. García-Mesa, M.D. Luque de Castro and M. Valcárcel, *J. Flow Injection Anal.*, 10 (1993) 262.
83 M. Valcárcel and M.D. Luque de Castro, *Non-Chromatographic Continuous Separation Techniques*, Royal Society of Chemistry, Cambridge (1991).
84 J.A. García-Mesa, M.D. Luque de Castro and M. Valcárcel, *Anal. Chem.*, 65 (1993) 3540.
85 M. Valcárcel and M.D. Luque de Castro, *Automatic Methods of Analysis*, Elsevier, Amsterdam (1988).
86 J.C. Boyd, R.A. Felder and J. Savory, *Clin. Chem.*, 42 (1996) 1901.
87 R.A. Felder, J.C. Boyd, K.S. Margrey, W. Holman, J. Roberts and J. Savory, *Anal. Chem.*, 63 (1991) 741 A.
88 G.L. Hawk and J. Strimaitis, *Advances in Laboratory Automation Robotics*, Zymark Co, Hopkinton, MA (1984–91).
89 J.M. Hawkins, *J. Autom. Methods Manage. Chem.*, 22 (2000) 47.
90 S. Peters, J. Hampsch, M. Cregor, C. Starrett, G. Gunaratna and C. Kissinger, *Curr. Sep.*, 18 (2000) 139.
91 S. Unger, *Biotech. Lab. Int.*, 5 (2000) 6.
92 D.A. Rudge, *Lab. Autom. Inf. Manage.*, 33 (1997) 81.
93 S.X. Peng, C. Henson, M.J. Strojnowski, A. Golebiowski and S.R. Klopfenstein, *Anal. Chem.*, 72 (2000) 261.
94 M. Traini, A.A. Gooley, K. Ou, M.R. Wilkins, L. Tonella, J.C. Sánchez, D.F. Hochstrasser and K.L. Williams, *Electrophoresis*, 19 (1998) 1941.
95 D. Gostick, *Lab. Update*, July–Aug. (2000) 8.
96 C. Schade, A. Wehren, F. Siegman, M. Collasius and D. Hauffe, *Lab. News*, Apr (2000) 14.
97 P.G. Schultz, *Chem. Eng.*, July 3 (2000) 27.
98 E.W.H. Jager, O. Inganäsm, U. Kybdström, *Science*, 288 (2000) 2335.
99 J. Ruzicka and T. Gubeli, *Anal. Chem.*, 63 (1991) 1680.
100 M. Guzmán and B.J. Compton, *Talanta*, 40 (1993) 1943.
101 M.F. Joyce, P.A. Langan and W.S. Reid, *Commun. Soil Sci. Plant Anal.*, 26 (1995) 19.

Index